1·29·03

Second Edition

CHEMISTRY OF THE ENVIRONMENT

Second Edition

CHEMISTRY OF THE ENVIRONMENT

RONALD A. BAILEY
HERBERT M. CLARK
JAMES P. FERRIS
SONJA KRAUSE
ROBERT L. STRONG

Department of Chemistry
Rensselaer Polytechnic Institute
Troy, New York

ACADEMIC PRESS
An Elsevier Science Imprint

San Diego   San Francisco   New York   Boston   London   Sydney   Tokyo

Sponsoring Editor	Jeremy Hayhurst
Production Managers	Joanna Dinsmore and Molly Wofford
Editorial Coordinator	Nora Donaghy
Cover Design	Amy Stirnkorb
Copyeditor	Brenda Griffing
Composition	Kolam Information Services
Printer	Malloy Lithographing

Cover photograph: Courtesy of NASA/JPL/Caltech.

This book is printed on acid-free paper. ∞

Academic Press
An Elsevier Science imprint
525 B Street, Suite 1900, San Diego, California 92101-4495, U.S.A.
http://www.academicpress.com

Academic Press
Harcourt Place, 32 Jamestown Road, London NW1 7BY, UK
http://www.academicpress.com

Library of Congress Catalog Card Number: 2001096794

International Standard Book Number: 0-12-073461-3

PRINTED IN THE UNITED STATES OF AMERICA
02 03 04 05 06 07 ML 9 8 7 6 5 4 3 2 1

CONTENTS

3

ENERGY AND CLIMATE

4

PRINCIPLES OF PHOTOCHEMISTRY

5

ATMOSPHERIC PHOTOCHEMISTRY

6

PETROLEUM, HYDROCARBONS, AND COAL

7

SOAPS, SYNTHETIC SURFACTANTS, AND POLYMERS

8

HALOORGANICS AND PESTICIDES

9

CHEMISTRY IN AQUEOUS MEDIA

10

THE ENVIRONMENTAL CHEMISTRY OF SOME IMPORTANT ELEMENTS

11

WATER SYSTEMS AND WATER TREATMENT

12

THE EARTH'S CRUST

13

PROPERTIES AND REACTIONS OF ATOMIC NUCLEI, RADIOACTIVITY, AND IONIZING RADIATION

14

THE NUCLEAR ENVIRONMENT

15

ENERGY

16

SOLID WASTE DISPOSAL AND RECYCLING

APPENDIX A

DESIGNATION OF SPECTROSCOPIC STATES   793

PREFACE

The predecessor to this book originated in the 1970s as notes to a junior/senior level course in "Chemistry of the Environment" that the authors developed at Rensselaer Polytechnic Institute in response to the interest in environmental concerns developing at that time. The present volume is an updated and expanded version of the original book published in 1978 and used ever since.

Major changes in environmental problems have occurred in the years since the publication of the first edition, including the ozone and global warming concerns, the nuclear power plant meltdown at Chernobyl in 1986, and the world's worst oil spills in Kuwait and the Persian Gulf in 1991. Although this second edition of the book refers to these and other disasters, and was written with current environmental topics in mind, it retains an emphasis on the principles of environmental chemistry, not on specific catastrophes. Significant alterations in coverage have been made in a number of topics to reflect these dynamics. For example, there is less discussion of the atmospheric chemical reactions of hydrocarbons in Chapter 6 because hydrocarbons have become a less serious environmental problem in the United States as a result of the mandated use of the catalytic converter on the exhaust systems of automobiles. The chapters on polymers and surfactants were combined into one chapter and follow the chapter on petroleum because the major pathways for the environmental degradation of these three classes of organics is by microorganisms and the reaction steps are similar in many instances. The separate chapter on pheromones was removed and an abridged version of this material has been added to the chapter dealing with pesticides because pheromones have not achieved the potential for insect control forecast for them in the 1970s. New material on environmental hormones has been included, and a new chapter on recycling has been added. The material dealing with the chemistry of particular elements of environmental concern has been expanded, and the material on nuclear chemistry redone. The latter we feel is particularly important in view of the possible resurgence of nuclear energy and the general (low) level of understanding of it. We have tried to avoid an excessive focus on U.S. concerns.

This book is intended as a text, but also as a reference book to provide a background to those who need an understanding of the chemical basis of many environmentally important processes. As a text, it is probably too extensive to be covered completely in a single course, but in our own use we have found that particular chapters and sections can be selected according to the topics desired. We have attempted to group chapters in a way that is suitable for this. In some cases, this has resulted in different aspects of some topics being covered in different chapters. We do not believe that this has led to either excessive fragmentation or redundancy. As either a text or reference, the reader will need some chemistry background; we have not attempted to write this book at a beginning level. Besides a good grounding in general chemistry, some knowledge of organic chemistry is needed for the sections dealing with organic compounds.

The objective of the book is to deal with the chemical principles and reactions that govern the behavior of both natural environmental systems and anthropogenic compounds important in the environment, although obviously it is a general coverage and does not pretend to go deeply into specialty areas such as marine chemistry or geochemistry. From time to time we may have stretched the scope of chemistry a bit—for example, we deal with atmospheric circulation because this is important to understanding some of the consequences of atmospheric chemistry—and we mention some nonchemical energy sources for completeness.

The book starts with a discussion of the atmosphere, leading into chapters on photochemistry and other atmospheric chemical reactions. Three chapters deal with the environmental chemistry of organic compounds of important types, including petroleum, detergents, pesticides, and plastics. Following sections deal mainly with water and the inorganic chemistry in natural water systems, including the environmental chemistry of selected elements, followed by a brief chapter on the lithosphere. Two chapters deal with nuclear chemistry in some detail. Other chapters deal with energy and recycling. Where appropriate, we have made reference to recent scientific publications, mainly in footnotes, and have suggested additional reading at the end of each chapter, but we have made no effort to give extensive literature coverage. We have included a few references to information on the Internet, which can be a valuable source of data if care is taken to use reliable sources such as governmental agencies and to *remember that much material is not reviewed*; some of it is just plain wrong. Exercises have been provided that can be used for self-study or for class assignments.

As in the original version, we have used the units that are commonly encountered in the literature dealing with particular fields, rather than rationalizing them to a standard such as SI units. We feel that this will better prepare the reader for further reading. We have included SI equivalents in many instances, and do provide a table of conversions and definitions that may help to make sense out of the many different units found.

R. A. Bailey
H. M. Clark
J. P. Ferris
S. Krause
R. L. Strong

ABOUT THE AUTHORS

Professor Ronald A. Bailey is an inorganic chemist whose research involves the synthesis of metal ion complexes and their spectroscopic and electrochemical behavior and structures in a variety of media. He became interested in environmental problems as a result of being asked about the complexing of metal ions in natural waters and sediments, and to provide assistance in explaining the chemistry involved in a proposed method for scrubbing stack gases. With this interest, he joined with other members of the Chemistry Department at Rensselaer Polytechnic Institute to develop a course for advanced undergraduates, "Chemistry of the Environment," that became the impetus for this book.

Herbert M. Clark (Ph.D., Yale University), Professor Emeritus of Chemistry at Rensselaer Polytechnic Institute, worked on the Manhattan Project before joining the faculty at RPI, where he established a course in nuclear chemistry and a laboratory course in radiochemistry. The areas of his research include solvent extraction of inorganic substances, radioactive fallout and rainout, adsorption of fission products from aqueous solution, and aqueous chemistry of technetium. He has been a member of the Subcommittee on Radiochemistry of the Committee on Nuclear Science of the National Academy of Sciences-National Research Council, and he has participated in major off-campus programs such as the evaluation of novel designs of nuclear power reactors along with methods for reprocessing their fuel and evaluation of methods for preventing the proliferation of nuclear weapons. His publications include three co-authored textbooks.

Professor James P. Ferris became interested in the Earth's current environment as a result of his research on the origins of life and his interest in the atmospheric chemistry of Titan, the largest moon of Saturn. His investigations of the origins of life have led him to consider the organic chemistry that took place in the environment of the primitive Earth about 4 billion years ago when life is believed to have originated.

Professor Sonja Krause is a physical chemist whose research deals with the thermodynamics and physical properties of solutions of synthetic and natural polymers. She is especially interested in the effects of electric fields on these mixtures and solutions. She became interested in the climate history of the earth when she learned that Hannibal crossed the Alps with elephants and wondered how that was possible. When she found out that the climate was much warmer in those days than at present, she learned as much as she could about the processes that control climate and began working on part of the undergraduate course at Rensselaer Polytechnic Institute that led to this book.

Robert L. Strong is Professor Emeritus of Chemistry at Rensselaer Polytechnic Institute, where he was a physical chemist with research in photochemistry. A native of southern California, he became interested in the nature of photochemical smog and the types of processes that could lead to the very rapid daytime buildup of pollutants in the lower atmosphere. He was involved in establishing the course "Chemistry of the Environment" at RPI and taught parts of it for several years before retiring in 1992.

1

INTRODUCTION

1.1 GENERAL

Humans evolved on earth, with its atmosphere, land and water systems, and types of climate, in a way that permits us to cope reasonably well with this particular environment. Being intelligent and inquisitive, humans have not only investigated the environment extensively, but have done many things to change it. Other living things also change their environment; the roots of trees can crack rocks, and the herds of elephants in present-day African game parks are uprooting the trees and turning forests into grasslands. We are beginning to understand that microorganisms play a major role in determining the nature of our environment, not only through their actions on organic material, but also in systems traditionally regarded as strictly inorganic—and not just at the earth's surface, but also at considerable depths. However, no other living things can change their environment in so many ways or as rapidly as humans. The possibility exists that we can change our environment into one that we cannot live in, just as the African elephant may be doing on a smaller scale. The elephants cannot learn to stop uprooting the trees they need for survival, but we should be capable of learning about our environment and about the problems we ourselves create.

One vital problem (among many) that we must solve is how to continue our technologically based civilization without at the same time irreversibly damaging the environment in which we evolved. This environment, which supports our life, is complex; the interrelations of its component parts are subtle and sometimes unexpected, and stress in one area may have far-reaching effects. This environment is also finite (hence the expression "spaceship earth"). That being so, there is necessarily a limit before significant environmental changes take place. Our modern technological society, coupled with largely uncontrolled population growth, places extreme stress on the environment, and many environmental problems come from prior failure to understand the nature of these stresses and to identify the limits. Many problems also come from refusal to accept that there are limits. Even for a component of our environment that is so large that for all practical purposes it can be taken as infinite, such as the water in the oceans, there are limits. We cannot destroy a significant amount of ocean water, but we can contaminate it.

Many environmental problems and processes are chemical, and to understand them, understanding the basic chemistry involved is necessary (but not necessarily sufficient). This is certainly true for those who wish to solve environmental problems, but also for those who need to make more general decisions that may have environmental consequences. It is important as well for those who would like to have some basis for understanding or evaluating the many, often contradictory, claims made regarding environmental problems. While the basic science underlying most environmental processes is well understood, many details are not. Too often, this lack of knowledge of the details of complex interactions is taken as an excuse to deny the validity of sound general conclusions, particularly when they run counter to political or religious dogma.

In this book, we shall use the term "environment" to refer to the atmosphere–water–earth surroundings that make life possible; basically this is a physicochemical system. Our total environment consists also of cultural and aesthetic components, which we will not consider here. Neither shall we deal extensively with ecology—that is, the interaction of living things with the environment—although some effects of the physicochemical system on life, and conversely, of living things on the physicochemical environment, will be included. Indeed, organisms play a major role in both organic and inorganic processes in the environment, and living things have made our environment what it is. Most obviously, the biological process of photosynthesis created the oxygen atmosphere in which we live.

1.2 POLLUTION AND ENVIRONMENTAL PROBLEMS

An immediate mental association with "environment" is "pollution." Yet what constitutes a pollutant is not easily stated. One tends to associate the term with materials produced by human activity that enter the environment with harmful effects especially on living organisms—for example, SO_2 from combustion of sulfur-containing fuels, or hydrocarbons that contribute to smog. However, these and other "pollutants" would be present even in the absence of humans, sometimes in considerable amounts. We will consider a "pollutant" to be any substance not normally present, or present in larger concentrations than normal. One type of source commonly associated with pollution consists of industrial plants, especially chemical, but also factories using organic solvents, heavy metals, and so on. Another source is mining, including waste rock dumps and smelters. Waste disposal dumps are yet another source; materials can be released from these dumps. Military bases, training, and testing facilities may be sources of pollution from heavy metals as well as explosives residues. Fuel storage facilities, airports, and on a smaller but more widespread scale, gasoline stations may pollute through spills or leaking tanks. Power plants and incinerators may emit pollutants to the atmosphere, while automobiles and other internal combustion engines provide a delocalized but abundant source.

Although any discussion of the environment must consider pollution, this will not be our primary aim. Rather, we will be concerned with some of the more important chemical principles that govern the behavior of the physicochemical environment and the interactions of its various facets.

Among the controversial topics as this book is written are global warming and ozone depletion. The chemistry of these phenomena is discussed in some detail later in this book (Chapters 3 and 5), but an introduction to them at this point may give a useful illustration of environmental problems. With respect to global warming, some aspects are beyond scientific dispute, some are highly probable, and some are still uncertain. There is no question that introduction of carbon dioxide and some other gases into the atmosphere will increase heat retention, and that the atmospheric concentration of carbon dioxide has been increasing. It is also well established that introduction of some (but not all) particulate material into the upper atmosphere will reflect solar energy and lead to cooling. What is in question is how much of the introduced materials will be removed by natural processes, what the relative heating and cooling effects will be, and other details that need to be understood before a quantitative prediction of the net result can be made. It is almost certain that the average temperature of the earth's surface has increased in the last few years, and likely that at least part of this increase is due to human activity, but natural variabilty makes the latter conclusions less certain. Also uncertain are collateral changes in atmospheric circulation, rainfall, and ocean currents that could accompany an increase in global average temperature. There are two primary responses to these uncertainties. One is the attitude that if there is any doubt, one can ignore the potential problem and continue business as usual. The other is to assume the worst and advocate massive changes at once, whatever the costs involved. An intermediate response, "Make the changes that have moderate cost but move in the right direction," is often overlooked.

Ozone destruction in the stratosphere by chlorofluorocarbons, as discussed in Chapter 5, was clearly shown to be inherent in the chemistry of the atmosphere long before its observed decrease caused responsive action to be taken. Even then, many people refused to accept that human activities could have a deleterious effect and used uncertainties in the details of the destruction mechanism to argue against the need for controls. Some still do.

Other responses to environmental problems are to focus on potential benefits, which is a perfectly valid thing to do if all the implications are understood. For example, it has been pointed out that global warming may have beneficial effects by permitting agricultural activity in the Arctic regions, but do the soils in these regions have the necessary characteristics to support domestic plant growth? Will the possible increase in productivity in the Arctic balance possible losses elsewhere? What about changes in patterns of precipitation that may accompany temperature changes? It is also pointed out that massive environmental changes, including higher temperatures, have occurred naturally, so that anthropogenic changes do not involve anything new. This ignores the rate of change: changes that human activity can bring about may take place much more rapidly than those arising from natural causes, although knowledge of details is generally inadequate to predict time scales. For example, people could adjust to the conversion of, say, Wisconsin or Provence to a desert over a thousand years, but the same change in a few decades would be disastrous to the people involved.

Two human activities that are strongly connected with environmental chemistry in general and pollution in particular are energy production and waste disposal, as discussed in Chapters 15 and 16, respectively. Aspects of the chemistry of these activities will be considered later, but some general comments are given here.

1.3 WASTE DISPOSAL

Waste disposal is a growing problem. The environmental system consists of a very large amount of material: about 5×10^{18} kg of air and about 1.5×10^{18} m^3 of water. In principle, comparatively large amounts of other materials can be dispersed in these, and consequently both the atmosphere and water bodies have long been used for disposal of wastes. This is sometimes done directly, and sometimes after partial degradation such as by incineration. In part, such disposal is based on the principle of dilution: when the waste material is sufficiently dilute, it is not noticeable and perhaps not even detectable. A variety of chemical and biological reactions may also intervene to change such input into normal environmental constituents (e.g., the chemical degradation of organic materials to CO_2 and H_2O).

Such disposal-by-dilution methods are not necessarily harmful in principle, but several factors must be considered in practice. The first is the problem of mixing. While the eventual dilution of a given material may be acceptable, local concentrations may be quite high because mixing is far from instantaneous. As the disposal of wastes from more sources becomes necessary because of population or industrial growth in a particular area, it becomes more difficult to avoid undesirable local excess either continuously or as natural mixing conditions go through inefficient periods. A second factor of long-term importance is the rate of degradation of waste material. It is obvious that if the rate of removal of any substance from a given sector of the environment remains slower than its rate of input, then the concentration of that material will eventually build up to an undesirable level. This is the case with many modern chemical products, including some pesticides and plastics that are only very slowly broken down into normal environmental components or rendered inaccessible by being trapped in sediments. Global contamination by such materials results from the dilution and mixing processes, and concentration levels increase while input continues. The examples of DDT and PCBs, discussed in Chapter 8, are well known. Other examples, such as the occurrence of pellets of plastics in significant amounts even in the open ocean, may ultimately be of equal concern. There is also the problem of natural concentration. It is well known that some toxic materials (e.g., DDT and mercury) are concentrated in the food chain as they are absorbed from one organism to another without being excreted. Some plants can concentrate toxic materials from the soil.

A third problem in disposal by dilution is the question of what concentration must be reached before harmful effects are produced. This is often difficult to answer, both because deleterious effects may be slow to develop and noticeable only statistically on large samples, and because often concentrations less than the parts-per-million range have to be considered. This is very low indeed, and often methods of detection and measurement become limiting in establishing the role of a particular substance. Finally, synergistic effects may come into play—that is, two or more substances existing at concentrations that separately are below the limits that are harmful may have highly deleterious effects when present together.

Disposal by burying is a process that does not depend on dilution. In general, however, it ultimately depends either on degradation to harmless materials, or alternatively on the hope that the material stays put. Such considerations are particularly important with highly toxic or radioactive waste materials. Leaching of harmful components from such disposal sites and consequent contamination of groundwater or extraction from soils and concentration in plants are obvious concerns associated with such disposal methods.

Disposal of insoluble materials in the oceans also depends upon the principle of isolation of the waste. Solid materials may be attacked by seawater, and harmful components leached out into the water, ultimately to enter the food chain. Barrels or other containers for liquid wastes may corrode and leak. Contamination of the biological environment is a possible result. Ocean currents may play a role in distributing such wastes in unforeseen ways, as with New York City sewage sludge dumped in the Atlantic Ocean that migrated a long way from its dump site into adjacent waters. An extreme example of this occurred in 1995, when 4500 incendiary bombs washed up on Scottish beaches, injuring a child.[1] These devices were believed to have come from the disturbance of an ocean dump of surplus munitions from World War II, dumped 10 km offshore in 1945, although the site had been used for munitions disposal from the 1920s up to the 1970s. Seven hundred antitank grenades and other weapons have washed up on beaches on the Isle of Man and Ireland from similar disposal sites.

Clearly, understanding of a disposal problem involves knowledge of physical processes such as mixing, chemical processes involved in degradation reactions, chemical reactions of the waste and its products with various aspects of the environment, and biological processes, meteorological processes, geology, oceanography, and so on.

1.4 RISK

The question of what constitutes a hazardous level of something—for example, radiation or a chemical that is carcinogenic or mutagenic—is not easily answered. In part this is because years may pass before the consequences become identifiable. Even for materials with a more immediate effect, establishment of toxic levels may be problematic if the effects are not highly specific. Moreover, individuals vary in response, and the same responses may be due to alternate causes. Data on hazardous exposure levels could be obtained from epidemiological studies on large groups of people, but this would require the exposure of people to conditions known to be harmful. Experimental studies must use other species than humans, with the accompanying problems of species specificity because of differing metabolisms, for example. (With our current knowledge, computer modeling or extrapolation from tests on simple organisms is not sufficient, although such exercises may be helpful.) To make the scale of such experiments reasonable, usually small numbers of individuals are tested at high levels of dose, and results extrapolated to lower levels to get an estimate of risk for low levels of exposure. Past practice has generally been to use a linear extrapolation, implicitly assuming that there is zero effect only

---

[1]R. Edwards, *New Sci.*, pp. 16–17, Nov. 18, 1995.

at zero concentration or exposure. Hence we had in the United States the Delaney clause prohibiting any detectable level of a carcinogenic additive in any food. There is abundant evidence, however, that organisms are often able to avoid or repair damage from substances up to a limit; that is, a threshold exists for many harmful effects, though that limit may be difficult to establish.

Even when hazardous levels can be identified, these are statistical; not all members of the exposed population will react the same way. Regulations designed to protect the public from exposure to harmful levels of substances involve some level of balance between cost and risk, although this practice may not be acknowledged explicitly at the time. Intelligent decisions about what this level should be involve many considerations, are often controversial, and are beyond the scope of this book.

1.5 ENERGY

Energy production is a major activity of modern society that causes many environmental problems. Some of these problems are associated with the acquisition of the energy source itself (e.g., acid drainage from coal mines as discussed in Chapter 10, oil spills as discussed in Chapter 6), others with the energy production step (e.g., combustion by-products such as CO, SO_2, partially burned hydrocarbons, and nitrogen oxides considered in Chapters 5 and 10), while still others arise from disposal of wastes (e.g., nuclear fission products, as will be seen in Chapter 14). In any device depending on conversion of heat to mechanical energy, (e.g., a turbine) the fraction of the heat that can be so converted depends on the difference between the temperature at which the heat enters, and that at which it leaves the device, as will be seen in Chapter 15. Practical restrictions on these temperatures limit efficiency and result in waste heat that must be dissipated, because making the exhaust temperature equal to ambient temperatures is not practical. This leads to the problem of thermal pollution. Such efficiency considerations also lead to the desire for ever higher input temperatures (i.e., the temperature in the combustion chamber). This in itself can produce secondary pollution problems, such as enhanced generation of nitrogen oxides.

1.6 POPULATION

Basic to many environmental problems is the problem of human population— its growth and the capacity of the earth to support it. Figure 1-1 shows the estimated human population from prehistoric times. Human population has increased by a factor of 4 from 1860 to 1961, while use of energy (other than

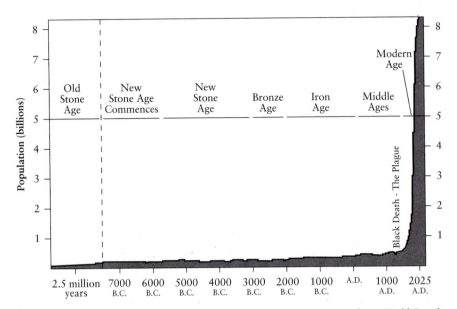

FIGURE 1-1 Human population trends from prehistoric times. Redrawn from *World Population: Toward the Next Century,* Copyright © 1994. Used by permission of the Population Reference Bureau.

human or animal power) increased more than 90-fold. This increase was accompanied by major increases in other resource needs as the technological base of society underwent drastic changes. Changes in resource use have not been uniform; a small fraction of the world's people, those in the developed nations, use most of the resources, although often these are obtained from lesser developed regions where the impact from mining, logging, and so on can be severe on local residents with little economic or political power to mitigate it. At the current rate of increase, 1.6% annually, the population will have doubled by the middle of the 21st century.[2] Many factors make predictions of the actual increase unreliable (a prediction from the U.S. Bureau of the Census is shown in Figure 1-2), but certainly the numbers are such that one must consider what the effective carrying capacity of the earth is. This also is hard to evaluate[3]; it depends upon the lifestyle assumed, and on technological

[2]Population increase is not uniform worldwide. Many industrialized nations have birthrates near or even below the replacement value, whereas population growth is largest in many of the least developed nations. The environment is deteriorating most severely in these less developed areas not from technological and industrial wastes and by-products, although these are increasing rapidly, but through environmental destruction driven by the needs of survival (e,g., deforestation in an effort to get firewood for cooking or to get land for growing food) or economic development.

[3]J. E. Cohen, *Science,* **269,** 341 (1995).

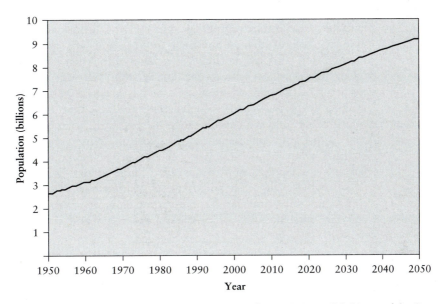

FIGURE 1-2 World population from 1950, projected to 2050. From U.S. Bureau of the Census, International Data Base.

solutions that may be developed for supplying food and other resources. Reliance on technological solutions yet to be developed to compensate for excessive population growth—which is the only alternative to population control if we are to maintain any reasonable standard of living—seems to be a dangerous act of faith in view of the consequences of the failure of workable solutions to materialize. Optimists point to increased crop yields per acre and to the increase in total food production without an increase in the land area harvested in arguing that technology will be able to keep pace with population growth. Even so, the food so produced is often not available to people who need it. It is also pointed out that most processes, including food and energy production, are not carried out, on average, close to the efficiency already demonstrated in the best operations, and far from the efficiency that is possible in principle. Thus, it can be argued that even with current technology, an increased population, and increased productivity to support it, are possible without increasing our impact on the environment. On the other hand, considering the uneven distributions of resources and population, and normal human tendencies for short-term self-interest, one can have little confidence that local disasters will be avoided.

All predictions of the effects of continued population growth and related activities based on extrapolation of existing conditions are rendered questionable because of the complicated interrelationships among environmental factors, and also by the increasing recognition that there is not always a linear relationship between the factors that influence a phenomenon and the level of the phenomenon itself. It has been proposed that many things, ranging from the spread of epidemics to the crime rate, exhibit a point of criticality: under a certain range of conditions, changes in the parameters that affect the phenomenon in question cause relatively proportional changes in it, but as these parameters approach some limiting values, small parameter changes begin to have very large effects. Is there such a critical point in the population–environment relationship? No one knows. Failure to account for nonlinear response to system perturbations is a major difficulty in understanding or predicting environmental consequences in many areas.

1.7 THE SCOPE OF CHEMISTRY OF THE ENVIRONMENT

Almost everything that happens in the world around us could come under the general heading "chemistry of the environment." Chemical reactions of all kinds occur continuously in the atmosphere, in oceans, lakes, and rivers, in all living things, and even underneath the earth's crust. These reactions take place quite independently of human activities. The latter serve to complicate an already complex subject.

To understand environmental problems, we must have knowledge not only of what materials are being deliberately or inadvertently released into the environment, but also of the processes they then undergo. More than this, we need to understand the general principles underlying these processes so that reasonable predictions can be made about the effects to be expected from new but related substances. We must also understand the principles that underlie natural environmental processes to anticipate human interferences. This chemistry-oriented book emphasizes the chemical principles underlying environmental processes and the chemistry of the anthropogenic components—the materials and changes that humans have introduced. But to put these into context, some topics that are rather distant from reactions and equations need to be discussed—as has been done in this chapter. In addition, knowledge in biological, meteorological, oceanographic, and other fields is equally important to the overall understanding of the environment. Indeed, although it is convenient to segment topics for study purposes, Barry Commoner's first law of the environment should always be kept in mind: "Everything is related to everything else."

Additional Reading

Alloway, B. J., and D. C. Ayres, *Chemical Principles of Environmental Pollution*, 2nd ed. Blackie, London, 1997.

Bunce, N. J., *Environmental Chemistry*, 3rd ed. Wuerz Publishing, Winnipeg, 1998.

Brown, L. R., and H. Kane, *Full House: Reassessing the Earth's Carrying Capacity*, Norton, New York, 1994.

Manahan, S. E., *Environmental Chemistry*, 5th ed. Lewis Publishers, Chelsea, MI, 1991.

O'Neill, P., *Environmental Chemistry*, 2nd ed. Chapman & Hall, London, 1993.

2

ATMOSPHERIC COMPOSITION AND BEHAVIOR

2.1 INTRODUCTION

There has been much discussion, both nationally and internationally, about the effects of various atmospheric pollutants on the earth's climate and on the earth's atmosphere. The United Nations Intergovernmental Panel on Climate Change (IPCC) report, data from the Goddard Space Science center of NASA, for example, shows that average global air temperatures have risen 0.6 to 0.7°C over the last century (Figure 2-1). The 1997 Kyoto Protocol is an agreement among more than 150 countries to attempt to limit global warming by putting limits on their emission of carbon dioxide and some other greenhouse gases (see Section 3.3.3) to limit global warming. Furthermore, other international treaties including the Montreal Protocol of 1988 and its amendments limit and even ban the use of chlorofluorocarbons and and the so-called halons in the United States and other countries at different times in order to stop the spread of the "ozone holes" in the stratosphere over the Arctic and Antarctic regions of the globe (see Section 5.2.3.5: Control of Ozone-Depleting Substances). Newspaper and magazine articles often make

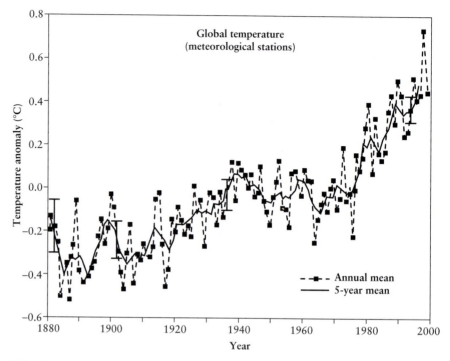

FIGURE 2-1   Near-global annual mean surface air temperature from 1880 to 1998 relative to the mean surface air temperature between 1951 and 1980 (~ 14°C). Data from the NASA Goddard Institute for Space Studies web site: http://www.giss.nasa.gov.

these problems seem relatively simple but, in fact, they raise a number of questions:

1. What criteria are used to tell us that the earth's climate is becoming warmer? How do we take account of the normal year-to-year variability of the climate in any one area on the earth's surface?

2. Assuming that the criteria for defining the stated changes in the earth's climate are sound, how do these twentieth-century climate changes compare with other variations in climate noted in historic and prehistoric times? In other words, is there a possibility that the climate changes noticed during the latter half of the twentieth century are due, at least partially, to natural causes? Computer models (see Section 3.3.5) detailing the influences of different factors on the earth's climate are being developed and may help answer this question.

3. Why should an increase in the carbon dioxide content of the atmosphere lead to a warming trend in the earth's climate? What other factors regulate the earth's climate?

4. What do we know about the general circulation of the atmosphere? After all, carbon dioxide and chlorofluorocarbons are released into the atmosphere at particular locations on the earth's surface. How far do they spread through the atmosphere?

5. Although the carbon dioxide released into the atmosphere is said to affect the total climate, the macroclimate, of the earth, we humans actually live in small areas where we worry about microclimate. What do we know, for example, about the special climates of cities as opposed to their surrounding countrysides? How do the pollutants produced by human activities in cities become trapped in the atmosphere above these cities, to become hazards to people's health?

We shall attempt to answer these questions in this and the following chapter, though not necessarily in the order in which they have been presented. The answers to most of these questions are complex and, in many cases, incomplete. It will thus be seen that questions and answers concerning the atmosphere and human effects on the atmosphere and climate are not simple and need a great deal of further study. This does not mean that action should not be taken, but it demonstrates the reason for the controversy surrounding these topics.

2.2 GASEOUS CONSTITUENTS OF THE ATMOSPHERE

Table 2-1 shows the constituents of clean, dry air near sea level. Usually, the atmosphere also contains water vapor and dust, but these occur in amounts that vary widely from place to place and from time to time. Carbon dioxide, the fourth constituent on the list, has, as we shall see, been increasing in concentration during this century, mostly through the burning of fossil fuels such as coal and petroleum products. The concentrations given in Table 2-1 are estimated global averages for 1989; some of the trace constituents, like methane, have also been increasing in concentration, as noted later (see Section 3.3.3 and Figure 3-14).

Trace constituents may vary in concentration in different parts of the atmosphere and in different parts of the globe. This is because many of these trace constituents are produced by different human and natural sources in different locations and are then removed from the atmosphere in different ways and by different means. For example, nitrogen oxides may be produced not only from human activities such as combustion in air but also from natural occurrences such as lightning discharges through the nitrogen and oxygen of the atmosphere and some photochemical reactions in the upper atmosphere (see Chapter 5, Section 5.2.1). Sulfur compounds arise not only from the combustion of sulfur-containing fuels but also from volcanic action. Methane

TABLE 2-1

Some Constituents of Particulate-Free Dry Air
near Sea Level

Component	Volume (%)
N_2	78.084
O_2	20.948
Ar	0.934
CO_2	0.0350
Ne	0.00182
CH_4	0.00017
Kr	0.00011
H_2	0.00005
N_2O	0.00003
H_2	0.00005
CO	0.00001
Xe	9×10^{-6}
O_3	$1{-}10 \times 10^{-6}$
NO_2	2×10^{-6}
NH_3	6×10^{-7}
SO_2	2×10^{-7}
CH_3Cl	5×10^{-8}
CF_2Cl_2	4×10^{-8}
$CFCl_3$	2×10^{-8}
C_2H_4	2×10^{-8}
H_2S	1×10^{-8}
CCl_4	1×10^{-8}
CH_3CCl_3	1×10^{-8}

generally comes from decaying organic matter while ammonia, H_2S, and some of the N_2O arise from decomposition of proteins by bacteria. The concentration of ozone, given in Table 2-1 at sea level, varies a great deal with altitude and somewhat with latitude. Section 2.5 discusses ozone concentration in greater detail.

Although oxygen is a very reactive gas, its concentration is presently constant in our atmosphere; this implies a dynamic equilibrium involving atmospheric oxygen. It is well known that photosynthesis in green plants adds large amounts of oxygen to the atmosphere during daylight hours. Although oxygen may also be produced in other ways, the amounts are much smaller than those produced by photosynthesis. It seems reasonable to suppose that the oxygen produced by photosynthesis makes up for the equally vast amounts of oxygen used up by the respiration of plants and animals, by weathering of rocks, by burning of fossil fuels, and by other oxidative processes that take place in the atmosphere. This balance, however, requires green plants that were

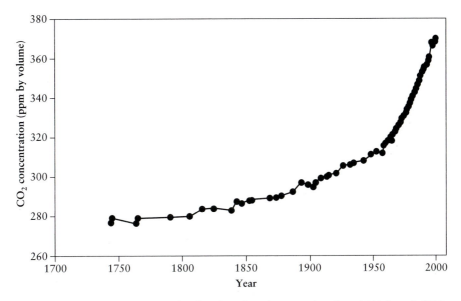

FIGURE 2-2 Historical variation of carbon dioxide in the atmosphere from 1744 through 1999. Data from the U.S. National Oceanic and Atmospheric Administration (NOAA) Climate Monitoring and Diagnosis Laboratory (CMDL), Carbon Cycle–Greenhouse Gases Branch; antarctic ice core samples and atmospheric measurements. http://www.cmdl.noaa.gov.

not present in the very distant past. The various theories that describe the earth as evolving from a primordial, lifeless state to its present condition, and the physical evidence that supports these theories, suggest that little or no oxygen was present on the primordial earth.

Carbon dioxide has been released into the atmosphere in significant amounts since the beginning of the industrial age through the burning of fossil fuels.[1] Furthermore, the destruction of forests in the tropics and temperate regions of the earth has reduced the removal of CO_2 from the atmosphere by green plants. Studies of carbon dioxide in ice cores from both Greenland and Antarctica have shown a large variability in the concentration of CO_2 in the atmosphere over the past 160,000 years (see Figure 3-13), but in more recent times it was about 280 ppm from 500 B.C. to about A.D. 1790. Figure 2-2 shows the average CO_2 concentration in the atmosphere from 1744 to 2000 as estimated from measurements made on Antarctic ice cores and in the atmosphere. Measurements made in the atmosphere at Mauna Loa, Hawaii, indicate that the average concentration of CO_2 was 315 ppm by 1958 and about 370 ppm by 2000. Figure 2-3 shows the data obtained at Mauna Loa between

[1]The carbon cycle, which includes carbon dioxide, is discussed in greater detail in Section 10.2.

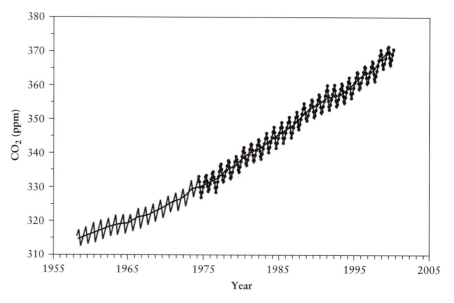

FIGURE 2-3 Monthly variation of atmospheric concentrations of carbon dioxide at Mauna Loa Observatory. Data prior to May 1974 are from the Scripps Institution of Oceanography; data since May 1974 are from the National Oceanic and Atmospheric Administration, Climate Monitoring and Diagnosis Laboratory Carbon Cycle–Greenhouse Gases Branch. http://www.cmdl.noaa.gov.

1958 and 2000. Mauna Loa is situated at a latitude of approximately 19°N. Note that there are seasonal variations in the carbon dioxide content of the atmosphere over Mauna Loa: less CO_2 is found in the summer, the growing season in the Northern Hemisphere. Hawaii may seem rather far south to feel the effects of this; more northerly areas like Scandinavia and northern Alaska show seasonal variations that have twice the amplitude of those over Mauna Loa. Seasonal variations in CO_2 concentration in the Southern hemisphere are much smaller.

At present, about 1.9×10^{10} metric tons of "new" CO_2 are released into the atmosphere every year, mostly from the burning of fossil fuels; this is estimated to be about 10% of the amount used by green plants in the same time. About 47% of this new CO_2 is absorbed by the oceans rather quickly; given enough time, it is possible that the oceans would absorb much more. More details on the distribution of CO_2 in nature and on the carbon cycle in general are given in Section 10.2. At this time, the atmospheric concentration of CO_2 appears to be increasing at the rate of 1.3 ppm per year. The possible effects of this increase on the earth's climate are discussed in Chapter 3.

Carbon monoxide as well as CO_2 is released to the atmosphere by the burning of fossil fuels, and also by volcanic and biological activity. It is possible that carbon monoxide is accumulating in the atmosphere above the Northern Hemisphere at this time, but the data are too variable for definitive conclusions. It is known, however, that carbon monoxide cannot be accumulating as rapidly as it is emitted into the atmosphere, and thus there has been much interest in the origin and fate of atmospheric CO. It appears that about 50% of the carbon monoxide entering the atmosphere each year comes from natural processes. Estimates of carbon monoxide entering the atmosphere each year made since 1990 are as follows: over 10^9 metric tons from natural causes such as the oxidation of methane arising from the decay of organic matter, the biosynthesis and degradation of chlorophyll, and release from the oceans, and over 10^9 metric tons from human activities. These figures indicate that removal of CO from the atmosphere must be occurring with great efficiency. Carbon monoxide reacts with hydroxyl radicals in the lower atmosphere (see Section 5.3.2), and there is evidence that fungi and some bacteria in the soil can utilize CO, converting it to CO_2. However, because of its toxicity, CO can be a major problem in areas in which its concentration is elevated. For example, although the worldwide concentration of CO varies between 0.1 and 0.5 ppm, its concentration in the air over large cities can be 50–100 times as great, up to 50 ppm. It has been difficult to quantify the toxic effects on humans due to the CO in the atmosphere above large cities because the CO concentration in automobiles in heavy traffic can be much higher and because smokers inhale large quantities of CO. However, in one study[2] it was shown that 20.2 ppm of CO above Los Angeles County in 1962–1965 resulted in 11 more deaths per day than occurred when the CO concentration was 7.3 ppm.

Sulfur and nitrogen compounds in the atmosphere are discussed at greater length in Chapter 10. At this point, we shall simply say that the sulfur compounds (mostly SO_2) that arise from the burning of fossil fuels come back to earth in the form of sulfuric acid dissolved in rainwater after undergoing further oxidation in the atmosphere (see Chapter 5). Some of the nitrogen oxides also dissolve in rainwater to form acids. Normal rainwater, which always contains CO_2 as well as some naturally occurring acids, may have a pH as low as 5.0 (see Section 11.4). In the last 50 years, however, rain and snow of much lower pH, that is, higher acidity, has been reported from many parts of the world. Rainfall in the northeastern United States now usually has a pH between 4.0 and 4.2, but individual rainstorms may have a pH as low as 2.1. Major industrial sources of atmospheric sulfur compounds are generally situated upwind of areas that have acid rain, and individual storms that have rain with very low pH have usually passed over large sources

[2] A. C. Hexter and J. R. Goldsmith, *Science*, **172**, 265 (1971).

of these compounds. Tracer experiments, in which a marker gas such as sulfur hexafluoride has been injected into a waste gas stream from a chimney and followed by an aircraft containing appropriate instrumentation, have shown that such gases do not disperse where they first appear but move with the prevailing winds. For example, gases released from the east coast of England have been followed to Scandinavia. The Scandinavian countries, which are downwind not only from England but also from Germany's heavily industrialized Ruhr Valley, have had rain with pH as low as 2.8. Even in South America's Amazon Basin, with very little nearby industrial activity, pH values as low as 4 have been recorded. Acid rains in Europe and elsewhere have led to rapid weathering of ancient structures and sculptures during the latter half of this century; some of these structures, such as the Parthenon in Athens, had stood virtually unchanged for hundreds and, in some cases thousands of years. Some of the chemistry involved is discussed in Section 9.2.2. In Germany and elsewhere, ancient forests are dying, presumably because of acid rain. The effects of acid rain on lakes, running waters, and fish are described in Section 11.4.

Methane is a trace gas that has received considerable attention in recent years. Studies of Antarctic ice cores indicate that the atmospheric methane concentration was as variable as the carbon dioxide concentration over the past 160,000 years (see Figure 3-13). However, from 25,000 B.C. to A.D. 1580, the methane content of the atmosphere was approximately constant; it began increasing at that time up to the present value of about 1.75 ppm. This increase was as large as 0.016 ppm/year in 1991; with some large oscillations, this rate of increase has dropped between 1991 and 1999. Natural sources of methane include anaerobic bacterial fermentation in wetlands and intestinal fermentation in mammals and insects such as termites, and these account for most of the methane entering the atmosphere. The "sinks" for atmospheric methane are not well understood, and the impact of human activities on methane formation and destruction has not been ascertained. The increase in methane content of the atmosphere can have important consequences for the earth's climate (see Section 3.3.3).

2.3 HISTORY OF THE ATMOSPHERE

2.3.1 Evidence and General Theory

Unfortunately, there is no agreement in the literature either on the composition or on the time of origin of the earth's original atmosphere. Evidence for the composition of the atmosphere at earlier times comes from a study of objects that were formed in contact with the atmosphere at those times. These objects may have changed since, but they still provide us with clues about the conditions existing when they were formed. The earliest such objects in existence are sedimentary rocks about 3×10^9 years old. It is believed that these rocks could not have been formed without prior atmospheric weathering, or without large amounts of liquid water. Therefore, water was probably present in the primitive atmosphere. On the other hand, the compositions of these and later rocks indicate that the atmosphere could not have contained much oxygen before 1.8×10^9 years ago. Also, the relative rarity of carbonate rocks such as limestone and dolomite among the oldest sedimentary rocks seems to indicate that bases such as ammonia cannot have been abundant in the primitive atmosphere. The presence of ammonia would have increased the pH of the primitive ocean and thus favored an abundance of carbonate rocks (see Chapter 9). Furthermore, in the absence of atmospheric oxygen and ozone, methane and ammonia would be rapidly dissociated after absorbing high-energy ultraviolet radiation. In this chapter, therefore, we shall assume the absence of large amounts of CH_4 and NH_3 from the primitive atmosphere. According to another school of thought, however, CH_4 and NH_3 must have been present in the primitive atmosphere, and evidence presented thus for for their absence has been insufficient. Moreover, experiments have been performed to show that an atmosphere dominated by CH_4 and NH_3 would be suitable for the evolution of complex organic molecules. Discussion of these conflicting viewpoints is beyond the scope of this book.

A combination of the geologic record and the present composition of the atmosphere, along with some reasonable conjectures, leads to the conclusion that the primitive atmosphere probably contained N_2, H_2O, and CO_2 in approximately their present amounts plus a small percentage of H_2 and a trace of oxygen. Volcanic gases contain H_2O, CO_2, SO_2, N_2, H_2, CO, S_2, H_2S, Cl_2, and other components. Of these, H_2O is the most abundant, followed by CO_2 and SO_2 or H_2S, depending on the volcano. Presumably, in the distant past, there was more volcanic action and thus the earth degassed somewhat faster than at present. The rate of "degassing" of the earth has varied throughout its history.

The earth at present is emitting mostly water vapor and carbon dioxide, some nitrogen and other gases, and no oxygen. These data must be reconciled with the composition of the atmosphere as given in Table 2-1. Calculations[3] indicate that the total quantity of water vapor released by the earth over all time is about $3 \times 10^5 \, g/cm^2$ averaged over all the earth's surface. Most of this water vapor has condensed and is presently in lakes and oceans. The oxygen now present in the atmosphere has probably arisen from photosynthesis after the evolution of green plants.

---

[3]F. S. Johnson in S. F., Singer, ed., *Global Effects of Environmental Pollution.* Springer-Verlag, New York, 1970.

2.3.2 Carbon Dioxide

Carbon dioxide has been emitted to the extent of about 5×10^4 g/cm^2 of the earth's surface over all time. Most of this CO_2 has dissolved in the oceans and much has precipitated from the oceans as calcium carbonate. Only about 0.45 g/cm^2 CO_2 remains in the atmosphere, while about 27 g/cm^2 averaged over all the earth's surface is dissolved in the oceans. There is 60 times as much CO_2 dissolved in the oceans as there is free in the atmosphere, but most of the carbon dioxide released by the degassing of the earth is now tied up in geologic deposits (i.e., carbonate rocks).

There are complex equilibria, discussed in Chapter 9, between the atmospheric CO_2 and that dissolved in the oceans. Recent experiments have shown that it takes about 470 days to dissolve half the CO_2 placed in contact with turbulent seawater. Complete equilibrium takes much longer, with estimates ranging from 5 to 500 years, but, given sufficient time, the oceans and rocks act as an enormous sink for any CO_2 that is added to the atmosphere. That is, most of the CO_2 added to the atmosphere eventually ends up in the oceans and in geologic deposits. However, we are now adding CO_2 to the atmosphere by burning fossil fuels in the amount of about 1.9×10^{10} metric tons per year (see Section 2.2); this adds to the 2.6×10^{12} metric tons of CO_2 already present in the atmosphere. Some of the added CO_2 will remain in the atmosphere even when atmosphere–ocean equilibrium has been achieved. The World Energy Council estimated that the amount of all recoverable fossil fuels was about 4.6×10^{12} metric tons[4] in 1993. This would allow 1.9×10^{10} metric tons to be burned each year for over 300 years. However, the provisions of the Kyoto Protocol may limit the amount of fossil fuels burned in the future so that the earth's supplies could last much longer than 300 years. The concentration of carbon dioxide in the atmosphere in the nineteenth century was about 290 ppm, while in 2000, as mentioned in Section 2.2, it was about 370 ppm. The increased concentration in this century is well documented and will be discussed later, together with its possible effects on climate. Much carbon dioxide is used by green plants in photosynthesis, but this carbon dioxide is returned to the atmosphere by decay and oxidation in living things.

Production of CO_2 from fossil fuels uses atmospheric oxygen, but this has little effect on the oxygen content of the atmosphere. For example, if all easily recoverable fossil fuels were burned at once, calculations indicate that only 10% of the atmospheric oxygen (a total of 5×10^{14} metric tons) would be used up.

---

[4]This is given in terms of the equivalent mass of petroleum, since fossil fuels include soft and hard coals as well as petroleum.

2.3.3 Nitrogen

Compared with CO_2, relatively small amounts of nitrogen have been released by degassing of the earth, only about $10^3\,g/cm^2$ of the earth's surface. But nitrogen is chemically inert, and at least 90% of this amount is still present in the atmosphere. The nitrogen no longer present in the atmosphere is mostly present in geologic deposits; the tiny fraction (10^{-8}) of the available atmospheric nitrogen that is "fixed" each year is soon returned to the atmosphere when plants and animals decay. The nitrogen cycle is discussed more fully in Section 10.3.

2.3.4 Oxygen

We have seen that atmospheric oxygen does not arise directly out of degassing of the earth but comes from photosynthesis in green plants and, possibly, from the photodissociation of water vapor in the upper atmosphere as shown in Equation 2-1.

$$H_2O\ (g) \xrightarrow{h\nu} H_2\ (g) + (1/2)O_2\ (g) \qquad (2\text{-}1)$$

Thus, during the photodissociation of water vapor, hydrogen gas is produced as well as oxygen gas. Hydrogen molecules, being much less massive than other atmospheric gas molecules, have a higher speed than these molecules at any temperature anywhere in the atmosphere, and many have a velocity greater than the escape velocity from earth's gravitational field. It has been calculated that hydrogen gas escapes at the rate of approximately 10^8 atoms/s per square centimeter of earth's surface. Over all time, about 10^{17} s, then, 10^{25} atoms/cm^2, about $17\,g/cm^2$, has escaped, which accounts for up to about $150\,g/cm^2$ of dissociated water vapor. This number corresponds to about 0.05% of the total water vapor released during the degassing of the earth; however, some of the escaped hydrogen was probably of volcanic origin. Furthermore, there is controversy over whether enough water vapor is transported to the upper layers of the atmosphere to allow photodissociation to take place.

The relative importance of photodissociation of water vapor versus photosynthesis is in dispute, with past estimates ranging from 100% photodissociation to 100% photosynthesis. In recent years, photosynthesis has been assumed to play the major role. In the present discussion, it is immaterial where all the oxygen now present in the atmosphere, in living things, and in oxidized rocks and minerals originated.

At what rate did oxygen accumulate in the atmosphere? There is some fossil evidence that the first anaerobic forms of life originated about 3.5×10^9 years ago. These life-forms may have released some oxygen into the atmosphere, adding to that produced by the photodissociation of water vapor. About 1% of the present amount of oxygen had accumulated in the atmosphere 6×10^8 years ago. At this point, respiration could begin in living things, as well as, ozone formation in the upper atmosphere, as will be discussed in Section 2.5. The atmospheric ozone screens certain ultraviolet wavelengths completely from the surface of the earth at present. When the oxygen level had reached 1% of its present value, screening of these harmful wavelengths was sufficient to permit the development of a variety of new life-forms in the oceans. By the start of the Silurian era, 4.2×10^8 years ago, the earth's oxygen level had reached about 10% of its present level, and sufficient ozone was present in the upper atmosphere to allow life to evolve on land. Most of the oxygen that has been produced in all this time has been used up in the oxidation of rocks and other surface materials on earth. However, more oxygen is still being produced, both by photodissociation and by photosynthesis.

2.4 PARTICULATE CONSTITUENTS OF THE ATMOSPHERE

Table 2-1 does not include the "particulate matter," dust, or, in general, the nongaseous matter that is usually present in the atmosphere. These are particles, which are $10^{-7} - 10^{-2}$ cm in radius, that is, from approximately molecular dimensions to sizes that settle fairly rapidly. Particulate matter in the atmosphere can be natural or man-made. Table 2-2 gives some estimates of the tonnage of particles with radii less than 2×10^{-3} cm (20 μm) emitted to the atmosphere each year.

Table 2-2 indicates that particles of human origin constitute between 5 and 45% of the total appearing in the atmosphere annually at the present time. Other estimates are as high as 60% These estimates cover a large range because some occurrences, like volcanic action, are inherently variable and because some of the other sources of particulate matter are very difficult to estimate.

Large particles can settle out of the atmosphere by sedimentation in the earth's gravitational field. Furthermore, it is estimated that approximately 70% of the particulate matter that enters the atmosphere is eventually rained back out, probably because the particles act as condensation nuclei for the water droplets that form rain clouds. The particles are accelerated by the gravitational force but retarded by friction with the gas molecules of the air, thus reaching a terminal sedimentation velocity which is related to the radius of each particle as shown in Table 2-3, which, by the way, refers to water droplets as well as particles. Water droplets with radii from 1 to 50 μm may be present in clouds, while droplets with radii of 100 μm generally fall as a drizzle. Larger droplets fall as rain.

TABLE 2-2

Estimate of Particles Emitted into or Formed in the Atmosphere Each Year

Particle sources	Quantity (megatons/year)
Natural	
Soil and rock debris	100–500
Forest fires and slash burning	3–150
Sea salt	300
Volcanic action	25–150
Gaseous emissions	
Sulfate from H_2S	130–200
NH_4^+ salts from NH_3	80–270
Nitrate from NO_x	60–430
Hydrocarbons from vegetation	75–200
Natural particle subtotal	773–2200
Human origin	
Direct emissions, smoke, etc.	10–90
Gaseous emissions:	
Sulfate from SO_2	130–200
Nitrate from NO_x	30–35
Hydrocarbons	15–90
Human origin particle subtotal	185–415
Grand total	958–2615

Source: Reprinted, with some modification, from *Inadvertent Climate Modification, Report on the Study of Man's Impact on Climate (SMIC)*, MIT Press, Cambridge, MA, 1971.

The smallest solid particles, many with radii of only 0.1–1.0 μm, have a maximum concentration about 18 km above the earth's surface. Some of these arise from volcanic eruptions, while others are particles of ammonium sulfate formed when H_2S and SO_2 in the atmosphere are oxidized and the resulting

TABLE 2-3

Sedimentation Velocity of Particles with Density 1 g/cm^3 in Still Air at 0°C and 1 atm Pressure

Radius of particle (μm)	Sedimentation velocity (cm/s)
< 0.1	Negligible
0.1 (10^{-5} cm)	8×10^{-5}
1.0	4×10^{-3}
10.0	0.3
100.0 (10^{-2} cm)	25

H_2SO_4 neutralized another atmospheric constituent, ammonia. The mean residence time of particulate matter in the lower atmosphere is between 3 and 22 days. In the stratosphere it is 6 months to 5 years, in the mesosphere from 5 to 10 years. (The meaning of "stratosphere" and "mesosphere" will be explained in the next section.) At any rate, the higher in the atmosphere—that is, the farther above the surface of the earth the particles have reached—the longer it takes them to come down.

Particles with radii from 0.02 to 10 μm may contribute to the turbidity of the atmosphere. We shall see later that the turbidity of the atmosphere is an important determinant of the earth's climate. The atmosphere has become more turbid over the past half-century, as shown by monitoring both in urban locations such as Washington, D.C., and in the mountains, such as Davos, Switzerland.

2.5 EXTENT OF THE ATMOSPHERE AND ITS TEMPERATURE AND PRESSURE PROFILE

Roughly 5.1×10^{18} kg of atmosphere is distributed over the 5.1×10^{14} m^2 of the earth's surface. This means that atmospheric pressure at sea level is about 10^4 kg/m^2, or 10^3 g/cm^2. This is approximately equivalent to one standard atmosphere of pressure (1013 mbar). Most of the atmosphere is fairly close to the surface of the earth. Fifty percent of the atmosphere is within 5 km (3 mi) of sea level, that is, at a height of somewhat more than 5 km above sea level, atmospheric pressure is 0.5 atm. There are quite a few mountains that are higher than this. One such mountain is Kilimanjaro in Tanzania, which rises nearly 6 km above sea level; Mount Whitney and the Matterhorn are just under 5 km in altitude.

Ninety percent of the atmosphere is within 12 km of sea level. By comparing 12 km with the radius of the earth, 6370 km, it is seen that 90% of the atmosphere exists in an extremely thin layer covering the earth's surface. The total extent of the atmosphere is much harder to specify, since there is no definite upper boundary. For most purposes, one can say that the atmosphere ends between 200 and 400 km above the earth's surface, but on occasion it is necessary to consider a few hundred or a thousand additional kilometers.

Nitrogen, oxygen, and carbon dioxide are reasonably well distributed at all altitudes in the atmosphere. Most of the water vapor and water droplets occur at altitudes below about 14 km, while most of the ozone is localized in the vicinity of 25 km above the earth's surface. There is, however, some water vapor at altitudes above 14 km, about 3 ppm in the stratosphere.

The exact manner in which temperature and pressure change with altitude in the atmosphere depends both on latitude and on the season of the year. Very close to the surface of the earth, temperature and pressure are extremely variable, even beyond the latitude and seasonal variations. For this reason, various national and international groups have proposed "standard atmospheres" at various times. For illustrative purposes (see Table 2-4), we shall use the U.S. Standard Atmosphere, 1962, which depicts idealized year-round mean conditions for middle latitudes such as 45°N, for a range of solar activity that falls between sunspot minimum and sunspot maximum. Standard information for other latitudes and for various times of year can be found in the Additional Reading at the end of this chapter. Table 2-4 shows that the pressure drops off continuously with increasing altitude, but the temperature goes through some strange gyrations. The temperature of the atmosphere decreases from sea level

TABLE 2-4

Temperature and Pressure of the Standard Atmosphere

Altitude (km)	P (mbar)	T (K)
0	1013	288 (15°C)
1	899	282
2	795	275 (2°C)
5	540	256
10	265	223 (−50°C)
12	194	217
15	121	217 (−56°C)
20	55	217
25	25	222
30	12	227
40	3	250
50	0.8	271 (−2°C)
60	0.2	256
70	0.06	220
80	0.01	180
90	0.002	180 (−93°C)
100	0.0003	210
110	7×10^{-5}	257
120	3×10^{-5}	350 (77°C)
130	1×10^{-5}	534
140	7×10^{-6}	714
150	5×10^{-6}	893 (620°C)
175	2×10^{-6}	1130
200	1×10^{-6}	1236
300	2×10^{-7}	1432
400	4×10^{-8}	1487
500	1×10^{-8}	1500 (1227°C)

Source: Adapted from *U.S. Standard Atmosphere*, U.S. Government Printing Office.

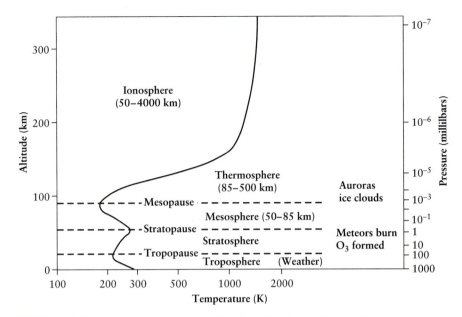

FIGURE 2-4   Temperature and pressure profile of the U.S. Standard Atmosphere.

up to 12 km, remains constant up to 20 km (at least in the U.S. Standard Atmosphere, 1962), increases from 20 to 50 km, decreases again from 50 to 80 km, remains constant to 90 km, and then increases asymptotically to about 1500 K above 300 km. These variations are probably less confusing as depicted in Figure 2-4, which also shows the change in pressure with altitude. Note that the temperature scale in Figure 2-4 is logarithmic, to make it easier to see the temperature variations in the first 100 km of the atmosphere.

Starting at the earth's surface, at sea level, the temperature of the standard atmosphere decreases steadily in the troposphere (sometimes referred to as the lower atmosphere) up to the tropopause. We shall see in Chapter 3 how temperature inversions, (i.e., regions where the temperature increases with increasing altitude) can sometimes occur in the troposphere. The troposphere contains most of the mass of the atmosphere, it is the only continuously turbulent part of the atmosphere, and it contains most of our weather systems. Atmospheric pressure drops off logarithmically with altitude near the earth's surface, but more slowly at higher elevations, and there are no pressure inversions.[5]

[5]The thermodynamic concept of an equilibrium temperature may not be valid at elevations above 350 km where the pressure is less than 10^{-7} mbar. The very few molecules present, seldom collide, and thus they probably are not at equilibrium. Thus, the temperature at altitudes above 300 km should be considered to be a mean kinetic temperature, having to do with the average kinetic energy of the molecules present.

Wherever the temperature rises with an increase in altitude in the atmosphere, there must be one or more exothermic (i.e., solar-energy-absorbing) chemical reactions involved. Without such processes, the temperature of the atmosphere would vary smoothly from warm at the earth's surface to very cold, close to absolute zero, at very high altitudes. Radiant energy from the sun is absorbed in various photochemical processes both in the stratosphere and in the thermosphere, thus increasing the temperature with increasing altitude as shown in Figure 2-4.

The photochemical processes of the thermosphere will be fully discussed in Chapter 5. At this point, let us simply note that atmospheric oxygen and nitrogen in the thermosphere above 100 km absorb virtually all the ultraviolet radiation having wavelengths below 180 nm (1800 Å). This circumstance is very fortunate for two reasons. First, this high-energy radiation would initiate chemical reactions in all complex molecules, with the result that life as we know it would be impossible if this radiation reached the earth's surface. Second, the large variations in the intensity of the sun's radiation at these short wavelengths have very little effect on the troposphere where our weather originates, since all these wavelengths are removed at much higher altitudes.

Radiation with wavelengths between 180 and 290 nm would also be destructive to life, but these wavelengths are absorbed in the stratosphere by oxygen and ozone. The maximum absorption of ozone is at 255 nm (see Section 5.2.3.1). The major reactions in the stratosphere, much simplified from the more complete discussion to be found in Section 5.2.3.1, are as follows:

$$O_2 \xrightarrow{h\nu} O + O \tag{2-2}$$

$$O_2 + O + M \rightarrow O_3 + M^* \tag{2-3}$$

$$O_3 \xrightarrow{h\nu} O_2 + O \tag{2-4}$$

$$O + O_3 \rightarrow 2O_2 + \text{kinetic energy} \tag{2-5}$$

Reaction (2-2) also occurs at higher altitudes, but does not lead to reaction (2-3) because the pressure is too low to allow termolecular collisions to occur with any reasonable frequency. A third atom or molecule M is needed for reaction (2-3), so that both energy and momentum can be conserved in the reaction. The symbol M^* refers to the increase in the energy of M while O_2 is combining with O to form O_3.

Ozone is formed down to about 35 km altitude, below which the ultraviolet radiation necessary for its formation ($\lambda < 240$ nm) has been fully absorbed. Most of the ozone in the atmosphere, however, exists at altitudes between 15 and 35 km, so that mixing processes must occur in the stratosphere as well as in the troposphere. Also, ozone concentrations are low over the equator, where, on the average, most of the sun's radiation impinges, and high near the poles, at latitudes greater than 50°, where less radiation comes in from the sun. Consequently, it is obvious that there are mechanisms for the transport of ozone from the equator to the poles. Furthermore, there are transitory losses of ozone over the Antarctic and (sometimes) Arctic regions during their early springtimes (see Sections 5.2.3.3 and 5.2.3.4, respectively). This ozone appears to be replenished during the Antarctic summer by transport from the stratosphere over the temperate regions of the southern hemisphere. Transport of gases through the tropopause and from one region of the stratosphere to another occurs readily; this transport has seasonal aspects and the mechanism is quite complicated and is still being studied.

Solar radiation with wavelengths greater than 290 nm (2900 Å) is transmitted to the lower stratosphere, the troposphere, and the earth's surface. We shall see that this remaining radiation, including UV with wavelengths exceeding 290 nm, visible and infrared, comprises more than 97% of the total energy from the sun.

2.6 GENERAL CIRCULATION OF THE ATMOSPHERE

Only very general patterns of atmospheric movement are within the scope of this chapter, where they are pertinent because we are interested in discussing general movements of pollutants and other constituents of the atmosphere. Atmospheric circulation will thus be considered in an extremely simplified manner. Books on meteorology, such as the ones given in the Additional Reading at the end of this chapter, may be consulted for further details.

There are two major reasons for the existence of a general circulation of the atmosphere. First of all, the equatorial regions on earth receive much more solar energy all year round than the polar regions. Hot air therefore tends to move from the equator to the poles, while cold air tends to move from the poles toward the equator. As we shall see, the surface of the earth absorbs a much larger proportion of the solar radiation than the atmosphere above it. Thus, in the troposphere, the air is heated by conduction, convection, and radiation from the ground. There is also some heating of the atmosphere when water vapor, recently evaporated from the earth's surface, condenses to form clouds.

If we recall that hot air is less dense than cold air and tends to rise above it (this will be discussed in Chapter 3 in connection with temperature inversions, Section 3.5), we shall expect hot air to rise near the equator and flow toward the poles at some elevation above the surface of the earth. The colder air from the polar regions is then expected to flow toward the equator nearer to the surface of the earth, as shown in Figure 2-5. This idea was first formulated by G. Hadley in 1735. Hadley's picture of atmospheric circulation also took account of the earth's rotation, but for the moment we should note the

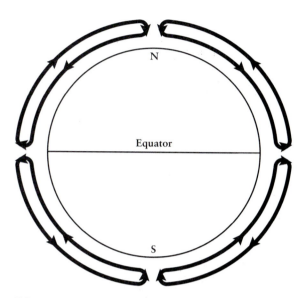

FIGURE 2-5 General circulation of the atmosphere as proposed by Hadley, 1735. These were the simplest forms of "Hadley cells," with hot air rising near the equator and moving toward the poles in the upper troposphere. Cold air, in this model, moves from the poles toward the equator near the surface of the earth.

following major oversimplifications in his model. First, this model showed a lot of air accumulating over the polar regions; second, no account was taken of unequal heating of land and sea; and, third, no account was taken of variations in solar radiation at various latitudes at different times of the year.

What happens when the rotation of the earth from west to east is taken into account? If we think of the earth as rotating around an axis that goes through the North and South Poles, we note that the surface of the earth at the poles does not rotate, while the surface of the earth at the equator rotates very quickly. The equatorial circumference of the earth is about 40,000 km, and one rotation occurs every 24 h, so that every point on the earth's equator is constantly moving at a velocity of 28 km/min in contrast to the stationary points at the poles. In the same way, the air above the poles does not rotate while the air above the equator rotates with the earth, and at the same velocity. Between the equator and the poles, points on the earth's surface and in the atmosphere above these points rotate at intermediate speeds, between zero and 28 km/min.

When air moves between latitudes in which the speed of rotation is different, it is subject to the Coriolis effect, which gives the wind speed an unexpected component with respect to an observer on earth. For example, when a parcel of air moves from one of the poles toward the equator, it is moving from an area where little or no rotation is taking place to an area where the earth is rotating quickly beneath it. This parcel of air tends to lag behind the rotating earth, that is, the earth is rotating away from this parcel of air, moving faster than the parcel of air from west to east. From the perspective of an observer who is rotating with the earth, this parcel of air seems to be moving backward (i.e., from east to west). In the Northern Hemisphere, where the parcel of cold air near the surface of the earth is moving north to south because of unequal heating, the overall result is to give this parcel of air a northeast-to-southwest motion with respect to someone on the earth's surface. We should therefore expect surface winds in the Northern Hemisphere to blow generally from the northeast. Similar reasoning would lead us to expect surface winds in the Southern Hemisphere to blow from the southeast.

A simplified version of the actual prevailing wind patterns on the earth's surface is shown in Figure 2-6. These prevailing surface wind patterns are most noticeable over the oceans and were of great practical importance in the days of large sailing ships, when the major wind systems received their names. The doldrums near the equator and the horse latitudes were much feared by sailors, because a ship could remain becalmed there for weeks at a time. The various wind belts have a tendency to shift north and south at times, and the winds sometimes reverse because of topographical features, weather patterns, or simply the greater variability of some of these wind systems. The trade winds and the prevailing westerlies are fairly reliable, at least over the larger oceans, but the polar easterlies barely exist. The prevailing westerlies in both

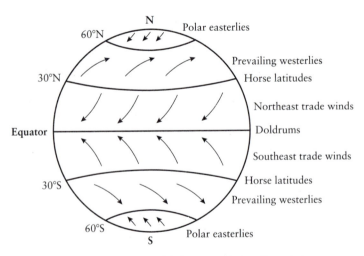

FIGURE 2-6   An extremely simplified version of prevailing wind patterns on the earth's surface. Polar easterlies, especially, are relatively insignificant.

hemispheres seem to be blowing in the wrong direction, at least according to Hadley's original ideas, moving from the southwest instead of the northeast in the Northern Hemisphere and from the northwest instead of the southeast in the Southern Hemisphere. Air is therefore flowing from the equator to the poles along the surface of the earth in these midlatitude regions, and is moving from a more rapidly rotating region of the earth's surface to a more slowly rotating region. Thus, these portions of the atmosphere are moving west to east faster than their destination, and the winds, which are a manifestation of this moving air, have a westerly component.

In summary, therefore, near the equator and near the poles, cold air is moving toward the equator near the surface of the earth with an apparent easterly (east-to-west) component, while, in the midlatitudes, warm air is moving from the equator toward the poles along the surface of the earth with a westerly (west-to-east) component. As stated earlier, these easterly and westerly components of winds moving between portions of the earth that rotate at different speeds are manifestations of the Coriolis effect.

The three main prevailing wind directions in each hemisphere have been explained in terms of a three-cell model (Figure 2-7) as distinguished from Hadley's original one-cell model (Figure 2-5). The current model has warm air rising at the equator and sinking near the poles just like the original Hadley model. The air circulation systems closest to the equator and to the poles move in the expected directions, with cold air traveling toward the equator near the surface of the earth, resulting in easterly winds. At latitudes 30°N and 30°S, these cells have downward moving air, while at 60°N and 60°S, they have upward moving air. To preserve these wind directions, the midlatitude cells have warm air moving from the equator toward the poles near the earth's surface, resulting in westerlies. The three-cell model, designed by Rossby in

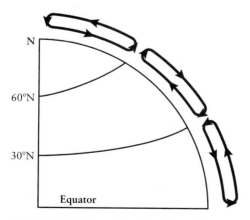

FIGURE 2-7   Three-cell model of general atmospheric circulation in the Northern Hemisphere.

1941, is also quite oversimplified because it gives the appearance that the cells are fixed in place and that there is almost no air circulation between cells. In fact, the boundaries of the cells move, and the mixing of parcels of air coming from different parts of the globe varies quite a bit with the seasons (see Section 5.2.3.3). Nevertheless, the three-cell model is a great improvement over the one-cell model because any model that has only easterly winds all over the globe is impossible. Such unidirectional winds would tend to slow down the earth's west-to-east rotation. In fact, easterly and westerly winds must balance on the average, since the atmosphere as a whole rotates with the earth. This fact was ignored in the single-cell model.

The simple scheme shown in Figures 2-6 and 2-7, although it explains the major prevailing wind systems on the earth's surface reasonably well, cannot adequately explain numerous other atmospheric phenomena. These include the so-called jet stream in the upper troposphere, which flows west to east, the systems of low and high pressure with circulating winds that constitute our weather systems, and the general west-to-east movement of these weather systems. It turns out that a good deal of energy is transmitted from the equatorial regions to the polar regions by these so-called cyclone and anticyclone weather systems. Further explanations are outside the scope of this chapter. Let us just note here the results of experiments with model liquid systems (i.e., suspensions of metal powders in liquids under conditions that simulate conditions in the earth's atmosphere) have shown, under certain conditions, flow patterns that correspond to the type of air circulation in cyclones and anticyclones and the slow drift of these flow patterns in the direction of motion of the fastest moving portion of liquid. The circulation experiments are carried out as follows. A liquid suspension is placed between two cylinders; the inner cylinder is cooled and stationary, while the outer cylinder is heated and rotating. The metal particles in the suspension can be seen in motion back and forth between the stationary cold cylinder wall and the rotating warm cylinder wall. Eddies sometimes form, and these can be seen moving in the direction of rotation of the outer cylinder, but much more slowly. Figure 2-8, taken from a report of one of these experiments, permits us to visualize the patterns that are observed. Both the steady waves and the irregular patterns in Figure 2-8 look somewhat like diagrams of the upper atmosphere jet stream seen with the North Pole at the center. The jet stream, observed in this way, has lobes going alternately north and south in which air travels generally west to east, in the direction of earth's rotation. The jet stream moves in the middle latitudes of the hemisphere.

In the continental United States, which is in the northern midlatitude zone, everything contributes toward a general west-to-east movement of the atmosphere. This is the region of prevailing westerlies, the jet stream moves west to east, and all major weather systems move west to east. This is why so many sulfur and nitrogen compounds from industrially polluted air seem to end up in

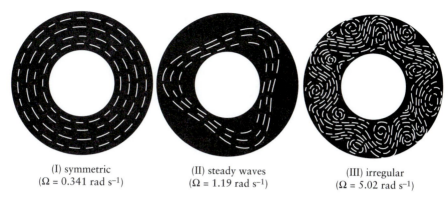

(I) symmetric
(Ω = 0.341 rad s^{-1})

(II) steady waves
(Ω = 1.19 rad s^{-1})

(III) irregular
(Ω = 5.02 rad s^{-1})

FIGURE 2-8 Drawings made from photographs illustrating three typical top-surface flow patterns of free thermal convection in a wall-heated rotating fluid annulus; inner wall at 16.3°C; outer wall at 25.8°C; working fluid, water. From R. Hide, "Some laboratory experiments on free thermal convection in a rotating fluid subject to a horizontal temperature gradient and their relation to the theory of global atmospheric circulation," In *The Global Circulation of the Atmosphere* (G. A. Corby, ed.). Royal Meteorological Society, London. Copyright © 1969. Used by permission of the Royal Meteorological Society.

rainfall over New England (northeastern sector), as mentioned earlier. Both the jet stream (in the upper troposphere) and our general weather systems contribute to the west–east motion of polluted air over the continental United States. There is also some north–south and south–north movement of polluted air at different times, especially between our industrial Midwest and south-central Canada. Furthermore, in Europe, polluted air seems to move generally southwest to northeast, causing problems for the Scandinavian countries which are thus in the path of pollutants transported from the United Kingdom and from central Europe.

At one time, people believed that virtually all the heat transfer from the equator to the poles occurred via the earth's atmosphere. It is now surmised that ocean currents like the Gulf stream perform about 50% of all heat transport between the equator and the poles; only half the heat is transported by the atmosphere.

2.7 CONCLUSION

In this chapter, we have examined the constituents of the earth's atmosphere and briefly discussed their origins. We have learned about the general temperature and pressure profile of the atmosphere and how and why, in general, it circulates around the earth. Up to this point, then, we have enough information to think about the answers to question 4 in Section 2.1. Winds, including the jet stream, occur mostly in the troposphere, but there must also be circulation patterns in the stratosphere to help distribute the ozone, which is produced mostly in the stratosphere above the tropics, to the regions above the poles.

Additional Reading (see also Chapter 3)

Calder, N., *The Weather Machine. How Our Weather Works and Why It Is Changing.* Viking, New York, 1974.

Deepak, A., ed., *Atmospheric Aerosols. Their Formation, Optical Properties, and Effects.* Spectrum Press, Hampton, VA, 1982.

The Enigma of Weather. A Collection of Works Exploring the Dynamics of Meteorological Phenomena. Scientific American, New York, 1994.

Finlayson-Pitts, B. J. and Pitts, J. N. Jr., *Chemistry of the Upper and Lower Atmosphere: Theory, Experiments, and Applications.* Academic Press, New York, 2000.

Inadvertent Climate Modification. A Report on the Study of Man's Impact on Climate (SMIC). MIT Press, Cambridge, MA, 1971.

Macalady, D. L., ed., *Perspectives in Environmental Chemistry.* Oxford University Press, New York, 1998.

Schneider, S. H., and R. Londer, *The Coevolution of Climate and Life.* Sierra Club Books, San Francisco, 1984.

Siskind, D. E., S. D. Eckerman, and M. E. Summers, eds., *Atmospheric Science Across the Stratopause,* American Geophysical Union, Washington, DC, 2000.

U.S. Standard Atmosphere, 1962; and *U.S. Standard Atmosphere Supplements,* 1966. U.S. Government Printing Office, Washington, DC.

Wayne, R. P., *Chemistry of Atmospheres,* 2nd ed. Clarendon Press, Oxford, 1991.

Walker, J. C. G., *Evolution of the Atmosphere.* Macmillan, New York, 1977.

EXERCISES

2.1. It is well known that the minimum altitude of the ionosphere is higher at night than during the day. Why might this be true?

2.2. Explain why the temperature of the earth's atmosphere increases
   (a) from about 15 to 50 km above the earth's surface, and
   (b) from about 85 km to the outer limits of the earth's atmosphere.

2.3. What are the three major influences that control earth's atmospheric circulation? Explain the effect of each of these.

2.4. The South Pole, during its midsummer, receives more solar energy during 24 h than any other place on earth, yet it remains extremely cold. Why might this be true?

2.5. A few air pollution experts have accumulated a small amount of evidence indicating that some of the smog particles from Los Angeles eventually come to earth in the vicinity of Albany, New York.

(a) Is this possible? Why or why not?

(b) Is this probable? Why or why not?

2.6. What compounds in the earth's atmosphere are considered to be responsible for acid rain? Why is rain with pH 5.5 not considered to be acid rain?

2.7. Make a sketch of the temperature profile of the earth's atmosphere without looking at Figure 2-4. The ordinate is altitude and the abscissa is temperature. Label the main parts of the atmosphere. Put in approximate temperatures in appropriate places.

2.8. There are several portions of the atmosphere in which the temperature change with altitude indicates that chemical reactions must be taking place. Name each of these parts of the atmosphere, and briefly describe the types of reaction taking place. Indicate why the reactions are different in different portions of the atmosphere.

3

ENERGY AND CLIMATE

3.1 ENERGY BALANCE OF THE EARTH

3.1.1 Incoming Radiation from the Sun

The discussion on the general circulation of the atmosphere in Chapter 2 presupposed that most of the heating of the earth's surface comes from solar radiation. The total solar energy reaching the surface of the earth each year is about 2×10^{21} kJ $(5 \times 10^{20}$ kcal$)$. Heat generated by radioactive processes in the earth and conduction from the core contribute 8×10^{17} kJ $(2 \times 10^{17}$ kcal$)$, and human activities contribute about 4×10^{17} kJ $(10^{17}$ kcal$)$ per year. This means that less than 0.1% of the total energy reaching the earth's surface each year comes from processes other than direct solar radiation. For the purposes of this chapter, therefore, we may treat the energy input to the earth as if it all came from the sun.

The sun is almost a blackbody radiator (with superimposed line spectra); a perfect blackbody is one that absorbs all radiation impinging on it; it also emits the maximum possible energy at any given temperature. The energy emitted from unit area of the blackbody per unit wavelength per unit solid angle in unit time, $B_\lambda (T)$, is given by Planck's blackbody equation:

$$B_\lambda(T)d\lambda = \frac{2hc^2 d\lambda}{\lambda^5[\exp(hc/\lambda kT) - 1]} \qquad (3\text{-}1)$$

where $k = 1.38 \times 10^{-23}$ J molecule^{-1} K^{-1} is Boltzmann's constant (the gas constant per molecule), h is Planck's constant, 6.63×10^{-34} J·s, c is the speed of light in vacuum, 3.00×10^8 m/s, λ is the wavelength of interest in meters, and T is the absolute temperature in kelvin.

Since the sun is emitting radiation mostly from its surface, and since we shall also wish to consider blackbody radiation from the surface of the earth, we shall rewrite equation (3-1) in the form

$$I(\lambda) = \frac{C_1}{\lambda^5[\exp(C_2/\lambda T) - 1]} \qquad (3\text{-}2)$$

where $I(\lambda)$ is the radiation intensity emitted by each square meter of surface of the blackbody at wavelength λ, T is the absolute temperature, $C_1 = 3.74 \times 10^{-16}$ W·m^2, and $C_2 = 1.438 \times 10^{-2}$ m·K (the symbol W refers to watts). The total intensity of radiation emitted by a blackbody at any temperature is given by the Stefan–Boltzmann law,

$$I = \sigma T^4 \qquad (3\text{-}3)$$

where $\sigma = 5.672 \times 10^{-8}$ W m^{-2} K^{-4} = 8.22×10^{-11} cal cm^{-2} min^{-1}K^{-4}.

The relative distribution of energy, proportional to equation (3-2), is plotted in Figure 3-1 as a function of wavelength for two temperatures. One may see that there is a wavelength of maximum emission λ_{max} that shifts to lower wavelengths as the temperature is increased. Most commonly, the wavelength of maximum blackbody emission is encountered in the infrared (e.g, the 3300 K curve in Figure 3-1 corresponds approximately to the output of a 200-W tungsten filament lamp). However, at 6000 K (roughly the blackbody temperature of the sun), λ_{max} is in the visible region (480 nm). Ninety percent of the solar radiation is in the visible and infrared, from 0.4 to 4.0 μm, with almost constant intensity (but see Section 3.3.1). The other 10% of the solar radiation intensity varies somewhat with time in the ultraviolet region and becomes extremely variable in the x-ray region of the spectrum. The wavelength at which maximum emission of radiation occurs at any temperature is given by Wien's displacement law,

$$\lambda_m T = 2897.8 \ (\mu\text{m·K}) \qquad (3\text{-}4)$$

Solar radiation is emitted in all directions from the sun, and very little of this reaches earth. In fact, earth is so far away from the sun that it picks up only about 2×10^{-9} of the total solar energy output. At a distance from the sun

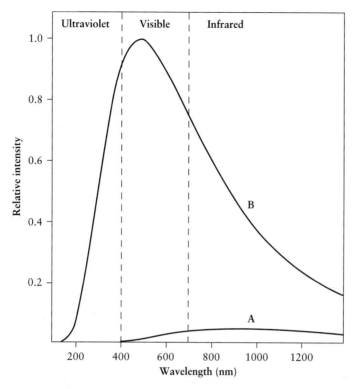

FIGURE 3-1 Distribution of energy from a blackbody radiator at 3300 K (curve A) and 6000 K (curve B).

equal to the average radius of the earth's orbit, the solar energy passing any surface perpendicular to the solar radiation beam is $1.367 \pm 4 \, \text{kW/m}^2$ $(1.95 \, \text{cal cm}^{-2} \, \text{min}^{-1})$; this is called the solar constant for the earth. Since the earth does not consist of a plane surface perpendicular to the path of the solar radiation but presents a hemispherical surface toward the sun, a recalculation of the solar constant must be made for radiation falling on this hemispherical surface. Actually, since we wish to use an average radiation intensity *averaged over the total surface of the earth*, we must compare the surface area of the whole earth, $4\pi r^2$, where r is the radius of the earth, with the area that the earth projects perpendicular to the sun's rays, πr^2 (see Figure 3-2). Thus, the earth's surface has four times the area of the circle it projects on a plane perpendicular to the sun's rays, if these rays are assumed parallel at such large distances from the sun. Therefore, the solar radiation that comes in toward the earth's surface is one-fourth of the solar constant, or approximately 343 W/m², averaged over the whole surface. These numbers indicate

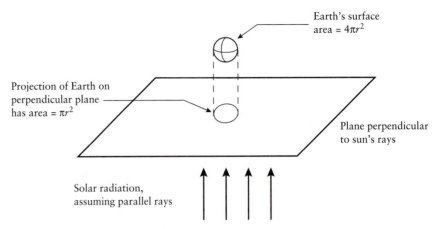

FIGURE 3-2 Comparison of earth's surface area with the area of earth's projection on a plane perpendicular to the sun's rays.

the total amount of solar radiation coming in to the top of the earth's atmosphere; some of this radiation is absorbed in the atmosphere and some is reflected.

The solar radiation that actually penetrates to the surface of the earth below the atmosphere no longer has the spectral distribution of black body radiation from a body at 6000 K (Figure 3-1). Figure 3-3 shows that no solar radiation with wavelength below 0.28 μm (280 nm) reaches the surface of the earth. We already knew this, from the discussion of absorption by oxygen in the thermosphere and by ozone and oxygen in the stratosphere (Section 2.5). The absorption of solar infrared radiation has not been discussed yet, but a large portion of the sun's infrared radiation is absorbed by water vapor, carbon dioxide, and trace gases in the atmosphere. Figure 3-4 specifically shows the absorption of solar radiation by water and carbon dioxide in the atmosphere; one can see that this absorption is complete at some wavelengths.

The absorption of solar infrared radiation by various constituents of the atmosphere is not very important because comparatively little of the solar radiation is in the infrared (Figures 3-1 and 3-3). These absorptions become much more important when we consider radiation emitted by the earth. It should be fairly obvious that the earth is emitting radiation because, on the average, the earth is in thermal equilibrium. It is in the path of 2×10^{21} kJ of solar radiation each year, and it is necessary that 2×10^{21} kJ per year leave the earth. If less energy than that leaves the earth, it will become hotter. Some of the solar energy, as we shall see, is immediately reflected, but some is absorbed and must be reemitted.

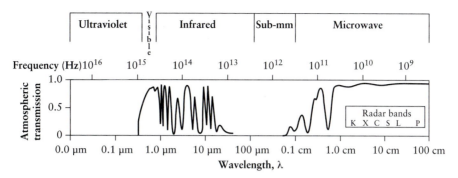

FIGURE 3-3 The transmission spectrum of the upper atmosphere: low transmission means high absorption. Redrawn from J. E. Harries, *Earthwatch, The Climate from Space*. Horwood, New York, Copyright © 1990.

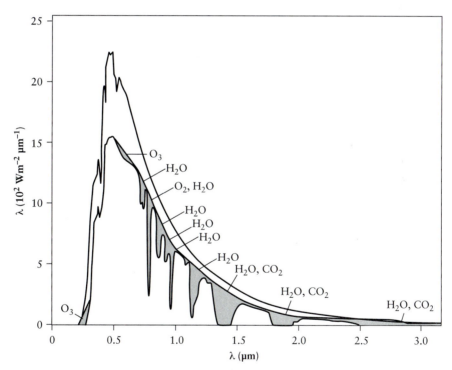

FIGURE 3-4 Spectral distribution of solar irradiation at the top of the atmosphere and at sea level for average atmospheric conditions for the sun at zenith. Shaded areas represent absorption by various atmospheric gases. Unshaded area between the two curves represents the portion of the solar energy backscattered by the air, **water** vapor, dust, and aerosols and reflected by clouds. Redrawn from J. P. Peinuto and A. H. Oort, *Physics of Climate*. American Institute of Physics, New York. Copyright © 1992 by Springer-Verlag GmbH & Co. Used by permission of the publisher.

3.1.2 The Greenhouse Effect

The present average temperature T of the earth's surface, taken at any one time over the whole surface of the earth, is close to 15°C (288 K). Thus, even though the temperature varies with time and place, the earth's surface must therefore radiate much like a blackbody at 15°C. Figure 3-5 shows the intensity of emission $I(\lambda)$ for blackbodies at 255 and 288 K. The wavelength for maximum radiation intensity is about 10 μm in this temperature range, that is, in the infrared region of the spectrum.

We should be able to find the average value of the earth's radiation temperature by calculation—by balancing incoming and outgoing radiation at the earth's surface. We shall call this mean radiation temperature T_r. Let us assume that energy in = energy out, ignoring radioactivity and human activities. We know from satellite measurements that the earth reflects about 30% of the total incoming solar radiation, so that only 70% is absorbed. Therefore, $I_{in} = I_{out} = 0.70 \times 343 \, \text{W/m}^2$. Using the Stefan–Boltzmann law, equation (3-3), we have

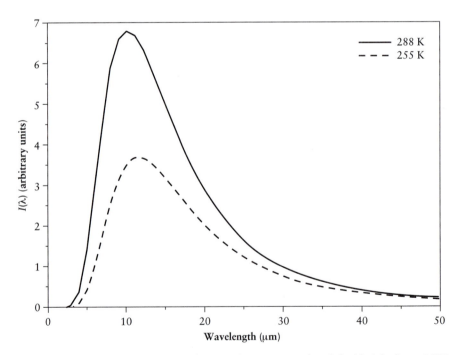

FIGURE 3-5 Radiation intensity (in arbitrary units) versus wavelength for black bodies at 255 K (dashed curve) and 288 K (solid curve).

$$T_r = \left[\frac{0.7 \times 343 \, \text{W/m}^2}{5.67 \times 10^{-8} \, \text{W m}^{-2} \, \text{K}^{-4}} \right]^{1/4} = 255 \, \text{K} = -18°\text{C} \qquad (3\text{-}5)$$

Therefore, to satisfy the law of conservation of energy, the earth–atmosphere system must be radiating energy into space at an average of $-18°$C. Therefore, the earth cannot be radiating energy out into space directly from its surface, at least on the average. Average temperatures in the neighborhood of $-18°$C are found in only two places from which radiation could logically occur: near the tropopause and in the region of the mesopause in the upper atmosphere. Since radiation cannot occur in the atmosphere unless many molecules are present to radiate, it is probable that most of the earth's radiation into space occurs from the atmosphere near the tropopause, probably the upper regions of the troposphere. This leaves us with two related questions: Why does the earth radiate to space mostly from the atmosphere and not from the surface, and why is the surface of the earth hotter than it should be, considering the energy balance calculations? That is, what traps heat between the surface of the earth and the tropopause?

Both these questions have to do with the absorption of infrared radiation by atmospheric water vapor, carbon dioxide, methane, and small amounts of other gases. Figure 3-5 shows that the earth's radiation is almost all in the infrared, whether we consider radiation from the surface (blackbody at $15°\text{C} = 288 \, \text{K}$) or from the atmosphere (blackbody at $-18°\text{C} = 255 \, \text{K}$). The absorption bands from 1.0 to 15.0 μm of CO, CH_4, H_2O, CO_2, and some of the other gases present in the earth's atmosphere are shown in Figure 3-6. The fraction of the earth's radiation absorbed at each wavelength depends on the amount of each gas present in the atmosphere and on the "strength" of each absorption band. Looking at Figure 3-6, one may observe that the gases normally in the atmosphere absorb in different regions of the infrared. In fact, the earth's infrared radiation is absorbed almost completely in many wavelength regions; this is true except in those wavelength regions that are called "atmospheric windows." These can be seen in Figures 3-3 and 3-6 as the regions in which the atmospheric transmission is close to 1.0 (Figure 3-3) or 100% (Figure 3-6). As we expect, Figure 3-3 shows that the atmosphere is almost completely transparent in the visible region of the spectrum, and that there are numerous complete and partial "windows" in the infrared. Both figures show that the region above 14 μm is almost completely opaque to infrared radiation, but the atmosphere is mostly transparent to the longer wavelength microwave radiation, especially the region that Figure 3-3 shows as radar bands.

Water vapor, which has more absorption bands than any of the other gases at almost all wavelengths shown in Figure 3-6, is the most important absorber of infrared radiation in the earth's atmosphere. After water vapor, CO_2 and O_3 are second and third in importance, respectively. The main atmospheric "window" (see Figures 3-5 and 3-6) is between about 8 and 12 μm except for a region

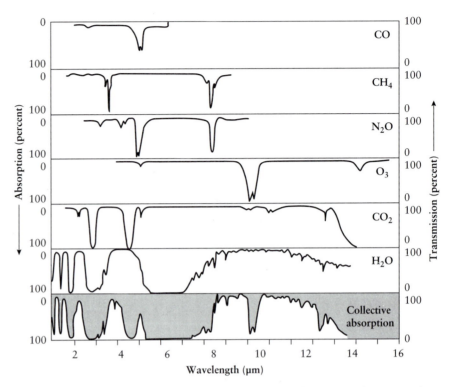

FIGURE 3-6 Absorption of some atmospheric gases in the infrared spectral region. Redrawn from M. L. Salby, *Fundamentals of Atmospheric Physics*. Copyright © 1996. Used by permission of Academic Press.

around 9.5 μm, in which O_3 is a strong absorber. We may note from Figure 3-6 that there are absorption bands of CO_2 in the atmospheric "windows" just below 5 μm and near 10 and 13 μm. CH_4 absorbs in the atmospheric "window" near 8 μm. This means that changes in the atmospheric concentration of these gases, and of others, including the chlorofluorocarbons discussed in Section 5.2.3.2, which can absorb more radiation in the infrared "windows," will be important in the discussion of global climate changes in Section 3.3.3.

When the sky is clear, terrestrial radiation in the wavelength region of the "windows" escapes directly to space. Clouds, when present, reflect this radiation back to the earth's surface and the radiation does not escape. However, on clear nights, especially in the winter, we have the phenomenon of "radiation cooling" of the earth's surface through the atmospheric "windows."

The infrared radiation that is absorbed in the troposphere and in the stratosphere would warm these regions of the atmosphere substantially if energy were not also radiated from these regions. This is the energy radiated at an average temperature of $-18°C$ (this average includes the higher temperatures at which radiation from the earth's surface escapes through the atmospheric windows) that leaves the earth. Thus, the atmosphere intercepts infrared radiation from the earth's surface and then reradiates the energy both back toward the earth's surface, thus warming it, and upward to space, thus cooling the earth so that the constant solar radiation does not slowly increase the earth's temperature. The earth's surface can therefore be much warmer than the average radiation temperature of the earth. This phenomenon is usually called the "greenhouse effect" because a related phenomenon is observed in a closed greenhouse. The glass or plastic windows of the greenhouse allow the visible solar radiation to enter the inside, but the upward-directed infrared radiation from the floor is absorbed or reflected by the same windows. In contrast to the "greenhouse effect" in the atmosphere, however, energy retention in the greenhouse is caused mostly by lack of convection (i.e., lack of mixing of the interior air with the surrounding atmosphere). The same effect occurs in a closed automobile that is left out in the sun. Both the greenhouse and the car interiors may thus become considerably warmer than the temperature of the surrounding atmosphere.

A greenhouse coefficient may be defined as the ratio of the surface temperature to the absolute radiation temperature of the whole system. For the earth, this coefficient is $288\,K/255\,K = 1.13$, while for the planet Venus it appears to be $743\,K/233\,K = 3.2$! Venus thus has a much larger greenhouse effect than earth; this is presently believed to be caused by carbon dioxide, sulfur dioxide, and possibly chlorine gas in the Venusian atmosphere. The Martian atmosphere contains mostly carbon dioxide but is so thin that the greenhouse coefficient is only $218\,K/212\,K = 1.03$. For quantitative discussions of the greenhouse effect on earth, it is convenient to define the greenhouse effect somewhat differently, as discussed and calculated in Section 3.3.1. Further discussions of the greenhouse effect on earth, and how this effect may vary with changes in the content of carbon dioxide and other trace gases in the atmosphere, are in Section 3.3.3.

3.1.3 Other Factors

The mean temperature of the earth's surface is certainly affected by the magnitude of the greenhouse effect, but it is also affected by the percentage of incoming solar radiation that is reflected back out to space, that is, by the earth's albedo. It has already been stated that approximately 30% of the incoming radiation is immediately reflected. This is an average value;

Table 3-1 shows the albedo of various major features on earth. Major changes in cloud cover, snow and ice, field and forest, and so on would change the average albedo of the earth. For example, a decrease in snow and ice cover would decrease the earth's albedo, thus leading to absorption of more solar radiation and, in the absence of other effects, a higher surface temperature for the earth. As it turns out, however, clouds also contribute to the greenhouse effect—they reflect some of the earth's infrared radiation back to earth instead of reradiating it like the atmospheric greenhouse gases. In the simplest picture, however, like the one considered in this book, clouds are considered only with respect to their reflectivity.

These and other factors in the energy balance of the earth are also important as shown schematically in Figure 3-7. The left-hand side of Figure 3-7 shows that 30% of the 343 W/m^2 that comes toward the earth from the sun is immediately reflected with no change, that water vapor and clouds in the troposphere and ozone and other ultraviolet absorbers remove an additional 20% of the total, and that the remaining 50% of the incoming radiation, about 172 W/m^2, is absorbed by the earth's surface. The central portion of Figure 3-7 shows the fate of the radiation from the earth's surface (at 15°C), about 390 W/m^2, which is 114% of the solar energy that enters the atmosphere (a result of the greenhouse effect, Section 3.1.2). Most of this energy is absorbed and reemitted (or reflected) by clouds and gases in the atmosphere. Other types of energy transfer also occur, as shown on the right-hand side of Figure 3-7. Latent heat transport refers to the heat of vaporization of water; water evaporates from the surface of the earth, thus cooling the surface, and then condenses in the atmosphere, thus heating the atmosphere. Sensible heat transfer refers to heat transfer by convection as heated air near the ground rises while cooler air higher in the troposphere sinks.

TABLE 3-1

Albedo of Various Atmospheric and Surface Features of the Earth

Albedo (% reflected)	Reflected from
30	Total (earth and atmosphere)
50–80	Clouds
40–95	Snow and ice
16–20	Grass
20–25	Dry, plowed fields
15–25	Green crops
5–20	Forests
18–28	Sand
14–18	Cities
7–20	Bare ground
7–23	Oceans

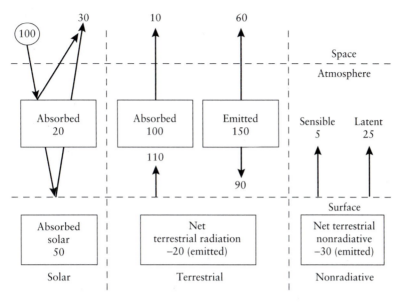

FIGURE 3-7 Energy–flux diagram showing radiative and nonradiative exchanges between the earth's surface, the atmosphere, and space. Units are percentages of global-average insolation (100% = 343 W/m). The flux at the surface exceeds 100% of the solar energy flux reaching the earth because of the absorption–emission cycle between the surface and the atmosphere. Redrawn from G. J. MacDonald and L. Sertorio, eds., *Global Climate and Ecosystem Change.* Plenum Press, New York. Copyright © 1990. Used by permission of Kluwer Academic/Plenum Publishers.

3.2 CLIMATE HISTORY OF THE EARTH

Before trying to decide whether humans are actually succeeding in changing the earth's climate at present, we must consider what sorts of climate change occurred before humans evolved, and what climate changes occurred before the major industrialization of the twentieth century. It is, unfortunately, beyond the scope of this chapter to discuss the various methods of calculating average temperatures at different times in the past; these methods are discussed in some of the references at the end of the chapter.

The variation of the mean surface temperature of the earth between the latitudes 40°N and 90°N as a nonlinear function of time from 500,000,000 B.C. to A.D. 1950 is shown in Figure 3-8. The time scale is arranged to emphasize recent years, for which more reliable data are available. More recent times are shown in more detail in Figure 3-9. Glacial periods occurred at intervals between 10,000 and 1,000,000 B.C. It looks, from Figure 3-8, as though there have also been major glacial episodes around 200,000,000 B.C.

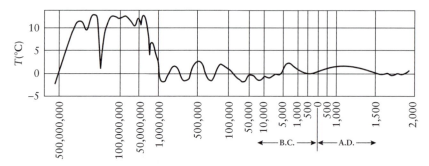

FIGURE 3-8   Temperature variations in the Northern Hemisphere (40°N – 90°N) as a function of time. Redrawn, by permission of the publisher, from J. E. Oliver, *Climate and Man's Environment.* Copyright © 1973, John Wiley & Sons, Inc.

and 500,000,000 B.C. In general, the average temperatures before 1,000,000 B.C. were warmer than those that have occurred since that time. Since 1,000,000 B.C., the interglacial periods, although warmer than either the glacial periods or the present, were much cooler than the eras before that time. Notice that the difference in average temperatures between glacial and interglacial periods was only about 5°C; that is, comparatively small differences in average temperature have large effects on the total climate. Notice also that variations up to 2°C have occurred since A.D. 1000. This means that average temperatures in the Northern Hemisphere have varied appreciably before human activities could have had any effect.

Figures 3-8 and 3-9 deal with very long-term variations in average temperature. If we are interested in climate changes possibly attributable to human intervention, we shall have to focus on much shorter term fluctuations. Let us, however, start with the period from 4000 to 3000 B.C. (6000–5000 years before the present), where Figures 3-8 and 3-9 show a temperature maximum, the "Holocene maximum." During this time many glaciers had melted, and sea level was much higher than it is today, and major civilizations existed in Egypt. Between 3000 and 750 B.C. the world generally become somewhat cooler, glaciers advanced, and sea level dropped. Canals were constructed in Egypt to take care of problems associated with the general drop in water levels. Another warming trend culminated in about A.D. 1200, the "medieval warm period," the warmest climate in several thousand years in the Northern Hemisphere. Glaciers retreated so far that the Vikings were able to travel to and settle in Greenland and Iceland. Southern Greenland had average temperatures 2–4°C above present temperatures while Europe had about 1°C higher average temperatures than at present.

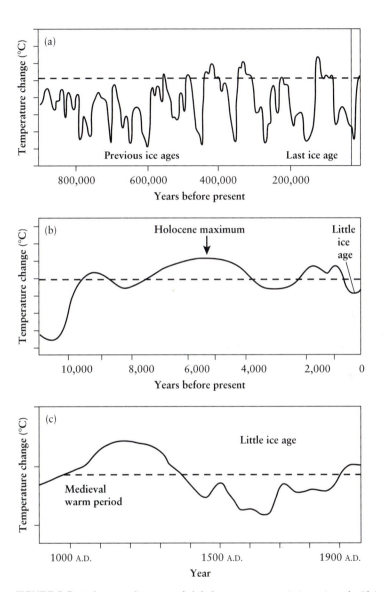

FIGURE 3-9 Schematic diagrams of global temperature variations since the Pleistocene on three time scales: (a) the last million years, (b) the last ten thousand years, and (c) the last thousand years. Dotted line nominally represents conditions near the beginning of the twentieth century. Redrawn from J. T. Houghton, G. J. Jenkins, and J. J. Ephraums, eds., *Climate Change. The IPCC Scientific Assessment.* Cambridge University Press, New York. Copyright © 1990. Used by permission of the Intergovernmental Panel on Climate Change.

Between A.D. 1300 and 1450, there was a cooling trend that made it harder for ships to travel between Greenland and Iceland and made Greenland cool enough to cause the Vikings to abandon their settlements. A warming trend between 1450 and 1500 was followed by the "Little Ice Age," which did not end in some areas until 1850. Glaciers advanced and winters were very cold in this period. Since 1850, there has been a distinct warming trend, with precipit-ous retreats of glaciers that ended temporarily around 1940 (see Figures 2-1 and 3-10). Figures 2-1 and 3-10 are very similar, but they refer to somewhat different time intervals and the temperature averages were obtained from somewhat different sources. Both figures show that the warming trend re-sumed about 1975. The retreat of the glaciers is easy enough to document, but these very short-term recent temperature trends are very hard to evaluate. First of all, while glaciers are retreating in one part of the world, they often are advancing elsewhere. What is the average trend? Furthermore, climate and temperatures are variable in any one place from year to year, and it is not easy to find a general trend among fluctuations. The problem in the evaluation of short-term trends lies in variability in climate that occurs both from place to place and from year to year. The place-to-place variability can be taken care of by averaging temperatures over a large enough area; that is why we see a global mean land temperature in Figure 2-1 and a combined land and sea surface temperature in Figure 3-10. The year-to-year variability can be seen very clearly in both figures. The averages, however, still do not tell the whole story; in this period, they represent mostly an increase in the number of warm months during each year, not an overall increase in actual temperature. Fur-thermore, although the overall temperature changes have been similar in the

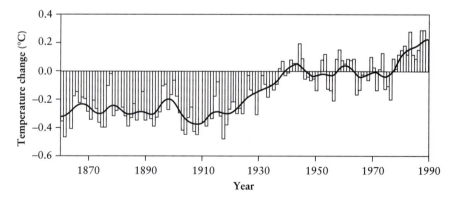

FIGURE 3-10 Global–mean combined land–air and sea–surface temperatures, 1861–1989, relative to the average for 1951–1980. Redrawn from J. Jäger and H. L. Ferguson, eds., *Climate Change: Science, Impacts and Policy.* Cambridge University Press, New York. Copyright © 1991. Used by permission of the Intergovernmental Panel on Climate Change.

Northern and Southern Hemispheres, the magnitudes have not been the same, and the warming and cooling trends did not necessarily occur in the same year, or, in some cases, in the same decade.

Note that the variations shown in Figure 3-10 are generally less than 1°C, but, as we have seen, it takes only about 5°C to change from a glacial to an interglacial period. The greatest variability in average temperature appears generally to be near the poles, not near the equator.

3.3 CAUSES OF GLOBAL CLIMATE CHANGES

3.3.1 General Comments

Climate is mainly effected by wind patterns over the earth's surface, and these changes in wind patterns are generally caused by changes in the earth's energy balance, as mediated, among other things, by changes in ocean currents and ocean temperatures. A well-known example is the El Niño/Southern Oscillation (ENSO), the eastward (toward South America) expansion of a pool of warm water usually found in the western Pacific Ocean; global temperatures generally rise during an ENSO.

The average surface temperature of the earth can be calculated in principle by employing the concepts discussed in Section 3.1. The starting point, as in Section 3.1, is $I_{in} = I_{out}$, where $I_{in} = (S/4)(1 - \alpha)$ and $I_{out} = I_0 - \Delta I$. In these expressions S is the solar constant, α is the earth's albedo, I_0 is the intensity of radiation emitted at the earth's surface, $I_0 = \sigma T^4$, T is the mean surface temperature of the earth; and ΔI is a term involving the greenhouse effect. Substituting for I_0 and solving for T, we have

$$T = \left[\frac{S(1 - \alpha)/4 + \Delta I}{\sigma} \right]^{1/4} \tag{3-6}$$

The variables in equation (3-6), S, α, and ΔI, are called "climate forcing agents" because changes in these variables "force" changes in climate.

If we take T, the average surface temperature of the earth, as an indicator of climate, then equation (3-6) allows us to determine the main factors that affect climate and even the magnitudes of possible effects. It can be seen that the solar constant, the earth's albedo, and the greenhouse effect all help to determine T.

It has always been tempting to attribute major climatic variations, like those between glacial and interglacial periods, to variations in the solar constant, that is, to variations in solar irradiation. There is considerable evidence for this hypothesis, more because of variations in the earth's orbit and in the inclination of different latitudes toward the sun during different seasons (see Section 3.3.2) than because of major changes in the emission of radiation from the sun. The sun is quite variable in its emission of x-rays and even some ultraviolet radiation, but the emission of visible and infrared radiation seems to be quite steady, increasing, however, about 0.12% from the minimum to the maximum number of sunspots in the (usually) 11-year sunspot cycle (see Figure 3-11). Small climate changes have been associated with the sunspot cycle during this century, but 11 years is a short time and there are many other influences on climate. Some extreme variations in sunspot activity have, however, been correlated with major peculiarities of climate. The coldest part of the "little ice age" appears to be concurrent with a 70-year minimum in sunspot activity during the late seventeenth and early eighteenth centuries, the "Maunder minimum." There is some evidence that the solar constant decreased by about 0.14% during the Maunder minimum.

We can use equation (3-6) to show that an 8% change in the solar constant could change the average surface temperature of the earth by 3°C. The values of S, α, and σ have been given in this chapter, and the present value of ΔI can be calculated from σ, the measured mean surface temperature of the earth T, and the mean radiation temperature of the earth T_r as found in Section 3.1:

$$\Delta I = \sigma\lfloor T^4 - T_r^4 \rfloor = 150 \text{ W/m}^2 \qquad (3\text{-}7)$$

The solar constant is presently being monitored by satellite measurements, and thus proper evaluations of its effect on climate will be possible in the future.

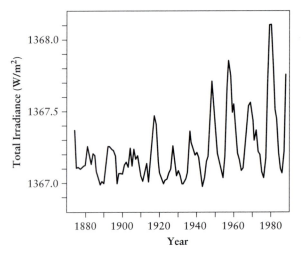

FIGURE 3-11 Reconstructed solar irradiance from 1874 to 1988. Redrawn from J. T. Houghton, G. J. Jenkins, and J. J. Ephraums, eds., *Climate Change, Science, Impacts, and Policy.* Cambridge University Press. New York. Copyright © 1990. Used by permission of the Intergovernmental Panel on Climate Change.

3.3.2 The Milankovitch Theory of Earth's Orbital Variations

The Milankovitch theory attempts to account for the existence of Ice Ages at various intervals (see Figure 3-9). This theory makes use of the dependence of the insolation, the total solar radiation impinging on the earth in different latitudes, on a number of factors involving the earth's revolution around the sun. There is quite a bit of controversy about the particular latitudes at which such variations in insolation are most important, but most workers feel that changes in insolation at high latitudes, closer to the poles than 60°N or 60°S, are the most important.

The first orbital variation, which affects all latitudes in a similar fashion, involves the change in the earth's orbit from almost circular to somewhat elliptical and back, with a somewhat variable period of 90,000–100,000 years. When the earth's orbit is at its highest ellipticity, the solar radiation intensity reaching the earth may vary up to 30% in the course of a year. Right now, the earth's orbit is not very elliptical and indeed is becoming more circular.

The second orbital influence on solar irradiation, which affects different latitudes differently, is the 21,000-year cycle in the precession of the earth's axis around the normal to the earth's orbit; this changes the season at which the Northern (or Southern) Hemisphere is closer to the sun. This influence is most important when the earth's orbit is even a little bit elliptical, as it is now. At present, for example, the earth is about 3×10^6 miles closer to the sun during winter in the Northern Hemisphere than during winter in the Southern Hemisphere. This means that there is about 7% less solar irradiation during summer in the Northern Hemisphere than during summer in the Southern Hemisphere.[1]

The third orbital influence on solar irradiation at different latitudes is the 40,000-year cycle in the tilt of the earth's axis with respect to the normal to the plane of the earth's orbit. This angle varies between 21.8 and 24.4 degrees; at present, it is 23.4 degrees.

These three types of orbital variation, with three different cycle lengths, have been used in various climate models to calculate average global temperatures in the absence of true solar constant variations and in the absence of changes in the greenhouse effect or earth's albedo for at least one million years before the present. Figure 3-12 is one of many comparisons between the calculations from the Milankovitch theory and actual measurements. (The "measurements" are actually reasonable guesses from various geological data.) The calculations were based on the assumption that the variation in insolation at latitude 50°N could be used if one further assumed that whenever

---

[1]This cycle is also connected with the change in the polestar, presently Polaris (almost); in 2600 B.C. the polestar was Alpha Draconis, about 25 celestial degrees away from Polaris.

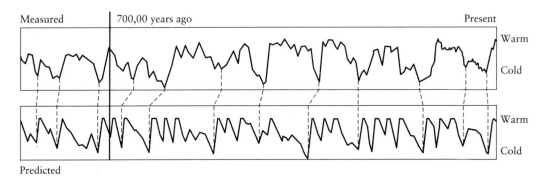

Measured   700,00 years ago   Present

Predicted

FIGURE 3-12   Comparison of warm and cold periods in the earth's climate as measured from geological data and as predicted from the Milankovitch theory. Redrawn from N. Calder, *The Weather Machine*. Viking Press, New York. London: BBC. Copyright © 1974. Used by permission of the author.

the summer sunshine at that latitude was at least 2% stronger than at present, glaciers tended to disappear, and that ice began to accumulate when the summer sunshine was less. Figure 3-12 shows reasonable agreement between the measured and calculated results. More complex assumptions have recently led to even better agreement between measured and calculated results. Therefore, we have a reasonable explanation for the existence of Ice Ages and interglacial periods. These orbital cycles exist and must interact with the other influences on the earth's climate.

3.3.3 Greenhouse Effect Changes

Small changes in the greenhouse effect ΔI can have appreciable effects on climate. These changes can occur when variations occur in the concentration of carbon dioxide, methane, and trace gases, especially those that can absorb radiation in the atmospheric "windows" as mentioned in Section 3.1.2. Figure 3-13 shows the close parallel between mean global temperature and the atmospheric carbon dioxide and methane concentrations as obtained from air trapped in Antarctic ice cores over the past 160,000 years. Although increased burning of fossil fuels is held responsible for the global increase in carbon dioxide concentration in the atmosphere during historic times, it appears that a major increase in carbon dioxide in the atmosphere started almost 20,000 years ago. Although this *could* be attributed to human activities, many of the other variations in carbon dioxide, and, for that matter, methane concentration, must have another cause. Figure 3-14 shows the recent (i.e., since 1840) increase in methane concentration in the atmosphere. In preindustrial times, the atmospheric concentration of methane was about 0.8 ppm by volume, increasing to 1.75 ppm in 1999 and still increasing, although the rate of increase appears to have decreased during the past two decades. The recent changes in CO_2 concentration in the atmosphere were shown in Figures 2-2 and 2-3. At any rate, there is definitely a positive correlation between the concentration of these gases and climate.

Because of the known correlation between the concentrations of carbon dioxide and methane and climate, even though the actual connection is quite complicated, the concept of relative global warming potentials (GWPs) has been developed for all the so-called greenhouse gases, those that absorb infrared radiation in the "atmospheric windows." This concept takes into account (1) the effect of an increase in concentration of the gas in the present

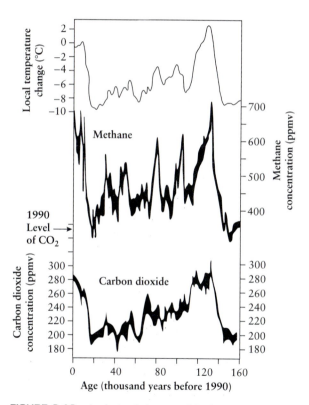

FIGURE 3-13 Analysis of air trapped in Antarctic ice cores shows that methane and carbon dioxide concentrations were closely correlated with the local temperature over the last 160,000 years. The concentration of carbon dioxide is indicated on the left ordinate. Redrawn from J. Jager and H. L. Ferguson, eds., *Climate Change: Science, Impacts and Policy*. Cambridge University Press, New York. Copyright © 1991. Used by permission of the Intergovernmental Panel on Climate Change.

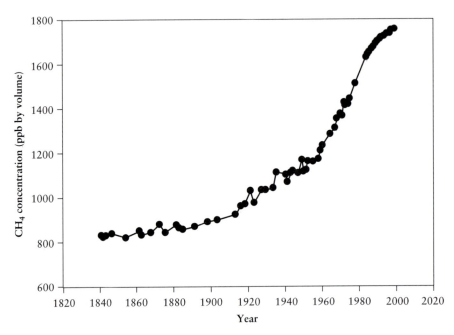

FIGURE 3-14   Change in methane concentration in the earth's atmosphere since 1840.

atmosphere both in absolute terms and relative to carbon dioxide, (2) the effect of the different times that the various gases remain in the atmosphere, and (3) some of the reactions that these gases undergo in the atmosphere (see Chapter 5) over long periods. The results of factor 3 are called "indirect effects" and are not always included in the calculations of the GWP. At any rate, the GWP is thus the time-integrated warming effect expected from the instantaneous release of 1 kg of a given gas in today's atmosphere, relative to that expected from an equal mass of carbon dioxide. Table 3-2 shows the total GWPs, based on both direct and indirect effects, calculated over 20, 100, and 500 years for some gases with respect to the earth's atmospheric composition in 1990 along with the atmospheric lifetimes of these gases as used in the calculations. Carbon dioxide has the lowest GWP in the table, but its contribution to the greenhouse effect, and thus potential global warming, is very large because this contribution depends on the product of the GWP and the number of kilograms of the gas actually emitted. The CFC's in Table 3-2 are discussed at greater length in Section 5.2.3.2 because of their ability to destroy the ozone in the stratosphere. It is important, by the way, not to consider the greenhouse effect and the GWP in isolation from the other possible effects on climate, because both positive and negative feedback influences are possible, as mentioned in Section 3.3.5.

TABLE 3-2

Atmospheric Lifetimes and Total Global Warming Potentials of Some
Greenhouse Gases (Based on the Atmospheric Composition in 1990)

Gas	Atmospheric lifetime (years)	Time horizon (years)		
		20	100	500
CO_2		1	1	1
CH_4	10.5	63	21	9
N_2O	132	270	290	190
CFC-11[a]	55	4500	3500	1500
CFC-12[a]	116	7100	7300	4500
HCFC-22[a]	16	4100	1500	510

[a]CFC's are chlorofluorocarbons while HCFC's are
hydrochlorofluorocarbons. The composition of each one can be
determined by using the Rule of 90 as follows: add 90 to the number in the
name of the gas. A number larger than 100 will be obtained, for example,
101 for CFC-11. The first digit is the number of C atoms, the second is the
number of H atoms, the third is the number of F atoms, and the number of
Cl atoms is obtained by assuming that the molecule contains no double or
triple bonds or rings. Thus, CFC-11 is $CFCl_3$, CFC-12 is CF_2Cl_2, and
HCFC-22 is CHF_2Cl.
  Source: J. T. Houghton, G. J. Jenkins, and J. J. Ephraums, eds.,
Climate Change: The IPCC Scientific Assessment. Cambridge University
Press, Cambridge, U.K., 1990.

Let us note that not all the carbon dioxide emitted into the atmosphere
remains there. For example, it has been calculated that in the period 1958–
1968, enough fossil fuels were burned to increase the carbon dioxide content
of the atmosphere by 1.24 ppm per year. The actual increase in CO_2 content
was 0.64 ppm/year. This is an indication of the enormous buffering effect of
the earth's oceans on the CO_2 concentration in the atmosphere. This buffering
effect was discussed in Chapter 2 and is considered further in Sections 9.2.2
and 11.2.

3.3.4 Albedo Changes

Changes in the earth's albedo have also been implicated in climate changes.
Although large-scale cultivation, irrigation, damming of rivers to form lakes,
and cutting down of forests all result in changes in the earth's albedo, these
causes have heretofore been considered minor. The possible results of volcanic
and other dusts in the atmosphere have, however, been investigated fairly
extensively.

Figure 3-15 shows an index of volcanic activity on earth between the years 1500 and 1970. Some of this volcanic activity was followed by worldwide cooling. For example, the "Little Ice Age" (1550–1850) occurred during a period of fairly high volcanic activity as well as low sunspot activity. Some of the highest peaks on Figure 3-15 are actually tremendous single volcanic explosions that were followed by noticeably cool years. The eruption of Asama in 1783 in Japan (the largest activity peak between 1700 and 1800 in Figure 3-15) was followed by three cool years, 1783–1785. The year 1816, called "the year without summer" (in New England, snow fell in June and there was frost in July), followed the eruption of Tambora, Sumbawa, in the Dutch East Indies in 1815. This eruption killed 5600 people and darkened the sky for days as far as 300 miles away. The large peak just before the year 1900 on Figure 3-15 is probably connected with the eruption of Krakatao in Indonesia in 1883. The eruption of Mount Agung in Bali in 1963, although not as spectacular as that of Tambora, occurred after accurate measurements of tropospheric temperature and aerosol measurements were being obtained, and is known to have resulted in a drop of 0.4 °C in the troposphere between latitudes 30°S and 30°N about one year later and lasting for about two years. It turns out that it is not the volcanic dust that contributes to tropospheric cooling, but aerosols consisting of submicrometer droplets of sulfuric acid derived from sulfur-containing volcanic gases. Thus, the eruption of Mount St. Helens in the northwestern part of the continental United States in 1980 did not cause noticeable cooling because the gases released by this volcano contained only a small amount of sulfur. Volcanic ash falls out of the atmosphere within a few

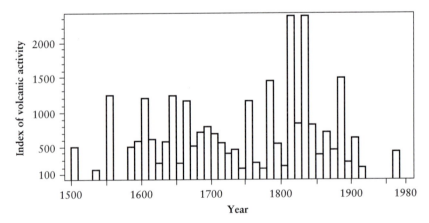

FIGURE 3-15 Index of volcanic activity, in arbitrary units, that is proportional to the relative amount of material injected into the stratosphere in various 10-year periods. Redrawn from *Inadvertent Climate Modification Report of the Study of Man's Impact on Climate (SMIC)*. Copyright © 1971. Used by permission of MIT Press.

weeks or months, but the sulfuric acid aerosol droplets can remain in the stratosphere for years. The 1982 eruption of El Chichón in Mexico apparently cooled the surface of the earth for several years because of the large amount of sulfuric acid aerosol formed afterward from sulfur-containing gases. The aerosols from the eruption of Mount Pinatubo in the Philippines and of Mount Hudson in Chile in 1991 apparently also caused global cooling for close to a year.

It has been postulated that a large amount of dust and soot would be released into the stratosphere during a major nuclear war. For various reasons, this dust and soot would probably remain aloft for about one year. This would then increase the earth's albedo to the extent that considerable cooling of the earth would occur over several years. Calculations indicate that the earth could cool up to 1°C for several years (recall that the difference between an ice age and an interglacial period is about 5°C). This scenario has been called "nuclear winter," and it includes major changes in precipitation and a probable depletion of the ozone in the stratosphere. Thus, this scenario includes a major disturbance of agriculture *all over the earth* in addition to depletion of the ozone layer that protects the surface of the earth from high-energy ultraviolet radiation.

3.3.5 Computer Models

We have seen in Table 2-2 that humans are presently contributing a large amount of particulate matter, and, certainly, aerosols, to the atmosphere. If this material, like that emitted by volcanoes, can contribute to a cooling of the earth, presumably by increasing the earth's albedo, then this may partially mitigate the effects from the simultaneous CO_2 emissions by humans. We have definitely been increasing both the carbon dioxide and the particle and aerosol content of the atmosphere, and these changes must have some influence on the greenhouse effect and the albedo, which have a further effect on climate. Human efforts are superimposed on natural trends such as the orbital variations considered in the Milankovitch theory, but it is not clear at present whether human changes or natural effects have the greater magnitude. Furthermore, the earth's climate is determined by a complex set of positive and negative feedback systems that no one, at present, understands completely. As an example of a negative feedback (or self-limiting) possibility, suppose that an increase in CO_2 content of the atmosphere caused an increase in temperature because of an increased greenhouse effect, which then caused an increase in evaporation of water from the oceans to form more cloud cover, which then reflected more of the sun's radiation to lower the temperature back toward its original value. It is an interesting exercise to dream up these self-limiting processes. The earth–atmosphere–ocean system is so complex that it is easy to imagine such self-limiting or even oscillating processes, but very difficult to develop models that produce results in which one has confidence. A positive feedback possibility involves the idea of a slightly colder than usual winter with extra snow cover at high latitudes as postulated in the calculations made for Figure 3-12. This would increase the earth's albedo, allowing less of the solar radiation to heat the earth and its atmosphere than before; the extra snow would not melt completely. This would cause an even colder winter (because of the extra snow cover in the fall) in the following year, and so on. Good climate models need to invoke both negative and positive feedback systems.

As we have seen, the earth's climate system is extremely complicated, so that computer models that can explore the consequences of any changes in the system must be quite simplified. Figure 3-16 shows schematically the components of the earth's climate system that may or may not be considered in the various computer models of climate. Some computer models involve only energy balance for the earth as a whole, some involve the vertical distribution

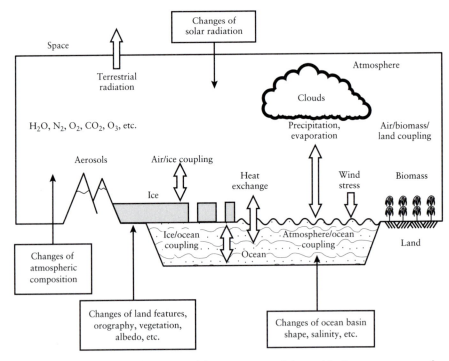

FIGURE 3-16 Schematic illustration of the components of the earth's climate system, together with some physical processes responsible for climate change. Redrawn from M. C. MacCracken, A. D. Hecht, M. I. Budyko, and Y. A. Izrael. eds., *Prospects for Future Climate*. Lewis Publishers, Chelsea, MI. Copyright © 1990. Used with permission.

of temperature for the earth's surface and atmosphere (again, as a whole), and some involve the general circulation of the atmosphere and the oceans, allowing for the calculation of the geographical distribution of temperature, precipitation, and other climate quantities. These models are constantly being refined and checked against each other and against past climates of the earth. For example, a model developed at the National Center for Atmospheric Research (NCAR) in Boulder, Clorado, has given an excellent simulation of the present climate by incorporating the effects of ocean eddies hundreds of kilometers in diameter. This model has been used to predict a global warming from the increasing CO_2 concentration in the atmosphere that is smaller than the amount anticipated by earlier models. Further comments are beyond the scope of this book.

3.4 SMALL-SCALE CLIMATE

As we have seen, it is very difficult to ascertain what effects human activities may have had or will have on large-scale climate. However, human effects on small-scale climate are well known and worth discussing for that reason. Every town, every reservoir, every plowed field changes the local climate to some extent. In this brief discussion of local climate, we shall confine ourselves to a consideration of urban climates, as distinguished from the surrounding countryside.

On the average, urban climates are warmer and have more precipitation than the surrounding countryside. Table 3-3 compares an average city with the surrounding countryside for a number of climate elements. Note that a city is dirtier, cloudier, foggier, has more drizzly rain (precipitation days < 5 mm total), more snow, less sunshine, less wind, and higher temperatures than the surrounding countryside. Wind speed is lowered in the city, except near very tall buildings, where appreciably higher wind speeds have been found in specific cases.

The differences in temperature between a city and its surroundings can be much more variable than the Table 3-3 averages might imply. In London on May 14, 1959, for example, the minimum temperature was 52°F (11°C) in the city's center and 40°F (4°C) at its edges. In the period 1931–1960, however, mean annual temperatures for London were 11.0°C (51.8°F) at the city's center, 10.3°C (50.5°F) in the suburbs, and 9.6°C (49.2°F) in the adjacent countryside.

Urban heat is generally assumed to arise from three causes. The most obvious source is the heat from combustion and general energy use in the city. A major cause of the warmer city nights is the release of heat stored in structures, sidewalks, streets, and so on in the daytime. A minor contributor to the warmth of cities is the back-reflection and reradiation of heat from pollutants in the atmosphere above the city.

TABLE 3-3

Microclimate of Urban Areas Compared with Surrounding Countryside

Climate element	Comparison with countryside
Particulate matter	10 times more
Emitted gases	5–25 times more
Cloud cover	5–10% more
Fog, winter	2 times more
Fog, summer	30% more
Total precipitation	5–10% more
Days with less than 5 mm	10% more
Snowfall	5% more
Relative humidity, winter	2% less
Relative humidity, summer	8% less
UV radiation, winter	30% less
UV radiation, summer	5% less
Sunshine duration	5–15% less
Annual mean temperature	0.5–1.0°C higher
Winter minimum temperature (average)	1–2°C higher
Annual mean wind speed	20–30% less

Source: Reprinted from *Inadvertent Climate Modification, A Report of the Study of Man's Impact on Climate* (SMIC). MIT Press, Cambridge, MA, 1971.

In the absence of winds, the pollutants in the atmosphere above a city often occupy a so-called urban dome (Figure 3-17), a sheath of air around the city that may be situated under a temperature inversion. The pollutants in an urban dome are reasonably well mixed, and the dome itself can often be distinguished from airplanes or from mountains near the city. Looking down on the urban dome of Los Angeles is something like observing a large pot of yellow pea soup! Urban domes have varying heights depending on conditions; an urban dome above a subarctic city in the winter may be only 100 m thick, while the dome over an oasis in the subtropics may be several kilometers thick. As we have noted, the urban dome tends to be warm; heat from combustion may supply up to 200 W/m², as in Moscow in the winter. When the wind blows, the urban dome becomes an urban plume, moving heat and pollutants down-wind (Figure 3-18).

FIGURE 3-17   The urban dome.

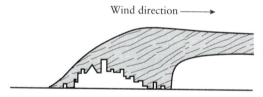

Wind direction ⟶

FIGURE 3-18   The urban plume.

The existence of a visible urban dome implies different atmospheric composition over the city and in the surrounding country. The air over a city contains smoke, dust, SO_2, and so on, in much higher concentrations than country air. The particles in the air above the city increase the albedo above the city, thus reducing the intensity of the incoming solar radiation. In some cities, on bad days, the intensity of the incoming solar radiation may be reduced up to 50%. This means that cities tend to have cooler days than the surrounding country; these cooler days do not, however, make up for the warmer nights. The added particulate matter over cities also increases the frequency of fogs, since the particles can act as condensation nuclei for water vapor.

The lower wind velocity in cities probably is caused by many obstacles (buildings) that are in the path of any wind in a city. The lack of standing water and the fast drainage of precipitation into storm drains result in a decrease of local evaporation and thus in the observed decrease of humidity in cities. However, when a city contains many cooling towers and there is generally a high water vapor output from burning of fossil fuels and industry, these factors, together with the extra condensation nuclei produced by the city, may cause an increased frequency of showers and thunderstorms downwind from the city and a weak drizzle within the city.

3.5 TEMPERATURE INVERSIONS

When pollutants are trapped for a long time over a city, it is usually because of a temperature inversion above the city. This may result in accumulations that can reach lethal proportions, at least for the sick and elderly. Temperature inversions occur when air at some elevation in the troposphere ceases to decrease smoothly in temperature with increasing altitude as is normal near the surface of the earth (see later: Figure 3-20). At a temperature inversion, the air temperature increases with increasing altitude for a distance; at higher altitudes, the normal decrease in air temperature with increasing altitude begins again and continues up to the tropopause. The ways in which such temperature inversions can occur will be discussed at the end of this section.

To see why a temperature inversion traps pollutants, it is necessary to know what happens in the absence of such an inversion. In the daytime, the ground is heated by solar radiation, and the air near the ground is heated by conduction from the ground. The air, if we assume that it is an ideal gas, obeys the equation of state for ideal gases:

$$PV = nRT \qquad (3\text{-}8)$$

where P is the pressure, V is the volume of n moles of gas, R is the gas constant, and T is the absolute temperature. For our heated air, the pressure P of the atmosphere above it remains constant, but T increases. If we consider a "packet" of n moles of this air, and rewrite equation (3-8) as $V = nRT/P$, then we see that upon heating, V must increase, and the density, n/V, decrease. Consequently, this packet of air will rise until it reaches a place in the troposphere where the surrounding air has the same density; that is, the same temperature and pressure. Both the temperature and the pressure of the rising air packet change continuously as it moves to regions of lower pressure and therefore expands. The expansion can be assumed to be adiabatic; no heat flows into or out of the packet during the process. Derivations of the equations for temperature and pressure changes in reversible adiabatic expansion of an ideal gas are given in most physical chemistry and many general chemistry textbooks. The result is

$$P_1/P_2 = [T_1/T_2]^{C_p/R} \qquad (3\text{-}9)$$

where C_P is the heat capacity at constant pressure of the gas, and subscripts 1 and 2 refer to properties of the gas before and after expansion, respectively. Qualitatively, this expression shows that if the gas is expanding, that is, if $P_2 < P_1$, then $T_2 < T_1$ and the temperature of the gas will drop. This equation, together with the known decrease of pressure with height, allows calculation of the adiabatic lapse rate of dry air. This rate is a decrease of 1°C for each 100 m, or 5.5°F for each 1000 ft of altitude. The rising air cools faster than the temperature lapse rate of the environment, which involves a decrease of only 0.65°C for each 100 m or 3.5°F for each 1000 ft of elevation. Dry air will eventually rise to a height where temperature and, therefore, the density matches that of the surrounding air. If the rising air contains a lot of moisture that condenses as the air rises and becomes cooler, then heat is given off during this condensation and the air cools more slowly with height; that is, its temperature lapse rate is less than that of dry air. This moist air will therefore rise higher than dry air and will probably form clouds as it rises.

These conclusions are illustrated in Figure 3-19 which shows what happens to heated dry air when the environment (the troposphere) is normal. The solid line shows the temperature of the lower atmosphere when the temperature of the surface of the earth is 24°C and there are no temperature inversions in the atmosphere. If the sun is shining, the surface of the earth absorbs heat and becomes warmer, heating the air just above it by conduction. The two broken lines in Figure 3-19 refer to two different "parcels" of air heated by the surface of the earth, to 25 and 26°C, respectively. Both parcels of air will begin to rise as just discussed; they rise with the temperature lapse rate of rising dry air. Since the temperatures of these two parcels of air are decreasing faster, 1°C per 100, than the temperature of the air through which they are rising, 0.65°C per 100 m, each parcel will eventually reach a height at which its temperature equals that of the air already present. Its pressure and density are then also equal to that of the surrounding atmosphere, and the parcel will stop rising. Figure 3-19 shows that the parcel that was heated to 25°C will rise to a height just below 300 m, while the parcel that was heated to 26°C by the surface of the earth will rise to a height just below 600 m. Hotter air rises to a higher altitude.

If a temperature inversion exists in the atmosphere at an altitude below that to which the heated air would normally rise, then the heated air may be trapped below the inversion, along with any pollutants that were present in the heated air. In Figure 3-20, which shows how heated air may be trapped when there is a temperature inversion in the troposphere, the solid line that

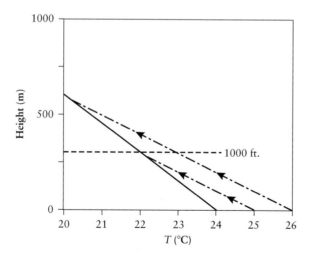

FIGURE 3-19 The fate of dry air heated by the earth's surface when the temperature lapse rate of the environment is normal: height vs temperature of the environment (solid line) and of rising, heated dry air (broken lines).

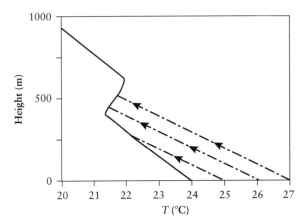

FIGURE 3-20   How a temperature inversion traps heated air and its associated pollutants: height vs temperature of the environment (solid line) and of rising, heated dry air (broken lines).

shows the variation of atmospheric temperature with altitude is not linear; it has a region in which temperature increases with increasing altitude between approximately 400 and 600 m altitude. This region of increasing temperature is a temperature inversion; above and below this region the air exhibits its normal lapse rate of 0.65°C per 100 m elevation. If the sun shines and the earth's surface is heated, the resulting ground-level heated air rises as shown by the broken lines in Figure 3-20. Three parcels of air are shown, one heated to 25°C, the second to 26°C, and the third to 27°C. The parcel of air heated to 25°C rises to the same altitude, just below 300 m, like the similar parcel in Figure 3-19; this happens because the temperature inversion occurs above this altitude and can affect only air that is attempting to rise through the inversion. Both the 26°C and the 27°C parcels of air are trapped by the inversion; each reaches an altitude at which its temperature equals that of the air already present within the temperature inversion zone. The 26°C parcel of air rises to less than 500 m, considerably lower than the 600 m it would ordinarily reach (Figure 3-19).

These temperature inversions that trap rising air are naturally occurring phenomena that may have two different causes. That is, they are of two different types, radiation inversions and subsidence inversions, explained in the following paragraphs. London tends to have radiation inversions in the winter; one of these trapped a highly polluted fog in 1952, producing the notorious London killer fog, which was considered responsible for over 3000 deaths. Los Angeles tends to have subsidence inversions that trap pollutants in the summer and fall.

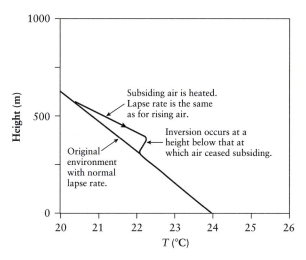

FIGURE 3-21 Mechanism by which a subsidence inversion occurs.

A radiation inversion generally occurs at night, when air near the ground is cooled by conduction from the cooled surface, especially on a clear night when radiation cooling occurs. Higher up in the troposphere, the air is not cooled appreciably, so, in the absence of winds, there is colder air underneath warmer air, an inversion by definition.

A subsidence inversion occurs in a region where high altitude air is pushed down nearer the ground because of topographical features in the path of prevailing winds or weather systems. This air is being pushed to regions of higher pressure and therefore is being compressed adiabatically. The same equation that held for rising air also holds for this descending air, but now $P_2 > P_1$ and $T_2 > T_1$; that is, the air warms up as it nears the ground. Figure 3-21 shows how this subsiding air creates an inversion; again horizontal winds must be absent for the inversion to be reasonably stable. The two mechanisms for forming an inversion have exactly the same results, trapping of heated air and any pollutants that may be present.

3.6 CONCLUSIONS

In this chapter, we have considered the climate history of the earth, the Milankovitch theory of the ice ages, the other factors that affect the earth's climate, the peculiarities of urban climate, and the formation and consequences of temperature inversions. It should now be possible to consider *all* the questions asked at the beginning of Chapter 2.

The information presented in Chapters 2 and 3 shows how complicated the earth's atmosphere and climate are. Many of the discussion questions asked at the end of this chapter have no easy answers. In spite of this, newspapers and magazines often contain articles that present the problems of the possible control of the earth's climate by humans as if they were amenable to simple solutions. These chapters should make it possible to consider these articles and problems with consideration of their inherent complexity.

Additional Reading

Berger, A., S. Schneider, and J. C. Duplessy, *Climate and Geo-Sciences. A Challenge for Science and Society in the 21st Century*. Kluwer Academic, Dordrecht, 1989.

Boeker, E., and R. van Grondelle, *Environmental Physics*. Wiley, New York, 1995.

Calder, N., *The Weather Machine. How Our Weather Works and Why It Is Changing*. Viking Press, New York, 1974.

Gribbin, J., *Forecasts, Famines and Freezes. Climate and Man's Future*. Walker New York, 1976.

Gribbin, J., *Future Weather and the Greenhouse Effect*. Delacorte Press, New York, 1982.

Gribbin, J., ed., *Climate Change*. Cambridge University Press, Cambridge, U.K. 5 1978.

Harries, J. E., *Earthwatch. The Climate from Space*. Horwood, New York, 1990.

Houghton, J. T., G. J. Jenkins, and J. J. Ephraums, *eds., Climate Change. The IPCC Scientific Assessment*. Cambridge University Press, Cambridge, U.K., 1990.

Hoyt, D. V., and K. H. Schatten, *The Role of the Sun in Climate Change*. Oxford University Press, New York, 1997.

Inadvertent Climate Modification. A Report on the Study of Man's Impact on Climate (SMIC). MIT Press, Cambridge, MA, 1971.

Jäger, J., and H. L. Ferguson, eds., *Climate Change: Science, Impacts and Policy. Proceedings of the Second World Climate Conference*. Cambridge University Press, Cambridge, U.K., 1991.

Krause, F., W. Bach, and J. Koomey, *Energy Policy in the Greenhouse*. Wiley, New York, 1992.

Lamb, H. H., *Climate History and the Future*. Princeton University Press, Princeton, NS, 1977.

MacCracken, M. C., A. D. Hecht, M. I. Budyko, and Y. A. Izrael, eds., *Prospects for Future Climate. A Special US/USSR Report on Climate and Climate Change*. Lewis Publishers, Chelsea, MI, 1990.

MacDonald, G. J., and L. Sertorio, eds., *Global Climate and Ecosystem Change*. Plenum, Press, New York, 1990.

Peixoto, J. P., and A. Oort, *Physics of Climate*. American Institute of Physics, New York, 1992.

Revkin, A., *Global Warming. Understanding the Forecast*. Abbeville Press, New York, 1992.

EXERCISES

3.1. How does the Greenhouse effect work (a) in a greenhouse and (b) in the earth's atmosphere?

3.2. The CO_2 and CH_4 content of the earth's atmosphere have been increasing steadily through this century.

(a) What effect do most people think this will have on the earth's climate? Why?

(b) Explain why your answer to part a might be wrong.

3.3. What are the three major factors that control earth's climate? Which of these factors may be affected by human activities? How?

3.4. What is an inversion? Explain how either a subsidence inversion or a radiation inversion occurs in the troposphere.

3.5. In 1997 some researchers made satellite measurements that indicated that the earth's polar regions are 0.55°C warmer during a full moon than during a new moon. The same article mentioned that another researcher cautions that the effect may be an artifact caused by reflection of lunar light off a piece of the spacecraft (the satellite doing the measurement). From what you have learned about earth's climate, explain why you feel either (a) that the measurements are probably correct or (b) that they are probably an artifact and cannot be relied on.

3.6. (a) Which items usually considered in the heat balance of the earth would change appreciably if the earth were dry (no water present in any form): snow, ice, liquid, or vapor? Explain the increase or decrease in the items that change.

(b) Would you expect a change in the mean temperature of the earth? Explain.

3.7. If all the earth's deserts were irrigated to produce crops of various kinds, what would be the effect, if any, on the earth's heat balance? Explain, and indicate whether you think the earth's mean temperature would increase, decrease, or remain the same.

3.8. Considering the increasing need for living space by the earth's ever increasing population, there have been proposals to try to make the northern latitudes more habitable by spreading many tons of soot or black dust on the Arctic ice sheet by cargo aircraft. This approach is expected to have the following results. Since the average thickness of this pack is about 3 m and it now shrinks by about 1 m during the summer melting season, the additional heat absorbed by a sooty surface might complete the melting of the ice in one or two seasons. The resulting open ocean would warm the lands on its border and also create more rainfall in an area that is now quite arid.

(a) Is this proposal feasible? Why or why not?

(b) Assuming that this proposal is feasible, what additional short-term results will it have?

(c) What are some possible additional long-term results of this procedure, if it works?

3.9 The temperature lapse rate of the troposphere is 0.65°C/100 m and the adiabatic lapse rate of dry air is 1°C/100 m. Explain, with a diagram, the consequences if the temperature lapse rate of the troposphere were 1.1°C/100 m.

4

PRINCIPLES OF PHOTOCHEMISTRY

4.1 INTRODUCTION

Visible and ultraviolet solar radiation is essential for virtually all life on earth as we know it, and photochemical reactions are some of the most important processes taking place in the human environment. Several examples of photochemical influences on our environment have already been given in Chapters 2 and 3. Direct utilization of solar energy as a possible alternative to fossil fuel combustion is discussed in Chapter 15, and its efficient usage is certain to be an increasingly important goal of research and development over the next few decades as we seek solutions for the worldwide energy problem. In Section 2.3.4 it is pointed out that atmospheric oxygen comes from photodissociation of water vapor in the upper atmosphere and from photosynthesis in the biosphere. Photosynthesis is a very important photochemical process that also leads to food production and storage of solar energy (Chapter 15). We discuss photochemical decomposition of petroleum, polymers, and PCBs and pesticides in Chapters 6, 7, and 8, respectively.

Photochemical processes occurring at high altitudes in the mesosphere and stratosphere are essential for the maintenance of the thermal and radiation balance at the surface of the earth (see Chapter 3), and they also provide phenomena such as night glow. The reactions following light absorption in the lower atmosphere that generate photochemical smog from atmospheric pollutants are less desirable! In association with living matter, light-induced reactions lead to such phenomena as vision, cyclic activities of plants and animals, and both cellular damage *and* repair. In a more prehistoric context, it has been suggested that the origin of life may have occurred via the photochemical synthesis of purines, essential building blocks of nucleic acids.

As we shall show later, absorption of light by a molecule results in excitation of the molecule to a higher electronic energy level. This is followed in many cases by reactions such as dissociation or interaction with other molecules, leading to the overall photochemical process. We shall see that the specific electronic state reached often determines the types of reaction that follow, so that the study of the absorption process is important to our understanding of photochemistry. In this chapter, we shall develop the basic principles of light absorption, electronic excitation, and subsequent photochemical and photophysical processes. As these principles are being developed, they will be applied to specific examples of photochemical processes occurring in our environment.

Photochemistry is the study of the interaction of a "photon" or "light quantum" of electromagnetic energy with an atom or molecule, and of the resulting chemical and related physical changes that occur. The energy necessary for the reaction (or at least for it to be initiated) is thus gained from the photon. This is in contrast to thermal reactions, in which the energy needed for the reaction is distributed among the molecules and among the internal vibrational and rotational motions of the molecule according to the Boltzmann distribution law (Figure 4-1).

As an example, consider the possible thermal dissociation reaction of molecular oxygen,

$$O_2 \rightarrow 2O \tag{4-1}$$

The fraction of molecules with thermal kinetic energy equal to or greater than a "threshold" energy E_c is approximately equal to the Boltzmann factor

$$\frac{N_c}{N} = e^{-E_c/RT} \tag{4-2}$$

where N_c is the number of molecules with energies equal to or greater than E_c, N is the total number of molecules, R is the gas constant, $8.314\,\mathrm{J\,K^{-1}\,mol^{-1}}$, and T is the temperature in kelvin. The energy required to break an oxygen bond is 5.1 eV, which is equivalent to 492 kJ/mol.[1] Equation 4-2 gives the fraction of O_2

---

[1]The electron-volt (eV) is the energy required to move an electron through a potential difference of one volt; it is equal to 1.602×10^{-19} J, or 96.49 kJ/mol.

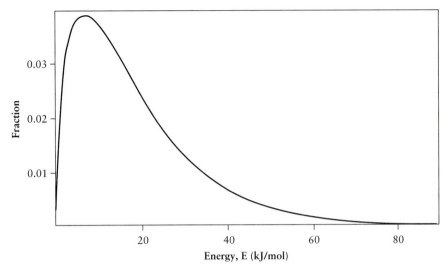

FIGURE 4-1 Distribution of kinetic energy for O_2 at 1500 K, expressed as the fraction of O_2 molecules having energies between E and (E + 1) kJ/mol.

molecules with energies E_c equal to or greater than 5.1 eV at 1500 K to be approximately 7×10^{-18}, which is too small to lead to even the most efficient thermal reactions involving oxygen atoms. Many reactions such as this one that are not feasible thermally can, however, be initiated by light, as will be shown.

4.2 ELECTROMAGNETIC RADIATION

4.2.1. Properties of Light

Monochromatic light is an electromagnetic wave of wavelength λ traveling at speed c ($= 3.0 \times 10^8$ m/s in a vacuum) and having a frequency ν equal to c/λ. One of the important characteristics or properties of an electromagnetic wave for photochemical purposes is that it can transport energy, as it does for example from the sun to the earth or from a lightbulb to an object. In addition, an electromagnetic wave can transport momentum, hence is capable of exerting a *radiation pressure*. Although so small that we do not ordinarily feel it, this radiation pressure has been measured under very carefully controlled conditions and is in complete agreement with classical theories of electromagnetic radiation.

Most of the experimental techniques used in photochemistry—polarization, prisms and gratings to produce monochromatic light, and so on—are based on the wave properties of light. It turns out, however, that many aspects of radiation, particularly its absorption and emission, are not adequately explained in this manner. Examples are the photoelectric effect and blackbody radiation, described in Section 3.1.1. The development of the Planck quantum hypothesis (1901), which was necessary to resolve these failures of the classical wave theory, led to the concept of the dual nature of matter and radiation and the theory of quantum mechanics.

In essence, quantum theory says that energy interacts with molecules only in discrete amounts rather than continuously, so that a light beam may be considered to be made up of a stream of discrete units (photons or quanta). Each photon has zero rest mass and possesses—and thus transports—a packet, or quantum, of energy ε given by the Einstein relationship

$$\varepsilon = h\nu = \frac{hc}{\lambda} \tag{4-3}$$

and momentum p

$$p = \frac{h}{\lambda} \tag{4-4}$$

where λ is the wavelength of the corresponding electromagnetic wave and h is Planck's constant, a universal natural constant equal to 6.63×10^{-34} J·s. Referring back now to the energetics of the oxygen dissociation reaction [equation (4-1)], it is easily shown from equation (4-3) that the bond dissociation energy of 492 kJ/mol corresponds to a wavelength of 243 nm:

$$\lambda = \frac{hc}{\varepsilon}$$

$$= \frac{(6.63 \times 10^{-34} \text{ J·s})(3 \times 10^8 \text{ m/s})(6.02 \times 10^{23} \text{ mol}^{-1})(10^9 \text{ nm/m})}{(492 \text{ kJ/mol})(1000 \text{ J/kJ})}$$

$$= 243 \text{ nm} \tag{4-5}$$

Thus when one photon of light having a wavelength of 243 nm is absorbed by an oxygen molecule, it will have sufficient energy to break apart or dissociate. Table 4-1 gives the energy ε of a photon and of an einstein of photons for several wavelengths and energy units.[2]

---

[2]An einstein is one mole (6.022×10^{23}) of photons.

TABLE 4-1

Photon Energies at Various Wavelengths

Wavelength	Photon energy, ε		
	J/photon	kJ/einstein	eV
1 m	2.0×10^{-25}	1.2×10^{-4}	1.24×10^{-6}
1 cm (10^{-2} m)	2.0×10^{-23}	0.012	1.24×10^{-4}
1 μm(10^{-6} m)	2.0×10^{-19}	120	1.24
1 nm (10^{-9} m)	2.0×10^{-16}	1.2×10^{5}	1.24×10^{3} (1.24 keV)
1 Å (10^{-10} m)	2.0×10^{-15}	1.2×10^{6}	1.24×10^{4} (12.4 keV)
1 pm (10^{-12} m)	2.0×10^{-13}	1.2×10^{8}	1.24×10^{6} (1.24 MeV)

4.2.2 Absorption of Light

The rather simple example just presented can be used to point out one of the basic principles of photochemistry (the Grotthus–Draper law, first formulated early in the nineteenth century): for light to be effective in producing photochemical transformations, not only must the photon possess sufficient energy to initiate the reaction, it must also be absorbed. Thus, even though we see from equation (4-5) that a 243-nm photon possesses enough energy to break the oxygen bond, in fact oxygen is *not* dissociated to any measurable extent when exposed to light of this wavelength because it is not absorbed.

A second important principle of photochemistry, which follows directly from quantum theory, is that absorption of radiation is a one-photon process: absorption of one photon excites one atom or molecule in the primary or initiating step, and all subsequent physical and chemical reactions follow from this excited species.[3] This principle is not strictly obeyed with the extremely high intensities possible from high-energy artificial light sources such as flash lamps or pulsed lasers. When such high-intensity radiation is used, the simultaneous absorption of two photons is possible. However, only single-photon absorption occurs in all environmental situations involving solar radiation.

In principle, electromagnetic radiation can extend essentially over an almost infinite range of wavelengths, from the very long wavelengths of thousands of miles and low energies [the relationship between the two given by equation (4-3)] to the very short wavelengths of 10^{-12} m, which is on the order of magnitude of nuclear dimensions. It is convenient to divide this electromagnetic spectrum into regions, as given in Table 4-2.

---

[3]As shown later, however, the overall effect from secondary reactions may be much greater than simply one molecule reacting per photon absorbed (see Section 4.4).

TABLE 4-2

The Electromagnetic Spectrum

Spectral region	Wavelength range	Energy range
Radio waves	> 10 cm	< 1.2 J/einstein
Microwave	10 cm–1 cm	1.2–12 J/einstein
Far infrared (IR)	1 cm–0.01 cm	12–1200 J (1.2 kJ)/einstein
Near IR	0.01 cm–700 nm	1.2–170 kJ/einstein
Visible	700 nm–400 nm	170–300 kJ/einstein
Near ultraviolet (UV)	400 nm–200 nm	300–600 kJ/einstein
Far (vacuum) UV	200 nm–100 nm	600–1200 kJ/einstein
X-rays and γ-rays	< 100 nm	> 1200 kJ/einstein

Photochemistry is generally limited to absorption in the visible, near-ultraviolet, and far-ultraviolet spectral regions, corresponding primarily to electronic excitation and (at sufficiently high energies) atomic or molecular ionization. Absorption in different spectral regions may lead to different electronic states of the absorbing molecule and electronically different intermediate species, as well as to different products of the photochemical reaction. It is therefore important to understand the designations of electronic states of atoms and molecules insofar as they relate to or indicate possible electronic transitions and transition probabilities. A summary of the recipes used in arriving at atomic and molecular state descriptions is given in Appendix A; the mathematical formulations of these principles are given in standard textbooks of quantum mechanics and spectroscopy, some of which are given in the Additional Reading for Appendix A.

Absorption in the low-energy radio, microwave, or infrared spectral regions results in no direct photochemistry unless very high-intensity laser radiation is used, the excitation for the most part only increasing the rotational and vibrational energies of the molecule and eventually being dissipated as heat. Some deleterious physiological effects do result from overexposure to microwave and infrared radiation (e.g., at radar installations or with improperly controlled microwave ovens), but these presumably are due to excessive localized heating. The chemical effects following illumination with very high-energy radiation (x-rays and γ-rays) are considered in Chapters 13 and 14.

It is worth noting that there is concern about the health effects of electric and magnetic fields (EMFs), primarily in two categories: ELF-EMFs, or extremely low frequencies (mainly 60 Hz) from power lines, household appliances, electric blankets, etc., and RF-EMFs, or radio frequencies transmitted by cellular phone antennas (800–1900 MHz, the frequency range between UHF radio/TV and microwave ovens). Although ELF-EMFs can have biological effects at very high intensities, electric fields are so greatly reduced by the human body that they are negligible compared with the normal body background electric fields; even the largest magnetic field normally encountered (~3000 milligauss) will not induce an electric current density comparable to that normally in the human body. Nevertheless, over the years several epidemiological studies have suggested fairly weak associations between exposures to ELF-EMFs, at levels found in typical residential areas, and some types of human cancers, particularly childhood leukemia.[4] On the other hand, there are other recent epidemiological studies that find no definitive link between typical ELF-EMF dosages and childhood leukemia.[5] Reviews in recent years have also arrived at conflicting conclusions. For example, one review strongly states that "there is no conclusive and consistent evidence that ordinary exposure to ELF-EMFs causes cancer, neurobehavioral problems or reproductive and developmental disorders."[6] However, an advisory panel to the National Institutes of Health concluded that although there is a "lack of positive findings in animals or in mechanistic studies," ELF-EMF is a "possible human carcinogen" and should be classified as such, and further strongly recommended that more research is needed.[7]

Many laboratory and epidemiological studies have also been carried out directed towards a possible link between RF-EMFs and brain cancer from handheld cellular phones where the antenna is placed near the head, even though they are operated at power levels (< 1W) well below which no known biological damage occurs. Most of these studies are controversial and inconclusive. However, the most recent (at the time of writing) reports—two case-control studies[8] and one cohort study of almost half a million cell telephone subscribers[9]—clearly find no association between cell phone usage and cancers of the brain. All agree, however, that risks associated with long-term exposures and/or potentially long induction periods cannot be ruled out. It is

[4]For example, a highly cited reference is M. Feychting and A. Ahlbom, Magnetic fields and cancer in children residing near Swedish high-voltage power lines, *Am. J. Epidemiol.*, **138**, 467–481 (1993).

[5]Such as M. S. Linet *et al.*, Residential exposure to magnetic fields and acute lymphoblastic leukemia in children, *N. Engl. J. Med.*, **337**, 1–7 (1997); M. L. McBride *et al.*, Power frequency electric and magnetic fields and risk of childhood leukemia in Canada, *Am. J. Epidemiol.*, **149**, 831–842 (1999), and correction, **150**, 223 (1999).

[6]National Research Council Committee on the Possible Effects of Electromagnetic Fields on Biologic Systems, *Possible Health Effects of Exposure to Residential Electric and Magnetic Fields*. National Academy Press, Washington, DC, 1997.

[7]NIEHS Report on Health Effects from Exposure to Power-Line Frequency Electric and Magnetic Fields, NIH Publication No. 99–4493. National Institutes of Health, Bethesda, MD, 1999.

[8]J. E. Muscat *et al.*, Handheld cellular telephone use and risk of brain cancer, *JAMA*, **284**, 3001–3007 (2000); P. D. Inskip *et al.*, Cellular-telephone use and brain tumors, *N. Engl. J. Med.*, **344**, 79–86 (2001).

[9]C. Johansen, J. D. Boice, Jr., J. K. McLaughlin, and J. H. Olsen, Cellular telephones and cancer—A nationwide cohort study in Denmark, *J. Natl. Cancer Inst.*, **93**, 203–207 (2001).

evident that continued long-term research is called for on the biological and biophysical effects from low-dosage exposure to ELF-EMFs and to RF-EMFs.

Direct physiological and biochemical responses occur from exposure to ultraviolet light. Because of the characteristics of atmospheric ozone absorption (see Section 5.2.3) and the degrees of deleterious biological effects resulting from absorption in this spectral region, the near-ultraviolet region (400–200 nm) is often divided into three bands: UV-A (400–320 nm), UV-B (320–290 nm), and UV-C (290–200 nm). Generally speaking, the UV-C band is virtually totally absorbed by atmospheric ozone and the UV-A band, although transmitted by ozone, is not carcinogenic at reasonable exposure levels,[10] so the UV-B band—partially transmitted by atmospheric ozone (see later: Figure 5-4) and directly absorbed by specific molecules, such as DNA or proteins—is for the most part the biologically active one.

The absorption of monochromatic electromagnetic radiation is given by the Beer–Lambert law, which says that the probability of light being absorbed by a single absorbing species is directly proportional to the number of molecules in the light path, which in turn is proportional to the concentration c (Beer's law) and the incremental thickness of the absorbing sample dx (Lambert's law)

$$\frac{dI}{I} = \text{probability of light of intensity } I \text{ being absorbed in thickness } dx$$

$$= \text{fraction that the intensity } I \text{ is reduced in thickness } dx$$

$$= -\alpha c \, dx \tag{4-6}$$

where α is the proportionality constant. The negative sign is included to account for the fact that the intensity is reduced by absorption, and thus dI is negative. Assuming that c and α are constant, integration of $-\alpha c \, dx$ from the incident light intensity I_0 at $x = 0$ to I at x

$$\int_{I_0}^{I} \frac{dI}{I} = \int_{I_0}^{I} d\ln I = -\int_{0}^{x} \alpha c \, dx \tag{4-7}$$

leads to

$$\ln\left(\frac{I}{I_0}\right) = -\alpha c x \tag{4-8}$$

or

$$\frac{I}{I_0} = e^{-\alpha c x} \tag{4-9}$$

[10]Radiation in the UV-A and visible regions has been implicated in malignant melanoma, possibly through energy or free-radical transfer from the broadly absorbing melanin. See R. B. Setlow, E. Grist, K. Thompson, and A. D. Woodhead, Wavelengths effective in induction of malignant melanoma, *Proc. Nat. Acad. Sci. USA*, **90**, 6666–6670 (1993).

The absorbance A is

$$A = \log(\frac{I_0}{I}) = \varepsilon cx \qquad (4\text{-}10)$$

where ε ($= \alpha/2.303$) is the extinction coefficient.[11] The fraction of light absorbed is given by

$$\frac{I_a}{I_0} = 1 - \frac{I}{I_0} = 1 - 10^{-\varepsilon cx} \qquad (4\text{-}11)$$

which holds for all concentrations. For small fractions of light absorbed—for example, at low concentrations of absorbers in the atmosphere—the fraction is directly proportional to the concentration of the absorbing species since $e^{-y} \cong 1 - y$ when y is much less than unity:

$$\frac{I_a}{I_0} = 1 - 10^{-\varepsilon cx} = 1 - e^{-2.303\varepsilon cx} \cong 2.303\varepsilon cx \qquad (4\text{-}12)$$

This relationship cannot hold at high concentrations as is shown by going to the extreme in the opposite direction, where $2.303\varepsilon cx$ is very large, so that essentially all the light is absorbed:

$$\frac{I_a}{I_0} \cong 1 - e^{-\infty} \cong 1 \qquad (4\text{-}13)$$

The light intensity I is normally expressed as the rate at which energy is transmitted through the cell or column of material, so that I_0 is then the light energy incident at the cell face per unit time—that is, it is the rate at which photons pass through the cell face. For reasons that will be apparent later, it is often convenient in photochemical systems to express the light absorbed, Ia, as the concentration of photons absorbed per unit time (photons volume^{-1} time^{-1}) or alternatively of einsteins absorbed per unit time (einsteins volume^{-1} time^{-1}). In these cases, I_0 is then the concentration of photons or einsteins passing through the cell face per unit time.

The units of α and ε are (concentration^{-1} length^{-1}), but obviously a variety of units may be used for concentration and length, leading to different values for α and ε. If c is given in molarity M (mol/dm^3) and x in centimeters, then α and ε have the units $M^{-1}cm^{-1}$ and ε is called the *molar extinction coefficient*, ε_M (sometimes referred to as the molar absorption coefficient or the absorptivity). Frequently, however, in gaseous systems such as the atmosphere it is convenient to express the concentration in pressure units. These are generally related to concentration units at low pressures by the ideal gas equation

---

[11]The constant α is also sometimes called the extinction coefficient. In usage here, however, we shall always refer to the extinction coefficient as the constant ε in the Beer–Lambert law in the decadic (base 10) form, equation (4.10), and to α as the proportionality constant in equation (4-6).

$P = (n/V)RT = cRT$, and therefore the temperature must be specified. If the pressure is to be expressed in atmospheres, $R = 0.08206 \, \text{dm}^3 \text{atm} \cdot \text{K}^{-1} \text{mol}^{-1}$.

If x is expressed in centimeters and the concentration of the absorbing species in molecules per cubic centimeter, as commonly used by atmospheric scientists, then the proportionality constant α in the base e form of the Beer–Lambert law, equation (4-8), has the units $\text{cm}^2/\text{molecule}$. This is called the *absorption cross section* and given the symbol σ. For very strongly absorbing species, at maximum absorption σ is roughly the order of magnitude of physical cross sections of molecules. In most cases we will use this absorption cross section and the symbol σ for expressing the absorption curves.

Table 4-3 summarizes and compares several of these Beer–Lambert law quantities for expressing the absorption of light by a single absorbing species.

As a comparison of these various ways of expressing the proportionality constants, consider a hypothetical gaseous species at a pressure of 1 matm (10^{-3} atm) and 25°C (298.15 K) that absorbs 40% of the incident light in 1 cm (i.e., is strongly absorbing). It follows from equations (4-9) and (4-12) that

$$\varepsilon_M = 5.43 \times 10^3 M^{-1} \text{cm}^{-1} = 222 \text{ atm}^{-1} \text{cm}^{-1}$$

$$\alpha = 1.25 \times 10^4 M^{-1} \text{cm}^{-1} = 511 \text{ atm}^{-1} \text{cm}^{-1}$$

and

$$\sigma = 2.1 \times 10^{-17} \text{cm}^2/\text{molecule}$$

Strictly speaking, the Beer–Lambert law applies only to monochromatic radiation, since ε is a function of wavelength. The extent to which use of nonmonochromatic light leads to significant error in the determination of concentration depends on the spectral characteristics of the absorbing and illuminating system. If the extinction coefficient and incident intensity are known as functions of wavelength, however, the total amount of light absorbed may be obtained by integrating over all wavelengths. Also, if there are i absorbing species or components present, the Beer–Lambert equation in base 10 form [equation (4-10)] becomes

TABLE 4-3

Beer–Lambert Law Quantities

Quantity	Name	Definition	Units
A	Absorbance	$\log_{10}(I_0/I)$	Unitless
α	Proportionality constant	$[\ln_e(I_0/I)]/cx$	Concentration^{-1} length^{-1}
ε	Extinction coefficient	$[\log_{10}(I_0/I)]/cx$	Concentration^{-1} length^{-1}
ε_M	Molar extinction coefficient	$[\log_{10}(I_0/I)/c(\text{mol}/\text{dm}^3)x(\text{cm})$	$\text{dm}^3 \text{mol}^{-1} \text{cm}^{-1}$
σ	Absorption cross section	$[\ln_e(I_0/I)]/c(\text{molecules}/\text{cm}^3)x(\text{cm})$	$\text{cm}^2/\text{molecule}$

$$\frac{I}{I_0} = 10^{-\sum_i \varepsilon_i c_i x} \tag{4-14}$$

where the exponential part of the equation represents the sum of the $\varepsilon c x$ terms for all the i absorbing species.

4.3 KINETICS OF THERMAL PROCESSES

Most chemical reactions are *kinetically complex*. That is, they take place by means of a series of two or more consecutive steps rather than by means of a single encounter of the reacting species. Nevertheless, the overall reaction between, say, two reactants A and B can be represented by the generalized equation

$$aA + bB \rightarrow cC + dD + \cdots \tag{4-15}$$

where the stoichiometry of the reaction is given by the coefficients $a, b, c, d, \ldots$ The *rate*, v, is then given by the differential equation

$$v = -\frac{1}{a}\frac{dn_A}{dt} = -\frac{1}{b}\frac{dn_B}{dt} = \frac{1}{c}\frac{dn_C}{dt} = \frac{1}{d}\frac{dn_D}{dt} = \cdots \tag{4-16}$$

where n_A is the number of moles of A, n_B is the number of moles of B, and so on, and dn_A/dt is the change in the number of moles of A per unit time. For reactions in which the volume of the system (V) is independent of time, the molar concentration of A is $[A] = n_A/V$ and $d[A] = V dn_A$. Similarly, $d[B] = V dn_B$, $d[C] = V dn_C$, and $d[D] = V dn_D$. It follows that the rate of the reaction per unit volume, r, usually referred to simply as the rate of the reaction, is

$$r = \frac{v}{V} = -\frac{1}{a}\frac{d[A]}{dt} = -\frac{1}{b}\frac{d[B]}{dt} = \frac{1}{c}\frac{d[C]}{dt} = \frac{1}{d}\frac{d[D]}{dt} = \cdots \tag{4-17}$$

In certain cases it is convenient to express the rate with respect to the rate of change in concentration of a single reactant or product:

$$r_A = \text{rate of change of A} = -\frac{d[A]}{dt}, \quad r_C = \text{rate of change of C} = \frac{d[C]}{dt}, \text{etc.} \tag{4-18}$$

However, it is important to remember that in general $r_A \neq r_C$, although the two rates are related by the proportionality factor c/a.

The expression for the experimentally determined rate of reaction in terms of the composition of the system can be quite involved. However, in many cases at constant temperature it is of the form

$$r = k[A]^\alpha [B]^\beta [C]^\gamma [D]^\delta \cdots [X]^\chi \cdots \tag{4-19}$$

where k is the temperature-dependent rate constant and X is neither a reactant nor a product—that is, it is a *catalyst*. The overall order of the reaction is the sum of the exponents $\alpha + \beta + \gamma + \delta + \cdots + \chi + \cdots$, and the order with respect to A is α, etc. For kinetically complex reactions there is in general no relationship between the exponents $\alpha, \beta, \gamma,$ and δ in equation (4-19) and the coefficients $a, b, c,$ and d in the stoichiometric equation (4-15).

A special situation for equation (4-19) is the *kinetically simple* reaction that involves only the single step of reactants coming together. In this case the rate of the reaction at a given temperature is proportional only to the concentrations of the reacting species—that is, it does not depend on the nature or concentrations of the products of the reaction. The overall order of the reaction is then the number of reactant molecules (called the *molecularity* of the reaction) that must come together in a single encounter for the reaction to occur. Such reactions are called *unimolecular, bimolecular,* or *trimolecular,* depending on the number of reactant molecules; this number can be one, two, or three, respectively.[12] Table 4-4 gives the rate expressions for the various possibilities of kinetically simple reactions.

The integrated solutions of the corresponding differential equations [defined by equation (4-17)] for the rates of these kinetically simple reactions, leading to reactant concentrations as a function of reaction time t, can be obtained by standard integration techniques.[13] These integrated forms are used to obtain reaction rate constants from experimental time-dependent concentration data.

A chemically complex reaction is the combination of two or more kinetically simple steps, giving the *mechanism* of the reaction. We will encounter complex reaction mechanisms in Chapter 5 in connection with many physical and chemical interactions occurring in the troposphere and stratosphere. Simple differential equations can be written for each kinetically simple step in a mechanism; however, integration of the new combined differential equation becomes difficult and in most cases impossible by the usual analytical techniques, since products of one kinetically simple step generally become reactants in one or more other steps in the mechanism.[14] Numerical methods

---

[12]The probability that more than three molecules may come together in a single collision is so small that such events are not considered in kinetic treatments.

[13]See C. Capellos and B. H. J. Bielski, *Kinetic Systems.* Wiley-Interscience, New York, 1972. Two examples: For the unimolecular (first-order) reaction $A \xrightarrow{k}$ product, we have

$$k = \frac{1}{t}\ln\frac{[A]_0}{[A]} \tag{4-20}$$

and for the bimolecular (second-order) reaction $A + B \xrightarrow{k}$ product, we write

$$k = \frac{1}{t([B]_0 - [A]_0)}\ln\frac{[A]_0[B]}{[A][B]_0} \tag{4-21}$$

[14]Examples of some solutions of complex reactions are given by Capellos and Bielski (see note 13).

TABLE 4-4

Kinetically Simple Rates of Reaction

I. Unimolecular (first-order)

$\quad$ A $\xrightarrow{k}$ products $\qquad r = k[A]$

II. Bimolecular (second-order)

$\quad$ 2A $\xrightarrow{k}$ products $\qquad r = k[A]^2$

$\quad$ A + B $\xrightarrow{k}$ products $\qquad r = k[A][B]$

III. Trimolecular (third-order)

$\quad$ 3A $\xrightarrow{k}$ products $\qquad r = k[A]^3$

$\quad$ 2A + B $\xrightarrow{k}$ products $\qquad r = k[A]^2[B]$

$\quad$ A + B + C $\xrightarrow{k}$ products $\qquad r = k[A][B][C]$

---

of integration are now possible for very complex mechanisms utilizing the tremendous computing power of present-day digital computers. Of course, justification of a mechanism (or model) by computer simulation of this type requires good rate constants and spatial and temporal (space and time) behavior of all pertinent species.

Two time quantities are sometimes used to describe the extent of a chemical reaction. One is the *half-life*, $t_{1/2}$, which is the time required for the reaction to be half-way completed; the other is the *mean-life*, τ, which is the average of the lifetimes of all the reacting molecules. In general $t_{1/2}$ and τ depend on the concentrations of the species involved in the reaction. A unique case (and one for which $t_{1/2}$ and τ are frequently used) is that of a first-order reaction A $\xrightarrow{k}$ products (such as radioactive decay; see Chapter 13), where it can be shown that $t_{1/2}$ and τ are independent of concentration:

$$t_{1/2} = \frac{\ln 2}{k} = \frac{0.693}{k} \tag{4-22}$$

and

$$\tau = \frac{1}{k} \tag{4-23}$$

The mean-life τ for a first-order reaction is the time required for the concentration of the reactant to decrease to $1/e\,(1/2.718)$ of its initial value, in contrast to $t_{1/2}$, which is the time required for the concentration of the reactant to decrease to half of its initial value.

The dependence of the rate on temperature is embodied in the rate constant k. The most useful relationship expressing this dependence is the empirical Arrhenius equation

$$k = Ae^{-E_a/RT} \tag{4-24}$$

or

$$\ln k = \ln A - \frac{E_a}{RT} \tag{4-25}$$

In this treatment the preexponential factor A and the activation energy E_a are assumed to be temperature independent, so that from equation (4-25) a plot of $\ln k$ vs $1/T$ should be linear, with a slope equal to $-E_a/R$. This behavior is found to be the case for many reactions, and this method involving the plot of $\ln k$ vs $1/T$ is the most common one for experimentally determining E_a. However, Arrhenius behavior will not necessarily be followed for kinetically complex reactions, and indeed reactions with very marked deviations from that predicted by equation (4-25) are encountered.

The simple collision theory of chemical kinetics, applicable to kinetically simple bimolecular gas-phase reactions between two unlike hard-sphere species 1 and 2, considers the rate of reaction $r\,(=kc_1c_2)$ to be proportional to the frequency of bimolecular collisions Z_{12}. The proportionality constant equating r and Z_{12} includes the Boltzmann factor [equation (4-2)], which in this case is the fraction of molecules in two dimensions defined by the trajectories of motion of two colliding particles with energies equal to or greater than a potential energy barrier E_c. Note that this is the same form of exponential term found in the Arrhenius equation (4-24); the difference between the two is that E_a is the experimentally determined empirical activation energy, whereas E_c is associated with a specific potential energy barrier that must be overcome for the kinetically simple reaction to occur. The rate is also proportional to a temperature-independent steric factor p, which allows for the possibility that not all colliding particles with sufficient energy to overcome the potential barrier do actually react. (For example, the molecules might not react unless they are in a particular orientation.) Thus,

$$r = pZ_{12}e^{-E_c/RT} = kc_1c_2 \tag{4-26}$$

so that

$$k = p\left(\frac{Z_{12}}{c_1c_2}\right)e^{-E_c/RT} = pZ'_{12}e^{-E_c/RT} \tag{4-27}$$

where Z'_{12} is the specific collision frequency and is proportional to $T^{1/2}$. The preexponential term in the Arrhenius equation (4-24) is therefore slightly temperature dependent, and this is in fact observed for reactions with low or zero activation energies, although any small temperature effect in this term is completely masked by the exponential term for reactions with large activation energies. For small species such as atoms and diatomic molecules, $Z_{12} \approx 2 \times 10^{-10}$ cm^3 molecule^{-1}s^{-1}. If the reaction goes at every collision (i.e., $E_c = 0, p = 1$), then $k = Z'_{12}$ and is called the *collisional* rate constant.

A kinetically simple *trimolecular* gas-phase reaction requires the simultaneous encounter of three molecules. Collisions of this type (*triple* collisions) are very rare in comparison to binary collisions and are usually encountered only in two-body gas-phase combination reactions where simultaneous conservation of energy and momentum requires the presence of a third body. Trimolecular atom or small-radical combination processes in the presence of inert third-body molecules typically have triple-collision rate constants of the order of 10^{-32} cm^6 molecule^{-2} s^{-1}. We will see several reactions of this type in Chapter 5.

4.4 KINETICS OF PHOTOCHEMICAL PROCESSES

The second principle of photochemistry, that absorption is a one-photon process, taken together with quantum theory, requires that each quantum that is absorbed bring about a change (excitation) in that one molecule. In essence, this may be considered to be a "bimolecular" process involving interaction of one photon and one molecule. What happens to the excitation energy, however, depends on the amount of energy absorbed and the nature of the excited molecule. For example, the excited molecule can "lose" all or part of its excitation energy by transferring energy to another species via a non-reactive collision (*collisional quenching*), or it can emit energy in the form of a photon of light (*fluorescence* or *phosphorescence*). Or, the excited molecule may undergo chemical processes (*isomerize, dissociate, ionize,* etc.), undergo chemical reactions with other species, or transfer its energy by collision to another molecule that ultimately results in a chemical transformation (*sensitization*). Thus, the overall number of molecules reacting chemically as a result of this absorption of a single photon may be virtually any value ranging from zero to a very large number ($> 10^6$). This latter is interpreted as arising from subsequent thermal chain reactions. We will encounter various reactions representing these physical and chemical processes in subsequent chapters.

The efficiency of a photochemical reaction is the *quantum yield*. The *overall* quantum yield Φ_J of a substance J, which may be a specific reactant or a stable product, is simply the total number of the specific reactant molecules lost or stable product molecules formed for each photon absorbed. Thus, for the generalized reaction (4-15):

$$\Phi_A = \frac{\text{number of A molecules reacted}}{\text{number of photons absorbed}}$$
$$= \frac{\text{number of moles of A reacted}}{\text{number of einsteins absorbed}} \qquad (4\text{-}28)$$

and

$$\Phi_C = \frac{\text{number of C molecules produced}}{\text{number of photons absorbed}}$$

$$= \frac{\text{number of moles of C produced}}{\text{number of einsteins absorbed}} \qquad (4\text{-}29)$$

Since r_A is the rate that A reacts, r_C is the rate that C is produced, and I_a by definition is the rate of light absorption, it follows that

$$\Phi_A = \frac{r_A}{I_a}, \quad \Phi_C = \frac{r_C}{I_a}, \text{ etc.} \qquad (4\text{-}30)$$

This overall quantum yield Φ_J represents the overall efficiency of the reaction initiated by the absorption of light, which does not have to be absorbed explicitly by a reactant molecule. As already pointed out, however, the initial excitation act of absorption of a photon of light by a molecule X,

$$X \xrightarrow{h\nu} X^* \qquad (4\text{-}31)$$

may be followed by a variety of physical and chemical primary processes (fluorescence, deactivation, sensitization, etc.) involving the excited molecule X^*. (Subsequent thermal chemical and physical reactions are considered secondary processes.) We can define a *primary quantum yield* ϕ_i as the fraction of X^* molecules undergoing a specific (*i*th) primary process. Thus, for fluorescence, f, in which light is emitted by an excited molecule,

$$\phi_f = \frac{\text{number of } X^* \text{ molecules emitting light}}{\text{number of photons absorbed by X}} \qquad (4\text{-}32)$$

$$= \frac{\text{fluorescence rate}}{\text{rate of light absorption}} = \frac{r_f}{I_a} \qquad (4\text{-}33)$$

Thus, in general, for the *i*th primary process the rate r_i is

$$r_i = \phi_i I_a \qquad (4\text{-}34)$$

The symbol $h\nu$ above the arrow in reaction (4-31) indicates that this is the light-absorbing reaction. Although this symbol should be used only for the kinetically simple light-absorbing step, it is often used even in the overall stoichiometric equation to signify a light-initiated reaction in which the overall mechanism consists of a complete set of primary and secondary processes. Since the energy of an absorbed photon has to go somewhere, to the extent that light absorption is a "one-for-one" event (the second principle of photochemistry), the maximum value for a single individual primary quantum yield is unity and the sum of all of the primary quantum yields for both physical and chemical processes must be unity:

$$\sum_i \phi_i = 1 \qquad (4\text{-}35)$$

where the summation includes all i modes of disappearance of the photoexcited species, including direct chemical reaction.

Additional Reading

Calvert, J. G., and J. N. Pitts Jr., *Photochemistry*. Wiley, New York, 1966.
Gilbert, A., and J. Baggott, *Essentials of Molecular Photochemistry*. CRC Press, Boca Raton, FL, 1991.
Logan, S. R., *Fundamentals of Chemical Kinetics*. Longman, Essex, U. K., 1996.
Moore, J. W., and R. G. Pearson, *Kinetics and Mechanism*. 3rd ed. Wiley, New York, 1981.
Suppan, P., *Chemistry and Light*. Royal Society of Chemistry, Cambridge, U. K., 1994.
Turro, N. J., *Modern Molecular Photochemistry*. Benjamin/Cummings, Menlo Park, CA, 1978.
Wayne, R. P., *Principles and Applications of Photochemistry*. Oxford University Press, Oxford, U. K., 1988.

EXERCISES

4.1. Ozone, O_3, is a weak absorber of electromagnetic radiation in the UV spectral region; its maximum absorption in the near ultraviolet region is at 255 nm.
   (a) Calculate the energy of a 255-nm photon in angstrom units (Å).
   (b) Calculate the energy of a 255-nm photon in joules per photon, in kilojoules per einstein, and in electron-volts.
   (c) Calculate the frequency of a 255-nm photon in reciprocal seconds and in reciprocal centimeters.
   (d) Calculate the momentum of a 255-nm photon in: kilogram-meters per second and in gram-centimeters per second.
   (e) How does the momentum of a 255-nm photon compare with the momentum of an ozone molecule at 25°C moving at a velocity of 400 m/s—that is, what is P_{photon}/P_{ozone}?

4.2. Give the spectral region in which each of the following photons might absorb:
   (a) a 5000-Å-wavelength photon
   (b) an 8-eV-energy photon
   (c) a 200-mile-wavelength photon
   (d) a 10^{13}-s^{-1}-frequency photon
   (e) a 5-cm-wavelength photon

4.3. Nitrogen dioxide (NO_2) is a major participant in photochemical smog (Chapter 5). It absorbs visible and ultraviolet light, with maximum absorption at 400 nm, where its absorption cross section σ is 6×10^{-19} cm^2/molecule.
   (a) Calculate its molar extinction coefficient ε_M at 400 nm and its extinction coefficient at 400 nm and 20°C in units of $atm^{-1}cm^{-1}$.

(b) Assume that the concentration of NO_2 in a polluted atmosphere above a city is constant at 2×10^{12} molecules/cm^3 to an altitude of 1000 ft and that the temperature is also constant at 20°C. Calculate the percent of sunlight absorbed by the NO_2 at 400 nm.

4.4. EDTA (ethylenediaminetetraacetic acid, MW 292.2) is an important industrial complexing agent (Section 9.5.6). It is poorly degraded in municipal sewage treatment plants and therefore is found in many surface waters. However, it forms a complex with the ferric ion (Fe^{3+}), Fe(III)-EDTA, which can be photochemically degraded [F. G. Kari, S. Hilger, and S. Canonica, *Environ. Sci. Technol.*, **29**, 1008–1017 (1995)]. At 366 nm, the molar extinction coefficient of Fe(III)-EDTA is 785 dm^{3}mol^{-1}cm^{-1} and the quantum yield of degradation is 0.034. In a sample of river water, the concentration of EDTA was found to be 280 μg/dm^3.
   (a) Calculate the fraction of light absorbed in 1-cm thickness (light path) of the river water sample at 366 nm if sufficient ferric ion is added to complex all the EDTA.
   (b) The intensity of 366-nm sunlight is approximately 10^{14} photons cm^{-3}s^{-1}. Calculate the time required to degrade all the Fe(III)-EDTA in one cubic centimeter of the sample in a cell 1 cm thick, assuming that the rate of the reaction does not change as the absorbing material is used up (a very poor assumption!).

4.5. We will see in Chapter 5 that the abstraction reaction of a hydrogen atom from an alkane molecule by an hydroxyl radical is an important initiating step in many atmospheric decomposition processes. This is a kinetically simple bimolecular reaction:

$$RH + \cdot OH \xrightarrow{k} R\cdot + H_2O$$

At 298 K (25°C), the rate constant $k_{298} = 6.9 \times 10^{-15}$ cm^{3}molecule^{-1}s^{-1} for RH = CH_4. Assume that you could have a closed flask containing initially methane and the hydroxyl radical at their estimated daytime concentrations in a typical polluted urban atmosphere: $[CH_4] = 4.2 \times 10^{13}$ molecules/cm^3 and $[\cdot OH] = 2.5 \times 10^6$ radicals/cm^3.
   (a) Calculate the initial rate of the reaction r,
   (b) Calculate the time required for half the ·OH radicals in the flask to be used up. (*Hint:* The initial concentration of methane is much greater than that of the hydroxyl radical, so the methane concentration can be assumed to be constant, giving an apparent first-order reaction.)
   (c) Assume now that all sources of methane in the troposphere were suddenly stopped, but that the hydroxyl radical concentration remained constant. Calculate how long it would take for half the methane molecules to be used up, assuming that the foregoing reaction is the only one leading to the disappearance of methane.

5

ATMOSPHERIC PHOTOCHEMISTRY

5.1 INTRODUCTION

We have already noted some of the important photochemical reactions occurring in the atmosphere that are important to the nature of our environment. In Section 2.3.4 we saw that photodissociation of water vapor in the upper atmosphere may be an important source of atmospheric oxygen. Molecular oxygen is photodissociated into oxygen atoms in the thermosphere, mesosphere, and upper stratosphere, and in so doing it absorbs most of the high-energy radiation below 180 nm. Ozone is produced in the stratosphere by the combination of an oxygen atom and an oxygen molecule. We saw in Section 2.5 that O_3 extends the earth's shield against lethal ultraviolet radiation from 180 nm to about 290 nm. The troposphere connects the biosphere (i.e., the earth's surface) to the stratosphere and therefore is closely involved in the chemistry of both. For example, anthropogenic emissions of oxides of nitrogen and hydrocarbons, from internal combustion automobile engines and stationary furnaces, are major contributors to photochemical reactions in the troposphere that drastically affect the quality of our environment. Strong vertical mixing occurs in the troposphere, allowing species like methane and water that are produced naturally, as well as anthropogenic substances such as oxides of nitrogen and chlorofluorocarbons, to be transported upward to contribute to photochemical processes occurring in the stratosphere.

It is pointed out in Chapter 4 that while direct photochemistry is primarily limited to initiation by *electronic* excitation of the absorbing molecule, different electronic states can result from absorption in different spectral regions, which in turn may lead to different intermediate electronic species and chemical products. A summary of the methods for generating the designations (or *term symbols*) for atoms and simple molecules is given in Appendix A. The following is a brief summary of these state designations.

The term symbol for an atom can be written in terms of atomic quantum numbers S, L, and J as

$$^{(2S+1)}L_J$$

where S represents the net (i.e., the resultant or vector sum) *spin* angular momentum (or simply the spin) of all the electrons of the atom; it can be zero or an integral number for an even number of electrons in the atom, or a half-integral number for an odd number of electrons. The quantity $(2S + 1)$ is called the multiplicity. The quantum number L describes the net electronic *orbital* angular momentum, and is given the symbols S, P, D, F, ..., for the allowed values of $L = 0, 1, 2, 3, \ldots$, respectively. J is a quantum number designating the total angular momentum of the atom—that is, the vector sum of the spin (S) and orbital (L) angular momenta. For example, for an oxygen atom (eight electrons) in its lowest (ground) electronic state, $S = 1$, $L = 1$, and $J = 2$, so the term symbol for the ground-state O atom is $^3P_2$.

For a diatomic molecule, the total term symbol is often given as

$$^{(2S+1)}\Lambda^{(+ \text{ or } -)}_{(g \text{ or } u)}$$

where S and $(2S + 1)$ are the same as for atoms, Λ represents the total orbital angular momentum along the internuclear axis (analogous to L, Λ is given the symbols $\Sigma, \Pi, \Delta, \Phi, \ldots$, for $\Lambda = 0, 1, 2, 3, \ldots$, respectively), and (g or u) and (+ or −) are associated with two types of symmetry operation that the diatomic molecule can undergo. States in which $S = 0$, 1/2, and 1 (multiplicity = 1, 2, and 3) are called *singlet*, *doublet*, and *triplet* states, respectively.

For polyatomic molecules, in many cases it is sufficient to designate an excited state in terms of the type of molecular orbital *from* which an electron is excited (the initial state) and the molecular orbital *to* which it is excited (the final state). In addition, the multiplicity $(2S + 1)$ of the molecule is included in the designation. Specific examples are given in Appendix A.

Radiative absorption leading to transitions between electronic states (in contrast to collisional energy transfer) are limited by *selection rules*. The following rules give the allowed transitions:[1]

1. For atoms:
   (a) $\Delta S = 0$ [Δ(multiplicity) = 0]; transitions where $\Delta S \neq 0$ are called *spin-forbidden*
   (b) $\Delta L = 0, \pm 1$
   (c) $\Delta J = 0, \pm 1$, but $J = 0 \rightarrow J = 0$ is not allowed
2. For diatomic and light-nuclei polyatomic molecules:
   (a) $\Delta S = 0$
   (b) $\Delta \Lambda = 0, \pm 1$
   (c) $u \longleftrightarrow g$;
   (d) $+ \longleftrightarrow +$ and $- \longleftrightarrow -$

Sometimes we will use the complete designations to provide specific information about the states. In other situations, where a specific state is not key to our understanding of the system or the process being given, involvement of an electronically excited state generally is represented simply by an asterisk (*). If no designation is used, the species is in its ground electronic state.

5.2 REACTIONS IN THE UPPER ATMOSPHERE

5.2.1 Nitrogen

We have seen that molecular nitrogen is the most prevalent species in the atmosphere. The reason that nitrogen is not so important photochemically is its large bond energy (7.373 eV, corresponding to a photon of wavelength $\lambda = 169$ nm), which limits its photodissociation chemistry to areas above the ozone layer. Molecular nitrogen absorbs only weakly between 169 and 200 nm, absorption in this region leading to a spin-forbidden excitation from ground-state N_2 to the $(^3\Sigma_u^+)$ excited state

$$N_2 \xrightarrow{h\nu} N_2(^3\Sigma_u^+) \tag{5-1}$$

---

[1]The selection rules are given here without proof or justification, which can be found in standard quantum mechanics and spectra reference books, such as M. D. Harmony, *Introduction to Molecular Energies and Spectra*, Holt, Reinhart, & Winston, New York, 1972, pp. 450–455. Strictly speaking, these selection rules are valid ("allowed" transitions) only for small molecules made up of light atoms; "forbidden" transitions can occur in larger molecules, particularly those containing heavy atoms, but they are generally of very low intensity.

that is a source of highly excited oxygen atoms by transfer of energy through collisions with ground-state O:

$$N_2(^3\Sigma_u^+) + O \rightarrow N_2 + O(^1S_0) \tag{5-2}$$

With photons of wavelengths below 100 nm, photodissociation occurs, possibly through photoionization and dissociative recombination involving ground-state O_2 and producing nitric oxide (NO)[2]:

$$N_2 \xrightarrow{h\nu\,(<\,100\,\text{nm})} N_2^+ + e^- \tag{5-3}$$

$$N_2^+ + O_2 \rightarrow NO^+ + NO \tag{5-4}$$

$$NO^+ + e^- \rightarrow N^* + O \tag{5-5}$$

An energetically excited nitrogen atom can be collisionally deactivated to the ground state by O_2, producing singlet oxygen (discussed in the following section),

$$N^* + O_2 \rightarrow N + O_2(^1\Delta_g) \tag{5-6}$$

or it can react with O_2 to produce NO:

$$N^* + O_2 \rightarrow NO + O \tag{5-7}$$

Nitric oxide might also be expected to form by direct combination of ground-state nitrogen and oxygen atoms,

$$N + O \xrightarrow{\;M\;} NO \tag{5-8}$$

where M can be any other atomic or molecular species such as N, O, N_2, or O_2. This is a kinetically simple trimolecular reaction that occurs at essentially every triple collision between an N, an O, and an M species with a rate constant of the order of 10^{-32} cm^6 molecule^{-2} s^{-1} (see Section 4.3). However, since photodissociation of N_2 only occurs at high altitudes above the ozone layer where the total concentration of M is very small, this recombination reaction is not a significant source of atmospheric NO.

As we will see in Sections 5.2.3, 5.3.2, and 5.3.3, NO plays a significant role in stratospheric photochemistry and it is also an important constituent in photochemical smog in the troposphere. At stratospheric heights it comes

---

[2]NO has an odd number of electrons (15), and therefore one of them must be unpaired. This unpaired electron occupies an antibonding π^* orbital and is spread over the whole molecule. Nitrogen dioxide (NO_2) also has an odd number of electrons (23) and thus also an unpaired electron, but in this case the electron is mostly localized on the N atom. Although NO and NO_2 are technically free radicals and are involved in many free-radical reactions, in this book we will not indicate the lone (unpaired) electron on each of them as a dot; this will also be the case for *atoms* with odd numbers of electrons, such as hydrogen or the halogens. The dot *will* be used on other free radicals in this book.

mainly from nitrous oxide, N_2O, by reaction with excited $O(^1D_2)$ oxygen atoms:

$$N_2O + O(^1D_2) \rightarrow 2NO \qquad (5\text{-}9)$$

It is generally accepted that the N_2O is not produced directly in the atmosphere, but rather in the biosphere, primarily by microorganism reactions in soils and oceans (roughly two-thirds natural, one-third anthropogenic). Nitrous oxide is a strong greenhouse gas (Table 3-2). It does not react with the hydroxyl radical ($\cdot OH$, the major initiator of removal of most trace gases in the atmosphere—see Section 5.3.2), nor does it absorb light above 260 nm, and therefore it is very stable in the troposphere. It is transported to the stratosphere primarily by large-scale movements of air masses within the troposphere. The concentration of N_2O in the atmosphere is now increasing above the preindustrial level at a rate of approximately 0.25%/year, due in large part to anthropogenic biomass burning and bacterial oxidation of fertilizer nitrogen (NH_4^+).[3] In addition to reaction (5-9), N_2O is destroyed in the stratosphere by photolysis. It absorbs light in a broad continuum starting at 260 nm, with a maximum at about 182 nm. The major photodissociation reaction is

$$N_2O \xrightarrow{h\nu\,(\leq\,260\,\text{nm})} N_2 + O(^1D_2) \qquad (5\text{-}10)$$

providing one stratospheric source of excited $^1D_2$ oxygen atoms. Other sources are discussed in the following section.

5.2.2 Oxygen

By far the most abundant photochemical reaction in the upper atmosphere is the photolysis of molecular oxygen. Figure 5-1 shows the potential energy curves for the various electronic states of O_2 of importance to us; Figure 5-2 is the absorption spectrum in the far-ultraviolet and vacuum ultraviolet regions (note logarithmic absorption cross section vertical axis). It was shown in Section 4.2.1 that the bond dissociation energy of oxygen (5.1 eV, or 492 kJ/mol) corresponds to a photon of wavelength 243 nm. Oxygen actually begins to absorb just below this wavelength, at 242.2 nm, in a spectral region known as the *Herzberg continuum*. Absorption is very weak ($\sigma \cong 10^{-23}$ cm²/molecule at 202 nm), but the spectrum is a continuum in this region, indicating that dissociation occurs with $\phi = 1$. Actually, excitation is to the weakly bound $^3\Sigma_u^+$ state (Figure 5-1, curve A) which rapidly dissociates into two ground-state atoms:

$$O_2\left(^3\Sigma_g^-\right) \xrightarrow{h\nu\,(<\,242.2\,\text{nm})} O_2\left(^3\Sigma_u^+\right) \rightarrow 2O \qquad (5\text{-}11)$$

[3]See Figure 10-6.

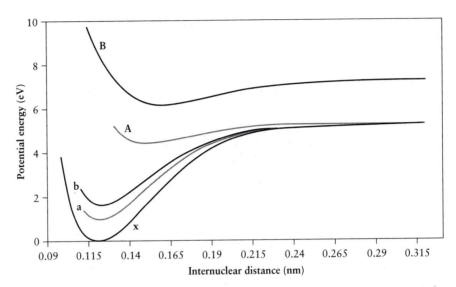

FIGURE 5-1 Potential energy curves for molecular oxygen. $X = {}^3\Sigma_g^-$, $a = {}^1\Delta_g$, $b = {}^1\Sigma_g^+$, $A = {}^3\Sigma_u^+$, $B = {}^3\Sigma_u^-$. Drawn from data of F. R. Gilmore, *J. Quant. Spectroscopy Radiat. Transfer*, 5, 369–390 (1965).

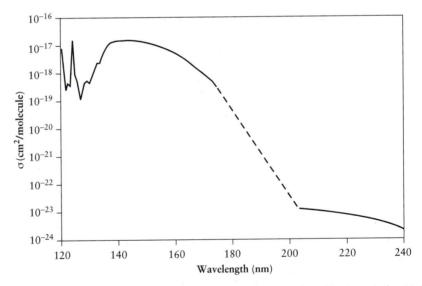

FIGURE 5-2 Absorption spectrum of molecular oxygen (note logarithmic vertical axis). The dotted line is the Schumann-Runge banded region. Drawn from data of M. Ackerman, in *Mesospheric Models and Related Experiments*, G. Fiocco, ed., D. Reidel, Dordrecht, 1971, pp. 149–159.

The very weak absorption shown in this case is a good example of the consequence of selection rule violation, the specific one in this case being that excitation from a negative (superscript $-$) to a positive (superscript $+$) state is a forbidden transition. The other selection rules given in Section 5.1 for diatomic molecules are obeyed, however.

Below 205 nm the absorption spectrum becomes much stronger and banded (the *Schumann–Runge bands*). The appearance of the banded spectrum rather than a continuum shows that absorption is now to a different potential energy curve from which dissociation no longer occurs at these energies. The Schumann–Runge bands get closer together as the wavelength decreases, and at 175 nm the bands blend together into a continuum known as the *Schumann–Runge continuum* which reaches maximum absorption at approximately 147 nm. Excitation in this Schumann–Runge region is to the bound $^3\Sigma_u^-$ state, curve B of Figure 5-1:

$$O_2\left(^3\Sigma_g^-\right) \xrightarrow{h\nu\,(<\,205\,\text{nm})} O_2\left(^3\Sigma_u^-\right) \qquad (5\text{-}12)$$

This is an allowed transition, hence has a large absorption cross section ($\sigma_{max} \cong 1.5 \times 10^{-17}\,\text{cm}^2/\text{molecule}$). The $^3\Sigma_u^-$ state dissociates at wavelengths shorter than 175 nm into one ground-state and one excited ($^1D_2$) oxygen atom:

$$O_2\left(^3\Sigma_g^-\right) \xrightarrow{h\nu\,(<\,175\,\text{nm})} O + O\left(^1D_2\right) \qquad (5\text{-}13)$$

The difference between the bond dissociation energy of O_2 into two ground-state oxygen atoms (492 kJ/mol) and the start of the Schumann–Runge continuum (175 nm, corresponding to a photon energy of 682 kJ/einstein) is 190 kJ/mol, which is the energy needed to excite an oxygen atom from its ground ($^3P_2$) state to the excited ($^1D_2$) state. This is a large excitation energy, making the $^1D_2$ atom a very reactive and important species in stratospheric and tropospheric photochemistry.

Below 130 nm there is sufficient energy to produce oxygen atoms in even higher electronic states, and some dissociation to the $^1S_0$ state also takes place,

$$O_2 \xrightarrow{h\nu\,(<\,130\,\text{nm})} O + O\left(^1S_0\right) \qquad (5\text{-}14)$$

and below 92.3 nm two excited atoms are produced

$$O_2 \xrightarrow{h\nu\,(<\,92.3\,\text{nm})} 2O\left(^1S_0\right) \qquad (5\text{-}15)$$

However, the ionization potential of O_2, 12.15 eV, corresponds to a photon of wavelength $\lambda = 102$ nm, and therefore a more likely reaction below 102 nm is photoionization

$$O_2 \xrightarrow{h\nu\,(<\,102\,\text{nm})} O_2^+ + e^- \qquad (5\text{-}16)$$

followed by dissociative recombination. Reactions (5-17) and (5-18) are two possible examples of this type of process:

$$O_2^+ + e^- \rightarrow O(^1S_0) + O + 269 \, \text{kJ/mol} \tag{5-17}$$

$$O_2^+ + e^- \rightarrow O(^1S_0) + O(^1D_2) + 80 \, \text{kJ/mol} \tag{5-18}$$

with (5-17) found to be the favored reaction.

Two other excited electronic states of O_2 are important in atmospheric photochemistry because of specific reactions to be discussed later with reference to their possible roles in the photochemical smog cycle, ozone photolysis, and photooxidation reactions with olefins. These are the $^1\Delta_g$ ("singlet oxygen") and $^1\Sigma_g^+$ states, with energies 94.1 and 157 kJ/mol, respectively, above the ground state (Figure 5-1). Direct excitation from the triplet ground state to either of these states is spin forbidden, and therefore oxygen absorbs only very weakly around 760 nm (157 kJ/mol), giving a banded spectrum called the *atmospheric bands*. The reverse radiative process is of course also spin forbidden and these states, when produced, are therefore relatively stable in the atmosphere, a factor contributing to their importance; the radiative lifetime of $^1\Sigma_g^+$ is 12 s, while that of $^1\Delta_g$ is 60 min. Furthermore, nonreactive deactivation by collision with other species, called *collisional quenching*, is also inefficient, the rate constants of deactivation by collision with ground-state O_2 being 5×10^{-16} and 2×10^{-18} cm^3 molecule^{-1} s^{-1} for the $^1\Sigma_g^+$ and $^1\Delta_g$ states, respectively.[4] In the absence of collisional quenching, the emissions from the normally "forbidden" radiative transitions

$$O_2\left(^1\Sigma_g^+\right) \rightarrow O_2 + h\nu(762 \, \text{nm}) \tag{5-19}$$

and

$$O_2(^1\Delta_g) \rightarrow O_2 + h\nu(1270 \, \text{nm}) \tag{5-20}$$

turn out to be two of the most intense bands in the atmospheric airglow and in auroral displays.[5] The $^1\Sigma_g^+$ state also relaxes to the $^1\Delta_g$ state by collisional deactivation.

Ground-state oxygen atoms can combine to form ground-state O_2 by the three-body recombination reaction

$$2O \xrightarrow{\text{M}} O_2 \tag{5-21}$$

[4]If deactivation occurred at every collision, the quenching rates would be approximately 4×10^5 and 10^8 times faster, respectively. See Section 4.3.

[5]*Airglow* is the faint glow of the sky produced by solar photochemical processes; it occurs at all latitudes. The *aurora* is produced by impact of high-energy solar particles; it is more intense than the airglow, but it is irregular and occurs near the poles.

where M can be another oxygen atom or any other atomic or molecular species. If reaction (5-21) is kinetically simple and therefore trimolecular, then

$$r = k[O]^2[M] \qquad (5\text{-}22)$$

The rate constant k for this reaction is very nearly temperature independent. In fact it has a slightly inverse temperature behavior, the rate decreasing with increasing temperature, implying a negative activation energy, which of course is physically impossible if activation energy is considered to be the height of a potential barrier.[6] The rate constant is also of the order of magnitude of the collisional frequency for triple collisions, indicating that the three-body recombination (5-21) occurs essentially at every encounter. Even so, at the very low pressures encountered in the upper atmosphere—for example, 3×10^{-7} bar at 100 km (see Table 2-4)—the rate of reaction (5-21) is small compared to the rate of O_2 photodissociation, and therefore above 100 km the primary oxygen species present is atomic oxygen.[7]

It has been seen that excited singlet ($^1D_2$ and $^1S_0$) atoms are formed from the photodissociation of molecular oxygen below 175 nm. These species also relax to lower electronic states with emission of light:

$$O(^1S_0) \rightarrow O(^1D_2) + h\nu(557.7\,\text{nm}) \qquad (5\text{-}25)$$

$$O(^1D_2) \rightarrow O(^3P_2) + h\nu(630\,\text{nm}) \qquad (5\text{-}26)$$

Again, both are "forbidden" transitions and therefore show only weak emission contributing to the airglow and aurora. Below 140 km, however, collisional deactivation to the triplet ground state with any atomic or molecular species M may become the dominant reaction, particularly with $O(^1D_2)$. If

---

[6]The reason for the negative activation energy is that oxygen atoms form an unstable intermediate complex with M, O·M, which then reacts with another O atom to produce O_2. As the temperature is raised, O·M becomes more unstable so that its concentration is decreased, thereby decreasing the rate at which O·M reacts with O, hence decreasing the overall rate of O_2 formation represented by reaction (5-21). Since reaction (5-21) is now kinetically complex, as pointed out in Section 4.3.1, the interpretation of the experimental activation energy as the height of a potential energy barrier is not valid. Assuming equilibrium between O, M, and O·M leads, however, to the same rate law as that for the trimolecular reaction (5-22). At *extremely* high pressures, well beyond those in any part of the atmosphere, the order of the reaction would decrease from third order to second order.

[7]It should be noted that radiative recombination processes are possible:

$$2O(^3P_2) \rightarrow O_2\left(^1\Sigma_g^+\right) + h\nu \qquad (5\text{-}23)$$

$$O(^3P_2) + O(^1D_2) \rightarrow O_2\left(^3\Sigma_g^-\right) + h\nu' \qquad (5\text{-}24)$$

However, these reactions are also governed by spin conservation and atomic radiative selection rules, and they also are negligible in comparison to O_2 photodissociation.

M is ground-state O_2, electronic energy is transferred and O_2 is excited to its $^1\Sigma_g^+$ state

$$O(^1D_2) + O_2 \rightarrow O + O_2\left(^1\Sigma_g^+\right) \qquad (5\text{-}27)$$

which decays collisionally to the $^1\Delta_g$ state:

$$O_2\left(^1\Sigma_g^+\right) \xrightarrow{\text{M}} O_2(^1\Delta_g) \qquad (5\text{-}28)$$

An important aspect of this energy transfer or sensitization is that spin is conserved. That is, the oxygen atom undergoes a singlet–triplet transition while the reverse triplet–singlet reaction occurs with molecular oxygen, so that total reactant and product spins are the same. This is a very efficient process under exothermic conditions, probably occurring within an order of magnitude of collisional frequency, and therefore reaction (5-28) is one of the major sources of "singlet oxygen" $O_2(^1\Delta_g)$ in the atmosphere.

5.2.3 Ozone

5.2.3.1 Ozone in a "Clean" Stratosphere

Below 100 km into the midmesosphere region, the three-body recombination of O atoms, reaction (5-21), becomes potentially important because of the increasing total third-body pressure. However, the O_2 concentration is also increasing as we go to lower altitudes, increasingly leading to absorption of the solar radiation below 243 nm. Photodissociation of O_2 is decreased, lowering the O atom concentration. Since the rate of reaction (5-21) varies as the square of the O atom concentration (5-22), this three-body recombination becomes insignificant below the stratopause at approximately 50 km. However, another three-body recombination between a ground-state oxygen atom and a ground-state oxygen molecule to produce ozone

$$O + O_2 \xrightarrow{\text{M}} O_3; \; r = k[O][O_2][M] \qquad (5\text{-}29)$$

does occur in this region [even though the trimolecular rate constant of (5-29) is less than that of reaction (5-21)] since it depends only linearly on the O atom concentration and the much higher O_2 concentration.[8]

Reaction (5-29) is the exclusive source of ozone in the stratosphere and mesosphere (as well as being the only source of ozone in the troposphere, although the source of O atoms is different: see Sections 5.3.2 and 5.3.3), and

[8] O and O_3 are often referred to as "odd oxygen," O_x.

is a major source of destruction of O atoms.[9] Despite its small concentration (it does not exceed 10 ppm anywhere in the atmosphere), ozone is one of the atmosphere's most important and remarkable minor constituents. It is found throughout the atmosphere, but particularly in the stratosphere in a well-defined layer between 15 and 30 km altitude. Since its production depends on the photodissociation of molecular oxygen by the sun, it is produced primarily in the stratosphere near the equator and transported toward the Arctic and Antarctic poles.[10] As pointed out in Chapters 2 and 3, not only does it provide the shield for the earth against damaging ultraviolet radiation, but also, through absorption of this radiation, it becomes the main energy reservoir in the upper atmosphere, hence a factor in climatic regulation. On the other hand, ozone is one of the most toxic inorganic chemicals known, and its presence at relatively high concentrations in lower atmosphere polluted air makes it a potentially dangerous substance for humans and other organisms.

The bond dissociation energy of ozone is only 101 kJ/mol, corresponding to a photon of wavelength of 1180 nm, and the primary quantum yield of dissociation is unity throughout the visible and UV regions. However, ozone absorbs only weakly in the visible region (*Chappuis bands*), leading to ground-state species

$$O_3 \xrightarrow{h\nu \text{ (visible)}} O_2 + O \tag{5-30}$$

The wavelength of maximum absorption in the visible is 600 nm, $\sigma_{max} \cong 5.6 \times 10^{-21} \text{ cm}^2/\text{molecule}$.

The major absorption of ozone, known as the *Hartley continuum* (shown in Figures 5-3 and 5-4), is in the ultraviolet commencing at 334 nm with $\lambda_{max} = 255.5$ nm, $\sigma_{max} = 1.17 \times 10^{-17} \text{ cm}^2/\text{molecule}$. This ultraviolet photolysis is quite complex, involving extensive secondary reactions. At wavelengths exceeding 310 nm, the predominant dissociative reaction is

$$O_3 \xrightarrow{h\nu \text{ (> 310 nm)}} O + O_2\left(^1\Delta_g \text{ or } ^1\Sigma_g^+\right) \tag{5-31}$$

but below 310 nm it is primarily

$$O_3 \xrightarrow{h\nu \text{ (< 310 nm)}} O(^1D_2) + O_2(^1\Delta_g) \tag{5-32}$$

[9]It has been proposed that the reaction between a highly vibrationally excited ground-state oxygen molecule and another ground-state O_2 molecule, and/or the reaction of a vibrationally *and* electronically excited $^3\Sigma_u^+$ oxygen molecule with a ground-state oxygen molecule, may also lead to the production of atmospheric ozone. See T. G. Slanger, Energetic molecular oxygen in the atmosphere, *Science*, **265**, 1817–1818 (1994).

[10]However, the ozone concentrations are about twice as large at the poles as at the equator because, until recently—see Section 5.2.3.2—there were very few ozone-destroying reactions occurring at the poles.

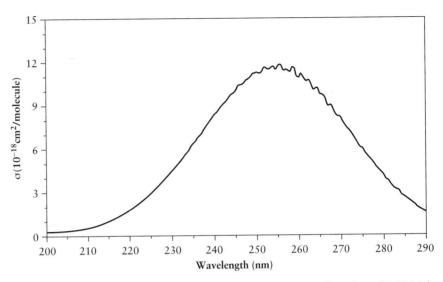

FIGURE 5-3 UV-C absorption spectrum of ozone at 298 K. Drawn from data of L. T. Molina and M. J. Molina, *J. Geophys. Res.*, **91**, 14,501–14,508 (1986).

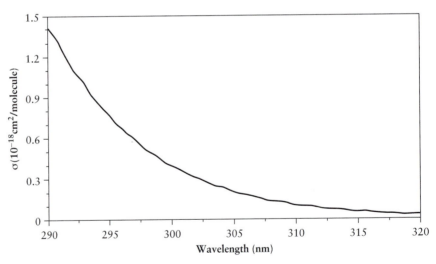

FIGURE 5-4 UV-B absorption spectrum of ozone at 298 K. Drawn from data of L. T. Molina and M. J. Molina, *J. Geophys. Res.*, **91**, 14,501–14,508 (1986).

and this is another primary source of singlet oxygen in the atmosphere.[11]

The several O_2 and O_3 photodissociation reactions just given may produce excited states of O and/or O_2 as dissociation products, depending on the absorbed wavelengths. However, collisional deactivation of the excited states in general is so rapid even at stratospheric pressures that only the ground-state allotropic forms of oxygen (O, O_2, and O_3) need to be considered in a natural or "clean" stratosphere.[12] In the following we will therefore write the photodissociation reactions for O_2 and O_3 simply as

5.2.3.2

$$O_2 \xrightarrow{h\nu\,(O_2)} 2O \qquad (5\text{-}36)$$

and

$$O_3 \xrightarrow{h\nu\,(O_3)} O + O_2 \qquad (5\text{-}37)$$

Total stratospheric formation of ozone is given by combination of reactions (5-29) and (5-36):

$$3O_2 \xrightarrow{h\nu\,(O_2),\,M} 2O_3 \qquad (5\text{-}38)$$

This represents direct conversion of solar energy to thermal energy and leads to stratospheric heating. A tremendous amount of solar energy is involved: the average rate of energy conversion for the whole earth is greater than $2 \times 10^{10}\ \mathrm{kJ/s}\ (kW)$, which is over three times the total human rate of energy use.

Ozone is destroyed by photolysis (5-37) and by reaction with ground-state O atoms

---

[11]Recent studies [A. R. Ravishankara *et al.*, *Science*, **280**, 60 (1998) and included references] indicate that $O(^1D_2)$ atoms can be formed at wavelengths at least as great as 330 nm by the photolysis of internally excited (IE) ozone molecules, $O_3(IE)$,

$$O_3(IE) \xrightarrow{h\nu\,(>\,310\,\mathrm{nm})} O(^1D_2) + O_2(^1\Delta_g) \qquad (5\text{-}33)$$

and, to a much lesser extent, by the "spin-forbidden" (see Section 5.1) reaction

$$O_3 \xrightarrow{h\nu\,(>\,310\,\mathrm{nm})} O(^1D_2) + O_2\left(^3\Sigma_g^-\right) \qquad (5\text{-}34)$$

[12]However, in a "dirty" atmosphere that includes species other than just the oxygen allotropes (e.g., water), singlet $O(^1D_2)$ atoms initiate many of the important free-radical reactions in the atmosphere, such as formation of the hydroxyl (·OH) radical:

$$O(^1D_2) + H_2O \to 2\,{\cdot}OH; \qquad k_{298} = 4.3 \times 10^{-12}\ \mathrm{cm^3\ molecule^{-1}\ s^{-1}} \qquad (5\text{-}35)$$

The importance of the $O(^1D_2)$—producing reactions above 310 nm given in footnote 11—is emphasized by the roughly threefold increase in solar intensity at sea level between 310 and 330 nm (Figure 3-4). Atmospheric reactions involving $O(^1D_2)$ atoms and the hydroxyl radical are covered in following sections.

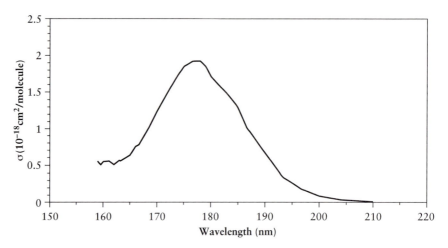

FIGURE 5-6   Absorption spectrum of CFC-12 (CF_2Cl_2). Drawn from data of C. Hubrich, C. Zetzsch, and F. Stuhl, *Ber. Bunsenges. Phys. Chem.*, **81**, 437–442 (1977).

foams, refrigerants, and electronic equipment cleansers. Their extreme inertness and insolubility in water also prevent them from being washed out of the atmosphere by rain, and for this reason as well, CFCs are very important in stratospheric ozone depletion. Long-term atmospheric monitoring of several CFCs, including the two most used ones, CFC-11 and CFC-12,[18] at a variety of "clean" sites around the earth, has shown that their rates of systematic increase have been consistent with their anthropogenic worldwide production rates. This indicates that no major sinks—that is, physical or chemical methods of destruction or removal from the atmosphere—exist for the CFCs within the troposphere. Thus, once released into the troposphere their only major loss is transport into the stratosphere by vertical movement of large air masses, which for the most part is independent of molecular weight. Once there, they may react with excited stratospheric oxygen atoms to produce reactive products such as Cl atoms and ClO·. More important, however, is that the CFCs absorb far-ultraviolet radiation, photodissociating with a quantum yield of unity into a chlorofluorinated radical and a chlorine atom. For CFC-12,

$$CF_2Cl_2 \xrightarrow{\;h\nu\;} CF_2Cl^{\cdot} + Cl \tag{5-47}$$

The ultraviolet absorption spectrum of CFC-12 is shown in Figure 5-6. Although radiation in this spectral region is completely prevented from reaching the troposphere by atmospheric oxygen and the ozone layer, it does partially penetrate into the stratosphere through the UV "window" between

[18]See Table 3-2, note *a*, for a way of determining the formula of each gas from its number.

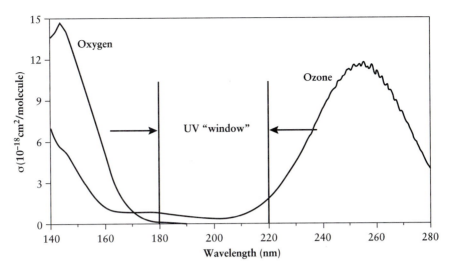

FIGURE 5-7 Solar UV window between oxygen and ozone absorption bands. Data for oxygen: 140–175 nm: M. Ackerman, in *Mesospheric Models and Related Experiments*, G. Fiocco, ed., D. Reidel, Dordrecht, 1971, pp. 149–159. 175–195 nm: K. Watanabe, E. C. Y. Inn, and M. Zelikoff, *J. Chem. Phys.*, **21**, 1026–1030 (1953). Data for ozone: $\lambda < 185$ nm: M. Ackerman, in *Mesospheric Models and Related Experiments*, G. Fiocco, ed., D. Reidel, Dordecht, 1971, pp. 149–159. $\lambda > 185$ nm: L. T. Molina and M. J. Molina, *J. Geophys. Res.*, **91**, 14,501–14,508 (1986).

approximately 180 and 220 nm formed by ozone and oxygen (Figure 5-7). Production of Cl atoms is then followed by the chain mechanism (5-45), with X = Cl:

$$Cl + O_3 \rightarrow ClO\cdot + O_2$$

$$ClO\cdot + O \rightarrow Cl + O_2$$

net: $$O_3 + O \xrightarrow{\ Cl\ } 2O_2 \tag{5-48}$$

If $\cdot OH$ or NO is also present, additional catalyzed chain mechanisms involving atomic chlorine are possible. For example, with $\cdot OH$,

$$Cl + O_3 \rightarrow ClO\cdot + O_2$$

$$\cdot OH + O_3 \rightarrow HO_2\cdot + O_2$$

$$ClO\cdot + HO_2\cdot \rightarrow HOCl + O_2$$

$$HOCl \xrightarrow{\ h\nu\ } \cdot OH + Cl$$

net: $$2O_3 \xrightarrow{\ Cl,\,\cdot OH,\,h\nu\ } 3O_2 \tag{5-49}$$

and with NO,

$$Cl + O_3 \rightarrow ClO^{\cdot} + O_2$$

$$ClO + NO \rightarrow Cl + NO_2$$

$$NO_2 + O \rightarrow NO + O_2$$

net: $$O_3 + O \xrightarrow{Cl, NO} 2O_2$$ (5-50)

or

$$Cl + O_3 \rightarrow ClO^{\cdot} + O_2$$

$$NO + O_3 \rightarrow NO_2 + O_2$$

$$ClO^{\cdot} + NO_2 \xrightarrow{M} ClONO_2$$

$$ClONO_2 \xrightarrow{h\nu} Cl + NO_3^{\cdot}$$

$$NO_3^{\cdot} \xrightarrow{h\nu} NO + O_2$$

net: $$2O_3 \xrightarrow{Cl, NO, h\nu} 3O_2$$ (5-51)

Note that mechanisms (5-49) and (5-51) do not involve atomic oxygen and therefore can occur at lower altitudes than (5-48) and (5-50). This is important in the massive destruction of stratospheric ozone over the earth's poles, discussed in Sections 5.2.3.3 and 5.2.3.4, as well as the important fact that HOCl and ClONO$_2$ serve as "reservoir" molecules in mechanisms (5-49) and (5-51), respectively, for the reactive Cl and ClO$\cdot$ species.

    For the anthropogenic chlorofluorocarbons to be major contributors to stratospheric ozone depletion, they must of course be the primary source of stratospheric chlorine atoms. Several "natural" sources, such as sea salt, burning biomass, and volcanoes, have been cited as depositing many times more chlorine into the atmosphere than the CFCs, suggesting that "natural" chlorine atoms have been continuously contributing to the ozone photostationary state existing in the stratosphere to a much greater extent than could possibly be attributable to perturbation by anthropogenic chlorine atoms. Several observations contradict this proposal. For one thing, natural chlorine-containing species are for the most part water soluble (unlike the CFCs), and therefore if they are emitted into the troposphere they will be rained out before they are transported into the stratosphere. Sodium is not detected in the stratosphere, also ruling out sea salt. Biomass burning produces methyl chloride,[19] but

[19]See Section 10.6.13.

recent estimates put the contribution of methyl chloride from biomass burning to the total amount of stratospheric chlorine at about 5%. Active volcanoes can discharge HCl (and, to a lesser extent, HF) into the lower atmosphere. However, in the troposphere volcanic HCl rapidly dissolves in water (which is in large excess) after it has condensed to a supercooled liquid, so that for the most part the HCl is washed out by rain with only about 0.01% of that emitted possibly getting to the stratosphere. Furthermore, the concentrations of HCl and HF would be expected to *decrease* with increasing height in the stratosphere if from volcanic sources; in fact, it is found that they actually *increase* at stratospheric altitudes[20] where concentrations of CFCs simultaneously decrease because they are in the process of photodecomposition. Of course, volcanoes undergoing *very* violent reactions do have the potential to deposit large quantities of HCl directly into the stratosphere, which bypasses washing out by rain. A very significant factor indicating that such large volcanic activity is not a major contributor to stratospheric chlorine is the series of global measurements that have been made since 1978 on the accumulation of HCl in the stratosphere. During that time there have been two major volcanic eruptions: El Chichón in Mexico in April 1982, and Mount Pinatubo the Philippines in June, 1991, with the former releasing about 1.8 megatons (1 megaton = 10^{12} g) and the latter 4.5 megatons of HCl into the atmosphere. In spite of these, stratospheric HCl has increased regularly with only a small (<10%) increase after the eruption in Mexico and even smaller after the eruption in the Philippines.

It should be pointed out, however, that volcanoes *can* influence the ozone layer in ways other than supplying chlorine.[21] Volcanic eruptions eject copious amounts of debris, including sulfur dioxide (SO_2)—approximately 20 megatons from Mount Pinatubo and 6 megatons from El Chichón, for example (although not all of this directly into the stratosphere).[22] Sulfur dioxide is oxidized to sulfur trioxide, SO_3, in the stratosphere by the ·OH radical and O_3:

$$SO_2 + \cdot OH \xrightarrow{\text{M}} HOSO_2\cdot \qquad (5\text{-}52)$$

$$\dot{H}OSO_2\cdot + O_2 \rightarrow SO_3 + HO_2\cdot \qquad (5\text{-}53)$$

SO_3 reacts with gaseous H_2O (although not in a simple combination reaction) to form sulfuric acid, H_2SO_4, producing a stratospheric sulfate layer of concentrated sulfuric acid aerosol droplets small enough to remain in the

---

[20]For example, at the equator the concentration of HF increases from zero at 16 km to > 1.0 ppb above 50 km.

[21]Volcanic eruptions can also temporarily affect the earth's climate; see Section 3.3.4.

[22]Even in the absence of volcanic activity, SO_2 is present in the stratosphere from oxidation of reduced compounds of sulfur from natural and/or anthropogenic sources (such as carbonyl sulfide, COS, which is sufficiently stable in the troposphere to reach the stratosphere).

stratosphere for years. Enhanced heterogeneous chlorine-catalyzed, ozone-depleting reactions can take place on these aerosols, as discussed in the next section in conjunction with the Antarctic ozone hole. It has been found, in fact, that stratospheric ozone depletion *is* enhanced after large volcanic activity, such as Mount Pinatubo, by these surface reactions, but not by ejected HCl. The volcanic debris also absorbs sunlight (e.g., the sunlight was attenuated between 15 and 20% at several Northern Hemisphere measuring sites at the peak of the effect following the Mount Pinatubo eruption), increasing the temperature of parts of the stratosphere and possibly altering the high-altitude winds that circulate tropical ozone around the globe.

Bromine atoms can also participate in the catalyzed mechanism (5-45):

$$Br + O_3 \rightarrow BrO\cdot + O_2$$
$$BrO\cdot + O \rightarrow Br + O_2$$

$$\text{net:} \quad O_3 + O \xrightarrow{\;Br\;} 2O_2 \tag{5-54}$$

Stratospheric bromine atoms come primarily from photolyses of methyl bromide (CH_3Br) and brominated CFCs (halons).[23] Methyl bromide (Section 8.6.2.3) occurs naturally in the atmosphere, mostly from oceanic marine biological activity and from forest and grass (biomass) fires.[24] It is also synthesized for agricultural uses as a soil fumigant, and it is emitted into the atmosphere from automobile exhausts. The anthropogenic halons, mainly H-1301 and H-1211, are used extensively for extinguishing or suppressing fires. Methyl bromide is rapidly degraded by soil bacteria; most of it released into the atmosphere is taken up by the oceans or is oxidized or photolyzed in the troposphere, but an appreciable amount does reach the stratosphere. The total concentration of bromine in the lower stratosphere is approximately 19 parts per trillion by volume (pptv). There are no known effective stratospheric reservoirs for Br or BrO·, unlike Cl and ClO· which have HOCl [mechanism (5-49)] and $ClONO_2$ [mechanism (5-51)], respectively. As a result, Br atoms are about 50 times more destructive of ozone than Cl atoms on a per-atom basis, although overall Cl atoms are more destructive because of the much greater concentration of Cl-containing compounds in the stratosphere than Br-containing compounds. In addition, ozone depletion by chlorine atoms is enhanced by bromine atoms through a synergistic interaction involving coupling of the BrO· and ClO· cycles:

---

[23]Halons are designated as H-wxyz, where w, x, y, and z are the number of carbon, fluorine, chlorine, and bromine atoms, respectively. Thus, CF_3Br is H-1301, and CF_2BrCl is H-1211.

[24]However most of these are deliberately set and thus are anthropogenic in origin.

$$BrO^{\cdot} + ClO^{\cdot} \rightarrow Br + ClO_2^{\cdot}$$

$$ClO_2^{\cdot} \xrightarrow{\text{M}} Cl + O_2$$

$$Br + O_3 \rightarrow BrO^{\cdot} + O_2$$

$$Cl + O_3 \rightarrow ClO^{\cdot} + O_2$$

net: $\qquad\qquad 2O_3 \xrightarrow{\text{Cl, Br}} 3O_2 \qquad\qquad$ (5-55)

It is estimated that at the time of writing approximately 25% of the global stratospheric ozone destruction is caused by bromine.

The depletion of ozone by a halogen-containing hydrocarbon depends on several factors, including the rate at which the hydrocarbon enters the atmosphere and its lifetime in the atmosphere (i.e., the time required for a given amount of a substance emitted "instantaneously" into the atmosphere to decay to $1/e$ of its initial value). The atmospheric lifetimes of fully fluorinated (perfluoro) gaseous compounds are very long (many centuries), since their loss results primarily from photolysis in the mesosphere, where pressures are very low and mixing with lower-altitude gases is extremely slow. Partial chlorination (CFCs) or bromination (halons) shortens the lifetime because of absorption and photolysis in the 180- to 220-nm UV window (Figure 5-7), as well as destruction by excited $O(^1D_2)$ atom reactions, both processes occurring in the stratosphere. Replacing some of the halogens with hydrogen or incorporating unsaturation in the molecule further greatly shortens the atmospheric lifetime because of reactions, such as with the $\cdot OH$ radical, in the troposphere. Atmospheric lifetimes of several halogenated gases are given in Table 5-1.

A factor that quantifies the ability of a gas to deplete stratospheric ozone is the *ozone depletion potential* (ODP), which is the steady-state depletion of ozone calculated for a unit mass of a gas continuously emitted into the atmosphere relative to that for a unit mass of CFC-11. Examples of ODPs are also given in Table 5-1. One-dimensional (1-D) and two-dimensional (2-D) models are used in the calculations of ODPs. (The ODPs in Table 5-1 are based on 2-D models.) Both types of model include modes of transport of atmospheric species, absorption and scattering of solar radiation, and approximately 200 chemical reactions involving about 50 reactive species; the difference between the two models is that the 1-D models consider only vertical distribution of species, assuming constant horizontal averages, whereas the 2-D models consider vertical and north–south (latitudinal) changes, but not east–west (longitudinal) effects. Major contributions to the uncertainties in the ODPs are the steady-state assumption and the lack of inclusion of the processes that produce the Antarctic ozone hole. Therefore the ODPs should be considered only as estimates to give a general picture of the effects of various constituents in halogenated gases. Table 5-1 lists the lifetimes and ODPs for

TABLE 5-1

Range of Lifetimes and Ozone Depletion Potentials (ODPs) of
Selected Halogenated Gases

Compound	Lifetime (years)	ODP
CFC-11	50	1.0
CFC-12	102	0.82
CFC-113	85	0.90
HCFC-141b	9.4	0.10
HCFC-142b	19.5	0.05
HCFC-22	13.3	0.04
HFC-134a	14	$< 1.5 \times 10^{-5}$
HFC-152a	1.5	0
H-1301	65	12
H-1211	20	5.1
CF_4	>50,000	0
CH_3Br	~ 1.3	~ 0.6
CF_3I	< 2 days	0
HFE-7100	4.1	0

Source: Scientific Assessment of Ozone Depletion: 1994,
World Meteorological Organization Global Ozone and
Monitoring Project, Report No. 37, U. S. Department of
Commerce (1995). Data for the hydrofluoroether HFE-7100
($C_4F_9OCH_3$—see Section 5.2.3.5) is from J. G. Owens, "Low
GWP Alternatives to HFCs and PFCs," presented at the Joint
IPCC/TEAP Expert Meeting on Options for the Limitations of
Emissions of HFCs and PFCs, Energieonderzoek Centrum
Nederland, Petten, The Netherlands, 1999.

selected CFCs, HCFCs, HFCs, and halons that together account for more than
95% of the organic chlorine and more than 85% of the organic bromine in the
atmosphere in 1995.[25] The very large ODPs for the halons are due primarily to
the synergistic coupled BrO· and ClO· cycle, mechanism (5-55).

CFCs were first developed in 1930, and as pointed out earlier, until recently
they (and the later developed halons) were used extensively in applications that
could benefit from their inertness, volatility, and safety. It is estimated that
approximately 90% of all the CFC-11 and CFC-12 produced to date has now
been released into the troposphere.

But has the release of CFCs and halons into the atmosphere already depleted
the stratospheric ozone layer to a measurable extent?

[25]S. A. Montzka *et al.*, Decline in the tropospheric abundance of halogen from halocarbons:
Implications for stratospheric ozone depletion, *Science*, **272**, 1318–1322 (1996).

To try to answer this question, we first consider how atmospheric ozone concentrations are experimentally measured. The Dobson spectrophotometer, in use since 1926, determines the total ozone contained in a vertical column passing through the earth's atmosphere. It does this by measuring the amount of radiation from the sun (or reflected from the moon) that is absorbed by the ozone at a pair of wavelengths near 300 nm. Currently there are more than 140 Dobson spectrophotometer operating stations positioned around the globe, forming the Global Ozone Observing System (GO$_3$OS). A similar ground-based instrument is the M-124 filter ozonometer; instead of a spectrophotometer, however, it uses two wideband optical filters, with maximum spectral sensitivities at 314 and 369 nm. Located at more than 40 sites (primarily in the former Soviet Union), these instruments also contribute to the GO$_3$OS network. Another ground-based or airborne system used to determine atmospheric ozone concentrations is differential absorption lidar (DIAL), which transmits light at two wavelengths, one that is at an ozone absorption line and is backscattered by ozone, and the other that is not at an ozone absorption (to correct for natural backscatter).[26]

In addition, ozone concentrations from the earth's surface to heights as high as about 30 km have been measured by electrochemical concentration cell ozonesondes, carried aloft by balloons. Total global coverage of ozone has also been measured since 1978 from above the atmosphere by a variety of instruments aboard polar- and precess-orbiting satellites. For example, the total ozone mapping spectrometer (TOMS) measures the ozone attenuation of six UV emissions from the earth between 312 and 380 nm. A similar UV-scanning satellite instrument is the solar backscattered ultraviolet (SBUV or SBUV/2) spectrometer. The microwave limb sounder (MLS), aboard the *Upper Atmospheric Research Satellite* (UARS) launched in 1991, determines stratospheric ozone concentrations from its microwave thermal emission lines at frequencies 183 and 205 GHz (1 GHz = $10^9 s^{-1}$). Satellites have also been used to determine the vertical atmospheric ozone concentration in the Stratospheric Aerosol and Gas Experiments (SAGE I, II, and III), and (since 1998) total ozone in the Global Ozone Monitoring Experiment (GOME).

First of all, measurements from these sources from as long ago as the 1950s to 1980 gave an average total "column" ozone concentration of about 312 Dobson units (DU) in the Northern Hemisphere and 300 DU in the Southern Hemisphere, with an overall global average of approximately 306 DU over this time period.[27] Superimposed on these averages, though, are periodic fluctuations

---

[26]Lidar is essentially radar that uses lasers as sources of pulsed electromagnetic radiation.

[27]The Dobson unit is a measure of the total integrated concentration of ozone in a vertical column extending through the atmosphere. It is the pressure in milliatmospheres (matm: 1 matm = 10^{-3} atm) that all the ozone in this column would have if it were compressed to one centimeter height. Thus, for 300 DU, the pressure of ozone in the hypothetical 1-cm column would be 300 matm. Assume now that the height of the column is increased to 30 km (3×10^6 cm) altitude

greater than 5%; these are due to seasonal variations (as much as 4% difference between a May–September maximum and a December minimum) and natural meteorological phenomena that include the tropical *Quasi-biennial Oscillation* (an approximately 26- to 27-month oscillation or cycle between easterly and westerly winds in the lower equatorial stratosphere), 11-year solar sunspot cycles and fluctuations, and the *El Niño–Southern Oscillation* of about 4 years (Section 3.3.1). Nevertheless, when the measurements over the 30-year period from 1964 to 1994 are treated and smoothed with a statistical model to filter out these natural components, they still show an overall global decline of approximately 4% in total ozone during the 13 years between 1978 and 1992, with however the change near the equator close to zero (Figure 5-8). About 64% of this total global change appears in the Southern Hemisphere.

In 1992, beginning in May, the global total ozone concentration dropped about 1.5% lower than expected from a linear extrapolation of the rate just given. In December 1992, the ozone levels in the northern midlatitudes over the United States were about 9% below normal, dropping to about 13% below normal in January and continuing well below normal into July, 1993, getting to as low as 18% lower at some sites. Similar record low stratospheric ozone levels were also observed over Canada and eastern and southern Europe, extending through August, 1993. Smaller unexpected midlatitude decreases appeared in the Southern Hemisphere. These extraordinary and unprecedented effects are believed to have resulted from the June 1991 eruption of Mount Pinatubo, which increased the stratospheric sulfate aerosol layer. However, volcanic aerosols had essentially disappeared by mid-1993, and in fact the annual decrease in total ozone had returned in 1994 to the pre-1991 values. This is as predicted, as the stratosphere recovered from volcanic debris.

---

at a constant temperature. (Over 90% of the ozone in the atmosphere is contained below 30 km.) Assuming ideal gas behavior, the average pressure of ozone within the column will now be 10^{-4} matm, since pressure is inversely proportional to volume (hence also inversely proportional to the height for a column of constant cross section) under these isothermal and ideal gas conditions. Note that the product Px, where P is the pressure and x the height of the cylindrical column of ozone, is independent of x. Thus, in this example,

$$Px = \left(10^{-4}\,\text{matm}\right)\left(3 \times 10^{6}\,\text{cm}\right) = (300\,\text{matm})(1\,\text{cm}) = 300\,\text{matm cm} = 300\,\text{DU}$$

so that

$$1\,\text{DU} = 1\,\text{matm} \cdot \text{cm} = 10^{-3}\,\text{atm} \cdot \text{cm}.$$

This is useful if the concentration c in the Beer–Lambert law (Section 4.2.2) is expressed in terms of pressure P. For example, the absorption cross section (σ) of ozone at 305 nm—midway in the UV-B band—is $2 \times 10^{-19}\,\text{cm}^2/\text{molecule}$, which converts to the base-10 extinction coefficient $\varepsilon = 2.14\,\text{atm}^{-1}\text{cm}^{-1}$ at 298 K. The fraction of light transmitted through the atmosphere at 305 nm is then given by Equation 4.11 as

$$\frac{I}{I_0} = 10^{-\varepsilon Px} = 10^{-\left(2.14\,\text{atm}^{-1}\text{cm}^{-1}\right)(300\,\text{DU})\left(10^{-3}\,\text{atm·cm/DU}\right)} = 0.228 = 22.8\%$$

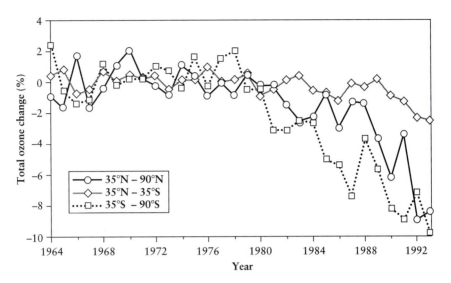

FIGURE 5-8 Percent change in total ozone relative to the 1964–1970 annual average for the Northern Hemisphere (35°N–90°N), the Tropics (35°N–35°S), and the Southern Hemisphere (35°S–90°S). The Mount Pinatubo eruption occurred in June, 1991. Drawn from data of K. Tourpali, X. X. Tie, C. S. Zerofos, and G. Brasseur, *J. Geophys. Res.*, **102**, 23,955–23,962 (1997).

5.2.3.3 The Antarctic Ozone Hole

The foregoing trends in global stratospheric ozone, although statistically significant, are admittedly small and obscured to a large extent by the seasonal variations that have been fairly regular for over 30 years. In striking contrast to this global behavior is the large depletion in total ozone above the Antarctic pole, first reported in 1985.[28] This depletion, called the *Antarctic ozone hole*, occurs in the Antarctic spring beginning in September and reaching a maximum (i.e., minimum total ozone) generally in early October. It disappears in November and early December as shifting springtime winds transport ozone-rich air from the middle and lower southern latitudes into the hole.

The depletion of ozone since 1975 has been very dramatic: the average ozone concentration above the Antarctic in October declined from approximately 310 DU in 1975 to 170 DU in 1987, and a record low concentration of 91 DU was measured on October 12, 1993, followed by lows of 102 DU and

[28]J. C. Farman, B. G. Gardner, and J. D. Shanklin, Large losses of total ozone in Antarctica reveal seasonal ClO_x/NO_x interaction, *Nature*, **315**, 207–210 (1985).

98 DU during October of the 1994 and 1995 Antarctic winters, respectively; a low of 95 DU was observed for the 2000 Antarctic winter, occurring on October 1, 2000. Most of the change occurs in the lower stratosphere between about 10 and 24 km, a zone in which the ozone concentration has reached essentially zero during the October minimum since the record low year of 1993. The area covered by the hole has also continued to increase since first observed in 1985; in 2000 it covered an area greater than 28×10^6 km^2, about the size of the North American continent and twice the size of Europe, extending to the tip of South America. Furthermore, after the hole breaks up, ozone-poor air drifts northward, resulting in increased ultraviolet radiation beyond Antarctica.

It is now accepted that the polar vortex, which is the strong cyclonic "whirlpool" with winds upwards of 100 m/s circling the pole, virtually isolating the gases within it from the middle latitudes, and the extremely cold Antarctic night are the distinguishing factors that lead to the formation of the large zone of temporary ozone depletion. Temperatures in the swirling vortex can reach lower than $-86°C$ (187 K), which is the temperature at which ice particles can form, producing clouds even in very dry stratospheric air (probably on sulfate aerosols) at altitudes between 10 and 25 km. These clouds of ice crystals are called *type II polar stratospheric clouds*, or type II PSCs. In addition to these type II PSCs, however, clouds also form at higher temperatures—as high as $-80°C$ (193 K). These are called type I PSCs.[29] Various possibilities have been suggested for the composition and phase or phases making up the type I clouds. These include crystals of nitric acid and water (particularly NAT, the nitric acid trihydrate $HNO_3 \cdot 3H_2O$) nucleated by frozen sulfate aerosols (sulfuric acid tetrahydrate, or SAT); ternary supercooled liquid solution droplets of water, nitric acid, and sulfuric acid; binary solutions of nitric acid and water; and a metastable water-rich nitric acid solid phase, as well as various combinations of these. The specific composition of type I PSCs may in fact be quite sensitive to the temperature and to the occurrence of freezing–thawing cycles during cloud formation.

Most of the Antarctic ozone depletion occurs below 20 km. In this region there are virtually no oxygen atoms present for two reasons. First, the relatively high third-body pressure maximizes the pressure-dependent $O + O_2$ recombination, reaction (5-29); and second, the lack of sunlight during the Antarctic night minimizes the O-atom-producing photodissociation of O_3, reaction (5-30). As a result, neither mechanism (5-45) nor (5-46) can be occurring to any significant extent. Nevertheless, observations clearly show that stratospheric chlorine species are involved in this dramatic ozone loss. The PSCs thus appear to be essential in formation of the Antarctic hole.

---

[29]Type I PSCs are sometimes further broken down into types Ia and Ib.

It is now known that one of the key roles played by PSCs is to provide a surface on which "active" chlorine, such as $ClO\cdot$, is converted to "inactive" chlorine by reactions such as

$$ClO\cdot + NO_2 \xrightarrow{\text{surface}} ClONO_2 \tag{5-56}$$

Chlorine nitrate, $ClONO_2$, is called a "reservoir" species because it traps "active" chlorine in an "inactive" form. Another reservoir species is HCl, which can be formed by "active" chlorine atoms reacting with methane:

$$Cl + CH_4 \rightarrow \cdot CH_3 + HCl \tag{5-57}$$

Neither of these reservoir molecules reacts with ozone, but the two do react together on PSC surfaces (a second key role for PSCs) by heterogeneous reactions such as

$$HCl + ClONO_2 \xrightarrow{\text{surface}} Cl_2 + HNO_3 \tag{5-58}$$

This reaction does not occur in the gas phase but is very fast on the ice surfaces and solutions making up the PSCs. Presumably an HCl molecule is adsorbed on the surface first, followed by collision of a $ClONO_2$ molecule and subsequent reactions represented by the overall reaction (5-58). Molecular chlorine, a "storage" molecule, accumulates and remains stored during the dark Antarctic night. When sunlight appears at the start of the Antarctic spring in September (beginning at the outer edge of the vortex and moving inward), molecular chlorine photodissociates into chlorine atoms

$$Cl_2 \xrightarrow{h\nu} 2Cl \tag{5-59}$$

and initiates the following cycle that leads to the destruction of ozone—at rates sometimes greater that 1% per day:

$$2(Cl + O_3 \rightarrow ClO\cdot + O_2)$$

$$2ClO\cdot \xrightarrow{M} ClOOCl \text{ (rate-determining step)}$$

$$ClOOCl \xrightarrow{h\nu} ClO_2\cdot + Cl$$

$$ClO_2\cdot \xrightarrow{M} Cl + O_2$$

$$\rule{10cm}{0.4pt}$$

net: $\qquad 2O_3 \xrightarrow{h\nu, M} 3O_2 \tag{5-60}$

This mechanism can also be bromine-enhanced by mechanism (5-55). Other "storage" molecules are HOCl and $ClNO_2$; they are produced by heterogeneous reactions like (5-58) and are also photolyzed in the Antarctic spring into Cl atoms, thereby further contributing to ozone destruction via (5-60).

For the observed large change in ozone concentration—that is, essentially its complete destruction between 10 and 24 km above the Antarctic pole—to occur by mechanism (5-60), the active ClO· concentration must remain relatively large for a month or so after the appearance of sunlight. Since ClO· is destroyed by (5-56), the concentration of NO₂ needs to be kept low. This occurs through the heterogeneous formation of HNO₃, reaction (5-58), on the PSCs. On the other hand, evaporation of the PSCs in the sunlight and with rising temperatures will liberate HNO₃, which photodissociates at wavelengths less than 320 nm, freeing NO₂:

$$HNO_3 \xrightarrow{\ h\nu\,(\leq 320\,nm)\ } \cdot OH + NO_2 \qquad (5\text{-}61)$$

We see, then, that PSCs not only lead to the containment of "storage" molecules and the eventual release (in early spring) of "active" molecules that catalytically destroy ozone; they also can lead to release of species, such as NO₂, that destroy "active" molecules and thus *prevent* extensive ozone destruction. It turns out that when the temperatures remain low enough (i.e., below the condensation temperatures of the PSC phases present) long enough, the particles making up the PSCs grow and sedimentation occurs. This leads to permanent removal of HNO₃ by transport to lower elevations, so that continued destruction of ozone occurs as a result of sustained high ClO· concentration. This process is called *denitrification*; it commences in the Antarctic during June and continues through the Antarctic winter and early spring.

In addition to the PSCs, the sulfate droplets in the stratospheric aerosol layer (see Section 5.2.3.2) also provide surfaces for heterogeneous reactions that contribute to ozone depletion. For example, dinitrogen pentoxide, N₂O₅, reacts readily with water on sulfate aerosols to form nitric acid:

$$H_2O + N_2O_5 \xrightarrow{\ sulfate\ aerosol\ } 2HNO_3 \qquad (5\text{-}62)$$

Since N₂O₅ is formed from NO₂ and NO₃,

$$NO_2 + NO_3 \xrightarrow{\ M\ } N_2O_5 \qquad (5\text{-}63)$$

the effect of reaction (5-62) is to tie up NO₂ so that less of the "inactive" chlorine nitrate is formed by reaction (5-56), leaving more "active" ClO· available for the ozone-destroying cycles. The stratospheric sulfate aerosol layer during the Antarctic spring is below 20 km and has a broad maximum between 10 and 16 km. It is now believed that this sulfate layer was responsible for the 1993 record low Antarctic ozone hole. The slight moderations in the minimum depletions of 1994 and 1995 may be due to disappearance of aerosol from the Mount Pinatubo eruption after 1993, although the 1996 hole appeared earlier than usual (maximum ozone depletion on October 5) and was as intense as during the preceding several years.

Direct observations of stratospheric ClO· by aircraft flying within the polar vortex conclusively show the involvement of CFCs and halons in the Antarctic ozone depletion process. For example, before the onset of the Antarctic spring and the development of the hole, the concentration of ClO· within the vortex from about 65°S has been more than 100 times greater than outside the vortex at lower latitudes. When the hole has fully developed in late September and early October, there is complete correlation between the ozone depletion and ClO· enhancement as a function of latitude and altitude within the vortex.

5.2.3.4 Depletion of Ozone above the Arctic

For several reasons, the formation of an Arctic ozone hole is much less likely to occur than the Antarctic ozone hole (which we saw in the preceding section has been increasing in strength since 1985):

The Arctic winters are generally warmer and rarely are cold enough for type II PSCs to form ($\sim -86°C$):

Warming toward the end of winter often occurs in February (equivalent to August in Antarctica) before there is enough sunlight to release active Cl and ClO·.

The polar vortex usually is more distorted and breaks up by the end of March.

In addition, during late winter and spring, ozone-rich air from the low latitudes above the equator is regularly transported northward at high altitudes to the North Pole region.[30] This "physical" ozone increase will tend to obscure any chemically induced depletion of ozone.

Nevertheless, as shown in Figure 5-8, total stratospheric ozone concentrations have in fact been significantly decreasing in the Northern Hemisphere since 1988, and it is possible that an unusually cold and prolonged Arctic winter could lead to ozone depletion effects and to the development of an Arctic hole similar to that in the Antarctic. During the mid-1990s the Arctic winters had below average stratospheric temperatures. For example, the Arctic winter of 1995–1996 was the coldest ever observed up to that time, reaching a low of $-88°C$ on March 3 and producing a vortex roughly equivalent in size to that of the Antarctic in a typical winter. However, final warming and breakup of the vortex began in early March, which was a month or two earlier than the usual Antarctic winter breakup. The Arctic winter of 1996–1997 was also extremely cold, but it was unique in that the polar vortex did not form until late December. Nevertheless, the vortex developed fast after that and remained very strong and symmetrical, generally centered near the North

---

[30]For example, this led, prior to 1988, to total late-spring ozone column concentrations greater than 450 DU above 70°N latitude—much above the year-round global average of about 306 DU.

Pole, into late April, isolating the polar region from the inward flow of ozone-rich air. Anomalous low temperatures—below that needed for type I PSC formation—continued through most of March. Concurrent high stratospheric concentrations of ClO· (as high as those in the Antarctic spring) in February indicated chemical ozone loss consistent with the halogen-catalyzed ozone depletion mechanism (5-60). The average Arctic ozone column concentration for the month of March (354 DU) was approximately 21% lower than pre-1988 values and reached a record low of 219 DU on March 26, 1997.

The Arctic winters of 1997–1998 and 1998–1999 were relatively warm, with low loss of ozone. However, the 1999–2000 Arctic winter was one of the coldest on record, with a low of −89°C on January 10 and temperatures below that needed for type I PSC formation (−80°C) continuing over a longer period (from about November 20 to March 10) and a larger area than in any previously observed winter. Atmospheric measurements made from November, 1999 to April, 2000 showed greater ozone depletion in early February (approximately 1% a day) than in the cold 1995–1996 winter, leading to a clear ozone minimum over the Arctic pole that extended during February and March over an area comparable to a typical Antarctic hole. About 70% of the ozone near 19 km above the Arctic was depleted by the end of March. In addition, studies of oxides of nitrogen loss in the polar vortex from mid-January to mid-March, 2000 (well after the breakup of the PSCs), showed the most severe Arctic denitrification ever observed.

In contrast, during the 2000–2001 Arctic winter a strong polar vortex formed in October and November, but mild stratospheric warming occurred after that with temperatures rising above the PSC-forming threshold by mid-February. This resulted in a heavy disturbance of the vortex and only a moderate total ozone loss for the season of approximately 20%.

5.2.3.5 Control of Ozone-Depleting Substances

Almost from the very beginning of the concern over potential stratospheric ozone depletion by nitrogen oxide or atomic chlorine from chlorofluorocarbons there have been repeated calls—mainly by the scientific community—for reduction and eventual elimination of the industrial uses of CFCs. In 1978, almost all uses of CFCs for aerosols (primarily CFC-11) were banned in the United States. However, it was not until it was unequivocally shown by the Antarctic ozone hole that ozone could actually be destroyed in the stratosphere by human releases of chemicals into the troposphere that very strong action was officially taken by governments: the *Montreal Protocol on Substances That Deplete the Ozone Layer*, signed in 1988 and amended several times since then. By January 1, 2001, 175 countries had signed this agreement, which banned production of all halons (the major fire-suppressant chemicals) as of January 1, 1994, and all CFCs, methyl chloroform (CH_3CCl_3), and carbon tetrachloride by January 1, 1996, for developed countries. The protocol has a 14-year grace period for developing countries (including production of small amounts by industrialized countries for use by developing countries), and also contains provisions allowing countries to petition for some exemptions based on lack of feasible chemical replacements, but it appears that relatively few such exemptions will be allowed. By 1993, the electronics industry in the United States had already almost completely stopped using CFCs (CFC-113) and methyl chloroform for precision cleaning, opting either for completely eliminating cleaning, or for cleaning with aqueous solutions. The United States capped consumption of methyl bromide (which accounts for about 10% of the ozone destruction) at 1991 levels in 1995, and a gradual phaseout started in 1999 with total consumption ban scheduled for 2005, in line with the 1997 amendments to the Montreal Protocol. The Japanese electronics industry phased out CFC use by 1995. There is some concern, however, about the possibilities for noncompliance with the provisions of the amended Montreal Protocol and with illegal trade of CFCs among developed and developing countries.

Many of the compounds being developed and used for replacement of the CFCs are the hydrochlorofluorocarbons, HCFCs, in which some of the CFC chlorine atoms have been replaced with hydrogen atoms that are readily abstracted by $\cdot$OH radicals in the troposphere.[31] As we saw Table 5-1, this greatly reduces their ozone depletion potentials, or ODPs.[32] For example, HCFC-141b (CH_3CFCl_2—see Table 3-2, note *a*, for determining compositions of CFCs, HCFCs, and HFCs from their names) has almost completely replaced CFC-11 as a blowing agent in the manufacture of polyisocyanurate foam insulation, and HCFC-124 ($CHClFCF_3$) is an alternative for medical applications. However, since the HCFCs still contain one or more chlorine atoms, they still possess finite ODPs (see Table 5-1). As a result, the HCFCs also had production caps imposed by 1996 and phaseout of HCFC-141b in the United States and Europe by 2003 has been mandated.

Replacement of all chlorine atoms with hydrogen atoms gives hydrofluorocarbons, HFCs, that have zero ODP. The hydrofluorocarbon HFC-134a (CH_2FCF_3) is a replacement for CFC-12 in automobile air conditioners, and by the end of 1993 all new automobile air conditioners produced in the United States and Canada were using it. A planned replacement for CFC-11 as a polyurethane foam blowing agent is HFC-245fa ($CF_3CH_2CHF_2$). HCFC-22

[31]J. S. Francisco and M. M. Maricq, Atmospheric photochemistry of alternative hydrocarbons, in *Advances in Photochemistry*, vol. 20, D. C. Neckers, D. H. Volman, and G. von Bünau, eds., pp. 79–163. Wiley, New York, 1995.

[32]The $\cdot CF_3$ radical, a fragment of HCFC-destroying reactions, is itself very stable and therefore potentially could also be involved in ozone-depleting catalytic cycles via $CF_3O\cdot$ and/or $CF_3OO\cdot$ radicals. However, laboratory studies and atmospheric models have shown that this is not the case—see Section 8.6.2.2.

($CHClF_2$), also a substitute for CFC-12, was widely used as a stationary air conditioner refrigerant, but it is being replaced by mixtures of various HFCs.

Other compounds without chlorine that are being developed for CFC replacements are the fluorinated ethers, HFEs. For example, HFE-7100 ($C_4F_9OCH_3$)[33] and HFE-7200 ($C_4F_9OC_2H_5$)[34] are now being used commercially for applications employing liquids such as cleaning electronic equipment, secondary refrigerants, and carrier fluids for lubricants. HFE-7100 and HFE-7200 are trade names for isomeric mixtures of the materials; for example, HFE-7100 is a mixture of 40% of the normal isomer n-$C_4F_9OCH_3$, $CF_3CF_2CF_2CF_2OCH_3$, and 60% of the iso isomer i-$C_4F_9OCH_3$, $(CF_3)_2CFCF_2OCH_3$. They are readily oxidized in the troposphere by ·OH radicals, giving short atmospheric lifetimes. The lifetime of 4.1 years given in Table 5-1 for HFE-7100 is a weighted average of 4.7 years for the normal isomer and 3.7 years for the iso isomer. Similarly, HFE-7200 has a lifetime of 0.8 years. Like the HFCs, they have zero ODPs.

The ideal replacement chemicals might seem to be the fully fluorinated perfluorocarbons (PFCs), since they are nontoxic and nonflammable, and have zero ODPs. Since, however, they are very unreactive, with no known atmospheric sinks below the mesosphere, they have extremely long atmospheric lifetimes—greater than 50,000 years for CF_4 (Table 5-1) and 10,000 years for C_2F_6. Since in general they absorb radiation in the near-infrared "window," they are very powerful greenhouse gases (see Sections 3.1.2 and 3.3.3). It is estimated, for example, that the impact of the PFCs on global warming may be greater than 5000 times than that of carbon dioxide 100 years after their release into the atmosphere. CFCs and HCFCs are also powerful greenhouse gases (i.e., large global warming potentials: see Table 3-2), although their atmospheric lifetimes (Table 5-1) are much less than those of the PFCs. Some HFCs such as HFC-23, a by-product in the production of HCFC-22, are also greenhouse gases. On the other hand, iodofluorocarbon compounds absorb strongly in the near-UV spectral region, leading to photodissociation (the C—I bond is weaker than the C—Br or C—Cl bond). Thus they have atmospheric lifetimes of only a few days and therefore have essentially zero ODPs and are not greenhouse gases. Trifluoromethyl iodide (CF_3I), which has an atmospheric lifetime of less than 2 days, is a promising replacement for H-1301 (CF_3Br) for controlling fires, but its toxicity makes it unsuitable for use in enclosed spaces.

[33]T. J. Wallington et al., Atmospheric chemistry of HFE-7100 ($C_4F_9OCH_3$): Reaction with ·OH radicals, UV spectra and kinetic data for $C_4F_9OCH_2$· and $C_4F_9OCH_2O_2$· radicals, and the atmospheric fate of $C_4F_9OCH_2O$· radicals, *J. Phys. Chem. A*, **101**, 8264–8274 (1997).

[34]L. K. Christensen *et al.*, Atmospheric chemistry of HFE-7200 ($C_4F_9OC_2H_5$): Reaction with ·OH radicals and fate of $C_4F_9OCH_2O(·)$ and $C_4F_9OCHO(·)CH_3$ radicals, *J. Phys. Chem. A*, **102**, 4839–4845 (1998).

For several decades prior to the discovery of the Antarctic ozone hole, the worldwide release of CFCs into the atmosphere exceeded a million tons per year. By 1993 this release of CFCs had dropped by 50%, and it is expected to dramatically decrease as the provisions of the revised and amended Montreal Protocol continue to be put into effect. As a result, the net chlorine and bromine concentrations in the troposphere peaked in 1994 and started to decline in 1995. Although tropospheric bromine continued to increase because of continuing release of halons, reactive stratospheric chlorine and bromine reached a maximum in 1999 or 2000. It is now calculated that if the provisions of the amended Montreal Protocol are adhered to, global stratospheric ozone should start to increase and the Antarctic ozone hole should start to shrink within 10 years, returning to 1979 levels by around the middle of the twenty-first century.

5.2.4 Water in the Atmosphere

We saw in Chapter 2 (Section 2.3.4) that the photolysis of water vapor is one of the two major sources of oxygen in our atmosphere. Reaction (2-1) represents the overall reaction leading to the formation of molecular oxygen and hydrogen. However, the precise mechanism of this reaction is not clear. The most likely path appears to be the photodissociation of water into H atoms, O atoms, and/or $\cdot OH$ radicals (depending on wavelength of light absorbed), followed by

$$H + \cdot OH \rightarrow H_2 + O \tag{5-64}$$

and the three-body atom recombination reaction (5-21). Combination of two $\cdot OH$ radicals

$$2 \cdot OH \rightarrow H_2 + O_2 \tag{5-65}$$

has also been proposed, however.

Water absorbs radiation below 186 nm. The absorption spectrum consists of a continuum region ($\lambda_{max} = 167.5$ nm, $\sigma = 5.2 \times 10^{-18}$ cm^2/molecule), a banded region between about 122 and 143 nm superimposed on a second continuum ($\lambda_{max} = 167.5$ nm, $\sigma = 9 \times 10^{-18}$ cm^2/molecule), and a strongly banded region to below 100 nm corresponding to transitions of H_2O^+. The presence of two continua indicates that two different excited states of H_2O are involved in the photochemistry of water.

The major (99%) primary process between 186 and 140.5 nm is dissociation into ground-state hydrogen atoms and hydroxyl radicals,

$$H_2O \xrightarrow{\;h\nu\,(186\text{–}140.5\,\text{nm})\;} H + \cdot OH \tag{5-66}$$

providing a source of radicals for reactions (5-64) and (5-65). Dissociation of water into H_2 and O may also be contributing in a minor way to absorption in this spectral region:

$$H_2O \rightarrow H_2 + O \text{ (ground state or excited }{}^1D_2 \text{ at } \lambda < 178 \text{ nm)} \qquad (5\text{-}67)$$

Between 140.5 and 124.5 nm the major reaction is still (5-66), but dissociations to excited OH* radicals

$$H_2O \xrightarrow{h\nu\,(140.5-124.5\,\text{nm})} H + \cdot OH^* \qquad (5\text{-}68)$$

and to ground-state O atoms

$$H_2O \xrightarrow{h\nu\,(140.5-124.5\,\text{nm})} 2H + O \qquad (5\text{-}69)$$

also occur. Below 124.5 nm (to 500 nm), minor products are excited H atoms and OH^+ ions in addition to the products from reactions (5-66), (5-67), (5-68), and (5-69).

Since water does not absorb dissociating radiation above 186 nm, it will not be photodissociated in the troposphere, which is protected from the high-energy photons by oxygen and the stratospheric ozone layer. Photodissociation does occur in the mesosphere and in the upper stratosphere as a result of partial transmission through the UV window (Figure 5-7). Absorption of nondissociating infrared radiation by water in the troposphere is of course an important contributor to the greenhouse effect (Section 3.1.2).

5.3 PHOTOPROCESSES IN THE TROPOSPHERE: PHOTOCHEMICAL SMOG

5.3.1 The General Nature of Smog

In polluted lower regions of the troposphere, particularly in urban environments (see Section 3.4), copious quantities of chemical compounds are released into the atmosphere, leading to a number of complex photochemical reactions producing a variety of eye and throat irritants, aerosols and reduced visibility, and other irritating or destructive species manifesting themselves in the general category of photochemical smog.

Unlike the notorious smogs (smoke plus fog) that were a part of London life as well as that of many other highly industrial urban areas from the transition to coal in the thirteenth century until massive cleanup efforts were undertaken in the past few decades, photochemical smog is of fairly recent origin. It was first detected through vegetation damage in Los Angeles in 1944 and is sometimes referred to as Los Angeles smog. However, it has now spread to virtually all major metropolitan areas and must be considered to be one of the most pressing actual or potential worldwide problems, with global aspects transcending national boundaries. Also in contrast to the poisonous so-called London killer smogs,[35] to date there do not appear to be major human disasters associated with photochemical smog. However, detailed epidemiological studies have shown that the concentrations of pollutants—particularly of ozone but also materials such as particulate matter, the oxides of nitrogen, and volatile organic compounds (VOCs)—are sufficiently high in photochemical smog to pose significant threats to human health, particularly among the very young, the elderly, those suffering from chronic respiratory illnesses, and cigarette smokers. Common symptoms are sore throat, mild cough, headache, chest discomfort, decreased lung function, and eye irritation. Ozone also injures vegetation, leading to reduction of agricultural yield and biomass production. Potentially, photochemical smog can have serious long-range global effects on the earth's radiation balance through submicrometer aerosol formation, and global distribution that may cause unfavorable changes in weather patterns in addition to its influence locally.

Photochemical smog requires rather specific meteorological and geographical conditions to be produced in large quantities. For example, the photochemical reactions must occur in a stable air mass that traps the pollutants, which can result when there is a temperature inversion (Section 3.5). The Los Angeles Basin is particularly suited for this to occur, and in fact temperature inversions occur there about 80% of the time in hot summer months. Even though California has the strictest pollution control programs in the United States, this basin is still one of the worst in the country with respect to photochemical smog air pollutants.[36]

There are other significant differences between "London" and "Los Angeles" smogs that suggest different remedial actions that might (or should) be taken in each case. London-type smogs occur mainly in the early morning hours in winter, where humidity is relatively high and virtually no solar photochemistry is taking place. Along with particulate matter, sulfur dioxide is the major contaminant, and this creates a chemically reducing atmosphere because of the ease of sulfur dioxide oxidation. On the other hand, photochemical smogs usually occur on warm sunny days (summer and winter) with initially clear skies and low humidity. Primary contaminants—those that are emitted directly into the atmosphere—are nitric oxide (NO), hydrocarbons, and carbon monoxide. Secondary contaminants, largely ozone and nitrogen dioxide, are produced by reactions of the primary contaminants in the atmosphere. They begin to build up during morning traffic hours and reach peak

[35]Over 3000 deaths were attributed to the four-day smog in London in December 1952; see Section 3.5.

[36]Apparently Houston, Texas has joined Los Angeles as having the worst air quality in the United States (Houston air is focus of $20 million study, *Chem. Eng. News*, Nov. 6, 2000, pp. 32–33).

intensity in midday, although hourly variations occur depending on automobile traffic/time distributions and even time variations for different constituents (Figure 5-9).

Natural and nonnatural sources of volatile organic compounds (VOCs) and oxides of nitrogen are covered in detail in Section 6.7. Naturally released methane and carbon monoxide can lead to photochemically induced ozone generation and formation of a variety of oxygen- and nitrogen-containing compounds if the nitrogen oxides concentration is sufficiently high. These reactions will be covered in the next section. The importance of the anthropogenic sources is that in an urban environment, with the appropriate climatic, geographic, and solar radiation conditions, they generate sufficiently high concentrations of the pollutants to undergo drastically more complex photochemical and secondary reactions, leading to photochemical smog. Hundreds of individual species are continuously discharged—at rates and light intensities that vary depending on the time of day and so on—into a complex moving air mass of variable size in which thousands of simultaneous and consecutive reactions take place. Complete quantitative mathematical modeling of photochemical smog requires reliable primary emission rates over a period of time and ambient concentrations of VOCs, physical three-dimensional

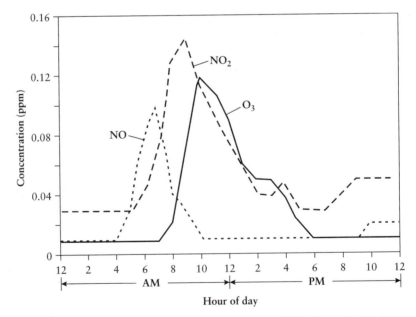

FIGURE 5-9 Photochemical smog buildup in Los Angeles, California, on July 10, 1965. Redrawn from K. J. Demerjian, J. A. Kerr, and J. G. Calvert, *Adv. Environ. Sci. Technol.*, 4, 1 (1974).

gas-phase (plume) transport behavior that includes topography, and accurate reaction rate data for the many hundreds of individual chemical processes taking place. Reasonable computer simulations generally lump the hundreds of individual VOC contaminants into a manageable smaller number of groups or classes of chemical species with similar structure and reactivity, such as alkenes or aldehydes. Each group is then represented by "surrogate" species.[37] In Section 5.3.3 we consider several reactions that are well established as primary steps in photochemical smog initiation in the presence of VOCs and associated generation of the accompanying obnoxious products.

5.3.2 Nitrogen Oxides (NO$_x$) in the Absence of Volatile Organic Compounds (VOCs)

As pointed out in Section 6.7, the main anthropogenic sources of nitrogen oxides in a polluted atmosphere are mobile and stationary petroleum and coal combustion chambers, where inside temperatures and pressures are high enough for the fixation of nitrogen to occur,

$$N_2 + xO_2 \rightarrow 2NO_x \qquad (5\text{-}70)$$

(NO$_x$ represents a mixture of oxides of nitrogen—mainly NO and NO$_2$) and where quenching to low temperatures outside the chambers is rapid enough to prevent the thermodynamically favored back-reaction (dissociation) from occurring. The major oxide of nitrogen produced is nitric oxide, NO, a relatively nontoxic gas, along with some nitrogen dioxide, NO$_2$. The concentration of NO in a polluted atmosphere is quite sensitive to automobile traffic patterns, in general being a maximum during the peak morning traffic hours (Figure 5-9). However, NO does not absorb radiation above 230 nm, and therefore it cannot be the primary initiator of photochemical reactions in a polluted lower atmosphere. On the other hand, NO$_2$ is a brown gas with a broad, intense absorption band absorbing over most of the visible and ultraviolet regions with a maximum at 400 nm ($\sigma_{max} \cong 6 \times 10^{-19}$ cm^2/molecule) as shown in Figure 5-10, and this gas is a major atmospheric absorber leading to photochemical smog. As we shall see later, there are other light-absorbing species present at very low concentrations in a polluted atmosphere that can also initiate the complex photochemical reactions.

---

[37]A mathematical modeling [R. A. Harley, A. G. Russell, and G. R. Cass, Mathematical modeling of the concentrations of volatile organic compounds: Model performance using a lumped chemical mechanism, *Environ. Sci. Technol.*, **27**, 1638–1649 (1993)] based on the Southern California Air Quality Study of the Los Angeles Basin on August 27–29, 1987 involved 35 species and 106 chemical reactions, with the atmospheric diffusion equation solved for each of the chemical species.

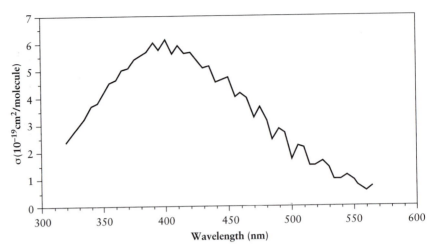

FIGURE 5-10 Absorption spectrum of NO_2 at room temperature averaged over 5-nm intervals. Drawn from data of M. H. Harwood and R. L. Jones, *J. Geophys. Res.*, **99**, 22,955–22,964 (1994).

The dissociation energy of NO_2 is 3.12 eV, which corresponds to a photon having a wavelength of 397.5 nm, and absorption below this wavelength leads to almost complete photodissociation ($\phi \approx 1$)[38]:

$$NO_2 \xrightarrow{h\nu\,(<\,397.5\,nm)} NO + O \qquad (5\text{-}71)$$

In the absence of other species, molecular oxygen may be formed by a reaction of an O atom with NO_2, reaction (5-43), and indeed the quantum yield of O_2 formation approaches unity below 380 nm. In the presence of ground-state molecular oxygen, however, ozone is produced by the three-body reaction (5-29) between O_2 and $O(^3P_2)$. Combination of reactions (5-71), (5-29), and (5-42) leads to the photostationary state

$$NO_2 + O_2 \xleftarrow{\quad h\nu\,(NO_2),\,M \quad} O_3 + NO \qquad (5\text{-}72)$$

and hence to a steady-state ozone concentration, $[O_3]_{SS}$. Since NO and NO_2 appear on different sides of reaction (5-72), the ozone concentration will depend on the ratio of nitrogen oxide concentrations: that is, $[O_3]_{ss}$ is

---

[38]Some dissociation even occurs at wavelengths greater than 397.5 nm—for example, the O_2 quantum yield is 0.36 at 404.7 nm—and thus energetically below the dissociation energy. But this dissociation is strongly temperature dependent and presumably is due to the contributions of internal rotational and vibrational energies.

proportional to $[NO_2]/[NO]$. However, since we have assumed a photostationary state, $[O_3]_{ss}$ will be a function of the light intensity.

Thus, NO_2 is the primary light-absorbing species in the atmosphere leading to photochemical smog and the ozone concentration is enhanced by increasing NO_2, but NO is the major oxide of nitrogen produced in high-temperature combustion chambers. Since the small amount of NO_2 produced directly would be completely depleted in a very short time in direct sunlight, at least one additional process leading to the oxidation of NO to NO_2 is required to produce the quantities of O_3 and other products generated in a polluted atmosphere. It is now agreed that the hydroxyl ($\cdot$OH) and hydroperoxyl ($HO_2\cdot$) radicals play key roles in the rapid oxidation of NO to NO_2 in the absence of VOCs.[39] The $HO_2\cdot$ radical is directly involved:

$$NO + HO_2\cdot \rightarrow NO_2 + \cdot OH \qquad (5\text{-}73)$$

However, we will see shortly that $HO_2\cdot$ radicals are formed in free-radical chain mechanisms in which $\cdot$OH radicals serve as chain carriers, so that $\cdot$OH and $HO_2\cdot$ are interconverted provided sufficient NO is present for reaction (5-73) to be dominant.

The major primary source of $\cdot$OH in the troposphere is the reaction of excited $O(^1D_2)$ atoms with water, reaction (5-35), following photolysis of ozone below 310 nm [see reaction (5-32)]. The global-averaged noon-time concentration of the hydroxyl radical in the lower troposphere is very low—of the order of 10^6 molecules/cm^3—with a lifetime of approximately one second. It does not react with the major tropospheric constituents (N_2, O_2, CO_2, H_2O), but it is the major initiator of oxidation in photochemical smog generation and in the removal of most trace gases in the atmosphere. The $HO_2\cdot$ radical can react with ozone, regenerating the $\cdot$OH radical

$$HO_2\cdot + O_3 \rightarrow \cdot OH + 2O_2 \qquad (5\text{-}74)$$

It can also self-combine to form hydrogen peroxide

$$2HO_2\cdot \rightarrow H_2O_2 + O_2 \qquad (5\text{-}75)$$

which dissolves in fog and clouds and is washed out of the troposphere by rain.

In the absence of VOCs, two species with major natural and anthropogenic sources that lead to $HO_2\cdot$ (hence to oxidation of NO to NO_2) are carbon

---

[39]Thermal oxidation of NO to NO_2 by O_2 is thermodynamically feasible but is not significant at normal NO pressures in the atmosphere. Similarly, the oxidation of NO by ground-state O atoms is unimportant at the low O-atom concentrations in the lower troposphere.

monoxide, CO, and methane, CH_4 (see Section 2.2). Carbon monoxide reacts rapidly (with very low activation energy) with the hydroxyl radical

$$CO + \cdot OH \rightarrow CO_2 + H; \qquad k_{298} \approx 2 \times 10^{-13} \, cm^3 \, molecule^{-1} \, s^{-1} \qquad (5\text{-}76)$$

followed by the very fast three-body (essentially triple-collision frequency—see Section 4.3) combination reaction

$$H + O_2 \xrightarrow{M} HO_2 \cdot \qquad\qquad (5\text{-}77)$$

Coupling these two reactions with reaction (5-73) gives the CO chain mechanism, with $\cdot OH$ the primary chain carrier and HO_2 the secondary chain carrier:

$$CO + \cdot OH \rightarrow CO_2 + H$$

$$H + O_2 \xrightarrow{M} HO_2 \cdot$$

$$NO + HO_2 \cdot \rightarrow NO_2 + \cdot OH$$

$$\text{net: } CO + NO + O_2 \xrightarrow{\cdot OH, M} CO_2 + NO_2 \qquad (5\text{-}78)$$

However, this CO chain does not contribute to the overall NO oxidation to the extent that methane does in a similar chain. The $\cdot OH$ radical is the major tropospheric "sink" for methane via the reaction

$$\cdot OH + CH_4 \rightarrow \cdot CH_3 + H_2O \qquad\qquad (5\text{-}79)$$

Reaction (5-79) results in a tropospheric lifetime for methane of about 12 years. The major fate of the methyl ($\cdot CH_3$) radical in the troposphere is the very fast (also of the order of a triple collision) three-body combination reaction with O_2 to form $CH_3O_2 \cdot$[40]:

$$\cdot CH_3 + O_2 \xrightarrow{M} CH_3O_2 \cdot \qquad\qquad (5\text{-}80)$$

The methylperoxy radical, $CH_3O_2 \cdot$, can react in the troposphere with a variety of species.

1. If the NO concentration is high enough, NO is oxidized to NO_2 in a fast reaction analogous to reaction (5-73), producing the methoxy ($CH_3O \cdot$) radical

$$CH_3O_2 \cdot + NO \rightarrow CH_3O \cdot + NO_2 \qquad\qquad (5\text{-}81)$$

---

[40]Actually, under tropospheric conditions the concentrations of $\cdot CH_3$ and O_2 and the lifetime of $CH_3O_2 \cdot$ are such that the reaction order varies between 2 and 3 depending on the overall pressure. That is, at very low pressures the reaction order approaches 3, whereas at high pressures it approaches 2.

followed by

$$CH_3O\cdot + O_2 \rightarrow H_2CO + HO_2\cdot \qquad (5\text{-}82)$$

The $\cdot OH$ radical is regenerated by reaction (5-73). Combination of these reactions gives the $\cdot OH$-initiated methane chain reaction for oxidation of NO:

$$\cdot OH \; + \; CH_4 \rightarrow \cdot CH_3 + H_2O$$

$$\cdot CH_3 \; + \; O_2 \xrightarrow{\;M\;} CH_3O_2\cdot$$

$$CH_3O_2\cdot + NO \rightarrow CH_3O\cdot + NO_2$$

$$CH_3O\cdot \; + \; O_2 \rightarrow H_2CO + HO_2\cdot$$

$$HO_2\cdot + NO \rightarrow NO_2 + \cdot OH$$

net: $\quad CH_4 + 2O_2 + 2NO \xrightarrow{\;\cdot OH, M\;} H_2CO + 2NO_2 + H_2O \qquad (5\text{-}83)$

[Reaction (5-75) is a possible chain-terminating step in this mechanism.] This gives an *increase* in ozone by the photostationary state (5-72).

2. On the other hand, if the tropospheric NO concentration is low (as in some very clean, unpolluted areas), instead of reacting with NO, the methylperoxy radical $CH_3O_2\cdot$ reacts with $HO_2\cdot$, forming methyl hydroperoxide:

$$CH_3O_2\cdot + HO_2\cdot \rightarrow CH_3O_2H + O_2 \qquad (5\text{-}84)$$

In effect reaction (5-84) serves as a "sink" for both $\cdot OH$ and $HO_2\cdot$, since the methane chain reaction (5-83) is now terminated and the interconversion of $\cdot OH$ and $HO_2\cdot$ is restricted, and this leads to a net *decrease* in ozone. The methylperoxy radical also rapidly combines with NO_2 in the three-body reaction

$$CH_3O_2\cdot + NO_2 \xrightarrow{\;M\;} CH_3O_2NO_2 \qquad (5\text{-}85)$$

but this reaction is generally not important in the troposphere because of the fast decomposition of methylperoxynitrate ($CH_3O_2NO_2$) back to the reactants.

Formaldehyde, H_2CO, is an overall product of the methane chain reaction (5-83), formed in the chain by the reaction of the methoxy radical with molecular oxygen (step 4).[41] It absorbs weakly in the near-ultraviolet spectral

---

[41]Formaldehyde is also a primary emission product of hydrocarbon combustion, and is produced in the photooxidation of a wide variety of higher hydrocarbons in VOC-containing atmospheres (next section); its average concentration in urban areas is approximately 3–16 parts per billion by volume (ppbv).

region. Its gas-phase absorption spectrum is strongly banded, with $\lambda_{max} \cong 305\,nm$ ($\sigma_{305} \approx 7 \times 10^{-20}\,cm^2/molecule$), and only very weak absorption near the visible region (e.g., $\sigma_{370} \approx 2 \times 10^{-21}\,cm^2/molecule$). This absorption forms an excited formaldehyde molecule, H_2CO^*, which dissociates either to H and the formyl radical $H\dot{C}O$,

$$H_2CO^* \xrightarrow{\;h\nu\;} H + H\dot{C}O \tag{5-86}$$

or to H_2 and CO:

$$H_2CO^* \xrightarrow{\;h\nu\;} H_2 + CO \tag{5-87}$$

Reaction (5-86) dominates below roughly 330 nm, corresponding to the dissociation energy of H_2CO, while (5-87) is the major dissociation above 330 nm. Thus, H_2CO is a source of $HO_2\cdot$ radicals by reaction (5-77) and by

$$H\dot{C}O + O_2 \rightarrow HO_2\cdot + CO \tag{5-88}$$

Formaldehyde also reacts with $\cdot OH$

$$\cdot OH + H_2CO \rightarrow H_2O + H\dot{C}O \tag{5-89}$$

and combines with $HO_2\cdot$ to form initially an alkoxy radical

$$HO_2\cdot + H_2CO \rightarrow HO_2CH_2O\cdot \tag{5-90}$$

which rapidly isomerizes to a peroxy radical

$$HO_2CH_2O\cdot \rightarrow \cdot O_2CH_2OH \tag{5-91}$$

However, excitation and subsequent dissociation steps, reactions (5-86) and (5-87), dominate in an atmosphere free from VOCs, leading to a noonday overall atmospheric lifetime of approximately 3 hours for formaldehyde.

A reaction not yet considered is the reaction between NO_2 and O_3:

$$NO_2 + O_3 \rightarrow NO_3\cdot + O_2 \tag{5-92}$$

The nitrate radical ($NO_3\cdot$) plays an important role in photochemical smog, particularly in nighttime reactions involving aldehydes and simple olefins. It reacts at close to collisional frequency with NO to produce NO_2:

$$NO_3\cdot + NO \rightarrow 2NO_2 \tag{5-93}$$

It combines with NO_2 to form dinitrogen pentoxide, N_2O_5, by the reversible reaction

$$NO_3\cdot + NO_2 \xrightleftharpoons{\;M\;} N_2O_5 \tag{5-94}$$

with a formation constant $K_{298} = 4.5 \times 10^{-11}\,\mathrm{cm^3/molecule}$. (Equilibrium is reached in about one minute under tropospheric conditions.) N_2O_5 also reacts with water to form nitric acid,

$$N_2O_5 + H_2O \rightarrow 2HNO_3 \qquad (5\text{-}95)$$

thus providing a sink for both $NO_3\cdot$ and NO_2. In the daytime $NO_3\cdot$ is rapidly photolyzed to NO and O_2

$$NO_3\cdot \xrightarrow{\ h\nu\ } NO + O_2 \qquad (5\text{-}96)$$

or to NO_2 and O

$$NO_3\cdot \xrightarrow{\ h\nu\ } NO_2 + O \qquad (5\text{-}97)$$

so that its daytime concentration is very low. However, at night it may be one of the most important tropospheric oxidizing agents. Additional reactions of $NO_3\cdot$ in the presence of methane and VOCs are covered next.

5.3.3 Reactions in Urban Atmospheres Containing Volatile Organic Compounds

We have just seen that the ozone concentration can be enhanced even in a VOC-free atmosphere containing oxides of nitrogen and methane—anthropogenic and/or natural—or, to a lesser extent, carbon monoxide. These ozone-producing processes are generally well understood now, and they give reasonable simulations in tropospheric computer modeling. In polluted urban areas, however, chemical reactions of VOCs dominate over those involving methane, leading to considerably more complexity because of the number (hundreds) of VOCs and their diverse chemistry. An excellent review by Atkinson[42] covers reactions under tropospheric conditions of several classes of organic compounds, such as hydrocarbons (alkanes, alkenes, alkynes, and aromatics and substituted aromatics) and oxygen- and nitrogen-containing compounds, and their degradation products. Although complete coverage of the photochemical smog mechanism is beyond the scope of this book, this section summarizes some of the more important types of these reactions.

One of the major tools used in determining the individual chemical processes taking place is the *environmental chamber*.[43] Also called smog cham-

[42]R. Atkinson, Gas-phase tropospheric chemistry of organic compounds, Monograph no. 2 of *The Journal of Physical and Chemical Reference Data*, J. W. Gallagher, ed. American Chemical Society and the American Institute of Physics for the National Institute of Standards and Technology, Gaithersburg, MD, 1994.

[43]A coverage in depth of environmental chambers is given in B. J. Finlayson-Pitts and J. N. Pitts Jr., *Chemistry of the Upper and Lower Atmosphere: Theory, Experiments, and Applications*, Academic Press, San Diego, CA, 2000.

bers, these vessels are often large (from tens to hundreds of cubic meters) receptacles in which reactant mixtures close to actual atmospheric conditions of concentration, pressure, relative humidity and temperature are illuminated with solar spectral distribution light. A typical smog chamber experiment for a mixture of NO, NO_2, and hydrocarbon (propene, C_3H_6) is shown in Figure 5-11. It is seen that the photostationary state projected for a VOC-free atmosphere [equation (5-72)] is destroyed by addition of the hydrocarbon. NO is rapidly oxidized to NO_2, O_3 is produced after an induction period, and the propene is oxidized to CO, CO_2, and a variety of oxygen- and nitrogen-containing organic compounds. This reaction mixture is of course an extreme simplification, containing only one hydrocarbon instead of the hundreds of VOC species emitted into a polluted urban environment. Note, however, that it qualitatively agrees with the actual photochemical smog buildup shown in Figure 5-9. That is, NO is rapidly oxidized to NO_2 in the presence of organic compounds (the VOCs emitted along with NO into the atmosphere starting around 6 A.M. from automobile exhausts) and the ozone builds up only after an induction period, during which the concentration of NO is lowered until O_3 is no longer destroyed by reaction (5-42).

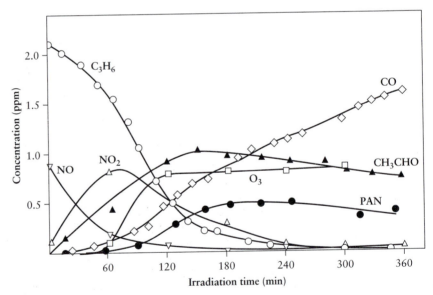

FIGURE 5-11 Typical photochemical smog chamber results for a mixture of hydrocarbon, NO, and air. Drawn from data of A. P. Altshuller, S. L. Kopczynski, W. A. Lonneman, T. L. Baker, and R. L. Slater, *Environ. Sci. Technol.*, **1**, 899 (1967); and K. J. Demergian, J. A. Kerr, and J. G. Calvert, *The Mechanism of Photochemical Smog Formation*, in *Advances in Environmental Science and Technology*, J. N. Pitts Jr., and R. L. Metcalf, eds., **4**, p. 1. Wiley, New York, 1974.

As we will see in this section, the organic alkyl ($R\cdot$), alkoxy ($RO\cdot$), and alkylperoxy ($RO_2\cdot$) radicals (in addition to $\cdot OH$ and $HO_2\cdot$) are important intermediates in the tropospheric degradation of VOCs and the formation of photochemical smog. Each is involved in many chain-propagating steps in which NO is converted to NO_2 before being terminated by reactions such as (5-75) or (5-103), shown later. We have seen that the methyl ($\cdot CH_3$) radical is produced in the chain-initiating H-atom abstraction by the $\cdot OH$ radical [reaction (5-79)] in the methane chain; similarly, this is the major mode of formation of alkyl and substituted alkyl radicals from several classes of compounds:

$$\cdot OH + RH \rightarrow R\cdot + H_2O \tag{5-98}$$

With alkenes, however, the major reaction of OH is addition to the carbon–carbon double bond to give β-hydroxyalkyl radicals,

$$RHC = CHR' + \cdot OH \xrightarrow{\quad M \quad} RH\dot{C} - C(OH)HR' \tag{5-99}$$

with only a small fraction of the $\cdot OH$ radicals involved in H-atom abstraction.

Radicals are also produced by photolysis of some VOCs. In Section 5.3.2, for example, we saw that the photolysis of formaldehyde below 330 nm generates H and $H\dot{C}O$ radicals [reaction (5-86)]. In a similar manner, higher aldehydes such as acetaldehyde, propanal, or butanal, which are formed in the tropospheric degradation of many VOCs, photodissociate into an $R\cdot$ radical and the formyl radical:

$$RCHO \xrightarrow{\quad h\nu \quad} R\cdot + H\dot{C}O \tag{5-100}$$

Smog chamber studies of hydrocarbon–NO mixtures have shown that addition of aldehydes does indeed enhance NO oxidation and production of secondary pollutants.

Most of the alkyl and substituted alkyl radicals very rapidly combine with O_2 to produce alkylperoxy radicals:

$$R\cdot + O_2 \xrightarrow{\quad M \quad} RO_2\cdot \tag{5-101}$$

For allyl, benzyl, substituted benzyl, or alkyls with more carbons than ethyl, reaction (5-101) is second order under atmospheric conditions (see footnote 40). The β-hydroxyalkyl radicals formed by the addition of $\cdot OH$ to the alkene double bond [reaction (5-99)] also rapidly add to O_2 forming β-hydroxy-alkylperoxy radicals, $RHC(\dot{O}_2)C(OH)HR'$.

All the alkylperoxy ($RO_2\cdot$) radicals react with NO with approximately the same rate constant. Both $CH_3O_2\cdot$ and $C_2H_5O_2\cdot$ primarily oxidize NO to NO_2, forming an alkoxy radical:

$$RO_2\cdot + NO \rightarrow RO\cdot + NO_2 \tag{5-102}$$

However, for the larger alkylperoxy radicals, addition to form organic nitrates also occurs,

$$RO_2{}^{\cdot} + NO \xrightarrow{\quad M \quad} RONO_2 \tag{5-103}$$

with reaction (5-103) becoming more important with increasing number of carbon atoms in R.

We thus see that the $HO_2{}^{\cdot}$ and $RO_2{}^{\cdot}$ radicals generated in an oxygen environment by the $\cdot OH$ radical in the degradation of many VOCs are at least qualitatively the major contributors to the enhanced NO-to-NO_2 oxidation—and therefore to buildup of O_3—in photochemical smog via reactions like (5-102) and chain reactions similar to the methane chain, (5-83).

Other radicals generated from VOCs react differently with O_2 other than by the addition reaction (5-101). For example, we saw in reaction (5-88) that the formyl (H$\overset{\cdot}{C}$O) radical from the photolysis of formaldehyde reacts with O_2 to give $HO_2{}^{\cdot}$ and CO. The α-hydroxy radicals also react with O_2 via H-atom abstraction to give the $HO_2{}^{\cdot}$ radical, as for example the simplest α-hydroxy radical, $\cdot CH_2OH$:

$$\cdot CH_2OH + O_2 \rightarrow HO_2{}^{\cdot} + H_2CO \tag{5-104}$$

This is the only process leading to the loss of $\cdot CH_2OH$ under atmospheric conditions. The vinyl radical $\cdot C_2H_3$ initially adds to O_2 across the double bond to form a $C_2H_3O_2$ adduct, which decomposes to H_2CO and H$\overset{\cdot}{C}$O giving the overall reaction

$$H\overset{\cdot}{C}=CH_2 + O_2 \rightarrow H_2CO + H\overset{\cdot}{C}O \tag{5-105}$$

The three major reactions of the alkoxy or substituted alkoxy (RO$\cdot$) radical from (5-102) in the troposphere are reaction with O_2, unimolecular decomposition, and (for four or more carbon atoms) unimolecular isomerization. However, all these reactions eventually lead to formation of $HO_2{}^{\cdot}$ and/or the oxidation of NO to NO_2 by reactions similar to (5-102).[44]

Aliphatic aldehydes and ketones are products of the atmospheric degradation of a wide variety of VOCs, primarily from reactions of alkoxy and alkylperoxy radicals. The $\cdot OH$ radical reacts with carbonyl compounds via H-atom abstraction; with the aldehydes, abstraction is mostly from the primary carbon to give the acyl radical and water:

$$\cdot OH + RCHO \rightarrow R\overset{\cdot}{C}O + H_2O \tag{5-106}$$

Acyl radicals are also produced by H-atom abstraction from aldehydes by the $NO_3{}^{\cdot}$ radical, giving HNO_3, and by other organic radicals, $R'\cdot$:

---

[44]For example, with the smallest alkoxy radical, methoxy ($H_3CO\cdot$), $H_3CO\cdot + O_2$ $\rightarrow H_2CO + HO_2{}^{\cdot}$.

$$R'\cdot + RCHO \rightarrow R'H + R\dot{C}O \tag{5-107}$$

Acyl radicals combine with O_2 to form peroxyacyl radicals $RC(O)O_2\cdot$

$$R\dot{C}O + O_2 \xrightarrow{\text{ M }} RC(O)O_2\cdot \tag{5-108}$$

which further contributes to NO oxidation, hence to photochemical smog, by the rapid reaction

$$RC(O)O_2\cdot + NO \rightarrow RC(O)O\cdot + NO_2 \tag{5-109}$$

followed by formation of another alkylperoxyl radical from the $RC(O)O\cdot$ (acyloxy) radical:

$$RC(O)O\cdot + O_2 \rightarrow RO_2\cdot + CO_2 \tag{5-110}$$

The peroxyacyl radical also combines with NO_2 in a three-body reaction to produce peroxyacyl nitrate, $RC(O)OONO_2$:

$$RC(O)O_2\cdot + NO_2 \xleftrightarrow{\text{ M }} RC(O)OONO_2 \tag{5-111}$$

The simplest and most abundant peroxyacyl nitrate is peroxyacetyl nitrate (PAN), $CH_3C(O)OONO_2$. PAN is an important component of photochemical smog. For example, it is highly *phytotoxic*: one of the most toxic compounds known for vegetation. It is also a potent lachrymator, and is perhaps the major eye irritant in photochemical smog. It is also a greenhouse gas in that it absorbs radiation in the tropospheric IR "window." It is unstable, decomposing thermally by the reverse of reaction (5-111) to the peroxyacetyl radical and NO_2; yet it is sufficiently stable in the absence of NO to aid in the transport of nitrogen oxides to initially unpolluted "downwind" areas. In fact, it has been detected even in remote parts of the troposphere. The next most abundant peroxyacyl nitrate is peroxypropionyl nitrate (PPN), $C_2H_5C(O)OONO_2$; although it may be more phytotoxic, its concentration is only about 10–20% that of PAN.

In addition to its key role in the NO-to-NO_2 oxidation, the hydroxyl radical is also the primary oxidant and remover of VOCs and other pollutants in the atmosphere. Briefly, its major reactions with VOCs, in addition to those already given are *addition*, across the carbon–carbon triple bond in alkynes and to the aromatic ring in monocyclic aromatics and in naphthalene and methyl- and dimethyl-substituted naphthalenes, and *H-atom abstraction*, from ethers, α, β-unsaturated carbonyls, and alkyl nitrates, and from both the C—H and O—H bonds in alcohols.

The hydroxyl radical can also initiate the oxidation of reduced forms of sulfur such as hydrogen sulfide to $\cdot$SH radicals:

$$\cdot OH + H_2S \rightarrow H_2O + \cdot SH \tag{5-112}$$

Further oxidation of ·SH leads to sulfur dioxide, SO_2. The hydroxyl radical is also involved in the continuing oxidation of SO_2 by the addition reaction to form the $HOSO_2$· radical, reaction (5-52), followed by oxidation to sulfuric acid, H_2SO_4, and sulfate aerosol droplets in the presence of water.[45] Large amounts of sulfur oxides are released into the troposphere over industrialized areas, and they become the major acidic component of acid rain (see Section 11.4).

While ozone is the major contaminant product and its concentration is taken as the measure of the intensity of photochemical smog, it also contributes to decomposition reaction of the VOCs. For example, the addition of O_3 to the double bond in alkenes is a major pathway in their atmospheric decomposition. An energy-rich ozonide is formed, which rapidly decomposes by breaking the C—C single bond and either of the two O—O single bonds to form a carbonyl and a biradical $RR'\dot{C}OO·$. This energy-rich biradical either decomposes, forming a variety of products including the ·OH radical, or reacts with aldehydes, H_2O, NO_2, and so on.

We have seen that the nitrate radical is formed by the reaction of NO_2 with ozone (5-92), that it is in equilibrium with NO_2 and N_2O_5 (5-94), that it reacts with NO to produce NO_2 (5-93), and that it is photolyzed to $NO + O_2$ (5-96) or to $NO_2 + O$ (5-97). It also reacts with alkanes via H-atom abstraction, similar to ·OH reactions:

$$NO_3· + RH \rightarrow R· + HNO_3 \qquad (5\text{-}113)$$

However, although the NO_3· concentration is greater at night than during the day because of its daytime photolysis, it is nevertheless much less important at nighttime than the ·OH radical is during the daytime as a source of alkane decomposition. NO_3· readily reacts with alkenes by addition to the carbon–carbon double bond, followed by rapid addition to O_2. The resulting β-nitratoalkylperoxy radicals can react with HO_2· and RO_2· radicals, and can also reversibly add NO_2 to give the unstable nitratoperoxynitrates, which may serve as temporary "sinks" for the peroxy radicals.

5.3.4 Summary of Photochemical Smog

Only a very small portion of the hundreds of reactions possible in polluted atmospheres have been presented, but these are representative of the types of mechanism being considered to account for photoinitiation, oxidation of NO and VOCs, and formation of noxious products associated with photochemical smog. Figure 5-12 shows a computer integration of the differential kinetic rate

---

[45]The oxidation of $HOSO_2$· to H_2SO_4 in the troposphere is not simply the addition of another ·OH radical. Rather, it is mostly by heterogeneous reactions taking place in cloud droplets, probably involving hydrogen peroxide, H_2O_2.

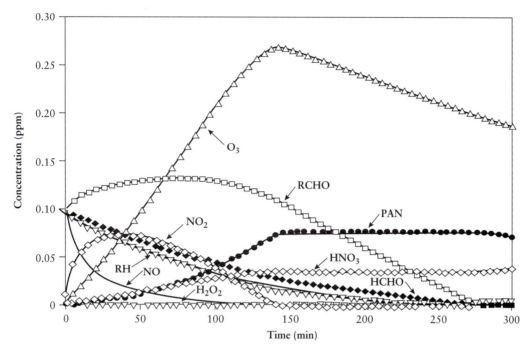

FIGURE 5-12 Computer simulation of a generalized mechanism representing photochemical smog (see text). Redrawn from J. H. Seinfeld and S. N. Pandis, *Atmospheric Chemistry and Physics: From Air Pollution to Climate Change*. Copyright © 1998. New York: John Wiley & Sons, Inc. Used by permission of John Wiley & Sons, Inc.

equations representing a generalized mechanism involving 20 steps—including many of the reactions and organic species given earlier—for a system initially containing equal concentrations (0.10 ppmv) of NO, H_2CO, and organic species groups RH (alkanes) and RCHO (aldehydes), and a small amount (0.01 ppmv) of NO_2.[46] Compare this computer simulation with the photochemical smog chamber results (that however did not have initial amounts of H_2CO and RCHO), where RH = C_3H_6, shown in Figure 5-11. Note the qualitative similarities between the two: the rapid conversion of NO to NO_2 and the subsequent loss of NO_2; the slower loss of RH (C_3H_6 in Figure 5-11) and buildup of ozone, after most of the NO has reacted; the buildup and decay of RCHO (CH_3CHO in Figure 5-11); and the induction period in the

[46]J. H. Seinfeld and S. N. Pandis, *Atmospheric Chemistry and Physics: From Air Pollution to Climate Change*, pp. 292–298. Wiley, New York, 1998.

production of PAN. Applicable trends are also evident on the measurements of actual atmospheric conditions (Figure 5-9).

Although so far there have not been any large disasters directly attributable to photochemical smog, the quality of the environment is seriously degraded by its presence, and it is now widespread in the industrial world. Ambient ozone is one of the most pervasive environmental problems facing urban and regional areas. In the next chapter we review many of the remedies and legislative actions being undertaken to reduce the emissions of particulates, VOCs, carbon monoxide, and oxides of nitrogen (primarily from automobiles), and to suppress or limit secondary pollutants. There continues to be a great need for reliable kinetic data for the many chemical processes occurring in a polluted environment and for accurate models to simulate their physical and chemical behaviors.

Additional Reading

Atkinson, R., Gas-phase tropospheric chemistry of organic compounds, Monograph no. 2 of *The Journal of Physical and Chemical Reference Data*, J. W. Gallagher, ed. American Chemical Society and the American Institute of Physics for the National Institute of Standards and Technology, Gaithersburg, MD, 1994.

Barker, J. R., ed., *Progress and Problems in Atmospheric Chemistry*. World Scientific, Singapore, 1995.

Biggs, R. H., and M. E. B. Joyner, eds., *Stratospheric Ozone Depletion/UV-B Radiation in the Biosphere*. Springer-Verlag, Berlin, 1994.

Birks, J. W., J. G. Calvert, and R. E. Sievers, eds., *The Chemistry of the Atmosphere: Its Impact on Global Change: Perspectives and Recommendations*. American Chemical Society, Washington, DC, 1993.

Brasseur, G. M., J. J. Orlando, and G. S. Tyndall, eds., *Atmospheric Chemistry and Global Change*. Oxford University Press, Oxford, U.K., 1999.

Brimblecombe, P., *Air Composition and Chemistry*, 2nd ed. Cambridge University Press, Cambridge, U.K., 1996.

Calvert, J. G., ed., *The Chemistry of the Atmosphere: Its Impact on Global Change*. Blackwell Scientific Publications, Oxford, U.K., 1994.

Finlayson-Pitts, B. J., and J. N. Pitts Jr., *Chemistry of the Upper and Lower Atmosphere: Theory, Experiments, and Applications*. Academic Press, San Diego, CA, 2000.

Kondratyev, K. Ya., and C. A. Varotsos, *Atmospheric Ozone Variability: Implications for Climate Change, Human Health, and Ecosystems*. Praxis Publishing, Chichester, U.K., 2000.

Makhijani, A., and K. R. Gurney, *Mending the Ozone Hole. Science, Technology, and Policy*. MIT Press, Cambridge, MA, 1995.

Moortgat, G. K., A. J. Barnes, G. LeBras, and J. R. Sodeau, eds., *Low-Temperature Chemistry of the Atmosphere*. Springer-Verlag, Berlin, 1994.

Newton, D. E., *The Ozone Dilemma: A Reference Manual*. ABC-CLIO, Santa Barbara, CA, 1995.

Seinfeld, J. H., *Rethinking the Ozone Problem in Urban and Regional Air Pollution*, Report of the Committee on Tropospheric Ozone Formation and Measurement, J. H. Seinfeld, Chairman. National Academy Press, Washington, DC, 1991.

Seinfeld, J. H., and S. N. Pandis, *Atmospheric Chemistry and Physics: From Air Pollution to Climate Change*. Wiley, New York, 1998.

Wang, W.-C., and I. S. A. Isaksen, eds., *Atmospheric Ozone as a Climate Gas. General Circulation Model Simulations.* Springer-Verlag, Berlin, 1995.

Warneck, P., *Chemistry of the Natural Atmosphere.* Academic Press, New York, 1988.

Wayne, R. P., *Chemistry of Atmospheres*, 3rd ed., Oxford University Press, Oxford, U.K., 2000.

Zellner, R., guest ed., *Global Aspects of Atmospheric Chemistry.* Steinkopff, Darmstadt, Germany, 1999.

EXERCISES

5.1. (a) Give the values for the molecular quantum numbers Λ and S for each of the five electronic states of molecular oxygen shown in Figure 5-1.

(b) For each of these five states, state whether (1) interchange through the center of symmetry, and/or (2) interchange through the plane of symmetry, leads to a change in sign of the electronic wave function. (See Appendix A.)

(c) Based on the selection rules for radiative absorption given in Section 5.1, which of the four electronic transitions between the ground state of molecular oxygen and the four excited states given in Figure 5-1 are allowed transitions? Justify your answer.

5.2. From Figure 5-2, determine the absorption cross section σ of O_2 at 160 nm. (Note that the vertical scale is logarithmic.)

5.3. Punta Arenas, Chile, is located at 53°S latitude, which is at the edge of the Antarctic ozone hole (Section 5.2.3.3). On October 4, 1995, the column ozone concentration was 325 DU; on October 13, it had dropped to 200 DU [V. W. J. Kirchkoff *et al.*, *J. Geophys. Res.*, **102**, 16109–16120 (1997)].

(a) Calculate the fraction of solar radiation intensity I/I_0 transmitted by atmospheric ozone at 297 nm on October 4 and on October 13. The spectrum of ozone is given in Figures 5-3 and 5-4.

(b) Calculate the percent increase in intensity at 297 nm from October 4 to October 13.

5.4. The thermodynamic equilibrium constant for the reaction $O + O_2 \xrightarrow{M} O_3$ is given by the expression

$$\ln K \left(\text{atm}^{-1} \right) = \frac{12,786}{T} - 15.32$$

(a) Calculate the equilibrium constant at the temperature of the atmosphere at an altitude of 25 km (Figure 5-5).

(b) The pressure of O_2 is about 0.005 atm at 25 km, and the concentration of ozone is given in Figure 5-5. Calculate the ratio of the equilibrium pressure of O atoms divided by the equilibrium pressure of O_2 (P_O/P_{O_2}), at 25 km.

(c) Would you expect this ratio to increase or decrease in the presence of sunlight? Why?

5.5. Using Figures 5-2, 5-3, and 5-5, calculate the percent of light transmitted at 205 nm by:

    (a) One meter thickness of molecular oxygen at an altitude of 25 km (where the pressure of O_2 is 0.005 atm) and (b) one meter thickness of ozone at an altitude of 25 km.

    (c) From the calculations in parts a and b, which of these two species would you infer contributes more to absorption in the UV window of Figure 5-7 at 205 nm? Make a rough estimate of the percent of solar radiation transmitted by the stronger absorber at 205 nm from the outer atmosphere to 25 km. (*Hint*: To calculate the column concentration, assume that its concentration decreases linearly from the value at 22 km to 0 at 50 km; therefore the average concentration from 22 km to 50 km is half the concentration at 22 km, and 0 above 50 km.)

5.6. Show that the appropriate combination of reactions (5-29), (5-36), (5-37), and (5-39) gives the Chapman cycle, equation (5-41).

5.7. Both mechanisms (5-48) and (5-49) lead to the catalytic destruction of ozone. What is the significance of one mechanism relative to the other?

5.8. (a) Devise a chain mechanism in which hydrogen atoms catalyze the destruction of ozone in the lower stratosphere.

    (b) Is water a reasonable source of H atoms in your mechanism? If so, why? If not, why not? (See Section 5.2.4.)

5.9. What is meant by the polar vortex? Why is it considered to be an essential feature of the Antarctic ozone hole?

5.10. What is the role of polar stratospheric clouds (PSCs) in the formation of the Antarctic ozone hole?

5.11. Explain in each case why the HCFCs and the PFCs have much lower ODPs than the CFCs.

5.12. What is the role of temperature inversion in photochemical smog?

5.13. Although the results of a photochemical smog chamber experiment shown in Figure 5-11 qualitatively agree in many ways with an actual case of photochemical smog (Figure 5-9), there are some significant differences between the two sets of data. In particular, note that the ozone concentration increases to a constant amount in Figure 5-11, whereas it decreases significantly after reaching a maximum in Figure 5-9. Discuss what might be the reason for this difference.

5.14. Calculate the percent of 397.5-nm light that would be absorbed by NO_2 in a polluted atmosphere that is 1000 ft thick and contains the maximum concentration of NO_2 given in Figure 5-9. Assume the pressure is constant at 1 atm and the temperature is 85°F (29.4°C).

5.15. (a) What are VOCs?

(b) What is the role of VOCs in the production of photochemical smog?

5.16. It is stated that NO_2, formed mainly from NO, is the main light absorber in the atmosphere leading to photochemical smog. Give several examples of reactions involving the oxidation of NO to NO_2 in an urban atmosphere.

5.17. We have seen that PAN, a component of photochemical smog, is thermodynamically unstable, decomposing into NO_2 and the peroxyacetyl radical $CH_3C(O)O_2\cdot$ [the reverse of reaction (5-111)]. This reaction is very sensitive to temperature changes; at 25°C the half-life of PAN is 27.5 mins, while at 100°C it is 0.18 s.

(a) Calculate the rate constants (in s^{-1}) at 25°C and at 100°C.

(b) Assuming that the empirical Arrhenius equation (4-24) is followed, calculate the activation energy E_a for the decomposition reaction.

6

PETROLEUM, HYDROCARBONS, AND COAL

6.1 INTRODUCTION

Organic compounds are abundant in the environment, both as the products of natural, mainly biological, activity and as waste materials and pollutants from industrial processes. Organic compounds generally may be considered to be substituted hydrocarbons that are made up of straight or branched chains, rings, or combinations of these structures. The hydrocarbon skeletons are modified by the presence of multiple bonds and substituents (oxygen, nitrogen, halogen, etc.) that markedly influence the reactivity of the molecules. Unsubstituted hydrocarbons are relatively inert, particularly toward the polar compounds in the environment. All hydrocarbons are thermodynamically unstable with respect to oxidation to CO_2 and H_2O, but the rate of this process is very slow except at elevated temperatures. However, the energy released by reaction with oxygen (about 12 kcal/g or 4500 J/g for saturated hydrocarbons) is very important for the generation of heat and electricity and for transportation in our industrial society.

Organic compounds are the basis for life on earth. The synthesis of organic compounds in living systems is catalyzed by other organic substances called enzymes. The degradation of organic compounds often follows a pathway that is the reverse of the synthesis, and the reactions are catalyzed by the same or similar enzymes. Many microorganisms utilize simple organic compounds for their growth, and their enzymes are responsible for the degradation of these substances. As a consequence, any compound that is synthesized by a living system is readily degraded biologically and is said to be "biodegradable." However, many of the carbon compounds that have been formed by geochemical reactions (crude oil, coal) or industrial chemical processes (polyethylene, DDT) are not readily degraded and are frequently regarded as pollutants. These compounds will accumulate in the environment unless their degradation is catalyzed by a microbial enzyme or they are rendered more susceptible to enzymatic degradation by some other environmental process discussed in Section 8.2. These environmental processes are effective only insofar as they render the pollutant more susceptible to biodegradation by introducing appropriate functional groups that can serve as points of microbial attack.

Since hydrocarbons are considered to be the parent compounds for organic materials, we shall consider them in this chapter. Coal, a potential source of hydrocarbons, is also discussed. Other organic compounds will be considered in later chapters. No attempt is made to deal with the general chemistry of the various classes of compounds and functional groups except as these reactions are important environmentally. It is assumed that the reader has some familiarity with the properties of hydrocarbons and their simple substituted derivatives.

6.2 THE NATURE OF PETROLEUM

Crude oil is the predominant source of the hydrocarbon compounds used for combustion and for industrial processes, although as the supply of crude oil decreases, coal, oil shale, and tar sands may become more important sources of liquid and gaseous hydrocarbons. Crude oil is also a major source of environmental pollution. It is estimated that 2–9 million metric tons (tonnes) of crude oil and hydrocarbons enter the environment each year.

Crude petroleum is a complex mixture of hydrocarbons that is separated into fractions of differing boiling ranges by the refining process shown schematically in Figure 6-1. The various fractions either are used directly as energy sources ("straight-run gasoline") or are modified chemically to yield more efficient energy sources such as high-octane gasoline. In addition, hydrocarbons may be converted to petrochemical compounds such as monomers, polymers, and detergents that are derived from petroleum.

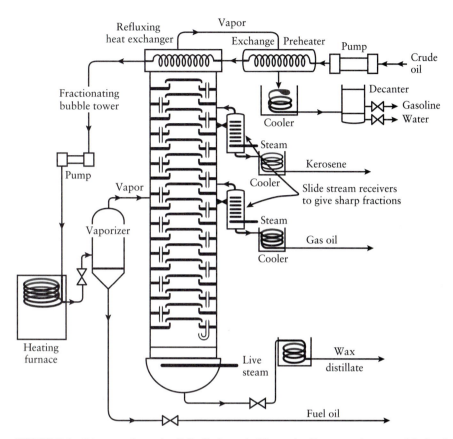

FIGURE 6-1 Diagram of a crude oil distillation unit. The crude oil enters at the upper right-hand corner of the diagram and the products are taken off below on the right. Redrawn with permission of Scribner, a Division of Simon & Schuster, Inc., from *The Chemistry of Organic Compounds*, Fourth Edition by James Bryant Conant and Albert Harold Blatt. Copyright © 1952 by Macmillan Publishing Company. Copyright renewed © 1980 by Grace R. Conant.

Sulfur, oxygen, nitrogen, and metal derivatives are also present in petroleum, and compounds containing these elements are emitted when petroleum is burned. Sulfur is the most important minor constituent because it is emitted into the air as sulfur oxides when petroleum is burned, and the sulfur compounds released poison the catalytic converter on automobiles (Section 6.7.3). The sulfur content of crude oil ranges from 0.1 wt% for low-sulfur to greater than 3.7 wt% for high-sulfur petroleum. The sulfur is present mainly in the form of thiophene or thiophene derivatives such as the following:

Thiophene   2,3-Benzthiophene   2-Ethyl-4,5-dimethylthiazole

Nitrogen from nitrogen compounds constitute a much smaller percentage ($\sim 1\%$ by weight) of petroleum than do sulfur compounds. The nitrogen is also present in the form of derivatives of the heterocyclic ring systems such as thiazole or substituted quinolines.

Thiazole   2,3,8-Trimethylquinoline

Metals are present in petroleum either as salts of carboxylic acids or as porphyrin chelates (Chapter 9). Aluminum, calcium, copper, iron, chromium, sodium, silicon, and vanadium are found in almost all samples of petroleum in the range of 0.1–100 ppm. Lead, manganese, barium, boron, cobalt, and molybdenum are observed occasionally. Chloride and fluoride also are commonly encountered. Some of these metallic components are probably introduced into the petroleum when it is removed from the ground. The use of emulsified brines and drilling muds (mixtures of iron-containing minerals and aluminum silicate clays) may account for the presence of some of these trace elements.

6.3 SOURCES OF CRUDE OIL AND HYDROCARBONS IN MARINE ENVIRONMENTS

6.3.1 Introduction

The major environmental effects of petroleum are due either to crude oil itself or to the automobile hydrocarbon emissions from the gasoline fraction of the crude oil. Oil spills are generally associated with the release of hydrocarbons to the oceans or other water bodies, where much of the oil floats and spreads over large areas of the water surface. Spills on land tend to be more easily contained to small areas, but they may result in the contamination of soil and local water supplies. This has been an important problem in cases of storage tanks that leak petroleum into the underground environment over long period of time. The primary environmental problems associated with spills in bodies of water

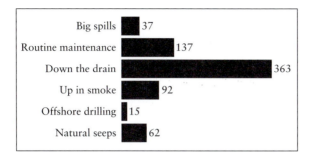

FIGURE 6-2 The annual input of petroleum hydrocarbons into marine environments annually in millions of gallons: Big spills, major oil tanker and oil well accidents; Routine maintenance, bilge cleaning and other discharges from oil tankers and other large ships; Down the drain, used automobile engine oil, runoff from land of municipal and industrial waste; Up in smoke, hydrocarbons from vehicle exhaust and other sources that are rained out in the oceans; Offshore drilling, Spill from offshore oil wells; Natural seeps, oil seeping into the oceans from natural, underground sources. Redrawn from http://seawifs.gsfc.nasa.gov/OCEAN_PLANET/HTML/peril_oil_pollution/html.

are the effects of the slowly degraded petroleum residues on aquatic life. Automobile emissions, on the other hand, cause atmospheric problems. Much spilled oil also enters the atmosphere by evaporation, but the problems associated with oil spills and automobile emissions are sufficiently different to warrant separate discussion in this chapter. The additional environmental problems of the surfactants, polymers, and pesticides synthesized from petroleum are discussed in subsequent chapters.

It is estimated that the crude oil input into the sea is 7×10^8 gal/year worldwide (Figure 6-2). The sources of the oil, with the exception of delivery from the atmosphere, are discussed in the subsequent sections of this chapter. The atmospheric input is too diverse and ill defined to permit a clear delineation of all the sources.

6.3.2 Natural Sources

Natural oil seeps led to the discovery of many of the major oil fields in the world. The first oil well in the United States was drilled in Titusville, Pennsylvania, because oil was observed seeping into the Oil River at that point. These oil seeps may be responsible for local environmental problems, but often the rate of seepage is so slow that the oil is effectively dispersed by environmental forces. A major oil seep, 3600 gal/day, is at Coal Oil Point near Santa Barbara on the coast of southern California. This seep produces an extensive oil slick on the ocean and often results in tar accumulation on the shore. The oil also supports a community of marine microorganisms that metabolize these hydrocarbons. Most seeps produce less than 40 gal/day and the environmental effect appears to be quite limited. The contribution of natural sources to the oil in the marine environment is 9% of all sources.

In 2000, earth-imaging satellites and radar discovered extensive natural seepage of oil into the Gulf of Mexico. It is estimated that the annual amount of oil released per year is twice that of the *Exxon Valdez* spill (2 × 11 million gallons: Section 6.4.1). This is a gradual release of oil, which wildlife is used to, and it forms a slick on the water 0.01 mm thick, which is of little danger to wildlife and is readily metabolized by bacteria. Some of these seeps have been occurring for thousands of years, and it has been proposed that there are food chains nurtured by the oil released by the seeps.

6.3.3 Offshore Wells

Leakage from offshore oil wells is responsible for only about 2% of the oil in the sea (Figure 6-2). Thus it is surprising that there is always major opposition from local communities when it is proposed to drill for oil along the East or West coasts of the United States. This mind-set is probably due to the images of the well that "blew" off Santa Barbara, California, in January 1969. This oil was pressurized by methane gas in the ground, and the pressure forced out the oil when the well was drilled. The high pressure forced oil out of the seafloor in the vicinity of the well and thus the flow could not be stopped by capping the well. It was necessary to drill additional wells to dissipate the gas pressure so that the oil flow could be controlled. Other examples include the North Sea blowout in 1977 and the Ixtoc 1 well, which continued for almost a year (from June 1979 to the spring of 1980) in the Gulf of Mexico off the coast of Campeche, close to the Yucatan Peninsula. The latter well spewed out 140 million gallons of crude oil (Figure 6-3). Now, however, unless the oil is pressurized, potential "blowout" sites can be identified by geologists, so the danger of major leaks from offshore oil wells appears to be minimal. There are 6000 wells off the coasts of Texas and Louisiana with no reports of major spills. One study has concluded that drilling offshore wells results in less environmental risk than transporting the oil in large tankers from the Gulf of Mexico or the Middle East because of the problems with tanker use discussed in Section 6.3.5.[1]

[1]W. B. Travers and P. R. Luney, *Science* **194**, 791–796 (1976).

FIGURE 6-3 The Ixtoc exploratory well, which blew out and caught fire June 3, 1979. It spilled 140 million gallons of oil into the Gulf of Mexico. From http://www.nwn.noaa. gov/sites/hazmat/photos/ships/08.html. Also see color insert.

6.3.4 Industrial and Municipal Waste

It is estimated that 51% of the oil in the marine environment is a result of industrial and municipal waste (Figure 6-2). About 1400 million gallons of used oil is generated each year in the United States. Most of this oil was used as a lubricant for automobile engines (crankcase oil) or as industrial lubricants. A large portion of this used oil is collected by commercial recyclers for direct use as fuel or for rerefining into new oil. It is estimated that 363 million gallons of this used oil ends up in the environment (Figure 6-2).

Automobile lubricating oil, changed by the vehicle owner, is a major source of the oil that is dumped or discarded in landfills. About 50% of the 190 million gallons of oil that is changed by do-it-yourselfers is brought to collection centers such as gasoline stations. The other 95 million gallons is either sent to landfills, dumped on the ground, or poured down storm sewers. This oil migrates through the soil and finally ends up in underground aquifers, rivers, lakes, and oceans. Until recently, used oil was sprayed on dirt roads to control dust but this practice was forbidden in 1992 because such oil contains toxic metals and toxic hydrocarbons that had been rapidly leached into water supplies.

California has an aggressive program of collecting and recycling oil. Of the 137 million gallons of used oil generated each year in the state, over half (83 million gallons) is collected. This program has resulted in noticeable decreases in the levels of oil present in soil and water samples in the state.

There have been major concerns raised concerning the burning of 800 million gallons of used oil each year to heat schools, hospitals, apartment buildings, and factories in the United States. The combustion of this oil is not regulated because it is not classified as hazardous. It is not labeled as hazardous because this would stigmatize the oil, causing people to be less willing to recycle it. Yet it contains lead, chromium, cadmium, and other toxic constituents, and many of these are emitted into the atmosphere during the combustion process. For example, in 1992 it was estimated that about 600,000 lb of lead compounds was released during the combustion process, an emission rate greater than any other industrial process. Oil cannot be legally combusted if it contains more that 100 ppm lead, but this standard is 10 times higher than the limit for other fuels imposed by the U.S. Environmental Protection Agency (EPA). Used oil that is to be burned must be tested first to determine that the levels of lead, chromium, cadmium, benzene, and other constituents are within the EPA standards for these substances. If not, the used oil is usually blended with other oil so that the mixture meets EPA standards.

6.3.5 Transportation

The major source of oil in the marine environment is spills resulting from crude oil transport. There are some 6000 oil tankers worldwide that transport crude oil, and these vessels are designed to ride low in the water when carrying a full load. After they unload their cargo they must take on seawater as ballast for their return trip to the oil fields. This ballast is pumped into holding ponds where the oil can be recovered. However, this ballast, along with the residual crude oil in their tanks, is pumped into the ocean before the next cargo of crude oil is loaded. It is estimated that 0.4% of the capacity of a tanker sticks to the tanker wall after the crude oil is discharged. If the tanker takes on ballast equivalent to one-third of its total capacity then about 1.6×10^4 gal of crude oil will be pumped out with the ballast of a tanker having a capacity of 12×10^6 gal. It is estimated that every year 137 million gallons of oil is introduced into the sea from "routine maintenance": ejection of tanker ballast and the many small spills resulting from crude oil transport (Figure 6-2). The sea lanes around the continent of Africa and the subcontinent of India are the most heavily oiled in the world as a result of the spills from the dense tanker traffic leaving the Persian Gulf (Figure 6-4).

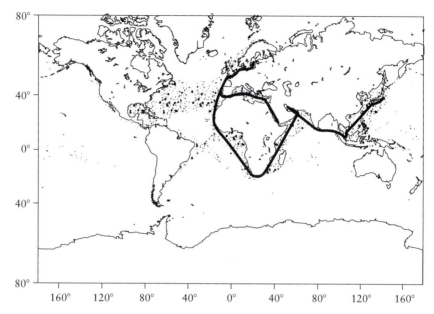

FIGURE 6-4 Positions at which visible oil slicks were observed between 1975 and 1978 by the Interglobal System Marine Pollution (Petroleum) Monitoring Project. Redrawn from E. M. Levy, M. Ehrhardt, D. Koknke, E. Sobtchenko, T. Suzuoki, and A. Tokuhiro, *Intergovernmental Oceanographic Commission*, UNESCO, 7 Place de Fontenoy 75700, Paris, France 1981. Used by permission of the United Nations.

6.3.6 Catastrophic Oil Spills

Catastrophic oil spills can have dramatic local environmental effects, and because of this they are well documented in the national press and television. Major oil spills became a problem on the east coast of the United States in World War II when German submarines sank over 60 oil tankers, releasing 6.5×10^5 metric tons of oil over a period of three years. Since the tankers were traveling close to the shore in an attempt to avoid the submarines, much of this oil eventually washed up on the beaches.

Today major oil spills occur each year as a result of oil tanker accidents (Figure 6-2). The huge tankers (Figure 6-5) release vast amounts of oil when accidents occur. These accidents are inevitable because of the steady stream of tankers traveling from the oil-producing centers in the Middle East and Alaska (Figure 6-4), just as automobile accidents are inevitable on busy highways. Catastrophic spills contribute only about 5% to the total amount of oil entering the marine environment as a result of transportation, but they have a

FIGURE 6-5 The "midsize" oil tanker *Ventiza*, built in 1986. It is 247 m (811 ft, or 2.7 U.S. football fields) long and 41.6 m (136 ft) wide. In 2000 the largest tanker in the world was the *Jahre Viking*, which is 458 m (1504 ft, or 5 U.S. football fields) long and 69 m (266 ft, or 0.89 U.S. football field) wide. From http://www.rigos.com/ventiza.html. Also see color insert.

major impact on the local environment if the oil reaches the shore. Catastrophic spills will occur less frequently if more double-hulled tankers (vessels having a 9- to 10-ft buffer area between the two hulls) are used. If an accident perforates the outer hull, there is a reasonable probability that the inner hull will not be breached and no oil will be released. Federal legislation passed in 1990 requires oil tankers proceeding between two destinations in the United States to have double hulls and mandates the removal from service of single-hull tankers over 25 years old. There has been an upsurge worldwide in the use of double-hulled tankers. This increase is probably due more to the potential for costly ligation resulting from an oil spill than to the laws passed by the U.S. Congress. There has been a decrease in the amount of oil spilled since 1984 (Figure 6-6; see also Table 6-1).

Oil tankers and other big ships have been identified as a major source of air pollution. These ships burn high-sulfur fuels that release large amounts of sulfur dioxide (SO_2) and nitrogen oxides (NO_x) as well as particulates into the air. The EPA plans to regulate these emissions, but international agreements will be required to enforce any such regulations.

As noted earlier, oil spills on land are more readily contained and are usually a less serious environmental problem than those at sea. One notable exception

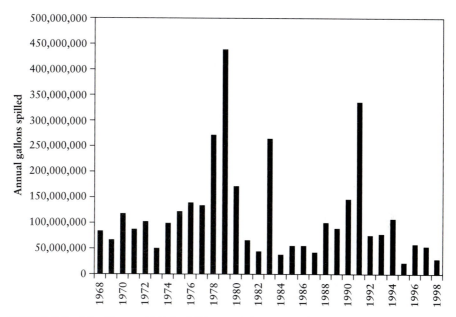

FIGURE 6-6 Oil spills over 10,000 gal (34 tonnes) since 1968. The major oil spills decreased in the 1984–1998 period. The 240 million gallons deliberately released in Kuwait during the Gulf war was the principal contributor to the total amount spilled in 1991. Redrawn from D. S. Etkin, *International Oil Spill Statistics: 1998*. Cutter Information Corp., Arlington, MA. Copyright © 1999. Used by permission of Cutter Information Corporation.

TABLE 6-1

Oil Spills In Order of Amount Spilled[a]

1. 26 January 1991: terminals, tankers; 8 sources total; sea island installations; Kuwait; off coast in Persian Gulf and in Saudi Arabia during the Persian Gulf War (240.0)
2. 03 June 1979: Ixtoc I well; Mexico; Gulf of Mexico, Bahia del Campeche (140.0)
3. 02 March 1990: Uzbekistan, Fergana Valley (88)
4. 04 February 1983: platform no. 3 well; Iran; Persian Gulf, Nowruz Field (80.0)
5. 06 August 1983: tanker *Castillo de Bellver*; South Africa; Atlantic Ocean, 110 km northwest of Cape Town (78.5)
6. 16 March 1978: tanker *Amoco Cadiz*; France; Atlantic Ocean, off Portsall, Brittany (68.7)
7. 10 November 1988: tanker *Odyssey*; Canada; North Atlantic Ocean, 1175 km northeast of St. Johns, Newfoundland (43.1)
8. 19 July 1979: tanker *Atlantic Empress*; Trinidad and Tobago; Caribbean Sea, 32 km northeast of Trinidad-Tobago (42.7)
9. 11 April, 1991: tanker *Haven*; Genoa, Italy (42)
10. 01 August 1980: production well D-103 (concession well); 800 km southeast of Tripoli, Libya (42.0)

(continues)

TABLE 6-1 (*continued*)

11. 02 August 1979: tanker *Atlantic Empress*; 450 km east of Barbados (41.5)
12. 18 March 1967: tanker *Torrey Canyon*; United Kingdom; Land's End (38.2)
13. 19 December 1972: tanker *Sea Star*; Oman; Gulf of Oman (37.9)
14. 23 February 1980: tanker *Irene's Serenade*; Greece; Mediterranean Sea, Pilos (36.6)
15. 07 December 1971: tanker *Texaco Denmark*; Belgium; North Sea (31.5)
16. 23 February 1977: tanker *Hawaiian Patriot*; United States; Pacific Ocean 593 km west of Kauai Island, Hawaii (31.2)
17. 20 August 1981: storage tanks; Kuwait; Shuaybah (31.2)
18. 25 October 1994; pipeline: Russia; Usinsk (in area that was closed to foreigners before collapse of Soviet Union) (30.7)
19. 15 November 1979: tanker *Independentza*; Turkey; Bosporus Strait near Istanbul, 0.8 km from Hydarpasa port (28.9)
20. 11 February 1969: tanker *Julius Schindler*; Portugal; Ponta Delgada, Azores Islands (28.4)
21. 12 May 1976: tanker *Urquiola*; Spain; La Coruña Harbor (28.1)
22. 25 May 1978: pipeline no. 126 well and pipeline; Iran; Ahvazin (28.0)
23. 05 January 1993: tanker *Braer*; United Kingdom; Garth Ness, Shetland Islands (25.0)
24. 29 January 1975: tanker *Jakob Maersk*; Portugal; Porto de Leisoes, Oporto (24.3)
25. 06 July 1979: storage tank tank no. 6; Nigeria; Forcados (23.9)

[a]Number in parentheses refers to millions of gallons spilled.
Source: excerpted from D. S. Etkin, *International Oil Spill Statistics: 1998*. Cutter Information Corp., Arlington, MA, 1999. http://www.cutter.com/oilspill

was the flooding caused by Hurricane Agnes in 1972, which resulted in the release of 6–8 million gallons of sludge and oil from storage tanks in Pennsylvania.

6.4 FATE OF AN OIL SPILL

The dispersal of an oil spill is illustrated in Figure 6-7. When oil is spilled, the lower boiling fractions evaporate and dissolve rapidly, with the result that 25–50% of a crude oil spill is soon lost from the surface of the sea. The extent to which these processes occur is determined by the temperature of water and the intensity of wind and wave action. The residue of higher molecular weight hydrocarbons is slowly degraded by microorganisms. This process may be slow because the other nutrients needed for growth, especially nitrogen, are quite limited in the ocean. In addition, the insoluble petroleum contains molecular species that are too large to be assimilated through the cell walls of microorganisms and will undergo biodegradation only after the petroleum has been broken down into smaller molecules by a combination of photochemical and oxidative processes. As a general rule, microorganisms cannot assimilate

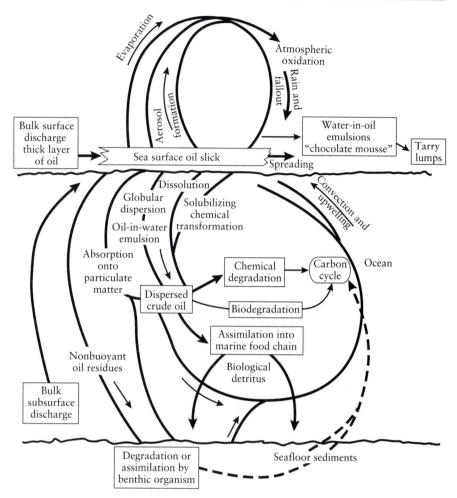

FIGURE 6-7   Fate of spilled oil. Redrawn with permission from R. Burwood and G. C. Speers, *Estuarine and Coastal Marine Science*, Vol. 2, pp. 117–135. Copyright © 1974. Used by permission of Academic Press.

organic molecules of molecular weight greater than 500 Da. Consequently the remaining higher molecular weight material is whipped up into a "mousse" of hydrocarbons and water by waves and winds, and this mousse slowly breaks up into tar balls. These float unless mixed with enough sand to give them a density that exceeds the density of seawater in which case they sink to the bottom. Continued action of light, oxygen, and waves breaks the material up into smaller and smaller particles and to molecules that can eventually be degraded by microorganisms.

6.4.1 Weathering of an Oil Spill

The polycyclic aromatic hydrocarbons in crude oil sensitize the photochemical formation of singlet oxygen (Section 5.2.2), which oxidizes the olefins present to cyclic dioxides and hydroperoxides:

$$\text{(structure)} + O_2 \, (^1\Delta_g) \longrightarrow \text{(structure)} \longrightarrow \text{(structure)} \tag{6-1}$$

$$\text{(structure)} + O_2 \, (^1\Delta_g) \longrightarrow \text{(structure)} \; OOH \tag{6-2}$$

These compounds can either rearrange directly [equation 6-1] or be cleaved photochemically to form hydroxyl radicals

$$\text{(structure)} \; OOH \xrightarrow{h\nu} \text{(structure)} \; O\cdot + \cdot OH \tag{6.3}$$

that attack the other unsaturated molecules present:

$$\text{(structure)} + \cdot OH \longrightarrow \text{(structure)} + H_2O \xrightarrow{O_2} \text{(structure)} \; OO\cdot \xrightarrow{RH}$$

$$\text{(structure)} \; OOH \longrightarrow \longrightarrow \longrightarrow \text{alcohols, phenols, carbonyl compounds, and acids} \tag{6.4}$$

$$+ \; R\cdot$$

The net result of these radical processes is the solubilization of the polycyclic aromatic hydrocarbons by conversion to phenols, alcohols, carbonyl compounds, and acids.

The breakup of the spill is slowed if it is washed up on the shore. Dispersal and degradation will continue if there is a strong wind and surf, but oil spills in quiet bays or lagoons may take 5–10 years to disperse in the absence of wave and wind action. The toxic effects on marine life likewise persist for 5–10 years in these environments.

The long-term effects of an oil spill are determined by the environment in which it occurs. Spills that never reach the shore have the least effect because most marine life tends to be near the land–sea interface, where food supplies are more abundant and fish travel to lay eggs and birds and amphibians raise their young. Thus a spill that remains at sea has less chance of damaging the eggs and rapidly developing young marine life, which are the organisms most susceptible to toxins of any sort.

6.4.2 Cleaning Up Oil Spills

One initial reaction to an oil tanker accident is to send another tanker or barge to the leaking vessel and try to offload the crude oil. Most tanker accidents, however, occur in stormy weather or as a result of hazardous conditions on the tanker. If this is the case offloading the crude is not feasible, and the first action is to put out a boom to contain the spill. Then skimmers, or boats with the capability to collect surface oil, are sent out. Skimmers work with small spills in calm water but have been less effective in large spills such as the 11-million-gallon spill by the tanker *Exxon Valdez* in Prince William Sound, near Valdez, Alaska, in 1989. The skimmers used there had a very low capacity and could not take on the oil as fast as it was coming out of the tanker. In addition, when the skimmers had filled their tanks they were 3 hours from shore, where the pumping facilities were, and the 6-hour round trip also limited their usefulness. Later, when wave action in Prince William Sound increased, the efficiency of the skimmers decreased markedly because they passed over the oil without removing it. (Booms used to control the spread of the oil are effective only in calm sea.) The fate of the oil from the *Exxon Valdez* released into Prince William sound is shown in Figure 6-8.

If the spill cannot be contained or collected, the next approach is to drop or spray dispersants on the oil to increase its miscibility with seawater. For example, dispersants were used to keep the toxic oil released by Ixtoc 1 in the Gulf of Mexico from reaching the shore. This treatment dispersed the oil in micelles. The dilution of the oil diminished the concentrations of the toxic hydrocarbons, and the dispersed oil underwent more rapid biodegradation than it would have as a separate layer on the surface. Studies of the toxic effects of this spill were limited, but there were no reported effects on the size or quality of the shrimp catch in that part of the Gulf of Mexico. The one danger with dispersants is their possible toxic effects on marine biota. For example, the phenolic dispersants used on the oil spilled from the *Torrey Canyon* in 1967 were more toxic than the crude oil. Recent studies have

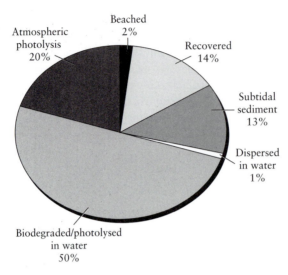

FIGURE 6-8 Fate of the oil from the *Exxon Valdez* spill of 1989 in Prince William Sound, near Valdez, Alaska. Redrawn from http://www.oilspill.state.ak.us/beaches/beaches/htm. Also see color insert.

identified detergents that are less toxic to marine life than crude oil but equally effective in dispersing the spill.

One treatment that is seldom used today is the addition of agents like chalk, which cause the oil to sink to the bottom. This approach has been abandoned because the precipitated petroleum has a long lifetime in the bottom sediments, where the dissolved oxygen content is low. There the persistent hydrocarbons have a toxic effect on marine life for many years. The hydrocarbons may also change an aerobic environment to an anaerobic one, which inhibits the degradation of the petroleum by aerobic micro-organisms.

Once the oil reaches the shore, a variety of techniques may be used to either collect the oil or wash it off the beaches. A spray of hot seawater was used to flush the oil from the *Exxon Valdez* off the rocks and beaches of Prince William Sound, and the oil released was collected from the surface of the water near the shore. This treatment may have been more detrimental than helpful. The number of microorganisms in the treated areas was found to be less than that present before treatment, suggesting that the treatment was more toxic to marine microbes than the oil.

Bioremediation, a different approach to cleaning oil spills, was also investi-gated for the removal of oil from the beaches in Prince William Sound. Fertilizers were sprayed on the oily beaches to enhance the growth of the oil-eating bacteria that occur in the natural environment. The fertilizer consisted of essential nutrients, such as nitrogen and phosphorus compounds that are present in only small amounts in the sea, mixed with emulsifying agents to solubilize the oil. These nutrients together with the carbon of the petroleum provide an environment in which petroleum-consuming bacteria can thrive. Inipol EAP 22, a mixture of oleic acid, lauryl phosphate, 2-butoxyethanol, urea, and water, appeared to be effective in enhancing the rate of loss of oil from beaches. In some instances a two- to fivefold increase in the rate of biodegradation of the oil was observed, and there was a marked improvement in the appearance of the treated beaches in comparison to control areas where no Inipol was used. However, in one case the visual improvement was misleading, for analysis showed no difference in the amounts of oil on the treated beach and in a control area that had not been treated.

Some studies were performed in an attempt to determine which of the components of the bioremediation mixture (Inipol EAP 22) were most effective in promoting bacterial degradation of oiled beaches in Prince William Sound. Nitrogen, which appears to be the limiting nutrient for hydrocarbon-metabolizing bacteria in the presence of an oil spill, is believed to be essential for bioremediation to be successful. The presence of oleic acid and lauryl phosphate stimulates the growth of other bacteria, including some nitrogen-fixing ones, which enhance the growth of the hydrocarbon-metabolizing bacteria. Oleic acid and lauryl phosphate are added only initially, since their continued addition mainly supports the growth of bacteria that do not consume oil. Large colonies of bacteria that do not metabolize oil will consume the other essential nutrients (e.g., nitrogen) necessary for the oil eaters to grow. Detergents that tend to disperse the oil also enhance the rate of bacterial consumption of oil because it is easier for the bacteria to ingest the dispersed hydrocarbons.

These factors have been evaluated qualitatively only because it is difficult to monitor the degradation of an oil spill in the ocean. Somewhat better data have been obtained for spills on beaches, but these data probably are not directly applicable to an oil slick on the ocean. At present there are no data showing that addition of known oil-metabolizing bacteria accelerates the degradation of the spill. The current view is that there is a bloom of oil-metabolizing bacteria as soon as the oil is spilled, and thus there is no need to add more bacteria. Addition of nitrogen- and phosphorus-containing fertilizers enhances the growth of the bacteria, which results in the rapid biodegradation of the oil. Too much fertilizer should be avoided, since this results in the proliferation of microorganisms that do not metabolize the oil but do deplete the molecular oxygen in the water. The oil-eating bacteria are aerobic, so the depletion of oxygen will slow their growth.

6.5 EFFECTS OF CRUDE OIL ON MARINE LIFE

The population of marine microorganisms that can metabolize soluble hydro-carbons increases dramatically in the vicinity of an oil spill. The growth of these microorganisms in the ocean is limited by the hydrocarbon content of the water, so they multiply at a much greater rate when there is a big increase in the hydrocarbon content of the seawater. As noted earlier, nitrogen is the growth-limiting nutrient in the presence of excess hydrocarbons.

Almost all forms of life absorb hydrocarbons and are affected by crude oil. The acute toxicity (an imminent health hazard) of crude oil is proportional to the percentage and structures of the aromatic compounds present. Soluble aromatic hydrocarbons containing two or more condensed rings have the highest acute toxicity. Ingestion of crude oil by birds and mammals inevitably leads to their death. The bulk of the oil ingested is a result of an animal's attempts to remove the oil from feathers or fur. Ingestion is also a consequence of eating oiled plants or animals. For example, seals can easily catch and eat oiled birds that are unable to fly. Plants and coral reefs are also susceptible to the toxic effects of crude oil. Sea grasses and mangrove forests have been killed by crude oil.

Birds and mammals are weakened by oil that binds fur or feathers together. This greatly decreases the insulating capacity of the fur or feathers, with the result that the animals lose heat at a much more rapid rate than before. Consequently they must eat much more food to maintain their body tempera-ture. This is a serious problem for mammals like polar bears and sea otters, which spend a lot of time in the sea and are unable to escape from the vicinity of the spill. These animals also experience eye irritation as a consequence of their surfacing in the oil and getting it in their eyes. Laboratory studies and other studies have demonstrated that crude oil is more toxic to eggs and developing young marine life than it is to adults. Such studies also indicate that marine life does not instinctively avoid oil spills.

The toxic effects of the crude oil appear to be short-lived. Most marine life repopulates a previously affected area within one to two years, and there is little visible effect on the ecosystem. The rate of recovery is strongly depend-ent on the activity of the ocean on the shoreline. In areas where there is active surf, the ocean constantly scrubs the beach and the rocks, and the oil is dispersed. Oil spilled in a protected area may persist for many years, so it takes correspondingly longer for the environment to return to its prespill condition.

The indirect effects of oil spills are not well known, but these are potentially as important as the direct effects just noted. It is known that fish and other forms of marine life are stimulated in their search for food, in their escape from predators, and in their mating by the organic compounds in the sea. The messages received by the fish may be masked by the presence of petroleum, or the compounds present in the crude oil may convey incorrect messages.

We now turn to some of the numerous laboratory studies to determine the effects of crude oil on various types of marine life as well as observations made at the sites of oil spills.

6.5.1 Fish

Laboratory studies have shown that dissolved hydrocarbons destroy the membranes in gills, which results in a diminished ability to absorb oxygen from the water. This loss of gill efficiency results in an enlarged heart. There is also less efficient utilization of food, an effect that may be related to diminished adsorption of food through the intestinal walls. The proposed decrease in food adsorption may also be due to damage to the lipid layers in the intestines. The diminished populations of bottom-feeding fish in the vicinity of some spills suggests that these may be affected most. This may be due to ingestion of oil adsorbed on the sediments.

6.5.2 Clams, Mussels, and Other Invertebrates

There is usually a sharp decrease in the populations of invertebrates in the vicinity of oil spills because, unlike fish, they do not have sufficient mobility to move away from the area of the spill and they live in sediments that contain adsorbed oil. These mollusks absorb oil into their systems, where it inhibits the production of eggs and sperm, causing an abrupt drop in their populations. Oil is also toxic to the adults, as shown by the large quantities of dead mollusks, crabs, and sea urchins that washed up on the shores of France after the *Amoco Cadiz* spill. The populations of these invertebrates gradually build up again over a period of several years.

6.5.3 Birds

The oiling of birds is usually one of the most dramatic visual effects of an oil spill. The toxicity of the oil ingested during a bird's attempts to remove the oil from its feathers usually results in the death of the animal. Very few of the birds recover, in spite of heroic efforts by rescue workers. There appear to be no long-term effects on the bird population; their numbers rebound rapidly within a year or so. For example 35,000 dead birds were retrieved as a consequence of the *Exxon Valdez* spill, and the total death toll was much higher–close to 200,000 birds, including about 200 adult bald eagles. However in one year's time it was observed that a significant number of birds occupied areas that had been oiled and, more specifically, 51% of the bald eagle nests in the oiled region were occupied and produced an average of 1.4 eaglets per nest, the usual ratio for this area.

It was demonstrated that ingestion of sublethal amounts of crude oil by herring gull chicks causes several sublethal effects that could impair the ability to survive. In this study the chicks suffered liver effects and ceased to grow in comparison to similar chicks that had not ingested oil. A similar study on young herring gulls and puffins showed oxidative damage to the hemoglobin of their red blood cells. Thus it appears unlikely that birds that have been subjected to a large amount of oil will survive for long even if the oil covering their feathers is removed. These birds will have ingested oil in their attempts to remove it from their feathers and consequently will be subject to its sublethal toxic effects.

6.5.4 Sea Otters and Other Marine Mammals:

Marine animals (like sea otters are very susceptible to the toxic effects of the oil mainly because they spend most of their time in the sea, swimming and surfacing through the oil. Exxon expended $8 million dollars for the helicopter capture of 357 otters ($80,000 per otter!), of which 234 survived. Many of the captured otters were not oiled or were only slightly oiled. Those that survived were cleaned and released, 45 with surgically implanted radio monitors. Only 22 of those with radio monitors were detected the following spring, suggestive of a high mortality rate. In addition, 878 dead otters were found, and it is likely that many more dead otters were never recovered. A report published in 2000 noted that sea otters born after the 1989 *Exxon Valdez* oil spill in Prince William Sound had a lower rate of survival than sea otters born before the oil spill. This suggests that the young are still experiencing the toxicity of the oil 8–10 years later.

These findings show that, as in the case of birds, it does not make sense to expend financial and human resources to try to rescue and resuscitate oiled mammals. Given the toxic effects of the oil, coupled with the stress the animals experience as a result of the capture, cleaning, and release process, there are very few survivors. Necropsies revealed severe emphysema (probably from the oil fumes) as well as liver, kidney, intestinal, and bone marrow abnormalities. It is likely that ingestion of crude oil will have the same effects on other mammals, including humans, since the same physiological effects were observed in laboratory studies on the effects of crude oil on rats. The one exception is oiled penguins (Figure 6-9). A survival rate, exceeding 90%, in comparison to the normal survival rate of nonoiled penguins, is reported for the African penguins oiled by spills at the tip of Africa.

FIGURE 6-9 Oil drips from the tail of a severely oiled African penguin. Photograph from http://www.uct.ac.za/depts/stats/adu/oilspill/pic02.htm. Copyright © L. G. Underhill, 2000. Used by permission of the University of Cape Town on behalf of L. G. Underhill. Also see color insert.

6.5.5 Sea Turtles

Laboratory studies on sea turtles were performed because the young hatch on the beach and must swim from the shore through the ocean. Since those that hatch on the African or Indian coasts must swim through the sea lanes where there are large amounts of crude oil (Figure 6-4), laboratory studies were carried out to determine the effects of sublethal doses of crude oil. Young turtles that had to surface to breath through an oil slick that was 0.05–0.5 cm thick developed pretumorous changes, lesions, and swelling over a period of 2–4 days. When there was the option of surfacing in the presence of the oil or in an area where there was no oil, the turtles did not avoid the oil. These findings suggest that there may be long-term toxic effects on the young as they swim through the sea lanes. They may not encounter the high concentrations of oil in the laboratory experiments, but they will be in contact with it for longer periods of time.

6.6 USE OF PETROLEUM IN THE INTERNAL COMBUSTION ENGINE

One of the major uses of petroleum as an energy source is in the automobile internal combustion engine. Most automobiles are driven by a four-stroke-cycle piston engine (Figure 6-10). Gasoline is mixed with air in the carburetor and drawn into the cylinder on the first stroke. The oxygen–gasoline vapor mixture is compressed on the second stroke and ignited by the spark plug. The expanding gases formed by combustion then drive the piston down on the third stroke, the stroke that provides the power to drive the car, and the combustion products are forced out of the cylinder on the fourth stroke.

The first automobiles had low-compression engines that could use the gasoline fraction distilled from crude oil (Figure 6-1), but as the automobile was improved, more powerful engines were developed. These high-compression engines required a more sophisticated fuel than straight-run gasoline. In high-compression engines, the hydrocarbons of simple petroleum distillates preignite before the top of the stroke of the compression cycle. Preignition decreases engine power because it results in an increase in the pressure on the piston before it has completed the upward stroke of the compression cycle. The preignition usually occurs explosively (knocking), which jars the engine and hastens its eventual demise.

The octane rating of gasoline is a measure of its tendency for preignition and detonation. The higher the research octane number (RON), the less likely that

   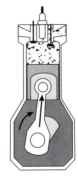

Stroke 1   Stroke 2   Stroke 3   Stroke 4

FIGURE 6-10   Schematic of a four-stroke-cycle engine.

knocking will occur. The octane rating is an empirical number that is measured on a standard test engine. It relates to the performance of the automobile engine when it is operating at low to medium speeds. The RON values of some representative compounds are listed in Table 6-2. The values of 0 for *n*-heptane and 100 for isooctane were established as arbitrary standards.

A variation of the octane scale designated motor octane number (MON) is also used to rate gasolines. The MON scale was established when it was observed that there was not good agreement between the RON determined in a test engine and that of the motor run under road conditions. The higher speeds and heavier weight of contemporary automobiles resulted in lower octane ratings in road tests than were observed in the test engine. The same test engine is used for both the MON and RON, but the operating conditions for MON determination give octane ratings that correlate more closely with those observed with an engine running at medium to high speed. The octane rating listed on gasoline pumps is an average of the RON and MON. It should be emphasized that octane ratings do not measure the energy released on combustion—all hydrocarbons release about the same amount of energy per gram when burned completely—but instead are a measure of the tendency for preignition and explosive combustion.

A number of methods have been developed to increase the RON rating (55–72) of straight-run gasoline. It was discovered in 1922 that the RON of straight-run gasoline could be increased to 79–88 by the addition of 3 g of "lead" per gallon. The "lead" or "ethyl fluid" that is added to gasoline is a mixture of about 60% tetraethyl lead [$Pb(CH_2CH_3)_4$], and/or tetramethyl lead [$Pb(CH_3)_4$], about 35–40% $BrCH_2CH_2Br$ and $ClCH_2CH_2Cl$, and 2% dye,

TABLE 6-2

Research Octane Numbers (Octane Ratings) for Some
Representative Hydrocarbons

Compound	RON
n-Octane	−19.0
n-Heptane	0[a]
n-Hexane	24.8
n-Pentane	61.7
2,4-Dimethylhexane	65.2
Cyclohexane	83
n-Butane	93.8
2,2,4-Trimethylpentane (isooctane)	100[a]
2,2,4-Trimethylpentene-1	103
Benzene	106
Toluene	120

[a]Standards for RON measurement.

solvent, and stabilizer. The lead prevents preignition by binding to hydroperoxyl free radicals, reactive intermediates with unpaired electrons, formed in the combustion process:

$$RH + O_2 \rightarrow R \cdot + \cdot OOH \tag{6-5}$$

This quenching decreases the combustion rate and prevents preignition. Leaded gasoline is no longer used in the United States and is discussed further in Section 6.7.4.

Catalytic re-forming, the conversion of aliphatic compounds to aromatic compounds, is the process of isomerization of linear aliphatic hydrocarbons to cyclic derivatives that are then dehydrogenated to aromatics as shown in reaction (6-6). The reaction pathway is determined by the nature of the binding of the hydrocarbons to the surface of a platinum or platinum-rhenium catalyst that is used. A product with a RON of 90–95 is obtained. In reaction (6-6), *n*-heptane is used to illustrate the cyclic compounds proposed as intermediates in the catalytic re-forming process.

Ethanol and other oxygenated organic compounds that enhance the octane hydrocarbon mixtures are discussed in Section 6.7.4. Apparently when compressed in the cylinder of the automobile engine, the substituted aromatics and oxygenated organics also react with hydroperoxyl and other free radicals to inhibit the preignition of gasoline vapors.

6.7 SOURCES OF HYDROCARBONS AND RELATED COMPOUNDS IN THE ATMOSPHERE

6.7.1 Introduction

Some hydrocarbons are natural components of the atmosphere. For example, most of the atmospheric methane is produced by anaerobic bacteria in swamps and bogs, and in the stomachs and intestines of cows, sheep, and termites; a much smaller percentage is released by natural gas wells. It has been proposed that many of the hydrocarbons found in the Gulf of Mexico may be due to the marsh plants indigenous to the Gulf Coast. Many plants, especially conifers and citrus, release terpenes into the atmosphere. It has been suggested that the

6-day air pollution alert in the Washington, DC, area in August 1974 was due mainly to the presence of terpenes originating from plants in the Appalachian Mountains. In the absence of any man-made pollution, there would be still parts-per-million concentrations of methane (Table 2-1) and parts-per-billion amounts of ethane, ethylene, acetylene, and propane in the atmosphere.

6.7.2 Hydrocarbons and Other Emissions from Automobiles

The automobile is a major nonnatural source of hydrocarbons in the atmosphere. The photochemical reactions of its gasoline and exhaust gases is an important source of an array of atmospheric oxidants that cause eye irritation and other problems for animals and plants (Section 5.3). Automotive air pollution is caused by the evaporation of gasoline or by tailpipe emissions. Gasoline evaporation occurs during the filling of the gas tank, and by the heating and cooling the automobile, as a result of diurnal temperature changes, and by operating the vehicle. In California and other areas of the country, vapor recovery systems connected to the gas tank collect the vapor displaced when it is filled. The displaced hydrocarbons are captured in canisters of activated carbon and later released into the engine during operation of the vehicle.

Tailpipe emissions from automobiles have been reduced significantly as a result of federal legislation, and this has resulted in a significant increase in air quality (Chapter 5). In 1968 the federal government decided to control automobile emissions by imposing standards for new cars (Table 6-3). It was assumed that older cars that did not meet the emission standards would eventually wear out and be taken off the road. The allowed emissions for new cars were continually lowered until 1981, and were unchanged during the 1981–1993 time period. Emission standards were further lowered starting in 1994. The legislated emissions from new cars were decreased by more than a factor of 10 between 1968 and 1994, and these reductions resulted in a decrease in air pollutants (Table 6-4).

The most difficult source of emissions to control is that from the exhaust pipe. This is because the internal combustion engine, as it is presently designed, will always emit some noncombusted hydrocarbons. The main source of these hydrocarbons is "wall quench" in the engine. The burning of the gasoline close to the walls of the cylinder is quenched because the zone next to the wall is much cooler than the combustion zone in the center of the cylinder. This relatively cold area slows the rate of combustion so that the hydrocarbons in this region are swept out by the piston before combustion is complete. Wall quench also occurs in the crevice above the piston rings between the cylinder wall and the piston (Figure 6-11).

There would still be hydrocarbon emissions from the automobile even if it were possible to eliminate wall quench. This is because there is insufficient

TABLE 6-3

Maximum Emissions from New Automobiles

	Hydrocarbons (g/mi)	CO (g/mi)	NO_x (g/test)	Evaporation (g/test)
Before controls[a]	10.60	84.0	4.1	47
1968	6.30	51.0	$(6.0)^b$	
1970–1971	4.10	34.0		
1972	3.00	28.0		
1973–1974	3.00	28.0	3.0	
1975–1976	1.50	15.0	3.1^b	2
1977	1.50	15.0	2.0	2
1978–1979	1.50	15.0	2.0^c	6
1980	0.41	7.0	2.0	6
1981–1993	0.41	3.4	1.0	2
1994	0.25	3.4	0.4	–
2004	0.125	1.7	0.2	

[a]Average values before legislation.

[b]Emissions of NO_x increased with control of HCs and CO. There was no NO_x standard at that time.

[c]Change in test procedure.

Source: Adapted from J. G. Calvert, J. B. Heywood, R. F. Sawyer and J. H. Seinfeld, *Science*, **261**, 37–44 (1993).

time for the complete combustion of gasoline to carbon dioxide and water. Addition of more air (oxygen) to the engine allows more complete combustion, but this results in an increase in the NO_x emissions (Figure 6-12) and a decrease in the power (Figure 6-13). The increase in NO_x is due to the higher

TABLE 6-4

Percent Decrease of Six Pollutants between 1975–1985 and 1986–1995

Pollutant	1975–1985	1986–1995
Pb	93	78
CO	47	36
SO_2	46	37
Particulates	20	17
NO_x	17	14
O_3	14	6

Source: Council on Environmental Quality, 22nd Annual Report, Washington, DC, March 22, 1992, p. 10, and *Chem. Eng. News*, pp. 22–23, Jan. 6, 1997.

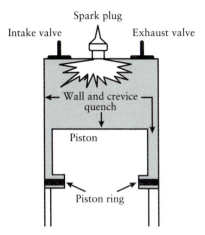

FIGURE 6-11 Regions of wall and crevice quench in the cylinder of a four-stroke cycle internal combustion engine.

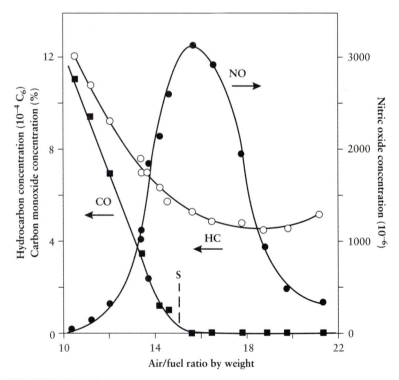

FIGURE 6-12 Effect of air/fuel ratio on hydrocarbon (HC), carbon monoxide (CO), and nitric oxide (NO) exhaust emissions; S, stoichometric operation. Redrawn from W. Agnew, *Proc. R. Soc. Ser. A.* 307, 153 (1968). Used by permission of The Royal Society of Chemistry.

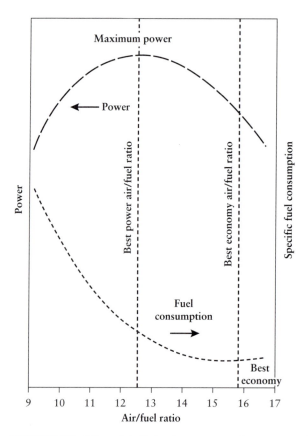

FIGURE 6-13   Effect of air/fuel ratio on power and economy. Redrawn from W. Agnew, *Proc. R. Soc. Ser. A.* 307, 153 (1968). Used by permission of The Royal Society of Chemistry.

temperature of the combustion zone in the engine in the presence of more oxygen.

Automobiles that travel farther on a gallon of gasoline are less polluting. In addition, they conserve valuable fuel resources. In 1978 the U.S. government instituted a program setting an average mileage goal for the fleet of cars manufactured by the major automobile companies. This corporate average fuel economy (CAFE) was gradually increased from 18 miles per gallon (mpg) in 1978 to 27.5 mpg in 1985. The U.S. Congress and presidents have been unwilling to increase the CAFE since 1985 because of concerns regarding the increased cost of the high mileage automobiles, and because the lighter cars that would have to be used might not protect their occupants in collisions as well as heavier vehicles. Environmental groups claim that the technology is available to produce a fleet of cars that average 45 mpg, but there is strong resistance to such a mandated increase from automobile manufacturers. The environmentalists were correct, since both conventional and hybrid gasoline–electric cars with mileage in the 40–50 mpg range were introduced in the United States in 2000.

6.7.3 The Catalytic Converter

Since the internal combustion engine, as presently designed, cannot operate without producing emissions, some device must be added to the automobile if both NO_x and hydrocarbon and carbon monoxide emissions are to be controlled. In 1973 recirculation of the exhaust gases was introduced to help control NO_x emissions. In addition, in some cars air was injected into the hot exhaust gases to facilitate continued oxidation of hydrocarbons. In 1975 in the United States, 85% of the new cars sold were equipped with a catalytic reactor in the exhaust stream to catalyze the oxidation of CO and hydrocarbons. Lead added to gasoline poisons the catalysts in the catalytic converter; therefore, starting with the 1975 models, all cars equipped with a catalytic converter were required to use lead-free gasoline.

The catalytic converter has evolved from a device that simply catalyzed the conversion of carbon monoxide and hydrocarbons to carbon dioxide, to a single unit that uses the reducing capacity of the gases emitted from the engine to first reduce NO_x to N_2 and then later to oxidize the reduced gases to carbon dioxide upon addition of air:

$$2HC + CO + 2NO_x + (3_xO_2) \rightarrow 3CO_2 + H_2O + N_2 \qquad (6\text{-}7)$$

$$2HC + CO + 3O_2 \rightarrow 3CO_2 + H_2O \qquad (6\text{-}8)$$

In 1979 the catalyst was a mixture of mainly platinum, rhodium, palladium, and ruthenium on a thin film of alumina coated on a honeycomb of ceramic or on ceramic pellets (Figure 6-14). Because of the expense of using platinum and rhodium, the catalyst evolved to being mainly palladium by 1997. This was made possible by higher engine temperatures, which minimize the deactivation of the palladium due to deposition of sulfur and phosphorus on its surface. The extent of oxidation of hydrocarbons and CO and the reduction of NO_x depends on the composition of the gases coming from the engine. If the engine is operating at an air/fuel ratio less than stoichiometric (reducing conditions, Figure 6-12), the efficiency for NO_x reduction is higher but the conversion of HC to CO_2 is less efficient. If the engine is operating at an air/fuel ratio higher than stoichiometric (oxidizing, Figure 6-12), the conversion of NO_x to N_2 is less efficient but the oxidation of hydrocarbons and CO is more efficient. There is a very small air/fuel window near the stoichiometric ratio for the optimal

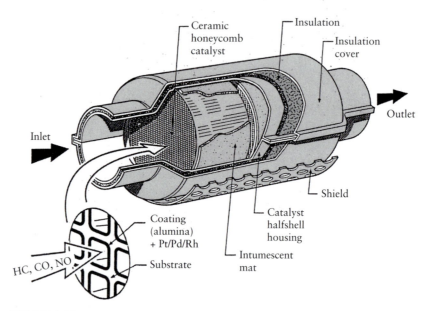

FIGURE 6-14 Cutaway view of an automotive catalytic converter. The honeycomb (enlarged, bottom left) contains 300–4000 square channels per square inch on which the metal catalysts are coated. Emissions pass through these channels and react on the catalyst surface. Provided by R. J. Farrauto, Engelhard Corp., NJ.

effectiveness of the catalytic converter. Ideally engines could be operated under oxidizing conditions that would permit efficient fuel consumption, resulting in the consumption of most of the hydrocarbons and CO. At the present time, however, there is no practical way to eliminate the NO_x formed in the absence of a reducing agent to convert it to N_2.

The 1990 amendments to the Clean Air Act mandated that emissions be reduced to even lower levels by 2004 and that the catalytic converter have a useful life of 100,000 rather than 50,000 miles. Emission standards more stringent than the national ones have been adopted by California and eleven northeastern and middle Atlantic states. A major step toward meeting the new emission standards will be the development of a catalytic converter that is operative when the engine is initially started. About 50% of the emissions from an automobile engine, as measured in a standard test procedure, are emitted during the first 100 s when the catalytic converter is being brought to its operating temperature of 400–600°C. Placement of the catalytic converter closer to the engine will result in a faster warm-up, but the converter will have to be reengineered to operate at about 1000°C, a temperature at which the present catalysts are inactivated and the ceramic core is not stable. In addition, the reduction of NO_x is less efficient at higher temperatures.

One possible solution is to electrically heat a portion of the catalyst when the car is started. The catalyst in a forward compartment of the converter would be coated on a metallic core that can be electrically heated to 400°C in 5 s by using power from the alternator. The remainder of the catalyst would be layered on a porous ceramic core as shown in Figure 6-14. Other possible solutions include the development of catalysts that are active at lower temperatures and the use of zeolites and other materials placed in the exhaust line before the converter, which adsorb the gaseous emissions when cold and release them when the operating temperature of the converter is reached. Meeting the requirement that the catalytic converter last for 100,000 miles will also require new catalyst formulations that are not poisoned by zinc, phosphorus, and sulfur, as are the present formulations. These elements are present in zinc dithiophosphate, a compound added to lubricating oil for the reduction of engine wear. Finally, there is active research on the development of better catalysts for the reduction of NO_x. This would make it possible to run the car engine at an air/fuel ratio of 18–21 instead of the stoichiometric conditions used currently (Figure 6-13). An air/fuel ratio of 18–21 results in the more complete combustion of gasoline to carbon dioxide and greater fuel economy as well.

6.7.4 Gasoline Additives: Tetraalkyllead, Ethanol, and Methyl *tert*-Butyl Ether

Tetraalkyllead compounds (mainly tetraethyl- and tetramethyllead), the antiknock compounds in lead fluid, the main antiknock additive in gasoline since the 1920s, are no longer used in the United States. The phaseout of lead in gasoline in the United States took place between 1975 and 1995. The Environmental Protection Agency initiated this phaseout mainly because of the adverse effect of lead on the catalytic converter, but also because of the possibility that the lead emissions from automobiles may be a health hazard (Section 10.6.10). In 1968, the combustion of leaded gasoline was responsible for 98% of the 185,000 tons (166,500,000 kg) of atmospheric lead in the United States. It is estimated that 7 million tons of lead was released to the environment from the use of this additive between 1923 and 1986.

Removing lead from gasoline has resulted in a major decrease in atmospheric lead (Table 6-4). For example, lead deposition in the White Mountains of New Hampshire decreased from 31 to 1–2 g/hectare in the 1975–1989 period (1 hectare $= 10^4 m^2$). In addition, there was a 76% decrease in the average blood lead level in U.S. residents from 1976 to 2000. Other studies have shown that up to 50% of the blood lead is due to the ingestion of atmospheric lead produced by leaded gasoline. It is well known that lead compounds are toxic, especially to children, so the removal of lead from gasoline should enhance the general health of the population. Unfortunately tetraethyllead is still used in the gasoline of many developing countries and in eastern Europe.

Aromatic organics, which have high octane ratings (Table 6-2), can be used in place of lead fluid to enhance the octane rating of straight-run gasoline. Being more toxic than aliphatic hydrocarbons, these aromatics constitute a health hazard when inhaled in large amounts, so other octane enhancers have also been used. The deleterious health effect due to aromatics in gasoline, however, is much less than the combined hazards of lead and organic emissions from cars without catalytic converters.

Ethanol is also used as an octane enhancer for petroleum; the 7–10% mixture in gasoline is often called gasohol. The action of ethanol as an octane enhancer was discovered in 1917, and at that time General Motors scientists proposed that it be used as a gasoline additive. A General Motors researcher discovered the antiknock effect of tetraethyllead in 1921, and the compound was selected as the approved additive by GM and Du Pont executives in 1923. They chose tetraethyllead instead of ethanol because the former, as a new material, it could be patented. The health hazards of lead compounds were known at the time, but tetraethyllead was chosen because money could be made on the patent. More recently gasohol was introduced because it can be produced by fermentation of biomass, mainly corn in the United States, and its incorporation into gasoline decreases the country's dependence on imported crude oil. Ethanol, derived from the fermentation of sugar, is not only used as a gasoline extender in Brazil, but is also used directly as a fuel. In 1991 ethanol constituted about 25% of the fuel used to power vehicles in Brazil and about one-third of these vehicles used only ethanol. The use of ethanol decreased in 1989–1991 mainly because of discoveries of crude oil in Brazil, which resulted in cheaper gasoline. The addition of ethanol to gasoline in the United States is subsidized by the government to aid midwestern farmers.

A disadvantage of ethanol is that it is hygroscopic and it picks up traces of water when gasoline that contains ethanol is shipped in pipelines. Since water is detrimental to automobile engines, ethanol must be shipped separately to local distributors, where it is blended with gasoline just before use. These two steps add to the cost of gasoline that contains ethanol.

More recently, another oxygenated organic compound, methyl *tert*-butyl ether (MTBE), has been used to enhance the octane rating of gasoline. It, together with ethanol, has the additional advantage of decreasing carbon monoxide (CO) emissions. As a consequence, the EPA has mandated the use of these or other oxygenated organics ("oxygenates") in U.S. cities in which the daily CO levels exceed the air quality standards prescribed by the Clean Air Act. The U.S. consumption of MTBE was 4 billion gallons a year in 1998, with one-third of the use in California. In the winter of 1992–1993, in cities where gasohol or gasoline containing 15% MTBE were used, there was an 80% drop in the number of days in which the CO emissions exceeded the national standard in comparison to the winter of 1991–1992.

Concerns with regard to the detrimental health effects of MTBE arose in the early 1990s. While MTBE has been used as an octane enhancer since 1979, when it was used in large amounts in 1992 in Fairbanks and Anchorage, Alaska, there were many reports of headaches, nausea, dizziness, and irritated eyes. In subsequent years, there were similar complaints from some of the residents of Colorado, California, New Jersey, North Carolina, Maine, and Wisconsin. More recently MTBE has been found in groundwater, where it imparts a noxious odor to drinking water when present in amounts greater than 3 ppb, although at the time of writing there are no definitive data on health hazards resulting from its ingestion. The rapid transport of MTBE in groundwater is due to its high solubility in water. It was estimated in 1999 that 250,000 gasoline stations in the United States had leaky underground storage tanks and that these are the source of the groundwater contamination. As of 2000, several states had banned the use of MTBE as a gasoline additive, and the EPA was planning a gradual phaseout of its use as an octane enhancer and as an additive for the reduction of carbon monoxide emissions. Ethanol appears to be the principal additive that will be used to reduce CO emissions in urban areas if the use of MTBE is banned. One can only imagine the health problems associated with lead and possibly MTBE that would have been avoided if ethanol had been chosen as the antiknock agent in 1923.

Another antiknock agent, methylcyclopentadienylmanganese tricarbonyl (MMT), has been used in Canada since about 1970. Long-term health effects have not been fully evaluated, and there is considerable debate concerning the possible deleterious effect of this additive on the engine monitoring equipment, which results in an increase in hydrocarbon emissions. The Canadian government had decided to discontinue the use of MMT because of possible neurotoxicity and interference with pollution-monitoring systems. In response to a lawsuit from the manufacturer based on the claim that the ban violated the North American Free Trade Agreement, however, the use of this additive was reinstated in 1998. Lawsuits also resulted in reapproval of MMT for use in the United States in 1995, although despite the lifting of the ban, addition of this antiknock agent to gasoline in the United States had not resumed as of 2000.

6.8 THE AIR POLLUTION PROBLEM IN URBAN CENTERS

In spite of the advances made in automobile emission control, air pollution is still a problem in urban centers. The daily ambient air levels of CO were exceeded periodically in 42 cities in 1991, and the corresponding ozone levels were exceeded frequently in 9 cities. Los Angeles continues to have high levels of photochemical smog and ozone, air pollution problems that are more severe than almost any other U.S. city. Currently the American city with the greatest air pollution is Houston, Texas. This continued urban pollution is initially surprising, considering that tailpipe emissions from new cars have been cut more than 10–fold over the 1968–1993 time period. Although the automobile emission standards for new cars have been lowered markedly, air sampling studies in highway tunnels, where other sources of polluted air are minimal, have shown that the average automobile emissions have not decreased as much as the mandated tailpipe emissions. For example, there has been only a three- to fourfold decrease in hydrocarbon and carbon monoxide and only a two- to threefold decrease in NO_x in the 1968–1993 time period (Table 6-3).

Tests on individual automobiles show that about 50% of the emissions come from 10% of the automobiles. Older automobiles have higher emissions. These cars were made when emission standards were more lenient, and they have catalytic converters with diminished or no ability to effect a decrease in tail pipe emissions. The higher emission rate of these older cars is partially offset by the lower number of miles they are driven annually in comparison to new cars. Surprisingly, about 20% of the new cars in every model year are high emitters. About 30% of these have had their emission controls intentionally disabled, and the emissions controls in the other 70% of such cars are not operating because they have not been properly serviced. It should be possible to solve these problems by providing incentives to replace old cars with newer models and mandating periodic checks of emission systems.

Others factors also contribute to the continued urban air pollution. One is that the miles driven in urban environments have doubled in the past 25 years. Another is emissions from stationary sources. For example, oil refineries, power plants, dry cleaning businesses, and auto repair shops emit hydrocarbon solvents, while restaurants and homes give off emissions from stoves, furnaces, fireplaces, lawn mowers, and grills. These sources, plus the natural emissions of terpenes from citrus and other plants, are becoming more important components of the photochemical smog in Los Angeles now that contributions from automobiles are decreasing.

6.8.1 Plans to Reduce Auto Emissions in Urban Centers

Several approaches are available to improve the quality of air in areas of high population density. Reformulated gasolines that have lower volatilities and use additives such as MTBE and ethanol to reduce CO emissions will probably be used. A catalytic converter that catalyzes the degradation of pollutants when the car is first started probably be mandated, as well. Alternative fuels, low-emission fuels such as natural gas, will be used in fleets of cars like taxis that operate in cities. In California in 1998, there was a requirement that 2% of all vehicles sold have zero tailpipe emissions. This percentage is mandated to increase to 10% by 2003. Other measures, such as stricter controls on the stationary sources mentioned earlier and limiting the number and types of vehicles in downtown areas, will also be implemented until the levels of carbon monoxide, ozone, volatile organics, and nitrogen oxide are decreased to the standards set by the federal EPA and by the state of California.

6.8.2 Low-Emission Automobiles

The adoption of vehicle emissions standards in California that are more restrictive than the national standards has forced the U.S. automobile companies to develop automobiles that have very low emissions and to produce zero-emission vehicles (ZEVs) as well. At present the only ZEVs are electrically powered, and there is a major emphasis in this area since the adoption by eleven eastern states of the California emission standards. The ZEVs must be an increasing number of the vehicles each auto-maker sells each year, starting in 2003. Strictly speaking, electrically powered cars do not emit pollutants during operation, but electricity from a power plant, which has emissions (see Chapter 15), is required to charge their batteries.

Studies carried out in 1992 show that the emissions resulting from the use of electric vehicles will be much lower than those from gasoline-powered automobiles even when the emissions from the power plants are taken into account. The net decreases in emissions for the electric car versus the gasoline-powered car are as follows: carbon dioxide, 2-fold; NO_x, 6-fold; volatile organics, 100-fold, and carbon monoxide, 200-fold. These decreases are possible because the emissions from power plants are controlled more readily than the emissions from thousands of automobiles. A small added benefit of the use of electric vehicles will be a decrease in consumption of oil used to power automobiles. Since oil is used to generate little of the electric power (5%) in the United States, the use of electric cars will not have significant impact on this need.

In 1999, however, the advantage of lower emissions from electric vehicles appeared to be decreasing in magnitude as new designs for catalytic converters for gasoline-powered automobiles resulted in lower emissions for these vehicles. Problems presently associated with electric vehicles include the short distance traveled, about 200 miles, before the battery needs charging, the long charging times, the weight of the bank of batteries required, the high cost of the vehicle in comparison to a gasoline-powered one of the same size, and the relatively short lifetime of the batteries (30,000–50,000 miles). In addition, if the electric vehicles are powered by lead batteries, there is a strong possibility of increased lead in the environment. None of these problems appear to be insurmountable at the present time. For example, it is possible in California to "fill up" an electric vehicle in parking lots, at "gas" stations, and at parking meters, and to pay by credit card. Two manufacturers introduced ZEVs in the United States in 2000. For the reasons just cited, however, they have not been popular. Vehicles with low and ultralow emissions are also prescribed in the California regulations. These will include hybrids, vehicles mainly powered by electricity but with the capability of also being powered by an internal combustion engine under certain driving conditions, such as steep hills or rapid acceleration. This engine, running in its most efficient mode, is also used to maintain the battery charge. Taxi fleets may be powered by natural gas and buses by fuel cells. Fuel cells generate power by the electrochemical oxidation of hydrogen or hydrocarbons, with up to 80% conversion of the chemical energy to electrical power. Very low emissions of carbon monoxide and NO_x are produced in these controlled oxidations (Section 15.9). Two manufacturers have introduced hybrid vehicles in the United States that use gasoline efficiently and generate very low emissions. These hybrids have gasoline mileage in the range of 40–50 mpg which is about the same as that of some conventionally powered vehicles of the same size.

Most large trucks and buses are powered by higher boiling diesel fuel (kerosene and light oil) (Figure 6-1), not gasoline. The diesel engine is similar to the internal combustion engine of the automobile except that the fuel is not ignited by a spark plug when compressed. Ignition proceeds as a result of the heat generated on compression of the air/fuel mixture. This combustion is not as fast as that resulting from the ignition of vapors by a spark, and uncombusted carbon, which is emitted as a black smoke containing particulates of carbon with adsorbed hydrocarbons, remains in the diesel cylinder. The 1990 amendments to the Clean Air Act mandated a sixfold decrease in these particulates in the 1991–1994 time period, along with reduction in the NO_x levels.

In 1997 the U.S. Environmental Protection Agency decided to extend its regulation of particulate emissions to particles 2.5 μm or greater in diameter. This was prompted by studies that noted a correlation between the level of atmospheric particulates and their associated sulfates with death rates. The studies noted an increase in heart and lung disease with the amounts of 2.5-μm particles in the air. This proposal drew a strong negative reaction from an industrial trade association but was supported by an independent study group. The EPA plans to obtain more definitive data on the distribution of these particles in the environment before it proposes specific regulations for their emissions. It is estimated that these proposals will be issued around 2010.

6.9 COAL

6.9.1 Coal Formation and Structure

Coal was probably the first of the fossil fuels to be used for energy. It was recognized as an energy source by the Chinese around 1100 B.C., while the ancient Greeks were probably the first of the western cultures to be aware of coal. The Romans reported that the "flammable earth" was being mined in Gaul when they captured that section of Europe. The first known coal mines in North America were operated by the Hopi Indians of Arizona some 200 years before Columbus.

Coal occurs mainly in the north temperate regions of the earth. Small deposits are located at the poles, but there are almost none in the tropics. The bulk of the world's deposits are located in Siberia in the former Soviet Union ($\sim 50\%$) while significant amounts are found in the United States ($\sim 25\%$) and China ($\sim 15\%$).

Coal was formed from the debris of giant tree ferns and other vegetation that grew in swamps and bogs 300 million years ago. When the plants died they were covered with sediment that prevented their oxidation by atmospheric oxygen. Further buildup of sediment and other geological processes subjected these materials to high pressures that resulted in their conversion to the solid, flammable substance we know as coal.

There are several classes of coal that reflect different stages in the metamorphosis from the carbohydrate and lignin of plants to the carbon of anthracite coal. Peat, a substance of a low heat value, is formed in the first stage of the process as a result of anaerobic microbial transformations of plant material. The lignin and microbial remains in the peat gradually lose H_2O and CO_2 as a result of high temperatures and pressures and are converted successively to lignite, bituminous coal, and anthracite. Anthracite, or hard coal, is about 85% carbon, with the remainder being mainly inorganic material and water.

The carbonization of plant carbohydrate results in the formation of a highly condensed aromatic ring structure similar to graphite. Aliphatic rings and chains as well as hydroxyl, carbonyl, carboxyl, and ether groupings are attached to these aromatic structures. Nitrogen, about 1% of the coal, is present as the cyclic amine function and sulfur as thiol (RSH), thioether (RSR), and disulfide (RSSR) groupings. Some of these structural elements are present in the postulated structure for bituminous coal (Figure 6-15). Oxidation of coal by aerobic bacteria removes some of the covalently bound organic sulfur.

Coal contains a variety of inorganic ions that ultimately end up in the ash residue after the coal is burned. Some 36 elements were detected in the ash resulting from the burning of West Virginia bituminous coal, with the major constituents being Na, K, Ca, Al, Si, Fe, and Ti. Toxic compounds of arsenic, selenium, and mercury (Section 10.6.7) are released during coal combustion. Pyrite (FeS_2), a major source of sulfur in coal, is not bound to the coal but occurs with it. It is formed by the action of anaerobic bacteria that reduce the sulfate in coal to sulfide [see reaction (11–21) in Chapter 11]. The sulfide reduces iron (III) to iron (II) with the corresponding oxidation of sulfide to S_2^{2-}.

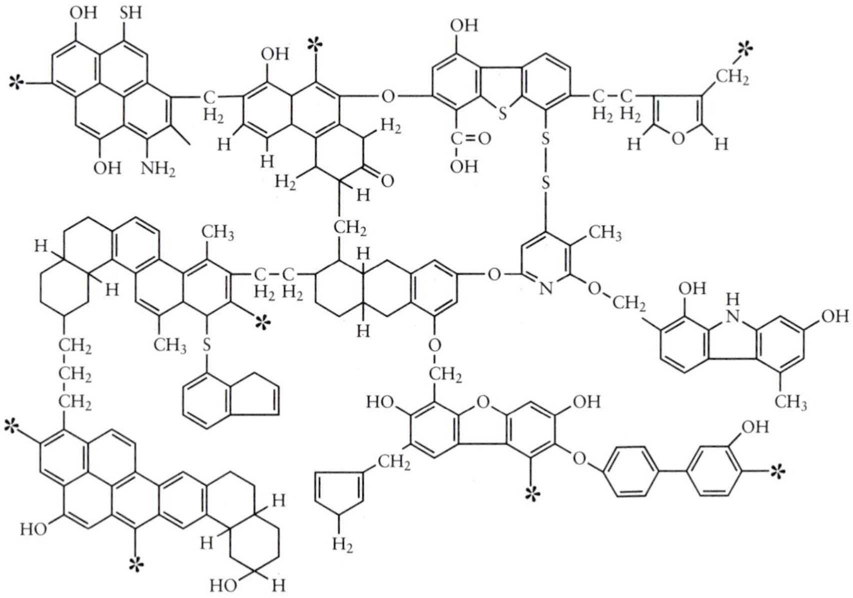

FIGURE 6-15 A postulated partial structure for bituminous (soft) coal. The stars indicate additional connecting points to the polymeric coal structure. Redrawn from N. Berkowitz, *An Introduction to Coal Technology*, 2nd ed. Academic Press, San Diego, CA. Copyright © 1994.

The combination of the two products yields pyrite (FeS_2). The total sulfur content of coal is about 0.6–3.2% by weight, with approximately half of that sulfur due to the pyrite present.

6.9.2 Coal Gasification

The gasification of coal was carried out extensively from about 1820 to 1920 for the production of coal gas, which was used for lighting (gaslights) and energy in the United States and elsewhere. However the discovery of abundant natural gas in the 1950s and the construction of pipelines to deliver the gas from the southern United States to the northeastern and western states resulted in the shift to the cleaner, cheaper natural gas as a fuel. As a consequence, the production of coal gas ceased and there were few new developments in coal gasification until the 1970s, when petroleum shortages prompted renewed interest in the use of coal, the most abundant fossil fuel in the United States.

Coal gasification, the first step toward the production of hydrocarbons from gasoline (Section 6.9.3.2), was used extensively by the Germans in World War

II when their access to petroleum was cut off. South Africa developed the technology further after World War II, and it was the source of 50% of their gasoline and related fuels. Gasification by the Lurgi process (Figure 6-16) is used in South Africa and in the Great Plains gasification facility in Beulah,

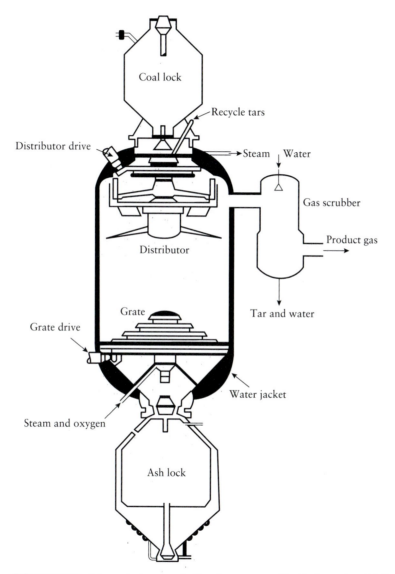

FIGURE 6-16 Schematic drawing of a Lurgi pressure gasifier. Redrawn from N. Berkowitz, *An Introduction to Coal Technology*, 2nd ed. Academic Press, San Diego, CA. Copyright © 1994.

North Dakota, a commercial plant that became operational in 1985. A number of other gasifiers that appear to be more efficient than the Lurgi reactor have been developed but the low cost of natural gas forestalled the construction of commercial units based on this new technology until 2001, when the cost of natural gas increased sharply. The Great Plains facility was selling syngas at a profit in 2001.

In the Lurgi process coal is added to the top of the reactor and steam and oxygen (not air) are injected into the bottom. The temperature in the reactor is 900–1000°C and the pressure is 3–3.5 MPa. This first step of the gasification process produces a mixture of carbon monoxide and hydrogen called "water gas":

$$4C(coal) + 2H_2O(steam) + O_2(hot\ air) \rightarrow 4CO + 2H_2 \qquad (6.9)$$

The "water gas" has a heating value of about 300 Btu per standard cubic foot (scf), or (11.2 mJ/m^3). In comparison, the heating value of natural gas (mostly methane) is 940 Btu/scf (35 mJ/m^3). The pieces of coal, which must be larger than a quarter-inch, migrate to the bottom of the reactor as they burn, and the inorganic ash that remains goes through the grate into the ash lock, while the gases are removed near the top of the reactor. A portion of the CO formed is reacted with steam in a second step, to generate hydrogen:

$$CO + H_2O \rightarrow CO_2 + H_2 \qquad (6.10)$$

This hydrogen, with that generated in reaction (6-9) is used in the third step to convert the remaining carbon monoxide to methane:

$$CO + 3H_2 \rightarrow CH_4 + H_2O \qquad (6.11)$$

Nitrogen and nitrogen compounds are converted to ammonia, which is removed by treatment with an acid wash and sold as fertilizer. Sulfur compounds are converted to hydrogen sulfide, which is oxidized to elemental sulfur or sulfate and sold. The operation of the Great Plains Gasification Facility is discussed in Section 15.3.5.

Considerable research has been performed on the gasification of coal in the ground, a process that avoids the costly, dangerous, and potentially environmentally destructive extraction of coal by underground or strip mining. Research on this topic was initiated in the 1930s in the Soviet Union, and the initial procedures were worked out there. Underground gasification was explored in the United States and Europe starting in 1945, but most of these projects were abandoned with the discovery of new sources of natural gas. In 1994 two operational facilities in the former USSR were producing gas for heating and power. One is in Angrensk, near Tashkent, and the other at Yuzhno-Abinsk near Novosibirsk. The basic process is shown schematically in Figure 6-17. The coal is ignited in the vicinity of the intake pipe(s) for air and

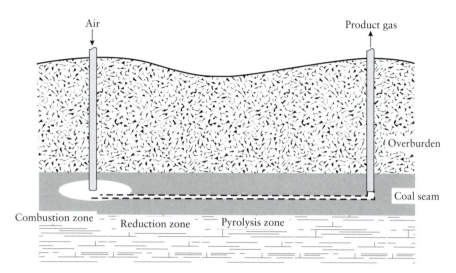

FIGURE 6-17 Schematic diagram of underground coal gasification. Redrawn from N. Berkowitz, *An Introduction to Coal Technology*, 2nd ed. Academic Press, San Diego, CA. Copyright © 1994.

steam, and the products of partial combustion and reduction (mainly carbon monoxide, carbon dioxide, and hydrogen) are collected from the outlet pipe. The success of the venture depends on the size of the coal seam and its permeability. Often the permeability needs to be enhanced by fracturing the coal seam by forcing high-pressure water or passing an electric current through it. If the coal seam is thicker than 1.5 m, it is possible to use this process to extract energy with better than average efficiency in comparison to mining and to gasify better than 75% of the coal. The discovery of large Siberian supplies of natural gas in the 1950s in the Soviet Union resulted in the closure of several underground gasification facilities and the cancellation of plans to build new ones.

6.9.3 Coal Liquefaction

6.9.3.1 Conversion of Mixtures of Carbon Monoxide and Hydrogen to Hydrocarbons

Reaction (6-12) gives the Fischer–Tropsch process for the hydrogenation of carbon monoxide in the presence of iron, cobalt, or nickel catalysts to form hydrocarbons.

$$CO + H_2 \xrightarrow{\text{Fe}} \text{hydrocarbons} \tag{6-12}$$

This process was investigated in the 1920s in Germany and was used in that country during World War II for the synthesis of fuel. The feedstock for the manufacture of gasoline was obtained by coal gasification (Section 6.9.2). This same approach to fuel formation was adopted in the 1950s by South Africa, which has abundant coal supplies but no petroleum. About half of the annual petroleum requirements for South Africa are produced from 6 million tons of coal. This method of hydrocarbon formation is basically inefficient, since all the bonds of coal are broken to form carbon monoxide, a partially oxidized product, and these carbon–carbon and carbon–hydrogen bonds are re-formed to produce hydrocarbons. It would be more efficient if the coal could be hydrogenated directly to hydrocarbons, a process we describe next.

6.9.3.2 Direct Conversion of Coal to Hydrocarbons

The first studies on the direct conversion of coal to hydrocarbons were performed in the laboratory of F. Bergius in Germany in 1918–1925. This chemist developed what is known as the Bergius process, in which pulverized coal is slurried in oil with an iron catalyst and heated with hydrogen at 450–500°C to form a mixture of oils having medium (180–325°C) and high (>325°C) boiling points. The medium boiling oil is sent to a second reactor, where it is catalytically cracked (broken into smaller hydrocarbons) and further hydrogenated at 470°C and a pressure of 240–300 atm to yield gasoline and kerosene. The high boiling fuel oil is slurried with another batch of pulverized coal, and the cycle is repeated. The first plant to carry out this process was constructed in Germany in 1927. Approximately a billion gallons of gasoline was produced in Germany by this process during World War II. The gasoline produced by the Bergius and related processes has a high octane rating because it consists of about 20–25% aromatic compounds, a reflection of the high proportion of aromatic structural units in coal.

Research on the direct conversion of coal to hydrocarbons was dormant from the 1950s until the oil embargoes of the 1970s because of the low price and high availability of crude oil from the Middle East. Interest was rekindled by the oil boycott of 1973, and a number of variations of the Bergius process have been developed. Incremental improvements in the process have been made, and now greater yields of hydrocarbon products can be had at lower temperatures and hydrogen pressures (Table 6-5). Development is continuing at a plant in Wilsonville, Alabama, where a consortium of power companies, petroleum companies, and the U.S. government is sponsoring research. Other processes, which differ from the original Bergius process, have also been investigated by a number of petroleum companies. Most of these projects have been terminated because of the continued availability of low-priced crude oil.

TABLE 6-5

Evolution of Coal Liquefaction Technology

Operating characteristics	1935	1945	1980	1986
Pressure (MPa)[a]	69	21	21	19
Max. temperature (°C)	480	465	455	440
Coal conversion[b]		94	94	94
Hydrogen consumption[b]	14	8	6	5.6
Hydrocarbons yield[b]	30	25	11	7
Distillable liquids yield[b]	54	54	51	65

[a]MPa = 10^6 pascals $\approx$ 10 atm.
[b]Wt % of moisture- and ash-free coal feed.
Source: Data from R. E. Lumpkin, *Science*, **239**, 873 (1988).

Additional Reading

Fuels Derived from Oil

Atlas, Ronald M., Slick solutions, *Chem. Bri.*, pp. 42–45, May 1996.

Courtright, Michael L., Improper disposal of used motor oil not only threatens the environment, it also wastes a valuable resource, *Machine Design*, **62**, 81–86 (1990).

Etkin, D. S., *International Oil Spill Statistics: 1998*. Cutter Information Corp., Arlington, MA., 1999, 66 pp.

Glaser, John A., Engineering approaches using bioremediation to treat crude oil–contaminated shoreline following the *Exxon Valdez* accident In Alaska. In *Bioremediation Field Experience*, P. E. Flathman, D. E, Jerger, and J. H. Exner, eds., pp. 81–103. Lewis, Boca Raton, FI, 1993.

Jackson, J. B. C., and 16 others, Ecological effects of a major oil spill on Panamanian coastal marine communities, *Science* **241**, 37–44 (1989).

Mielke, J. E., *Oil in the Ocean: The Short-and Long-Term Impacts of a Spill*. Congressional Research Service, Library of Congress, 90–356 SPR, July 24, 1990.

Prince, Roger C., Petroleum spill bioremediation in marine environments, *Crit. Rev. Microbiol.*, **19**, 217–242 (1993).

Smith, J. E., *"Torrey Canyon" Pollution and Marine Life*. Cambridge University Press, London, 1968.

Steering Committee for the Petroleum in the Marine Environment *Update, Oil in the Sea. Inputs, Fates and Effects*. National Academy Press, Washington DC, 1985, 601 pp.

Steinhart, C. E., and J. S. Steinhart, *Blowout. A Case Study of the Santa Barbara Oil Spill*. Duxbury, Belmont, CA, 1972.

Tanker Spills: Prevention by Design. National Research Council, National Academy Press, Washington DC, 1991.

Thayer, A. M., Bioremediation: Innovative technology for cleaning up hazardous waste, *Chem. Eng. News*, pp. 23–44, Aug. 26, 1991.

U.S. Congress, Office of Technology Assessment, Bioremediation for Marine Oil Spills— Background Paper OTA-BP-O-70. U.S. Government Printing Office, Washington DC, May 1991.

Using Oil Spill Dispersants on the Sea. National Research Council, National Academy Press, Washington, DC, 1989.

Vehicle Emissions

Calvert, J. G., J. Heywood, R. F. Sawyer, and J. H. Seinfeld, Achieving acceptable air quality: Some reflections on controlling vehicle emissions, *Science*, **261**, 37–44 (1993).

Farruto, Robert J., Ronald M. Heck, and Barry K. Speronello, Environmental catalysis, *Chem. Eng. New*, p. 34, Sept. 27, 1992.

Coal

Berkowitz, Norbert, *An Introduction to Coal Technology*, 2nd ed. Academic Press, San Diego, CA, 1994.

Crawford, Don L., ed., *Microbial Transformations of Low Rank Coals*. CRC Press, Boca Raton, FL, 1993.

Lumpkin, Robert E., Recent progress in the direct liquefaction of coal, *Science*, **239**, 873–877 (1988).

Schobert, Harold H., *Coal: The Energy Source of the Past and Future*. American Chemical Society, Washington, DC, 1987.

Also see the references given at the end of Chapter 2.

EXERCISES

6.1. A supertanker with a large hole in its hull is sinking in 50 ft of water in the Mediterranean near the Italian Riviera. The immediate principal concern is whether the tanker will break up and release all its cargo before the oil can be pumped out of the wreck. The Mediterranean is generally calm at this time of year, and there are no storms in the forecast. You are in charge of minimizing the damage to the beaches and wildlife at this expensive resort area.

(a) How will you remove the oil from the wreck?

(b) How will you minimize the damage from the oil spilling into the Mediterranean?

(c) How will you protect the beaches from the oil. How effective will your approach be?

(d) Predict the short- and long-term effects of this spill on marine life in the area.

6.2. Two major tanker spills of the same amounts of Kuwait crude oil have occurred, one in the middle of the North Atlantic, far from land, where there are frequent storms, and the other in the calm water of Long Island Sound, close to highly populated areas in New York and Connecticut. Compare these spill with regard to the following:

(a) Ease and feasibility of cleanup by means of floating booms to control the spread of the oil and ships to "vacuum" it up

(b) Rate of natural dispersal of the crude oil

    (c) Extent of the immediate toxic effects on marine life

    (d) Possible short- and long-term toxic effects to residents (human and other) in the area of the spill.

6.3. A brilliant geology student at your university proposed a new theory for the formation of crude oil and predicted that there is a huge deposit under the school's football field. Ten years after your graduation, preliminary drilling confirmed the student's prediction. In spite of the pleas by the football coach and threats of bodily harm to the president by the football team, the university administration decides to proceed with drilling a well on the 50 yard line.

    As a prominent graduate from your university you are asked to outline the possible health and other hazards of this operation (drilling, pumping, and transporting) so that proper precautions could be taken to minimize the health and other hazards to students, staff, and faculty. Give your recommendations and discuss the scientific bases for them.

6.4. What are the most likely first steps in the air oxidation of an olefin present in oil spilled in the environment? Use the following generic formula for an olefin in your proposed reactions.

$$RCH_2CH = CHR'$$

6.5. Predict the fate of each of the following compounds if introduced into the ocean as part of an oil spill. If the compound reacts chemically or is degrade by microorganisms under these conditions, give the equations(s) for the reaction in the environment.

    (a) Styrene

    (b) *n*-Butane

    (c) Benzo[*a*]pyrene

    (d) Asphalt

    (e) Kerosene

    (f) *n*-Propylbenzene

6.6. Write the chemical reactions for the initial steps in the combustion of isooctane (2,2,4-trimethylpentane) in the internal combustion engine.

6.7. Why can't an internal combustion engine be "tuned" (adjusted) to prevent it from emitting hydrocarbons, CO and NO_x?

6.8. The catalytic converter of present-day automobiles removes hydrocarbons (HC), CO, and NO_x from emissions by both oxidative and reductive processes.

    (a) Briefly explain how oxidation and reduction of these compounds proceed in the same device.

    (b) Write balanced chemical equations for both the oxidation and reduction processes.

    (c) Why do you smell the emission of hydrocarbons from many cars equipped with functioning catalytic converters when their engines are first started?

6.9. The addition of catalytic converters to the tailpipes of automobiles has resulted in a 10-fold or greater decrease in the emissions from new automobiles over the past 25 years. Give reasons for the much smaller decrease in the amounts of these gases in the atmospheres of urban areas.

7

SOAPS, SYNTHETIC SURFACTANTS, AND POLYMERS

7.1 INTRODUCTION

Soaps, surfactants and polymers are discussed together, following the discussion of petroleum, because most of the polymers and the surfactants in detergents are made mainly from chemicals derived from petroleum. Natural fats and oils are also used in the manufacture of surfactants (Figure 7-1). Soaps and surfactants contain segments of linear or lightly branched hydrocarbon chains that are, in the main, broken down to acetate up on metabolism by microorganisms in the environment. The biodegradation of petroleum occurs by the same pathway once a terminal carbon has been oxidized to a carboxyl grouping. The rate of degradation by other environmental reagents such as sunlight, oxygen, or water is much slower than that of microbial degradation.

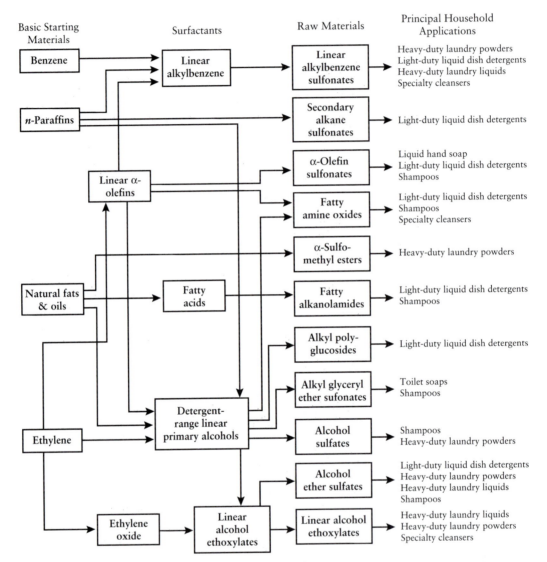

FIGURE 7-1 Surfactants for household detergents: petrochemical raw materials and uses. Redrawn from "Soaps and Detergents," Susan Ainsworth, *Chem. Eng. News*, Jan. 24, 1994. Used by permission of SRI Consulting. Also see color insert.

7.2 SOAPS AND SYNTHETIC SURFACTANTS

Soaps and synthetic surfactants have similar structures, and their mechanisms of cleansing are similar. Both contain a hydrophobic portion (usually a hydrocarbon chain) to which a hydrophilic (polar) group is attached. Such materials are surface active: that is, they concentrate at the surface of an aqueous solution or form aggregates called micelles (Figure 7-2) within the solution. Surface-active materials lower the surface tension of the water so that the water better penetrates the surface and interstices of the object being cleaned. If the binding of the micelles of soap or synthetic surfactant to the substance being cleaned is greater than the binding of the soiling agent, the latter will then be transferred into the micelles upon agitation. Oils and greases will also be adsorbed by the hydrophobic end of the surface-active agent. The soiling agent can be washed away because the affinity of the polar group of the surfactant for water keeps the micelle–dirt complex suspended in the water.

The foam covering many otherwise scenic rivers and streams was a source of much public concern for many years. This environmental problem was caused by slowly degrading synthetic surfactants and was probably the first modern environmental problem to be solved as a result of public opinion.

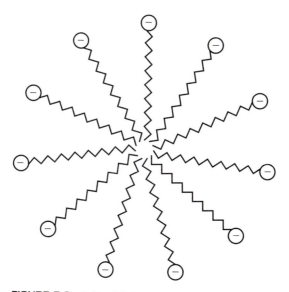

FIGURE 7-2 A Simplified two-dimensional representation of a three-dimensional micelle. The hydrophobic hydrocarbon chains associate in the center of the micelle. Nonpolar substances associate with the hydrocarbon chains of the micelle, become solubilized, and are removed from the fabric.

Some indication of the nature of the problem was apparent as early as 1947 when the aeration tanks at the sewage disposal plant at Mount Penn, Pennsylvania, were inundated with suds. This spectacular effect was the result of a promotional campaign during which free samples of a new detergent had been distributed at local stores. Everyone used the new product at about the same time, but unfortunately the microorganisms residing at the disposal plant metabolized the synthetic surfactant in this detergent very slowly. Despite this early warning, slowly degrading surfactants were manufactured up to 1965 before the detergent industry responded to public pressure and voluntarily switched to the manufacture of surfactants that degrade more rapidly in the environment.

7.2.1 Soaps

Soaps are prepared by the saponification (base hydrolysis) of animal or vegetable fats as shown in reaction (7-1) (note, however, that the R groups are usually different). Hydrolysis converts the triglycerides in the fats to the sodium salts of carboxylic acids (soaps) and glycerol. The carboxylate group is the polar, water-soluble end of the soap.

$$
\begin{array}{cccc}
\overset{\displaystyle O}{\overset{\displaystyle \|}{R C}}-O-CH_2 & & & HO-CH_2 \\
\overset{\displaystyle O}{\overset{\displaystyle \|}{R C}}-O-CH + 3\,NaOH \longrightarrow & \overset{\displaystyle O}{\overset{\displaystyle \|}{3\,R C O^-}}\,Na^+ & + & HO-CH \qquad (7\text{-}1) \\
\overset{\displaystyle O}{\overset{\displaystyle \|}{R C}}-O-CH_2 & \text{Soap} & & HO-CH_2 \\
& & & \text{Glycerol}
\end{array}
$$

The sodium salt of stearic acid, $CH_3(CH_2)_{16}CO_2H$, is the major product released upon hydrolysis of animal fat, while the sodium salt of oleic acid, $CH_3(CH_2)_7CH=CH(CH_2)_7CO_2H$, is the main product from the hydrolysis of olive oil. Hydrolysis of palm oil yields approximately equal amounts of the salts of oleic acid and palmitic acid, $CH_3(CH_2)_{14}CO_2H$.

These naturally occurring fatty acids contain linear hydrocarbon chains, which are synthesized biochemically from the acetate ion ($CH_3CO_2^-$). They are also readily degraded back to acetate by microorganisms in the environment. Since long-chain ("fatty") acids occur naturally, it is not surprising to find some foam and suds at waterfalls, the surf at the beach, and other places of turbulence in streams, rivers, and oceans owing to the presence of natural soaps in the water.

The main disadvantage of the use of salts of carboxylic acids as cleansing agents is the insoluble precipitates that form with Ca^{2+} and Mg^{2+}, which when present in water make it hard. This curdy, gray precipitate leaves a deposit on clothes or "ring" around collars or bathtubs. Not only are the deposits undesirable, but greater amounts of soap must be used to make up for that lost by precipitation. Another disadvantage of soaps is that they are salts of weak acids that protonate in mildly acidic solutions. The polarity of the "head" of the detergent decreases when protonated and, as a consequence, the product's effectiveness at solubilizing oils and greases is diminished.

7.2.2 Detergents

Detergents are a mixture of synthetic surfactants, builders, bleaches, enzymes, and other materials designed to enhance the specific cleaning power of this mixture. Millions of pounds of these materials are produced in the United States each year (Table 7-1). The main advantage of synthetic surfactants over soaps is that the former do not precipitate in hard water. Since, however the cleansing power of synthetic surfactants is also decreased markedly by Ca^{2+} and Mg^{2+}, it is necessary to add builders that enhance the cleansing action of synthetics by inactivating the Ca^{2+} and Mg^{2+} in hard water. The builders are polycarboxylic acids, silicates, zeolites (inorganic aluminosilicates), or polyphosphates (e.g., $Na_5P_3O_{10}$; Section 10.5), which bind Ca^{2+} and Mg^{2+} thus preventing them from binding to the surfactant. Polyphosphates, which were used extensively as builders in the 1970s, were replaced in large part by zeolites because it was believed that phosphate from polyphosphates initiated algal blooms on lakes (Section 9.1.2), with the resulting eutrophication. Since the phosphate in detergents is easily removed by treatment with Fe^{3+} at sewage disposal plants, a process that has been in use in Sweden for many years, it may have been premature to ban the use of polyphosphates. It should also be noted

TABLE 7-1

U.S. Detergent Chemical Use in 1996

Use	Amount (lb $\times 10^6$)
Builders	3805
Surfactants	2000[a]
Bleaches, brighteners, and enzymes	190
Fragrances and softeners	128
Other	260

[a]Surfactant use in 2001 projected to be 2400×10^6 lb.
Source: E. M. Kirschner, *Chem. Eng. News*, p. 39, Jan. 26, 1998.

that detergents account for only 20–25% of the phosphate in lakes and rivers, while the bulk of the remainder comes from fertilizer and animal waste runoff from farms. Consequently, the phosphates in detergents are not the main cause of lake eutrophication. In accordance with the view that phosphates have a minor role in lake eutrophication, these compounds have been classified as ecologically acceptable for use in home laundry detergents in Europe. Currently 25% of the home laundry detergents sold in Europe contain phosphates.

It is reported that zeolites are not as effective as polyphosphates in enhancing the cleansing power of surfactants and that they have environmental problems as well. Studies suggest that zeolites promote surface algal blooms in the Adriatic Sea. In natural waters they bind naturally occurring humic acids (Section 9.5.7) to form particles that float. Microorganisms that bind to these particles grow and produce a foul-smelling algal bloom.

7.2.3 Surfactants

7.2.3.1 The General Nature of Surfactants

It has been possible to synthesize a wide variety of surfactants by varying the structure of both the polar and nonpolar ends of the molecule. The polar end of the linear alkylsulfonates is the anionic sulfonate group RSO_3^-. The linear alkylsulfonates [see later: reaction (7-7)] and alcohol sulfate

$$[CH_3(CH_2)_n \overset{\overset{\displaystyle CH_3}{\displaystyle |}}{C}HOSO_3^- Na^+$$

and the related alcohol ether sulfates

$$RO(CH_2CH_2CH_2)_nCH_2CH_2OSO_3^- Na^+$$

are the major laundry surfactants used in the United States (Table 7-2). The nonionic surfactants, alcohol ethoxylates $[RO(CH_2CH_2)_nCH_2CH_2OH]$, and alkylphenol ethoxylates [see later: reaction (7-11)], which have alcohol groups as the polar entity, are also produced in large amounts (Table 7-2). Smaller amounts of cationic surfactants $[CH_3(CH_2)_nCH_2N(CH_3)_3Cl^-]$ are also manufactured. These are not as efficient as the anionic and nonionic cleaning agents but are used as germicides and fabric softeners.

7.2.3.2 Synthesis of Linear Alkylsulfonates (LAS)

Alkyl benzene sulfonates (ABS), the original slowly degrading surfactants, and linear alkylsulfonates are prepared by similar procedures. The alkyl chain in both is prepared by the oligomerization (formation of short polymers) of

TABLE 7-2

Laundry Surfactant Use in the United States in 1991

Surfactant	Amounts ($lb \times 10^6$)	Percent
Linear alkylsulfonates	740	40
Alcohol sulfates	220	12
Alcohol ether sulfates	260	14
Alcohol ethoxylates and alkylphenol ethoxylates	390	21
Other anionics	200	11

Source: Modified from A. M. Thayer, *Chem. Eng. News*, p. 33, Jan. 25, 1993.

propene (propylene), but with the ABS derivatives a branched oligomer is formed in the following acid-catalyzed process:

$$\begin{aligned}
CH_3CH{=}CH_2 + H^+ &\longrightarrow CH_3\overset{\underset{|}{CH_3}}{CH}{}^+ \xrightarrow{CH_3CH=CH_2} (CH_3)_2CHCH_2\overset{\underset{|}{CH_3}}{CH}{}^+ \\[2mm]
\xrightarrow{2CH_3CH=CH_2} &(CH_3)_2CH(CH_2\overset{\underset{|}{CH_3}}{CH})_2CH_2\overset{\underset{|}{CH_3}}{CH}{}^+ \longrightarrow \\[2mm]
&(CH_3)_2CH(CH_2\overset{\underset{|}{CH_3}}{CH})_2\overset{\underset{|}{CH_3}}{CH}CH{=}CH_2 + H^+
\end{aligned}$$

(7-2)

It is not possible to synthesize linear hydrocarbon molecules by the straightforward acid-catalyzed oligomerization of olefins because the more substituted secondary carbocation is formed, with the formation of a branched chain hydrocarbon as shown in reaction (7-2). The branched-chain oligomers were used because they were readily prepared in acid-catalyzed reactions.

Linear hydrocarbons can be synthesized or separated from petroleum. They can be synthesized from olefins by use of catalysts such as triethylaluminum to give either alkanes or alkenes. In the Alfol process, linear alcohols are formed by the air oxidation and subsequent hydrolysis of the oligomers formed initially [reactions (7-3) and (7-4)]. Olefins can be formed by pyrolysis of the trialkylaluminum adduct in the presence of ethylene in reaction (7-5).

$$CH_3CH_2{-}\overset{\underset{|}{CH_2CH_3}}{Al}{-}CH_2CH_3 + nCH_2{=}CH_2 \longrightarrow {-}\overset{|}{Al}{-}(CH_2CH_2)_nCH_2CH_3$$

(7-3)

$$\begin{aligned}
{-}\overset{|}{Al}{-}(CH_2CH_2)_nCH_2CH_3 &\xrightarrow{O_2} {-}\overset{|}{Al}{-}O(CH_2CH_2)_nCH_2CH_3 \\[2mm]
&\xrightarrow{+H_3O} CH_3CH_2(CH_2CH_2)_nOH + Al(OH)_3
\end{aligned}$$

(7-4)

$$-\overset{|}{Al}-(CH_2CH_2)_nCH_2CH_3 \xrightarrow[\Delta]{CH_2=CH_2}$$

$$CH_3CH_2(CH_2CH_2)_{n-1}CH=CH_2 + Al(CH_2CH_3)_3 \qquad (7\text{-}5)$$

Urea crystallizes in a unique fashion from hydrocarbon solutions, and this is the basis of a method of separating linear and branched hydrocarbons in petroleum fractions. Linear hydrocarbons may separate from the branched-chain kind in petroleum fractions by encapsulation in the urea crystals. The urea crystallizes as helices, with the linear hydrocarbons encased in the central channel of the tubes formed by these helices. Branched-chain hydrocarbons do not fit into the 5-Å-diameter tube, so the linear hydrocarbons can be separated by merely filtering out the urea crystals. "Lightly branched" hydrocarbons with more than 10 carbon atoms will also form urea inclusion compounds and these are also present in the mixture of "linear" hydrocarbons separated by this procedure.

The use of molecular sieves has provided a cheaper means for the separation of linear and branched hydrocarbons than the use of urea. Molecular sieves are semipermeable zeolites that have a channel structure similar to that of urea when it is crystallized from hydrocarbons. The molecular sieves with a diameter of 5 Å (0.5 nm) are used to selectively screen linear hydrocarbons from petroleum.

Virtually the same procedures are used to complete the synthesis of both the ABS and linear alkylsulfonate detergents. The propene or other olefin oligomer is attached to benzene in an acid-catalyzed Friedel–Crafts reaction (7-6), the alkylated benzene is sulfonated with sulfur trioxide, and then, in reaction (7-7), the sulfonic acid product is neutralized with NaOH to generate the anionic surfactant.

$$R'CH=CH_2 \xrightarrow{HF} R'HC-\overset{CH_3}{\overset{|}{\bigcirc}} \qquad (7\text{-}6)$$

$$R'HC-\overset{CH_3}{\overset{|}{\bigcirc}} \xrightarrow[2.\ NaOH]{1.\ SO_3} R'HC-\overset{CH_3}{\overset{|}{\bigcirc}}-SO_3^-Na^+ \qquad (7\text{-}7)$$

7.2.3.3 Synthesis of Nonionic Alcohol Ethoxylates

The alcohol and alkylphenol ethoxylates, the largest group of nonionic detergents currently in use, are prepared from the anion of the alcohol or alkylphenol and ethylene oxide.

Reactions (7-8)–(7-11) ($n = 5$–10) outline the synthesis of the nonylphenol ethoxylate, one of the major nonionic detergents manufactured in the United States.

$$C_9H_{19}\!\!-\!\!\left\langle\bigcirc\right\rangle\!\!-\!\!OH \xrightarrow{OH^-} C_9H_{19}\!\!-\!\!\left\langle\bigcirc\right\rangle\!\!-\!\!O^- + H_2O \qquad (7\text{-}8)$$

$$C_9H_{19}\!\!-\!\!\left\langle\bigcirc\right\rangle\!\!-\!\!O^- + \;\triangle\!\!O \longrightarrow C_9H_{19}\!\!-\!\!\left\langle\bigcirc\right\rangle\!\!-\!\!OCH_2CH_2O^- \qquad (7\text{-}9)$$

Ethylene oxide

$$C_9H_{19}\!\!-\!\!\left\langle\bigcirc\right\rangle\!\!-\!\!OCH_2CH_2O^- + {}_n\triangle\!\!O \longrightarrow$$

$$(7\text{-}10)$$

$$C_9H_{19}\!\!-\!\!\left\langle\bigcirc\right\rangle\!\!-\!\!O(CH_2CH_2)_nCH_2CH_2O^-$$

$$C_9H_{19}\!\!-\!\!\left\langle\bigcirc\right\rangle\!\!-\!\!O(CH_2CH_2)_nCH_2CH_2O^- \xrightarrow{H^+}$$

$$(7\text{-}11)$$

$$C_9H_{19}\!\!-\!\!\left\langle\bigcirc\right\rangle\!\!-\!\!O(CH_2CH_2)_nCH_2CH_2OH$$

7.2.4 Microbial Metabolism of Hydrocarbons, Soaps, and Synthetic Surfactants

Microorganisms can utilize the energy of oxidation released by some hydrocarbons, soaps, and surfactants as well as some of the organic fragments released during their growth. The metabolism of these compounds is affected by enzymes—high molecular weight proteins that catalyze a broad range of chemical transformations in living systems. Enzymes are characterized by their extreme catalytic efficiency (the catalyzed reactions typically proceed about 10^9 times as fast as the uncatalyzed reaction) and specificity. (There is usually a specific enzyme for each chemical transformation that takes place in a cell.) Many enzymes require a small, nonprotein prosthetic group, called a coenzyme, for catalytic activity. Coenzymes are bound to the enzyme during the course of the reaction, and sometimes they combine with one or more of the reacting chemical species.

Fatty acids (including protonated soaps) occur widely in biological systems and are therefore readily degraded by microorganisms in enzyme-catalyzed reactions. Linear hydrocarbons and linear detergents are similar in structure to soaps, and these are degraded in the same or similar pathways.

$$RCH_2CH_2CO_2H + CoASH \longrightarrow RCH_2CH_2\overset{\overset{\displaystyle O}{\displaystyle \|}}{C}SCoA \qquad (7\text{-}12)$$

$$RCH_2CH_2\overset{\overset{\displaystyle O}{\|}}{C}SCoA + FAD \longrightarrow RCH=CH\overset{\overset{\displaystyle O}{\|}}{C}SCoA + FADH_2 \qquad (7\text{-}13)$$

$$RCH=CH\overset{\overset{\displaystyle O}{\|}}{C}SCoA + H_2O \longrightarrow R\overset{\overset{\displaystyle OH}{|}}{C}HCH_2\overset{\overset{\displaystyle O}{\|}}{C}SCoA \qquad (7\text{-}14)$$

$$R\overset{\overset{\displaystyle OH}{|}}{C}HCH_2\overset{\overset{\displaystyle O}{\|}}{C}SCoA + NAD^+ \longrightarrow R\overset{\overset{\displaystyle O}{\|}}{C}CH_2\overset{\overset{\displaystyle O}{\|}}{C}SCoA + NADH + H^+ \qquad (7\text{-}15)$$

$$R\overset{\overset{\displaystyle O}{\|}}{C}CH_2\overset{\overset{\displaystyle O}{\|}}{C}SCoA + H_2O \longrightarrow R\overset{\overset{\displaystyle O}{\|}}{C}OH + CH_3\overset{\overset{\displaystyle O}{\|}}{C}SCoA \qquad (7\text{-}16)$$

The steps in the oxidative metabolism of fatty acids are outlined in reactions (7-12)–(7-16). Each of these reactions is catalyzed by an enzyme even though the enzyme is not specified in the reaction. Coenzyme A (CoASH) is a complex structure that contains a thiol group (—SH) as the reactive center. The acetyl coenzyme A [$CH_3C(O)SCoA$] produced in reaction (7-16) may be used by the microorganism for the synthesis of the specific fatty acids needed for growth by a pathway that is essentially the reverse of the one shown here. Alternatively, the acetyl CoA may be oxidatively degraded to carbon dioxide and water to provide the energy needed to drive microbial metabolic processes. Flavin adenine dinucleotide (FAD) and nicotinamide adenine dinucleotide (NAD$^+$) are the oxidized forms of hydrogen transfer coenzymes, while FADH$_2$ and NADH are the respective reduced forms.

Originally it was believed that branching in the alkyl chain of alkylbenzene sulfonate (ABS) detergents impeded specific steps in reactions (7-12)–(7-16) and that this was the reason for their slow environmental degradation. It was proposed that the branched hydroxyester reaction formed in (7-17) would not be oxidized further to the keto derivative [see reaction (7-15)] and bio-degradation would be halted at this roadblock. However, it has been found that microorganisms are able to get around these simple barriers either by oxidatively removing the methyl groups or by cleaving off propionyl CoA [$CH_3CH_2C(O)SCoA$] in place of acetyl CoA. Chain branching merely slows the degradative process.

$$\underset{\underset{\displaystyle CH_3}{|}}{R}C=CH\overset{\overset{\displaystyle O}{\|}}{C}SCoA + H_2O \longrightarrow \underset{\underset{\displaystyle CH_3}{|}}{R}\overset{\overset{\displaystyle OH}{|}}{C}CH_2\overset{\overset{\displaystyle O}{\|}}{C}SCoA \qquad (7\text{-}17)$$

The biodegradation of surfactants has been studied in considerable detail, but their structural diversity, even within specific types, makes it difficult to formulate guidelines for biodegradation based on structure. These studies are further confused by the need to "acclimate" the bacteria population with the surfactant to ensure that consistent results are obtained. The need for acclimation suggests that all the members of the population are not able to metabolize the surfactant and acclimation promotes the growth of strains of microorganisms that are able to degrade it. The following guidelines have been developed to facilitate understanding of the microbial degradation of ABS and LAS detergents.

1. Highly branched ABS detergents degrade slowly in part because the alternative pathways required to cleave the side chain methyl groups proceed more slowly than the route shown in reactions (7-12)–(7-16).

2. The ease of cleavage of the surfactant increases, the further the polar sulfate or sulfonate grouping is from the alkyl terminus of the chain. Consequently, surfactants with branched chains that are shorter than those with linear chains containing the same number of carbon atoms degrade more slowly. This rule may be related to the ease of binding of the surfactant to a degradative enzyme if the site for binding its polar group is in a hydrophilic region of the enzyme and the site for binding its hydrophobic end is in a hydrophobic region of the enzyme.

3. The rate of degradation of surfactants is slower in concentrated solutions than in dilute solutions. This finding suggests that minor constituents of the surfactant mixture inhibit the microbial degradative enzymes. Their concentrations are not high enough in dilute solutions to inhibit all the microbial enzyme present.

These guidelines also apply to the alkyl substituent of the alkyl- and alkylphenol ethoxylates. The alkyl chain will degrade slowly if it is highly branched, and then the ethoxylate grouping will degrade first. The rate of degradation of the alkyl chain decreases as the number of ethoxylate groups increases. For example, the extent of degradation of the ethoxylate chain of an alkylphenol ethoxylate decreases as the number of ethoxylate groupings increase from 5 to 20. This decrease may reflect the increased difficulty of transport of the more hydrophilic detergent (with more ethoxylate groupings) through the hydrophobic cell membrane of microorganisms.

The alkylphenol ethoxylates degrade more slowly than the alcohol ethoxylates. In some cases it has been possible to isolate what appear to be intermediates in the degradation of the alkylphenol ethoxylates in which one or two ethoxylate groupings remain attached to the alkylphenol. These compounds have little surfactant action or solubility and precipitate with the other sludge at the waste disposal plant (Section 11.5.2). Their anaerobic (nonoxygenic) degradation product, nonylphenol, has been found in sewage sludge [see reaction (7-18)].

$$C_9H_{19} - \bigcirc - OCH_2CH_2OCH_2CH_2OH \xrightarrow{[H]} C_9H_{19} - \bigcirc - OH \qquad (7\text{-}18)$$

Nonylphenol

[H] = microbial degradation under reducing conditions

Nonylphenol is toxic to fish and other marine life, and it may be an environmental hazard when it is leached from sewage sludge that is dumped on land or in the ocean. It is also has estrogenic hormonal activity in mammals (Section 8.5.1). It is not clear whether the nonylphenol released by the degradation of nonionic surfactants is present in sufficient quantities in the environment to cause estrogenic effects.

While the biochemical pathway for the degradation of ethoxylate surfactants has not been determined, some insight into the mechanism can be obtained from studies on the degradation of glycols and simple ethoxylate oligomers shown in reaction (7-19).

$$ROCH_2CH_2OH \xrightarrow{[O]} RO\overset{\overset{\displaystyle HO}{|}}{C}HCH_2OH \xrightarrow{[O]} RO\overset{\overset{\displaystyle HO}{|}}{C}H\overset{\overset{\displaystyle O}{\parallel}}{C}H \xrightarrow{[O]} RO\overset{\overset{\displaystyle HO}{|}}{C}H\overset{\overset{\displaystyle O}{\parallel}}{C}OH \longrightarrow$$

$$ROCH_2OH + CO_2 \xrightarrow{[O]} RO\overset{\overset{\displaystyle O}{\parallel}}{C}H \xrightarrow{H_2O} ROH + H\overset{\overset{\displaystyle O}{\parallel}}{C}OH \qquad (7\text{-}19)$$

R = $HOCH_2CH_2OCH_2CH_2O^-$, where [O] indicates oxidation.

The microbial breakdown of detergents and petroleum requires the terminal oxidation of the hydrocarbon chain to a carboxylic acid group to initiate the β-oxidation procedure outlined in reactions (7-12)–(7-16). This terminal oxidation is a well-documented process, and in some instances it has been established that the corresponding alcohol and aldehyde are reaction intermediates. Molecular oxygen is the oxidizing agent, and the oxygen is activated by binding to the iron-containing enzyme cytochrome P-450. The reaction follows the following sequence:

$$RCH_2CH_2CH_3 \xrightarrow{[O]} RCH_2CH_2CH_2OH \xrightarrow{[O]} RCH_2CH_2CO_2H \xrightarrow{[O]}$$
$$R_2CO_2H + CH_3CO_2H \qquad (7\text{-}20)$$

As discussed in Section 6.4, these reactions are carried out in microorganisms, but are slow, especially with high molecular weight materials that do not readily pass through the microbial cell wall.

The microbial oxidation of the cyclic hydrocarbons and aromatic compounds present in detergents and petroleum is also catalyzed by the enzyme cytochrome P-450 in microorganisms. It has been established in some cases that arene oxide intermediates are formed, as in the following reaction sequence for the oxidation of benzene.

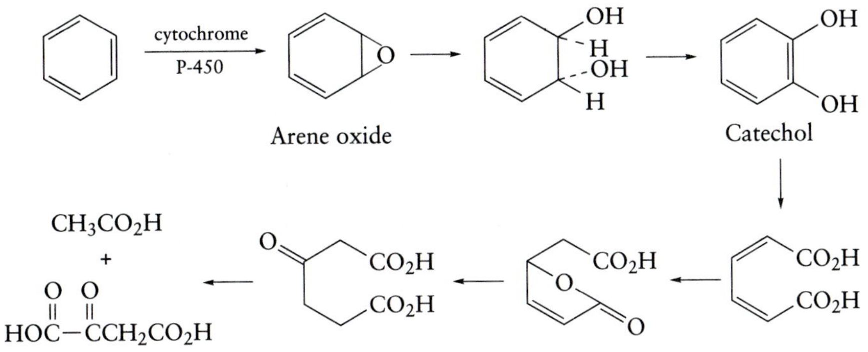

Arene oxide   Catechol

$$(7\text{-}21)$$

The microbial oxidation of the more highly condensed aromatics such as naphthalene or of the alkyl-substituted benzenes proceeds via salicylic acid, which is then oxidized to catechol in accordance with reaction (7-22) and finally to oxaloacetate.

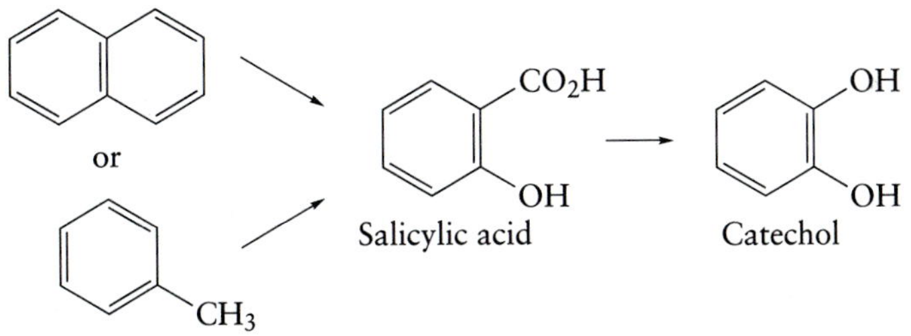

Salicylic acid   Catechol

$$(7\text{-}22)$$

The sulfate esters in alcohol sulfates are readily hydrolyzed by microbial sulfatase enzymes. Sulfonates present in the surfactants are subject to rapid oxidative elimination by a variety of aquatic bacteria:

$$RCH_2SO_3^- \xrightarrow{[O]} R\overset{\overset{\displaystyle OH}{|}}{C}HSO_3^- \longrightarrow RCHO + HSO_3^- \qquad (7\text{-}23)$$

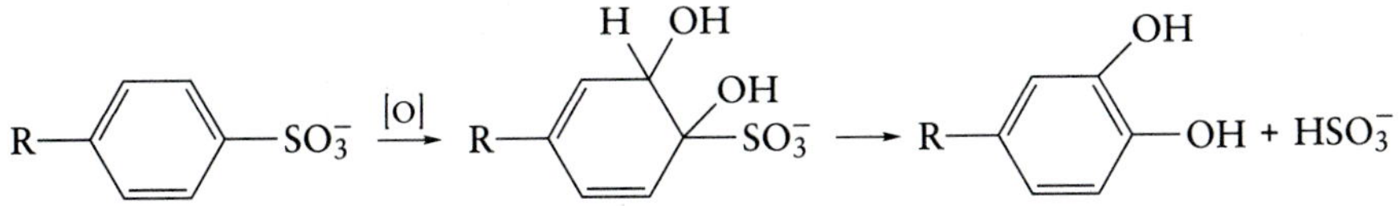

$$(7\text{-}24)$$

Therefore, these sulfonate groups do not slow the environmental degradation of the alcohol sulfates. The bisulfite formed is rapidly oxidized to bisulfate by microorganisms.

7.2.5 Environmental Effects of Biodegradable Organic Compounds: Biological Oxygen Demand

The presence of biodegradable organic compounds in lakes and rivers can result in an environmental problem. Oxygen is required for the microbial degradation of these compounds, and relatively small quantities of organic material can deplete the supply of dissolved oxygen. The amount of oxygen consumed in the microbial degradation of organic compounds can be estimated if the crude assumptions are made that the organic compounds are pure carbon and that they are all converted to carbon dioxide.

$$C + O_2 \rightarrow CO_2 \qquad (7\text{-}25)$$

From reaction (7-25) and the ratio of the molecular weight of oxygen to the atomic weight of carbon, it can be seen that 32/12, or about 2.7 g of oxygen, is required to oxidize 1 g of carbon. Since water can dissolve only about 10 ppm of oxygen, this means that in the absence of further dissolution of atmospheric oxygen, only 4 ppm of carbon compounds can deplete all the dissolved oxygen in a body of water. This calculation is an oversimplification because microbial oxidation does not proceed instantly and the oxygen is always being replenished. However, the oxygen near the bottom of deep lakes can be depleted by dissolved organics because of the slow rate of oxygen transport to these low levels (Chapter 9). Fish have difficulty breathing when the oxygen level drops to 5 ppm (one-half the maximum level).

Biological oxygen demand (BOD), a measure of the biodegradable organic material in water, is determined by the amount of oxygen required for the microbial oxidation of the dissolved organic content of a water sample. The analysis measures the amount of dissolved oxygen lost from a water sample kept in a sealed container at 2°C over a 5-day period. Chemical oxygen demand (COD) is another measure of dissolved organic compounds. It is determined by measuring the equivalents of acidic permanganate or dichromate necessary for the oxidation of organic constituents. Neither BOD nor COD measures the total oxidizable carbon (TOC) content. Total oxidizable carbon is measured by the amount of CO_2 formed upon combustion after the water and the carbonate in the sample have been removed. Biodegradable and chemically degradable compounds reduce the oxygen level in lakes and rivers. However, the major sources of dissolved organics in lakes and rivers are the effluent from sewage disposal plants, manure from animal feed lots, industrial wastes, and decomposing plants and algae.

7.3 POLYMERS

Polymers are macromolecules, that is, molecules containing many atoms that are of very large size and high molecular weight (ranging from 10 KDa to over 100 kDa). They are made up of a small number of repeating groups. Many macromolecules are components of living systems; examples are proteins, which are polymers of amino acids, and starch and cellulose, which are polymers of cyclic polyhydroxy compounds (sugars). Since biological polymers are built up in living systems, they can typically be broken down by these systems (i.e., they are biodegradable). Biodegradable substances will not cause long-term environmental problems, although the decomposition of large amounts of such materials can be a problem in the short term through their effect on the BOD. In aqueous solution, their degradation will contribute to the biological oxygen demand and use up the dissolved oxygen (see Section 7.2.5).

The latex obtained from the sap of rubber trees is apparently an exception to the rule that naturally occurring polymers are biodegradable. The latex is modified by cross-linking and other processes during its formulation into tires, gloves, and other products. These chemical processes increase the strength and durability of the material and also its resistance to biodegradation.

Synthetic polymers are made up of a great number of simple units (monomers) joined together in a regular fashion. Some examples are given in Figures 7-3 and 7-4. Many plastic items are not formulated from the polymeric material alone. They may contain lower molecular weight compounds called plasticizers that increase their flow and processibility. Polychlorinated biphenyls (PCBs) were formerly used for this purpose, as mentioned in Chapter 8. Esters of phthalic acid are other examples of plasticizers. Fillers, such as wood flour, cellulose, starch, ground mica, asbestos, or glass fiber, are often added to increase the strength of the plastic or to provide bulk at low cost. Heavy metal compounds, including those of tin, cadmium and lead (Sections 10.6.7 and 10.6.10), may be present as plasticizers or to confer resistance to degradation. Plasticizers may be leached or evaporated from polymeric materials in use, while both plasticizers and fillers may contribute to problems of environmental disposal and degradation. The release of PCBs upon incineration of chlorinated polymers is referred to in Chapter 8. Fillers may be inert, but if released as dust during incineration of wastes, they can contribute to atmospheric particulate material. Asbestos particles are health hazards, in view of the known connection between asbestos dust and lung cancer.

In August 2000 an independent panel appointed by the U.S. National Toxicology Program reported "serious concern" that when vinyl polymers are used, the "exposure to intensive medical procedures may have adverse effects on the reproductive tract of male infants." The basis for this concern is that di(2-ethylhexyl)phthalate (DEHP) may be leached from vinyl tubing by

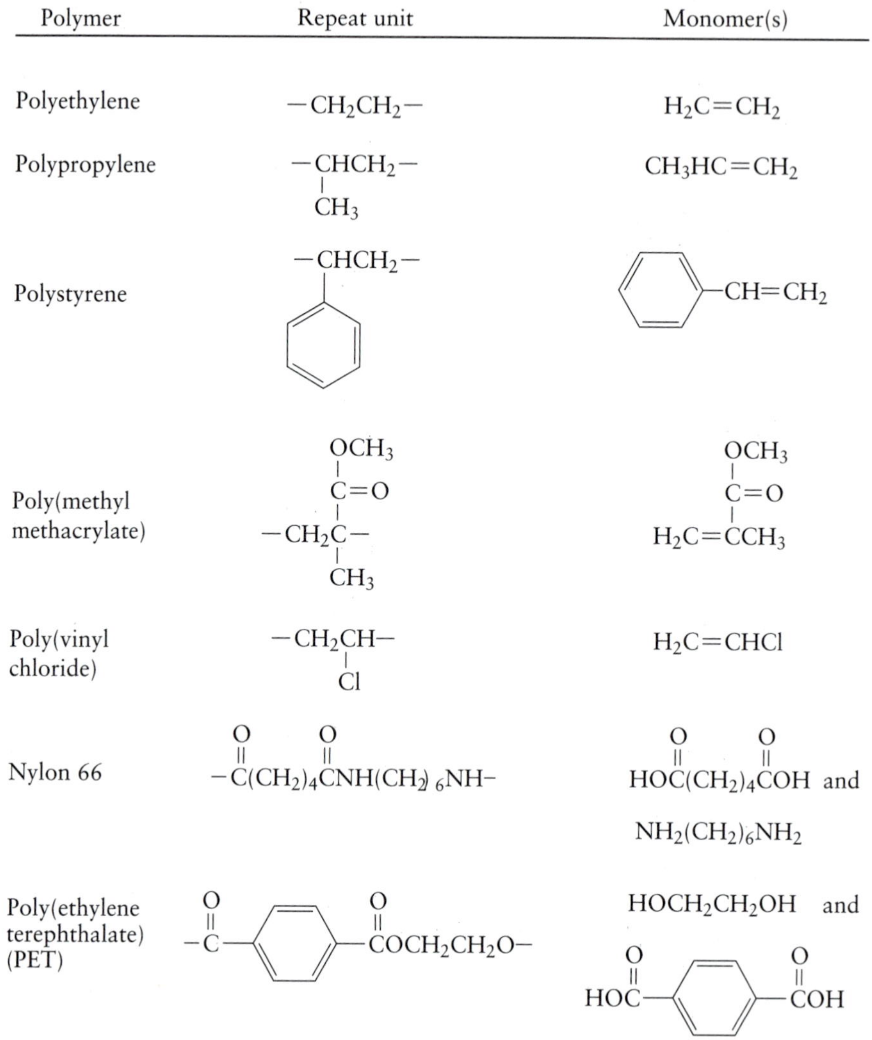

Polymer	Repeat unit	Monomer(s)
Polyethylene	$-CH_2CH_2-$	$H_2C{=}CH_2$
Polypropylene	$-CHCH_2-$ $\quad$	$\quad$ CH$_3$	$CH_3HC{=}CH_2$
Polystyrene	$-CHCH_2-$ (phenyl)	(phenyl)$-CH{=}CH_2$
Poly(methyl methacrylate)	$-CH_2C-$ with OCH$_3$, C=O, CH$_3$	$H_2C{=}CCH_3$ with OCH$_3$, C=O
Poly(vinyl chloride)	$-CH_2CH-$ $\qquad$Cl	$H_2C{=}CHCl$
Nylon 66	$-\overset{O}{\overset{\|}{C}}(CH_2)_4\overset{O}{\overset{\|}{C}}NH(CH_2)_6NH-$	$HO\overset{O}{\overset{\|}{C}}(CH_2)_4\overset{O}{\overset{\|}{C}}OH$ and $NH_2(CH_2)_6NH_2$
Poly(ethylene terephthalate) (PET)	$-\overset{O}{\overset{\|}{C}}-\text{(benzene ring)}-\overset{O}{\overset{\|}{C}}OCH_2CH_2O-$	$HOCH_2CH_2OH$ and $HO\overset{O}{\overset{\|}{C}}-\text{(benzene ring)}-\overset{O}{\overset{\|}{C}}OH$

FIGURE 7-3 Some typical nonbiodegradable polymers and the monomers from which they are formed.

fluid running through it to an infant with a medical problem. The "serious concern" was directed to the use of vinyl polymers that contained DEHP (Figure 7-5) as the plasticizer. Six other commonly used phthalate ester plasticizers were rated as generating only "minimal" or "negligible" concern. The panel also raised the "concern" that the "exposures of pregnant women to

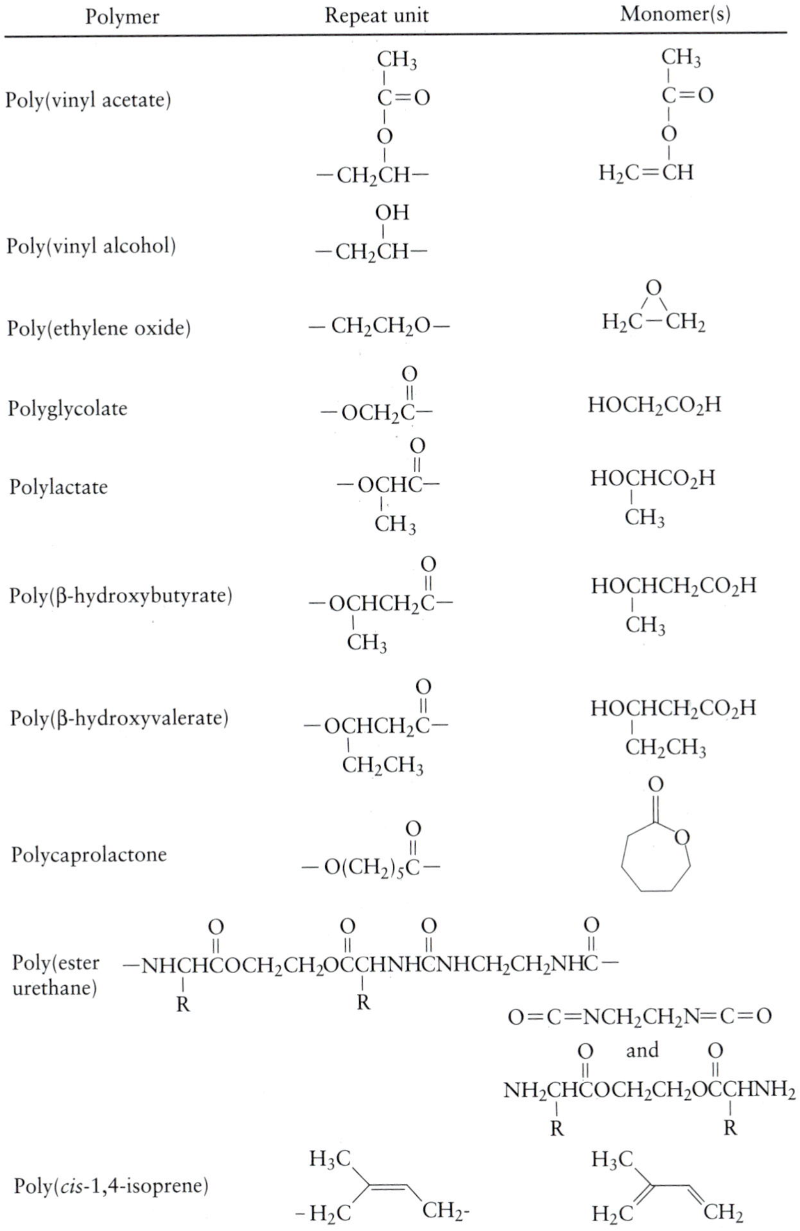

Polymer	Repeat unit	Monomer(s)
Poly(vinyl acetate)	$\begin{array}{c}\mathrm{CH_3}\\|\\\mathrm{C{=}O}\\|\\\mathrm{O}\\|\\\mathrm{-CH_2CH-}\end{array}$	$\begin{array}{c}\mathrm{CH_3}\\|\\\mathrm{C{=}O}\\|\\\mathrm{O}\\|\\\mathrm{H_2C{=}CH}\end{array}$
Poly(vinyl alcohol)	$\begin{array}{c}\mathrm{OH}\\|\\\mathrm{-CH_2CH-}\end{array}$	
Poly(ethylene oxide)	$\mathrm{-CH_2CH_2O-}$	$\mathrm{H_2C{-}CH_2}$ (epoxide)
Polyglycolate	$\mathrm{-OCH_2\overset{O}{\overset{\|}{C}}-}$	$\mathrm{HOCH_2CO_2H}$
Polylactate	$\mathrm{-OCH\overset{O}{\overset{\|}{C}}-}$ $\;\mathrm{CH_3}$	$\mathrm{HOCHCO_2H}$ $\;\mathrm{CH_3}$
Poly(β-hydroxybutyrate)	$\mathrm{-OCHCH_2\overset{O}{\overset{\|}{C}}-}$ $\;\mathrm{CH_3}$	$\mathrm{HOCHCH_2CO_2H}$ $\;\mathrm{CH_3}$
Poly(β-hydroxyvalerate)	$\mathrm{-OCHCH_2\overset{O}{\overset{\|}{C}}-}$ $\;\mathrm{CH_2CH_3}$	$\mathrm{HOCHCH_2CO_2H}$ $\;\mathrm{CH_2CH_3}$
Polycaprolactone	$\mathrm{-O(CH_2)_5\overset{O}{\overset{\|}{C}}-}$	(caprolactone ring)
Poly(ester urethane)	$\mathrm{-NHCHCOCH_2CH_2OCCHNHCNHCH_2CH_2NHC-}$ with O (×4) and R groups	$\mathrm{O{=}C{=}NCH_2CH_2N{=}C{=}O}$ and $\mathrm{NH_2CHCOCH_2CH_2OCCHNH_2}$ with R groups
Poly(cis-1,4-isoprene)	$\begin{array}{c}\mathrm{H_3C}\\\diagdown\\\mathrm{-H_2C}\diagup{=}\diagdown\mathrm{CH_2-}\end{array}$	$\begin{array}{c}\mathrm{H_3C}\\\diagdown\\\mathrm{H_2C}\diagup{=}\diagdown\mathrm{CH_2}\end{array}$

FIGURE 7-4 Some typical polymers that are biodegradable under aerobic conditions only and the monomers from which they are formed. Note that a poly(vinyl alcohol) monomer is not formed by the polymerization of the vinyl alcohol but rather by hydrolysis of poly(vinyl acetate).

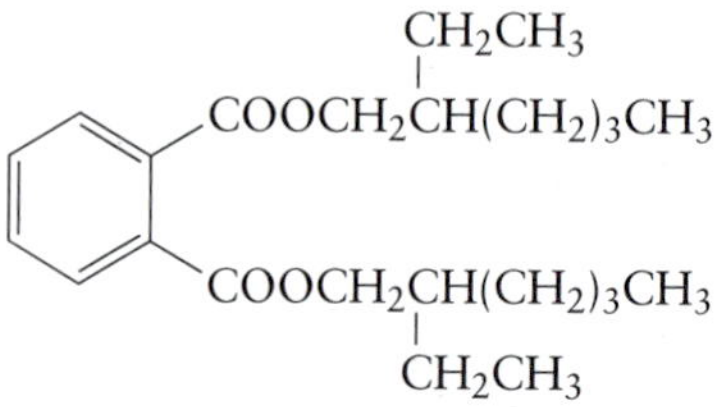

$$\text{[benzene ring]} \begin{matrix} \overset{\displaystyle CH_2CH_3}{\underset{|}{COOCH_2CH(CH_2)_3CH_3}} \\ \\ \underset{\displaystyle CH_2CH_3}{\overset{|}{COOCH_2CH(CH_2)_3CH_3}} \end{matrix}$$

FIGURE 7-5 The structure of the phthalic acid plasticizer di(2-ethylhexyl)phthalate.

DEHP might affect development of their offspring." The concern arose from rodent studies in which juvenile rats that ingested DEHP had lower testicular weight, testicle degeneration, and reduced sperm counts in comparison to controls. From these results, the possibility was raised that DEHP might be leached from vinyl surgical tubing by fluid running through it to infants with serious medical problems. In addition the report led many toy manufacturers to stop making toys with plastics that contained phthalate plasticizers. While there are few expensive or non toxic plasticizers that can replace phthalates in vinyl polymers, there are plastics that can replace vinyl in medical devices. DEHP is also a potential carcinogen.

Phthalates are abundant in the environment because they are used extensively (10×10^9 lbs produced worldwide in 1999) and because they are degraded at a moderate rate by microorganisms that use a variation on the oxidative pathway illustrated in reaction (7-24).

The bulk of the monomers used in polymer synthesis are synthesized directly or indirectly from petroleum. Therefore, our present dependence on plastic materials is ultimately a dependence on sources of crude oil.

7.3.1 Polymer Synthesis

Polymers may be formed from monomers in chain reaction or step reaction polymerization. The chain reaction process is illustrated by the free-radical polymerization of ethylene or its derivatives. The specific example of the polymerization of acrylonitrile to Orlon is given in reactions (7-26)–(7-30).

$$ROOR \longrightarrow 2RO\cdot \qquad\qquad (7\text{-}26)$$

$$RO\cdot \; + \; H_2C{=}\underset{\displaystyle CN}{\overset{|}{CH}} \longrightarrow ROCH_2\underset{\displaystyle CN}{\overset{|}{CH}}\cdot \qquad (7\text{-}27)$$

$$\underset{\underset{CN}{|}}{ROCH_2CH\cdot} + \underset{\underset{CN}{|}}{H_2C=CH} \longrightarrow \underset{\underset{CN}{|}\quad\underset{CN}{|}}{ROCH_2CHCH_2CH\cdot} \qquad (7\text{-}28)$$

$$2\,\underset{\underset{CN}{|}\quad\underset{CN}{|}\quad\underset{CN}{|}}{ROCH_2CH(CH_2CH)_nCH_2CH\cdot} \xrightarrow{\text{combination}}$$

$$\underset{\underset{CN}{|}\quad\underset{CN}{|}\quad\underset{CN}{|}\;\underset{CN}{|}\quad\underset{CN}{|}\quad\underset{CN}{|}}{ROCH_2CH(CH_2CH)_nCH_2CH-CHCH_2(CHCH_2)_nCHCH_2OR} \qquad (7\text{-}29)$$

$$2\,\underset{\underset{CN}{|}\quad\underset{CN}{|}\quad\underset{CN}{|}}{ROCH_2CH(CH_2CH)_nCH_2CH\cdot} \xrightarrow{\text{disproportionation}}$$

$$\underset{\underset{CN}{|}\quad\underset{CN}{|}\quad\underset{CN}{|}}{ROCH_2CH(CH_2CH)_nCH=CH} + \underset{\underset{CN}{|}\quad\underset{CN}{|}\quad\underset{CN}{|}}{CH_2CH_2(CHCH_2)_nCHCH_2OR} \qquad (7\text{-}30)$$

In free-radical polymerizations, a small amount of initiator, such as a peroxide, is added to the monomer to start the polymerization. The initiator decomposes to form radicals (7-26), which add to the monomer in the chain-initiating step (7-27). Further addition of monomer takes place in the chain-propagating step (7-28). Polymer growth stops in the chain-terminating steps, which require the reaction of two radicals such as the combination and disproportionation reactions shown in (7-29) and (7-30).

Ionic processes may also be involved in chain reaction polymerization. Cationic polymerization of vinyl monomers is initiated by acids. The synthesis of propene oligomers from propene by this process has already been discussed (Section 7.2.3.2). Anionic polymerization is often initiated by strong bases such as butyllithium ($CH_3CH_2CH_2CH_2^-\,Li^+$). In these chain reaction polymerization processes, the molecular weight of the final polymer is dependent on the maintenance of the active species, be it the radical, cation, or anion formed initially. In step reaction polymerization, monomers and oligomers react to form polymers. The formation of a nylon illustrates this process:

$$\underset{}{\overset{O}{\overset{\|}{}}\quad\overset{O}{\overset{\|}{}}} \\ HOC(CH_2)_4COH + NH_2(CH_2)_6NH_2 \xrightarrow{-H_2O} \overset{O\quad\;\;O}{\overset{\|\quad\;\;\|}{HOC(CH_2)_4CNH(CH_2)_6NH_2}}$$

$$\xrightarrow[-H_2O]{NH_2(CH_2)_6NH} NH_2(CH_2)_6\overset{O}{\overset{\|}{NHC}}(CH_2)_4\overset{O}{\overset{\|}{CNH}}(CH_2)_6NH_2 \qquad (7\text{-}31)$$

$$\xrightarrow[\text{between oligomers already formed}]{\text{continued addition of monomers and reaction}} \underset{\text{a Nylon}}{\left[\overset{O}{\overset{\|}{C}}(CH_2)_4\overset{O}{\overset{\|}{C}}NH(CH_2)_6NH\right]_n}$$

The polyester poly(ethylene terephthalate) (PET), (Dacron or Terylene) is prepared in a similar fashion from ethylene glycol and terephthalic acid (reaction 7-32). These step reaction polymerizations are examples of condensation polymerizations. A small molecule, water here, is eliminated in each step. Many of the biodegradable polymers are aliphatic polyesters prepared by step reaction condensation reactions as shown earlier (Figure 7-4).

$$HOCH_2CH_2OH + HO_2C-\langle\!\!\!\!\bigcirc\!\!\!\!\rangle-CO_2H \longrightarrow$$

$$\left[OCH_2CH_2O\overset{\displaystyle O}{\overset{\|}{C}}-\langle\!\!\!\!\bigcirc\!\!\!\!\rangle-\overset{\displaystyle O}{\overset{\|}{C}} \right]_n \quad (7\text{-}32)$$

PET

7.3.2 Paper

Paper is composed mainly of cellulose, a polymer made up of β-glucose units linked through oxygen atoms. Paper is generally produced by the chemical removal of lignin, resins, and other components of wood to leave cellulose fibers (pulp), which is compacted to produce the paper sheet. Mechanical grinding can also be used to facilitate the removal of the lignin, but this yields a lower quality paper. The kraft pulping process hydrolyzes the lignin with a sodium hydroxide–sodium sulfide solution, producing organic sulfur compounds and other organic material in what is called black liquor. This can be a considerable source of pollution, but most of it is treated and recycled for reuse of the reactants. The sulfite process, used for high-quality papers, uses calcium and magnesium sulfites produced from SO_2 and the corresponding carbonate.

The pulp, especially from the kraft process, contains oxidized organic compounds that produce a brown color and must be bleached. Traditionally, this has required chlorine to produce the actual bleaching agent employed, chlorine dioxide (Section 11.5.1). Paper manufacturing thus has been a heavy user of chlorine and sodium hydroxide, and paper plants have often operated their own chloralkali plants with their associated environmental problems (Section 11.5.1). Recent practice is to replace the chlorine-based bleach with an alternative process that employs oxygen as the bleaching agent,

Most paper products contain various additives—fillers such as clay or titanium dioxide, resins to improve wet strength, sizings, coloring, and so on. These differ depending on the type of paper and add to the complications of paper recycling (Section 16.7.2).

7.3.3 The Fate of Polymers after Use

About 108 million tons of plastics was produced in the world in 1996, with 24% of the production in the United States, 25% in western Europe, 9% in Japan, and 25% in the rest of Asia. In the same year western Europe discarded 19.3 million tons of plastic waste; 10% of this waste was recycled, 14% was used for energy, and the rest was placed in landfills. In 1990 25 million tons of plastics was manufactured in the United States, while during the same time period 16 million tons was placed in landfills (this does not include about 1.8 million tons of tires also placed in landfills). Only about 0.4 million tons of the plastic was reused or recycled. Plastic material constitutes about 10% by weight and 20% by volume of the material in landfills, with paper being the principal constituent at 32% by weight and by volume. As discussed in Section 16.2, relatively little decomposition takes place in landfills, so very little of this plastic material disappears with time.

The plastic that is not placed in landfills, recycled, used for energy, or incinerated, is scattered either on land or in the sea. These materials, which have been designed to be strong and stable, persist for a long time. For example, it is estimated that the six-pack plastic strap made of high-density polyethylene has a lifetime of 450 years in the sea. It is estimated that 6 million tons of garbage is dropped in the oceans each year by boats. The amount of plastic material in this garbage has not been estimated, but it includes commercial fishing nets, fishing line, ropes, bags, lids, six-pack rings, disposable diapers, gloves, tampon applicators, bottles, plastic foam, and pellets to name a few.

The principal concern associated with plastic materials scattered on land is one of aesthetics, not toxicity or harm to ecosystems. Plastic articles dumped in oceans are not aesthetically pleasing when they wash up on beaches, but of even greater importance is their danger to marine life. It has been estimated that plastics kill or injure tens of thousands sea birds, seals, sea lions, and sea otters each year, and hundreds of whales, porpoises, bottlenose dolphins, and sea turtles. A large portion of the reported fatalities are due to the entanglement of birds and mammals in the driftnets of commercial fishing boats. These nets, up to 30 km long, entangle seals, bottlenose dolphins, porpoises, and birds. For example, the 4–6% annual decline in the population of the northern fur seal (20,000–40,000 seals), which live in the Bering Sea off Alaska, is attributed to entanglement in drift nets. Most of these drift nets are in use and cannot be classified as a disposal problem, but free-floating drift nets, which have broken loose or have been discarded and continue to trap sea life, are an environmental problem. The most dramatic victims are whales that are unable to dive and feed because they are entangled in large piece of drift net. Dying, beached whales wrapped in drift nets have been found on both the east and west coasts of the United States.

Other plastic items also endanger sea life. For example, seal pups play with plastic items floating in the sea. If pups happen to stick their heads through plastic collars, as they grow the plastic will sever their neck arteries or strangle them. Of perhaps even greater danger is the ingestion of plastic items by some marine life. A floating plastic bag can look like a tasty jellyfish to a sea turtle. When turtles consume a lot of plastic it blocks their intestines, preventing them from assimilating food. Sea birds, which often mistake plastic pellets for fish eggs or other marine food, also suffer from intestinal blockage if they consume large quantities of plastic. The use of biodegradable polymers will help to reduce the level of plastic in the ocean, but extensive research will be necessary to develop polymers that not only are biodegradable but also have the strength and other desirable properties of the nonbiodegradable polymers in use today. Moreover, after such materials have been developed, the consumer will have to be convinced to purchase biodegradable products even though they will slowly degrade and therefore may have a shorter useful life than the corresponding nonbiodegradable product.

7.3.3.1 Biodegradable Polymers

A goal of the plastics industry is to design biodegradable polymers that can be substituted for materials that are difficult to collect and reuse. These include water-soluble polymers, polymers designed for use for fishing or other marine applications that end up in the marine environment, and polymers employed with other materials such as coatings on cups, diapers, and sanitary products. Polymers that tend to be used only once such as six-pack plastic straps, plastic cutlery, and hospital waste should also be degradable. Microorganisms use degradable polymers for their growth, and thus the polymer will be converted to compounds that can be utilized in the natural cycle of life on earth. As noted already, these polymers will break down only where there is abundant molecular oxygen, so the designation "biodegradable" does not apply to the anaerobic environment in a landfill.

7.3.4 Environmental Degradation of Polymers

7.3.4.1 Nonbiodegradable Polymers

Many hydrocarbon polymers have chemical reactivity similar to that of high boiling petroleum fractions and, as a consequence, these compounds are very persistent in the environment (Section 7.2.4). Until recently, the goal of the polymer chemist was to design a polymer that degrades very slowly in the environment. Each new polymer was tested extensively to be sure that it did not break down rapidly in its particular use. For example, a material designed for use in vinyl siding for homes would be tested for its resistance to degradation by oxygen, water, and sunlight. Some products made from vinyl polymers appear to degrade in the environment, but this is mainly due to the loss of their phthalate ester plasticizer, which results in a change of the mechanical properties of the polymer. They vinyl products become brittle and crack, but the actual degradation of the the polymer proceeds more slowly.

7.3.4.2 Photooxidation

Polymers will be degraded if they absorb sunlight at wavelengths greater than that of the radiation that is not absorbed by stratospheric ozone (290 nm: see Section 5.2.3). Since many polymers do not have functionality that results in light absorption at wavelengths greater than 290 nm, direct photolysis is not usually a major degradative pathway. Indirect photochemical processes, such as oxidation by photochemically generated singlet oxygen (Sections 5.2.2 and 6.4) can also result in polymer degradation.

The peroxides formed by polymer processing may trigger photochemical polymer degradation. Mechanical shearing of polymers during processing results in radical formation as a result of chain breaking [reaction (7-33)], and these radicals react with molecular oxygen to form hydroperoxides, as shown in reaction (7-34) and (7-35).

$$R - R \xrightarrow{\text{mechanical shearing}} 2R^{\cdot} \qquad (7\text{-}33)$$

$$R^{\cdot} + O_2 \rightarrow ROO^{\cdot} \qquad (7\text{-}34)$$

$$ROO^{\cdot} + RH \rightarrow ROOH + R^{\cdot} \qquad (7\text{-}35)$$

Hydroperoxides absorb light at wavelengths greater than 290 nm and are cleaved to hydroxyl radicals and alkoxy radicals with a quantum yield close to one, according to reaction (7-36).

$$ROOH \xrightarrow{h\nu} RO^{\cdot} + {\cdot}OH \qquad (7\text{-}36)$$

The radicals react further as follows to form alcohols:

$$RO^{\cdot} + RH \rightarrow ROH + R^{\cdot} \qquad (7\text{-}37)$$

Alternatively, hydroperoxides cleave to aldehydes or ketones [reaction (7-38)], and these undergo further photochemical chain scission as outlined next in Section 7.3.4.3.

$$ROOH \xrightarrow{h\nu} R'(CO)R'', R'''(CO)CH_3 \qquad (7\text{-}38)$$

The same radical processes may be triggered by impurities in the polymers that form peroxides on irradiation in the presence of oxygen. For example, iron(III) and titanium(IV) present in a polymer initiate peroxide formation from molecular oxygen. Exposure to the ozone, formed as a result of air pollution (Section 5.3), also leads to peroxide formation. Antioxidants, which react with free radicals, are added to many polymers to slow their rate of degradation.

7.3.4.3 Photodecomposition Triggered by Carbonyl Groups

Carbonyl groups that are present in polymers as a result of the decomposition of peroxides or are incorporated into the polymer during synthesis absorb UV light at 300–325 nm and initiate reactions leading to the cleavage of the polymer backbone. Two processes can occur, called Norrish type I and Norrish type II. In the type I reaction, bond breaking proceeds by a radical pathway (7-39), while a concerted electron shift occurs in the type II reaction (7-40):

$$
\underset{R-C-R'}{\overset{O}{\parallel}} \xrightarrow{h\nu} R\cdot \ + \ \cdot\underset{C-R'}{\overset{O}{\parallel}} \quad \text{and} \quad R-\underset{C\cdot}{\overset{O}{\parallel}} \ + \ \cdot R' \tag{7-39}
$$

$$
CO + \cdot R' \qquad R\cdot + CO
$$

$$
\begin{array}{c} O \ \ H \\ R-C \quad CHR \\ H_2C-CH_2 \end{array} \xrightarrow{h\nu} \left[\begin{array}{c} O \ \ H \\ R-C \quad CHR \\ H_2C-CH_2 \end{array} \right] \longrightarrow R-C \overset{OH}{\underset{CH_2}{\diagdown}} \ + \ \underset{CH_2}{\overset{CHR}{\parallel}} \tag{7-40}
$$

Many polymers contain small amounts of carbonyl groups that are formed by the decomposition of peroxides, which initiate the slow decomposition of the polymer. For example, unstabilized 2.5-mil-thick films of polyethylene fragment in less than 90 days in the summer Texas sun. It should be noted that this polyethylene is only "fragmented," not totally broken down. Polymers have been prepared in which carbonyl groups were purposely incorporated into the chain to facilitate their extensive photochemical degradation. Polystyrene cups in which 1% of the styrene units also contained a ketone grouping broke down to a wettable powder after standing outside in the sun for 3 weeks. Currently a copolymer of ethylene and carbon monoxide is being used to make the plastic strap that holds beverage cans in a six-pack because it undergoes photochemical decomposition in sunlight. This addresses the problem of wildlife getting caught in this strap as well as its general lack of environmental aesthetics.

7.3.4.4 Hydrolysis

Polyesters, polyamides, and polyurethanes undergo slow hydrolytic degradation to form the corresponding acid and amine or alcohol. This random process does not usually result in the loss of the strength of the polymer.

However, the rate of cleavage increases with time because the carboxyl and amine or alcohol groups that are formed serve as catalysts to cleave the ester or amide groupings tethered in their vicinity. These catalyze chain scission and loss in the strength of the polymer. The extent of chemical cleavage is also dependent on the physical properties of the polymer. Polymers with a high degree of crystallinity are less susceptible to cleavage than amorphous ones. The crystalline phase is less accessible to penetration by water, so there is less chance that hydrolysis will occur. Biodegradation, to be discussed next, often involves an enzymatic acceleration of the hydrolytic process.

7.3.5 Biodegradation

Since biodegradation requires metabolism by microorganisms present in the environment, most of the biodegradable polymers contain functional groups that are subject to attack by microbial enzymes. This means that these polymers contain ester or amide groups that can be hydrolyzed or linear chains that can be oxidatively cleaved by the process described for linear surfactants (Section 7.2.4). A major problem is the inability of microorganisms to ingest high polymers through their cell walls; it is generally not possible for a microorganism to ingest a molecule that has a molecular weight greater than about 500 Da. Some microorganisms that degrade polyesters secrete esterases, enzymes that catalyze the hydrolysis of ester groups, which break the polymer into smaller fragments that can then be ingested. As noted earlier, polyesters also undergo slow hydrolysis at the ester bond in the environment, which will provide smaller, ingestible fragments. Other processes of chemical degradation may also provide fragments small enough to be biodegraded.

Synthetic polymers in commercial use that undergo biodegradation in the presence of molecular oxygen were listed in Figure 7-4. These constituted about 2% of the total U.S. plastics production in 1992 and are mainly polyesters. Poly(ethylene glycols) are prepared from ethylene oxide by a process similar to that outlined for nonylphenol ethoxylates (Section 7.2.3.3). Poly (ethylene terephthalate) (Figure 7-3), is a polyester manufactured in large amounts that does not undergo biodegradation, apparently because its aromatic rings do not bind in the catalytic sites of the microbial esterases.

Poly(β-hydroxybutyrate) (PHB) and the copolymer of β-hydroxybutyrate and β-hydroxyvalerate (PHBV) (Figure 7-4) are unique in that they are synthesized by microorganisms. These polymers are prepared and stored in the cell in granules. When grown under conditions where nitrogen is a limiting nutrient, the microorganisms produce 30–80% of their cell weight as granules that contain mainly PHB. The genes for the enzymes required to produce PHB have been cloned and expressed in *E. coli* and in a plant.

PHB is a brittle polymer that may be used in plastic soft drink bottles but is difficult to mold because its melting point and decomposition point are almost the same. It has been possible to induce the microorganisms to prepare the copolymer PHVB by adding some β-hydroxyvalerate to their growth medium. The PHVB formed is more malleable than PHB and has physical properties similar to those of polypropylene. Now manufactured in Great Britain, it is used for making products that are marketed on the basis of their "environmentally friendly" or "green" image (e.g., shampoos, razors, writing pens). In 1999 it was reported that a genetically engineered plant (oilseed rape) produced PHVB directly. This process is not economically feasible at present, but it does suggest the possibility of the future preparation of polymers using renewable resources.

A polyethylene containing 5–20% starch granules is used in the manufacture of shopping bags that are claimed to be biodegradable. The starch granules are degraded by fungi and bacteria. Biochemical degradation of the starch granules from the plastic film enhances its permeability to other reactants. Other additives such as transition metals may also be added to the polyethylene to enhance the photodegradation of the polyethylene via hydroperoxides (Section 7.3.4.2). It is not clear whether all the polyethylene is eventually completely degraded to biomolecules like acetate (Section 7.2.4).

7.4 CONCLUSIONS

The environmental problems associated with nonbiodegradable surfactants have been solved with the design of biodegradable replacements. A potential problem with biodegradable surfactants is the eutrophication of lakes due to the buildup of organics resulting from the microbial degradation process. Since, however, the principal cause of most lake eutrophication is runoff of organic waste (manure) and fertilizer from farms, the role of surfactants in this problem appeare to be minor.

Polymers derived from petroleum will continue to predominate into the twenty-first century. It is likely that new technology will be developed to recycle a greater proportion of the materials. The extent of the recycling may depend on whether "cradle to grave" responsibility is mandated for the company that initially synthesizes the polymer. Not all synthetic polymers will be used in a way that permits efficient recycling. Research in progress suggests that it may be possible to devise biodegradable polymers for these uses as long as they can be subjected to decomposition in an aerobic environment such as that present in composting. Although progress has been made in the synthesis of biodegradable polymers, these materials cost more and, in general, their properties are not as useful as those prepared from petroleum feedstocks.

This is a rapidly developing area, so it appears likely that biodegradable polymers will assume an increasingly larger percentage of the polymer market.

Additional Reading

Soaps, Detergents, and Synthetic Surfactants

Ainsworth, Susan J. Soaps and detergents, *Chem. Eng. News*, p. 34, Jan. 24, 1994. (A summary of industrial trends in the soap and detergents is published annually in the January issue of *Chemical & Engineering News*.)

Davidsohn, A. S., and B. Milwidsky, *Synthetic Detergents*, 7th ed. Longman Scientific and Technical, New York, 1987.

Degradation of Synthetic Organic Molecules in the Biosphere. National Academy of Sciences, Washington, DC, 1972.

Meloan, C. E., Detergents, soaps, and syndets, *Chemistry* 49(7), 6 (1976).

Swisher, R. D., *Surfactant Biodegradation*, 2nd ed., revised and expanded. Dekker, New York, 1987.

Polymers

Aguado, J., and D. Serrano, *Feedstock Recycling of Plastic Wastes*. Royal Society of Chemistry, Cambridge, U.K. 1999 192 pp.

Alexander, Martin, *Biodegradation and Bioremediation*. Academic Press, San Diego, 1994, pp. 278–280.

Ching, Chauncey, David L. Kaplan, and Edwin L. Thomas, eds., *Biodegradable Polymers and Packaging*, Technomic, Lancaster, PA, 1993.

Dawes, Edwin A., ed., *Novel Biodegradable Microbial Polymers*. Kluwer Academic, Dordrecht, 1990.

Hamid, Halim, S. Mohamed B. Amin, and Ali G, Maadhah, eds., *Handbook of Polymer Degradation*. Dekker, New York, 1992.

Satyanarayana, D., and P. R. Chatterji, *Biodegradable polymers: Challenges and strategies*, J. Macromol. Sci. Rev. Macromol. Chem. Phys., **C33**, 349–368 (1993).

Vert, M., J. Feijen, A. Albertsson, G. Scott, and E. Chiellini, eds., *Biodegradable Polymers and Plastics*. Royal Society of Chemistry, Cambridge, U.K., 1992.

EXERCISES

7.1. (a) What structural units are cleaved from a soap when it decays in a sewage disposal plant or in the environment? Illustrate by equations the cleavage of one unit from the salt of octadecanoic acid $[CH_3(CH_2)_{16}COOH]$.

(b) Identify the agent(s) in a sewage plant or environment responsible for this degradation.

      (c) Why were some surfactants slow to break down in a sewage plant and in the environment?

7.2. (a) Surfactants and some hydrocarbons are metabolized by certain microorganisms in the environment. Write one or more reaction(s) to illustrated the overall process by which one fragment is cleaved by these microorganisms from hydrocarbon with the formula $CH_3(CH_2)_8CH_3$ and from surfactant of general formula $CH_3(CH_2)_{16}CH_2OSO_3^{2-}$

    (b) Explain with words and/or equations why the following compound is resistant to microbial degradation. Be specific in your answer.

$$CH_3CHCH_2CHCH_2CHCH_3$$
$$|||$$
$$CH_3CH_3CH_3$$

7.3. Outline the steps in the free-radical polymerization of styrene.

7.4. (a) Which of each of the following pairs of compounds degrades more rapidly in the environment? $CH_3(CH_2)_8CH_3$ and $CH_3(CH_2)_{48}CH_3$, polyethylene and polyhydroxybutyrate, branched-chain nonylbenzenesulfonate and linear nonylbenzenesulfonate, n-decane and a linear hydrocarbon of formula $C_{50}H_{102}$.

    (b) Give one reason for your answer based on the structure of the compound.

    (c) Use chemical equations to illustrate the first step by which each of the more environmentally reactive compounds degrades in the environment.

7.5. Chemists have devised synthetic polymers with a wide array of uses. Some are stronger than steel and many are much less expensive than paper.

    (a) What problems are associated with the disposal of these polymers on land and in the ocean?

    (b) Many polymers, even the so-called degradable ones, don't breakdown in these environments. Why not?

    (c) What problems are associated with recycling synthetic polymers?

7.6. Materials constructed of the following polymers were thrown away in a forest by slovenly campers.

    (a) Which of them will have been degraded by environmental in a year's time? Explain your answers.

    (b) Give the reaction pathway by which degradation occurred for those that did break down ($n = 1000$ in every example).

Polylactate $\left[OCHC \right]_n$
$\overset{O}{\underset{\|}{}}$
CH_3

Polyalanine $\left[HNCHC \right]_n$
$\overset{O}{\underset{\|}{}}$
CH_3

Polyacrylonitrile $\left[CH_2CH \right]_n$
CN

Polydichlorostyrene $\left[CHCH_2 \right]_n$
with dichlorophenyl group (Cl, Cl)

(c) Write the reaction pathway for the polymers that decay at a significant rate in the environment.

7.7. Of the garbage that is dumped in the ocean by ships mainly the plastic ware (bottles, spoons, cups, etc.) ends up on beaches. Why aren't the other items present in garbage (paper, food, etc.) also washed up on the beaches?

8

HALOORGANICS AND PESTICIDES

8.1 INTRODUCTION

Haloorganic compounds have many uses, such as pharmaceutical agents, fibers, building materials, agricultural chemicals, solvents, and cleaning agents. Many of these compounds are inexpensive to manufacture and are used in large quantities. Many were manufactured and deliberately distributed in the environment as pesticides to control plant and insect pests. Some were allowed to be released in the environment because their long-term toxicity and environmental hazards were not understood. Yet others reached the environment because they are volatile or because they were discarded in open dumps and landfills.

The scope of the environmental problems associated with the extensive use of slowly degrading haloorganics has been known for many years, and steps have been taken to remedy some of these problems. The environmental stability of some of these compounds is high, and the manufacture of the more stable compounds has been banned in the United States and some other countries. The toxicity and environmental problems associated with some of these compounds are appreciable. Examples include chlorofluorocarbons, polychlorinated biphenyls (PCBs), and organochlorine pesticides.

The reaction proceeds via the stable allylic free radical formed by hydrogen abstraction with a chlorine atom (8-7). This radical reacts further with Cl_2 to give allyl chloride (8-8), that in turn adds Cl_2 to yield 1,2,3-trichloropropane (8-9).

$$CH_2{=}CHCH_3 + Cl\cdot \longrightarrow CH_2{=}CHCH_2^{\cdot} + HCl \qquad (8\text{-}7)$$

$$CH_2{=}CHCH_2^{\cdot} + Cl_2 \longrightarrow CH_2{=}CHCH_2Cl + Cl\cdot \qquad (8\text{-}8)$$

$$CH_2{=}CHCH_2Cl + Cl_2 \longrightarrow \underset{\underset{Cl}{|}}{ClCH_2CHCH_2Cl} \qquad (8\text{-}9)$$

Elimination of HCl from 1,2,3-trichloropropane yields 1,3-dichloropropene. An 87% yield of approximately equal amounts of the cis and trans isomers of 1,3-dichoropropene is obtained in this reaction, along with a 12% yield of 3,3-dichloropropene (8-10).

$$CH_2{=}CHCH_2Cl + Cl_2 \longrightarrow CH_2{=}CHCHCl_2 + HCl \qquad (8\text{-}10)$$

The degree and position of halogenation of hydrocarbons depends on the ease of hydrogen abstraction (or stability of the corresponding radical) at each carbon atom and the reaction conditions, such as the ratio of chlorine to hydrocarbon.

More vigorous conditions are required to chlorinate aromatic compounds than aliphatic compounds. The aromatic chlorination reaction proceeds by an ionic pathway. For example, polychlorinated biphenyls are prepared from biphenyl by direct chlorination using ferric chloride as a catalyst (8-11). The mechanism involves the initial formation of a chloronium ion [reaction (8-12)] that undergoes an electrophilic addition to the aromatic ring system as illustrated in reaction (8-13) with benzene.

$$\text{(biphenyl)} + Cl_2 \text{ (excess)} \xrightarrow{FeCl_3} \text{(chlorinated biphenyl)} + HCl \qquad (8\text{-}11)$$

(One of many products)

$$Cl_2 + FeCl_3 \rightleftharpoons FeCl_4^- + Cl^+ \qquad (8\text{-}12)$$

$$Cl^+ + \text{(benzene)} \longrightarrow \text{(intermediate)} \longrightarrow \text{(chlorobenzene)} + H^+ \qquad (8\text{-}13)$$

Chlorination of some organics takes place under conditions used to disinfect water with chlorine gas. The chlorine is converted to hypochlorous acid

(HOCl), which is the oxidizing and chlorinating reagent discussed later [Section 11.5.1] reaction (11-11)]. Compounds with aliphatic double bonds or aromatic compounds containing phenolic hydroxyl groups are chlorinated under these conditions. But aromatic compounds such as biphenyl are not chlorinated.[1]

8.3.2 Reactions

The aliphatic organochlorine compounds are generally more reactive than the corresponding aromatic derivatives. Two limiting reaction pathways (displacement and elimination) are observed for aliphatic chloro derivatives in the presence of nucleophilic reagents.

8.3.2.1 Displacement Reactions

Displacement may proceed by unimolecular (S_N1) [reactions (8-14) and (8-15)] or bimolecular (S_N2) [reaction (8-16)] reaction pathways. The rate-limiting step of the S_N1 reaction is ionization of the alkyl halide to the planar carbocation (8-14), while the rate of the S_N2 reaction is proportional to the concentrations of both alkyl halide and nucleophile (8-16).

$$(8\text{-}14)$$

$$(8\text{-}15)$$

$$(8\text{-}16)$$

Transition state

[1]R. A. Larson and Eric J. Weber, *Reaction Mechanisms in Environmental Organic Chemistry*, Chapter 5, pp. 275–358. CRC Press, Boca Raton, FL, 1994.

The mechanism of the displacement reaction is determined by the solvent, the displacing agent, and the structure of the chloroorganic molecule. The S_N1 (carbocation) mechanism is favored by polar solvents that stabilize the ionic intermediates or when the displacing agent is a poor nucleophile such as water. Carbon compounds with sterically hindered halogens resistant to direct (S_N2) displacement or containing tertiary, allylic, and benzylic halogens (which ionize to stabilized carbocations) usually react by an S_N1 pathway.

The ease of ionization of a carbon-halogen bond increases in the order $F < Cl < Br < I$. This trend reflects the decreasing strength of the carbon-halogen bond (Table 8-1). The C—F bond is so strong that ionization is never observed under environmental conditions; hence carbon–fluorine bonds present in fluorocarbons are not cleaved in the environment.

8.3.2.2 Elimination Reactions

Two limiting reaction mechanisms are available for the elimination reaction: a unimolecular (E1) pathway [reactions (8-14) and (8-17)] and a bimolecular (E2) pathway [reaction (8-18)]. The extent of olefin or alcohol formation is determined by steric effects at the carbocation and the stability of the olefinic product.

$$\begin{array}{c} CH_3 \\ \diagdown \overset{+}{C}-CH_3 \\ CH_3 \diagup \end{array} \longrightarrow \begin{array}{c} CH_3 \\ CH_2=C \diagup \\ \diagdown CH_3 \end{array} + H^+ \qquad (8\text{-}17)$$

$$\begin{array}{c} Base \\ + \\ CH_3CH_2CH_2Cl \end{array} \longrightarrow \left[\begin{array}{c} CH_3 \qquad\qquad Cl \\ \diagdown \overset{+}{CH} \text{---} CH_2 \diagup \\ H \\ Base \end{array} \right] \longrightarrow \begin{array}{c} CH_3 \\ \diagdown C=CH_2 \\ \diagup \\ H \end{array} + H\ \overset{+}{Base} + Cl^- \qquad (8\text{-}18)$$

$$\text{Transition state}$$

The factors that favor the S_N1 reaction also favor E1 elimination because the same carbocation intermediate is involved (8-14). The relative proportion of elimination to displacement is determined by the relative nucleophilicity and basicity of the group attacking the carbocation.

8.3.2.3 Predominant Chemical Reaction Pathways in the Environment

When haloorganics are in the environment, and not absorbed by a microorganism, plant, or animal they are usually in the presence of water, that has a moderate pH (7 ± 2) and very low concentrations of other reactants. Consequently S_N2 or E2 reactions will not be observed because the concentration of nucleophiles, acids, and bases is very low. Ionization reactions are favored

TABLE 8-1

Bond Dissociation Energies and Electronegativities of CH_3X

Compound	Bond dissociation energies		Atomic electronegativity of X
	kcal/mol	kJ/mol	
CH_3—H	104	425	2.1
CH_3—F	108	452	4.0
CH_3—Cl	84	352	3.0
CH_3—Br	70	293	2.8
CH_3—I	56	234	2.5

Source: Adapted from R. T. Morrison and P. N. Boyd, *Organic Chemistry*, 6th ed. Prentice Hall, Englewood Cliffs, NJ, 1992.

in water because its solvation of ions and high dielectric constant stabilizes the ionic intermediates. Thus S_N1 and E1 reactions are usually observed.

The E2 elimination is observed with strong bases in nonpolar solvents and is not likely to take place under environmental conditions. For example, the dehydrochlorination of DDT and DDT analogues with sodium hydroxide proceeds by an E2 mechanism in 92.6% ethanol [see reaction (8-19)]. The second-order rate constant decreases by 50% in going from 92.6% ethanol to 76% ethanol. The higher polarity of 76% ethanol–water favors the carbocation (E1) reaction pathway. The elimination proceeds mainly by an E1 mechanism at neutral pH in aqueous solution:

$$(8\text{-}19)$$

Nucleophilic displacement of aryl and vinyl halides is a very slow process that does not occur readily under environmental conditions. The S_N1 or E1 reactions are not observed because the sp^2 hybrid carbon atom of the aryl or vinyl halide is more electronegative than the sp^3 hybrid and cannot stabilize a positive charge and become a carbocation. Furthermore, the lone pairs of electrons on the chlorine participate in the bonding and strengthen the bond as shown in the following resonance structures.

Thus the ground states of the vinyl and aryl halides are more stable than the corresponding ground states of aliphatic halides.

This results in a greater energy of activation (E_A) for carbocation formation and a slower rate of reaction. This is shown qualitatively in Figure 8-1.

The reactivity of aryl halides is enhanced by electron-withdrawing groups. For example, a nitro substituent greatly accelerates the rate of displacement of aryl-bound halogens:

(8-20)

This is due to the formation of an intermediate that is stabilized by the electron-withdrawing substituents. Halide ion displacement from an aromatic ring containing electron-withdrawing groups might thus occur under environmental conditions.

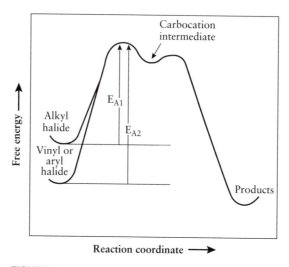

FIGURE 8-1   Reaction coordinate for the conversion of halogenated organics to products by S_N1 or E1 processes.

8.4 MICROBIAL DEGRADATION OF ORGANICS

Biodegradation is the key to the destruction of organic compounds in the environment (Section 8.2). Proteins and nucleic acids of dead plants and animals are degraded rapidly because they are metabolized by microorganisms prevalent in the environment. Xenobiotics—compounds that occur rarely in the environment, such as crude oil, nylon or polychlorinated biphenyls—either are not degraded or are degraded slowly because the population of micro-organisms available to metabolize them is small. The presence of xenobiotics in the environment may promote the growth of strains of microorganisms that do metabolize these compounds. Microorganisms mutate rapidly, so in some cases there is a mutant that can live by metabolizing the xenobiotic. This is why microbiologists who are looking for a population of microorganisms that degrade a particular xenobiotic search the soil samples in the vicinity of a spill of that compound.

The degradation of xenobiotics is usually carried out by a cooperating group of microorganisms called a consortium. Different steps in the degradation are carried out by different microorganisms. For example, the degradation of 3-chlorobenzoate is performed by three different groups of bacteria, as follows:

$$
\text{(structure)} \xrightarrow[\text{Step 1}]{H_2} \text{(structure)} \xrightarrow{\hspace{1cm}} \xrightarrow[\text{Step 2}]{} \xrightarrow{\hspace{1cm}} CH_3COO^- + H_2 \xrightarrow[\text{Step 3}]{} \begin{array}{c} CH_4 \\ + \\ CO_2 \end{array} \qquad \text{(8-21)}
$$

The hydrogen generated in step 2 is utilized by the group of microorganisms doing the reductive dehalogenation in step 1 in the process outlined in reaction (8-21).

The degradation of organic materials to inorganic compounds of C, H, N, P, the process of mineralization, is the ultimate goal of the degradation of xenobiotics. Mineralization is usually measured by the conversion of carbon to CO_2. But biodegradation may result in the intermediate conversion of the carbon to new microorganisms, not achieving mineralization until these intermediate products die. This is not a concern, since the realistic goal of biodegradation is the conversion of a xenobiotic compound to a biotic one that is readily metabolized by a large population of microorganisms, hence will not accumulate in the environment. These biological compounds will eventually be degraded to CO_2. For example, the acetate formed in the degradation of 3-chlorobenzoate is readily metabolized by both anaerobic and aerobic microorganisms. The crucial step in the degradation of 3-chlorobenzoate is the reductive elimination of chloride ion, a reaction that is carried out by relatively few microorganisms.

Both anaerobes and aerobes degrade xenobiotics. Anaerobes often use reductive processes to degrade organics, and aerobes use oxidative processes. Both carry out other processes in which there is no change in the oxidation state of the xenobiotic.

8.4.1 Reductive Degradation

The anaerobic dehalogenation of a pesticide was discovered during an investigation of the loss of potency of γ-hexachlorocyclohexane [Lindane: see Section 8.9, reaction (8-40)] in a cattle tick bath. The cattle were made to swim through the bath containing hexachlorocyclohexane to kill the ticks. Since the chemical is expensive, the same bath was used for several years. The potency of the bath decreased slowly with time, but when fresh pesticide was added to it, the rate of loss of the hexachlorocyclohexane dropped more rapidly. The rapid loss in activity was traced to microorganisms, which evolved in the bath over several years. These microorganisms metabolized the hexachlorocyclohexane by removal of its chlorines as chloride.

Bacterially mediated reductive dehalogenation is an important pathway for the destruction of unreactive haloorganics. The reaction usually proceeds in anaerobic environments such as marine sediments and landfills with the replacement of halogens with hydrogens. The principal pathway for the reductive dechlorination of 2,3-dichlorobenzoate is shown in reaction (8-22). The benzoate formed is readily degraded by other microorganisms, as shown in Section 6.4.

$$\tag{8-22}$$

Reductive dechlorination is an important first step in the degradation of heavily chlorinated organics like pentachlorophenol (PCP). PCP is rapidly converted to tetrachlorophenols, but the subsequent loss of chloride in the process of forming tri-, di-, and monochlorophenols proceeds much more slowly than the initial loss of chloride ion. The faster initial rate reflects the higher oxidation potential of PCP, which makes it easier to reduce than its more lightly chlorinated metabolites.

The reductive dehalogenation of aliphatic compounds has also been observed. For example, 1,1,1-trichloroethane (methylchloroform) can be reduced to the corresponding dichloro derivative:

$$(Cl)_3CCH_3 + H^+ + 2e^- \longrightarrow (Cl)_2CHCH_3 + Cl^- \tag{8-23}$$

8.4.2 Oxidative Degradation

The energy derived from the oxidation of xenobiotics with atmospheric oxygen drives the process of oxidative degradation. These oxidations occur mainly with lightly chlorinated chloroorganics because they require the attack of an electrophilic oxygen on an electron-rich center such as an aromatic ring or a double bond. The delocalization of these electrons by electronegative chlorine atoms decreases the electron density on the aromatic ring; hence these compounds are less readily oxidized by molecular oxygen.

The overall oxidative degradation of chlorobenzene is shown in reaction (8-24).

$$(8\text{-}24)$$

In step a, a microbial dioxygenase enzyme catalyzes the formation of a *cis*-dihydrodiol, which is then oxidized (step b) to a halogenated catechol. Enzyme-catalyzed oxidative ring cleavage takes place either between the two phenol groups (step c) or adjacent to one of the phenol groups (step d) to give dicarboxylic acids, which are readily degraded to simpler compounds and eventually to $CO_2, H_2O,$ and Cl^-.

8.4.3 Hydrolytic Degradation

Hydrolytic cleavage of halogens, esters, amides, and other groupings are catalyzed by both aerobes and anaerobes. Dichloromethane (methylene chloride), a compound that decomposes very slowly in the presence of water, is hydrolyzed by microorganisms to formaldehyde. The enzyme-catalyzed reaction usually proceeds by the S_N2 displacement of chloride ion by the thiol peptide glutathione (G-SH). The resulting intermediate is hydrolyzed to glutathione and formaldehyde:

$$G\text{-}SH + CH_2Cl_2 \longrightarrow G\text{-}SCH_2Cl \longrightarrow {}^+G\text{-}S{=}CH_2 \xrightarrow{H_2O} G\text{-}SH + CH_2O + H^+$$

$$+ \qquad\qquad + \qquad\qquad\qquad (8\text{-}25)$$

$$HCl \qquad\qquad Cl^-$$

The microorganisms that catalyze this reaction utilize formaldehyde as an energy source.

8.5 TOXICITY

Acute or immediate toxicity to humans and animals is not usually a problem for most commercial haloorganic compounds. The toxicity associated with some haloorganics is of the chronic type, where the deleterious effect appears 2–30 years after the initial exposure. For example, vinyl chloride, the monomer polymerized to poly(vinyl chloride) (Figure 7-3), evidently caused cancer in factory workers making the polymer twenty years after they started working with it. Since these workers all contracted an unusual type of liver tumor that was not observed in the general population, it was concluded that vinyl chloride was the causative agent.

Most commercial halocarbons are nonpolar compounds that are removed from the bloodstream by the liver. There, their presence induces the synthesis of the enzyme cytochrome P-450, which catalyzes the oxidative degradation of the nonpolar organics in the liver. However, if the compounds are heavily chlorinated, they may have high oxidation potentials and are oxidized slowly, or not at all, by cytochrome P-450 (Section 7.2.4). The high levels of this enzyme that are induced in the liver catalyze the oxidation of other nonpolar organics such as steroids. Since other steroid hormones are produced by cytochrome P-450 oxidation, the presence of excess cytochrome P-450 can change the normal hormonal balance.

Some of these haloorganics have many deleterious health effects other than causing cancer. Low levels can cause endocrine, immune, and neurological effects. Haloorganics vary dramatically in chronic and acute toxicity according to structure, and examples of specific problems are presented when selected types of compound are discussed in the sections that follow.

8.5.1 Environmental Hormones

The possible role of industrial chemicals in human health was debated extensively in the 1990s. Some chronic effects on humans have been anticipated since the observation of the decline in the number of eagles and hawks in the 1960s and 1970s, that correlated with the buildup of chloroorganics in the environment. There was some reassurance that humans might not be affected by these compounds because of the resurgence of the birds in the 1980s following the ban of the use of DDT in 1972. However, there were still reports of the decrease in populations of some birds and the apparent feminization of males in which enhanced levels of chlorinated organics were found.

The observation of abnormal sexual development of a number of forms of wildlife and the detection of higher levels of chloroorganics in the same animals led to the proposal that some industrial chemicals have hormonal activity that mimics or inhibits the activities of the sex hormones. Estrogen, a mixture of three steroid hormones, has essential roles in both males and females. A specific ratio of estrogen to androgen (male hormones) is required during prenatal and postnatal development for sex differentiation and for the development of reproductive organs. In females estrogen also prepares the uterus to accept the fertilized egg, helps with lactation, and lowers the risk of heart attacks, but stimulates the development of uterine and breast cancer. The presence of too much estrogen in males inhibits sperm production and the development of the testes.

In a number of instances industrial chemicals or their breakdown products have been shown to have estrogen-like activity that interferes with the sex hormones in wildlife. For example, the feminization of seagulls was attributed to the higher levels of chloroorganics in these birds. PCBs appear to exert the same effect in terns living near a toxic waste site containing these chemicals. PCBs also demonstrate estrogen-like activity in turtles. Laboratory studies have established that o,p-DDT, p,p'-methoxychlor [structure given in reaction (8-35)], kepone, and some PCBs and the phenols derived from them are

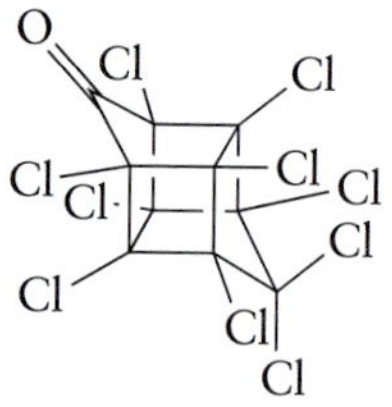

Kepone

estrogen mimics. Nonchlorinated organics have also been shown to be estrogen mimics. Examples include bisphenol A, a degradation product of polycar-

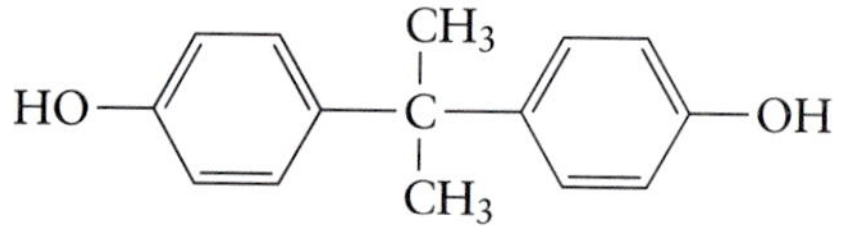

Bisphenol A

bonate polymers, and alkylated phenols such as p-nonylphenol, a degradation product of polyethoxylate detergents (Section 7.2.2.3).

The interpretation of the effect of a mixture of environmental compounds found in humans on their health is complicated by the varied biological responses triggered by the presence of the compounds. Some industrial compounds have been shown to have antiestrogenic effects. These include 2,3,7,8-tetrachlorodibenzo-p-dioxin [TCDD: see Section 8.10.3, reaction (8-58)] and polychlorinated benzofurans [Section 8.7.3, reaction (8-30)]. In principle, these compounds could negate the effects of the environmental estrogens.

Another problem with TCDD is that it may cause endometriosis, a disorder in women in which tissue migrates from the uterus to the abdomen, ovaries, bowels, and bladder. This can result in internal bleeding and infertility. The possible role of TCDD in endometriosis was suggested by a long-term study of the reproductive effects of TCDD in monkeys. TCDD was included in the diet of 24 female monkeys for four years and then no more was administered. In the 6-to 10-year period after the initial administration of the TCDD, 3 of the monkeys died of endometriosis. Investigation of the remaining monkeys revealed that 75% of those given a high dose of TCDD had moderate to severe cases of endometriosis, while only 42% of those given a weak dose had the ailment. None of the control monkeys had endometriosis.

More recently it has been discovered that DDE, a breakdown product of DDT, has antiandrogen activity. Antiandrogens compete with the male sex hormones for binding at androgen receptor sites and thus inhibit the activity of the androgens. The net result of this inhibition is comparable to the presence of compounds with estrogen activity. It appears likely that the feminization of newborn alligators living in a lake containing high levels of DDE should really be called demasculinization resulting from the antiandrogenic effect of the DDE.

There is general agreement on the assignments of estrogenic, antiestrogenic, or antiandrogenic properties to the compounds listed in Table 8-2. Disagreements arise when these data are used to explain recent health problems in humans. For example, the increases in breast cancer observed in older women over the past 50 years have been tentatively attributed to long-term exposure to environmental hormones. The assumption is that since estrogen is a cause of breast cancer, environmental hormones will also cause cancer. Correlations have been reported between the PCB levels in tissue and DDE present in serum in women who have breast cancer, but there is no evidence of a direct link between breast cancer and these compounds. It has also been proposed that the 40% decline in sperm counts in men over the past 50 years is due to exposure to estrogen-like and antiandrogen compounds in the environment. In addition,

TABLE 8-2

Proposed Hormonal Activity of Environmental Compounds

Estrogen-like	Antiestrogens	Antiandrogen
PCBs	2,3,7,8-Tetrachlorodioxin	DDE
o,p-DDT	Polychlorinated benzofurans	
Methoxychlor	Benzo(a)pyrene	
Kepone		
Toxaphene		
Dieldrin		
Endosulfan		
Atrazine		
Bisphenol A		
Nonylphenol		

it has been proposed that there is a correlation between the increase in testicular cancer and the presence of estrogen-like compounds in the tissue of males. As in the case of breast cancer, there is no unambiguous link between the medical observations and environmental compounds.

An important argument against the important role of environmental chemicals as estrogens in humans is the observation that the cruciferous vegetables (brussel sprouts, cauliflower, kale, etc.) contain bioflavonoids that exhibit estrogenic activity.[2] It is calculated that the estrogenic potential of the compounds in these vegetables is 10^7 times higher than that of the environmental chemicals because they constitute such a high proportion of our diet. Scientists on both sides of the debate agree that more data are required before conclusive statements can be made concerning the importance of environmental estrogens on humans.

Many of the persistent organic pollutants (POPs) that are believed to be environmental hormones are still being produced and used in the world. These compounds are carried from where they are manufactured or used around the globe by weather patterns. The United Nations sponsored and adopted a treaty in 2000 intended to end the production of these and other persistent chlorinated compounds throughout the world and to destroy their stocks. This is the first attempt to control these compounds globally. The "dirty dozen" compounds include the insecticides aldrin, chlordane, DDT, dieldrin, endrin, heptachlor, hexachlorobenzene, mirex, and toxaphene plus the industrially used polychlorinated biphenyls and the industrial by-products, dioxins and benzofurans. The treaty took effect upon ratification by 122 nations in December, 2000. The use of DDT for the control of malaria is the one exception to the ban, since the insecticide is still used for this purpose in Africa, Asia, and South America. The structures and discussions of most of these compounds are given in subsequent sections of this chapter.

8.6 CHLOROFLUOROCARBONS, HYDROFLUOROCARBONS, AND PERHALOGENATED ORGANICS

Chlorofluorocarbons (CFCs) and perhalogenated organics were used for a long time as aerosol propellants, refrigerants, solvents, and foaming agents, but this use has been curtailed. So-called halons, which contain bromine as well as fluorines, are still used to extinguish fires in critical situations such as on aircraft. Most of these compounds are nontoxic, stable, colorless, and nonflammable: ideal substances for use in a variety of industrial applications. The high stability of these compounds, which made them so attractive to industry, also explains

---

[2]S. E. Safe, Environmental and dietary estrogens in human health: Is there a problem? *Environ. Heath Perspect.*, **103**, 346–351 (1995).

why they are implicated in two major environmental problems, global warming (Section 3.3.3) and the destruction of the ozone layer (Section 5.2.3.2).

8.6.1 Structural Types

The chlorofluorocarbons that were banned from production in the industrialized nations in 1996 (Section 5.2.3.5) are highly halogenated derivatives that do not contain a C—H bond (e.g., CCl_3F, $C_2F_2Cl_4$, and $C_3F_2Cl_6$. The perchlorinated compound carbon tetrachloride (CCl_4), which had been used extensively as a degreasing and cleaning solvent, is as deleterious to the environment as the CFCs. Methylchloroform (1,1,1-trichloroethane, CCl_3CH_3), which was also used for degreasing, although it has one-tenth the ozone-destructive power of CFCs, was nevertheless a major contributor to the destruction of the ozone layer because it was used in large amounts in industry.

Bromofluorocarbons (halons), such as $CBrF_3$, used to extinguish fires in military and commercial aircraft, are believed to act in much the same way that tetraethyllead prevents preignition in internal combustion engines (Section 6.7.4), that is, by combining with free radicals and stopping the oxidative chain reaction central to the combustion process. The production of halons that destroy the ozone layer (Section 5.2.3.2) was banned in 1994 in the industrialized countries but will continue in developing nations. There are still large supplies available that, when put to use to fight fires, will eventually end up in the ozone layer. The level of halons in the atmosphere was still increasing in 1998.

It is ironic that the same person, Thomas Midgely Jr., invented both chlorofluorocarbons and tetraethyllead. These compounds, which were highly acclaimed when discovered in the 1920s, made possible the widespread use of refrigerators and automobiles, staples of the American culture and economy. The manufacture and use of leaded fuels in the United States continued until 1995, while the manufacture of chlorofluorocarbons was halted in 1996. The phaseout in the use of these compounds started some 10–15 years earlier than the final cutoff dates.

8.6.2 Atmospheric Lifetimes[3]

8.6.2.1 Chlorofluorocarbons and Perhalogenated Organics

Carbon tetrafluoride (CF_4), the most stable of the perhalogenated organics, is estimated to have a lifetime in the environment that exceeds 50,000 years—

---

[3]The lifetimes referred to in this section are the reciprocal of the first-order rate constant for their degradation. See A. R. Ravishankara and E. R. Lovejoy, Atmospheric lifetime, its application and determination: CFC-substitutes as a case study, *J. Chem. Soc. Faraday Trans.*, **90**(15), 2159–2169 (1994).

about 10 times longer than the history of human civilization. It is formed as a by-product of the production of aluminum by the electrolysis of aluminum oxide in a melt of cryolite (Na_3AlF_6) (Section 12.2.5). The CF_4 is formed by the reaction of fluorine with the carbon electrodes at the high temperatures ($1000°C$) and voltages used in the electrolytic process. About 30,000 tons of CF_4 is released globally into the atmosphere each year by this process.

The presence of strong C—F and C—Cl bonds and the absence of C—H bonds are central to the environmental stability of the CFCs and perhalogenated organics. The trends in the energy of dissociation of the carbon–halogen bond of halomethanes show the C—F bond to be the strongest (Table 8-1). The large energy of dissociation of the C—F bond is a consequence of the greater overlap of the bonding orbitals because of the similarity in size of the carbon and fluorine atoms. In addition, because of its high electronegativity (4.0), the fluorine atom reduces electron density on carbon and deactivates the carbon atom for reaction by S_N1 or E1 processes (Section 8.3.2). Because the C—Cl bond is weaker than the C—F bond, a CFC will react to form a carbocation by ionization of the chlorine atom as follows:

$$
\begin{array}{ccc}
\underset{Cl}{\overset{Cl}{\diagdown}}\!\!\!\!C\!\!\!\!\overset{F}{\diagup} & \longrightarrow & \underset{Cl}{\overset{Cl}{\diagdown}}\!\!\!\!\overset{+}{C}\!\!\!\!\overset{F}{\diagup} \; + \; Cl^-
\end{array}
\tag{8-26}
$$

The postulated carbocation intermediate in reaction (8-26) has a very high energy. Consequently, the chloride atom ionizes from the carbon to form the carbocation only at high temperatures. While CCl_4 is more reactive than CF_4, the relatively high strength and electronegativity (3.0) of the C—Cl bond (Table 8-1) explains the environmental stability of CCl_4.

CFCs and perhalogenated organic compounds are stable in the earth's atmosphere because they do not contain C—H groups and therefore are resistant to attack by hydroxyl radicals (Section 5.2.3.5). Since abstraction by a hydroxyl radical ($\cdot OH$) of a chlorine atom from a CFC to form ClOH is not energetically favored, CFCs only react after diffusion up to the ozone layer, where solar UV radiation initiates the dissociation of the chlorine atom (Section 5.2.3.2). The chlorine atom catalyzes the dissociation of ozone. Since CF_4 does not contain chlorine or bromine, it does not cause the destruction of the ozone layer, but its global warming potential is several thousand times greater than that of a comparable amount of carbon dioxide (Section 3.1.2).

Methylchloroform, an industrial solvent, is a compound whose production was curtailed as a result of the 1987 Montreal Protocol agreement (Section 5.2.3.5). It is not as inert as the chloroflurocarbons, with a lifetime of 5.2 years in the Northern Hemisphere and 4.9 years in the Southern Hemisphere. Its atmospheric concentration leveled off in 1991 and started to decrease in 1995. The rapid decay of methyl chloroform is due to its reaction with atmospheric hydroxyl radicals to form $\cdot CH_2CCl_3$ (Section 5.2.3.5). It was possible to calculate from these lifetimes that the concentration of hydroxyl radicals in the atmosphere of the Southern Hemisphere is $15\pm10\%$ higher than in the Northern Hemisphere.

The annual production of CFCs decreased by 90% in the 1986–1995 time period. Since the lifetimes of many CFCs in the atmosphere are in the range of 50–150 years, limitations on their production will not result in a decrease in their atmospheric concentrations until about 2001–2010 (Figure 8-2). However the global agreement to halt the production and use of chlorofluorocarbons made in the Montreal Protocol in 1987, and its subsequent modifications, have resulted in halting the increase in buildup of stratospheric chlorine in 1994 at one-third the level that might have formed if no controls had been put in place until 2010. It is predicted that the ozone levels will decrease in 2050 to those observed in the time period when the Antarctic ozone hole was discovered (1980–1984: see Section 5.2.3.3).

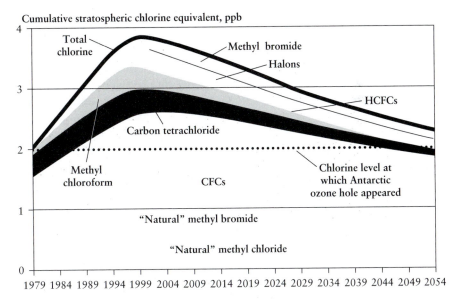

FIGURE 8-2 Projected trend in levels of chlorine, CFCs, HCFCs, and haloorganics in the atmosphere as a consequence of The Montreal Protocol. Redrawn from, "Looming ban on production of CFCs, halons spurs switch to substitutes," *Chem. Eng. News*, p. 13, Nov. 15, 1993. Used by permission of E.I. du Pont de Nemours and Company. Also see color insert.

8.6.2.2 Hydrochlorofluorocarbons and Hydrofluorocarbons

Hydrochlorofluorocarbons (HCFCs) and hydrofluorocarbons (HFCs) have been designed to take the place of CFCs, which were banned from production in industrialized nations on January 1, 1996. Many times, a chemist who has prepared a useful but unstable compound is asked to come up with a compound that has similar properties but does not break down. The assignment in the case of the CFCs, however, was to make a less stable substitute that breaks down more rapidly in the environment. The approach used was to include C—H bonds in these substitutes for the CFCs so that they would be susceptible to oxidation by hydroxyl radicals in the atmosphere (Section 5.2.3.5). Two examples of the hundreds of HCFCs and HFCs prepared are CH_3CHF_2 (HFC-152a) and CF_3CH_2F (HFC-134a). Both have no ozone destruction potential, but both are greenhouse gases. In fact, CF_3CH_2F has about five times the global warming potential of CH_3CHF_2. The global warming potentials of CFCs, like $CFCl_3$, are about 30 times greater than that of CH_3CHF_2.

The atmospheric lifetimes of CH_3CHF_2 and CF_3CH_2F are estimated to be 2 and 14 years, respectively, values that are considerable shorter than the 50-to 100-year lifetimes of the CFCs. In addition, since these compounds contain no bromine or chlorine, they will not destroy the ozone layer. There was a concern that the CF_3 radical formed in the atmospheric degradation of CF_3CHF_2, and other CFC substitutes that contain the CF_3 group, would catalyze the degradation product of ozone, via the formation of $CF_3OO\cdot$ and $CF_3O\cdot$ by reaction with molecular oxygen and ozone, respectively. Laboratory studies and atmospheric models showed that these oxygenated trifluoromethyl radicals would react in the lower stratosphere, where they would have a negligible effect on the level of atmospheric ozone. There is also the concern that the trifluoroacetic acid, an important atmospheric degradation product of CFC substitutes containing the CF_3 group, would accumulate at toxic levels in ecosystems where water evaporates to form concentrated solutions of organic and inorganic compounds. While it has been demonstrated that trifluoroacetic acid is not very toxic to humans, plants, and animals, and is degraded by some anaerobic bacteria, there is still concern that it may accumulate in the ecosystems and be toxic to the life there. These conclusions were reached on the basis of modeling (calculations) and have not been verified by experimental studies.

The HCFCs and HFCs are not expected to be permanent replacements for chlorofluorocarbons. Rather, their use is planned until 2020, when they are expected to be replaced by even more benign substitutes. For example, while compounds that contain the CF_3 group (e.g., CF_3CHF_2), do not destroy the ozone layer, they are degraded to CHF_3, which has a 400-year atmospheric lifetime and 8000 times the global warming potential of carbon dioxide. Policy makers expect that chemists will find readily degradable compounds that neither destroy the ozone layer, cause global warming, nor degrade to sub-

stances that are toxic to any forms of life on earth. These compounds should
have the desirable properties of CFCs so that all humans can continue to enjoy
all the benefits of modern civilization without an increase in the cost of living.
It remains to be seen whether these desirable characteristics can be achieved.

8.6.2.3 Other Compounds

Progress toward the goal of producing a new generation of halons for use as
fire retardants has not been as successful as the search for effective replacements for CFCs. The current plan is to carefully husband the available supplies
of CF_3Br and $CBrClF_2$ by reserving their use for situations in which other
retardants cannot be used while the search for effective replacements continues. There was hope that CF_3I would be a direct replacement of CF_3Br,
but the toxicity of the iodide prevents its use in aircraft and other closed
environments. The production of halons in 1995 decreased to almost zero in
both industrialized countries and in all developing countries except China,
where the production increased 2.5-fold between 1991 and 1995.

Another ozone-destroying chemical, methyl bromide (Figure 8-2), is used in
large amounts as a soil and grain fumigant. It is an excellent agent for the
removal of fungi, bacteria, viruses, insects, and rodents from soil prior to
planting crops. Unlike many other pesticides, it does not persist in the soil
but is decomposed in a few weeks. Even though its atmospheric lifetime is
short, its use in large amounts results in the transport of some of the pesticide
to the ozone layer. It was proposed in 1992 that methyl bromide is responsible
for 10% of the depletion of the ozone layer. The photochemically generated
bromine atoms destroy ozone about 50 times more rapidly than do chlorine
atoms (5.2.3.2). California, the largest agricultural user of methyl bromide of
any U.S. state, released 8×10^6 kg in 1988. It was proposed in 1992 that the
use of methyl bromide in developed nations be phased out in 2005. There is
great concern about this proposed ban in the agricultural community because
there is no obvious replacement.

In 1992 it was found that methyl chloride, methyl bromide, and more highly
halogenated methane derivatives are formed in the environment. These compounds are produced by microorganisms living in salt marshes, coastal lands,
decaying vegetation and in beds of kelp (seaweed) growing on the ocean floor.
This is the principal source of atmospheric methyl chloride, since the compound is not manufactured in large amounts. Methyl chloride is a source of the
stratospheric chlorine that destroys the ozone layer (Section 5.2.3.2).

Since these compounds have been in the environment for a long time, it is
likely that there are microorganisms that metabolize them for their carbon and
energy. Indeed, in a field that was fumigated with methyl bromide, microbiologists found an anaerobic microbe that uses this compound as its sole energy
source. Other microorganisms were discovered in the environment, both

aerobes and anaerobes, which metabolize these halomethane derivatives. Aerobes oxidize them to carbon dioxide and halide ions. Anaerobes use them to methylate the sulfide ion formed by the reduction of sulfate to generate the volatile dimethyl sulfide. These metabolic pathways proceed more rapidly than the simple hydrolysis of methyl halides to methanol but more slowly than the rate of formation of these haloorganics.

These new developments have resulted in extensive study of the sources and sinks of methyl bromide in the environment. Anthropomorphic sources of methyl bromide include its release during fumigation, biomass burning, and combustion of leaded gasoline in automobile engines (Section 6.6). These sources, together with production in the environment, are estimated to put 150×10^6 kg of methyl bromide into the atmosphere each year. The use of methyl bromide for fumigation releases 41×10^6 kg a year, or about 30% of the total amount released into the atmosphere. It is possible, however, to reduce the amount emitted during soil fumigation by a factor of 10 by covering the fields with a sheet of nonpermeable polymer while the methyl bromide is injected into the soil. The polymer sheet prevents the escape of methyl bromide to the atmosphere. If the field is covered for several weeks, the methyl bromide decomposes and is not released when the sheets of plastic are removed. If this practice is shown to be applicable on a large scale, it may not be necessary to ban the use of methyl bromide as an agricultural fumigant. At present there are no plans to rescind the ban on the production of methyl bromide in industrialized countries.

8.7 POLYCHLORINATED AND POLYBROMINATED BIPHENYLS

Polychlorinated biphenyls (PCBs) were first used industrially in the 1930s, but their use was not widespread until the 1950s. Their extreme stability made them especially effective in many applications. For example, they are resistant to acids, bases, heat, and oxidation. This stability had prompted their use as heat exchange fluids and dielectrics in transformers and capacitors. They were also used in large amounts as plasticizers and solvents for plastics and printing inks and in carbon paper. It is estimated that in the 50-year period of their use, over 1.4 billion pounds was manufactured and several hundred million pounds released in the environment.

PCBs were first available commercially in the United States in 1929. The last major manufacturer of PCBs was Monsanto, under the trade name of Arochlor. These products are identified by a four-digit number; the first two digits indicate if it is pure or a mixture, and the last the second two digits designate the weight percent chlorine present. The annual sales of Arochlors increased from 40 million pounds in 1957 to 85 million pounds in 1970. The cumulative U.S. production is estimated at one billion (10^9) pounds. World production was 200 million pounds in 1970. Because of the increasing environmental concerns, Monsanto stopped selling Arochlors in 1970 for use where they could not be recovered and terminated manufacture of them completely in 1977 when the EPA recommended that PCBs not be used as heat transfer fluids in the production of foods, drugs, and cosmetics.

PCB production has been banned in most of the industrial nations of the world. Russia, the principal exception, claims that because all its transformers were built on the use of PCBs as the dielectric, it must continue to manufacture PCBs for these transformers until 2005. The government has pledged to destroy all the unused stock of PCBs in 2020. Thus it appears likely that Russia will be a major source of PCBs in the environment in the twenty-first century. PCB manufacture also continues in some nonindustrialized countries in the world.

8.7.1 Environmental Problems

The stability of PCBs, which makes them so useful commercially, also results in their persistence in the environment. The first report that these compounds were prevalent and were being ingested by aquatic life came in 1966, when Jensen detected PCBs in fish. Since the production of PCBs has been banned in the United States, the current sources in the environment are evaporation from landfills, incineration of household trash in backyard barrels, volatilization from lakes and other repositories, and synthesis and use elsewhere in the world. For example, since PCBs have been dumped into and washed into the Great Lakes for many years, these lakes are now one of the major sources of atmospheric PCBs. PCBs bound to dust particles have been carried over the globe by atmospheric circulation and are present in wildlife from the Antarctic to isolated Pacific atolls. Their atmospheric concentrations are highest near urban centers in the United States and Europe, where they are present in higher quantities than in rural areas.

8.7.2 Toxicity

Early concerns regarding the toxicity of PCBs were fueled by the "Yusho incident" in Japan in 1968, when more than a thousand persons ate rice oil contaminated with PCBs that had leaked from a heat exchanger used to process the rice oil. Persons who ate 0.5 g or more (average consumption was 2 g) developed darkened skin, eye damage, and severe acne. It is not certain whether the subsequent deaths of some of the patients were due to the PCB poisoning. Recovery was slow, with symptoms still present after three years. Several infants were born with the same symptoms, demonstrating that PCBs can readily cross the placental barrier. Similar effects were observed on Taiwan in 1979, when cooking oil contaminated with PCBs was used by many people.

While the concentration of PCBs in lakes, rivers, and oceans is low because of the compounds' low solubility in water, concentrations in animals that feed almost exclusively on fish can build up to 100 million times the concentration in water. Animals at the end of a food chain, such as those that eat fish, consume vast amounts of fish in their lifetimes, with the result that even though an individual fish may contain a low level of PCBs, the cumulative amount consumed is very large. The PCBs remain dissolved in the livers and other fatty tissues of these animals because they are so insoluble in water and because they are degraded very slowly by cytochrome P-450 in the liver. For example, herring gulls, eagles, bottlenose dolphins, seals, killer whales, and beluga whales all have been found to contain high levels (from 10 to > 200 ppm) of PCBs. The normal PCB concentration in humans is 1–2 ppm.

It is known that mink are very sensitive to PCBs; only 5 ppm causes complete reproductive failure. In contrast, there is little decrease in survival of offspring in rats containing 100 ppm. There have been several episodes of chicken poisoning by feed containing PCBs. As little as 10–20 ppm causes enlarged livers, kidney damage, decreased egg production, and decreased hatchability of the eggs produced, as well as developmental defects in the chicks that do hatch. Similar effects have been observed in the birds living on the shores of the Great Lakes. Some species of shellfish, shrimp, and fish also exhibit sensitivity to PCBs.

There is also some concern that chronic toxic effects may occur in the animals at the end of food chains that contain high levels of PCBs. It is difficult to assign such effects unambiguously to PCBs because the animals also contain other stable chloroorganics in their fatty tissues. In addition, the biological effects vary with the degree of chlorination of the biphenyls and with the polychlorinated benzofuran [see Section 8.7.3, reaction (8-30)] content of the PCBs. Nor is it clear that all the toxic effects observed in humans who ingested PCB-containing cooking oil in Japan and Taiwan were due to the PCBs, since chlorobenzofurans present might have been implicated as well. PCBs and benzofurans have environmental half-lives of about 15 years, so significant amounts will be in the environment well into the twenty-first century.

It was reported in 2000 that PCBs and their dioxin and benzofuran impurities may inhibit the development of children. The study showed that young breast-fed children appear to have a weakened immune system in comparison to those fed formulas. The mothers' milk contained PCBs, while the formulas did not. The effect on the children's immune systems was small, but the breast-fed children were eight times more likely to get chicken pox and three times more likely to have at least six ear infections. This study will have to be confirmed by independent investigators before the conclusions can be generally accepted.

8.7.2.1 Polybrominated Biphenyls (PBBs)

Polybrominated biphenyls, first sold commercially in 1970, were produced in much smaller amounts (5×10^6 lb./year) than PCBs. They were used as flame-retardant and fire proofing materials in the polymers in typewriters, calculators, televisions receivers, radios, and other appliances. Their chemical properties are similar to those of PCBs in that they are very stable. They are an environmental problem in the Pine River, in St. Louis, Michigan, where about a quarter-pound was spilled daily during manufacture. Their deleterious effect to livestock and fowl was observed in 1973 when PBBs, sold under the commercial name Firemaster, were mistakenly substituted for Nutrimaster, a dairy feed supplement, in animal feed. Animals fed PBBs were not fit for human consumption, and 20,000 cattle, pigs, and sheep had to be destroyed along with 1.5 million domestic fowl. The PBBs caused weight loss, decreased milk production, hair loss, abnormal hoof growth, abortions, and stillbirths. The manufacture of PBBs in the United States was banned in 1977.

8.7.3 Chemical Synthesis and Reactivity

PCBs are manufactured by the direct chlorination of biphenyl in the presence of Fe or $FeCl_3$, as shown earlier in reaction (8-11). A mixture of isomeric products is obtained, the structures of which depend on the reaction temperature and the ratio of Cl_2 to biphenyl. Since the biphenyl is usually contaminated with polyphenyls such as tetraphenyl, the chloro derivatives of these products are also obtained in reactions analogous to (8-11).

PCBs are among the most persistent compounds in the environment. As noted previously, they are resistant to both hydrolysis and oxidation and, as we discuss shortly, undergo slow microbial degradation. It is not surprising that biphenyls are so stable, since the aryl halide functional group is exceedingly resistant to hydrolysis (see Section 8.3.2.3).

Highly chlorinated PCBs are photodegraded with the long-wavelength UV light (> 290 nm), which is not absorbed by the ozone layer. The initial step in the photodegradation process involves the dissociation of a chlorine atom, and the radicals that form abstract a hydrogen atom from another organic molecule, as shown in reactions (8-27) and (8-28). When the photolysis proceeds in the presence of water, phenols are obtained as products [reaction (8-29)] by the reaction of water with the triplet excited state of the PCB.

$$\text{(Ph-C}_6\text{H}_4\text{-Cl)} \xrightarrow{h\upsilon} \text{(Ph-C}_6\text{H}_4\text{·)} \; + \; Cl^- \qquad (8\text{-}27)$$

$$\text{(biphenyl)}\cdot \;+\; RH \;\longrightarrow\; \text{(biphenyl)}-H \;+\; R\cdot \qquad (8\text{-}28)$$

$$\text{(biphenyl)}-Cl \;\xrightarrow{h\nu}\; \left[\text{(biphenyl)}-Cl\right]^{*} \;\xrightarrow{H_2O}\; \qquad (8\text{-}29)$$

$$\text{(biphenyl)}-OH \;+\; HCl$$

Photochemical reactions are not a major pathway for the environmental destruction of PCBs because only the highly chlorinated derivatives absorb at wavelengths greater than 290 nm and the quantum efficiency of the reactions is low. Lightly chlorinated biphenyls are not degraded photochemically in the environment because they do not absorb light at wavelengths greater the 290 nm; they absorb at the shorter wavelengths, which do not penetrate the ozone layer.

PCBs have low water solubility (~ 50 ppb for tetrachlorobiphenyls and 1 ppb for hexachlorobiphenyls) and are more dense than water. When discharged into lakes and rivers they sink to the bottom, where they are adsorbed on hydrophobic organic sediments. PCBs, as noted earlier, have low volatility (10^{-4}–10^{-11} atm), with the largest amounts in the atmospheres of urban areas (0.2–20 ng/m^3) and the lowest in marine air (0.02–0.34 ng/m^3). The atmospheric reservoir of PCBs is less than 0.1% of the total amount in the environment. In spite of these low volatilities, atmospheric transport is the principal way PCBs are spread globally as a gas or bound to particles. The Great Lakes of North America have become a major source of atmospheric PCBs owing to the revolatilization of the PCBs present there. PCB concentrations are high in urban areas as a result of their volatilization from municipal landfills and waste disposal sites, where they were dumped prior to 1978, when their disposal was regulated.

The destructive disposal of PCBs requires extreme reaction conditions because they are such stable compounds. Since they are resistant to air oxidation at temperatures below 700°C, incineration at lower temperatures, for example, while burning trash, results in partial oxidation to chlorobenzofurans:

$$\xrightarrow{O_2} \qquad (8\text{-}30)$$

A polychlorinated
dibenzofuran

Reductive removal of the chloro groups proceeds with strong reducing agents such as $LiAlH_4$, sodium, or titanium(III), or by catalytic hydrogenation to generate biphenyl, a compound that can be oxidatively destroyed by combustion. Complete hydrolysis can be attained by treatment with NaOH in methanol at 300–320°C and pressure of $180 \, kg/cm^2$.

High-temperature combustion (~1000°C) is used most frequently for complete destruction (> 99.9999% of the PCBs can be destroyed). Cement kilns, with their high temperatures (1500°C) and basic conditions, are ideal sites for the combustion of PCBs. In this process a slurry of crushed limestone and silicate-containing rock, containing 30–40% water, is fed in one end of a 500-ft rotating tube, and this slurry slowly migrates by gravity to the other end of the tube, where oil or coal and the PCBs are burning to give the high temperature. The PCBs are pyrolyzed and oxidized to carbon dioxide while burning, and the chlorine is released as HCl. The HCl reacts with the basic compounds present in the limestone to form calcium and magnesium chloride as it passes through the cement and slurry. Cement formation is a slow process, so the PCBs have a long residence time in the kiln where the temperatures are maintained at 1370–1450°C in the presence of molecular oxygen. Almost all of the PCBs are destroyed, and the only emissions of chlorinated compounds observed are those detected when no PCBs were added to the kiln.

8.7.4 Microbial Degradation

Soil samples containing PCBs have been found to contain microorganisms that metabolize the PCBs. Different degradative pathways have been observed which proceed in the presence and absence of molecular oxygen.

8.7.4.1 Aerobic Biodegradation

A variety of microbial types, most of which are members of the genus *Pseudomonas*, degrade PCBs in the presence of oxygen. They attack the more lightly chlorinated PCBs (mono- to tetrachlorinated) to give phenolic products by a pathway that is similar to the microbial oxidation of nonchlorinated aromatic compounds. The metabolic pathways used by these diverse microorganisms for PCB destruction are similar (Figure 8-3). This suggests that the degradation is catalyzed by the same group of enzymes and thus has the same genetic basis in all the microorganisms. Since it is unlikely that the same degradative pathway evolved separately in each genetic type, it appears likely that these unrelated microorganisms obtained the ability to degrade PCBs by gene transfer. These mobile genes have made it possible for microorganisms living in the presence of PCBs to gain a selective advantage over the others by utilizing the energy and carbon present in PCBs for their own metabolic processes.

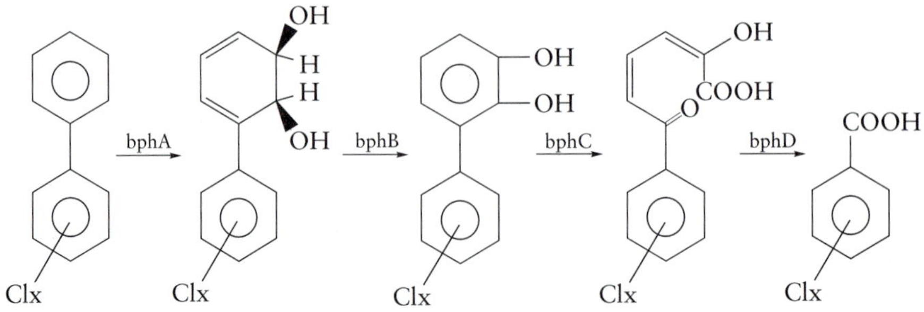

FIGURE 8-3 Degradation of biphenyl (bph) and chlorobiphenyls (clx) by the 2,3-dioxygenase pathway in *Pseudomonas* strain LB400. Redrawn from F. J. Mondello, *J. Bacteriol.*, 171, 1725 (1989). Used by permission of the American Society for Microbiology.

The genes encoding the enzymes responsible for the aerobic degradation of PCBs have been isolated, cloned, and incorporated into microorganisms such as *E. coli*. These genetically engineered microorganisms have the same ability to degrade lightly chlorinated PCBs as the ones found in the soil samples. It may be possible to design microorganisms that degrade PCB's faster than those that occur naturally.

8.7.4.2 Anaerobic Biodegradation

The bulk of the PCBs present in the environment are present in the sediments of lakes, rivers, and oceans where there is little or no oxygen. In some lakes, like Lake Ontario, where there is little homogenization of sediments, there are close correlations between the U.S. sales of PCBs and the concentrations of PCBs in sediment cores (Figure 8-4). Since sediments are the principal repository for PCBs, anaerobic degradative pathways are likely to be of greater importance than aerobic ones.

Studies of the degradation of PCBs in the sediments of the Hudson River in New York resulted in the discovery of microorganisms that degrade PCBs in the absence of oxygen. These microorganisms degrade the more highly chlorinated PCBs (> tetrachloro) to the mono-, di-, and trisubstituted isomers. This reductive dehalogenation results in the substitution of a hydrogen for the chloro group on the aromatic ring. Similar reductive dechlorinations have been observed in the environment with other chlorinated aromatic compounds, as shown earlier in reaction (8-23). Anaerobic microorganisms reduce the chloro groups from the highly substituted biphenyls to give lightly chlorinated biphenyls, and the aerobic microorganisms destroy the lightly substituted biphenyls. In one proposed bioremediation procedure, PCB-containing soil is

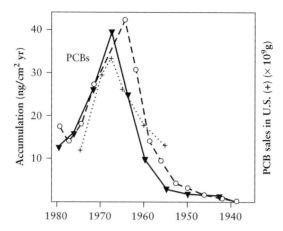

FIGURE 8-4 Correlation of PCBs in Lake Ontarico core samples with PCB sales in the United States: E-30 (triangles) and C-32 (circles) are cores from two different sites in Lake Ontario. Redrawn from D. L. Swockhamer and S. J. Ehrenreich, in *Organic Contaminonts in the Environment. Environmental Pathways and Effects*, K. C. Jones, ed., Elsevier Applied Science, London. Copyright © 1991.

to be incubated first with anaerobic microorganisms in the absence of oxygen and then with aerobes in the presence of oxygen. This should result in the destruction of the PCBs in sediments, while requiring a minimal energy input to the process.

8.8 DDT AND ITS DEGRADATION PRODUCTS

The insecticidal effects of DDT were discovered by the Swiss chemist Paul Müller in 1940, and DDT was used extensively in World War II for the control of diseases spread by flies, mosquitoes, and other insects. Initially, it was very effective in killing the mosquitoes that carry the microorganism responsible for malaria. It was used directly on humans to control head lice without apparent toxic effects. In 1948 Müller was awarded the Nobel Prize in chemistry for his discovery.[4] DDT was used in even greater amounts in the 1950s and 1960s to control insects that attack crops, in addition to its use for fly and mosquito control. There was a general feeling at that time that we would be able to conquer all insect pests with DDT and other pesticides.

[4]The initial beneficial effects of DDT are described in M. Gladwell, "The Mosquito Killer," *The New Yorker*, July 2, 2001, pp. 42–51.

8.8.1 Synthesis

DDT [1,1,1-trichloro-2,2-bis(*p*-chlorophenyl)ethane] (also called dichlorodi-phenyltrichloromethane, hence DDT), is prepared by reaction (8-31) between trichloroacetaldehyde (chloral) and chlorobenzene. The mechanism of the reaction involves the electrophilic addition of a carbocation to the aromatic ring; a Friedel–Crafts type of reaction. The mechanism is shown in reactions (8-31)–(8-34).

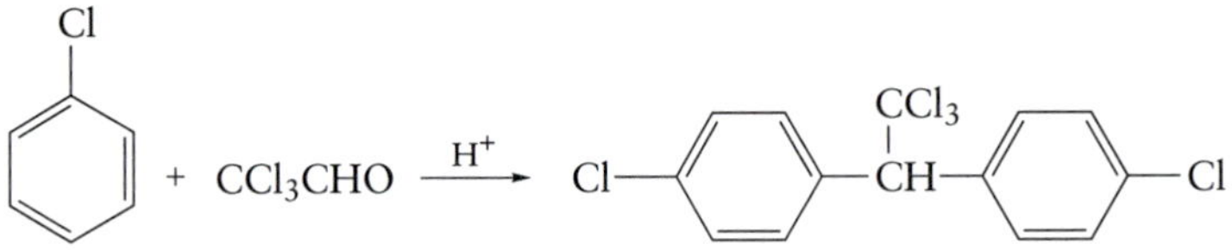

$$\text{(8-31)}$$

p, p'- DDT

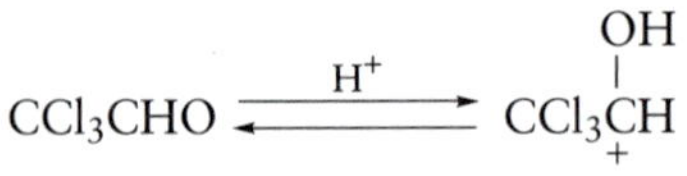

$$\text{(8-32)}$$

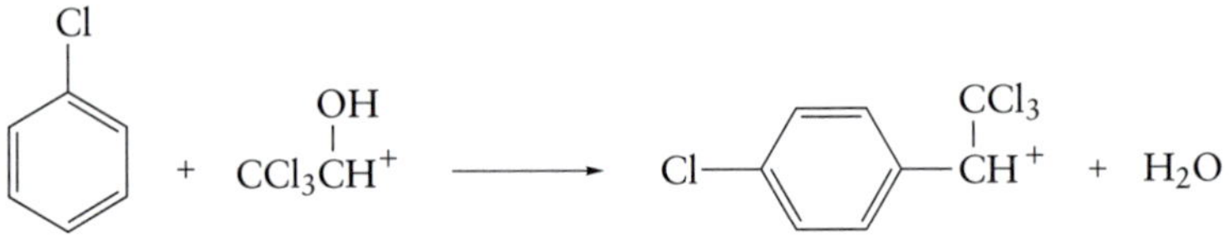

$$\text{(8-33)}$$

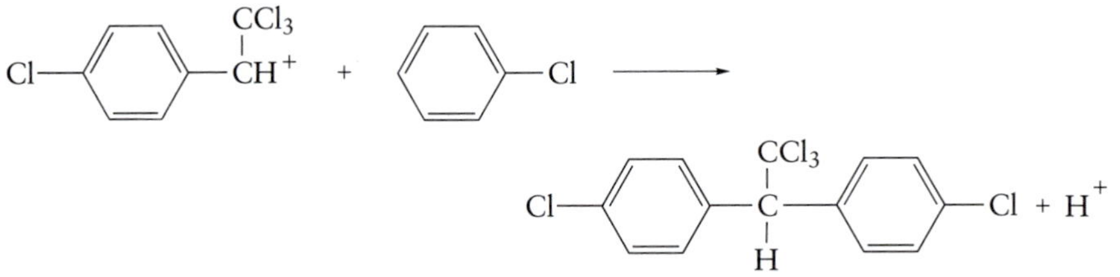

$$\text{(8-34)}$$

The *p,p'*-isomer shown in reaction (8-31) is not the only reaction product. Appreciable amounts of *o,p'*- and *o,o'*-DDT are also obtained. This synthetic approach has been applied for the preparation of a great variety of DDT analogues. One important example, methoxychlor, is used in place of DDT in many applications because it is biodegradable.

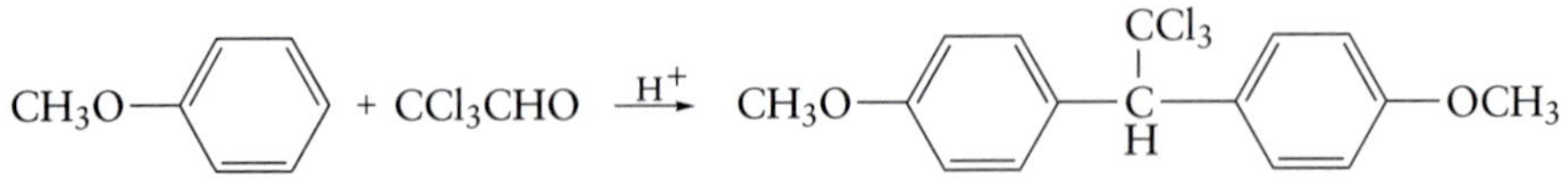

$$\text{(8-35)}$$

Methoxychlor

Methoxychlor in less effective as an insecticide, however, and about three times as costly to prepare, since anisole (methoxybenzene) is a more expensive starting material than chlorobenzene. The estimated LD_{50} (lethal dose for 50% of the population) of methoxychlor in rats is 5000–7000 mg/kg (VS 250–500 mg/kg for DDT).

8.8.2 Environmental Degradation

The first step in the degradation of DDT is the relatively rapid elimination of HCl to form DDE (1,1-dichloro-2,2-bis(p-chlorophenyl)ethylene) (reaction 8-37) when the DDT is heated in water. However, the subsequent hydrolysis of DDE to DDA is extremely slow, as indicated in reaction (8-37), because there are only unreactive vinyl and aryl chlorides in DDE.

DDD (also known as TDE), a major microbial metabolite of DDT formed by reductive dechlorination, has also been used as an insecticide (Rothane). Because of reactions (8-36) and (8-37), both DDD and DDE accompany DDT and DDA in the environment, so when the mixture of DDT and its degradation products is reported in environmental samples, it is often referred to as DDT.

Methoxychlor and other DDT analogues that do not contain the p,p'-chloro group are much less persistent than DDT in the environment. This is due to their metabolism by soil microorganisms to phenols that are readily degraded to acetate in accordance with reactions (8-38) and (8-39). o,p-DDT (8-38) is known to have estrogenic activity (Section 8.5.1).

$$(8\text{-}38)$$

$$(8\text{-}39)$$

The enzymes involved in these oxidations are probably of the cytochrome P-450 type mentioned earlier (Section 8.5). Cytochrome P-450 enzymes are also present in insects. The lower toxicity of methoxychlor (vs DDT) may reflect the effective oxidative detoxification of methoxychlor by some insects.

8.8.3 Mechanism of Action of DDT and DDT Analogues as Pesticides

Insects sprayed with DDT exhibit hyperactivity and convulsions consistent with the disruption of the nervous system by DDT. Many theories have been suggested for the toxic effect of DDT, and the exact mechanism is not known. A theory that appears to be plausible and has some experimental basis suggests that the DDT molecules are of the correct size to be trapped in the pores of the nerve membranes, which are thus distorted, allowing sodium ions to leak through and depolarize the nerve cell so that it can no longer transmit impulses. This theory states that the toxicity of DDT is not due to its chemical reactivity but rather to its size and geometry, which allow for the blockage of the pores of the nerve membranes. This theory is supported by the observation that a variety of quite different compounds that are stereochemically similar to DDT exhibit DDT-like activity.

This "special fit" hypothesis is supported by the observation that the biological activity of methoxychlor analogues decreases rapidly when R in the following formula contains five carbon atoms or more.

$$RO-\underset{}{\bigcirc}-\overset{CCl_3}{\underset{H}{C}}-\bigcirc-OR$$

Methoxychlor analogue; R is substituted for CH_3

Presumably the longer chain methoxychlor analogues are not able to fit into the nerve pores.

Insects are especially susceptible to DDT because it is readily absorbed through the insect cuticle. DDT is not appreciably absorbed through the protein skin of mammals and does not exhibit acute toxicity. This is why it could be safely applied directly to humans to kill head lice. If the polarity of the DDT analogue is increased by introducing $-NO_2$, $-CO_2H$, $-CO_2CH_3$, and $-OH$ substituents into the aromatic ring, the toxicity to insects is lost. This loss is probably due in part to decreased absorption through the exoskeleton and in part to decreased adsorption on the nonpolar nerve membrane.

8.8.4 Chronic Toxicity of DDT and Related Compounds

The large-scale use of DDT as an insecticide entailed the release of large amounts into the environment, where it and its degradation products persist for decades. These compounds are so volatile that, like PCBs, they were spread worldwide in small amounts by the global winds. Yet while the use of DDT was increasing, it was noted that there was a decrease in many species, especially those at the end of food chains, on earth. These included especially eagles, hawks, and falcons, birds that feed mainly on fish.

Correlations were observed between the decreasing number of young hatched and the amounts of DDT and its degradation products in the eggs that did not hatch. A greater number of broken eggs than usual were found in the nests, and all the eggs, both whole and broken, had thinner shells than normal. In addition, the adults tended to mate later than usual in the summer, giving their progeny less time to mature before winter arrived. Thus fewer of the immature adults survived to have offspring the following summer.

All these factors resulted in a precipitous drop in the populations of these fish-eating birds. The presence of both DDT and its degradation products and PCBs in the eggs that did not hatch suggested that both chloroorganics may have been responsible for the collapse of the populations of these birds. In the United States, the manufacture and use of DDT was stopped in 1972, and the manufacture of PCBs was halted in 1977. Within 10 years after the DDT ban, there was a marked increase in the number of young hatched by the affected birds and a corresponding decrease in the amounts of DDT and its degradation products in the eggs that did not hatch. In addition, there was a corresponding increase in the thickness of the egg shells. During this time period there was no change in the levels of PCBs in the eggs, suggesting that DDT and its derivatives were responsible for the initial decrease in these bird populations. Independent studies have shown that low concentrations of PCBs have little effect on the thickness of avian eggshells.

There are two proposals to explain the connection between eggshell thickness, mating behavior, and the DDT. In one theory the connector is the monooxygenase enzyme cytochrome P-450. As noted in Section 8.5, high levels of liver cytochrome P-450 are induced by some chlorinated organics, levels that remain high if the chloroorganics are heavily chlorinated and thus difficult to oxidize. DDT residues induce cytochrome P-450 isozymes, structurally different P-450s which have similar catalytic activity and cause changes in the level of the steroid hormone estradiol, which controls mating behavior.

Estradiol

The production of eggshells is also influenced by estradiol. The shell of an egg is not formed until the last day before laying. About 60% of the calcium needed to form the egg is obtained from the bird's food intake, and the remainder comes from calcium stored in bones. Since the level of calcium deposited in the bones is regulated by levels of estradiol in the blood, a low level of estradiol results in a low level of stored calcium. The absence of stored calcium could be a significant factor in egg survival, since a 20% decrease in shell thickness results in extensive egg breakage. A 20% decrease in shell thickness is correlated with as little as 25 ppm DDT or its breakdown products observed in the egg.

In addition, it has been observed that some pesticides inhibit the formation of eggshells even in the presence of an adequate supply of calcium. Presumably these pesticides interfere with the enzyme carbonic anhydrase, which is responsible for the conversion of carbon dioxide to carbonate. The latter is required to combine with calcium to form the calcium carbonate of the shell. It has been established that DDE interferes with the formation of the shell in this way.

The connector between DDT, DDE, and thin eggshell and delayed mating of birds in the second proposal is the induction of hyperthyroidism in birds. This proposal has been questioned by Moriarity.[5] It has been observed that when

[5]F. Moriarity, ed., *Organochlorine Insecticides: Persistent Organic Pollutants*. Academic Press, New York, 1975.

these chemicals are administered to some strains of birds, there is a doubling of the size of the thyroid gland. Increased metabolism, pulse, and heart size are indicative of hyperthyroidism. Studies on birds with hyperthyroidism have shown that they delay mating, and lay lighter eggs with thinner shells. Male birds with hyperthyroidism have significantly lower fertility than those that do not. It is not clear at the time of writing if either of these proposals is correct or whether there is yet another explanation for the biological effects of DDT and its degradation products on fowl.

After DDT was banned, its levels in the environment slowly decreased over a 10-year period (the half-life of pure DDT is about 10 years, but its degradation products last much longer), and consequently its levels in fish and the animals that consumed the fish also decreased, and the steroidal hormonal balance gradually swung back closer to the pre-DDT levels. The result of the ban is that most of the eagles, hawks, pelicans, and falcons in the lower 48 states of the United States have returned to close to their pre-DDT levels either naturally or, in the case of some eagles, with the help of their reintroduction from Alaska and Canada (Figure 8-5). Another example is the dramatic

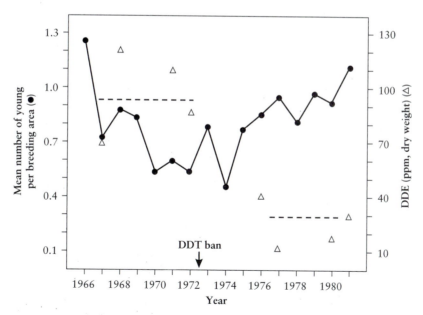

FIGURE 8-5 Summary of average annual bald eagle reproduction and DDE residues in addled (spoiled) eggs in north western Ontario, 1966–1981. Dashed lines indicate weighted mean concentrations of DDE residues in clutches before (94 ppm) and after (29 ppm) the ban of DDT. Means for the 16-year period are 57 ppm DDE (weighted mean) and 0.82 young per breeding area. Redrawn with permission from James W. Grier, *Science,* **218** 1232–1236. Copyright © 1982 American Association for the Advancement of Science.

TABLE 8-3

Brown Pelican Breeding in Coastal Southern California (U.S.) and Northern Baja California (Mexico) and the Content of DDT and Its Breakdown Products in Anchovies and Pelican Eggs

Year	Nests built	Young fledged	DDE, DDT, and DDD[a]	
			Anchovies (ppm, fresh wet weight)[b]	Pelican eggs (ppm, lipid mass)[c]
1969	1125	4	4.27	1055
1970	727	5	1.40	
1971	650	42	1.34	
1972	511	207	1.12	>221
1973	597	134	0.29	183
1974	1286	1185	0.15	97

[a]About 95% DDE.
[b]A major food for the brown pelican.
[c]Addled and broken eggs.
Source: Adapted from D. W. Anderson, R. W. Risebrough, L. A. Woods, Jr., L. R. Deweese, and W. G. Edgecomb, Brown pelican: Improved reproduction of the southern California coast, *Science*, **190**, 806–808 (1975).

increase in the successful hatching of the brown pelican after 1970, when effluent from a DDT manufacturing facility in Los Angles was diverted from the Pacific Ocean to a landfill (Table 8-3).

Although the manufacture and use of DDT has been banned in Europe and North America, it is still used in large amounts in India, Africa, the Middle East, Southeast Asia, and Central and South America. One of its major uses is for the control of the anopheles mosquito, which transmits malaria to humans. The total DDT use in developing countries at the turn of the century was comparable to that used in the developed countries in the 1960s. The volatility and stability of DDT and its degradation products, which have an approximate half-life of 60 years in temperate climates, suggests that these compounds will continue to be an environmental problem for many more years unless all the nations of the world decide to ban the manufacture and use of the pesticide.

8.9 OTHER CHLOROORGANIC PESTICIDES

8.9.1 Lindane

8.9.1.1 Synthesis

Lindane, the γ-isomer of 1,2,3,4,5,6-hexachlorocyclohexane, is prepared by the photochemical initiated addition of chlorine to benzene, as shown in reaction (8-40).

$$\text{(8-40)}$$

In this reaction four of the eight possible isomers of 1,2,3,4,5,6-hexachlorocyclohexane, are formed in appreciable amounts. These isomers differ in the relative orientations of the chloro substituents. Axial (a) groups are perpendicular to the cyclohexane ring and are labeled in the α-isomer in reaction (8-40). Equatorial groups are in the equatorial plane of the ring as shown in the β-isomer. Only one of these isomers, the γ-isomer, exhibits significant insecticidal action. Initially, the total reaction mixture was used as the insecticide. However, that practice was discontinued because it resulted in the distribution into the environment of large amounts of potentially toxic chlorocarbon compounds, which do not have insecticidal activity.

Initially the use of lindane was banned in the United States, but that ban was lifted, and it is now imported for some specific uses. For example, lindane is used as a soil insecticide and on cotton and forest crops. Lindane is excreted rapidly, and its acute mammalian toxicity is low.

The β-isomer tends to accumulate in mammalian tissues and exhibits chronic toxicity. The volatility of the γ-isomer is high, which makes it useful as a soil insecticide, but this volatility results in its distribution far from the point of use. 1,2,3,4,5,6-Hexachlorocyclohexane accumulates in plankton and fish (presumably the β-isomer), along with DDT residues and PCBs, with the highest concentrations being found in bottlenose dolphins and other animals at the end of the food chain.

8.9.1.2 Environmental Degradation

Lindane is an inert chloroorganic that undergoes a very slow elimination of HCl in aqueous solution in path (a) of reaction (8-41). Presumably this is an E1 reaction [see reactions (8-14) and (8-17)]. Lindane is also degraded by microorganisms by a combination of reductive and elimination pathways to 1,2,4-trichlorobenzene [see path (b) of reaction (8-41)].

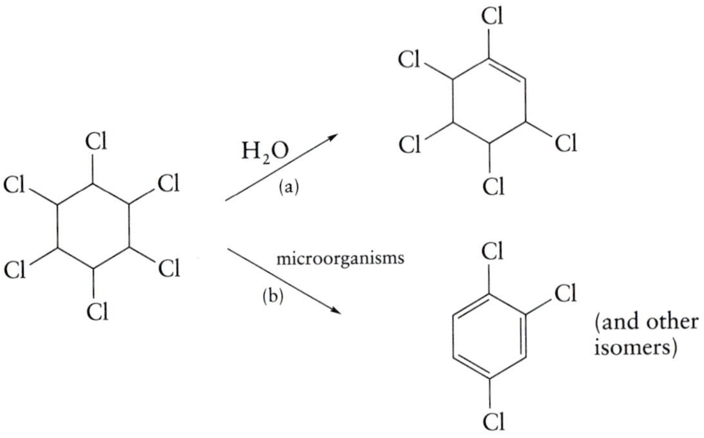

(8-41)

8.9.2 Polychlorinated Cyclodienes

8.9.2.1 Synthesis

A series of bicyclic insecticides, mirex, chlordane, heptachlor, aldrin, and dieldrin, can be synthesized by the Diels–Alder addition of hexachlorocyclopentadiene to olefins, as follows:

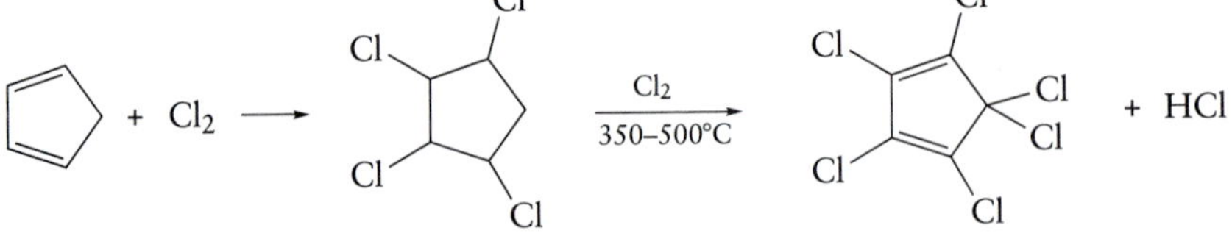

hexachlorocyclopentadiene

(8-42)

The preparation of these compounds is shown in reactions (8-43) and (8-44). It is possible to use the Diels–Alder reaction to prepare a whole series of potential insecticides simply by varying the diene and olefin.

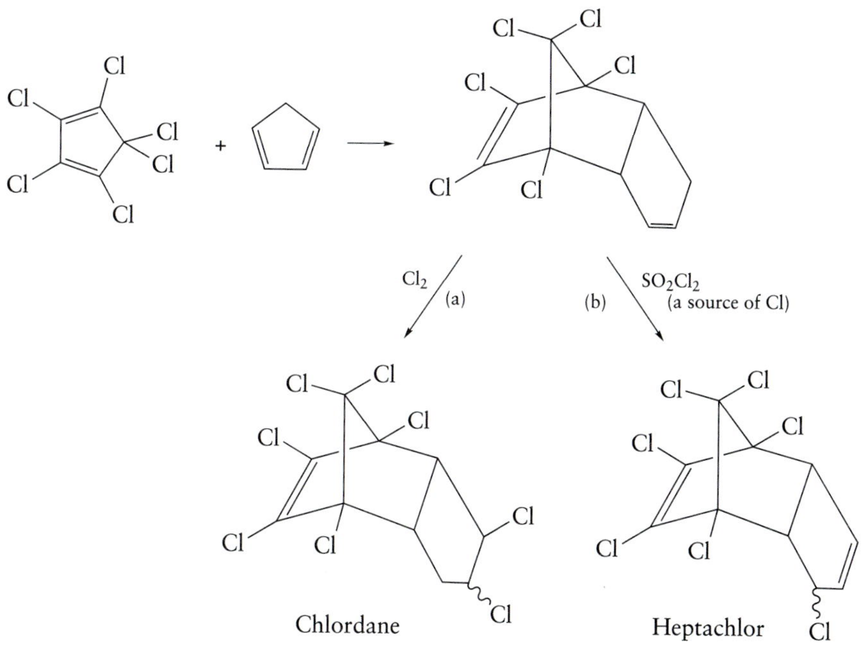

Chlordane

Heptachlor

(8-43)

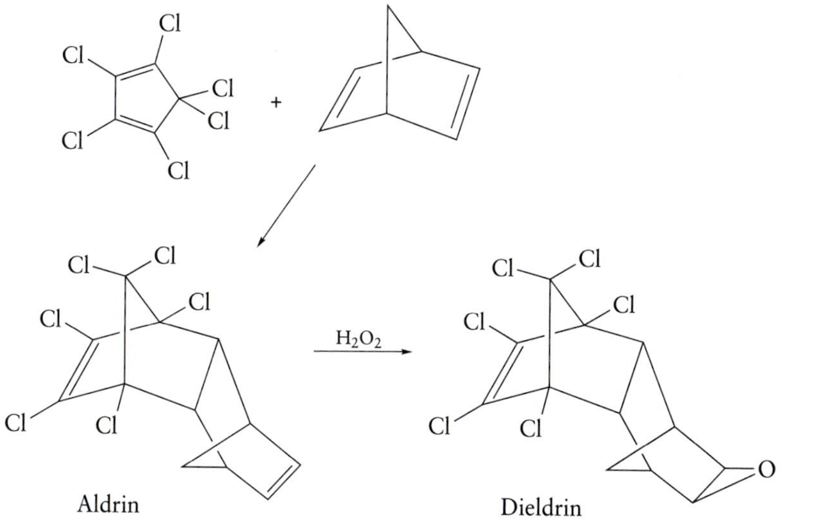

Aldrin

Dieldrin

(8-44)

Mirex, another insecticide, is prepared by the aluminum chloride catalyzed cyclodimerization of hexachlorocyclohexadiene, as in reaction (8-45). It is one of the most stable known organic compounds and has a melting point of 485°C.

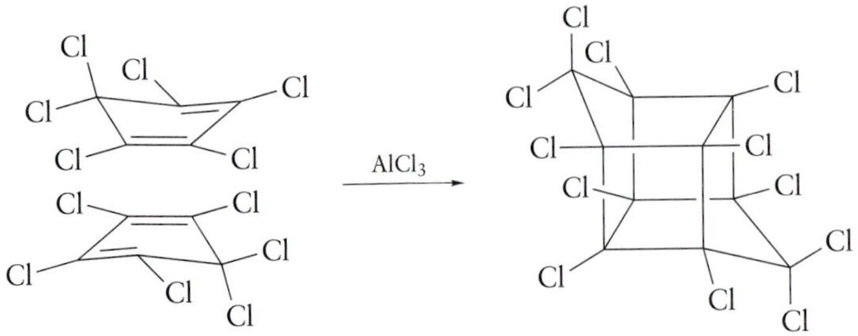

(8-45)

Mirex

(8-46)

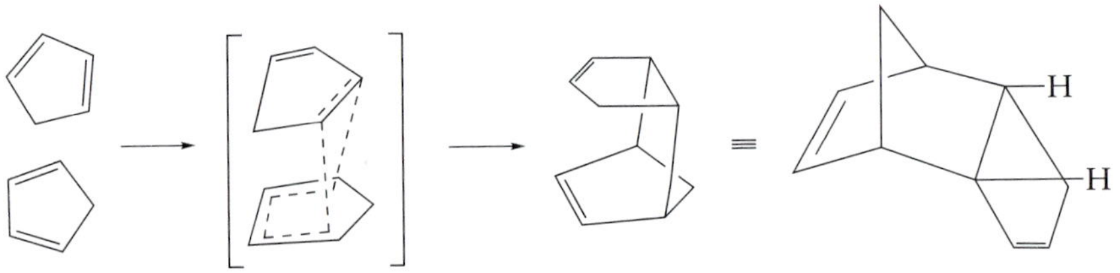

Endo (predominant product)

(8-47)

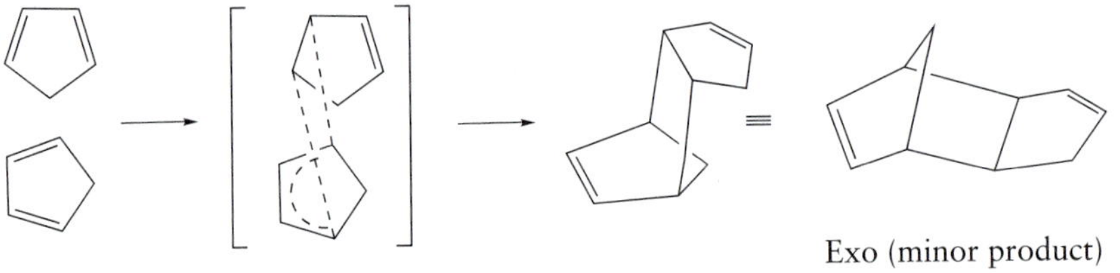

Exo (minor product)

The endo adduct shown in reaction (8-46) is the major product resulting from the Diels–Alder addition of a diene and an olefin. No intermediate has been detected, but the reaction product can be rationalized as having originated from a transition state in which there is maximum interaction between the π bonds of the olefin and diene. This is illustrated in reactions (8-46) and (8-47), in which the upper cyclopentadiene may be regarded as the olefin, and the lower cyclopentadiene, the diene.

8.9.2.2 Environmental Degradation

The cyclodiene insecticides are very resistant to environmental degradation. One might expect that the tertiary, allylic chloro groups at the bridgehead carbons would be especially susceptible to S_N1 hydrolysis or E1 elimination [see reaction (8-14)]. However, the rigid geometry of the bicyclic ring system prevents carbocation formation, as shown in reaction (8-48).

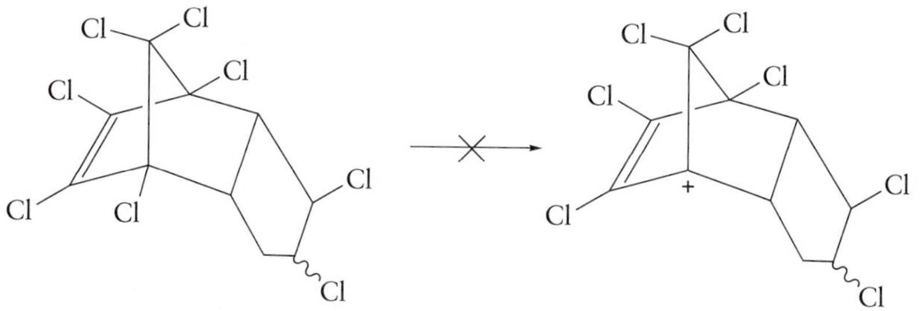

(8-48)

A carbocation is more stable if the positive central carbon atom and the three substituents attached to it are in the same plane. It is not possible to obtain a planar carbocation intermediate at the bridgehead carbon atom of these compounds without introducing extraordinary strain in the bicyclic system. Consequently, the chloro groups at the bridgehead position are quite unreactive.

Some of the cyclodiene insecticides are susceptible to oxidative attack by soil microorganisms, plants, and animals, and by UV. However, the resulting products, such as the one illustrated in reaction (8-49), are often more toxic than the starting pesticide.

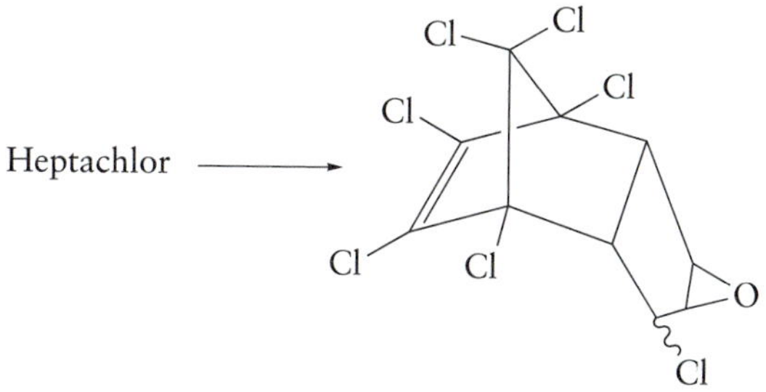

Heptachlor ⟶

(8-49)

Photodieldrin, shown in reaction (8-50), is formed by the action of ultraviolet light (UV-B, 290–320 nm) as well as by the action of oxidase enzymes on Dieldrin. The photoproduct of heptachlor is more resistant to environmental degradation than heptachlor itself.

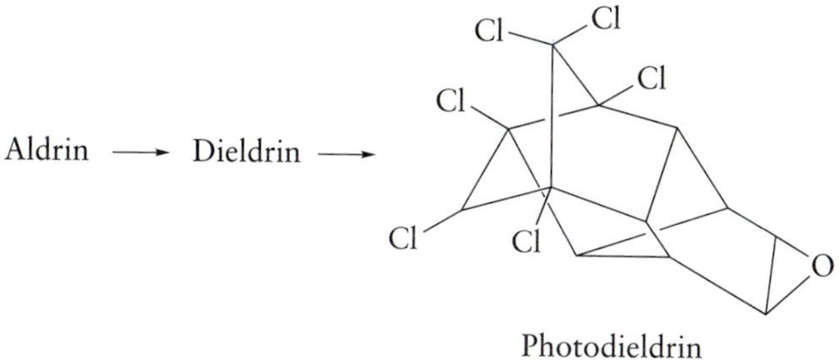

Aldrin ⟶ Dieldrin ⟶

(8-50)

Photodieldrin

Aldrin, dieldrin, heptachlor, and chlordane have been used to kill grubs and other soil pests and were applied mainly to corn and cotton. Dieldrin has also been used to control the anopheles mosquito.

The mechanism of action of these bicyclic organochlorine compounds is similar to that of DDT. However, their toxic effects as insecticides suggest that the site of action lies in the ganglia of the central nervous system rather than the peripheral nerves that are affected by DDT.

Mirex has a long lifetime in the environment and a half-life in mammals of about 10 years. It was used extensively in the southern United States in the 1950s and 1960s in an attempt to control the fire ant, an accidentally introduced species from South America. Fire ants live in large colonies in mounds they construct in pasture land. Humans and livestock that step in the mounds are aggressively attacked and bitten by hordes of fire ants coming out of the mound. At one point attempts were made to control the fire ants by using World War II bombers to drop a mix of ant food and mirex on pastures. This killed not only the fire ants but also all the other ants and many other insects as well. When the plan failed to eradicate the fire ant, the Mirex treatment was terminated. Since the fire ant has a more efficient mechanism of recovery than the indigenous ants, it took over the fields that had been treated with Mirex when the pesticide was washed away. So the Mirex eradication plan actually enhanced the spread of the fire ant by killing the indigenous ants that were the barrier to its rapid spread throughout the South.

Because of their potential danger as carcinogens and because they accumulate in fatty tissues of birds and mammals, the Environmental Protection Agency has terminated the use of mirex (1971), aldrin and dieldrin (1974), and heptachlor and chlordane (1976). Exemptions for use in special instances have been granted.

An example of an exemption gone bad resulted in the contamination of milk in Hawaii, where a request for the use of heptachlor to control aphids (mealybugs) on pineapple plants in was granted in 1978. The aphids cause "mealybug wilt," which withers pineapple roots. The aphids are carried to the pineapple plants by ants which, in return, collect the "honeydew" produced by the aphids. The aphids were controlled by killing the ants with Heptachlor. At the same time heptachlor was being used, agricultural scientists were working on a procedure to use pineapple plants, after the pineapples had been harvested, as cattle feed. The plants were not supposed to be fed to cattle until one year after the last application of heptachlor, but in 1982 this procedure was not followed. Consequently, high concentrations of the hydrophobic heptachlor ended up in the milk of the cows that were fed the pineapple plants, and the contaminant was passed on to those drinking the milk, mainly children. These high levels were detected in the semiannual screening of pesticides in milk by the Hawaii Health Department.

8.10 HERBICIDES: 2,4-DICHLOROPHENOXYACETIC ACID AND 2,4,5-TRICHLOROPHENOXYACETIC ACID

8.10.1 Introduction

2,4-Dichlorophenoxyacetic acid (2,4,-D) and 2,4,5-triphenoxyacetic acid (2,4,5-T) are herbicides that are especially effective in the control of broad-leafed plants, yet they have little or no effect on grasses. Hence they are used to control weeds and other unwanted vegetation where only grasses are desired. These compounds mimic the plant growth hormone (auxin) indoleacetic acid.

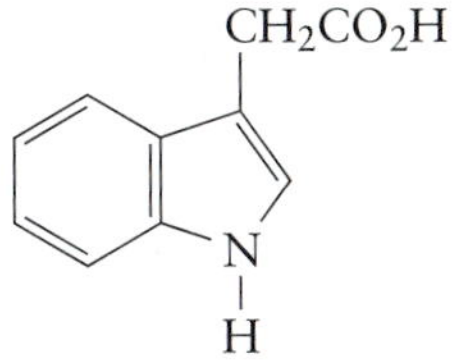

Indoleacetic acid

However, the large amounts used (in comparison to the natural auxin) cause abnormal growth and eventual death of the plant. Specific results include little or no root growth, abnormal stem growth, and leaves with little or no chlorophyll.

8.10.2 Synthesis

2,4-D and 2,4,5-T are manufactured by the S_N2 displacement of the chloro group of chloroacetate with a phenoxide anion as in reactions (8-51) and (8-52) or by the chlorination of phenoxyacetic acid, as in reaction (8-53).

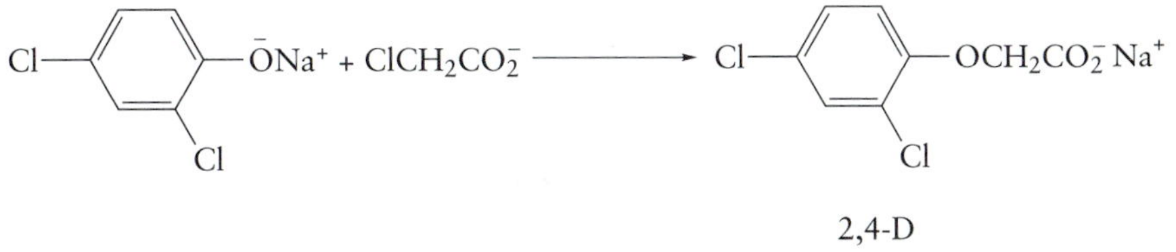

$$(8-51)$$

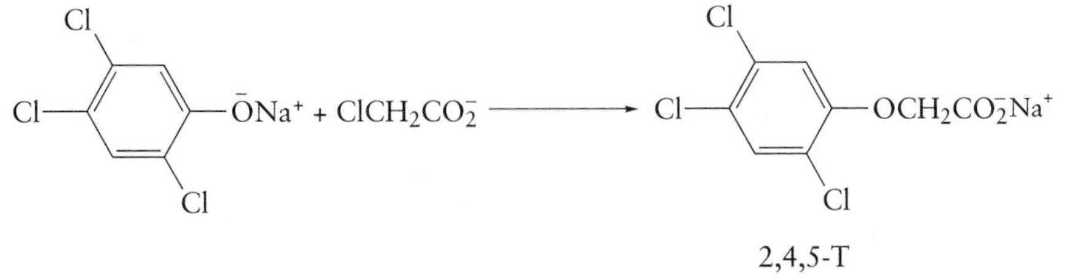

2,4-D

$$(8-52)$$

2,4,5-T

$$\text{C}_6\text{H}_5-\text{OCH}_2\text{CO}_2^-\text{Na}^+ + \text{Cl}_2 \longrightarrow \text{Cl}-\text{C}_6\text{H}_3(\text{Cl})-\text{OCH}_2\text{CO}_2^-\text{Na}^+ \qquad (8\text{-}53)$$

2,4-D

A variety of closely related 2,4-D analogues also show herbicidal activity. The carboxyl group can be replaced by an amide, nitrile, or ester group. In addition, a longer side chain can be used in place of the acetic acid moiety. However, this side chain must contain an even number of carbon atoms to be active. This result is consistent with β-oxidation of the side chain by the plant (Section 7.2.4) to form the biologically active 2,4,-D.

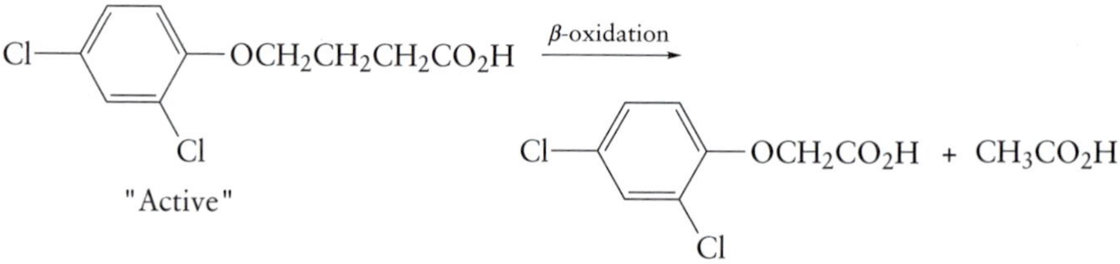

$$\text{Cl}-\text{C}_6\text{H}_3(\text{Cl})-\text{OCH}_2\text{CH}_2\text{CH}_2\text{CO}_2\text{H} \xrightarrow{\ \beta\text{-oxidation}\ } \text{Cl}-\text{C}_6\text{H}_3(\text{Cl})-\text{OCH}_2\text{CO}_2\text{H} + \text{CH}_3\text{CO}_2\text{H} \qquad (8\text{-}54)$$

"Active"

$$\text{Cl}-\text{C}_6\text{H}_3(\text{Cl})-\text{OCH}_2\text{CH}_2\text{CO}_2\text{H} \xrightarrow{\ \beta\text{-oxidation}\ } \text{Cl}-\text{C}_6\text{H}_3(\text{Cl})-\text{OH} + \text{CO}_2 + \text{CH}_3\text{CO}_2\text{H} \qquad (8\text{-}55)$$

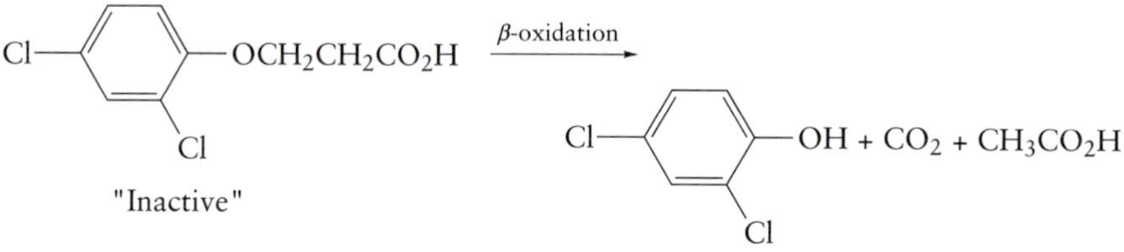

"Inactive"

8.10.3 Environmental Degradation

2,4-D and 2,4,5-T are subject to rapid environmental degradation. These compounds are cleaved by microorganisms to the corresponding phenol, which is readily degraded further. Presumably, carbon dioxide and formic acid are the other reaction products:

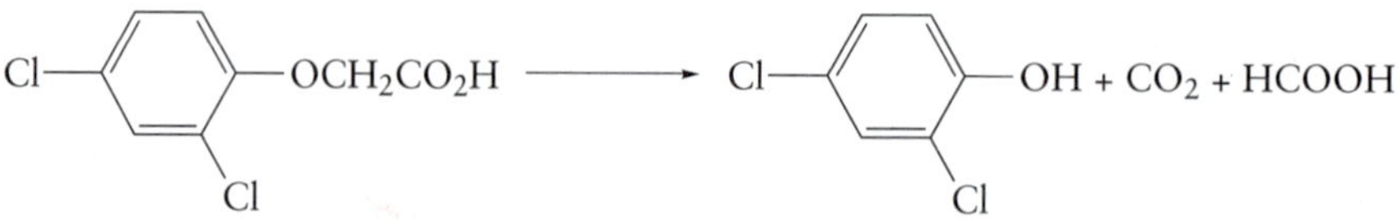

$$\text{Cl}-\text{C}_6\text{H}_3(\text{Cl})-\text{OCH}_2\text{CO}_2\text{H} \longrightarrow \text{Cl}-\text{C}_6\text{H}_3(\text{Cl})-\text{OH} + \text{CO}_2 + \text{HCOOH} \qquad (8\text{-}56)$$

In addition, the photochemical cleavage of the acetic acid side chain and the photochemical hydrolysis of the aryl chloro groups have been demonstrated. These findings suggest that the long-term environmental impact of these compounds is small.

The environmental problem that has been associated with these compounds is due to 2,3,7,8-tetrachlorodibenzo-*p*-dioxin (TCDD) impurities produced during their manufacture. The synthesis of 2,4-dichlorophenol and 2,4,5-trichlorophenol (8-57) results in the formation of small amounts of TCDD (8-58), which are carried along with the final product.

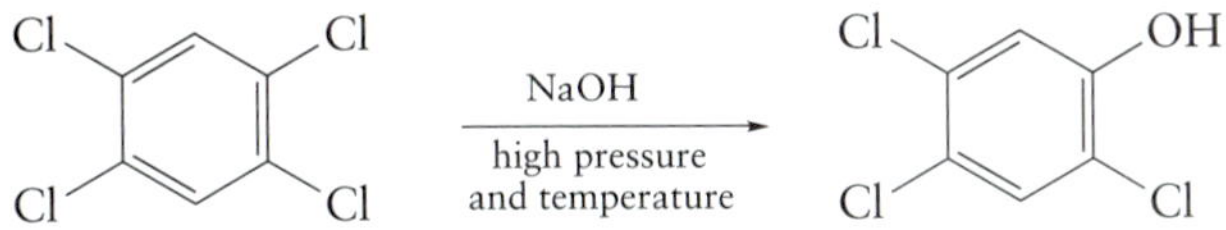

(8-57)

2,4,5-Trichlorophenol

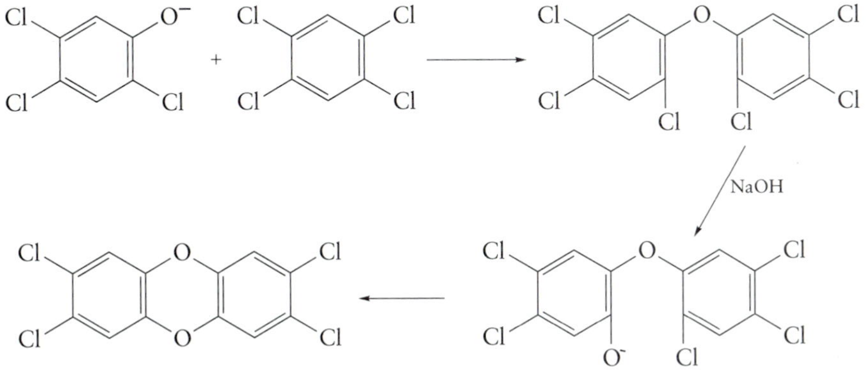

(8-58)

2,3,7,8-Tetrachlorodibenzo-*p*-dioxin (TCDD)

TCDD is formed because of the vigorous conditions that are required to hydrolyze the aryl chloro groups. The presence of TCDD in 2,4,5-T led to the ban of its use in the United States in 1979. Related toxic five-membered ring derivatives [polychlorinated dibenzofurans: reaction (8-30)], formed in the manufacture of PCBs, may be also responsible for some of the toxicity of these compounds. It is generally considered that 2,4-D is safe, but it has been established that it belongs to the class of compounds called peroxisome proliferators. Cell structures containing bundles of enzymes are called peroxisomes. Compounds that induce the formation of peroxisomes are also known to cause tumors and testicular damage in rodents. It has also been suggested as the cause of non-Hodgkins lymphomas in the Vietnam veterans who handled Agent Orange (discussed shortly) and as an immune system suppressor.

TCDD and other dibenzofurans, substituted with different amounts and different patterns of substituents, as well as polychlorinated benzofurans, are formed during the incineration of municipal waste. The polychlorinated dibenzodioxins and polychlorinated dibenzofurans are present in the waste after incineration in amounts 2–100 times greater than the amounts present before incineration. Their formation is not due to pyrolysis of chloroorganics [e.g., poly(vinyl chloride)] present. Rather, it has been established that the polychlorinated dibenzofurans and polychlorinated dibenzodioxins are synthesized from the carbon and chlorine formed in the combustion process in reactions catalyzed by the fly ash formed at 250–400°C in the postcombustion zone in the incinerator. Fly ash is the fine particulate inorganic material formed from the inorganics that constitute about 25% of municipal waste. This ash consists of an array of metal salts, any of which could serve as a catalyst for the synthesis of TCDDs and polychlorinated benzofurans. Isotope labeling studies have shown that these compounds form directly from carbon or from the chlorinated phenols that are produced in the combustion process.

The level of the toxicity of TCDD (Section 8.7.2) varies dramatically in animals, and it is not possible to estimate its toxicity to humans (Table 8-4). It was listed as "known to be a human carcinogen" by the U.S. National Toxicity Program in January 2001. The toxicity to humans and a wide array of mammals was, however, demonstrated when dioxins were present in some oil applied to the soil in a horse arena in Missouri. Contact with only 32 µg of dioxins per gram of soil was sufficient to kill the birds, cats, dogs, and horses that used the arena. Humans were less sensitive, but one child developed hemorrhagic cystitis and several other skin lesions. The acnegenic effect has been observed in workers involved in the manufacture of 2,4,5-T.

Similar toxic effects were noted in 1976 when 1–4 lb of TCDD was accidentally released over a 123-acre area near the northern Italian town of Seveso. At a plant where 2,4,5-trichlorophenol was manufactured, a reactor had overheated, blowing out a plume that contained TCDD and trichlorophenol. Many animals were killed by the TCDD, but the main immediate problem

TABLE 8-4

Variation in TCDD Toxicity

Animal	LD_{50} (mg/kg)
Guinea pig	0.6–2.5
Mink	4
Rat	22–330
Monkey	<70
Rabbit	115–275
Mouse	114–280
Dog	>100–<3000
Hamster	1150–5000

Source: Adapted from D. J. Hanson, Dioxin toxicity: New studies prompt debate, regulatory action, *Chem. Eng. News*, p. 8, Aug. 12, 1991.

observed in humans was skin irritation caused by the trichlorophenol. High levels of TCDD were measured in the human population of that area, and the people have been monitored regularly since their exposure. No birth defects were observed, but in 1993 it was reported that those exposed to the plume appeared to have a higher risk for certain types of cancer than did a control group that was not exposed. It is believed that these cancerous tumors are due to TCDD and not the trichlorophenol, because the tumors reported are of the type induced by TCDD.

The toxicity of TCDD is at the center of a long-running debate in the United States concerning the toxicity of Agent Orange, a 1:1 mixture of 2,4-D and 2,4,5-T that was used to defoliate the jungles of Vietnam. About 17.6 million gallons of defoliant, mainly Agent Orange, were sprayed from airplanes from 1965 to 1971 in an attempt to make it more difficult for enemy guerilla forces to hide from U.S. troops. The 2,4,5-T used in the Agent Orange was contaminated with TCDD, and it was estimated from analysis of blood samples that 2 ppm was the median level of TCDD in each soldier who handled Agent Orange.

There have been a number of studies of the servicemen who handled Agent Orange to determine whether some of the symptoms they exhibited were due to TCDD, since the median value of TCDD in their bodies was about three times that of a control group that did not handle the defoliant. Initial studies suggested that those who handled Agent Orange were not at higher risk for cancer than the rest of the population, but in 1993 a National Academy of Sciences panel determined from epidemiological studies that soft tissue carcinoma, non-Hodgkins lymphoma, and Hodgkins disease were associated with handling Agent Orange. In addition, they concluded that there was suggestive evidence of an association with multiple myeloma, as well as respiratory tract and prostate cancer. The possibility of an association between handling Agent Orange and adult-onset diabetes was reported in 2000. The increased danger was described as small. It was not possible to determine whether the TCDD in the Agent Orange was the source of the health problem. The NAS study group did suggest that 2,4-D was the likely cause of the non-Hodgkins lymphomas.

There have been few studies of the effect of Agent Orange on the citizens of Vietnam who are living in the area where the herbicide was sprayed. Such research is expected to increase if the relations between the United States and Vietnam continue to improve. Anecdotal data suggest that children of parents living in the sprayed region have higher levels of birth defects and disease than children born to parents living in regions that were not sprayed. Adults in the sprayed areas have higher levels of dioxin in their tissue than those living in other parts of Vietnam. Consumption of fish containing high levels of dioxin is a likely source of the dioxin.

There have been a number of sites in the United States made uninhabitable by the presence of high levels of TCDD and other chloroorganics. In 1980 part of the community of Love Canal, New York, was closed and the residents relocated because hazardous chloroorganics, including TCDD, were leaking into the basements of homes built on top of a waste disposal site. The whole town of Times Beach, Missouri, was purchased and closed by the federal government in 1984 because of the high levels of TCDD and other chloroorganics dumped in the soil there. The contaminated soil (240,000 tons) in Times Beach was incinerated to remove the chloroorganics, and the site where the town once was will be turned into a nature preserve. This decontamination was completed in August 2001.

The toxic effects of TCDD in mammals are triggered by its strong binding to a protein called the Ah (aryl hydrocarbon) receptor. Once TCDD has bound to the receptor, a second protein, the translocating factor, facilitates the transfer of the receptor–TCDD complex into the cell nucleus. When a sufficient number of the receptor–TCDD complexes are inside the nucleus, they bind and distort the DNA at a specific locus, which prompts the translation of the information at that point into proteins. One of the proteins formed is a cytochrome P-450 (Sections 8.4, 8.7.4), which oxidizes some aryl hydrocarbons to reactive carcinogens. TCDD binds the most strongly of all the chloroorganics to the Ah receptor, but other compounds that bind less strongly include other chlorodibenzo-*p*-dioxins, chlorobenzofurans, PCBs, and some halogenated naphthalenes. It is estimated that about one-third of the TCDD-like toxicity is caused by the much higher concentrations of PCBs present in humans.

8.11 ORGANOPHOSPHORUS INSECTICIDES

8.11.1 Introduction

Organophosphorus insecticides were used for insect control because the chloroorganics exhibited chronic toxicity and were persistent in the environment. Three examples of the many types of organophosphorous insecticide will be discussed; parathion, methyl parathion, and malathion [shown later in reactions (8-64) and (8-65)]. These organophosphorus compounds exhibit high insecticidal action against a wide variety of species. This is not really an asset because beneficial as well as harmful insects killed by these broad-spectrum insecticides. In addition, since the compounds break down rapidly in the environment, frequent applications are necessary for total pest control. The organophosphorus compounds are also highly toxic to vertebrates as well as to insects and, as a consequence, are much more dangerous to the workers using them than are chloroorganics. The LD_{50} of parathion is 6.4 mg/kg, while that of methyl parathion is 15–20 mg/kg when measured in rats, typical values for the very toxic organophosphorus insecticides. Malathion is atypical in that its LD_{50} is 1200 mg/kg; this low toxicity allows its use where there is an appreciable human presence. Malathion is detoxified by hydrolysis of its ester groups. These ester groups are rapidly hydrolyzed by esterase enzymes present in mammals and absent in insects. Whereas organophosphorous pesticides are not an environmental problem, their broad toxicity makes them dangerous to use. Their induction of resistant strains of insects illustrates one of the problems resulting from the exclusive use of pesticides for insect control.

At the time of writing, the U.S. Environmental Protection Agency was discussing a ban of organophosphate pesticides. Some groups believe that the trace residues of these compounds in foods can have neurotoxic effects on the development of the brain in infants and children.

8.11.2 Synthesis

The phosphoric acid esters used in the synthesis of the organophosphorus insecticides are prepared industrially by the reaction of phosphoryl chloride with alcohols:

$$\underset{\underset{Cl}{|}}{\overset{\overset{O}{\|}}{Cl-P-Cl}} + 2ROH + 2R_3N \longrightarrow \underset{\underset{OR}{|}}{\overset{\overset{O}{\|}}{RO-P-Cl}} + 2R_3N \cdot HCl \qquad (8\text{-}59)$$

Under suitable conditions, diester chlorides (dialkyl phosphorochloridates) may be prepared. The same reaction pathway may be used for the preparation of corresponding sulfur analogues (dialkyl phosphorothiochloridates):

$$\underset{\underset{Cl}{|}}{\overset{\overset{S}{\|}}{Cl-P-Cl}} + 2ROH + 2R_3N \longrightarrow \underset{\underset{OR}{|}}{\overset{\overset{S}{\|}}{RO-P-Cl}} + 2R_3N \cdot HCl \qquad (8\text{-}60)$$

Both the phosphoric acid and thiophosphoric acid derivatives may also be prepared by the action of chlorine on the corresponding derivative of phosphorous acid:

$$\underset{\underset{RO}{|}}{\overset{\overset{O(S)}{\|}}{RO-P-H}} + Cl_2 \longrightarrow \underset{\underset{OR}{|}}{\overset{\overset{O(S)}{\|}}{RO-P-Cl}} \qquad (8\text{-}61)$$

Most nucleophiles attack phosphorochloridates at the phosphorus atom and displace the chloro group. This reaction [i.e., reaction (8-62)] is analogous to the S_N2 displacement on carbon (Section 8.3.2).

$$HS^- + RO-\overset{\overset{O}{\|}}{\underset{\underset{OR}{|}}{P}}-Cl \longrightarrow HS \cdots \overset{\overset{\delta^-}{\overset{O}{\|}}}{\underset{RO\diagdown OR}{P}} \cdots \overset{\delta^-}{Cl} \longrightarrow HS-\overset{\overset{O}{\|}}{\underset{\underset{OR}{|}}{P}}-OR \qquad (8\text{-}62)$$

Transition state

The transition state for the reaction is shown in reaction (8-62).

Most of the organophosphorus insecticides are triesters of phosphoric or thiophosphoric acid that are prepared by the attack of a nucleophile on the phosphorochloridates or phosphorothiochloridates, as shown in the examples that follow. Since the ester group is more difficult to displace than the chloro group, the triesters exhibit greater stability than the phosphorochloridates. As a consequence of the slow displacement of the ester, a reaction pathway different from that just illustrated is sometimes observed. For example, in neutral and acid solutions water can attack the α-carbon atom of the phosphate ester grouping with resultant cleavage of the carbon–oxygen bond:

$$H_2O + R-O-\overset{\overset{O}{\|}}{\underset{\underset{OR}{|}}{P}}-OH \longrightarrow H_2\overset{+}{O}R + \overset{-}{O}-\overset{\overset{O}{\|}}{\underset{\underset{OR}{|}}{P}}-OR \qquad (8\text{-}63)$$

It has been shown by isotope tracer studies using ^{18}O that the P—O bond is not broken. Reaction (8-63) is more likely under environmental conditions than the direct displacement on phosphorus shown in reaction (8-62).

The synthesis of parathion and methyl parathion also involves an S_N2-like displacement:

$$(RO)_2\overset{\overset{S}{\|}}{P}-Cl + {}^-O-\hspace{-4pt}\left\langle\hspace{-4pt}\bigcirc\hspace{-4pt}\right\rangle\hspace{-4pt}-NO_2 \longrightarrow (RO)_2\overset{\overset{S}{\|}}{P}-O-\hspace{-4pt}\left\langle\hspace{-4pt}\bigcirc\hspace{-4pt}\right\rangle\hspace{-4pt}-NO_2 + Cl^-$$

$$(8\text{-}64)$$

$$R = CH_3CH_2-\text{ , parathion}$$

$$R = CH_3-\text{ , methyl parathion}$$

The synthesis of malathion involves a nucleophilic addition of a thioacid to a double bond that is conjugated to a carbonyl group as in reaction (8-65), a Michael addition.

$$(CH_3O)_2\overset{\overset{S}{\|}}{P}-SH + \overset{\overset{CHCO_2CH_2CH_3}{\|}}{CHCO_2CH_2CH_3} \longrightarrow (CH_3O)_2\overset{\overset{S}{\|}}{P}-S\overset{\overset{CH_2CO_2CH_2CH_3}{|}}{CHCO_2CH_2CH_3} \qquad (8\text{-}65)$$

Malathion

8.11.3 Toxicity

The toxic effect of the organophosphorus insecticides is due to their interference in the transfer of nerve impulses from one nerve cell to the next. When a nerve impulse reaches the end of a nerve cell, it triggers the release of a minute amount of the compound acetylcholine:

$$(CH_3)_3\overset{+}{N}CH_2CH_2O\overset{\overset{\displaystyle O}{\|}}{C}CH_3 \xrightarrow[\text{cholinesterase}]{H_2O} (CH_3)_3\overset{+}{N}CH_3CH_2OH + CH_3CO_2H \qquad (8\text{-}66)$$

$$\text{Acetylcholine} \qquad\qquad\qquad\qquad \text{Choline} \qquad \text{Acetic acid}$$

The acetylcholine activates a receptor on an adjacent nerve cell, causing it to carry the impulse to the next nerve cell. The acetylcholine is then hydrolyzed to choline and acetic acid by the enzyme cholinesterase as in reaction (8-66). Other enzymes then convert the choline and acetic acid back to acetylcholine on the nerve endings.

The organophosphorous pesticides bind chemically to the cholinesterase so that it can no longer catalyze the hydrolysis of acetylcholine. The resulting excess of acetylcholine hyperstimulates nerves and produces convulsions, irregular heartbeat, and choking in vertebrates.

8.11.4 Environmental Degradation

The organophosphorus insecticides undergo a very rapid hydrolysis in the environment. There is no buildup of residues; however, frequent application of the pesticide is required. In the typical hydrolytic process given in reaction (8-67), the nitro group of parathion is reduced by soil microorganisms to give the nontoxic aminoparathion, which is stable and remains bound to the soil.

$$(RO)_2\!-\!\overset{\overset{\displaystyle S}{\|}}{P}\!-\!O\!-\!\!\left\langle\!\!\bigcirc\!\!\right\rangle\!\!-\!NO_2 \xrightarrow{H_2O} HO\!-\!\overset{\overset{\displaystyle S}{\|}}{\underset{\underset{\displaystyle OR}{|}}{P}}\!-\!O\!-\!\!\left\langle\!\!\bigcirc\!\!\right\rangle\!\!-\!NO_2 +$$

$$(8\text{-}67)$$

$$ROH + RO\!-\!\overset{\overset{\displaystyle S}{\|}}{\underset{\underset{\displaystyle OR}{|}}{P}}\!-\!OH \quad + \quad HO\!-\!\!\left\langle\!\!\bigcirc\!\!\right\rangle\!\!-\!NO_2$$

It is not known whether this "inactive" form of parathion constitutes an environmental problem.

Organophosphorus insecticides sometimes undergo conversion to more toxic substances when released into the environment. Reaction (8-68) shows how parathion is readily oxidized by atmospheric oxygen or in an enzymatically catalyzed process to a derivative that is four times as toxic as parathion itself.

$$(CH_3CH_2O)_2\overset{S}{\overset{\|}{P}}-O-\!\!\!\bigcirc\!\!\!-NO_2 \xrightarrow{O_2} (CH_3CH_2O)_2\overset{O}{\overset{\|}{P}}-O-\!\!\!\bigcirc\!\!\!-NO_2 \quad (8\text{-}68)$$

Since parathion exhibits no anticholinesterase activity itself, its insecticidal action may be due to its conversion to this oxygenated analogue in the insect.

8.12 RESISTANCE: THE BUGS FIGHT BACK

The discovery of the general toxicity of DDT to insects in the 1940s and its apparent lack of toxicity to other life-forms suggested that the problem of insect control had been solved. But by 1947, it was not possible to control flies without a considerable increase in both the amount of DDT used and the number of sprayings. This approach increased the environmental burden of DDT markedly. Eventually DDT ceased to be a useful control agent for flies, mosquitoes, and many agricultural pests. Often the DDT would "knock down" flies, which would stagger around for a while as if drunk, and then suddenly fly away again. It was later discovered that the survivors had an enzyme, named DDT dehydrochlorinase, that catalyzes the loss of HCl from DDT to form the nontoxic DDE. Presumably the "drunken" behavior occurred during the few minutes needed for the enzyme to deactivate the DDT so that it could no longer disrupt nerve impulses in the fly.

The problem of insect resistance is a general one, which occurs with the large scale use of a particular pesticide for insect control (Figure 8-6). A dramatic example of this is the chronology of the Colorado potato beetle resistance shown in Table 8-5. It should be noted that not all insects develop resistance to pesticides, and some that are resistant in one part of the country may not be resistant in other areas.

The usual approach for dealing with resistant strains is to switch to another pesticide. But as shown for the Colorado potato beetle, insects that are resistant to one pesticide often develop resistance to another. Moreover, new pesticides are not being marketed as fast as resistant strains are developing, and the new pesticides are usually more costly than the old ones.

Insects possess the biochemical machinery to combat insecticides. Initially it seemed very unlikely that this would be the case because past generations of these insects never encountered synthetic chemicals. But the resistance to pesticides is part of their genome. In addition, the mechanism of detoxification is a general one, which can detoxify pesticides of different types. Presumably the

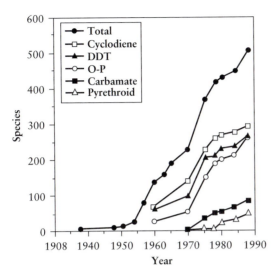

FIGURE 8-6 Chronological increase in the number of insect and mite species resistant to at least one type of insecticide (total), and species resistant to each of the five principal classes of insecticides. Not discussed in this chapter were polychlorinated cyclodiene (cyctodienes), pyrethrums (pyrethroid), and carbamates. From G. P. Georghiou, in *Managing Riointance to Agrochemicals*, ACS symposium series 421, p. 20. American Chemical Society, Washington, DC, 1990.

biochemical pathways to deal with synthetic pesticides developed over the past 250 million years as a result of warfare between plants and insects. Most plants don't just sit there allowing themselves to be chewed up by worms and beetles. For example, the members of the chrysanthemum family produce pyrethrums, a mixture of compounds toxic to many insects. (The pyrethrums extracted from the dried flowers of the chrysanthemum family are used as insecticides.)

The insect strains that survived these wars were the ones that evolved a biochemical apparatus to deal with the natural pesticides, and they use the same apparatus to deal with the synthetic agents. Another example is that birch trees generate a varying array of chemicals throughout the growing season to overcome the genetic ability of insect predators to adapt to the birch tree "insecticides." Resistance to DDT and the cyclodiene insecticides is controlled by genes that initiate the synthesis of proteins that bind these compounds. The protein–insecticide complex then induces the synthesis of detoxifying enzymes. In the case of DDT, the products are glutathione transferases, which catalyze the loss of HCl from DDT, and cytochrome P-450s, which deactivate by addition of a hydroxyl group to the DDT, are produced.

Application of the same or similar pesticides to a field for several years may result in the development of resistant strains of insects. Since the insects that

TABLE 8-5

An Abbreviated Chronology of Colorado Potato Beetle Resistance to Insecticides in Long Island, New York

Insecticide	Year introduced	Year first failure detected
Arsenicals	1880	1940s
DDT	1945	1952
Dieldrin	1954	1957
Endrin	1957	1958
Carbaryl	1959	1963
Azinphosmethyl	1959	1964
Monocrotophos	1973	1973
Phosmet	1973	1973
Phorate	1973	1974
Disulfoton	1973	1974
Carbofuran	1974	1976
Oxamyl	1978	1978
Fenvalerate[a]	1979	1981
Permethrin[a]	1979	1981
Fenvalerate + p.b.[a]	1982	1983
Rotenone + p.b.[a]	1984	?

[a] M. Semel, New York State Agriculture Experimental Station, Riverhead, NY, personal communication, 1984: p.b. = piperonyl butoxide.

Source: G. P. Geovghiou, in *Pesticide Resistance: Strategies and Tactics for Management* Natl. Acad. Science, Washington, DC, 1985, Table 6, pg. 33.

are not resistant to the pesticide are killed off, the farmer is helping to select and build up a population of resistant insects that are not killed by the pesticide. Since most pesticides are not specific for an insect pest, they will usually also kill off insects that are natural predators of the harmful insects. Consequently, when a resistant insect strain develops, it causes even more damage because the number of natural predators has been reduced by the pesticide.

8.13 BIOCHEMICAL METHODS OF INSECT CONTROL

The development of insect resistance has been an important factor prompting the search for methods of insect control that are not totally dependent on the use of pesticides. These methods are outlined briefly because their adoption by farmers will help both to minimize the amounts of synthetic pesticides in the environment and to slow the development of resistant strains of insects. It should be noted that the majority of U.S. farmers use synthetic pesticides, but the amounts used per acre have decreased as biological methods of control are used increasingly in conjunction with the application of synthetic pesticides.

8.13.1 Pheromones

Pheromones are chemicals secreted by one species to affect the behavior of another of its own species. Insect pheromones are used mainly to indicate the location of food or to attract a mate (sex pheromones) (Figure 8-7). In most instances the female secretes sex pheromones. Insect sex pheromones have been studied extensively with the goal of controlling the insect by disrupting mating behavior or by attracting the males to traps where a toxic agent kills them. Pheromones are usually species specific: that is one can selectively control a specific insect pest. Presumably, the insects will not become resistant because pheromones, being produced by their own species, should not trigger the production of protective enzymes.

Thirty years ago pheromones held great promise as species-specific, biodegradable agents for the control of insects which would not promote the development of resistant strains. Initially, this promise was not realized. It was

FIGURE 8-7   From *The Scientist*, p. 3, March 30, 1998. Copyright © 2001 by Sidney Harris. Used with permission.

not possible to devise low-cost, large-scale syntheses of these compounds so that they could be manufactured in sufficient quantities for large-scale use. The supply problem was aggregrated by the pheromones rapid rate of degradation in the environment, making their effectiveness short-lived.

There has been a gradual resurgence in the use of pheromones because the problems associated with their manufacture and application were addressed. The problem of their rapid degradation in the environment was solved by encapsulation in capillary tubes from which the pheromone slowly diffuses into the field or orchard. Political and economic factors drove these innovations because of the desire to limit the use of pesticides, which persist in the environment, leave residues on fruits and vegetables, and lead over time to the development of resistant strains of insects. Pheromones are now an important part of the integrated pest management (IPM) approach (Section 8.14) for the control of insect predators on major crops.

Since pheromones are present only in trace amounts in insects, special techniques are required to determine the structures of these compounds. For example, only $900\,\mu g\,(9 \times 10^{-4}\,\text{gm})$ of pure male sex attractant was isolated from 135,000 virgin female fall armyworm moths. The isolation and analysis of the pheromone is even more difficult if it consists of a mixture of compounds that are effective only when present in the proper ratio. The cotton boll weevil pheromone "grandlure" is an example of a pheromone that consists of four substances in the following proportions:

The chemical constitution of the boll weevil pheromone "grandlure"

This pheromone is unusual in that it is released by the male to attract females. The pheromone of the pink boll worm moth, an insect whose larvae eat cotton, consists of two substances:

$$\underset{\substack{\| \\ O}}{CH_3\overset{O}{C}OCH_2(CH_2)_5CH=CH(CH_2)_2CH=CH(CH_2)_3CH_3}$$

$$\text{cis} \qquad\qquad \text{cis or trans}$$

The pink bollworm moth pheromone "gossyplure"

Once the structure of the pheromone was determined, an efficient method of chemical synthesis was devised to make a sufficient amount available for large-scale agricultural use.

8.13.1.1 Use of Pheromones in Insect Control

Pheromones have been used in at least four different approaches to insect control.

1. *To monitor the population of a specific insect*. The pheromone is placed in traps, and the number of insects trapped is recorded. The population density of the insect can be monitored to measure the effectiveness of the insecticides being used or the insects' migration into new areas. The traps must be designed to be compatible with the behavior of the target insect, which means that design and color can be chosen only after appropriate field tests. The migration, in a wind tunnel, of a male moth to a source of the pheromone produced by the female has been monitored and is shown in Figure 8-8.

2. *To trap males (or females in some instances)*. A large number of traps containing the pheromone are used to attract the males in an area. To be effective, almost all the males must be trapped, so that they are not available for mating. If as few as 10% are not trapped, the local population will build up again within a few generations, which for insects may only be a matter of weeks.

3. *For male (female in some instances) confusion*. Capillary tubes are used to distribute large amounts of the pheromone in the field, causing the air to be permeated with the female sex pheromone. The male is surrounded by the attractant, which saturates the insect's pheromone receptors, preventing the male from locating the female. This techniques and technique 2 are effective only if there is a low population of the insects to be controlled. For

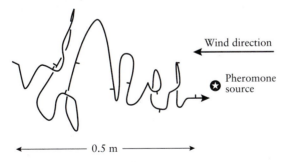

Wind direction

Pheromone source

← 0.5 m →

FIGURE 8-8 Track of male moth flying toward a source of female heromone in a wind tunnel. Marks on track at one-second intervals. Redrawn with permission from W. Booth, *Science*, **239**, 136. Copyright © 1988 American Association for the Advancement of Science. Source: *The Insects* by R. F. Chapman.

example, male confusion is not effective if the population of insects is high enough for the male to see the female.

4. *To attract (aggregate) or disperse both males and females.* Some pheromones attract both sexes to a particular area, resulting in a population large-enough to form colonies. This was observed with beetles, which require a certain number of conspecifics to launch a mass attack on a tree. Once that number has been reached, a dispersant pheromone is emitted that diverts potential new colony members. The aggregation pheromone is used in insect control to bring the target organisms together in one location, where they can be efficiently killed by a pesticide. The dispersant is used to divert the insects away from the area in which the crop is growing. These attractants and dispersants are species specific and have no effect on other, some potentially beneficial, insects.

The male confusion approach has proved to be useful in the control of insects that feed on grapes, apples, pears, nectarines, and currants. Studies are in progress on the application of the male confusion technique to insect control on about 10 other crops. One example is the control of the grape berry moth, the only insect pest that attacks Concord grapes. The pheromone is encapsulated inside thin, hollow polyethylene tubes that are tied to the wires used to hold up the grapevines. The release of the pheromone is controlled by the length of time it takes for its vapor to diffuse down this capillary tube into the atmosphere. Initial studies show that this approach keeps the damage to grapes at or below the 2% level, much lower than that observed with use of synthetic insecticides.

The juice from concord grapes is used in making wine, juice, and jelly, but poorer product is obtained if more than 2% of the grapes are damaged. The vineyards tend to be near homes, and thus synthetic pesticides cannot be used. Since the grape arbors are small, it is economically feasible to use the pheromone, which is a 9:1 mixture of two compounds, to confuse the male moths.

$$\underset{CH_3}{H}\diagdown\!\!=\!\!\diagup\underset{CH_2(CH_2)_6CH_2OCCH_3}{H} \overset{O}{\overset{\|}{}} \quad \text{9 parts}$$

$$\underset{CH_3}{H}\diagdown\!\!=\!\!\diagup\underset{CH_2(CH_2)_8CH_2OCCH_3}{H} \overset{O}{\overset{\|}{}} \quad \text{1 part}$$

Grape berry moth sex pheromone

8.13.2 Insect Attractants

Insect attractants, chemicals that attract insects to traps, are used in many of the same ways as insect pheromones. The attractants are usually substances that lead the insects to food. A simple example is the use of narrow-necked bottles containing fruit juice or beer to trap wasps. The volatile fatty acid esters in the beer or fruit juice attract the wasps to the trap. Trimedlure, a mixture of isomers of the isobutyl esters of chloromethylcyclohexane carboxylic acid, was used successfully in eliminating Mediterranean fruit flies from Florida by attracting them to insecticide-containing traps. An improved version of this attractant in which the chloro groups are replaced with iodo groups has been developed.

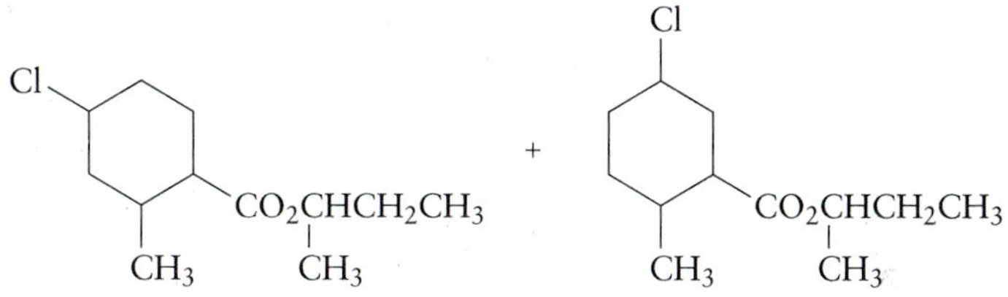

Trimedlure

8.13.3 Hormones

8.13.3.1 Introduction

Insect hormones regulate growth and maturation from the larva to the adult. The juvenile hormone controls the development of the immature larva through successive growth stages. If, however, it remains present during the metamorphosis of the larva to the adult, a deformed larva results or the adult that is formed soon dies. The molting hormone, ecdysone, is required for the differentiation to the adult. If the immature larva is treated with the molting hormone, it passes through its life cycle at a rapid rate and dies prematurely. Insect growth and development could be controlled if these hormones, or compounds with similar biological activity, could be applied at critical stages in the development of the larva to the adult. For example, compounds have been developed that bind to ecdysone receptors and accelerate the growth of the larvae of the tobacco hornworm, an insect that eats tobacco plants.

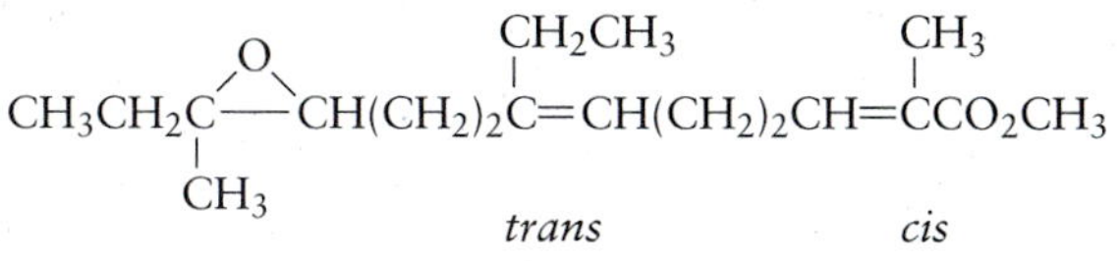

Juvenile hormone

α-Ecdysone
(molting hormone)

8.13.3.2 Antijuvenile Hormones from Plants

The continuing battle between insects and plants led to the synthesis by the common ageratum plant of two compounds that block the action of juvenile hormones. Compounds with this general biological effect are called antiallatrotropins. The particular compounds isolated from ageratum were named precocene I and precocene II because they caused premature metamorphosis of the larvae of a number of insects.

Precocene I   Precocene II

The adult females that develop after treatment with the precocenes are sterile, and only a few of the males are capable of successful mating with normal females. In addition, treatment of newly developed adult females with precocenes prevents the development of ovaries and, in the case of the cockroach, prevents the secretion of sex pheromones. Treatment of the adult potato beetle with precocenes caused it to prematurely stop feeding and go into a dormant stage. Precocene II also prevented the development of the eggs of the milkweed bug and the Mexican bean beetle.

The wide range of biological activity exhibited by the precocenes suggests that the antiallatrotropins could be very effective agents for insect control. They have a decided advantage over juvenile hormones in that their toxic properties are effective on the adult and on the egg as well as on the larva. In addition, their effect on the larval stage is to accelerate development to the adult, thereby shortening the larval stage, which is generally the most destructive.

8.13.4 Viruses and Bacteria

8.13.4.1 Introduction

The natural enemies of insects are obvious agents for insect control. These organisms are species specific, and it will probably be a long time before the insect develops defense mechanisms to ward off the predators. Viruses, bacteria, fungi, and protozoa have been suggested as possible natural control agents, although only viruses and bacteria are currently in use.

8.13.4.2 Viruses

Viruses and bacteria have two major advantages over other methods of pest control: they are highly infectious, and they spread rapidly. Often they become established in the environment, hence need to be applied only once. Therefore, in the ideal case it will not be necessary to mass-produce the virus but rather to rely on its reproductive ability in the presence of the insect host. This ideal was achieved in the case of the European spruce sawflies in Canada. This insect has been virtually eliminated by a virus predator.

Some viruses must be applied each year because the natural population does not build up rapidly enough to prevent significant crop damage. The production of viral pesticides is difficult because the viruses can be grown only in the tissues of their hosts. Consequently, the production of viruses also requires the growth of large numbers of host insects.

The use of genetic engineering techniques to get plants that produced protective viruses was reported in 1996. The full genome of the virus that attacks the larvae of the cotton bollworm and the tobacco budworm was inserted into the DNA of tobacco plants, causing the plants also to produce viruses when they grow. The viruses expressed by the plants protected them from attack by the cotton bollworm, while those without the viral genome were not protected. Concerns about the possible spread of the viral genome to other plants will have to be addressed before this approach can be field-tested.

8.13.4.3 Bacteria

The first successful application of a bacterium for the control of insects was the use of the spores of *Bacillus popilliae* against the Japanese beetle in 1939. When this bacterium invades the beetle grub, it multiplies rapidly and causes the grub to turn an abnormal white color, hence the name "milky disease."

This bacterium, together with other predators, succeeded in reducing the Japanese beetle to a minor pest in the United States by 1945.

There has been a recent emphasis on the use of *Bacillus thuringiensis*, which is effective against over a hundred species of caterpillars including the cabbage looper and the tomato hornworm. The bacterium, which paralyzes the gut of larvae and inhibits feeding, requires a pH of 9 or higher for effective growth and is therefore specific for high pH caterpillars. *B. thuringiensis* must be used annually because the bacterium becomes established in the fields too slowly to prevent heavy crop damage after only one application.

In an approach designed to take advantage of the effects of the toxic protein produced by *B. thuringiensis*, the gene responsible for the production of this protein has been cloned and inserted in the genome of cotton, tomato, potato, soy beans, corn and other crop plants. The gene is passed on to the next generation of plants via its seeds. Many fewer applications of synthetic pesticides are required on fields where plants containing the *B. thuringiensis* gene plants are grown. It is estimated that U.S. cotton farmers have reduced the amounts of insecticides used by 1.2 million kg/year. The decreases with two other major crops, corn and soy beans, have not been as impressive because they use a different farming methodology.

The results obtained from these transgenic plants have been impressive, but unfortunately, insects have been shown to develop resistance to the *B. thuringiensis* toxin in the same way they resisted pesticides. This toxin will have to be used sparingly or else the transgenic plants will have only a small window of usefulness. Strategies are being developed to monitor crops to detect the development of resistance in one area and to keep resistant strains from being transferred to other areas. To slow the development of resistant strains, plots of nontransgenic plants are also grown to maintain a competing population of nonresistant insects.

An additional concern with the use of transgenic plants is that the gene could be transferred to wild relatives and thereby increase their resistance to insect pests. While cotton, corn, and potatoes have no wild relatives in the United States, the incorporation of a gene for insect resistance or some other trait into a plant that is cross-pollinated with a wild plant could result in the generation of a weed population that is much more difficult to control. A thoughtful discussion of the potential ecological risks and benefits of genetically engineered plants and a discussion of how little scientific data we have in this area is given by Wolfenbarger and Phifer.[6]

Finally, there has been much public concern in Europe and an increasing concern in the United States about potential health problems associated with the consumption of genetically modified food. While there are no data to

[6]L. L. Wolfenbarger and P. R. Phifer, The Ecological risks and benefits of genetically engineered plants, *Science*, **290**, 2088–2093 (2000).

demonstrate that problems exist, there appears to be a feeling that problems will arise at a future time, much as the chronic toxicity of some pesticides became apparent only some 20 years after their initial use. So far the companies that developed the genetically modified foods have not been successful in convincing the public that their products are safe to eat and will benefit the environment because lower amounts of pesticides are used in growing.

8.14 INTEGRATED PEST MANAGEMENT

The present strategy is to manage the insect pests by use of all possible techniques in an integrated program rather than by relying on a single approach. The goal is to minimize the use of synthetic pesticides and related agents while maintaining crop yield, quality, and profitability. This strategy limits the buildup of pesticides in the environment, the presence of pesticide residues on fruit and vegetables, and the development of resistant insect strains. More emphasis is placed on the use of scientific farming and natural predators for insect management. Crop rotation prevents the buildup of one particular pest. Growing plants that the insect prefers next to the desired crop is effective in diverting the interest of the pest to these plants and slows the development of resistant strains. Beneficial insects such as the ladybug, praying mantis, and lacewing (aphid lion) may also be used to control insects that destroy crops.

Integrated pest management requires close monitoring of the crop and the insect population, but it can be cost efficient because the required number of applications of synthetic insecticides is far lower than that needed with narrower approaches. Traps containing pheromones provide data on the population of the insect to be managed. It will be important to have local weather information available to estimate the extent of crop and insect development in the fields. This information must be available online so that every farmer can consult the data daily by computer.

There have been a number of successful applications of the integrated pest management (IPM) approach, and the history of the growth of cotton in southern Texas is one example.[7] In the 1930s the boll weevil, the pink bollworm, and the cotton flea hopper were the principal cotton pests. Early planting and early destruction of crop residues controlled the pink bollworm, while the boll weevil was controlled by the use of calcium arsenate and the flea hopper by sulfur dust. These pesticides did not kill many other insects and thus there were few outbreaks of the secondary cotton predators, the tobacco budworm and the pink bollworm, since other insects controlled these. At

---

[7]P. L. Adkisson, G. A. Niles, L. K. Walker, L. S. Bird, and H. B. Scott, Controlling cotton's insect pests: A new system, *Science*, 216, 19–22 (1982).

the end of World War II chlorinated pesticides became available, and all the insects, favorable and unfavorable, were killed as a result of the use of 10–20 DDT applications per year. When the boll weevil became resistant to DDT, the organophosphorus insecticide methyl parathion [Section 8.11, reaction (8-64)] was applied along with the DDT. DDT was required to control the pink bollworm and the tobacco budworm because methyl parathion killed their natural insect enemies. But by 1960 the budworm and bollworm became resistant to DDT and other insecticides. This problem was partially solved by the use of greater amounts of methyl parathion. This almost eliminated the boll weevil as a pest, so the two secondary insect pests, the bollworm and the tobacco budworm, which had developed resistance, became the major cotton predators. By 1968 it became impossible to control these secondary pests with 15–20 applications of mixtures of many types of insecticides (Table 8-6), and the production of cotton dropped precipitously and was not profitable.

Since the bollworm and the tobacco budworm were a problem only when their natural insects enemies were killed off, new ways had to be devised to control the boll weevil and the cotton fleahopper without overkilling these natural enemies. A plan for managing the boll weevil was devised from a study of its life cycle in cotton plants. In the fall and winter, the weevils go into a dormant phase in the cotton stalks. It was found to be effective to spray the fields just before the start of the dormant phase and then grind up the stalks and plow them under. This kills many of the adults that would otherwise winter over in the ground, and also destroys the refuges in the cotton plants of any that survived the insecticide. In the spring, insecticides were used to kill the adult

TABLE 8-6

Increase in the Resistance of the Bollworm and Tobacco Budworm to Certain Organochlorine and Carbamate Insecticides in Southern Texas between 1960 and 1965

| | Median lethal dose (mg/g larva) | | | |
| | Bollworm | | Tobacco budworm | |
Compound	1960	1965	1961	1965
DDT	0.03	1000+	0.13	16.51
Endrin	0.01	0.13	0.06	12.94
Carbaryl	0.12	0.54	0.30	54.57
Strobane and DDT	0.05	1.04	0.73	11.12
Toxaphene and DDT	0.04	0.46	0.47	3.52

Source: P. L. Adkisson, G. A. Niles, L. K. Walker, L. S. Bird, and H. B. Scott, Controlling Cuttun's Insect Pests: A New System, *Science*, **216**, 19–22 (1982).

weevils that survived before they had a chance to reproduce. Insecticides were applied early in the growing season to minimize the damage of the flea hopper. The timing of the insecticide applications was crucial to prevent killing the natural predators of the bollworm and budworm. Finally, early-flowering strains of cotton were used to permit the cotton to be harvested before boll weevil and leafhopper damage became serious. This approach greatly decreased the use of costly synthetic pesticides and thus the cost of growing cotton. As a result, the crop increased sufficiently to become profitable to grow in southern Texas once more.

The boll weevil still caused $200 million in crop losses in the United States in 1999, and $70 million is spent each year to control it. The integrated pest management approach has been used effectively to eliminate the boll weevil from large areas of the United States and adjacent Mexico, but it is still a major problem in other regions. The Southwest Boll Weevil Eradication program has eliminated this pest in southern California, western Arizona, and northwest Mexico. This area is continually monitored by means of traps baited with the pheromone grandlure to detect migrating weevils. The number of females detected in traps illustrates the effectiveness of the eradication program: it decreased from 4,030,356 in 1988 to none in 1992. One female was detected in 1996, but no others could be found after more traps were placed in the area. It was concluded that the lone weevil had hitched a ride on a train that passed through the area near the trap. No more weevils had been detected up to 1999.

The eradication program is still in progress in other parts of the United States. It has been successful in an area encompassing Virginia and part of South Carolina. It was still in progress in 1999 in sections of North Carolina, South Carolina, Georgia, Florida, Alabama, Mississippi, Louisiana, and Texas. The goal of the program is to eradicate the boll weevil in the United States and Mexico by 2010. A similar approach is under way to eradicate the pink bollworm in the northwestern corner of Arizona.

8.15 PROPOSED PHASEOUT OF ALL CHLORINE-CONTAINING COMPOUNDS

Concern that the health hazards due to chloroorganics described in this chapter may be impacting on the health and reproductive ability of terrestrial life-forms has resulted in the call by environmental groups for a phaseout of the manufacture of chlorine-containing compounds in the United States.[8] As noted in this chapter, some chloroorganics have already been banned from production in the United States, and a worldwide ban on the production and use of 12 chlorinated organics was enacted by the United Nations in December 2000.

[8] J. T. Thornton, *Pandora's Poison: Chlorine, Health, and a New Environment*. MIT Press, Cambridge, MA, 2000.

The groups proposing the ban on chlorinated organics suggest that among the approximately 15,000 such compounds in commerce, there may be others that are also health hazards. Consequently, they believe it would be prudent to phase out the use of all chlorinated compounds. The association of manufacturers of chlorinated compounds, the Chlorine Institute, suggests that compounds be evaluated on a case-by-case basis, since it is not known which ones are health hazards.

A complete phaseout of the manufacture of chloroorganics could also include a ban on the use of chlorine in water purification, as a bleach in the production of paper, and as an ingredient in chlorinated pharmaceutical drugs. Local bans on the use of chloroorganics already exist in some cities in Europe. The use of poly(vinyl chloride) in pipe and public buildings materials is banned because it releases TCDD and benzofurans when burned. Chlorinated pesticides, including some compounds that are readily degraded, have been banned or restricted. Several large U.S. industrial firms have developed long-range plans to deal with such a phaseout, should it occur, and they have taken steps to cut back on the use of chlorinated organics.

This phaseout would be very costly for the economy of most countries. Non-chlorine-containing substitutes would have to be found for many of the 15,000 commercial compounds that contain chlorine.

One study, by a consulting firm commissioned by the Chlorine Institute, estimates an annual cost of $102 billion for such a phaseout in the United States. Even if this estimate is inflated by as much as a factor of 10, it is still an exceedingly expensive proposal. Since many of the chloroorganics that are believed to be the sources of health problems are manufactured in countries around the world, it would not be prudent for any country to take this step without first implementing a global plan for phasing out these compounds as was done for the chlorofluorocarbons (Section 8.6). The global circulation of volatile chloroorganics means that the United States can still be impacted by compounds produced in other countries.

Additional Reading

General

Jones, K. C., ed., *Organic Contaminants in the Environment. Environmental Pathways and Effects.* Elsevier Applied Science, London, 1991, 338 pp.

Larson, R. A., and Eric J. Weber, *Reaction Mechanisms in Environmental Organic Chemistry.* CRC Press, Boca Raton, FL, 1994.

Schwarzenbach, R. P., P. M., Gschwend, and D. M. Imboden, *Environmental Organic Chemistry.* Wiley, New York, 1993.

Thornton, J., *Pandora's Poison: Chlorine, Health and a New Environment*, MIT Press, Cambridge, MA, 2000, 539 pp.

Chlorofluorocarbons

Ravishankara, A. R., and E. R. Lovejoy, Atmospheric lifetime, its application and determination: CFC-substitutes as a case study, *J. Chem. Soc. Faraday Trans.* **90**(15), 2159–2169 (1994).

Polychlorinated Biphenyls

Alexander, M., *Biodegradation and Bioremediation*, Chapter 13, Academic Press, Orlando Fl, 1994.

Bunce, N. J., *Environmental Chemistry*, 3rd ed. Wuerz, Winnipeg, Canada, 1998.

Waid, J. S., ed. *PCBs and the Environment*, Vol. I–III. CRC Press, Boca Raton, FL, 1986

Polychlorinated Dibenzodioxins and Dibenzofurans

Altwicker, E. R., Formation of PCDD/F in municipal solid waste incinerators: Laboratory and modeling studies, *J. Hazardous Mater.*, **47**, 137–161 (1996).

Bumb, R. R., and eleven others, Trace chemistries of fires: A source of chlorinated dioxins, *Science*, **210**, 385–390 (1980).

Biodegradation of Chloroorganics

Boyd, S. A., M. D. Mikesell, and J.-F. Lee, Chlorophenols in soils, in *Reactions and Movements of Organic Chemicals in Soils*, B. L. Sawhney and K. Brown, eds. Soil Science Society of America, Madison, WI, 1989.

Kuhn, E. P., and Suflita, J. M., Dehalogenation of pesticides by anaerobic microorganisms in soils and ground water—A review, in *Reactions and Movements of Organic Chemicals in Soils*, (B. L. Sawhney and K. Brown, eds)., pp. 111–180 Soil Science Society of America and American Society of Agronomy, Madison WI, 1989.

Schwarzenbach, R. P., P. M., Gschwend, and D. M. Imboden, *Biological Transformation Reactions*, in *Environmental Organic Chemistry*, Schwarzenbach, Gschwend, and Imboden, eds., pp. 485–546. Wiley, New York, 1993.

Singh, V. P., Biotransformations: Microbial biodegradation of health-risk compounds, in *Progress in Industrial Microbiology*, Vol. 32. Elsevier, Amsterdam, 1995.

Wackett, L. P., Dehalogenation reactions catalyzed by bacteria, in *Biological Degradation of Wastes*, A. M. Martin, ed., pp. 187–203. Elsevier Applied Science, London, 1991.

Toxicity of Chloroorganics

Hansen, L. G., Halogenated aromatic compounds, in *Basic Environmental Toxicology*, L. G. Cockerham and B. S. Shane, eds. pp. 199–230. CRC Press, Boca Raton, FL, 1994.

Jefferies, D. J., The role of the thyroid in the production of sublethal effects by organochlorine insecticides and polychlorinated biphenyls, in *Organochlorine Insecticides: Persistent Organic Pollutants*, F. Moriarity, ed., pp. 131–230. Academic Press, New York, 1975.

Moriarity, F., *Ecotoxicology*, pp. 109–115. Academic Press, New York, 1983.

Peakall, D., *p, p'*-DDT: Effect on calcium metabolism and concentration of estradiol in the blood, *Science* **168**, 529 (1970).

Environmental Hormones and Related Topics

Davis, D. L., and H. L. Bradlow, Can environmental estrogens cause breast cancer? *Sci. Am.* pp. 166–172, October 1995.

Colburn, T., D. Dumanski, and J. P. Myers, *Our Stolen Future. Are We Threatening Our Fertility, Intelligence and Survival? A Scientific Detective Story.* Dutton, New York, 1996.

Hileman, B. Environmental estrogens linked to reproductive abnormalities, cancer, *Chem. Eng. News*, pp. 19–23, Jan. 31, 1994.

McLachlan, J. A., and S. A. Arnold, Environmental estrogens, *Am. Sci.*, **84**, 452–461 (September–October 1996).

Safe, S. E., Environmental and dietary estrogens and human health: Is there a problem? *Environ. Health Perspect.* **103**, 346–351 (1995).

Pesticides

Büchel, K. H., ed., *Chemistry of Pesticides*. Wiley, New York, 1983, 517 pp.

Duke, S. O., J. J., Mann, and J. R., Plimmer, eds., *Pest Control with Enhanced Environmental Safety*, ACS Symposium Series 524. American Chemical Society, Washington, DC 1993.

Heitefuss, R., *Crop and Plant Protection*. Ellis Horwood, Chichester, U.K., 1989.

Somasundaram, L. and J. R., Coats, eds., *Pesticide Transformation Products*, ACS Symposium Series 459. American Chemical Society Washington, DC , 1990.

Insect Resistance to Pesticides

Brattsten, L. B., C. W. Holyoke Jr, J. R. Leeper, and K. F. Raffa, *Insecticide resistance: Challenge to pest management and basic research*, *Science* **231**, 1255–60 (1986).

Green, M. B., H. M. LeBaron, and W. K. Moberg, eds., *Managing Resistance to Agrochemicals*, ACS Symposium Series 412. American Chemical Society, Washington, DC, 1990.

National Research Council, *Pesticide Resistance*, National Academy Press, Washington, DC, 1986.

Insect Pheromones

Cardé, R. T., and A. K. Minks, Control of moth pests by mating disruption: Successes and constraints, *Ann. Rev. Entomol.*, **40**, 559–585 (1995).

Jutsum, A. R., and R. F. S. Gordon, eds., *Insect Pheromones in Plant Protection*. Wiley-Interscience, New York, 1989.

Nordlund, D. A., R. L. Jones, and W. J. Lewis, eds., *Semiochemicals: Their Role in Pest Control*, Wiley, New York, 1981, pp. 306.

Biochemical Control of Insects

Duke, S. O., J. J. Menn, and J. R. Plimmer, eds., *Pest Control with Enhanced Environmental Safety*, ACS Symposium Series 524. American Chemical Society, Washington DC, 1993.

Genetically Engineered Plants

Howse, P., I. Stevens, and O. Joves. *Insect Pheromones and Their Use in Pest Management*. Chapman and Hall, London, 1998.

Wolfenbarger, L. L., and P. R. Phifer, The ecological risks and benefits of genetically engineered plants, *Science*, **290**, 2088–2093 (2000).

Integrated Pest Management

Luttrell, R. G., G. P., Fitt, F. S. Ramahlo, and E. S. Sugonyaev, Cotton management: Part 1. A worldwide perspective, *Ann. Rev. Entomol.*, **339**, 517–526 (1994).

EXERCISES

8.1. (a) Which one of each of the following pairs of compounds degrades more rapidly in the environment? Give a reason(s) for your answer which is based on the structure of the compounds.

(b) Illustrate the pathway(s) by which the more reactive compound degrades in the environment using an equation(s) which show reaction intermediates.

1. $CH_3CH_2CH_2CH = CHCl$ and $CH_3CH_2CH = CHCH_2Cl$
2. $(CH_3)_3CCl$ and $CH_3CH_2CH_2CH_2Cl$
3. Dieldrin and parathion
4. DDT and DDE
5. $CFCl_3$ and CH_3CHF_2
6. $(CH_3)_2CFCH_2CHFCH_2Cl$ and $(CH_3)_2CClCH_2CHFCH_2Cl$
7. CH_3CCl_3 and CCl_4
8. Chordane and heptachlor 9

9. and

8.2. A recently discovered potential problem is that of "environmental estrogens."

(a) Briefly explain what "environmental estrogens" are and give their sources. Do any occur naturally in the environment?

(b) What are their proposed health effects in humans and other mammals.

(c) What are the structures of the proposed environmental estrogens? Are these structures similar to each other or to that of the estrogenic hormones? Is it possible devise a theory for how these compounds exert their estrogenic effect in mammals and birds based on their structures

8.3. Bisphenol A is believed to be an environmental estrogen, yet it is not directly dispersed in the environment. How does it get there?

8.4. The manufacture of PCBs is prohibited in all the industrial countries of the world except Russia. The manufacture of PCBs was stopped in the United States in 1977. In spite of these bans, PCBs are still raining out in the Great Lakes and are found in other locations worldwide. Give reasons for the continued presence of these pollutants in the environment so long after their manufacture was stopped in most of the world.

8.5 (a) Some microorganisms degrade chlorinated aryl hydrocarbons. Write two different possible pathways for the microbial degradation of 2,3-dichloronaphthalene.

(b) What environmental conditions are required for each degradative process.

8.6. Part of your campus was constructed on a landfill that was used to dispose of chlorinated organics during World War II. Recently, these organics have started to escape from the landfill, and their fumes are wafting through the adjacent law school. Analysis of the material in the landfill reveals that the main compounds present are DDT, DDE, CCl_4, $CHCl_3$, chlorobenzene, 4-chlorophenol, CH_3CCl_3 and 1,2,4,5-tetra-chlorobenzene. Small amounts of 2,3,7,8-tetrachlorobenzo-p–dioxin were also detected.

(a) What are the acute and chronic toxicity problems for humans, faculty, and other mammals associated with these compounds?

(b) What are the environmental dangers of these compounds?

(c) The following proposals have been made to stop the release of these compounds on your campus. Discuss the feasibility and each proposal. Include the advantages and disadvantages, the methods that should be used, and the potential long-term consequences of each plan.

1. Removing the chemicals from the landfill and trucking this mixture of liquids to another site.

2. Removing the chemicals from the landfill and incinerating these compounds on site.

3. Capping the landfill with clay and installing drainage areas and monitoring stations to minimize and keep track of emissions.

4. Leave it alone to reduce the environmental impact of an overpopulation of lawyers.

5. Suggest other options.

8.7. Organophosphorus compounds kill a wide array of insects and are readily degraded in the environment. Why do they have only limited use in insect control?

8.8. Insect pheromones would appear to be ideal for insect control because they are species specific and biodegradable. Why do they have only limited use in crop protection?

8.9. The resistance of insects to some pesticides is a well-known phenomenon.

(a) Explain how insects become resistant.

(b) What is the molecular basis (in the insect DNA) of this resistance?

(c) Not all strains of insects become resistant to a particular insecticide. Postulate why this is the case.

8.10. Define integrated pest control and compare it with the conventional spraying techniques to control insect pests on crops. Include the following points in your discussion.

(a) Potential impact on the ecosystem in the field being cultivated and in the general vicinity of the farm

(b) The use of pheromones

(c) The development if insect resistance

(d) Use of special strains of plants

9

CHEMISTRY IN AQUEOUS MEDIA

9.1 WATER IN THE ENVIRONMENT

9.1.1 The Water Cycle

Much of environmental chemistry takes place in aqueous systems. Water itself takes part in a cyclic process of evaporation, precipitation, and transport, shown schematically in Figure 9-1. Approximately $5 \times 10^{14} \, \text{m}^3$ $\left(5 \times 10^{17} \, \text{kg}\right)$ of liquid water evaporates and precipitates annually, with most of this being over the oceans. About $4 \times 10^{13} \, \text{m}^3$ $\left(4 \times 10^{16} \, \text{kg}\right)$ more water falls on land as precipitation than evaporates from it, and this is the annual global volume of freshwater runoff.

Less than $2 \times 10^{13} \, \text{m}^3$ of water is permanently present in the atmosphere as vapor or cloud. The bulk of all water, more than 97%, is found in the oceans, where it contains much dissolved inorganic material and small but important amounts of organic substances. For many purposes ocean water can be considered in two categories: surface water, in which there is relatively rapid mixing, and a larger amount of deep water, with mixing on much longer time scale. Surface freshwaters (lakes, streams, etc.) are of highly variable

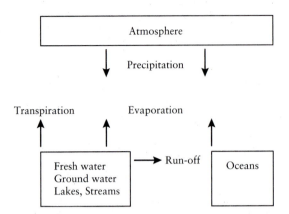

FIGURE 9-1   The water cycle. Volcanoes also add a small contribution to the atmosphere.

composition; usually, however, they contain small amounts of dissolved substances, biological materials, and frequently much suspended matter. These water sources are most heavily used for agricultural, industrial, and domestic purposes. Ice, in the Arctic and Antarctic ice caps, makes up the largest reservoir of fresh water.

Subsurface or ground waters, held in sand, gravel, or porous or fractured rock, are normally quite free of suspended and organic materials because of their percolation through soil and rock formations, but may contain dissolved inorganic material. These waters also are used extensively in some locations; about 20% of the U.S. freshwater usage comes from this source. A great deal of groundwater is used for irrigation in the western states, and over half the population relies on groundwater as a drinking water source. The term "aquifer" refers to a permeable geological formation that is saturated with water and capable of producing a significant flow at a well or spring.[1] One type of aquifer formation consists of fractured rock: blocks of more or less solid rock separated by cracks and fissures. Much of the water in this type of aquifer is in the fissures, but some is in pores in the rock itself. The latter typically is released very slowly in comparison to the former. A second type of aquifer consists of more finely divided material such as sand; flow in this will follow the rules of flow through porous media.

[1]The water-saturated region of soil or porous rock is referred to as the phaedose zone; the upper bound of this is called the water table. The region above this, where the pores are unsaturated, is the vadose zone. In the region between them (the capillary zone), water rises upward through capillary action.

FIGURE 6-3 The Ixtoc exploratory well, which blew out and caught fire June 3, 1978. It spilled 140 million gallons of oil into the Gulf of Mexico. From http://www.nwn.noaa.gov/sites.hazmat/photos/ships/08.html.

FIGURE 6-5 The "midsize" oil tanker *Ventiza*, built in 1986. It is 247 m (811 ft, or 2.7 U.S. football fields) long and 41.6 m (136 ft) wide. In 2000 the largest tanker in the world was the *Jahre Viking*, which is 458 m (1504 ft, or 5 U.S. football fields) long and 69 m (266 ft, or 0.89 U.S. football field) wide. From http://www.rigos.com/ventiza.html.

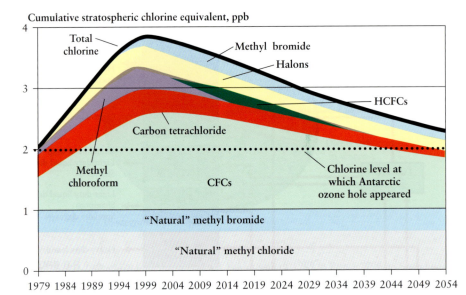

Cumulative stratospheric chlorine equivalent, ppb

FIGURE 8-2   Projected trend in levels of chlorine, CFCs, HCFCs, and haloorganisms in the atmosphere as a consequence of The Montreal Protocol. Redrawn from "Looming ban on production of CFCs, halons spurs switch to substitutes," *Chem. Eng. News*, p. 13, Nov. 15, 1993. Used by permission of E.I. du Pont de Nemours and Company.

Freshwaters are classified as hard when they contain significant amounts of calcium or magnesium salts in solution. Hardness is easily demonstrated through the precipitate formed with soap (Section 7.2).

Cycling between the atmosphere and the reservoirs of liquid water obviously involves evaporation and condensation (precipitation), but also transpiration through plants. Vegetation can play a significant role in this transport, either increasing or decreasing water loss from land surfaces, depending on the circumstances.

Cycling involving the groundwater reservoir is much more complex. Some aquifers (confined aquifers) are isolated by relatively impervious rock or clay so that input from rain may be extremely slow. Since extensive withdrawal of water from these is not always matched by recharge, a groundwater supply, unlike surface waters, may not be a renewable resource on the time scale of interest to the users. One estimate suggests that these aquifers are being depleted by 160 billion cubic meters of water annually, most extensively in India, North Africa, China and the United States. Removal of water may be accompanied by ground settling (e.g., Mexico City, Venice, Bangkok) or, if near an ocean, inflow of salt water. The Ogallala aquifer that underlies a large portion of the Great Plains of the United States is one example of a major groundwater source that is heavily used for domestic, industrial, and agricultural purposes. Indeed, much of the agriculture in this area has depended on this source of irrigation water. In many locations, withdrawal has exceeded recharge by two or three orders of magnitude, with the result that flow rates are decreased, deeper wells are needed, and pumping expenses are higher. There is real concern about the continued availability of water from this source, but little positive conservation action.

Contamination of a deep aquifer is, for all practical purposes, an irreversible event. Such contamination may arise from leaching from waste dumps or agricultural activities if there is exchange with surface waters, or from deep-well disposal that inadvertently penetrates an impervious layer. Other aquifers (unconfined) are near the surface and are readily recharged by rainfall. Because they are accessible, they are also easily contaminated.

Cleanup of contamination from a localized pollution source may be possible in some cases because distribution of the contaminants is normally slow and follows the flow of the water in the porous medium. Therefore, if caught in time, the contamination may be restricted to a relatively small volume of the total aquifer. Pumping out of water, destruction of the contaminants *in situ* by biological or other means, or construction of impermeable barriers around the contaminated zone are examples of remedial methods that may be applicable in particular cases (see Chapter 12). None is cheap.

9.1.2 Structure and Properties of Water

Chemical interactions in aqueous systems make up an important area of environmental chemistry, and the properties of water determine the behavior observed in solution. The water molecule is angular and, owing to the large difference in electronegativity of oxygen and hydrogen, quite polar. Consequently, water molecules will have strong electrostatic interactions with one another, and with dissolved ionic materials or other highly polar molecules. Because the electronegativity of oxygen is large enough that the bonding electrons have little effect in screening the hydrogen nucleus from interaction with centers of negative charge on other molecules, hydrogen bonding of water to molecules with highly electronegative atoms is important. The energy of typical hydrogen bond interactions is significantly larger than that of ordinary dipole–dipole interactions of polar molecules, although much weaker than most covalent bonds (Table 9-1).

Hydrogen bond interactions, to a first approximation, may be regarded as a special case of an unusually strong dipole–dipole attraction. This leads to molecular association, may determine details of molecular structure, and in appropriate cases produces stronger hydration of dissolved species than would have been the case based on a simple dipole–dipole interaction. Hydrogen bonding is of greatest importance in aqueous solutions of molecules with oxygen and nitrogen groups, and particularly with OH or NH groups that not only are hydrogen-bonded by the hydrogen atoms of the water molecules but can themselves hydrogen-bond to the water oxygen.

The structure and density properties of water are largely determined by the hydrogen bonding of the water molecules with one another. As is illustrated in Figure 9-2, in ice each water molecule is surrounded tetrahedrally by four others; it is H-bonded twice through its own hydrogen atoms, and twice more via the oxygen through to the hydrogens on other water molecules. This gives a coordination number of 4, which is not very effective in filling space. Consequently, the structure is very open, with low density. Melting destroys the regular structure and results in an increase in the packing density, although the greatest part of the hydrogen bonding remains intact. With further

TABLE 9-1

Energies of Different Types of Bonds

Type of bond	Typical bond energy (kJ/mol)
Covalent	200–500
Dipole–dipole	5–20
Hydrogen bond	20–30

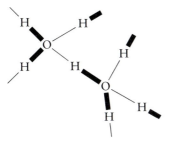

FIGURE 9-2 Hydrogen bonding between two water molecules of ice. The longer, thin lines represent the hydrogen bonds.

warming, increased hydrogen bond breaking allows the packing efficiency to increase, while greater thermal motion acts to reduce the density. The interplay of these counteracting tendencies results in the variation of density illustrated in Figure 9-3. The maximum density occurs at 4°C.

The facts that liquid water has a maximum density at 4°C and is more dense than ice have important consequences in natural water systems. As an unstirred body of water is cooled from the surface, the upper levels sink as their density becomes greater than that of the warmer, deeper water. If cooling continues below 4°C, the density decreases, and this water remains on the surface. Upon freezing, the still lower density ice floats. This property is of obvious significance in preventing the complete freeze-up of bodies of water,

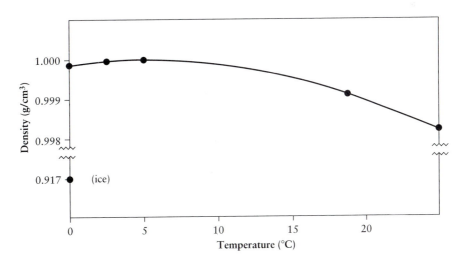

FIGURE 9-3 Variation of the density of water with temperature.

but the density gradients in the liquid are also responsible for stratification effects of considerable importance. Figure 9-4 illustrates stratification in a typical deep lake. In summer, the upper level, the epilimnion, consists of comparatively warm water. Photosynthesis takes place in this region, and the oxygen content is likely to be high. The thermocline represents a region of rapid temperature change. The lower level, the hypolimnion, consists of colder water. Although oxygen is more soluble in cold than in warm water, the only source of oxygen here is diffusion from higher levels. Since oxygen is used up by decomposing organic matter that has settled to the bottom, the oxygen content of the hypolimnion is likely to be low and reducing conditions may exist. A typical lake of this sort will "turn over" in the fall, and often again in the spring, but otherwise there will be little mixing. "Turnover" refers to displacement of the bottom water by surface water as the latter reaches the temperature of maximum density. Nutrients taken up by organisms in the epilimnion are largely released when the organisms decay in the hypolimnion. Thus this "turn-over" is important for biological productivity, because it brings nutrients back to the levels at which photosynthesis takes place.

Lakes are often described as oligotrophic, eutrophic, or dystrophic. An oligotrophic lake has comparatively pure water, is low in nutrient materials, and has low biological production. This would be a geologically young lake. A eutrophic lake has a high nutrient content and high biological production. The final stage, a dystrophic lake, is shallow, marshy, contains much organic material, and has a high biological oxygen demand (Section 7.2.5). Since most of the oxygen is used up by decaying vegetation, actual production is low. These stages form a natural sequence in the developmental history of a

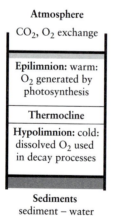

FIGURE 9-4   Schematic representation of stratification in a deep lake.

lake. The sequence of events can be speeded up artificially by introduction of nutrients to enhance algal growth. The term "eutrophication" is commonly used to describe this process.

The ocean is an example of an environmentally important system that contains a large amount of dissolved salts that increases the density of seawater compared with that of fresh-water; the density is greater the higher the salinity. Salinity is expressed as grams of salt per thousand milliliters; a typical ocean value is 35‰ (3.5%). The salt content affects other properties of water—for example, the freezing point is decreased. Water of 35‰ salinity freezes at $-1.9\ °C$. Since the ice produced is pure water (except for such brine as may be trapped in inclusions), the salinity of the unfrozen water increases. The variation of density with temperature also differs from that of freshwater; there is no maximum density before the freezing point but a continuing density increase until the freezing point is reached.

Salinity adds a second factor influencing stratification in saltwater systems, because higher salinity also gives a higher density. Low-salinity water—for example, river runoff, ice melt or abundant rainfall—will tend to float on higher salinity water. Freezing in polar regions or evaporation in the tropics will increase the salinity of the surface layer. Density differences, along with wind effects, lead to ocean current circulation patterns that are analogous to circulation in the atmosphere, although obviously constrained by landmasses (Figure 9-5). For example, in the North Atlantic, warm tropical water (the Gulf stream) flows west and north toward the Arctic, where it turns eastward. As evaporation increases the salinity and density, dense Arctic water sinks and flows along the bottom, back toward the equator. Eventually these bottom

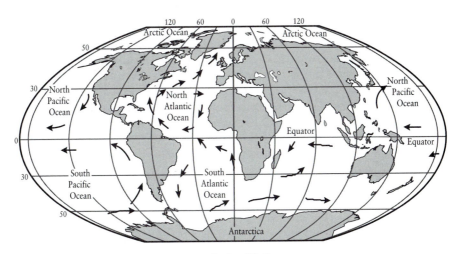

FIGURE 9-5 Surface ocean currents (greatly simplified).

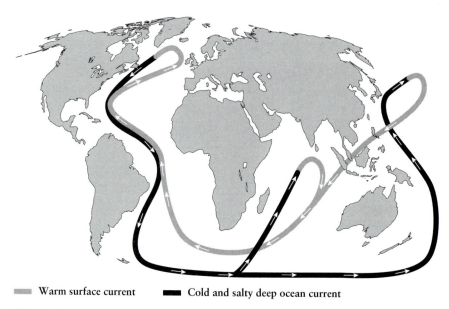

■■ **Warm surface current** ■ **Cold and salty deep ocean current**

FIGURE 9-6 Thermohaline circulation and the "global conveyor belt." From TOPEX/Poseidon project office at the Jet Propulsion Laboratory, Pasadena, California. http://geosuml.sjsu.edu/~dreed/130/lab10/30.html.

waters, which are rich in nutrients, return to the surface. Areas of extensive upwelling are important for marine life.

Besides the North Atlantic Gulf stream, similar equator-to-pole surface flows exist in all the earth's major oceans, producing clockwise surface circulation in the Northern Hemisphere and counterclockwise circulation in the Southern. (See the discusion of Coriolis effect in Section 2.6.) While the surface currents are largely wind driven, the return deep-water flows are driven by density differences that arise from differences in salinity and temperature and are called thermohaline circulation. Together, the surface and deep currents are referred to as a "global conveyor belt," carrying heat from the tropics to the polar regions. These flows have an important influence on climate, keeping the climate of northern Europe milder than that of eastern North America at the same latitude, for instance.

The relationship between global climate and these circulation patterns is clearly important, although not fully understood. One proposal suggests that increased global warming could lead to additional polar ice melt that could result in overrunning of the saline water by a layer of lighter, low-salinity water, causing the original flow pattern to be disrupted and further changing the climate. Recent studies suggest that shutdown of present ocean circulation by global warming could perversely cause European temperatures to drop several degrees[2] and that severe short-term temperature swings could result. There is evidence that this has occurred in the past. It is complexities such as these that make prediction of the effects of an increase in global average temperature difficult.

Besides the unusual density properties that water exhibits as a consequence of its hydrogen-bonding interactions, water also has high values for several thermodynamic properties (e.g., specific heat and enthalpy of vaporization) that are important factors in climate. Specific heat refers to the amount of heat energy required to change the temperature of a unit amount of substance by one degree. Its value for water is $4.184 \, \mathrm{J \, g^{-1} \, {}^{\circ}C^{-1}}$ (it varies slightly with temperature) compared to about $0.22 \, \mathrm{J \, g^{-1} \, {}^{\circ}C^{-1}}$ for air at constant pressure. Consequently, passage of cool air across a body of warmer water can noticeably warm the air with only a small change in water temperature. Ocean currents can convey large amounts of energy, which is why they are so important to climate.

The heat of vaporization of water—that is, the energy required to convert the liquid to the vapor, or the energy released in the reverse process—is 2.266 kJ/g. This energy will be absorbed from a body of water through evaporation, with consequent cooling of the water. It is also released to the air when water vapor condenses, thus warming the air, and is the energy source for cyclonic storms. The heat of fusion of water—that is, the energy required to melt ice, or that released when the liquid freezes—is 0.334 kJ/g, not a particularly large value in comparison to other liquids, since many of the hydrogen bond interactions responsible for interactions in the solid remain in the liquid.

Variation of density with temperature means that the ocean volume, and consequently the sea level, will change with temperature. Warming will cause expansion, although this is a small effect, especially if warming were limited to the surface layer. Melting of floating sea ice (e.g., the Arctic Ocean, where major decreases in ice thickness have been noted in recent years) would not in itself have much effect either, because floating ice displaces a volume of water equal to its mass. (This is not to say that low-lying regions would not be impacted by relatively small changes.) The unlikely melting of land-based ice (e.g., much of the Antarctic, and to a lesser degree the Greenland ice caps) would be much more significant, as this would add more water to the oceans. Sea levels are increasing slowly.

---

[2]R. A. Kerr, Warming's unpleasant surprise: Shivering in the greenhouse? *Science*, **281**, 158 (1998). For more technical discussions, see F. Joos, G.-K. Plattner, T. F. Stocker, O. Marchal, and A. Schmittner, *Science*, **284**, 464 (1999); R. A. Wood, A. B. Keen, J. F. B. Mitchell, and J. M. Gregory, *Nature*, **399**, 572 (1999); and W. S. Broecker, *Science*, **278**, 1582 (1997).

9.1.3 Solubility

The solubility of a substance in a liquid is determined by the interactions that take place between the molecules, and by entropy effects. That is, for a solution process to take place, the free energy of the system must decrease. The free energy change depends upon the enthalpy change, which at constant pressure is the heat associated with changes in the various interactions involved (bond energies, dipole–dipole interaction energies, etc.), and upon the entropy change that can be regarded as a measure of the change of order, or degrees of freedom. The processes involved in dissolving one substance in another are disruption of solute–solute interactions, at least partial disruption of solvent–solvent interactions caused by the presence of solute molecules, and formation of solvent–solute interactions. The first two of these involve absorption of heat (endothermic). The change in enthalpy will be favorable for solution only if the solvent–solute interactions, which are exothermic, are larger in magnitude than the other interactions. In the case of water, large solvent–solute interactions can arise only through H-bonding, coordination, ion–dipole attractions, or perhaps dipole–dipole attractions. Solute molecules not able to interact in one of these ways will not interact strongly with the water molecules. Because of these factors, there is a general tendency for like substances to dissolve in one another, and for unlike substances not to do so. Water provides an example of this tendency; it is generally a good solvent for ionic and highly polar (especially hydrogen-bonding) substances, but poor for nonpolar substances such as hydrocarbons. The increase in entropy on mixing is always a driving force toward solubility, although it may be largely outweighed by energy effects in some cases.

The foregoing considerations are valid as a qualitative guide to the type of solubility behavior to be expected. However, a prediction of poor solubility does not necessarily mean no solubility at all. Substances may be important environmentally at very low levels of solubility, so that even if a material is described as "insoluble," its presence in aquatic systems cannot be dismissed. Concentrations in the parts-per-million (ppm) or even parts-per-billion (ppb) range may be significant for some pollutants, and solubilities to this level will be found for most substances. Further, the distinction between dissolved and suspended matter is not always clear cut. Very finely divided particulate matter (colloids) that settle or coagulate very slowly may pass through most filtering systems. The surface areas per unit mass of such material will be very high, so that its reactivity may be much greater than that of the bulk solid.

The solubility of a gas in a liquid depends upon the partial pressure P of the gas. Normally, Henry's law applies at low solubilities, so that for a gaseous substance A, the equilibrium constant for the process

$$A_{(gas)} \rightleftharpoons A_{(solution)} \tag{9-1}$$

can be written as

$$K = P_{A_{(gas)}}/C_{A_{(solution)}} \qquad (9\text{-}2)$$

where C represents concentration in appropriate units, such as moles per liter. Solubility of a gas is usually small unless a reaction with the solvent is involved, and in most cases it decreases with increasing temperature.

The pressure dependence of gaseous solubility is well known from the release of carbon dioxide from pressurized carbonated beverages when the container is opened. It is also well known from this experience that release of the gas is not instantaneous, but is accelerated by agitation, which aids in the nucleation step in the formation of the gas bubbles. An environmental analogue is found in some deep crater lakes in Cameroon. Since these lakes are tropical, there is no seasonal turnover to mix the upper and lower levels and they can be highly stratified. The lower levels of some of these lakes contain concentrations of carbon dioxide, either from decaying vegetation or, more likely, from geological seepage, that are far greater than equilibrium values at surface pressures. In August 1986 about 1700 people were killed by suffocation from carbon dioxide released when one of these lakes, Nyos, "turned over" in response to some physical stimulus[3] (perhaps earth movements, perhaps wind effects, although a volcanic eruption also was proposed). Thirty-four people in the same country died in 1984, when a similar phenomenon occurred in Lake Monoun.

The equilibrium solubility of a slightly soluble substance that ionizes in solution (e.g., a salt) is expressed in terms of the solubility product. For the solution reaction of the pure solid M_aX_b,

$$M_aX_b = aM^{b+}_{(solution)} + bX^{a-}_{(solution)} \qquad (9\text{-}3)$$

the equilibrium expression[4] is

$$K_{sp} = [M^{b+}]^a[X^{a-}]^b \qquad (9\text{-}4)$$

The equilibrium solubility depends on the concentration of the ions from all sources and will be modified by competing reactions that involve these ions.

---

[3]R. A. Kerr, Nyos, the killer lake, may be coming back, *Science*, **244**, 1541 (1989). Also, see Volume 39 (1989) of *Journal of Volcanology and Geothermal Research*, which contains a number of papers on the Lake Nyos event. More recently, experiments aimed at degassing the lake by inserting a 200-m-long pipe to the bottom were carried out. A self-powered jet of water results that releases 10 liters of CO_2 per liter of water. Expansion of this process is planned, in spite of some concerns that it might destabilize the lake sufficiently to cause another sudden overturn.

[4]It will be the convention in this chapter to use square brackets to represent the equilibrium concentration of a species, while C, with a suitable subscript, will be used to represent analytical or total concentrations of substances that exist in solution as one or more species.

Nonionic solids or liquids will also have an equilibrium solubility, but these are not generally treated by the equilibrium constant approach. Solubilities in general may range from very small to infinite. In environmental situations, normally one is dealing with low concentrations, and consequently, equilibrium considerations will be of concern only with slightly soluble substances.

The equations used in equilibrium expressions in this book are written in terms of concentrations rather than activities. The activity a_i of species i in solution of concentration C_i is

$$a_i = \gamma_i C_i \tag{9-5}$$

where γ_i is the activity coefficient of the substance. In effect, γ_i is the factor by which the actual concentration must be multiplied to maintain the validity of the thermodynamically exact equations. The activity coefficient γ_i may be greater or less than unity, depending on both the concentration C_i and the nature and concentration of other species in the solution. Because $\gamma_i \to 1$ as $C_i \to 0$, use of concentrations in dilute solutions is acceptable for most purposes and necessary in most environmental problems, where activities generally are not known. Some care must be exercised, however, as with seawater, which contains roughly 0.5 M NaCl in addition to other components and cannot be considered to be a dilute solution in this context. The presence of a relatively large concentration of an "inert" salt such as NaCl produces nearly constant values for the activity coefficients of ions present in small amounts, such as H_3O^+, although these values may be far from unity. Equilibrium constant expressions will give correct results in this case, but only for an equilibrium constant having a value different from that applicable to the dilute solution case. Equilibrium constants valid for a solution of a particular overall composition are sometimes called stoichiometric equilibrium constants.

Although these thermodynamic relationships will permit calculation of equilibrium concentrations, it must be remembered that many natural systems will not be at equilibrium because of kinetic factors. Supersaturation may occur because of difficulty in forming the initial nuclei on which the particles may grow. Surface ocean water, for example, is supersaturated in calcium carbonate. The process of dissolution also may be slow, so that undersaturation even in the presence of excess undissolved material may persist. Finally, in natural systems, interactions with other materials in solution may have large effects.

9.1.4 Acid–Base Behavior

Another property of water is that of autoionization: that is, the process

$$2H_2O \rightleftharpoons H_3O^+ + OH^- \tag{9-6}$$

The species H_3O^+ represents the hydrated proton. Because of its small size and consequent high charge density, a proton will not exist alone in solution but will interact with the electron clouds of solvent molecules. In water, it will share one of the lone pairs from an oxygen atom. The actual species present probably is more complex than H_3O^+ (perhaps $H_9O_4^+$), but this is not particularly important for our purposes. This proton transfer process, however, is a very important type of reaction in aqueous chemistry, and acid–base behavior is widespread. Typical waters have a hydrogen ion concentration between 10^{-6} and 10^{-9} M (pH 6–9; pH $= -\log[H^+]$). This is a comparatively narrow range, which is determined by the balance of acidic and basic substances dissolved in the waters. Absolutely pure water at 25°C would have a hydrogen ion concentration of 10^{-7} M, from the equilibrium expression

$$K_w = [H^+][OH^-] = 10^{-14} \qquad (9\text{-}7)$$

Both biological processes and purely inorganic processes, such as dissolution or precipitation of minerals, are dependent upon pH. We are interested in what determines the pH of natural waters, and equally important, how resistant the pH value is to natural or human influences.

The oxygen atom of a water molecule possesses two pairs of nonbonding electrons (unshared pairs). These may be donated to empty, low-energy orbitals in another atom to form a coordinate bond, and such activity is the basis of Lewis acid–base behavior. This has particular importance with metal ions, as discussed in more detail later.

9.2 ACID–BASE REACTIONS

9.2.1 General Behavior

The most useful definition of acids and bases in aqueous chemistry is the Brønsted–Lowry definition: an acid is a proton donor; a base is a proton acceptor. This definition may be illustrated by an equation,

$$HCl + H_2O \rightleftharpoons H_3O^+ + Cl^- \qquad (9\text{-}8)$$

Here, HCl is acting as the acid (proton donor) and H_2O as the base (proton acceptor) as the reaction goes from left to right. Such reactions are very rapid and reversible, although, as here, the reaction may go very far to the right. It may be noted that if this reaction were reversed, H_3O^+ would be the acid, and Cl^- the base. $HCl-Cl^-$ and $H_3O^+-H_2O$ make two sets of acid–base pairs (conjugates).

Another example of an acid–base reaction is

$$NH_3 + H_2O \rightleftharpoons NH_4^+ + OH^- \qquad (9\text{-}9)$$

Here, H_2O acts as the acid, and this illustrates the amphoteric character of water: that is, its ability to act either as acid or base, depending on the substance with which it is reacting. It is inherent in the Brønsted approach that the term "acid" or "base" has meaning only with respect to a reaction like (9-9).

The strength of an acid or a base is given by the equilibrium constant for its ionization process. That is, for the reaction of a weak acid HA with water,

$$HA + H_2O \rightleftarrows H_3O^+ + A^- \qquad (9\text{-}10)$$

$$K_a = \frac{[H_3O^+][A^-]}{[HA]} \qquad (9\text{-}11)$$

An analogous expression could be set up for a base, with the constant usually called K_b. In place of K_a and K_b, we often find pK_a or pK_b, where $pK = -\log K$.

The larger the value of the equilibrium constant K_a, the stronger the acid. A so-called strong acid is one that is completely dissociated (equilibrium far to the right) and, in fact, all the available protons exist as H_3O^+. All acids above a certain strength appear equally strong in water, since the acid actually present is solely H_3O^+ in those cases. Strong acids may not be equally strong in other solvents. Similar comments hold for bases, with OH^- being the strongest base in water.

Brønsted acids fall into three major classes: binary acids, oxo acids, and hydrated metal ions. Binary acids contain hydrogen and one other element. One naturally occurring example is HCl, which is emitted in large amounts by volcanic activity. HCl also may be formed on incineration of chlorine-containing organic substances. Most of the HCl from these sources is dissolved in rainwater and washed out of the atmosphere in a short time (see Chapter 5 in connection with ozone destruction). A second example is H_2S, which also has volcanic origins and is a decomposition product of organic material such as proteins under oxygen-poor conditions. Strengths of binary acids increase with increasing bond polarity and with decreasing bond strength.

Oxo acids are those in which the ionizable hydrogens exist as OH groups attached to a not-too-electropositive third element. Environmental examples include H_2CO_3 (formed from CO_2 in water), H_2SO_4 (a result of sulfur oxide emissions, important in acid rain and in influencing climate through effects of sulfate aerosols), and H_3PO_4 (originating from phosphates in detergents and fertilizers). These particular examples are polyprotic acids. Strengths of oxo acids increase with the number of nonhydroxo oxygen atoms attached to the third element.

Hydrated metal ions may give acidic solutions. Many metal salts on solution in water become strongly hydrated to form complexes such as

$[Fe(H_2O)_6]^{3+}$, and such species may be ionized through reaction such as the following:

$$[Fe(H_2O)_6]^{3+} + H_2O \rightleftarrows [Fe(H_2O)_5(OH)]^{2+} + H_3O^+ \qquad (9\text{-}12)$$

More than one proton may be lost. Processes of this sort, which are important in the behavior of metal salts in water and in the hydrolysis of these salts to form insoluble products, are further treated in Chapter 10.

Bases, like acids, may be weak or strong. Some examples of bases are NaOH, which in solution ionizes to give OH^- and is a strong base; NH_3, a weak base whose action depends on taking a proton to form NH_4^+; and Na_2CO_3, a salt that also accepts protons via reactions

$$CO_3^{2-} + H_2O \rightleftarrows HCO_3^- + OH^- \qquad (9\text{-}13)$$

and

$$HCO_3^- + H_2O \rightleftarrows H_2CO_3 + OH^- \qquad (9\text{-}14)$$

All Brønsted bases have unshared pairs of electrons that can be donated to the 1s orbital of a proton. This picture of acid–base behavior leads to the Lewis definition of acids and bases. A base is an electron pair donor; an acid is an electron pair acceptor.

Acid–base equilibrium constants (ionization constants) may be used for calculations, such as that of hydrogen ion concentration or pH. Simple examples are discussed in most general chemistry and basic analytical texts, and the details are not dealt with here. However, as an alternative to individual calculations, the variation with pH of the concentration of the various species in equilibrium in a weak acid or base solution may conveniently be expressed graphically in a way that gives a rapid approximation of behavior. For a given total concentration of dissolved acid, $C_T = [HA] + [A^-]$, one can rewrite equation (9-11) for K_a in the form

$$[HA] = \frac{C_T[H_3O^+]}{K_a + [H_3O^+]} \qquad (9\text{-}15)$$

or

$$[A^-] = \frac{C_T K_a}{K_a + [H_3O^+]} \qquad (9\text{-}16)$$

In the limit that $K_a \ll [H_3O^+]$ (pH $\ll$ pK), $[HA] = C_T$ and $\log[HA] = \log C_T$; that is, $\log[HA]$ is independent of pH. Thus, a plot of $\log[HA]$ versus pH has zero slope as long as the pH is small. In this same region, $\log[A^-] = \log C_T + \log K_a + $ pH. The slope $d(\log[A^-])/d(pH) = 1$. If pH $\gg$ pK_a, similar arguments give the slopes of the plots of $\log[A^-]$ versus pH and $\log[HA]$ versus pH as 0 and -1, respectively. When pH = pK_a, $[HA] = [A^-]$.

We can use these approximations to show the concentration of the species present as a function of pH graphically. Figure 9-7 is such a plot for a hypothetical monobasic acid of $pK_a = 6$ at $C_T = 10^{-3}$ M. Point 1 corresponds to a solution of HA in water, point 3 to a solution of a salt (e.g., NaA), and point 2 to equal concentrations of HA and A^-. The scale of the ordinate will depend on the value of C_T. The pH (labeled H_3O^+) and pOH (labeled OH^-) lines in Figure 9-7 are plotted from pH + pOH = 14. For a solution of only HA in water, the electroneutrality requirement gives $[H_3O^+] = [A^-] + [OH^-]$, and pH thus is easily evaluated. In fact, since $[OH^-]$ is so small, $[H_3O^+] = [A^-]$ here. A solution of the salt NaA of the same total concentration would be partially hydrolyzed to form HA and OH^- in equal concentrations. The pH is given where these lines cross. Electroneutrality requires $[Na^+] + [H_3O^+] = [A^-] + [OH^-]$.

A titration curve showing the variation of pH with equivalents of base added to the HA solution discussed can be calculated and is illustrated in Figure 9-8. The pH values given by pure HA and by the completely neutralized system (equivalent to the NaA solution) have been discussed (points 1 and 3). The half-neutralized pH (point 2) also follows easily, since $[A^-] = [HA] = 1/2C_T$ here. Values of pH outside points 1 and 3 (dashed lines) can be achieved if additional strong acid or base is present. It can be seen that in certain regions, the pH is comparatively insensitive to added acid or base, while in others it is quite sensitive. The term "buffer capacity" is used to denote the rate of change of pH with added strong base or acid. (Buffer capacity is

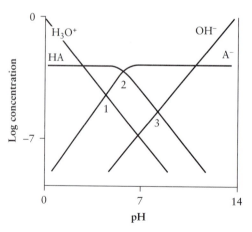

FIGURE 9-7 Log concentration versus. pH for a weak acid HA of pK 6 and $C_T = 10^{-3}$ M in water.

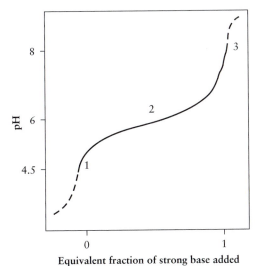

FIGURE 9-8 Titration curve for a weak acid of pK_a about 6. Broken lines indicate the effect of excess strong acid (on the left) or excess strong base (on the right).

defined as the number of equivalents per liter of strong acid or base required to change the pH by 1 unit.) The pH changes rapidly in regions of equivalent or neutralization points such as 1 and 3.

9.2.2 Carbonic Acid

The most important environmental acid is carbonic acid, which is produced in natural waters by the dissolution of CO_2 gas. The dissolution of gaseous CO_2 is comparatively slow; the average lifetime of a CO_2 molecule in the atmosphere is about seven years. The air–solution transfer process may be catalyzed by the enzyme carbonic anhydrase in waters containing biological materials. When CO_2 enters solution, two steps are actually involved:

$$CO_2(g) \rightleftarrows CO_2(aq) \tag{9-17}$$

$$CO_2(aq) + H_2O \rightleftarrows H_2CO_3(aq) \tag{9-18}$$

The species $CO_2(aq)$ (molecular CO_2 associated with water of hydration) and $H_2CO_3(aq)$ are not normally distinguished, so that an overall reaction can be written:

$$CO_2(g) + H_2O \rightleftarrows H_2CO_3^* \tag{9-19}$$

where

$$[H_2CO_3^*] = [CO_{2(aq)}] + [H_2CO_{3(aq)}] \tag{9-20}$$

$$K_{H_2CO_3^*} = \frac{[H_2CO_3^*]}{P_{CO_2}} = 2.8 \times 10^{-2} \tag{9-21}$$

Here, the Henry's law constant is in units of moles per liter per atmosphere (mol liter^{-1}atm^{-1}). At a partial pressure of CO_2 in the atmosphere of 3×10^{-4} atm, this gives an equilibrium concentration of $H_2CO_3^*$ of 1.03×10^{-5} mol/liter.

The ionization equilibria, which involve fast reactions, are

$$H_2CO_3^* + H_2O \rightleftharpoons H_3O^+ + HCO_3^- \tag{9-22}$$

$$K_{a_1} = \frac{[H_3O^+][HCO_3^-]}{[H_2CO_3^*]} = 4.5 \times 10^{-7} \tag{9-23}$$

[the ionization constant for $H_2CO_{3(aq)}$ alone is 3.16×10^{-4}; the value in equation (9-23) is smaller because only a part of the dissolved CO_2 is in the form of H_2CO_3] and

$$HCO_3^- + H_2O \rightleftharpoons H_3O^+ + CO_3^{2-} \tag{9-24}$$

$$K_{a_2} = \frac{[H_3O^+][CO_3^{2-}]}{[HCO_3^-]} = 4.8 \times 10^{-11} \tag{9-25}$$

We can obtain the following expressions from the equilibrium constants and C_T $(C_T = [H_2CO_3]^* + [HCO_3^-] + [CO_3^{2-}])$:

$$[H_2CO_3^*] = \frac{C_T}{1 + K_{a_1}/[H_3O^+] + K_{a_1}K_{a_2}/[H_3O^+]} \tag{9-26}$$

$$[HCO_3^-] = \frac{C_T}{[H_3O^+]/K_{a_1} + 1 + K_{a_2}/[H_3O^+]} \tag{9-27}$$

$$[CO_3^-] = \frac{C_T}{[H_3O^+]/K_{a_1}K_{a_2} + [H_3O^+]/K_{a_2} + 1} \tag{9-28}$$

We can find the concentration of any species as a function of pH at any value of C_T, or find slopes of the log concentration plots in various regions as illustrated in the general case just presented. Figure 9-9 is a plot of the fraction of C_T that exists in the form of each of the species as a function of pH. Note that at the pH values of most natural water systems, almost all the carbonate is in solution as the bicarbonate ion.

The various equilibrium constants on which Figure 9-9 is based are for pure water containing only CO_2. In seawater, the much higher ionic strength affects activities, and if concentration expressions are used, the corresponding

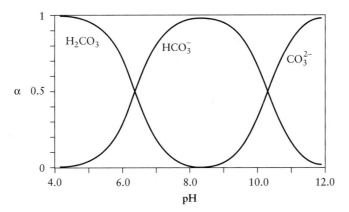

FIGURE 9-9 Plot of the carbonate species present in solution in pure water as a function of pH; α is the fraction of the total for each of the species.

stoichiometric equilibrium constants are different (see the list of values in Table 9-2). The net effect is that in seawater all the curves in Figure 9-9 would be shifted by 0.5–1 pH unit to the left, for the same C_T. The concentration of species in a CO_2 solution as a function of pH is illustrated in Figure 9-9 for a fixed total carbonate concentration C_T ($C_T = [H_2CO_3]^* + [HCO_3^-] + [CO_3^{2-}]$) in pure water.

The pH of a solution containing only CO_2 as the carbonate source would be given from $[H_3O^+] = [HCO_3^-] + 2[CO_3^{2-}] + [OH^-] \simeq [HCO_3^-]$ (the concentration of all of the cations, only H_3O^+ in this case, must be electrically balanced by the anions). This gives a pH of about 5.6 for water in equilibrium with the atmosphere. Natural waters typically are more basic; for example, seawater has a value near 8.1. This is the result of reactions of CO_2 with

TABLE 9-2

Equilibrium Constants of Some Carbonate Reactions

Reaction	"Pure" water		Seawater	
	0°C	25°C	0°C	25°C
$CO_2 + H_2O \rightleftharpoons H_2CO_3^*$	8×10^{-2}	2.8×10^{-2}	6.5×10^{-2}	3×10^{-2}
$H_2CO_3^* + H_2O \rightleftharpoons HCO_3^- + H_3O$	2.6×10^{-7}	4.5×10^{-7}	7.2×10^{-7}	9×10^{-7}
$HCO_3^- + H_2O \rightleftharpoons CO_3^{2-} + H_2O$	3×10^{-11}	4.8×10^{-11}	4×10^{-10}	8×10^{-10}
$CaCO_3(\text{calcite}) \rightleftharpoons Ca^{2+} + CO_3^2$		4.2×10^{-9}		6.4×10^{-7}

bases from rocks. Indeed, a pH of 8 is remarkably near the equivalence point for HCO_3^- given in Figure 9-10. It is evident from the figure that an isolated sample of seawater is not buffered, since it is near an equivalence point and holds its pH of about 8.1 rather precariously. The buffer capacity is very small.

Natural water not only can react with sediments, but it is in equilibrium with atmospheric CO_2 that will serve to keep the concentration of $H_2CO_3^*$ fixed according to the solubility relationships [equations (9-1) and (9-3)]. If P_{CO_2} (g) is constant, then $[H_2CO_3^*]$ and not C_T is constant within the time scale permitted by the rate of the gas–solution exchange reaction and mixing. Figure 9-11 gives the log concentration versus pH curves for this case, with $H_2CO_3^*$ fixed by a partial pressure of CO_2 of 3×10^{-4} atm. The pH of a solution of only CO_2 in water is not much different from the situation of the total concentration C_T being constant, but the titration curve is very different, as shown in Figure 9-12. This system at pH 8 is strongly buffered toward added base, much less so toward acid. At this pH, most of the dissolved CO_2 is present as HCO_3^-, and the amount of this in the oceans exceeds the amount of gaseous CO_2 in the atmosphere.

As noted, the pH of the ocean and most other natural waters is far from that predicted by CO_2 alone, and additional base must be present. This may be provided by soluble carbonate salts. In addition, equilibria with insoluble carbonates also contribute to establishment of the pH. The most obvious of these carbonates is $CaCO_3$, since calcium is a very common element in aqueous systems, although others such as $MgCO_3$ are important as well.

If $CaCO_3$ precipitation controls the carbonate ion concentration through the Ca^{2+} concentration of the water, it in fact controls the pH and acts as a buffer system for changes in carbon dioxide pressure. Increased $[H_2CO_3^*]$

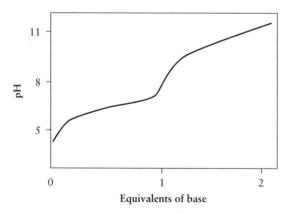

FIGURE 9-10   The titration curve for carbon dioxide.

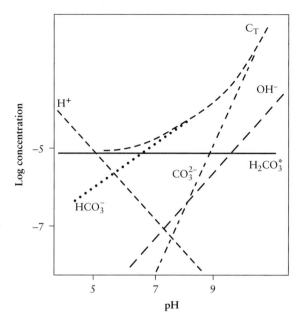

FIGURE 9-11 Log concentration versus pH for carbonate species in equilibrium with a fixed atmospheric pressure of CO_2: 3×10^{-4} atm.

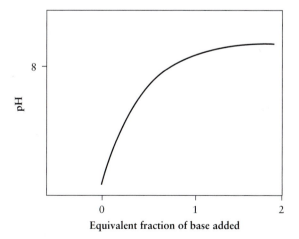

FIGURE 9-12 Titration curve for a solution of CO_2 in equilibrium with a fixed atmospheric pressure of CO_2 in freshwater.

arising from a higher CO_2 pressure would first cause the pH to drop as HCO_3^- and H_3O^+ form. This lower pH would also lower $[CO_3^{2-}]$, since the reaction $CO_3^{2-} + H^+ \rightarrow HCO_3^-$ is favored at lower pH values. The carbonate ion in turn is replaced in the solution as $CaCO_3$ dissolves; that is, the extra H_2CO_3 is titrated by the basic carbonate. The sum of these reaction is

$$CO_2 + H_2O + CaCO_3 \rightleftarrows Ca^{2+} + 2HCO_3^- \qquad (9\text{-}29)$$

The relation of these reactions to the carbon cycle and the increase in atmospheric carbon dioxide is discussed further in Section 10.2.

The pH and calcium ion values in many freshwater systems are consistent with $CaCO_3$ acting as a CO_2-buffering system. However, there are a number of complications in the quantitative treatment of calcium carbonate equilibria, such as the existence of different crystalline forms (calcite and aragonite) with slightly different solubility constants, and the problem of supersaturation; that is, the solubility product concentrations may be exceeded without precipitation taking place. Surface ocean waters are often supersaturated with calcium carbonate, while deep waters often are undersaturated. There is evidence that supersaturation may be encouraged by organic materials, which hinder the nucleation process that is essential for initiating formation of the solid phase. Supersaturation is a metastable condition. That is, it is not one of thermodynamic equilibrium, but the rate at which the system moves toward equilibrium is slow.

Variation of carbonate solubility with water CO_2 content is responsible for various deposits encountered in freshwater systems. Many soil waters have higher than normal concentrations of dissolved CO_2, because they absorb this gas as it is formed from reactions of microorganisms in soils at partial pressures exceeding the normal atmospheric value and ground-water often is under pressure. $CaCO_3$ is then more soluble (it is also more soluble under pressure, see Section 9.3). When the water comes to the surface, however, the pressure is released, excess CO_2 leaves the solution, and $CaCO_3$ can deposit. This result is illustrated by stalactite and stalagmite formation in caves in limestone regions. Carbonic acid in rainwater also contributes to the weathering of buildings and monuments made of carbonate rock; this is discussed further in Section 12.4.

Waters containing Ca^{2+} (or Mg^{2+}) and HCO_3^- are called temporarily hard waters, since heating will drive CO_2 from solution and cause the soluble calcium bicarbonate to convert to the insoluble carbonate in the reverse of reaction (9-29). Permanent hardness, caused by other Ca^{2+} or Mg^{2+} salts such as sulfate or chloride, is not eliminated by heating.

An important property often measured for waters to be used industrially or domestically is alkalinity, which is a measure of water quality. Titration of a water sample with acid to a pH of 8 (the phenolphthalein end point) converts all carbonate to bicarbonate, and the amount of acid required for this is a measure of the carbonate or hydroxide content (carbonate and caustic alkalinity, respectively). Large amounts of carbonate alkalinity, and any caustic alkalinity, are undesirable because such waters are unpalatable and may be corrosive in some uses. Titration to pH 5 (methyl orange end point) converts bicarbonate to CO_2, and is taken as a measure of bicarbonate content. Caustic alkalinity is present whenever the amount of acid required to reach pH 8 (the phenolphthalein end point) exceeds that needed to go from pH 8 to pH 5 (the methyl orange end point). Waters with excess acidity, rather than alkalinity, are encountered occasionally, usually in polluted systems such as those associated with acid mine drainage (Section 10.6.3).

9.2.3 pH and Solubility

From the preceding discussion of the solubility of $CaCO_3$ in the presence of acid, it is clear that an increase in acidity will lead to an increase in solubility of a sparingly soluble salt when the anion is the conjugate base of a weak acid. Heavy metal carbonates (e.g., $PbCO_3$) are examples of compounds for which the solubility increase with decreasing pH can be of environmental concern. Hydroxides and oxides are other common materials that will be affected similarly. It does not follow that increasing pH will not be of concern. Amphoteric substances may exhibit increased solubility in both acidic or basic conditions, and in many cases there is a particular pH that gives a minimum solubility. This is illustrated by alumina as discussed in Section 12.3.1. Figures 9-13 and 9-14 illustrate the solubility behavior of calcium carbonate

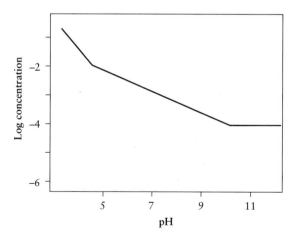

FIGURE 9-13   Solubility of $CaCO_3$ as a function of pH.

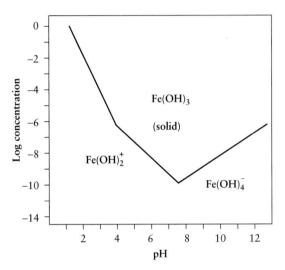

FIGURE 9-14 Solubility of amorphous $Fe(OH)_3$ as a function of pH. Increased slope of the solubility plot at pH values below about 4 is due to the formation of $Fe(OH)^{2+}$ and other species rather than $Fe(OH)_2^+$ at high acidity.

(nonamphoteric) and ferric hydroxide (amphoteric). In the latter case, solubility increases at high pH as soluble $Fe(OH)_4^-$ forms. At low pH, solubilty increases through the formation of soluble cationic species.

Lead carbonate can be used as an example to show how the solubility depends on pH in a quantitative way.[5] The solubility product of $PbCO_3$, given from

$$PbCO_3 \rightleftarrows Pb^{2+} + CO_3^{2-}; \qquad K_{sp} = 7.4 \times 10^{-14} \qquad (9\text{-}30)$$

gives the solubilty of Pb^{2+} as 2.72×10^{-7} M. However, in the presence of acid, the reaction is

$$PbCO_3 + H_3O^+ \rightleftarrows Pb^{2+} + HCO_3^- + H_2O \qquad (9\text{-}31)$$

This is just the difference of reactions (9-30) and (9-24), and the equilibrium constant for reaction (9-31) is

$$K = \frac{[Pb^{2+}][HCO_3^-]}{[H_3O^+]} = \frac{K_{sp}}{K_{a_2}} = \frac{7.4 \times 10^{-14}}{4.8 \times 10^{-11}} \qquad (9\text{-}32)$$

[5]The equations necessary for an exact calculation of the solubility of carbonates have been discussed by M. Bader and by S. Roo, L. Vermeire, and C. Görller-Walrand, *J. Chem Educ.*, **74**, 1160 (1997).

At a pH of, say, 5, and in the absence of other sources of bicarbonate, the solubility of Pb^{2+} is now 1.2×10^{-4} M. In natural waters, this would be further modified by other bicarbonate sources.

9.3 EFFECTS OF TEMPERATURE AND PRESSURE ON EQUILIBRIUM

Equilibrium constants vary with temperature and pressure in a way that can be predicted thermodynamically. The equilibrium constant is related to the standard free energy change of the process:

$$\Delta G^\circ = -RT \ln K = \Delta H^\circ - T\Delta S^\circ \tag{9-33}$$

where ΔG° is the standard free energy change, T the absolute temperature, R the gas constant, ΔH° the standard enthalpy change, and ΔS° the standard entropy change, for the reaction. In simple terms, ΔH° is a measure of the change in energy (e.g., bond energies and other interaction energies), and ΔS° a measure of the change in the degrees of freedom. The change in $\ln K$ with temperature at constant pressure is easily seen by differentiating the expression for $\ln K$ (assuming that ΔH° and ΔS° are independent of T, as is normally a good approximation over reasonably small temperature ranges). That is, at constant pressure,

$$\left[\frac{d \ln K}{dT} \right] = \frac{d}{dT} \left[\frac{-\Delta H^\circ}{RT} + \frac{\Delta S^\circ}{R} \right] = \frac{\Delta H^\circ}{RT^2} \tag{9-34}$$

Thus, the temperature variation of K depends on whether the enthalpy change of the process is negative (exothermic) or positive (endothermic). An exothermic heat of solution results in a decrease in solubility with increasing temperature. This is the case with $CaCO_3$, for example, which is less soluble in hot than in cold water. This property has consequences for carbonate solubility behavior in lakes and oceans. Values of the equilibrium constants used in this book refer to 25°C unless otherwise indicated.

The pressure variation of the equilibrium constant can be derived from the effect of pressure on ΔG° at constant temperature:

$$\frac{d}{dP}(\Delta G^\circ) = \Delta V^\circ \tag{9-35}$$

and

$$\frac{d}{dP}(\ln K) = \frac{-\Delta V^\circ}{RT} \tag{9-36}$$

where ΔV° is the standard volume change on reaction. For condensed (solid or liquid) systems, ΔV° is small; hence pressure effects are also small. However, pressures at the ocean's floor may exceed 200 atm, and this can have significant effects. For the dissolution of $CaCO_3$, $\Delta V° \approx -60\,cm^3/mol$ (i.e., the volume decreases because of improved packing of water molecules around the ions), and thus $CaCO_3$ should be more soluble in deep water than at the surface at the same temperature.

These effects can be predicted qualitatively from Le Chatelier's principle: when a system at equilibrium is perturbed by a change in concentration, temperature, or pressure, the equilibrium will shift in a way that tends to counteract the change.

9.4 OXIDATION–REDUCTION PROCESSES

The oxidation state of an atom in a molecule is a concept helpful in keeping account of electrons. In many cases it is an indication of the number of electrons involved in bonding, but this is not always true and the oxidation state (or oxidation number) need not have any physical meaning. The oxidation state is equivalent to the charge on a positive or negative ion in an ionic substance, while for a covalent compound it is arrived at arbitrarily by assigning the electrons in a bond to the most electronegative atom. Electrons forming bonds between two atoms of the same electronegativity are shared equally in the assignment.[6]

Reactions that involve changes of oxidation states, called oxidation–reduction (redox) reactions, are governed by the oxidation potentials of the systems involved. The potential E of the redox couple or half-cell, M/M^{n+}, is given a numerical value based in principle on an electrochemical cell in which M is in equilibrium with its ions:

$$M^{n+} + ne^- \rightleftarrows M \tag{9-37}$$

The potential is measured against a reference couple, the primary one being H_2/H^+, which is defined to have a standard potential value of exactly zero volts. The potential E is related to the free energy change of the process;

$$\Delta G = -nFE \tag{9-38}$$

where F is the Faraday constant, 96,500 coulombs (C), and n is the number of electrons transferred. Positive potentials and negative ΔG values are characteristic of spontaneous processes.

Standard potentials $E°$ refer to values where the reactants are at unit concentration (strictly, unit activity). Values are tabulated as the electromotive

---

[6]Oxidation state is conventionally expressed in parenthetical roman numerals after the symbol or name for the element [e.g., Fe(II)].

series, usually as reduction potentials, that is, for the process as written in reaction (9-37). The potential varies with concentration according to the Nernst equation. For the process given in reaction (9-37), we have

$$E = E^\circ - \frac{RT}{nF} \ln \frac{1}{[M^{n+}]} = E^\circ + \frac{0.059}{n} \log [M^{n+}] \qquad (9\text{-}39)$$

at 25°C (R is the gas constant, F is the Faraday constant, and T is the absolute temperature.) Potentials on this scale are sometimes given the symbol E_H to indicate that values are on the hydrogen scale.

It is often desirable to express the oxidizing or reducing properties of a system in terms of the concept of pE, which is analogous to pH. That is, in a formal sense, pE = $-$ log (concentration of electrons).[7] Although there are no free electrons in typical chemical systems, the value of pE provides a measure of the oxidizing or reducing properties of a solution. Alternatively, pE serves as a quantitative indication of the oxidizing or reducing conditions necessary for a given oxidation state of an element to be stable. The quantity pE gives a unitless scale analogous to pH, while E_H is a scale measured in volts. In practice, pE is evaluated from the expression

$$\text{pE} = \frac{E}{(RT/F)\ln 10} = \frac{E}{0.059} \quad \text{at } 25\,°C \qquad (9\text{-}40)$$

The pE value for oxygen in aqueous media can be evaluated for illustration. The oxygen half-cell reaction in acid solution is[8]

$$O_2(g) + 4H^+ + 4e^- \rightleftarrows 2H_2O; \quad E^\circ = +1.229\,V \qquad (9\text{-}41)$$

$$E = E^\circ + \frac{0.059}{4} \log\left([H^+]^4 P_{O_2}\right) \qquad (9\text{-}42)$$

Under conditions of equilibrium with the atmosphere, where the partial pressure of O_2 is 0.21 atm, we have

$$\text{pE} = \frac{1.229}{0.059} + \frac{4(-\text{pH}) + \log 0.21}{4} \qquad (9\text{-}43)$$

At a pH value of 8, in the range reasonable for natural waters, pE = 12.6. This will vary with pH and the oxygen partial pressure. In the absence of other oxidizing agents, the oxidizing power in a body of water is in fact determined by the concentration of dissolved oxygen. If the water is not saturated with

---

[7]More properly, pH = $-$ log (activity of H$^+$); then pE = $-$ log (activity of electrons).
[8]In basic solution, the reaction is

$$O_2(g) + 2H_2O + 4e \rightleftarrows 4OH^-; \quad E^\circ + 0.401\,V$$

The acid and base reactions are related by the ion product of water, and either can be used under all conditions.

oxygen, equation (9-42) should have the value of the oxygen partial pressure that would be in equilibrium with the actual O_2 concentration according to Henry's law.

When used to describe a set of conditions or an environment, "oxidizing" and "reducing" are relative terms; that is, they have meaning only with respect to the reaction or compound being considered. Conditions that exhibit low values of pE will favor reductions of many elements and compounds, and the lower the value, the greater the number of materials that can be reduced. High pE values will favor oxidation similarly.

Oxidizing or reducing power in aqueous solution is limited by the fact that water itself can be oxidized or reduced according to the reactions

$$2H_2O \rightarrow O_2 + 4H^+ + 4e^- \tag{9-44}$$

and

$$2H_2O + 2e^- \rightarrow H_2 + 2OH^-; \qquad E° = -0.828 \text{ V} \tag{9-45}$$

respectively. [Reaction (9-44) is just the reverse of (9-41).] The pE values of these reactions set the limits of strength for oxidizing or reducing substances that can be stable in water. The values are dependent on the pH and on the oxygen or hydrogen concentrations (or equilibrium partial pressures).

An example of the use of pE is shown by consideration of the equilibrium between sulfate S(IV) and sulfide S(-II) [reaction (9-46)], where the conditions under which SO_4^{2-} or sulfide predominates in solution can be related to the pE value.

$$SO_4^{2-} + 9H_3O^+ + 8e^- \rightleftharpoons HS^- + 13H_2O \tag{9-46}$$

Under most conditions of pH that are of interest environmentally, the predominant form of sulfide is HS^-. Other sulfur species (e.g., sulfite, SO_3^{2-}, and elemental sulfur) will be ignored for simplicity; in much, but not all, of the aqueous chemistry of sulfur compounds, they are not important. For reaction (9-46), $E°$ is 0.25 V. Therefore,

$$\begin{aligned} pE &= \frac{E°}{0.059} + \frac{1}{8} \log \frac{[SO_4^{2-}]\,[H_3O^+]^9}{[HS^-]} \\ &= 4.24 + \frac{1}{8} \log \frac{[SO_4^{2-}]}{[HS^-]} - \frac{9}{8} pH \end{aligned} \tag{9-47}$$

One can consider how the ratio of $[SO_4^{2-}]$ to $[HS^-]$ varies with pE at a particular pH. At a pH of 8, we have

$$pE = -4.76 + \frac{1}{8} \log \frac{[SO_4^{2-}]}{[HS^-]} \tag{9-48}$$

From this, it can be seen that the concentrations of SO_4^{2-} and HS^- will be equal at $pE = -4.76$. On the other hand, at the pE of oxygen-saturated water, 12.6, the ratio $[SO_4^{2-}]/[HS^-]$ is very large (log $[SO_4^{2-}]/[HS^-] = 138.9$), and hence SO_4^{2-} is the stable form. Plots can be made for $[SO_4^{2-}]$ and $[HS^-]$ as a function of E or pE (the slopes of the plots are easily established, as was the plot of pH vs log concentration for weak acids) at a particular total concentration of SO_4^{2-} plus HS^- to show the behavior graphically (see Section 10.4).

Plots of potential E vs pH that show the regions in which particular species are stable are called Pourbaix or predominance area diagrams. An example for sulfur is given in Figure 9-15. Each region represents the conditions of E and pH necessary for a particular species to be stable. There is a small region in which elemental sulfur is favored; this is not included in the foregoing discussion. A vertical line on this diagram represents an equilibrium reaction that is independent of potential, that is, an acid–base reaction. It is the value of pK. A horizontal line represents a redox reaction that is independent of pH, and is simply the $E°$ value, while sloping lines represent redox reactions that are pH dependent; the slope depends on the number of electrons and the number of hydrogen ions involved in the reaction. A line representing equilibria between soluble species marks conditions at which concentrations are equal: one species or the other predominates on either side of the line. If only one species is soluble, it is necessary to specify what solution concentration will be necessary for "predominance"—in other words, what level of concentration

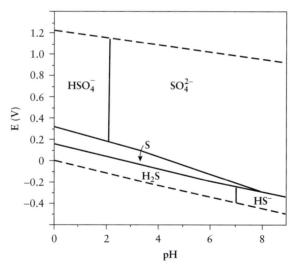

FIGURE 9-15 Pourbaix diagram for the sulfur system. The dashed lines are the stability limits for water. Adapted from G. Faure, *Principles and Applications of Inorganic Geochemistry*. Copyright © 1991. Reprinted by permission of Pearson Education, Inc. Upper Saddle River, NJ 07458.

will be deemed important—and draw the plot accordingly. The upper and lower limits in this diagram are the stability limits for water; they are not limits on the existence of particular states in the absence of water.

Potentials are based on thermodynamics and represent equilibrium situations; they say nothing about how fast equilibrium will be reached. Many redox reactions have multistep and complex mechanisms. Kinetic limitations may make reactions slow even when they are strongly favored thermodynamically. This book would be carbon dioxide and water now if it were not for kinetics! Many electrode reactions require a larger potential than calculated to overcome a kinetic barrier (the overpotential). Redox reactions involving oxo anions often take place slowly, the more so the higher the oxidation state of the central element, and the smaller it is. Thus, the perchlorate ion, ClO_4^- is reduced more slowly than chlorate, ClO_3^-, and chlorate more slowly than iodate, IO_3^-. Reactions involving many diatomic molecules such as O_2 also tend to be slow.

There are two general mechanisms for redox reactions. In the *outer sphere* mechanism, the reactants must approach each other to allow the electron to be transferred, but no direct bonding takes place between them. In the *inner sphere* mechanism, a bonded intermediate must form first (although it may be transient and unstable). This process involves the same needs for activation, and so on as the kinetics of other chemical reactions. Even outer sphere processes may require an activation of the reactants before the electron can be transferred. Many redox reactions do not involve transfer of electrons as such, but rather transfer of an atom. The oxidation state changes because of our definition. An example is the reaction between the common oxidizing agent hypochlorous acid and the nitrite ion,

$$ClOH + NO_2^- \rightarrow NO_3^- + Cl^- + H^+ \tag{9-49}$$

where the oxygen atom originally on the Cl in ClOH is transferred to the nitrogen. In other reactions involving oxo ions, the first step is protonation of one of the oxygens to form water, which dissociates to leave an electron-deficient species that can bond to a lone pair on the reducing agent Not only are the potentials of such reactions affected by pH according to the Nernst equation, but so are the rates at which they can occur.

9.5 COORDINATION CHEMISTRY

9.5.1 General Aspects of Coordination Chemistry

A metal ion invariably is a species with vacant, low-energy orbitals. Though obviously not a metal ion, H^+ may be taken to be a simple analogue of one.

Just as this species never exists in solution as such but always acquires some electron density by sharing an electron pair from another atom, so metal ions generally make some use of their empty orbitals by means of coordinate bonding of this type. The extent to which this is done depends on the energy of the orbitals that are accepting the electrons, and, for representative metals, is roughly parallel to the electronegativity of the metal. Coordinate bond formation is very slight for the alkali metals, and it increases across a period in the periodic table. This type of bonding is also greater for metal ions with small sizes and with high oxidation states. Transition metal ions—that is, those with empty inner orbitals—are usually particularly strong complex formers. The picture of complex formation, [i.e., donation of a nonbonding pair of electrons on an ion or molecule, the ligand into an empty orbital on the metal] falls into the Lewis classification of acids and bases, with the metal ion as the acid. The metal ion–donor atom bond may be highly polar, but is essentially a covalent interaction in typical systems. In extremes, however, such as with the alkali metals, it may equally well be regarded as an ion–dipole interaction.

The common donor atoms are the halide ions, the atoms O, N, S, and P in their compounds and C in a few examples such as CO and CN^-. The overall ligand may be neutral or anionic (cationic in a few cases), and the complex as a whole may be anionic, cationic, or neutral depending on the sum of the charges on the metal ion and ligands. While there is no general correlation between the strength of ligand behavior and charge, many of the common ligand molecules are anions: for example, the carboxylate group—COO^-, inorganic oxo anions, and the halide ions. Coordination tends to facilitate proton loss by neutral ligands through changes in the electron density of the bonds, as has already been illustrated by the aqua complexes (Section 9.2.2). For strong Bronsted bases (i.e., ligands that form weak acids), there will be a competition between protons and the metal ion for the ligand, and complexing will be influenced by pH. Even with weak Brønsted bases (e.g., Cl^-, where HCl is a strong acid) there may be a pH dependence, since in basic media competition from OH^- ligands for the metal ions may become important.

9.5.2 Coordination Number and Geometry

The number of donor atoms linked to a metal ion is the coordination number of the complex. The most common coordination number is 6, with the donor atoms lying at the corners of a regular or distorted octahedron as illustrated in Figure 9-16. A 4–coordinate, tetrahedral structure also is common, and a variety of other structures are found in specific examples.

The coordination geometry (i.e., the arrangement of the donor atoms around the central metal ion), favored by a given element may vary with its oxidation state, with the nature of the donor atoms, and with other factors.

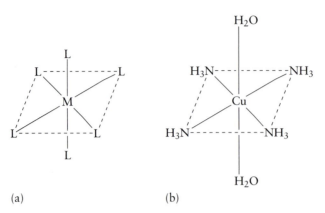

FIGURE 9-16 (a) An octahedral complex. (b) A common distortion of an octahedron: two bonds are longer than the other four.

Comparatively few metal ions form complexes exclusively of a single geometry, although most have a limited number of preferences. The chemical behavior of metal ions in systems in which geometry is critical (e.g., biological activity) may be related to this property. As has been stated, an octahedral, 6-coordinate geometry is the one most commonly encountered, and most of the aqua complexes can be expected to exhibit this. Not all octahedral complexes are regular, even if all ligands are identical. Hexacoordinate complexes of Cu(II) are somewhat extreme examples of this, since the usual configuration has four "normal" bonds in a plane, with the remaining two ligands attached by long, weak bonds above and below this plane. Indeed, many Cu(II) complexes are often referred to as square planar, although in fact the additional two weak interactions are usually present, especially in solution, where solvent molecules can take up these positions. The striking deep-blue complex formed when ammonia is added to a solution of a Cu(II) salt has this highly distorted octahedral shape, with four NH_3 ligands in a plane, but two H_2O molecules still weakly attached (Figure 9-16b).

With ligands more highly polarizable than water or ammonia (e.g., halide ions), the ions of the first transition series metals tend to form tetrahedral, 4-coordinate complexes. Truly 4-coordinate square-planar compounds are found for some ions, chiefly those that have eight d electrons [e.g., Pt(II), Au(III)]. This is related to the particular stabilization possible with this unique number of d electrons for this geometry in favorable cases. A few other ions [e.g., Ag(I), Hg(II)] are encountered commonly, but not exclusively, in linear, 2-coordinate complexes. Again, this is related to a particular electronic configuration: 10 d electrons. Other coordination numbers and geometries are less common in environmental systems.

A donor atom with more than one available nonbonding pair of electrons may donate these simultaneously to two or more metal ions, acting to bridge them together. Water molecules can do this, although it is more common with OH^- ligands. A ligand with two or more separate donor atoms may bridge, but often these ligands are found attached to one metal ion through both donor atoms. This behavior is called chelation, and leads to more stable complexes. An example of a chelating ligand with two donor atoms (bidentate) is ethylenediamine, $H_2NCH_2CH_2NH_2$. Each N possesses one nonbonding electron pair, as in NH_3. The molecule is flexible and can bend to fit two adjacent coordination sites in most geometries. Ligands also are known with 3 (tridentate), 4 (tetradentate), 5 (pentadentate), and 6 (hexadentate) donor groups. Some are illustrated later in this chapter. A ligand with a single donor atom is called unidentate. Chelation obviously results in a ring structure.

9.5.3 Metal–Ligand Preferences

The various metals show differing complexation tendencies and preferences for donor atoms. Ions of the alkali metals (Li, Na, K, Rb, Cs) have little electron pair acceptor tendency, although in aqueous solution they are hydrated by predominantly ion–dipole interactions. This tendency is greater for the smaller ions, which also have the lowest energy orbitals and greatest covalency in their interactions. Complexing is weak for this family. The alkaline earth ions, especially Mg(II) and Ca(II), are very common in the environment. Owing largely to their higher charge densities, they show a larger tendency to attract the ligand electrons than do the alkali metals, and complexing is important in their behavior. Much of the interaction can be ascribed to Coulombic forces, and since the metals are quite electropositive, buildup of electron density on them is unfavorable. Their strongest complexes are formed with strongly electronegative donors such as F^- and oxygen ligands, which form highly polar bonds. Much more covalent in its bonding with ligands than the other alkaline earths, Be(II) forms stronger complexes than the other members of its family, and in biological systems it may act as a poison because it preferentially ties up coordination groups that are normally involved more weakly with Mg(II) or Ca(II). Chelation is very important for the formation of stable complexes with this family. The greater stability of chelate complexes in comparison to those with analogous unidentate ligands lies not so much with stronger metal–ligand bonding, but rather with an entropy effect. The number of degrees of translational freedom, and therefore the entropy, is increased by replacing several unidentate ligands (e.g., water molecules) by a polydentate one, and so increasing the number of independent species (see Section 9.5.5).

Ions of charge greater than 2+, and transition metal ions generally, are strongly complexed by a variety of ligands. The fluoride ion, and ligands with oxygen donors, are favored by ions of electropositive metals such as Al and those of first three or four transition element families and also most metal ions in high oxidation states. Nitrogen donors become favored by the mid–transition element families in their lower oxidation states [e.g., Co(II), Fe(II)] although higher states still prefer F^- and oxygen donors. Ions of some of the heaviest and the least electropositive metals [Pt(II), Au(III), Hg(II)] favor larger, more polarizable (more covalent) ligands such as I^- or Cl^- in preference to F^-, and sulfur or phosphorus ligands in preference to oxygen or nitrogen donors. For example, the gold in seawater is believed to be present as $AuCl_4^-$. However, it may be noted that the large amount of chloride in seawater for the most part is not involved in complexing because the predominant cations prefer oxygen donors such as water or carbonate.

Metal–ligand interactions are examples of Lewis acid–base behavior, and the preference of a metal ion for a particular ligand can be correlated with the concept of hard and soft acids and bases. Lewis acids and bases can be ranked according to an empirical property of hardness or softness, which is determined by such factors as size, charge, polarizability, and the nature of the electrons available for interaction. A hard acid typically is small, with a high positive charge and tightly held, chemically inactive electrons, while a soft acid is large, with a small or no positive charge and electrons in orbitals that are readily influenced by other atoms. Hard bases are small, and of high electronegativity, while soft bases are large, with easily distorted electron clouds. (Primarily, these are properties of the acceptor or donor atoms in polyatomic acids and bases. However, the properties of an individual atom in a molecule are often strongly influenced by the rest of the molecule, especially when there is extensive electron delocalization.) The examples in Table 9-3 illustrate these classes. There is a continuous range of this property, but no numerical values can be given.

TABLE 9-3

Some Lewis Acids and Bases on the Hard–Soft Scale

Hard acids	Intermediate acids	Soft acids
$H^+, Li^+, Na^+, K^+, Be^{2+}, Mg^{2+}, Ca^{2+},$ $Al^{3+}, Cr^{3+}, Fe^{3+}$, high oxidation state metal ions	$Fe^{2+}, Co^{2+}, Ni^{2+}, Cu^{2+},$ $Zn^{2+}, Pb^{2+}, Sn^{2+}$	$Cd^{2+}, Pd^{2+}, Pt^{2+}, Hg^{2+}, Cu^+,$ Ag^+, Tl^+
Hard bases	Intermediate bases	Soft bases
$F^-, OH^-, H_2O, NO_3^-, CO_3^{2-},$ carboxylic acid anions ($RCOO^-$), NH_3, aliphatic amines (RNH_2)	Br^-, NO_2^-, aromatic amines (e.g., $C_6H_5NH_2$), pyridine	I^-, CN^-, CO, H^-, organic sulfur compounds (R_2S), phosphorus and arsenic compounds (R_3P, R_3As), π-electron donors such as C_2H_4 and C_6H_6, R^-

The utility of the hard–soft concept is simply the rule that like prefers like; that is, interaction among species of similar hardness or softness is preferred over interactions involving a large difference in hard–soft properties. Thus, a metal such as Hg(II), which is soft, prefers to complex with sulfur rather than oxygen donors. This is not to say that complexing between hard and soft species will not take place, only that it is normally less favorable than if these properties are matched.

9.5.4 Water as a Ligand

In aqueous systems, water is the most abundant ligand. Distortion of the electron distribution on a water molecule as a result of coordination causes the protons to be more readily ionizable, as already mentioned. The result may be the formation of hydroxo (HO^-) or oxo (O^{2-}) complexes, a process called hydrolysis. That is, the aqua complex[9] $[M(H_2O)_x]^{n+}$ (x is usually 6) may be partially or even completely deprotonated, depending on the nature of M^{n+} and the pH. This behavior is illustrated by Fe^{3+} or Al^{3+}. One has loss of protons in, for example, the following reactions:

$$[Al(H_2O)_6]^{3+} + H_2O \rightleftharpoons H_3O^+ + [Al(H_2O)_5OH]^{2+} \qquad (9\text{-}50)$$

$$[Al(H_2O)_5OH]^{2+} + H_2O \rightleftharpoons H_3O^+ + [Al(H_2O)_4(OH)_2]^+ \qquad (9\text{-}51)$$

Ultimately this may lead to oxo species through loss of two protons from one O, and finally to oxo anions such as $[AlO_3]^{3-}$ if the medium is basic enough. There may be polymerization—for example, as follows:

$$2[Al(H_2O)_5OH]^{2+} \rightleftharpoons OH^- + [(H_2O)_5Al - (OH) - Al(H_2O)_5]^{5+} \qquad (9\text{-}52)$$

where the hydroxo ligand bridges two metal ions. With further deprotonation and polymerization, repulsion among ligands usually results in a change in coordination number from octahedral 6 to tetrahedral 4. After aging, compounds with compositions such as $FeO(OH)$ and $AlO(OH)$, which have bridging ligands around each metal ion to give a three-dimensional cross-linked structure, result. The reactions are all pH sensitive and often reversible, but once the final highly polymerized products have formed, they may be quite resistant to dissolution upon increase of acidity.

The extent of these hydrolysis reactions can be correlated roughly with size and charge of the metal ion, being more extensive for the smallest and most highly charged ion at a given pH. In general, one has the following:

---

[9]It is the convention to write the formula for a complex ion or molecule in square brackets to define the coordination sphere. This practice should not be confused with the use of square brackets to denote concentrations.

1+ *ions*: form simple aqua complexes over most of the pH range.

2+ *ions*: aqua complexes predominate in acidic media, but hydroxo species occur in basic media. However, larger alkaline earths remain aqua complexes even in strongly basic media, while very small ions such as Be^{2+} are hydrolyzed under neutral or mild acidic conditions.

3+ *ions*: chiefly present as hydroxo complexes in the pH range of natural waters.

4+ *ions*: most of these form hydroxo or oxo complexes in all but highly acidic solutions.

The highly polymerized hydroxo precipitates that frequently form upon hydrolysis of the 3+ and 4+ ions are often of a gelatinous nature, resulting from the irregular cross-linked network that is built up. These usually rearrange and become more crystalline with time. The formation of such precipitates can be used in flocculation processes for water purification as discussed in Chapter 11.

9.5.5 Complex Stability and Lability

The formation of a complex $[ML_6]^{n+}$ in solution is part of an equilibrium process that can be represented by a series of steps, each described by its own equilibrium constant (often called formation constant):

$$[M(H_2O)_6]^{n+} + L \rightleftharpoons [M(H_2O)_5L]^{n+} + H_2O \tag{9-53}$$

$$K_1 = \frac{[[M(H_2O)_5L]^{n+}]}{[[M(H_2O)_6]^{n+}][L]} \tag{9-54}$$

$$[M(H_2O)_5L]^{n+} + L \rightleftharpoons [M(H_2O)_4L_2]^{n+} + H_2O \tag{9-55}$$

$$K_2 = \frac{[[M(H_2O)_4L_2]^{n+}]}{[[M(H_2O)_5L]^{n+}][L]} \tag{9-56}$$

(The ligand L has been written as a neutral species for simplicity; it may equally well be an anion.)

The overall process is

$$[M(H_2O)_6]^{n+} + 6L = [ML_6]^{n+} + 6H_2O \tag{9-57}$$

and the overall equilibrium constant is the product of the stepwise constants:

$$K_{overall} = \prod_i K_i = \frac{[[ML_6]^{n+}]}{[[M(H_2O)_6]^{n+}][L]^6} \tag{9-58}$$

These equilibria must hold along with ionization or solubility equilibria.

Complex formation changes the solubility of an insoluble metal compound by removing the metal ions from the equilibrium as they are tied up in the complexed form. For example, in the presence of a ligand L, the reaction for the solution of lead carbonate becomes

$$PbCO_3 + L \rightleftharpoons [PbL]^{2+} + CO_3^{2-} \tag{9-59}$$

which is the sum of the solubility equation giving K_{sp} and the complex formation equation giving K_f. The equilibrium constant is just the product of these two constants,

$$K = \frac{[[PbL]^{2+}][CO_3^{2-}]}{[L]} = K_{sp}K_f \tag{9-60}$$

An increased concentration of L in solution leads to an increase in the concentration of soluble lead. If the ligand also takes part in acid–base equilibria, as many do (e.g., the anions of Brønsted acids), the complexation process will be pH dependent:

$$PbCO_3 + HL \rightleftharpoons [PbL]^+ + CO_3^{2-} + H^+ \tag{9-61}$$

$$\frac{[[PbL]^+][CO_3^{2-}][H^+]}{[HL]} = K_{sp}K_fK_a \tag{9-62}$$

Of course, $[CO_3^{2-}]$ and $[H^+]$ are not mutually independent; they are related by the carbonate equilibria.

A large value for the formation constant means that the formation equilibrium lies far to the right. In the presence of excess ligand, for example, the amount of "free" metal ion would be small in this case. However, it is important to distinguish between thermodynamically stable systems in the foregoing sense and those that are kinetically inert. Complexes in solution normally undergo continuous breaking and remaking of the metal ligand bonds. (This is true of many other systems also. For example, a weak acid that is largely undissociated according to its ionization constant nevertheless exchanges protons at a very fast rate with solvent water molecules.) If the bond-breaking step is rapid, and consequently ligand exchange reactions are rapid, the complex is called labile. If the ligand exchange reactions are slow, the complex is said to be inert. Lability or inertness has no direct relation to thermodynamic stability, although many inert complexes are thermodynamically stable. Many metal ions complexed by unidentate ligands will undergo complete exchange in a time of fractions of a second: some are much slower, hours or days. These inert complexes are associated with particular electronic configurations of the metal ion. Exchange rates are lower with multidentate ligands. Inert complexes in particular have properties different from both the free metal ion and the ligands. An example is $[Fe(CN)_6]^{3-}$, in which the cyanide groups do not exhibit the toxicity of free CN^-.

Complex stability can be correlated with several properties of the metal ion and the ligand. The following are most important:

Size and oxidation state of the metal. Smaller size and higher positive oxidation state lead to stronger complexing. This can be understood in terms of the stronger bonding that arises with shorter metal–ligand distances and the greater Coulombic contribution from a greater charge.

Chelation. The free energy change upon complex formation depends on both the enthalpy and the entropy changes of the reaction. The enthalpy change is made up largely from the metal–ligand bond energies, while the entropy change involves the change in the degrees of freedom of the system. The number of independent molecules increases when unidentate ligands are replaced by a polydentate one, resulting in a positive contribution to the entropy change and a larger negative free energy change than for a one-to-one exchange. This is a widely accepted explanation for the observed greater stability of chelated systems compared with those of analogous unidentate ligands: the greater the degree of chelation, the greater is the stability. The size of the chelate ring formed is of prime importance. Five-membered rings (including the metal atom) are most favored, with six-membered rings a close second, and other sizes of minor importance.

The electronic configuration of the metal ion. The degeneracy of the d orbitals of a transition metal ion is destroyed when the ion is surrounded by ligands. The d orbitals are said to be split by the ligand field. The splitting pattern depends on the geometry of the complex, while the magnitude of the splitting depends on the nature of the ligand and the metal ion. Occupancy of the lower energy orbitals contributes to the stability of the system (the ligand field stabilization energy) although this is counteracted if the higher energy orbitals are also filled. Thus, this contribution to stability will depend on both the electronic configuration and the magnitude of the d orbital splitting. With a given metal ion, the common ligands usually have the following relative splitting effects: $I^- < Br^- < Cl^- < F^- < H_2O$ (and most other oxygen donors) $< NH_3$ (and amines) $< CN^-$. With octahedral complexes, the electronic configurations leading to a large ligand field stabilization energy involve 3, 6 (with strongly splitting ligands), and 8 d electrons (d^3, d^6, and d^8), while d^0, d^5 (with weakly splitting ligands), and d^{10} have no stabilization from this source.[10] Trends in the stability of complexes of different metals with a given ligand, for example, $Mn(II) < Fe(II) < Co(II) < Ni(II) < Cu(II) > Zn(II)$, can be traced to the variation of ligand field stabilization energy with elec-

---

[10] Any intermediate or advanced level textbook on inorganic chemistry will contain a detailed discussion of this topic.

tronic configuration, coupled with the normal trends in bond energies with size.

The hard–soft match of the metal and ligand, as already discussed.

The donor strength of the ligand. One measure of this characteristic is given by the strength of the ligand's conjugate acid. For example, if a ligand is the anion of a strong acid, this implies that the ligand is not an effective donor toward a proton, and in such cases it is generally not a good donor to a metal ion either. The parallelism is rough, because the metal ion acceptor orbitals are quite different from the 1s orbital of the proton and there are some notable exceptions (e.g., CO).

9.5.6 Some Applications of Complexing

The formation of complexes may tie up a metal ion to such an extent that solubility is increased considerably: that is, the concentration of free metal ions required to achieve saturation according to the solubility product requirements is less easily reached. Solids that would otherwise precipitate are thus kept in solution. This is often desirable when waters containing dissolved minerals are employed for industrial or domestic use, and is one reason for the use of phosphates as "builders" in detergent formulations, as discussed in Chapter 7. The calcium and magnesium ions in hard water form insoluble salts with soaps and interfere with the action of most detergents. The soap precipitates have a curd like character and are very difficult to wash away. Hard waters also require larger amounts of the surfactant for adequate cleaning. The Ca(II) and Mg(II) ions can be held in solution by complexing, and certain phosphate ions were widely used for this purpose, most commonly, sodium tripolyphosphate $Na_5P_3O_{10}$. The simplest form of phosphate, orthophosphate PO_4^{3-}, is tetrahedral. Polyphosphates are based on this structure, but with shared oxygens. The tripolyphosphate ion has the following structure:

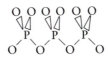

Unlike orthophosphate, this ion, can form chelates having six-membered rings, and it is potentially tridentate. The extra stability introduced is illustrated by the following equilibrium constants:

$$Ca^{2+} + HPO_4^{2-} \rightleftharpoons [CaHPO_4]; \qquad K = 5.02 \times 10^2 \qquad (9\text{-}63)$$

$$Ca^{2+} + P_3O_{10}^{5-} \rightleftharpoons [CaP_3O_{10}]^{3-}; \qquad K = 1.25 \times 10^8 \qquad (9\text{-}64)$$

Besides its complexing role, the polyphosphate ion protonates to produce a basic solution that improves the cleaning action of soaps and detergents by ensuring complete ionization of the hydrophilic anionic functional group. The polyphosphate is slowly hydrolyzed to simple phosphate species.

As discussed in Chapter 7, with low cost and no apparent toxicity, phosphates received extensive use in detergent formulations until the 1970s, with typically 35–50% of the detergent formulations being sodium tripolyphosphate. Since phosphate is an essential nutrient and an effective fertilizer, extensive increases in the growth of algae in freshwater systems and consequent increase in biological oxygen demand (Section 7.2.5) and eutrophication were ascribed to detergents entering the waters through waste discharge. As a result, phosphates were banned in home detergents in many areas, including much of the United States and Canada, although it now appears that this was an overreaction.

From the general coordination properties discussed for Ca(II), we can predict that a satisfactory complexing agent for replacement of phosphate in detergents would have to be a chelating agent, since only then would adequate stability of the complex be likely. Oxygen should be the chief donor atom, since only this donor atom is hard enough to interact strongly with Ca(II) while at the same time being part of a more elaborate molecule. There are also requirements of nontoxicity, biodegradability, and low cost. An example of a very effective chelating agent for calcium is ethylenediaminetetraacetic acid, EDTA:

$$HOO-CH_2 \diagdown \qquad \diagup CH_2-COOH$$
$$N-CH_2-CH_2-N$$
$$HOO-CH_2 \diagup \qquad \diagdown CH_2-COOH$$

This is potentially hexadentate. It coordinates as the anion, with an oxygen from each ionized carboxyl group and the two nitrogens being the donor atoms. (It may be added parenthetically that while EDTA is potentially hexadentate, it may not be so in all its complexes. For example, a water molecule may occupy the sixth coordination site, leaving a carboxyl group free.) EDTA is widely used in analytical chemistry and for other purposes that call for the sequestration of metal ions in soluble form. A related substance, nitrilotriacetic acid (NTA),

$$\diagup CH_2COOH$$
$$N-CH_2COOH$$
$$\diagdown CH_2COOH$$

was suggested as a phosphate replacement, and large-scale plans for its introduction were under way in about 1970. Again, an oxygen from each ionized carboxylate group and the nitrogen can act as the donor atoms. Because of concerns (now believed to be unfounded) that NTA could enhance the toxic effects of heavy metals (e.g., cadmium), plans for its use in the United States were canceled, although NTA is used in Canada and Finland, for example.

Sequestration of metal ions to prevent them from interfering with useful processes or materials is not limited to the laundry. Trace amounts of metals can have undesirable catalytic activity that can be prevented by the presence of complexing agents. For example, a salt of EDTA is sometimes added to canned foods to protect them from undesirable effects from trace metal components. Some forms of metal poisoning have been treated by a strong ligand (e.g., EDTA), which removes the metal ions from binding sites in the body and leads to its excretion as a soluble complex. (Such treatment is not a simple matter; the complexing agent also removes essential metals.) The unavailability of essential trace metal nutrients in soils has been overcome by addition of complexing agents that extract the needed metal ions from insoluble minerals to produce soluble forms to aid plant growth.

There is a flip side to the benefits of increased solubility of metal ions from complexing. That is the possibility that the solubility of toxic elements such as heavy metals could be increased through the formation of soluble complexes from otherwise insoluble salts.

9.5.7 Complexing in Natural Systems

Most natural water systems contain an excess of Ca(II) compared with the available strong ligands. Consequently, although calcium complexes are rather weak, the amount of ligand left over to complex with trace metals is often small, perhaps often negligible. In some situations, however, complex formation is believed to be important in natural waters—for example, when the concentration of dissolved metal ions is higher than expected in the solubilization of metal ions for transport in soils and plants, and in the sometimes unexpected chemical behavior of trace metals. The enrichment of aqueous trace metals by plants tends to follow the normal stability sequence for metal ion complexes [this is, in part, that the complex stability with a given ligand that follows the order Mn(II) < Fe(II) < Co(II) < Ni(II) < Cu(II) > Zn(II)]. This suggests that trace metal uptake by plants is related to complex formation. There is evidence that unpleasant outbreaks of algal growth (e.g., the Florida "red tide") are linked to the availability of soluble metal complexes.

Most naturally occurring organic ligands are decomposition products or by-products of living organisms. There are several main types.

Humic acid, fulvic acid, humin, etc. Soils and sediments normally contain organic matter derived from decaying vegetation. Much of this organic matter is made up of fulvic or humic acid materials, or humin. These are members of a related series distinguished primarily on the basis of solubility differences. Fulvic acid is soluble in both acid and base, humic acid in base only, and humin is insoluble. The molecular weights increase in the same sequence, ranging from near 1000 to perhaps 100,000 Da. They are not definite simple compounds, but polymeric materials. Their exact origin is uncertain: One theory suggests that they are oxidized and decomposed derivatives of lignin; another that they are formed by polymerization of simple decomposition products from plant materials like phenols and quinones. The structures contain aromatic rings with oxygen, hydroxyl, and carboxyl groups, and some nitrogen groups, all of which have potential coordination abilities. Typical units are as follows.

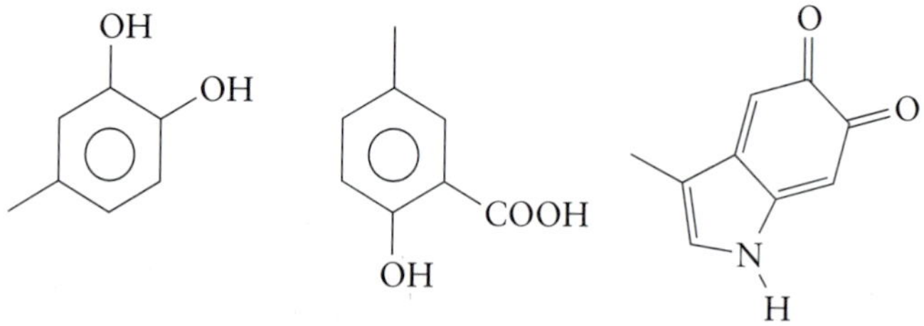

It is well established that strong interactions between these materials and metal ions take place; humic substances have a high capacity for binding metal ions. They form a potential reservoir of trace metals in soils. The availability of trace metals in soil to plants evidently depends on complex formation. Most soil sources of essential metals, which include V, Mo, Mn, Co, Fe, and Cu, are insoluble, and the small amounts that can be made available through acidic attack of the oxides, carbonates, silicates, and so on in which they are present appear to be inadequate for normal plant needs. Large amounts may be available through a reservoir of coordinated metal held in humic acids in the soil, which may be released on further decomposition, or by an attack from agents such as citric or tannic acids, or other substances generated by plant roots. Humic substances are often associated with color in water, and are sometimes accompanied by a high iron content through complex formation.

Polyhydroxy compounds: carbonates, sugars. The following grouping,

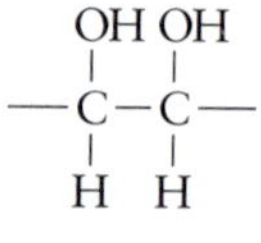

often present in sugars, can chelate, although it does not normally form strong complexes. These substances are widespread in metabolic products and in carbohydrate decomposition products.

Amino acids, peptides, and protein decomposition products. A simple amino acid, such as the following

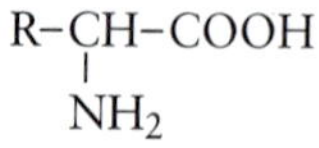

$$R-CH-COOH$$
$$|$$
$$NH_2$$

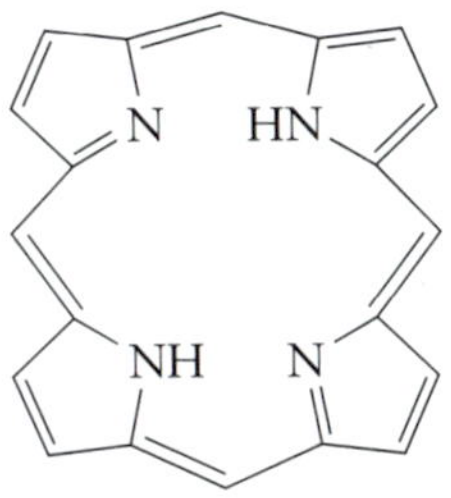

FIGURE 9-17 The porphin structure.

is an effective chelating agent. Such molecules can be products of protein hydrolysis, although there is no definite evidence for their importance in natural water systems. Peptides and related substances are more important in this respect; for example, the ferrichromes are very stable, soluble iron complexes. These ligands, which have high specificity for Fe(III), are common in microorganisms and are present as decomposition intermediates. They seem to be useful in transporting iron into cells and making it available to iron-containing enzymes. The iron is bound in an octahedron of oxygen atoms from groups such as the following.

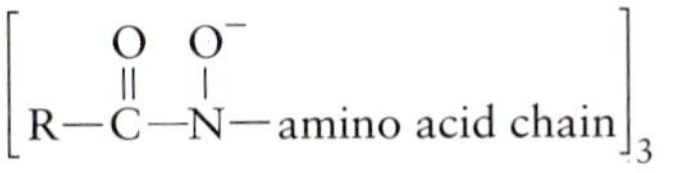

$$\left[\begin{array}{c} \overset{\displaystyle O}{\underset{\displaystyle \|}{}} \ \overset{\displaystyle O^-}{\underset{\displaystyle |}{}} \\ R-C-N-\text{amino acid chain} \end{array}\right]_3$$

Porphyrins and related compounds. Compounds based on the tetrapyrrole nucleus make up an important class of naturally occurring complexing agents. The parent compound is porphin,[11] the structure for which is given in Figure 9-17. Considerable scope for electron delocalization is possible in this ring structure (i.e., resonance forms exist), and partly because of this, these compounds exhibit a high stability. Substituents on the porphin skeleton, and minor changes to the ring structure, are found in the various naturally occurring compounds of this class. The porphyrin structure is planar, with sufficient room in the center of the ring system for binding a metal ion, as shown in Figure 9-18. This is an example of a macrocyclic ligand.

The electron delocalization just referred to may involve orbitals and electrons from the metal as well as those of the organic molecule. Metal ions that tend to form octahedral complexes may have ligands above and below the porphyrin plane; the bonds are also influenced by the electron delocalization in the plane. Examples of compounds of this kind are the iron-containing hemoproteins, including hemoglobin and myoglobin; cobalt-containing vitamin B_{12} and related enzymes (based on a slightly modified ring called the corrin structure); and chlorophyll, with magnesium as the metal.

[11] A substituted porphin is called a porphyrin.

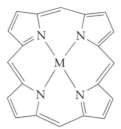

FIGURE 9-18 A porphin ring containing a metal ion.

The structure of the metal ion portions of hemoglobin and myoglobin, used in biological oxygen transport and storage, is shown in Figure 9-19. In addition to the four porphyrin nitrogen atoms in the coordination sphere of the Fe(II) ion, a fifth nitrogen from a histidine group in the amino acid portion of the molecule is attached to one of the remaining coordination sites. The sixth coordination site of the Fe(II), which is the active site, may have a water molecule replaced reversibly by an oxygen molecule (dioxygen). That is, O_2 is taken up when its partial pressure is high, and released when it is low. Other ligands may be attached to the sixth site in place of water or dioxygen. If such ligands are not replaced readily, as in the case with CO, which bonds strongly to the Fe(II) in hemoglobin, the oxygen transport function is blocked and death can result. This illustrates one process by which the biological function of a species containing a metal ion can be disrupted.

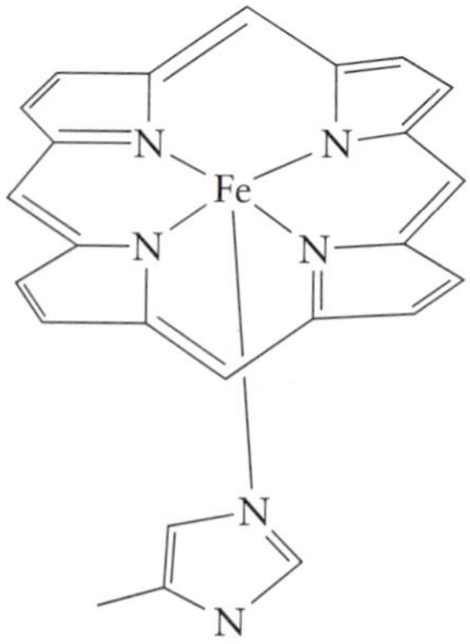

FIGURE 9-19 The iron center in hemoglobin and myoglobin. Groups attached to the porphin ring are not shown.

The exact behavior of the metal ion in this and analogous systems is determined not only by the immediate, coordinate environment that influences the electron density on the metal, but also by the extensive protein structure that is part of the molecule. This protein structure may have indirect effects on electron density and may exhibit important steric effects that control the overall reactions. Thus, iron-containing enzymes with direct coordination structures similar to hemoglobin, but different structures of the protein chains, may not exhibit reversible oxygen uptake. Rather, they may act as electron transfer agents through the conversion of Fe(II) to Fe(III).

Complex formation with organic materials also may influence mineral weathering reactions, largely through solubilizing a weathering product that otherwise would be insoluble, or through the effect of complexing on the stability of different oxidation states of an element. For example, Fe(II) is usually oxidized to Fe(III) only slowly; the reaction may be faster if it can proceed as follows:

$$\text{Fe(II)} + \text{organic ligand} \rightarrow \text{Fe(III)} - \text{organic complex} \qquad (9\text{-}65)$$

Ligands such as tannic acid, amino acids with SH groups, or phenols are particularly effective. We may also encounter more complicated cycling reactions in which the Fe(III) complex oxidizes the organic ligand, reverting to Fe(II) again. Organic materials in groundwater lower pE and result in solubility and stability of Fe(II) and Mn(II). [Normally, these elements are present as insoluble Mn(IV) and Fe(III) oxide species.] The presence of either of these ions is an indicator of organic contamination of the groundwater source.

Complexation may provide a buffering of free metal ion concentration. To maintain the equilibrium conditions required by the complex stability constant, excess metal may be tied up as a complex, or released from a complex form. A regulation of this sort probably occurs in organisms and may be present in other parts of the environment; for example, organic-rich sediments may have some action of this kind in water bodies.

Complex formation in seawater is significant, but the predominant cations, Ca(II), Mg(II), Na(I), and K(I), are largely present as the aqua species. About 10% of the alkaline earth ions and 1% of the alkali metal ions in seawater are complexed, mostly by the sulfate ion. Relative concentrations are such that this takes roughly half of the sulfate ion concentration. Carbonate and bicarbonate ions are present in much lower concentrations, but the amounts present are extensively complexed with the more abundant cations. This complexing is often compensated for in equilibrium calculations through the use of "stoichiometric" constants valid for seawater conditions (Section 9.1.3).

9.6 COLLOIDAL MATERIAL

Although we have been discussing concentrations of dissolved substances, a good deal of material may be present in water as suspended particles. If the particles are small enough, they may pass through filters and travel in aqueous systems very much like material in true solution. Such suspensions of small particles are called colloids. Typically, colloids are particles in the size range of 1–10 μm, although some suspended material in natural water systems may be larger than this. Mineral materials (clays) such as silica, silicates, aluminosilicates, metal oxides or hydroxides, and polymeric organic material from the excretions or decay of organisms are common sources. Particles will settle out at a rate that depends on the density of the particle and the square of its diameter (Stokes' law). The density depends on the composition. The size, because of the dependence on the square of the diameter, is most important. Small particles will settle very slowly and are easily kept in suspension by agitation, but coagulation or coalescence into larger particles will lead to much more effective settling. Figure 9-20 shows the size relationships of materials in aqueous systems.

Colloidal material has a high surface area per unit mass; clay particles, for example, may have a specific surface area in excess of $50\,\mathrm{m^2/g}$. A large surface area results in high activity, particularly for adsorption. All surfaces tend to adsorb other substances from their surroundings. Polar surfaces, which are typical of the materials that make up natural colloids, will adsorb polar molecules or ions. A result of these surface charges is that coagulation into larger particles that can more readily settle out is hindered by the Coulombic repulsions. While small amounts of electrolytes stabilize a colloidal suspension through contribution to this surface charge, addition of large concentrations of

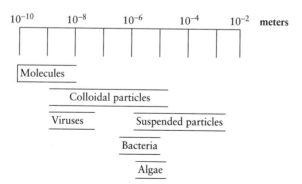

FIGURE 9-20 Size spectrum for particles in water. Adapted from W. Stumm, *Environ. Sci. Technol.*, **11**, 1066 (1977).

electrolytes may aid in the aggregation of colloidal material into larger particles. Part of the effect is due to adsorption of ions of opposite charge, which reduces the overall charge on the particle. In addition, with more ions available in solution, the solution layer above the surface that must contain counterions to give electrical neutrality can be thinner, allowing closer approach of the particles. The repulsive forces from the charge on the particles depend on their separation distance squared. However, the attractive forces that can hold them together when they coagulate, van der Waals forces, depend on the separation to the fourth power. Consequently, only when close approach is possible will the attractive forces predominate.

The activity of surfaces in adsorbing other substances is of significant importance in modifying concentrations, reactivities, and movement of such substances in the environment. This applies not only to colloidal suspensions in water, but also to the finely divided clays and humic substances in soils, and to the liquid–gas surfaces in foams and even to the surfaces of water systems themselves. Trace amounts of materials can be built up on the surfaces of particles to levels that are much greater than indicated by bulk analytical values. This represents a potential process for the concentration of toxic materials. Also, it is well established that concentrations at the surface of a water body can be different from the concentrations in the bulk; a surface film perhaps 50 µm thick may have concentrations several hundred times the bulk value. Although the total amount in such a thin film is small, it is this film that becomes spray. The composition of the film can affect properties such as surface tension and these surface effects can be important in phenomena such as energy transfer or gas transfer across water–air interfaces.

9.7 ADSORPTION AND ION EXCHANGE

The behavior of elements present in large amounts in an aquatic system is controlled by the rules of solubility, acid–base equilibria, complexation, and so on, but the behavior of trace substances is largely controlled by adsorption on solid surfaces. Material will be partitioned between the surface and the bulk liquid, depending on the strength of the adsorption. Adsorption can be approximated by the equation

$$K_d = C_s/C_{aq} \qquad (9\text{-}66)$$

where C_{aq} is the concentration in the aqueous phase, C_s is the quantity adsorbed per unit amount of solid, and K_d is the partition or distribution coefficient. If adsorption becomes more difficult as the amount already adsorbed increases (i.e., if the adsorption sites available vary in activity), the relation between C_{aq} and C_s may not be linear. In addition, there is a limit to how much material can be adsorbed. Consequently, this expression is most applicable to low levels of adsorption.

The usual environmental particles in water systems, sediments, and soils are oxides of metals or metalloids. Since surface metal atom will adsorb OH^- or H_2O from the water, the entire surface can be regarded as oxide. In a complex solid, all sites may not have the same affinity for the adsorbate, and thus adsorption may not show a linear relation between amount adsorbed and concentration in solution. At some point the surface will be saturated. Competition may occur between different adsorbates, and adsorption may be pH dependent.

A typical oxide surface site may be represented as follows

$$\equiv X{-}O^- \qquad \equiv X{-}OH \qquad \equiv X{-}O\!\!\diagup^{\textstyle H^+}_{\textstyle H}$$

with an acid–base (and thus a pH-dependent) equilibrium ($\equiv X$ represents an atom to which a surface oxygen atom is attached). The surface may be positive, negative, or neutral, depending on the pH. The isoelectric point is the pH at which the surface is neutral. The isoelectric point for silica and many silicates is less than 3, so that under most conditions in natural waters and soils the particles carry a negative charge. Values for iron oxides tend to be near pH 7, while those for aluminum oxide and hydroxide materials are strongly basic. While actual values of neutrality will vary with electrolytes, these give some idea of the types of ions likely to be absorbed by a colloid of a given kind.

If a metal ion M^{2+} is present, possible interactions include the following.

$$\equiv X{-}O{-}M^+ \qquad \equiv X{-}O\!\!\diagup^{\textstyle H}_{\textstyle M^{2+}} \qquad {\equiv X{-}O \atop \equiv X{-}O}\!\!\diagup\!\!\!\diagdown M \qquad \equiv X{-}O{-}M{-}OH$$

The O—M linkage may be primarily electrostatic with an ion such as Na^+, or it may have more covalency with transition metal ions. In any case, the strength of the interaction will be greater for smaller and more highly charged ions. A problem associated with this correlation, however, is knowing whether the free ion or the hydrated ion size is important. A small or highly charged ion may hold the water molecules that hydrate it strongly enough to ensure that they are retained upon adsorption, while a larger or lower charged ion may be bound without its waters of hydration and thus appear smaller.

The last of the foregoing structures illustrates the possibility that anions (OH^- in this case) can be adsorbed by an attached cation; more generally, $\equiv X{-}O{-}M{-}Y$ (Y is any anion). Another form of possible anion binding is

$$\equiv X{-}O\!\!\diagup^{\textstyle H^+}_{\textstyle H}\cdots Y^-$$

The $\equiv X\!-\!O$ group can also adsorb oxo anion species by replacing another group as follows, for example,

$$\equiv X\!-\!O\!-\!\overset{\displaystyle OH}{\underset{\displaystyle O}{\overset{|}{\underset{\|}{P}}}}\!-\!OH$$

or ionized forms.

In any of these processes, one ion can be exchanged for another; for example, H^+ is replaced by a metal ion, or M^{2+} can be replaced by another metal ion, or Y^- by Y'^-. That is, the adsorption of ions by the surface can lead to ion exchange behavior, an important process for nutrients in soils, for example (see Section 12.3.1). The pH behavior of adsorption is species specific, but in general, metal ions tend to be held most strongly at high pH and released at low pH, where H^+ can replace them. Anions behave differently: Phosphate adsorbed by alumina, for example, is held most strongly at low and especially neutral pH, being released at high pH, where OH^- controls the behavior.

The ability of a material such as a clay to absorb ions is measured by its ion exchange capacity, determined by how much Na^+ can be replaced by another ion, and is usually expressed as milliequivalents per 100 g. Values for mineral particles may exceed 100. A clay with an ion exchange capacity of only 10 mequiv/100 g could absorb 1.04 g (5 mmol) of Pb per 100 g, for example. Thus retention capabilities may be quite significant. In a real situation, of course, there will be competition among the various ionic species present.

Neutral organic molecules, if polar, can be strongly adsorbed on oxide surfaces. In many natural water systems, the amount of dissolved organic carbon (e.g, soluble humic compounds) is sufficient to saturate the suspended inorganic particles and completely cover their surfaces. These organic surface films may still have exposed functional groups that can attract cations, but the organic layer may also adsorb hydrophobic organic materials. In this way, trace organic materials as well as heavy metals can be significantly concentrated in sediments as the suspended matter settles.

9.8 CONCLUSION

In this chapter, we have endeavored to survey the main principles that underlie the chemical behavior of "natural" and "unnatural" (i.e., human-influenced) environmental processes occurring in aqueous media. Some of these principles provide a basis for quantitative predictions, although in reality the complexity of many environmental systems makes calculations that work in laboratory situations difficult to apply with great certainty to the real world. In any event, they show the factors that must be included in an understanding of the aqueous environment.

Additional Reading

Liss, P. S., and R. A. Duce, eds., *The Sea Surface and Global Change*. Cambridge University Press, Cambridge, U.K., 1997.

Morel, F. M. M., and J. G. Hering, *Principles and Applications of Aquatic Chemistry*. Wiley, New York, 1993.

Pankow, J. F., *Aquatic Chemistry Concepts*. Lewis Publishers, Chelsea, MI, 1991.

Schnoor, J. L., *Environmental Modeling: Fate and Transport of Pollutants in Water, Air and Soil*. Wiley, New York, 1996.

Stumm, W., *Chemistry of the Solid–Water Interface: Processes at the Mineral–Water and Particle–Water Interface in Natural Systems*. Wiley, New York, 1992.

Stumm, W., ed., *Aquatic Surface Chemistry: Chemical Processes at the Particle–Water Interface*. Wiley, New York, 1987.

Stumm, W., and J. J. Morgan, *Aquatic Chemistry: Chemical Equilibrium and Rates in Natural Waters*, Wiley, New York, 1996.

EXERCISES

9.1. What weight of water would have to evaporate from 1 m^3 of water to produce a temperature decrease of 1 degree celsius? How much energy would be released to the atmosphere when this vapor condensed?

9.2. What would be the temperature decrease required in 1 M^3 of water to increase the temperature of an equal volume of air by 1 degree celsius? Assume $25°C$ where the density of air is 1.1843×10^{-3} g/ml, and no evaporation.

9.3. Henry's law constant for O_2 in water at $25°C$ is 1.28×10^{-3} mole $\text{liter}^{-1}\text{atm}^{-1}$.

   (a) What is the concentration of dissolved oxygen in moles per liter? In grams per liter? In ppm?

   (b) How many grams of organic material $(CH_2O)_n$ could be converted to carbon dioxide per liter of air-saturated water?

9.4. What is the concentration of Cd^{2+} in water in contact with cadmium carbonate with a bicarbonate ion concentration of 1×10^{-3} mole/liter at a pH of 4? $K_{sp}, CdCO_3 = 5.2 \times 10^{-12}$.

9.5. What is the predominant species in a carbonate solution at pH 2? At pH 7? At pH 11?

9.6. List some acids that are important in the environment. Why is each important? What are the sources?

9.7. Explain the cause of stratification in a lake.

9.8. The pH expected for rainwater under unpolluted conditions is around 5.6. Explain why this value is expected.

9.9. The pK_a of hypochlorous acid is 7.5; that of HF is 3.18. Which is the stronger acid? Which acid contains the stronger conjugate base?

9.10. If a liter of water is saturated with CO_2 at 25°C and 5 atm pressure, how many grams of CO_2 would be released if the pressure were reduced to 1 atm. and the system equilibrated?

9.11. The solubility product for cadmium sulfide is 8×10^{-27}, assuming the reaction is $CdS \rightarrow Cd^{2+} + S^{2-}$.
   (a) Calculate the concentration of dissolved Cd^{2+} in pure water.
   (b) Since S^{2-} is a strong Brønsted base, what will the actual process be?
   (c) Find the equilibrium constant for the actual process from the equilibrium constants that are available, and from this find the Cd^{2+} concentration at pH 7 and 4.

9.12. Would you expect coal to be able to undergo ion exchange reactions? Explain. (See Chapter 6.)

9.13. In a solution containing the following metal ions and ligands, match up the most likely metal–ligand combinations: Ca^{2+}, Cu^{2+}, Pb^{2+}, NH_3, HS^-, $H_2PO_3^-$.

9.14. (a) Write the equation for the reaction for the dissolution of CdS in water containing the ligand EDTA.
   (b) From the standard expressions for the solubility product, the formation constant and the ionization constants of the acid EDTA, derive an expression and the value for the equilibrium constant for the reaction in part a.
   (c) Use the results of part b to explain the pH dependence of the solubility of CdS under these conditions.

9.15 The pK value for the formation constant of $[AuCl_2]^+$ is 9.8. What fraction of Au(III) in a solution that is 10^{-2} M Au(III) and 0.05 M Cl^- is present as the complex, assuming no other species are found?

9.16. At what value of pE and E_H would NO_3^- and NO_2^- be in equilibrium at pH 5?

9.17. Discuss how metal ions could be complexed by humic acid.

9.18. (a) A water sample, pH 6, contains 10^{-4} M Mn(II). What values of E_H and pE are required for this to be stable with respect to MnO_2?
   (b) What value of P_{O_2} would this correspond to?
   (c) Would you expect aqua complexes of iron to be soluble under these conditions? If so, in what state?

9.19. If a gram of a solid substance of density $1\,g/cm^3$ were divided into colloidal particles that were $10\,\mu m$ sized cubes, what would the surface area be?

9.20. Figure 9-13 shows the variation of solubility of calcium carbonate with pH. Explain the changes in slope of the plot of pH vs log C.

9.21. A 100-g sample of a montmorillonite clay was found to be able to absorb 1.85 g of sodium ions at saturation. What is its ion exchange capacity? What weight of Cd could be held (as Cd^{2+}) by the clay?

9.22. MnO_2 (solid) can be reduced to soluble Mn(II).

(a) Give the equation for the half-reaction.

(b) What potential would be required to give a concentration of Mn(II) of 1×10^{-4} at pH 7?

(c) What equilibrium oxygen pressure would this correspond to?

10

THE ENVIRONMENTAL CHEMISTRY OF SOME IMPORTANT ELEMENTS

10.1 INTRODUCTION

Some 90 naturally occurring elements exist in the environment, each with its own chemistry. Those elements present in high abundance determine the nature of the environment as a whole through the properties and behavior of themselves or their compounds. These, and some of lesser abundance, have important biological roles. This chapter will consider how the chemistry of some of the most important elements relates to the properties of the environment and to biological effects.

The elements believed to be essential for the growth and development of organisms are listed along with others of environmental importance in Table 10-1. Many essential elements are present in organisms in trace amounts (micronutrients), and it is possible that others will be found that are essential at the trace level. Eliminating the ingestion of very low levels of virtually any element by an organism, as must be done to prove that absence of that element is deleterious to the organism's development, is extremely difficult. The presence of an element in an organism does not signify that it is necessarily playing an active role. Of more immediate concern is the prevention of the ingestion of

TABLE 10-1

Elements That Are Biologically Essential, Especially Toxic, or of Other Environmental Interest
(in order of atomic number)

Element	Biological role	Toxicity in animals	Environmental sources and comments
Hydrogen, H	Component of organic molecules.		
Lithium, Li	None known, but suspected.	Slight toxicity.	Used in medicine; lithium batteries.
Beryllium, Be	None.	Very toxic (beryllosis).	Coal combustion is the primary source; amount released globally is small. A hazard to metalworkers.
Boron, B	Essential for plants.	Slight toxicity.	Used in detergents, glass.
Carbon, C	Component of all organic molecules.	Depends on chemical form (e.g., CO, CN^- are very toxic). Cannot generalize for organics.	CO_2, CO from fossil fuels, biological activity. CN^- used to leach gold and silver ores and in metal electroplating.
Nitrogen, N	Component of proteins.	Depends on chemical form (e.g., nitrogen oxides are toxic).	Nitrogen oxides are combustion by-products; contribute to acid rain. NO_3^-, NH_3 from sewage and fertilizers cause water pollution problems. NO is a nerve transmitter in the body.
Oxygen, O	Required for respiration for most organisms.	O_3 (ozone), peroxides are toxic.	
Fluorine, F	Essential trace element (2.5 ppm); reduces tooth decay.	Excess causes discolored teeth, other toxic effects at higher levels.	A contaminant of some phosphorus fertilizers, an additive to drinking water, toothpaste, etc. Released along with some fluorocarbons from aluminum smelters. These sources could add to natural food sources to give excessive levels in some locations.
Sodium, Na	Electrolyte component of biological fluids, nerve function.	Related to some forms of hypertension; otherwise low toxicity.	
Magnesium, Mg	Generally essential element; enzymes, chlorophyll.	Low.	
Aluminum, Al	Possible essential trace element.	Moderately toxic to plants, some animals. Uncertain role in Alzheimer's disease.	May be made soluble by acid (e.g., acid rain) but is normally in insoluble forms and not accessible; aluminum sulfate used in water treatment through precipitation of $Al(OH)_3$.

(continues)

TABLE 10.1 (*continued*)

Element	Biological role	Toxicity in animals	Environmental sources and comments
Silicon, Si	Essential trace element for some animals; structural role in some plant and animal species.	Certain compounds are toxic as a result of physical form.	Asbestos, silicate dusts can affect the lungs (cancer, silicosis).
Phosphorus, P	Essential in metabolic processes, bones, teeth, phospholipids.	Low in inorganic phosphates but phosphate esters (insecticides), PH_3 toxic.	Phosphate in fertilizers and sewage can contribute to water pollution.
Sulfur, S	Component of proteins.	SO_2, SO_3, H_2S toxic.	SO_2 and SO_3 from combustion and smelting lead to acid rain; oxidation of sulfides produces H_2SO_4 in mine wastes; anaerobic decay releases H_2S.
Chlorine, Cl	Electrolyte component of biological fluids.	Low as Cl^-; high as Cl_2 and oxidized forms.	Some organochlorine compounds are carcinogenic or toxic. Stratospheric Cl destroys ozone. Industrial and natural sources.
Potassium, K	Essential for most organisms; nerve action.	Low.	Possible water pollutant from fertilizers.
Calcium, Ca	Essential; bones, shells, enzymes.	Low.	
Titanium, Ti	None known.	Nontoxic; not available in soluble forms.	TiO_2 common in paints and other products.
Vanadium, V	Essential trace element in many organisms; enzymes and metabolism.	Toxic; particulates may cause lung disease.	Combustion of fossil fuels.
Chromium, Cr	Essential trace element; enzyme in sugar metabolism.	Toxic and carcinogenic in Cr(VI) form, Cr(III) less toxic.	Metal plating and antirust treatment of wastes; Cr(III) forms are insoluble.
Manganese, Mn	Essential trace element; many enzymes.	Moderately toxic.	Coal burning, mining, smelting, and metal processing.
Iron, Fe	Essential in significant amounts; oxygen transport, respiration, and other enzymes.	Low toxicity except when intake is excessive.	Widespread natural occurrence; release to aqueous systems often associated with mines for coal and for other metals.
Cobalt, Co	Essential trace element for many organisms; enzymes, vitamin B_{12}.	Toxic, especially to plants.	Deficiency is more often a problem than excess; low soil values lead to inadequate levels in plant foods. Metal industries, fly ash are local sources.

(*continues*)

TABLE 10.1 (*continued*)

Element	Biological role	Toxicity in animals	Environmental sources and comments
Nickel, Ni	Essential trace element.	Toxic; carcinogenic in some forms. Some people develop dermatitis from skin contact with nickel-containing metals.	Metal smelting and refining, burning fossil fuels.
Copper, Cu	Essential trace element; redox enzymes.	Moderately toxic to mammals, more so to other organisms.	Mining, metal industries; sometimes used as algicide in the form of cupric sulfate.
Zinc, Zn	Essential trace element; many enzymes.	Slightly toxic.	Burning fossil fuels, metal plating.
Arsenic, As	Essential trace element.	Very toxic to animals. Reported carcinogen.	Mining wastes, released on burning coal, used in insecticides and herbicides. Present in some groundwater sources. Domestic wastewater contains large amounts at low levels.
Selenium, Se	Essential trace element; enzymes, protective against heavy metals.	Highly toxic to animals.	Can be concentrated in plants and taken up by animals.
Bromine, Br	Probable essential trace element.	Nontoxic as Br^-; organobromine compounds similar to Cl analogues.	Br reaching the stratosphere destroys ozone; some (halons, e.g., $CBrF_3$, $CBrClF_2$) have fire-extinguishing applications; others used as flame retardants. Biological as well as industrial sources of organic compounds.
Molybdenum, Mo	Essential trace element; enzymes (including those involved in nitrogen fixation).	Moderately toxic; excess can cause copper deficiency by interfering with Cu uptake.	Industrial smoke; particulate emissions may cause lung disease.
Silver, Ag	None.	Highly toxic to fish, much less so to humans.	Photographic wastes (reduced by economic considerations); mining (especially cyanide treatment of low-level ores).
Cadmium, Cd	Possible essential trace element in some organisms	Toxic; a cumulative poison in mammals.	Metal smelting and refining; impurity in Zn, metal plating, batteries; mining wastes caused "itai-itai" disease in Japan.
Tin, Sn	Essential trace element.	Low toxicity but concern about some organotin compounds.	Organotin compounds used as wood preservatives and antifouling agents; stabilizer in plastics; atmospheric release from burning oil.

(*continues*)

TABLE 10.1 (*continued*)

Element	Biological role	Toxicity in animals	Environmental sources and comments
Antimony, Sb	None known.	Possible carcinogen.	Used in fire retardants, some solders, ceramics, lead storage batteries, electronics; in metal ores with arsenic.
Iodine, I	Essential trace element (thyroid function).	Low toxicity as I^-.	Organic iodides (e.g., CH_3I) released by marine organisms. Stratospheric iodine is an effective ozone destroyer, but compounds usually photolyzed at low altitudes.
Barium, Ba	Some evidence that it is an essential element.	Moderately toxic in soluble forms.	Drilling fluid in oil and gas wells, used in paints and other products; solubilization from natural sources by increased acidity.
Tungsten, W	Component of enzymes in organisms living near oceanic hydrothermal vents.	Usual forms insoluble, nontoxic.	
Mercury, Hg	None.	Highly toxic in soluble forms or as vapor; a cumulative poison.	Organomercury fungicides, chlorine – alkali manufacturing (now much reduced), chemical processes (wide-spread mercury poisoning in Minamata, Japan, from fish contaminated by Hg pollution from a chemical plant), Hg release in processing gold-containing ores; release to atmosphere in coal combustion.
Lead, Pb	None.	Toxic; a cumulative poison.	Formerly lead additives in gasoline were a major source, lead-based pigments in paints (mainly old paint), mining, batteries, lead pipes in old plumbing, solder, pottery glazes.
Radon, Rn	None.	Carcinogen.	An α-emitting decay product of uranium found in many rocks; a non reactive gas, it can diffuse into buildings and be ingested through breathing.

(*continues*)

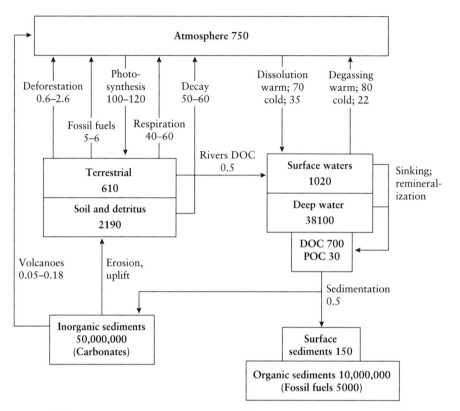

FIGURE 10-2 The carbon cycle: Quantities in 10^{15} grams of carbon, fluxes in 10^{15} grams of carbon per year. DIC, and dissolved inorganic carbon; DOC, dissolved organic carbon; POC, particulate organic carbon. Data from Post *et al.*, *Am. Sci.*, **78**, 310 (1990), and Houghton *et al.*, eds., *Climate Change 1995*, Cambridge University Press, Cambridge, U.K., 1996.

Most of the ocean carbon is in the deep water, and some small fraction of this is transferred to sediments, where it is sequestered virtually permanently. About 3×10^{15} g of C is estimated to be in marine biota.

IPCC estimates of annual perturbations to the natural cycle averaged over the 1980–1989 decade, indicate that a total of 7.1×10^{15} g of C is released by anthropogenic sources; 5.5×10^{15} g from fossil fuels and cement production, and 1.6×10^{15} g from changes in tropical land use, primarily destruction of forests. Of these emissions, the ocean takes up about 2×10^{15} g of C, Northern Hemisphere forest regrowth takes up another 0.5×10^{15} g, and other terrestrial sinks (increased plant growth from fertilization and other effects) 1.3×10^{15} g. About 3.3×10^{15} g of C is left in the atmosphere.

Take-up by the biosphere is dependent on the partial pressure of CO_2, since many (but not all) plants increase their growth at higher CO_2 pressures. This suggests that some buffering of atmospheric increases may come from the biosphere. The extent of this such action unknown, particularly in as much as other nutrients often are limiting. Experiments on some forest trees have shown that they respond to an increase in carbon dioxide with a period of rapid woody growth, but this ends relatively quickly. Arguments pointing out various advantages of a CO_2 increase have been proposed—primarily reduced water transpiration and increased food production. Countering these possible benefits are the disadvantages that might come from associated effects such as climate change.

As mentioned, many uncertainties remain about the sizes of some reservoirs in the carbon cycle and about turnover rates. The estimates concerning terrestrial carbon are confused by uncertain land use data. In addition, there are indications that dissolved organic carbon in the oceans has been underestimated, and many of the transport rates are known very poorly. For example, the amount of carbon as CH_4 trapped in clathrates as discussed later in this section is uncertain. These uncertainties are one of the factors that make greenhouse and other climate predictions difficult (Section 3.3).

From what has been discussed here and in Chapter 9, it is clear that the increase in atmospheric carbon dioxide the earth is now undergoing should be offset in part by the dissolution and neutralization reactions that it can undergo. Archer et al.[4] have modeled the processes to estimate the fate of a sudden injection of CO_2 into the atmosphere, considering the rates of the various processes. Their results suggest that the ocean should absorb 70–80% of such an input in 200–450 years. Reactions of seafloor $CaCO_3$ should take up another 10–15% after 5500–6800 years. The same investigators estimate that 5–8% would then remain in the atmosphere, but that too would be taken up by reactions with other basic rocks in weathering reactions of the types discussed in Section 12.3.1[5] after 200,000 years. These time periods could change if climate changes alter ocean circulation and mixing patterns, or with changes in the biological activity that is largely responsible for mixing in the sediments that influence reaction with the solid carbonates, for example. At any rate, these times are long in human terms.

With growing concerns about the increase of atmospheric carbon dioxide, proposals have been made for sequestering some of this. Sequestration may involve removal of CO_2 from the atmosphere after it has been released, or separating CO_2 from an exhaust gas stream and dealing with it in concentrated form. Five proposals are summarized briefly.

---

[4]D. Archer, H. Kleshgi, and E. Maier-Reimer, *Geophys. Res. Lett.*, **24**, 405 (1997).
[5]Essentially, reactions with Ca and Mg silicates to give the carbonates.

1. Increase forest growth, with the assumption that the wood produced will remain intact for some long period of time. Forests can be carbon dioxide sinks, but under natural circumstances will reach an equilibrium with CO$_2$ released upon decay (or combustion) equal to that absorbed. Only wood that is prevented from decaying can be a long-term trap, but in the short term, significant CO$_2$ might be removed from the atmosphere by increased forested areas.

2. Increase soil carbon through incorporation of vegetation. As with forest growth, equilibrium will eventually be reached.

3. Increase ocean biological uptake by plankton. As discussed later (Section 10.6.3), the growth of plankton can be significantly increased in some parts of the ocean by suitable fertilization. Some of the carbon taken up in their growth would be incorporated into marine sediments, but it is difficult to estimate how much, or to judge what other environmental consequences might result from the growth and decay of high concentrations of these organisms. Fertilization would have to be an on-going process over large areas of the ocean.

4. Injection of separated CO$_2$ into locations where it would be isolated and stable. There are at least three possibilities.

    (a) Exhausted petroleum fields. CO$_2$ is already injected into oil fields to enhance recovery by dissolving in the oil and reducing its viscosity. Much of this CO$_2$ is released, but a significant fraction remains. If no oil recovery is attempted, none would be released.

    (b) Very deep saline aquifers, where the CO$_2$ would be above critical pressure. About 600,000 metric tons removed from natural gas is being injected into an aquifer under the North Sea. It is estimated that this aquifer alone has the capacity to store 400 years' worth of CO$_2$ production from all the European power stations.

    (c) Deep sea locations, also under supercritical conditions, where the CO$_2$ would remain an insoluble liquid or solid.[6] Below 3000 m in the ocean, liquid CO$_2$ is more dense than water and sinks; under these conditions, some of it also forms a solid hydrate, similar to methane hydrate discussed shortly. There seem to be few data on potential effects of slow release of CO$_2$ at high local concentrations from such sinks.

5. Not sequestration, but recycling of carbon to reduce the rate of increase of CO$_2$ in the atmosphere is another alternative. Use of renewable fuels (e.g., alcohol derived from plants as a motor fuel), is one such example. A second possibility is the use of bioengineered bacteria to convert CO$_2$ to useful fuel such as methane. Nonbiologically, catalysts that could economically allow reduction of CO$_2$ to more useful compounds could achieve the same result.

[6]P. G. Brewer, G. Friederich, E. P. Peltzer, and F. M. Orr Jr., *Science*, **284**, 943 (1999).

Essentially all the transformations in the carbon cycle involve CO_2 or, in aqueous media, carbonate or bicarbonate ions. However, incomplete combustion of reduced forms of carbon produces small amounts of carbon monoxide, especially from internal combustion engines, but much is also produced from oxidation of methane and other hydrocarbons, and some from natural microorganisms and vegetation.[7] The natural level of CO in the atmosphere is uncertain, but is probably under 0.1 ppm (see Table 2-1). This can be exceeded locally, but the lifetime of CO in the atmosphere is short. A variety of chemical and photochemical reactions, but primarily reaction with the hydroxyl radical, convert CO to CO_2, and soil organisms provide another sink.

The toxicity of carbon monoxide is well known and is the cause for the concern over high carbon monoxide concentrations that can develop in areas of heavy automobile traffic. The effects of chronic, sublethal CO levels on health are not fully understood. Higher concentrations such as can build up in a closed room through a faulty space heater, or in a closed car through a leaking exhaust system, can quickly be fatal. Carbon monoxide is toxic through attachment to the coordination site of the iron atom in hemoglobin (see Section 9.5.7).

Biological processes release more complex organic molecules to the atmosphere. Some of these can have significant environmental consequences, but in terms of the carbon budget they are not important. Other organic molecules may be dissolved in seawater (referred to as dissolved organic carbon, DOC). Undissolved organic materials (particulate organic carbon, POC) play a role in carbon transport processes between surface and deep waters as the particles sink. Biological pumping refers to the process in which photosynthesis in marine organisms in surface waters results in the production of biomass (and also inorganic carbonates such as foraminifera shells) that eventually falls to lower depths. Much of the organic material decays and is recycled, but some fraction may enter the sediments.

The cyanide ion CN^-, occurring as hydrogen cyanide (hydrocyanic acid, HCN), and salts such as KCN, is also unimportant in the overall carbon cycle but has local significance. The cyanide ion is a very strong ligand for many metal ions, often forming soluble complexes with them. It does have some natural sources (e.g., apricot pits), but it is more important in industrial applications such as its use in mining. Several major releases of cyanide-containing water to rivers have taken place with disastrous consequences on wildlife and are discussed further in Section 12.2.6. The toxic cyanide ion is quickly oxidized to less toxic products, mostly cyanate, CNO^-, so the direct effects from it are short term. Indirect effects from the heavy metals that

---

[7]Global atmospheric CO is estimated to amount to between 1400 and 3700×10^{12} g/yr, with less than 25% each from biomass and fossil fuel burning and about half from oxidation of other hydrocarbons.

FIGURE 10-3   The structure of methane hydrate.

accompany such releases may be longer lasting. Deliberate use of cyanide compounds to incapacitate fish by some third world fishermen is another cause of environmental damage.

Methane is a small but potentially important component of the atmosphere because it is a greenhouse gas, and its concentration in the atmosphere, although small, has been increasing (Section 2.2). Approximately one-third of the total methane emissions are anthropogenic, with the decreasing order of importance being rice production, cattle growing, waste systems, natural gas losses, burning of biomass, and coal mining. Natural sources are dominated by release from anaerobic processes in swamps, sediments, and other anoxic locations.

One potentially large reservoir of methane exists in the form of gas hydrates in ocean sediments and permafrost.[8] The open structure of ice was described in Section 9.1. In the presence of small molecules such as methane, water can crystallize in a structure of linked pentagons containing cavities in which the molecule is trapped. Such combinations are called clathrates, or cage compounds. The solid, which resembles ordinary ice, can form at a few degrees above 0°C if the pressure is high enough. This structure of solid water, based on hydrogen bonding as in normal ice, owes its stability to the weak van der Waals interactions of the guest molecule with the water molecules in the cage and is not stable if the cavities are empty. The structure of a cavity is shown in Figure 10-3. The overall structure is based on 46 water molecules that link to form six such cavities with pentagonal faces, and two slightly larger cavities having two hexagonal faces; the overall stoichiometry is $8CH_4 \cdot 46H_2O$ (almost 1:6) if all the cavities are filled, which they may not be—these systems are nonstoichiometric. Other small molecules may form clathrates with the same structure (e.g., ethane, H_2S, CO_2), while other clathrate structures exist with larger gas molecules.

[8]K. Krajick, *Nat. His.*, **106**(4), 28 (1997).

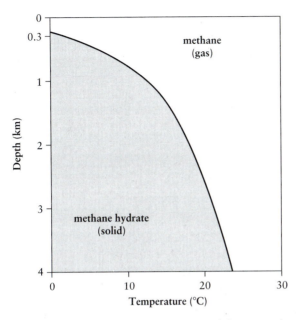

FIGURE 10-4 Ocean depth–temperature stability range of methane hydrate.

Figure 10-4 shows the pressures and temperatures at which methane hydrate is stable. (Pressure increases by about 1 atm for each increase in depth of 10 m). In ocean sediments at depths of 1000 ft (300 m) or more, or in permafrost below about 400 ft (120 m), conditions may be suitable for methane clathrate formation from biologically generated methane, and deposits have been found in such regions. Methane hydrates have also been observed in natural gas pipelines, clogging and sometimes damaging them. The amount of gas hydrate deposits in nature is unknown, but estimates of the methane in clathrates or trapped below layers of them range up to twice the carbon in all the oil, coal, and ordinary natural gas deposits on earth. As yet, plans for tapping this methane as an energy source are highly preliminary, although at least one Siberian gas field, Messoyakha, apparently has produced natural gas at least partially from such a source since the 1970s.

Recognition of the existence of methane hydrate reservoirs has led to expressions of concern over possible release of methane if global warming allows the decomposition of some hydrates as permafrost warms, or as warmer ocean water gets into the depths through changes in circulation patterns. Because of the strong greenhouse property of methane, as discussed earlier, this could have a reinforcing effect. Interestingly, there are some indications that such releases may have occurred in past glacial periods when sea levels fell enough to release the pressure sufficiently. This might have been responsible for warming periods which occurred then. At the present time, knowledge of these systems is inadequate to make definite predictions.

10.3 NITROGEN

Nitrogen exists in nature in several oxidation states: N(–III) as in NH_3, NH_4^+, and various organic compounds; N(III) in nitrites, NO_2^-, and N(V) in nitrates, NO_3^- as well as N(0) in N_2 in addition to other formal oxidation states in oxides of which nitrous oxide (N_2O), nitric oxide (NO), and nitrogen dioxide (NO_2), are most important environmentally.

The elemental form N_2 contains a triple bond with a large bond energy (946 kJ/mol). Consequently, reactions that require the N—N bond to be broken are likely to take place with difficulty, even if the overall energy change of the reaction is favorable. As a result, N_2 is relatively inert. Some of its most important environmental reactions are produced by microorganisms, which can provide a reaction mechanism of low activation energy to convert N_2 to ammonia and amines.

The aqueous redox chemistry of nitrogen involves primarily NO_2^-, NO_3^-, and NH_4^+; these take part in oxidation–reduction processes, which are expressed in the following equations:

$$NO_3^- + 2H^+ + 2e^- \rightarrow NO_2^- + H_2O \tag{10-1}$$

$$NO_2^- + 8H^+ + 6e^- \rightarrow NH_4^+ + 2H_2O \tag{10-2}$$

$$NO_3^- + 10H^+ + 8e^- \rightarrow NH_4^+ + 3H_2O \tag{10-3}$$

These reactions depend on pH, and the equilibrium composition of a nitrogenous system depends on this as well as on the redox potential pE of the system (Section 9.4). Nitrite has a comparatively narrow range of pE over which it can exist at significant concentrations (Figure 10-5); most commonly the stable forms of nitrogen are NH_4^+ or ammonia in reducing environments, and NO_3^- in oxidizing ones. At pH ranges near neutrality, the NO_2^- stability region lies near pE 6.5.

The essential features of the nitrogen cycle are shown in Figure 10-6. There are several processes of importance.

1. Nitrogen fixation refers to the conversion of atmospheric N_2 to another chemical form, most frequently N(–III). In nature, formation of N(–III), amine nitrogen, is a microbial process, most importantly involving bacteria that have a symbiotic relationship with the roots of certain plants. Legumes such as clover, peas, and alfalfa that are associated with *Rhizobium* bacteria are best

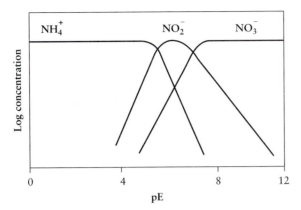

FIGURE 10-5 The relative concentrations of the nitrogen species in solution as a function of pE at pH 7.

known in this respect, but some aquatic bacteria and blue-green algae are important in marine systems, and some trees such as alder (with bacteria of the *Frankia* genus) contribute in forest regions. This complex biochemical process depends on a large metal-containing enzyme, nitrogenase, in which

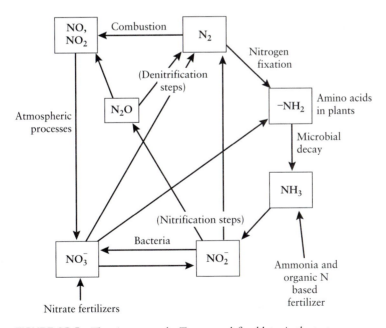

FIGURE 10-6 The nitrogen cycle. Terms are defined later in the text.

both iron and molybdenum take part. The process leads to amine nitrogen, which is incorporated into amino acids in the plant. Plants generally, however, absorb nitrogen from the soil in the form of the nitrate ion. Other natural fixation processes—for example, formation of nitrogen oxides in the atmosphere by lightning discharges—also contribute to a lesser degree. Much nitrogen is fixed industrially, as discussed later.

2. Nitrification is the conversion of amine nitrogen to nitrate. Decay of protein material produces NH_3, which is oxidized through nitrite to nitrate. This is also a bacterial process. Nitrification is reversible, under anaerobic conditions, since reduction of nitrate can provide an energy source for bacteria through the net change

$$4H^+ + 4NO_3^- + 5CH_2O \rightarrow 2N_2 + 5CO_2 + 7H_2O \qquad (10\text{-}4)$$

(CH_2O represents the approximate composition of the organic material that is used up in this process.)

An intermediate in the reduction of nitrate is nitrite. The nitrite ion NO_2^- is relatively toxic because of its interaction with hemoglobin. Nitrite in the blood results in the oxidation of the Fe(II) of hemoglobin to Fe(III), forming methemoglobin, which has no oxygen-carrying ability; this disease is called methemoglobinemia. While cases of direct nitrite poisoning are rare, the nitrate ion can be reduced to nitrite in the stomach, and for this reason food and water with high nitrate contents are dangerous. This reduction is especially possible in the stomachs of infants, where the low acidity allows growth of nitrate-reducing microorganisms. Some evidence exists that nitrites in the body can react with organic amines to form possibly carcinogenic nitrosamines,

$$\begin{array}{c} R \\ {}_{\diagdown} \\ N\text{--}N{=}O \\ {}^{\diagup} \\ R' \end{array}$$

Sodium nitrate and nitrite (the latter often produced from nitrate *in situ*) are common additives in cured meats, where they act to prevent the growth of bacteria, but nitrite is also added to produce flavor and an attractive color through the formation of nitrosylmyoglobin. Concerns about the risks from nitrate and nitrite have led to reductions in their use.

3. Denitrification is the formation of gaseous N_2 from nitrate to return nitrogen to the atmosphere. Primarily it is carried out by soil bacteria and involves reduction by organic carbon compounds with the production of CO_2. This process can be used in an anaerobic process to remove nitrates from wastewater by addition of a reducing agent such as methanol as food for the bacteria:

TABLE 10-3

Global Nitrogen Reservoirs

Nitrogen in	Amount (g $\times 10^{15}$)
Atmosphere	4×10^9 (99.99% N_2; 99% of the rest is N_2O)
Biomass	1×10^4 (land); 8×10^2 (ocean)
Ocean	2×10^7 (dissolved N_2); 6×10^5 (dissolved inorganic); 2–5×10^5 (organic)
Soils	6×10^4 (organic); 1×10^4 (inorganic)
Rocks; sediments	6×10^8

Source: Data from P. M. Vitousek, J. D. Aber, R. W. Howarth, G. E. Likens, P. A. Matson, D. W. Schindler, W. H. Schlesinger, and D. G. Tilman, *Ecol. Appl.*, 7, 737 (1997).

$$5CH_3OH + 6NO_3^- + 6H^+ \rightarrow 5CO_2 + 3N_2 + 13H_2O \qquad (10\text{-}5)$$

Some nitrogen is returned to the atmosphere as N_2O generated bacterially in soils. Nitrous oxide is comparatively inert, but it is rapidly destroyed by processes that are not entirely clear. It can be photolyzed to N_2 at high altitudes, where radiation exists at wavelengths short enough to be absorbed by N_2O (see Chapter 5), but it does not seem likely that this reaction can be responsible for the low N_2O levels usual at low altitudes. Other processes for N_2O removal must exist at or near the earth's surface. It is a greenhouse gas, and its concentration in the atmosphere has been increasing recently by a few tenths of a percent per year from the effects of fertilizers, land clearing, and industrial processes, although the details are not well understood.

Some ammonia is also released naturally from decay of organic materials and is present in air either as NH_3 gas or as an ammonium salt aerosol. These are removed from the atmosphere in rain. Ammonia is the main basic material in the atmosphere.

Table 10-3 gives estimates of the amount of nitrogen contained in the various environmental reservoirs. The atmosphere contains most of the earth's nitrogen as N_2. Inputs of "fixed" nitrogen, oxidized or reduced forms, are given in Table 10-4. There are widely varying estimates for some of these quantities; for example, estimates of nitrogen oxides produced by lightning discharges vary from 2 to 20×10^{15} g of N per year, and fossil fuel combustion from 14 to 28×10^{15} g of N per year.[9] As can be seen from these data, anthropogenic inputs exceed natural ones. The cycle is unbalanced as the amount of fixed nitrogen is continually increasing.

[9]See, for example, G. Brasseur, J. J. Orlando, and G. S. Tyndall, eds., *Atmospheric Chemistry and Global Change*, Oxford University Press, Oxford, U.K., 1999, Chapter 5, for some of these estimates.

TABLE 10-4

Input of Fixed Nitrogen into the Environment

Nitrogen input	Amount ($g \times 10^{12}$ / yr)
Natural	
Fixation	90–140
Lightning	5–10
Anthropogenic	
Fertilizer	80
Crops (legumes, etc.)	32–53
Fossil fuel combustion	20
Biomass combustion	40
Land use changes	30
Total anthropogenic	~210

Source: Data From D. A. Jaffe, in *Global Biogeochemical Cycles*, S. S. Butcher, R. J. Charlson, G. H. Orians, and G. V. Wolfe, eds., Academic Press, San Diego, CA, 1992.

Fixed nitrogen includes gaseous compounds such as NO_x, N_2O and NH_3 as well as nitrates and so on. NO_x released to the atmosphere annually amounts to about 64×10^{12} g of N, about half from natural sources; NH_3 is about 53×10^{12} g of N, over half of which has an anthropogenic origin, and N_2O, about 15×10^{12} g of N. There is some interconversion of these species as one is oxidized or reduced to another.

Fertilizers make up the largest source of anthropogenic input of nitrogen to the environment (see also Section 10.5). Because nitrogen is an essential nutrient for plants, and often a limiting one, large-scale use of nitrogen-containing fertilizers has become commonplace. These may take the form of nitrates, which are immediately available to the plant, or ammonia or organic nitrogen-based fertilizers. These must be converted to inorganic nitrate before use by plants, and since this conversion takes place over a period of time, they can provide a longer lasting source of nitrogen than nitrate fertilizers. Nitrates tend to be quite soluble and weakly held by ion exchange forces, and so may easily be leached from the soil and wasted. Organic nitrogen fertilizers may or may not be soluble, but soluble materials can be formulated to resist dissolution (e.g., pellets may be coated with sulfur and wax). However, it should be emphasized that movement of materials through the soil is highly complicated by absorption and ion exchange processes; even readily soluble materials can be retained for long periods in some types of soil (Section 12.3). Thus ammonia itself, although a very water-soluble gas, is an effective fertilizer, as are its aqueous solutions. It is held in the soil as the ammonium cation by ion exchange. Other widely used compounds are NH_4NO_3, $(NH_4)_2SO_4$, ammonium phosphates, and urea. Note that since nitrogen is generally assimilated by plants as the nitrate ion, the original source of the nitrogen does not affect the plant. Rather, choice of the form a fertilizer should take is based on the rate of release of NO_3^-, extent of leaching (which results in waste and ultimately water pollution), and other economic factors. Estimates are that half of the nitrogen applied as fertilizer is never taken up by plants but enters into runoff water.

Industrial nitrogen fixation to produce fertilizers (and other nitrogen compounds) is based primarily on the Haber process:

$$1/2N_2 + 3/2H_2 \xrightarrow[\text{catalyst}]{\text{elevated } T, P} NH_3 \qquad (10\text{-}6)$$

The source of hydrogen is normally natural gas or petroleum, either of which react with steam under the action of a catalyst in a process such as

$$CH_4 + H_2O \rightarrow CO + 3H_2 \qquad (10\text{-}7)$$

$$CO + H_2O \rightarrow CO_2 + H_2 \qquad (10\text{-}8)$$

Consequently, fertilizer production and cost are closely linked to the supply of these fossil fuels. Ammonia is the starting material for most other industrial nitrogen compounds; for example, nitrates are produced through reactions that include oxidation of ammonia by air in the presence of a platinum catalyst.

Oxides of nitrogen in the atmosphere are a cause of concern with respect to air pollution problems, as discussed in Chapter 5. The main compounds of concern are NO and NO_2, which are by-products of combustion processes. They are formed from the reaction of N_2 in the air with O_2. The primary product is NO, but some NO_2 is produced as well. Production of nitrogen oxides (often called NO_x) is directly related to the temperature in the combustion zone of a furnace or engine and can be reduced by operating at lower temperatures. This generally reduces the efficiency of the device, however. Reduction of the amount of excess air in the combustion chamber also will reduce NO_x emission, but at the expense of an increase in the amount of incompletely burned fuel. A two-stage combustion, the first at high temperature with an air deficiency, followed by completion of the combustion at a lower temperature, can be quite effective in reducing NO_x emission. Figure 10-7 shows NO_x equilibrium concentrations as a function of temperature at one particular fuel/air ratio, equivalent to a 20% excess of oxygen. The reaction between oxygen and nitrogen to form the oxides is kinetically complex, requiring first the reaction of oxygen with another molecule to generate oxygen atoms, and reaches equilibrium in a finite time only at high temperatures. Once formed, the NO and NO_2 concentrations will by "frozen in" when the temperature drops to a value at which the rate becomes insignificant.

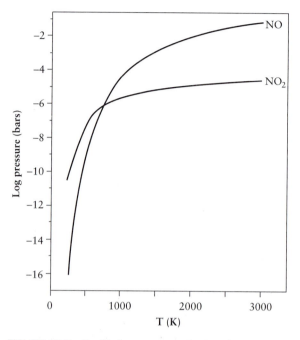

FIGURE 10-7 Equilibrium pressures of NO and NO$_2$ in air with a 20% excess of oxygen over fuel. Redrawn from S. S. Butcher and R. J. Charleson, *An Introduction to Air Chemistry*, Copyright © 1972. Used by permission of Academic Press.

If production of NO$_x$ is not reduced, absorption of NO$_x$ from the combustion gases is an alternative process to reduce pollution levels. Such a practice is used for SO$_2$ removal, as will be discussed in Section 10.4, but is more difficult for NO$_x$. Various methods have been proposed for both acid and alkaline scrubbing of stack gases. Alkaline scrubbers can remove SO$_2$ and NO$_x$ simultaneously, but neither NO nor NO$_2$ is efficiently absorbed by a simple basic solution. However, a mixture of NO and NO$_2$ can be absorbed through an equilibrium reaction,

$$NO + NO_2 \rightleftarrows N_2O_3 \qquad (10\text{-}9)$$

Although this equilibrium lies far to the left under normal conditions, it is shifted to the right if the N$_2$O$_3$ is absorbed into a basic solution; this occurs readily, since N$_2$O$_3$ is the anhydride of nitrous acid,

$$N_2O_3 + H_2O \rightarrow 2HNO_2 \qquad (10\text{-}10)$$

The nitrite salts formed in a basic solution can be oxidized to nitrate. This process, however, requires equimolar amounts of both NO and NO$_2$ in the gas stream, or recycling of the excess.

Catalytic reduction of NO_x is possible; reaction with CO or CH_4, for example, can produce N_2:

$$NO + CO \rightarrow CO_2 + 1/N_2 \qquad (10\text{-}11)$$

This reaction is favored thermodynamically but is very slow under normal circumstances. Various metals (e.g., Ag, Cu, Ni, Pd) are effective catalysts but are subject to poisoning in practical use and are sometimes expensive. Some oxides (e.g., copper chromite [$Cu_2Cr_2O_4$] and Fe_2O_3) are also effective catalysts for this process. Commercial systems for catalytic NO_x removal from flue gases use ammonia as the reducing agent in the following reactions:

$$4NO + 4NH_3 + O_2 \rightarrow 4N_2 + 6H_2O \qquad (10\text{-}12a)$$
$$6NO_2 + 8NH_3 \rightarrow 7N_2 + 12H_2O \qquad (10\text{-}12b)$$

Only a few such systems are in use , but up to 80% removal of NO_x is claimed for a 1000-MW coal-fired plant in Japan.

The nitrogen cycle, like the carbon cycle, is not balanced because of human activities (Table 10-4). Although less well known than the carbon dioxide problem, the increase in fixed nitrogen compounds can have comparably significant environmental effects. Industrial nitrogen fixation plays a major role in this imbalance, but also involved is release of fixed nitrogen from soils and biomass from land clearing and wetland draining, and burning of biomass and fossil fuels. "Natural" fixation has increased through planting of more leguminous and other plant crops. Increased fixed nitrogen in soils increases plant growth, which in fact is helpful regarding the carbon dioxide problem, as more carbon is sequestered in biomass. However, this is coupled with more release of nitrous oxide to the atmosphere in denitrification processes as a consequence of more nitrogen compounds available to soil bacteria. Atmospheric nitrogen compounds contribute to acid rain and to soil acidity; this, along with the acidity increase that results from the bacterial activity, encourages the solubilization of other elements.

Excessive nitrogen fertilization has several direct consequences. One is a change in the number of plant species, which occurs because those that thrive on the nitrogen drive out those that prefer low nitrate conditions. Major ecosystem changes can result. A second is disruption of the soil chemistry. As mentioned earlier, soil acidity changes can change the availability of other nutrients, and with excessive nitrogen, one or another of these can be growth limiting. Conversion of these other nutrients to forms that can be leached away can ultimately decrease soil fertility. Another serious concern is the runoff of the excess nitrogen from soils. Nitrate salts are soluble, and not strongly held by ion exchange processes. When they enter freshwater systems, they can increase algal growth and eutrophication, although in most freshwaters, phosphate is limiting. The major effect is felt in estuaries and coastal areas of the ocean, where nitrogen is normally limiting and additional input can lead to harmful algal blooms and damage to important marine life. Consequences of the nitrogen overload are most obvious in northern Europe and to a lesser extent in the United States, where fertilizer use has been highest and nitrogen inputs to forests and freshwater systems have increased to 10–20 times the natural level.

10.4 SULFUR AND THE SULFUR CYCLE

Sulfur is an important, relatively abundant, essential element. As is true of many elements, it takes part in a biogeochemical cycle discussed shortly. It is a major component of air pollution, particularly in industrialized areas, although natural sources of sulfur also contribute. Several oxidation states are encountered in environmental systems; the most stable under aerobic conditions is S(VI) as in SO_3 and sulfates. The reduced form S(–II) is encountered in organic sulfides, including some amino acids, in H_2S, and in metallic sulfides. It is a reduction product of sulfates under anaerobic conditions. Oxidation of sulfides produces chiefly SO_2 [S(IV)] as the immediate product. Sulfur dioxide and sulfites, the salts produced when SO_2 reacts with base, are reducing agents and are used as antioxidants in some foods (e.g., cut fresh fruits and vegetables, some shellfish, wine). While sulfite is generally not considered harmful, it causes asthmatic reactions, sometimes severe, in individuals who are sensitive to it. Elemental sulfur [S(0)] also occurs in nature, as well as some intermediate forms such as S_2^{2-}.

Figure 10-8 shows the sulfur species that are in equilibrium in solution at pH 10 as a function of pE. At this pH, S(II) (as HS^-) predominates below a pE of −7, and SO_4^{2-} above; the crossover shifts to larger pE values as the pH decreases; in more acidic media, elemental sulfur can be a stable intermediate.

The most important interrelationships of these states of sulfur are shown in the sulfur cycle (Figure 10-9). A large amount of sulfur is released to the atmosphere as SO_2 produced from the combustion of sulfur-containing fuels. The sulfur may be present in the fuel as organosulfur compounds, or as inorganic sulfide contaminants such as FeS_2. In coal the amounts of each are comparable. More than twice as much SO_2 has been produced from coal combustion as from the burning of petroleum in the United States in recent years; coal and petroleum together make up the source of close to 90% of the total SO_2 emissions, with ore smelting being third. The contribution from coal can be expected to increase if the use of high sulfur coals becomes necessary to replace low-sulfur petroleum, especially in power generation. Other sources of sulfur oxides are volcanic activity, chemical processing, and incineration. Biomass burning can contribute small amounts of OCS, H_2S, and other sulfur compounds in addition to SO_2. A small amount (about 5%) of the sulfur

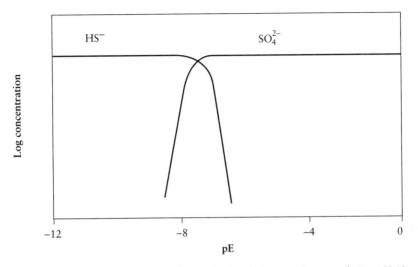

FIGURE 10-8 Concentration of sulfur species in solution as a function of pE at pH 10.

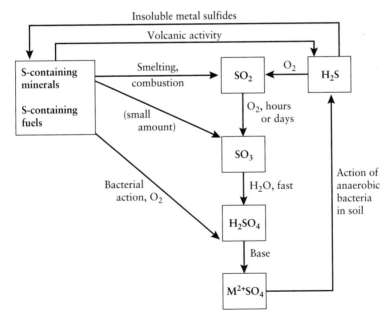

FIGURE 10-9 The sulfur cycle.

oxides produced in combustion processes is emitted as SO_3 rather than SO_2. Furthermore, catalytic exhaust converters on automobiles convert much of the sulfur emissions from automobile exhausts to SO_3. Sulfur dioxide is also formed from H_2S, which is released to the atmosphere from volcanoes and from anaerobic decomposition of organic matter in soils and sediments.

Organic sulfur compounds released to the atmosphere through biological processes include H_2S, CS_2, OCS, CH_3SCH_3, CH_3SH, and CH_3SSCH_3. The first three are emitted chiefly from anaerobic water systems, while the fourth, dimethyl sulfide, is the product of ocean phytoplankton, and is the largest biogenic source of sulfur compounds in the atmosphere. Table 10-5 gives some estimates of the amount of sulfur emitted to the atmosphere per year; many of these numbers are highly approximate and, in the case of volcanic activity, sporadic.

The residence times of gaseous sulfur species in the atmosphere are generally short: on the order of 2 days for SO_2 and only a little longer for SO_3, which depends on washout by precipitation. The organic compounds undergo oxidative reactions with oxygen atoms, ozone, and free radicals such as the hydroxyl radical, and contribute to the SO_2 content; their lifetimes also are no more than a few days. However, dimethyl sulfide is converted to methane sulfonic acid, CH_3SO_3H, which is a component of the particulate material that makes up cloud condensation nuclei and so plays a role in climate control. Another exception is COS, which is resistant to further oxidation and has a residence

TABLE 10-5

Estimated Annual Sulfur Emissions[a,b]

Source	Sulfur emissions (g $\times 10^{12}$/yr)
Anthropogenic	
Combustion, smelting	113
Chemical industry	29
Volcanic	28
Biogenetic	
Land	18
Marine	39

[a]Dust and sea-spray sources not included.
[b]Estimates differing from those in the cited source can be found in G. Brasseur, J. J. Qrlando, and G. S. Tyndall, eds., *Atmospheric Chemistry and Global Change*, Oxford University Press, Oxford, U.K., 1999, Chapter 5.
Source: Data from R. J. Charleson, T. L. Anderson, and R. E. McDuff, in *Global Biogeochemical Cycles*, S. S. Butcher, R. J. Charlson, G. H. Orians, and G. V. Wolfe, eds., Academic Press, San Diego, CA, 1992.

time of about a year. Because of this longer life, COS is more uniformly distributed in the atmosphere than the other sulfur compounds and can reach the stratosphere, where it is photolyzed and oxidized to SO_2 and sulfate, making up the major source of stratospheric sulfate particles except on the rare occasions when volcanic activity injects products to these altitudes. Some hydrogen sulfide generated in soils and sediments undergoes reoxidation reactions to S, SO_2, or SO_4^{2-}, plus precipitation reactions, to form insoluble metal sulfides. Many of the processes are biochemical. Little H_2S or other reduced sulfur is evolved from localized sources.

Sulfur dioxide is thermodynamically unstable with respect to the higher oxide SO_3 under natural conditions. However, in the atmosphere the reaction of SO_2 with O_2 is relatively slow. The rate is influenced by photochemical processes and by catalysts. The most important oxidation process involves catalysis by metal salts present in water droplets or on dust particles. In fog or cloud, SO_2 reacts with water to form sulfurous acid H_2SO_3. This is followed by oxidation:

$$H_2SO_3 + 1/2O_2 \rightarrow H_2SO_4 \tag{10-13}$$

In the presence of the hydroxyl radical in polluted air (see Section 5.3), the bisulfate radical is formed and then reacts with water and oxygen to form sulfuric acid and the hydroperoxyl radical.

Sulfur trioxide itself is extremely hygroscopic and immediately reacts with water vapor to form sulfuric acid; water droplets in air containing sulfur oxides are in fact a dilute solution of H_2SO_4. This acid may be neutralized to sulfate salts by basic substances such as ammonia that may be present from industrial or natural biological processes.

Sulfuric acid is responsible for much of the corrosiveness of air in industrial localities. The effects can be noticed in the deterioration of some construction materials (H_2SO_4 on carbonate causes decomposition to CO_2, Section 12.4.1) and of metals, paint, and so on.

There is good evidence that atmospheric sulfur compounds provide a cooling influence on climate. Incident solar radiation is reflected directly by sulfate salt aerosols, and cloud formation is encouraged by the additional nucleation sites that can be provided by high atmospheric sulfur levels. It appears that increased atmospheric pollution over the last few decades may have reduced the global warming effects of the greenhouse gases by a significant amount. Natural volcanic as well as anthropogenic sources contribute to this. Atmospheric sulfates are washed out with rain, but very violent volcanic release can result in their being transported to stratospheric levels, where they can remain for periods of years and participate in the chemical reactions of the upper atmosphere.

Rainwater containing H_2SO_4 is acidic and contributes to the acidity of lakes and streams. As discussed in Section 11.4, rainwater in the northeastern United

States has shown pH values near 4 in recent years. This is well below the value expected for water in equilibrium with atmospheric CO_2. Not all this acidity is necessarily caused by sulfur oxides; nitrogen oxides also play a part. In addition, the pH of rainwater depends not only on the acidic oxides, but also on basic materials emitted to the atmosphere from the same or independent sources. Thus, pH may be influenced by such factors as the nature of the fuels used for heating or power generation (e.g., high or low sulfur content), the combustion temperatures in widespread use (this can influence the formation of nitrogen oxides, see Section 10.3), the nature of particulate emissions, and extensive industrial or natural emissions of other kinds.

Because emission of SO_2 presents a major pollution problem, considerable attention has been given to efforts to reduce these emissions. Worldwide combustion of fossil fuels produces the order of 10^8 tons of SO_2 annually, and increasing use of high-sulfur fuels requires some means of reducing the amount of SO_2 evolved to control pollution problems. Pretreatment of fuels represents one approach. For example, coal can be separated from its FeS_2 component on the basis of different densities. Removal of organic sulfur from coal or oil requires more elaborate and costly chemical treatment. For example, the crushed coal may be extracted with a heated sodium hydroxide solution under moderate pressure to remove nearly all of the inorganic sulfur, a large portion of the organic sulfur, and many of the trace metal components.

Removal of SO_2 from exhaust gases is another approach to the pollution problem, and a variety of techniques has been proposed. Since SO_2 is an acid, it can be removed by reaction with a base such as calcium carbonate.

$$CaCO_3 + SO_2 \rightarrow CaSO_3 + CO_2 \qquad (10\text{-}14)$$

The calcium sulfite can be converted to sulfate (gypsum)

$$2CaSO_3 + O_2 + 4H_2O \rightarrow 2CaSO_4 \cdot 2H_2O \qquad (10\text{-}15)$$

which has some commercial value and at any rate is easier to handle. The base most likely used in such processes is calcium carbonate or calcium magnesium carbonate, both of which available cheaply as limestone and dolomite, respectively. In one approach, the crushed carbonate is added with the fuel and the sulfate salts are precipitated by electrostatic precipitators, or washed from the gas stream by a scrubber. Alternatively, the stack gases may be scrubbed by a slurry of $CaCO_3$. This method can be very effective, but it produces large amounts of sludge containing $CaSO_4$, $CaSO_3$, unreacted carbonate, and ash, which may contain heavy metals. Disposal of this sludge raises problems, since landfill is its most likely fate.

Other bases can be used to scrub the gases. One alternative process uses MgO as the base because $MgSO_3$ can be decomposed at reasonably low temperatures to produce SO_2 as a feedstock for sulfuric acid production, with the MgO regenerated for reuse. Sodium hydroxide or sodium carbonate scrubbing will produce sodium sulfite and bisulfite, as well as some sulfate. These materials remain in solution, but eventually have to be disposed of. Catalytic oxidation, using a solid catalyst bed after removal of particulates, will serve to convert the SO_2 to SO_3 that is dissolved in water to produce a solution of sulfuric acid with commercial value. Other approaches that can give useful products to partially offset costs have been proposed but are not in general use. Although gypsum, sulfuric acid, or sulfur itself, to which the products of these adsorption processes can be converted, have large-scale use, they are available very cheaply from other sources and do not have much economic value as by-products. Calcium carbonate adsorption is the predominant approach at this time.

To be practical, such processes must operate in very large scale systems and must do so reliably with a minimum of maintenance. The volumes of gas to be handled are enormous—for example, on the order of 10^8 ft^3/h from a 750-MW coal-fired generating plant. Many by-products such as $CaSO_4$ have little value, and indeed their disposal may add considerable expense. The capital equipment requirements are likely to be large; economics is a major consideration in applying such treatments. Despite these problems, stack gas scrubbers seem to be the most practical means of reducing power plant SO_2 emissions at present.

Dry adsorption processes for both SO_2 and NO_x in stack gases are available and have been applied to some facilities. Adsorption on activated coke at 100–200°C produces adsorbed H_2SO_4. Injection of ammonia results in catalytic decomposition of the NO_x to N_2 and water as well as neutralization of the sulfuric acid. The coke is then heated to 500°C, which decomposes the ammonium sulfate back to nitrogen, water, and sulfur dioxide (which can be converted to useful chemicals) and regenerates the absorber.

10.5 PHOSPHORUS, FERTILIZERS, AND EUTROPHICATION

In solution, the mono- and dihydrogen phosphate ions, HPO_4^{2-} and $H_2PO_4^{-}$ are the predominant forms of phosphate at the usual pH values (since H_3PO_4 is a strong acid), and it is in these forms that phosphate is taken up by organisms. The orthophosphate ion PO_4^{3-} will exist in significant concentration in solution only at very high pH values.

Polyphosphates can be formed from the heat-induced condensation polymerization of simple orthophosphate units, for example,

$$2H_3PO_4 \rightarrow H_2O + H_4P_2O_7 \qquad (10\text{-}16)$$

which occurs readily with pure orthophosphoric acid. The product here is pyrophosphoric acid, which is made of two tetrahedra with a shared oxygen making a common corner (Figure 10-10).

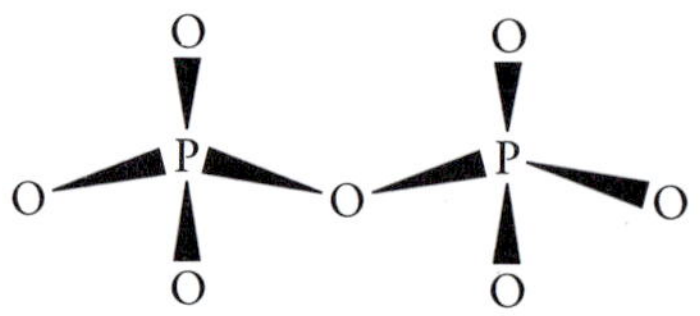

FIGURE 10-10 Pyrophosphoric acid.

Many polyphosphate compounds can be prepared by heating simpler phosphates in reactions analogous to reaction (10-16). Other polyphosphates may be cyclic such as trimetaphosphate (Figure 10-11).

The linear tripolyphosphate used in detergents was discussed in Section 9.5.6. Other condensed phosphates that have been used include pyrophosphate, tetraphosphate, and hexametaphosphate. In dilute solutions, polyphosphates are hydrolyzed to the orthophosphate, although the reactions are not always fast. Inorganic phosphate concentration is limited by solubility, since many metal phosphates are insoluble. For example, K_{sp} for $FePO_4$ is 10^{-23}. Aluminum phosphate, $AlPO_4$, and hydroxyapatite, $Ca_5OH(PO_4)_3$, are also very insoluble. Ligands that form strong complexes with the metal, or protonation of the phosphate ions, can increase the effective solubility, however. Typically, maximum availability of phosphate in soils is around pH 7.

Natural waters may contain organophosphorus compounds that can make up a significant fraction of the total soluble phosphorus content. These compounds are generally of unknown composition and are derived from biological products or possibly P-containing insecticides such as the phosphate esters discussed in Chapter 8, or phosphono compounds such as phosphonomethylglycine [Glyphosate: $(OH)_2P(O)CH_2NHCH_2CO_2H$] used as a herbicide.

Although phosphorus compounds in a variety of other oxidation states are known in chemistry [especially P(–III) in PH_3 and derivatives, P(III) in phosphorous acid, PCl_3 etc.], the inorganic forms of these oxidation states have no general importance under environmental conditions. The natural phosphorus cycle consists of weathering and leaching of phosphate from rocks and soils, and runoff to the oceans, which serve as a sink. Biological cycling intervenes in

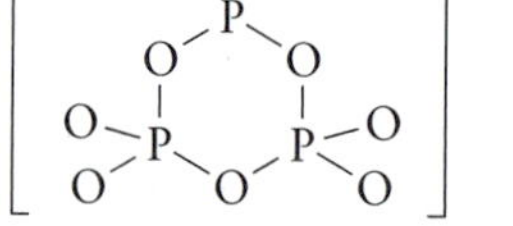

FIGURE 10-11 Cyclic trimetaphosphate.

this process, but there are no atmospheric steps except as particulates. The low phosphate levels in water systems are increased easily by anthropogenic processes such as the use of fertilizers and phosphate-containing detergents. Runoff from cattle feedlots is another important source of environmental phosphate as well as nitrogen compounds.[10]

Phosphorus is an essential nutrient material for plants and animals, being required in biological synthesis and energy transfer processes. The overall photosynthesis reaction in aquatic organisms (70% of all photosynthesis takes place in the ocean) results in the eventual formation of biological material that has an overall C:N:P ratio of approximately 100:16:1. The N:P ratio in ocean water is about 13:1 (Table 11.1). In freshwater lakes, the ratios vary considerably, but N is usually in excess; typical ranges are N, 10^{-4} to 10^{-5} M; P, 3×10^{-5} to 3×10^{-7} M. (There is evidence that natural processes act to make phosphorus the most important algal nutrient in any event.[11]) One milligram of phosphorus results in production of 0.1 g of organic material if N is in excess, and in systems where phosphorus is the limiting nutrient, additional phosphorus input can greatly increase the biological yield. The 0.1 g of organic material produced requires more than 0.1 g of oxygen for decomposition, and if this is not replaced, the water becomes depleted of dissolved oxygen, with consequent decrease in the abundance of the higher life-forms and production of undesirable products of anaerobic reactions: NH_3, H_2S, and CH_4. The enhanced growth and decay of algae caused by increased phosphorus levels is the cause of the concern over phosphate pollution and its effects on eutrophication processes in freshwater lakes (Section 9.1). Large anoxic regions in the oceans, for example, the Gulf of Mexico, the Black Sea, and the Kattegat Strait, have been linked to agricultural fertilizer use, but nitrate rather than phosphorus.[12]

Phosphate fertilizers are produced commercially from insoluble phosphate rock, which has the formula $Ca_2(PO_4)_2 \cdot CaX$, where X can be CO_3^{2-}, $2(OH)^-$, and others, but is usually $2F^-$. Soluble phosphates are produced by an acid displacement reaction:

$$Ca_3(PO_4)_2 \cdot CaX + 3H_2SO_4 \rightarrow Ca(H_2PO_4)_2 + 3CaSO_4 + H_2X \quad (10\text{-}17)$$

If fluoride is present, the HF produced[13] will react with silica in the rock to produce SiF_4 and fluorosilicate salts (M_2SiF_6). Residual fluoride impurities may remain in the product, along with other elements such as As or Cd that may be present in the ore.

---

[10]M. A. Mallin, Impacts of industrial animal production on rivers and estuaries, *Am. Sci.*, **88**, 26 (2000).

[11]D. W. Schindler, Evolution of phosphorus limitation in lakes, *Science*, **195**, 260 (1977).

[12]D. Ferber, Keeping the Stygian waters at bay, *Science*, **291**, 968 (2001).

[13]If X = $2F^-$, H_2X in reaction (10-17) represents 2HF.

The $Ca(H_2PO_4)_2$— $CaSO_4$ mixture is often sold as such as "superphosphate." The sulfate itself is a useful component, since sulfur is also an important nutrient. A higher phosphate content can be obtained in so-called triple superphosphate, produced by the process

$$Ca_3(PO_4)_2 \cdot CaX + 6H_3PO_4 \rightleftarrows 4Ca(H_2PO_4)_2 + H_2X \qquad (10\text{-}18)$$

Treatment of superphosphates with ammonia solution produces ammoniated phosphate fertilizers. Phosphoric acid is formed from excess H_2SO_4 on $Ca_3(PO_4)_2$, and can serve as the source material for other phosphate chemicals through neutralization and condensation reactions.

As with nitrogen, phosphate fertilizers are subject to leaching from soils, adding to the nutrient content of runoff water and lakes, and contributing to eutrophication. Because of its higher charge, however, phosphate is held more strongly than nitrate by ion exchange, as well as through the formation of insoluble phosphate salts.

Besides phosphorus and nitrogen, a third major component of fertilizers is a soluble salt of potassium. Most common lawn fertilizers consist of a mixture of these three materials, and the composition specifications (e.g., 6.10.6) refer to the percentage composition by weight in terms of N, P_2O_5, and K_2O.

Figure 10-12 shows the use of commercial fertilizers over time; the large increase in nitrogen fertilizer use since 1960 is notable. Fertilizer use

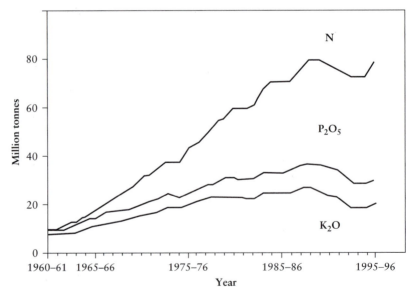

FIGURE 10-12 Global fertilizer use. Used by permission of International Fertilizer Industry Association (IFA); http://www.fertilizer.org.

worldwide has decreased in recent years. European use began to decrease in the 1970s for economic reasons and with changing agricultural practices; there has been a large decrease in use in the former Soviet Union, but demand in the United States and developing countries continues to increase.

In addition to phosphorus, potassium, and nitrogen, other fertilizer elements are needed in particular cases. Important examples of these are calcium, magnesium, and sulfur. Special fertilizers are produced to meet the need of some crops for available trace metals, such as zinc. Unfortunately, in some cases waste materials such as dusts from cement kilns or steel recycling furnaces are used in the production of these fertilizers. These may include other, toxic, heavy metals including cadmium, lead, and mercury which may further add to the buildup of these materials in soils. At this time, there are no national standards for heavy metal content of fertilizers in the United States although there is in some other nations.[14]

10.6 BIOLOGICALLY IMPORTANT METALS AND OTHER ELEMENTS

10.6.1 Introduction

In addition to the elements already discussed, a variety of others have important biological effects. The biological availability of an element, and its ultimate effect on an organism, may depend on the chemical form in which the element is encountered. Thus, solubility properties, the oxidation state, or the bonding characteristics in a particular compound may determine whether it will be inert, poisonous, or detoxified by normal metabolic pathways. Toxic compounds occurring naturally are very often degraded by organisms to nontoxic compounds, but excessive concentrations, or new compounds, may overcome these metabolic capacities.

Table 10-1 listed, among other elements, some of the metallic elements released to the environment by industrial and other processes that may pose concern as health hazards in trace amounts. Many of the heavy elements not listed in this table are either very rare or very insoluble in forms encountered in the environment. Any element, if ingested in large enough amount, will cause undesirable physiological changes. We are using "toxic" to imply the danger of harmful effects with amounts that could realistically be ingested inadvertently or as contaminants in food, water, or air.

The importance of the chemical form of an element is illustrated by nickel subsulfide, Ni_2S_3, and nickel carbonyl, $Ni(CO)_4$, both of which are potent carcinogens (nickel carbonyl is also highly toxic in other ways). Most other nickel compounds have either no or much weaker effects. Both Ni_2S_3 and

---

[14]B. Hileman, Fertilizer concerns prompt new standards, *Chem. Eng. News*, **78**(19), 24 (1998).

surfaces. Under these conditions a more or less stable suspension or colloid results. It is not uncommon for the particles of ferric hydroxide to be extremely small and not easily distinguished from material in true solution.

As discussed earlier, Fe(II) may form under anaerobic conditions. Considerable amounts of ferrous sulfide exist, for example, as pyrite FeS_2 (involving the S_2^{2-} species). This is often found in coal areas, where presumably it was formed under oxygen-free conditions, but it is also common in other mines that contain sulfide minerals. Exposure of pyrite to air and moisture leads to the formation of so-called acid mine drainage: highly acidic water that is the cause of considerable damage to natural water systems in coal mining areas. The following reactions are involved:

$$FeS_2(s) + 7/2O_2 + 3H_2O \rightleftharpoons Fe(II)(aq) + 2SO_4^{2-} + 2H_3O^+ \qquad (10\text{-}28)$$

In this step, which typically takes place through the action of bacteria, the S_2^{2-} is oxidized, forming sulfuric acid, while the now soluble Fe(II)(aq) is free to react further, but does so more slowly:

$$Fe(II)(aq) + 1/4O_2 + H_3O^+ \rightleftharpoons Fe(III)(aq) + 3/2H_2O \qquad (10\text{-}29)$$

When the acidity is low enough, Fe(III) hydrolyzes, giving as an overall reaction

$$[Fe(H_2O)_6]^{3+} \rightleftharpoons Fe(OH)_3 + 3H_3O^+ \qquad (10\text{-}30)$$

[$Fe(OH)_3$ may not be a real substance; this formula is used here to refer to the poorly defined precipitate that upon drying would be largely FeO(OH).]

One mole of FeS_2 eventually gives 4 mol of H_3O^+: two from oxidation of S_2^{2-} and two from the oxidation of iron. This process deposits insoluble ferric hydroxide that coats stream beds, as well as contributing to the acidity. Since the formation of Fe(III) from Fe(II) is slow, this may take place over a considerable length of the stream.

If the Fe(III)(aq) is formed in contact with pyrite, an additional process is involved:

$$FeS_2 + 14Fe(III)(aq) + 24H_2O \rightarrow 15Fe(II)(aq) + 2SO_4^{2-} + 16H_3O^+ \quad (10\text{-}31)$$

The oxidation of reduced forms of sulfur by Fe(III) can also take place with other metal sulfides that may be present, including those of toxic heavy metals such as lead or cadmium. Consequently, the acid mine liquids can contain important concentrations of these elements. Much of the heavy metal content may be transferred to the sediments by adsorption on the ferric hydroxide particles as they precipitate.

The role of bacterial processes in these apparently inorganic reactions is very significant. Various thiobacilli, for example, *Thiobacillus ferrooxidans*, can oxidize inorganic sulfides to sulfate as a source of energy. Besides iron, a variety of other metal sulfides, including those of copper, nickel, arsenic, and zinc, can be used. The bacteria are in a sense a catalyst for these reactions, but unlike true catalysts, which remain unchanged in the process, the bacteria grow. Bacteria can also gain energy from the oxidation of Fe(II) to Fe(III).

We may represent the stability relationships in a system such as that of iron as a function of pE (or E_H; see Section 9.4) and pH diagrammatically. Figure 10-13, which illustrates this, shows that iron metal will be oxidized (will rust) unless the oxygen content is low enough to give a pE below −10. The ferrous ion is soluble in slightly acidic reducing solutions, but the carbonate will precipitate near neutrality if CO_2 is present. Fe(III) is soluble only at pH values below 4 under oxidizing conditions. Under environmental conditions, other components, especially carbonate and sulfide phases, might also be involved in the equilibria. Examples may be found in the reference given in the figure caption.

Another major species in the iron system is Fe_3O_4. This oxide, called magnetite, contains iron in both the (II) and (III) states. The exact stability relationships of this compound are unclear, but it has been suggested that the reaction

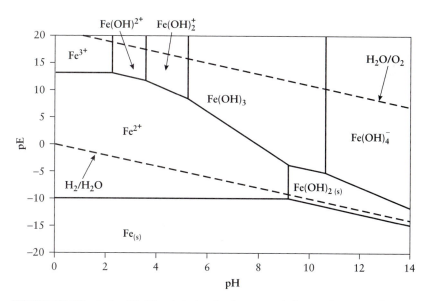

FIGURE 10-13 A simplified Pourbaix (predominance area) diagram for iron in the presence of water and oxygen. Redrawn from J. G. Farmer and M. C. Graham, in R. M. Harrison, ed., *Understanding Our Environment*. Cambridge: Royal Society of Chemistry. Copyright © 1999. Used by permission of The Royal Society of Chemistry.

$$12FeO(OH)(s) \rightleftarrows 4Fe_3O_4(s) + 6H_2O + O_2 \qquad (10\text{-}32)$$

may have acted to regulate the O_2 pressure in the primitive atmosphere.

Iron is a very common constituent of fresh waters, where it is sometimes present in the Fe(III) state either as colloidal material or in solution as a complex with organic material. In large amounts it imparts a characteristic rust color. In some groundwaters that are anaerobic, Fe(II) may be present, and since compounds of this oxidation state are soluble, comparatively high concentrations can be reached (often in the range of 1–10 mg/liter). These Fe(II) compounds of course are oxidized under ordinary conditions to Fe(III) and ultimately are the source of the colloidal material. Although this oxidation of Fe(II) upon aeration may be moderately slow, significant occurrence of iron in this oxidation state in surface waters requires either the presence of organic material or a high acidity (see Figure 10-13), and in either case is suggestive of pollution.

Iron in water is not toxic, but it does have some annoying effects. It can cause "rust" stains on clothing, porcelain, and so on, as well as giving the water itself an undesirable color and taste. The U.S. Public Health (USPH) standard for maximum iron content in drinking water is 0.3 mg/liter, and this is based on color and taste effects rather than on toxicity. (Long-term ingestion of excess iron can be toxic, but this condition is unusual.)

10.6.4 Chromium

Chromium is known as an inert metal most familiar for its use for decorative and protective plating, but also with other industrial applications. It is an essential element in trace amounts, being a component of an enzyme involved in sugar metabolism, but it is also one that can cause serious health effects in some forms. The two common oxidation states environmentally are Cr(III) and Cr(VI). Chemically, Cr(III) behaves somewhat like Fe(III) in that it is readily hydrolyzed to an insoluble hydroxide. In this form it is comparatively harmless as long as it remains immobilized. Because of its electronic configuration, compounds of Cr(III) tend to be chemically inert—that is, reactions such as exchange of ligands tend to be quite slow.

One widespread use of Cr(III) salts is in leather tanning; hides are soaked in chromium solution to chemically alter components of them to produce durability. Considerable amounts of chromium are released in tannery wastewater (about 0.4 kg of Cr per 100 kg of hides), and where this is untreated, river and soil pollution results. This is a serious problem in parts of India, for example, where tanning is a major industry.

Chromium(VI) compounds are encountered as oxo species, which are good oxidizing agents. One such application is treating steel to resist corrosion.

Solutions of the chromate ion, CrO_4^{2-}, are effective in "passivating" steel, presumably by reacting to form a protective oxide layer on the surface. Wastes from such operations are a prime source of chromium pollution. In acid, chromate is in equilibrium with dichromate:

$$2CrO_4^{2-} + 2H_3O^+ \rightleftharpoons Cr_2O_7^{2-} + 3H_2O \qquad (10\text{-}33)$$

Chromium(VI), a possible carcinogen, is highly toxic (USPH limit for chromium in water, 0.05 mg/liter), causing skin lesions upon excessive exposure. Chromate dusts have been associated with lung cancer.

Although organic materials may reduce Cr(VI) to the less toxic Cr(III), redox potentials predict that Cr(III) may be oxidized to Cr(VI) by oxygen in well-aerated systems. This, and other possible reactions, make soils contaminated by Cr(III) quite hazardous. Stabilization of Cr(III)—for example, by converting it to the very insoluble and unreactive sulfide—can reduce this hazard.

10.6.5 Manganese

Manganese is an essential element, and one of moderate toxicity. In its highest oxidation state [Mn(VII)] as permanganate, MnO_4^-, it is a strong oxidizing agent. Its more stable states are Mn(IV), which is encountered in natural systems chiefly as insoluble MnO_2, or as compounds of Mn(II). The chemistry of these states resembles that of Fe(II) and Fe(III); Mn(II) salts are soluble in water, but the dioxide is insoluble, although often producing a colloidal or gelatinous and highly absorbent precipitate. Mn(II) is encountered only under reducing conditions in natural waters, as is Fe(II). Sources of manganese are industrial pollution and acid mine drainage. Long-term inhalation of MnO_2 produces neurological problems. An organometallic manganese compound, methylcyclopentadienyltricarbonylmanganese, has been used for some years in Canada as an alternative to tetraalkyllead as an antiknock agent in gasoline and is more toxic than inorganic manganese compounds. As mentioned in Section 6.7.4, attempts by Canada to prohibit its use were prevented by legal action on the grounds that the prohibition violated the North American Free Trade Agreement (NAFTA).

10.6.6 Copper

Copper is an essential metal for many organisms, including humans. It is used in enzymes that modify redox reactions and in some oxygen-carrying systems. Its function in these is associated frequently with its ability to exist as both Cu(I) and Cu(II). The most common state is Cu(II). Like many essential metals, large amounts are toxic, and the USPHS limit in drinking water is set at 1 mg/liter. It is particularly toxic to lower organisms and has been used as an algicide in lakes (as $CuSO_4$) and is also used in wood preservatives, either associated with other metals such as chromium or arsenic, or complexed with organic ligands (e.g., 8-quinoline, dimethyldithiocarbamate). Other sources are erosion of copper pipes, industrial wastes, and weathering of rocks. The carbonate and hydroxide are of low solubility, so the natural level in water is low in the absence of complex-forming substances.

10.6.7 Zinc, Cadmium, and Mercury

Zinc, cadmium, and mercury make up one family of the periodic table. These three elements are representative metals possessing two valence electrons. Their position in the periodic classification immediately follows the transition series, and in keeping with the usual periodic trends, they have comparatively high electronegativity values for metallic elements and form bonds with nonmetals of significant covalent character. The covalent properties are emphasized on going down the family from zinc to mercury.

A common metal, zinc is released into the environment through mining and industrial operations. It is comparatively nontoxic, as suggested by USPHS maximum limits of 5.0 mg/liter in drinking water. Its simple compounds are readily hydrolyzed, while oxygen and nitrogen are the favored donor atoms. It is a constituent of enzymes—for example, carboxypeptidase A, which is active in the hydrolysis of a terminal peptide linkage of a peptide chain. The active site in this enzyme is a Zn(II) ion coordinated by two imidazole nitrogens and two oxygens from a bidentate carboxylate group from amino acid units in the protein portion of the enzyme, and a water molecule (Figure 10-14). The zinc can accept a bond from an oxygen in the substrate. The shift in electron distribution in the peptide resulting from this coordination facilitates the hydrolysis step.[16] Stereochemistry and hydrogen-bonding interactions of the protein chain assist in this process. Carbonic anhydrase, which catalyzes the conversion of carbon dioxide to bicarbonate, is another zinc enzyme. In this, the zinc is in a 4-coordinate site involving three imidazole nitrogens and a water molecule.

A primary use of zinc is in metal plating, where it is used extensively in galvanizing iron to prevent rust. Corrosion or rusting of iron is an electrochemical process involving reduction of oxygen [reaction (9-41)] and oxidation of the

---

[16] The process is not simple. Coordination of the substrate may cause the carboxylate to change to unidentate coordination to maintain the coordination number. The coordinated water is deprotonated to act as a hydroxyl group. Structures and mechanisms of zinc and other enzymes are discussed in S. J. Lippard and J. M. Berg, *Principles of Bioinorganic Chemistry*, University Science Books, Mill Valley, CA, 1994.

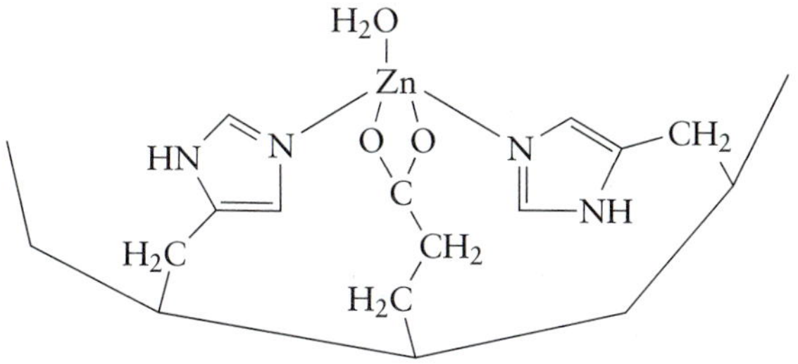

FIGURE 10-14 The active site in the zinc enzyme carboxypeptidase A. The configuration of the protein chain that is important for the reaction is not shown.

metal to Fe(II). The Fe(II) is further oxidized to Fe(III), which forms a porous coating of hydrated oxide, often increasing further corrosion. A more active metal in contact with the iron will be oxidized preferentially, serving to protect the latter. This is the case with zinc. In contrast to a protective coating of paint, which will permit corrosion to start in cracks or pinholes, an electrochemically protective process will function even if the coating is not intact. The zinc on galvanized iron is sacrificed as it performs its function.

Cadmium, the second member of the zinc family, resembles zinc in its chemistry in many respects. Indeed, obtaining zinc free from a cadmium impurity is difficult. A major chemical difference is a tendency of cadmium to form more covalent bonding than does zinc, and more stable complexes. It is extremely toxic and a very hazardous heavy metal; kidney damage, and at high levels bone damage, are the main effects. Inhalation of vapors can lead to lung damage. Its limit in drinking water is 0.01 mg/liter. The reason for the high toxicity may in part lie in its similarity to zinc; it can replace the latter in enzymes, for example, but because of stronger bonding and perhaps stereochemical differences, the function of the enzyme is disrupted. It is cumulative; ingested cadmium remains in the body for many years, so that small but continued intakes can lead to significant body burdens over time. However, much of this cadmium appears to be bound irreversibly to thionein, a small sulfur-rich protein, the main function of which appears to be to tie up heavy metals as a protective mechanism.

Evidence for a cadmium role in a marine plankton carbonic anhydrase enzyme, suggests that this may be another example of a toxic element that has an essential biological function.

Cadmium also is a frequent material in industrial waste discharges and has been introduced into water systems through mining operations. It has also been employed in metal plating, where it is used as a sacrificial coating to prevent rust on steel in the same way as zinc. This use has been largely stopped except for special applications. Some early cases of cadmium poisoning came

TABLE 10-6

Major Uses of Cadmium

Batteries	Rechargeable NiCd batteries use about 70% of Cd production
Metal plating	For corrosion resistance, especially marine and aerospace applications
Television phosphors	Blue and green as well as black and white phosphors
Pigments	Orange or yellow pigments (CdS, CdSe, and combinations) for plastics, paints, ceramics, glasses; have high-temperature stability
Plastic stabilizer	Used in poly(vinyl chloride) to protect against sunlight and heat degradation (usually as long-chain carboxylates)
Photovoltaic cells/semiconductors	CdS, CdTe
Solder/alloys	Especially fusible metals (e.g., in sprinkler systems, fuses)
Fungicides	Various compounds have limited use, mostly on golf courses

from dissolution of the element from cadmium-plated utensils by acidic foods such as fruit juices. Cadmium sulfide and selenide are used in some red and orange paint pigments. Although these are very insoluble, this does not guarantee biochemical inertness, as it is well known that metals can be solubilized by bacterial attack on solid inorganic sulfides. While this process has not been reported for cadmium sulfide pigments, it does suggest the danger of assuming harmless behavior on the basis of in vitro chemical properties alone. Some uses of the element are summarized in Table 10-6.

The most notable widespread example of cadmium poisoning occurred along the lower Jinzu river in central Japan, where approximately 200 people developed a disease called "itai-itai"[17] after eating rice grown in paddies watered from the river, which was contaminated with cadmium released from lead and zinc mine wastes. The disease, which affects the bones, was first noticed before World War II and became more prevalent during and after the war. Many years passed before the cause was officially recognized.[18]

Mercury, the third member of the zinc and cadmium family, is a very toxic, cumulative poison, having its chief effects on the nervous system. The currently recommended maximum daily intake is $0.1\,\mu g$ per kilogram of body

---

[17] A rough translation is "ouch-ouch"; this is an extremely painful brittle bone disease.

[18] Jun Kobayashi, Pollution by cadmium and the itai-itai disease in Japan, in *Toxicity of Heavy Metals in the Environment*, Part I, F. W. Oehme, ed., p. 199. Dekker, New York, 1978. Also E. A. Laws, *Aquatic Pollution*, 2nd ed., p. 391. Wiley, New York, 1993.

weight. Mercury tends to be still more covalent in its bonding character than cadmium, especially with heavy elements such as sulfur, which coordinates very strongly with Hg(II), a soft Lewis acid (see Section 9.5.3). Indeed, mercuric compounds bind strongly to the amino acid sulfur atoms contained in many protein and enzyme structures, thereby disrupting their normal physiological processes. The Hg—C bond has comparatively low polarity and organomercury compounds are quite stable in aqueous media, in contrast to the more polar organozinc and -cadmium compounds that are easily hydrolyzed. Another feature of mercury is its ability to exist in the Hg(I) state. This state is stable in solution over a very limited pE range, and under most circumstances would be important only as insoluble solid compounds or in the presence of elemental mercury. The elemental form is liquid at normal temperatures, with a significant vapor pressure. Because of its volatility, Hg is found universally at very low levels, and cycles through the atmosphere.

Table 10-7 gives some of the major uses and sources of mercury. Many consumer uses are being phased out or reduced. For example, mercury batteries are no longer available, and the mercury additives in standard dry cells are being replaced. Mercury is still found in fluorescent lights, but new designs reduce the amount from the order of 20 mg to under 5 mg per tube. Industrial applications also are being minimized, with past applications of mercury compounds as catalysts and in paints and pigments, for example, now ended in the United States. Total use in the United States has declined from over 2000 metric tonnes annually in the 1970s to less than 500 tons in 2000. At the same time, however, its cost has dropped and use in developing nations has increased.[19]

The environmental chemistry of mercury depends on both organic and inorganic forms, which differ considerably in their toxic effects, although all are hazardous. The vapor from the elemental form has severe physiological effects if inhaled, since in this form it can be transported to the brain and enter brain cells, where it causes irreversible neurological problems. The long-term limit is a billionth of a gram (1 ng) per liter of air. This is easily exceeded in a closed room containing exposed mercury (e.g., a laboratory), although the rate of evaporation is usually slow, and reasonable ventilation will keep enclosed areas safe. Since the liquid metal is relatively inert, as are many of the insoluble inorganic mercury compounds, oral ingestion of the liquid metal (e.g., dental amalgams) is less hazardous. Nevertheless, concern has been raised about the release of mercury as the vapor from dental fillings, with suggestions that some individuals are sufficiently sensitive to mercury that they should have such fillings removed. It is generally considered that these concerns have no scientific basis.

---

[19]See J. Johnson, The mercury conundrum, *Chem. Eng. News*, p. 21, Feb. 5, 2001, for a discussion of current trends in use and marketing of mercury.

TABLE 10-7

Some Applications of Mercury and Sources from Which It Is Released to the Environment

Items containing mercury	
As the metal	
Batteries	Mercury batteries for cameras, etc. (now discontinued); additive to dry cells to improve performance (now replaced).
Lights	Mercury vapor lights for street and commercial lighting, fluorescent and neon lights.
Switches	Used in relays, thermostats, etc.
Thermometers	
Barometers	
Dental amalgams	Alloyed with gold or silver.
As organomercury	*(Uses in these categories are now generally restricted.)*
Fungicides	Agricultural use, especially on seed grain.
Slimicides	Especially in pulp and paper mills.
Antifouling paints	
Antimildew paints	
As inorganic mercury	
Medicinal products	Especially as topical skin treatments.
Pigments	Often as mixed Hg-Cd-S-Se compounds.
Applications	
Mining	As metal; amalgamation method for separating gold.
Chlor-alkali plants	Metallic electrode material.
Chemical industry	Mercury compounds as catalysts (especially acetaldehyde, vinyl chloride synthesis).
Sources of release	
Power plants	Coal contains some mercury, which is released on combustion; oil releases a very small amount.
Incinerators	Municipal and other incinerators release Hg from waste.
Soil/plants	A considerable amount of mercury is released to the atmosphere in natural processes from the soil and through plant transpiration.
Estimated U.S. demand in 2001[a]	
Electrical switches, thermostats etc.	30%
Chlor-alkali production	23%
Dental amalgams	22%
Lighting	14%
Instruments	11%
Total demand	220 tons

[a] J. Johnson, *Chem. Eng. News*, 21, February 5, 2001.

Acute mercury poisoning through exposure to metallic mercury and its vapor has been of concern mainly to miners and to some industrial workers, but a notable incident involving others occurred in Arkansas in late 1997[20]

[20]S. C. Gwynne, The quicksilver mess, *Time*, Jan. 26, 1998.

when teenagers found a large amount of mercury metal in an abandoned neon light factory and, being unaware of its hazards, treated it as a toy—even dipping cigarettes in it before smoking them. Several people were hospitalized as a result of exposure, and a number of homes and other buildings were contaminated as the finders distributed the mercury to friends. There is also concern about the use of mercury in religious rituals by people of some Caribbean and Latin American areas, where mercury-containing candles are burned and other mercury-containing materials that can release the vapor are employed.

The natural volatility of elemental mercury means that there is a mercury cycle in the environment. Mercury ore deposits normally produce a considerable mercury vapor pressure that results in a low level, worldwide distribution of mercury, and recent studies have shown mercury vapor release from plants and soils far from mercury contamination sites.[21] The element has been found in ice samples that predate industrial use of the metal. Atmospheric release from anthropogenic sources is of some concern; the U.S. EPA is planning to require reduction of mercury release from power plants.[22] Coal-burning power plants in the United States release about 50 tons (33% of the total U.S. release; 1995 figures) of mercury per year to the atmosphere, mostly as elemental mercury from low-sulfur coals and as Hg(II) compounds from high-sulfur coals. In the latter case, existing scrubbers are effective in removing much of the mercury from the exhaust gas before it is released to the atmosphere. With elemental mercury, however, additional treatment is required. This can include addition of activated carbon to absorb the mercury before the dust collectors, or use of an oxidizing agent in a wet scrubbing system. Since the mercury concentrations are the order of $1-10\ \mu g/m^3$ in a gas flow of as much as a million ft^3/min in a 250-MW plant, mercury removal is both technically challenging and expensive. Industrial boilers in the United States contribute another 28 tons (18%) to the atmosphere annually. Other anthropogenic sources of release to the atmosphere are incinerators: 30 tons (18%) from municipal waste incinerators, 16 tons (10%) from medical waste, and 7 tons (4.4%) from hazardous waste incinerators. Regulations are already in place to limit this release. Industrial sources, mainly cement and chlor-alkali plants, contribute another 10%.

Inorganic mercury compounds [chiefly Hg(II)] can cause intestinal, liver, and kidney damage, but unlike the metallic vapor cannot easily enter the brain cells, and effects may be reversible. Some mercury compounds have been used medicinally; in particular, calomel, Hg_2Cl_2, has been used as a skin lotion. While this is very insoluble and inert, it is not without risk. In 1999 about 60 people in Arizona were found to have elevated mercury levels and some with

[21]M. Rouhi, *Chem. Eng. News*, p. 36, Feb. 5, 1996.
[22]J. Johnson, *Chem. Eng. News*, p. 18, Jan. 1, 2001.

early symptoms of mercury poisoning from the use of a Mexican beauty creme that contained calomel.

Organomercury compounds are much more toxic; as early as 1870, several chemists had been killed or permanently injured from working with the then new compounds of methyl- or ethyl-mercury.[23] These compounds are lipid soluble, and since higher organisms do not decompose and excrete them effectively, small amounts in the environment tend to be concentrated in organic tissues in the food chain. Two general classes need to be considered: organomercurium cations, RHg^+, which form ionic bonds to anions such as nitrate or more covalent ones to softer anions such as chloride, and species with two organic groups, R_2Hg. These compounds, especially dimethylmercury, are readily transferred into cells and even through the placental barrier to the fetus. Both forms can be found in water systems, as discussed shortly, and are concentrated in biological materials.

Besides originating from natural sources, mercury pollution may come from organomercurials (e.g., methyl, ethyl, and phenyl compounds) used as fungicides in the treatment of seed grains, and from industrial discharges of organomercurials (e.g., phenyl mercuric acetate fungicides from pulp and paper mills), from inorganic mercury compounds, or from the metal. In North America, large amounts of the metallic mercury were discharged from chloralkali plants in which mercury cathodes were used in the electrolysis of NaCl solutions in the productions of Cl_2 and NaOH. Such inorganic mercury discharges were long considered to be inert and harmless. Unfortunately, bacterial action can convert inorganic mercury in sediments to soluble methylmercury and dimethylmercury species. Conversion to diaphragm-type cells to eliminate mercury release has largely overcome this source of pollution.

The inorganic forms of mercury undergo interconversions mediated by microorganisms; for example, insoluble HgS is converted to soluble Hg(II) by bacterial oxidation, and soluble Hg(II) is reduced by some bacterial enzymes to Hg(0). This forms a process by which mercury can be eliminated from the bacterium through volatilization. If Hg(II) and Hg(0) are present together under appropriate conditions, they will enter into an equilibrium reaction to produce Hg(I); the latter exists as the diatomic ion Hg_2^{2+}.

Alternatively, Hg(II) is eliminated by some bacteria through methylation. This method appears to proceed through transfer of a CH_3^- group mediated by vitamin B_{12}, a cobalt-containing coenzyme, that participates in this reaction. Vitamin B_{12} contains cobalt in a porphyrin-like environment of four planar nitrogens, with a fifth nitrogen from a dimethylbenzimidazole group also attached (see Section 9.5.7). A methyl group on the active site of the Co(I) may be transferred as CH_3^- to a suitable acceptor such as Hg(II), and replaced

---

[23]In 1997 a research chemist died as a result of the spill of a few drops of dimethylmercury, which penetrated her latex gloves and were absorbed through the skin.

by H_2O. The cobalt is oxidized from the unusual Co(I) state to the more common Co(II) state in this process. The methylmercurium ion CH_3Hg^+ can be further methylated to $(CH_3)_2Hg$ in a similar process. The actual product that predominates depends on the pH of the medium, because the dimethyl compound decomposes to the more stable monomethyl derivative in slightly acidic solution. Methylmercury compounds also can be reduced bacterially to Hg(0). There also is evidence that Hg(II) can be methylated by water-soluble methylsilicon, methyltin, and methyllead compounds if these should be introduced from industrial wastes.

The methylmercurium ion CH_3Hg^+ produced in sediments is partially extracted into the water phase and taken up into the food chain. Dimethylmercury can volatilize to the atmosphere, where it is photolyzed to Hg(0) and methyl radicals. The reactions are summarized in Figure 10-15:

These reactions of mercury occur naturally but are increased by anthropogenic input into the cycle. The resulting accumulation in the aquatic food chain leads to high mercury levels in predatory species of fish; in the ocean, these are especially swordfish, sharks, and large tuna. Typical values of Hg in swordfish reach 0.8–0.9 ppm Hg, and it is not unusual to find values over 1 ppm, the maximum level considered safe by the U.S. Public Health Service. For this

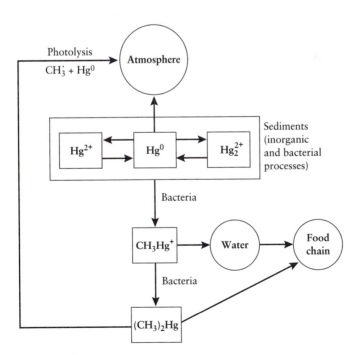

FIGURE 10-15 The mercury cycle.

reason it is recommended that swordfish not be eaten more frequently than once per week, or once per month by women who could become pregnant because of the added risk to neural development of the fetus.

The mechanism of formation of methylmercury in the environment is applicable to other elements (e.g., tin, palladium, platinum, thallium) that are heavy metals that form covalent, relatively nonpolar metal–carbon bonds. Some of these elements are employed industrially and could be discharged in wastes. Thallium in particular is toxic in its inorganic as well as organic forms, and it bears other resemblances to mercury in its behavior. On the other hand, the more electropositive metals including zinc, cadmium, and lead are not methylated this way (at least, not to any significant extent), and their alkyls are not stable in aqueous solution.

The environmental hazards of organomercury compounds are well documented. Serious and extensive poisoning of the population of Minamata, Japan, which occurred during the 1960s, was caused by the consumption of fish that had been contaminated by discharges of inorganic mercury compounds and methylmercuric chloride from a local industry, Chisso Corporation, which manufactured vinyl chloride and other organic chemicals. Actually, discharges began as early as 1932, and toxic effects were noticed in the early 1950s. Discharges continued until 1968, even though the cause of the disorder was known by 1960. More than 700 persons died, and over 2000 were seriously affected. Laws summarizes the irresponsible behavior of both the company and the authorities in this case.[24] Other cases of mercury poisoning from industrial discharges have occurred in Japan and China, while still others have been traced to seed grain treated with organomercury fungicides, a practice that was conducted for many years. Although never intended for direct consumption, this grain did find its way into food supplies. This has happened even indirectly, through feeding treated grain to animals shortly before their slaughter. In 1969 several members of a farm family in New Mexico were poisoned by eating meat from a pig that had been fed organomercury-treated grain. The pig itself had showed no sign of illness before slaughter; tests have shown that these animals can consume and retain considerable amounts of organomercury compounds without showing overt symptoms of toxicity. In two instances in Iraq, in 1956 and 1961, large numbers of people were poisoned by flour made from treated wheat. Other incidents occurred in Guatemala in 1963–1965, when treated seed grain again was consumed by local farmers. Dangers of consuming fish that had concentrated organomercury from contaminated rivers and lakes are well known, and consumption of fish from many freshwater sources has been banned in numerous cases in North America because of their mercury content. Although no

---

[24]E. A. Laws, *Aquatic Pollution: An Introductory Text*, 2nd ed., Wiley, New York, 1993, p. 372.

major incidents of poisoning have been confirmed, health effects have been found in communities that make heavy use of freshwater fish from contaminated water in their diets; examples are people of northern Ontario, northern Quebec, and peoples living along rivers in the Amazon Basin of Brazil.

10.6.8 Boron

Boron in nature is encountered as boric acid or its salts. Boric acid $B(OH)_3$ is quite weak, and this is the form expected in aqueous systems. The $B(OH)_3$ $\rightleftharpoons B(OH)_4^-$ equilibrium in the ocean acts as a buffer analogous to the carbonate system, but at a much lower concentration. Boron compounds (e.g., borax, which is a hydrated form of sodium tetraborate,[25] $Na_2B_4O_7$), are widely used industrially. Sodium perborate widely used as a bleach in detergents and other applications, and consequently boron is entering water systems. Sodium perborate written variously as $Na_2B_2O_4(H_2O_2) \cdot 2H_2O$ or $NaBO_3 \cdot 4H_2O$, is actually $Na_2[B_2(O_2)_2(OH)_4]$, in varying stages of hydration and purity in which two B atoms are linked by two peroxo groups:

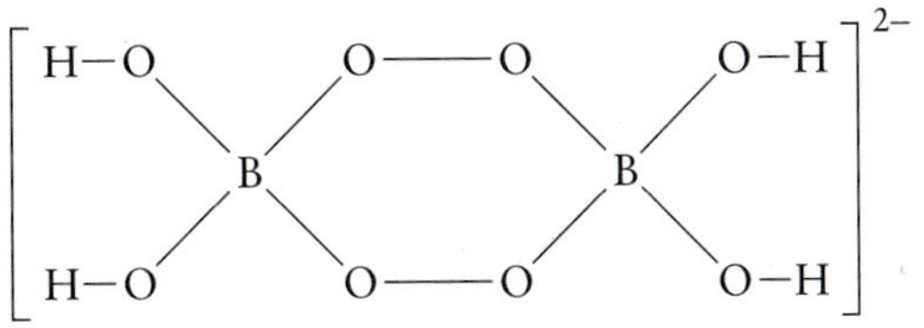

To complete the tetrahedral coordination, there are two hydroxo groups on each B. Under basic conditions in water, sodium perborate decomposes to release hydrogen peroxide, which is the active bleaching agent. Although necessary for some plants, including algae, boron is also known to be toxic to plants at higher concentrations. Toxicity generally is not very well understood, but a maximum limit of 1 mg/liter has been set by the USPHS for drinking water.

10.6.9 Aluminum

Aluminum is a strongly oxophilic element; that is, it combines strongly and preferentially with oxygen. The naturally occurring forms of Al are, consequently, oxides or oxo anion species. Many minerals are aluminosilicates: network structures in which aluminum and silicon are associated with oxygen

---

[25]Borax should be written $Na_2[B_4O_5(OH)_4] \cdot 8H_2O$, where the B_4O_5 unit can be regarded as a B_4O_4 ring, with the fifth O atom bridging two B atoms across the ring.

in insoluble, relatively inert materials. These are discussed in more detail in Chapter 12. Aluminum metal itself is highly reactive, but an impervious layer of oxide that forms upon exposure to air renders it inert. The chemical stability of these oxo compounds leads to the general assumption that Al could be considered a nontoxic element. However, it is now recognized that aluminum can be converted to soluble forms—for example, under acidic conditions such as those produced by acid rain—and that these soluble forms are toxic to plants, fish, and possibly to other animal life as well.

10.6.10 Tin and Lead

Tin and lead are the last two members of the carbon family but, in keeping with the general tendency for metallic character to increase with atomic number in a family, they show typical metallic properties. However, they have comparatively weak electropositive characteristics and strong electron acceptor properties. Bonds to nonmetals such as carbon have considerable covalent character and make up an important aspect of their chemistry.

Both tin and lead form compounds in which they have the oxidation states (IV) and (II). The former state is found in the organometallic compounds that have considerable environmental significance. Overall, lead compounds show much greater toxicity than those of tin, and environmental lead poses a more significant hazard than does tin.

Organotin compounds are used for a variety of purposes, such as wood preservatives, marine antifouling paints (especially tri-*n*-butyltin compounds), fungicides, and stabilizers for poly(vinyl chloride). The organotin compounds have the general formula R_nSnX_{4-n}, where X is a suitable anion; the most important systems are those for which $n = 2$ or 3. The toxicity of these organotin compounds is greatest for trialkyl compounds with short carbon chains. Dioctyltin compounds are sufficiently nontoxic to be used in plastics employed for food packaging. However, extraction of mono- and dialkyltin compounds into drinking water from plastic pipes is of some concern, as is accumulation of organotin compounds in some marine environments; health effects are not well established at this point. Microbial and photochemical reactions are known to degrade some organotin compounds now in use fairly rapidly, although the behavior with respect to degradation of many of these compounds has not been established fully. On the other hand, microbial processes can also form methylated tin species.

Tributyl tin has been widely used in marine anti-fouling paints, which are designed to kill marine organisms that attach to and foul boat hulls. The tributyl tin group is attached to an organic polymer coating through an Sn-O linkage to an ester group. Slow hydrolysis by sea-water releases the tributyl tin as the chloride (n-Bu$_3$SnCl) or oxide [(n-Bu$_3$Sn)$_2$O], killing the organisms that cause fouling but also allowing the compound to accumulate in the ocean where it is toxic to a variety of marine life in heavily used locations such as harbors. It is particularly toxic to shellfish, crustaceans, and juvenile fish, and is bioaccumulated. Tributyl tin can survive in an aerobic medium for months, and for much longer in anaerobic sediments. Many nations, including the United States, have imposed severe restrictions on the use of this substance.

The final degradation product of organotin compounds is an inorganic tin oxide. Unlike the case with lead, inorganic forms of tin are relatively nontoxic. Tin metal is used to coat steel to make cans used for packaging food and other materials. It also finds use in lead-free solders and pewter.

The toxicity of lead in the environment has caused extensive concern in recent years. The current U.S. limit for lead in water is 0.015 mg/liter. Like mercury, Pb(II) forms comparatively covalent bonds with appropriate donor groups in complexes, generally favoring sulfur and nitrogen over oxygen donors, and it may owe some of its physiological action to replacement of other metals in some enzymes. Low levels have subtle effects on the nervous system, while higher levels can lead to many symptoms, such as severe effects on the nervous system, including loss of sight and hearing, as well as symptoms of gout, headache, insomnia, anemia, kidney damage, diarrhea, stomach pains, intestinal paralysis, and eventually death.

A great deal of the environmental lead, and most of the airborne lead, has come from organolead compounds widely used as the tetraalkyls added to gasolines (Section 6.6). These were adopted in the 1920s through largely U.S. developments to allow an increase in automotive engine compression ratios, even though they were known to be toxic and an alternative, methanol, was widely proposed as an effective alternative.[26] Upon combustion the organolead compounds are converted to elemental lead, lead oxide, or a lead halide. These products are not volatile, and atmospheric lead is essentially all particulate. Much of this will settle near the source, and soils near areas with heavy vehicular traffic often have elevated lead levels. As discussed in Section 6.7.4, lead additives have been removed from gasoline in much of the world, in part because of concerns about lead toxicity, but also to permit the use of catalytic converters.

Although atmospheric lead is primarily particulate, it is still distributed worldwide. Analysis of Greenland ice cores show a relatively constant, low lead content up to about 1750, but with a period of increased concentration at levels corresponding to the Roman period, when it is known that mining and use of lead increased (Figure 10-16). Values increased after 1750 as the Industrial Revolution encouraged continued increases in lead use. Lead concentrations in the ice then went up much more rapidly after about 1940, corresponding to widespread use of lead in gasoline.

[26]An extensive "exposé" was published by J. L. Kitman in *The Nation*, March 20, 2000; available at http://www.thenation.com.

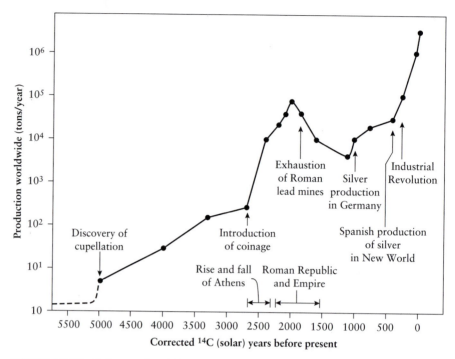

FIGURE 10-16 Estimates of world lead production over time. Note that cupellation is a process for purifying precious metals by melting the impure metal in a cupel (a flat, porous dish) and then directing a blast of hot air onto it. The unwanted metals are oxidized and partly vaporized and partly absorbed into the pores of the cupel. Redrawn with permission from D. M. Settle and C. C. Patterson, *Science*, **207** 1167–1175. Copyright © 1980 American Association for the Advancement of Science.

Many common lead salts, such as the carbonate, hydroxide, and sulfide, are extremely insoluble in water; this normally limits the dissolved lead content in a body of water, although in principle complexing agents in the water may considerably increase the effective solubility.

Various technological sources of lead (Table 10-8) can serve as origins of ingested lead in humans and consequent lead poisoning. As with other heavy metals strongly bound by biological complexing agents, lead is a cumulative poison and can act through long-term ingestion of relatively small quantities. Lead can be retained in the body for long periods, especially in bones, where it can replace some of the calcium. Lead poisoning in children from ingestion of paint fragments is well known. This is primarily a hazard with older paints, in which lead pigments were widely used. Basic lead carbonate, $(PbCO_3)_2$ $Pb(OH)_2$, known as white lead, is an example of a white pigment, while lead chromate is an orange or yellow one. The oxides, sulfate, and a variety of other lead compounds also were employed. Large amounts of white pigment is used even in colored paints to give opacity; lead content does not relate to color. These lead materials are largely absent in modern paint formulations, because of both the availability of superior materials (TiO_2) and legal restrictions arising from recognition of the hazard. It is interesting that the problem of lead poisoning in children from the lead in paint was recognized as early as 1900, and white lead in indoor paint was prohibited in France in 1909; international treaties to eliminate it were proposed in 1922, but the United States did not ban lead in interior paints until 1971. Even then, lead remained legal in paints for some exterior uses.

In spite of the toxicity of lead, and occasional calls for its use to be decreased, annual use in the United States has increased, from 1230 metric tons in 1970 to about 1600 metric tons in 1998, even though its use in gasoline decreased. Most of the increase (to well over 80% of the total, most of which is recycled) has been in batteries.

Less common now, but of possible significance in the past, is the extraction of lead into drinking water from lead pipes used in plumbing systems. (One may note the similarity between "plumbing" and *plumbum*, the Latin word for lead.) Lead pipes are not now used in drinking water systems (although they

TABLE 10-8

Some Applications and Uses of Lead

As metal	
Batteries	Lead–acid storage batteries are most widely used.
Water pipes	No longer used; some may still exist in old construction
Solder	As an alloy; also now removed from drinking water applications
Structural use	Has been used for roofing, glass mounting, etc.
Radiation shielding	High density makes it a good absorber of radiation
Shot	An alloy, also contains antimony and arsenic
Antifriction metal	Some bearing compositions
Type metal	
As inorganic lead	
Pottery glazes	
Glass	Increases index of refraction; gives sparkle to crystal
Insecticides	No longer used
Paints	White pigment and base, also colored pigments; now removed from indoor and most outdoor paints
Stabilizers in plastics	Heat stabilizer in poly(vinyl chloride) (often as organic acid salt)
Putty, caulk, etc.	
Matches	
As organolead	
Gasoline additives	Generally phased out except in undeveloped nations

may remain in some old constructions), but lead-containing solder has been common in copper piping systems, and in some older water system fixtures, including drinking fountains, one can find both solder and brass components that contain lead. Hot water, and water that has been in contact with the plumbing for some time, can dissolve significant quantities of lead from these sources. Therefore, it has been recommended that water be allowed to run for a few minutes before being used for drinking, and that water from hot water lines not be consumed. Many measurements have shown higher (sometimes much higher) levels of lead in first-draw water in the morning from household supplies than are found after the water has been allowed to run for a few minutes. In addition, the low-solubility lead salts that form as coatings on such plumbing surfaces can be further solubilized if the acidity of the water increases. Although lead-based solders and other components are not now used in drinking water systems in the United States, only recent installations are likely to have lead-free solder.

A more subtle source of ingested lead is pottery utensils. Lead salts have been a common constituent of some glazes used since antiquity to decorate pottery, and while apparently inert, some lead may in fact be leached out under acidic conditions, particularly if the glaze is improperly fired. Acidic foods (e.g., orange juice) can provide these conditions. One theory for the decline of the Roman Empire proposes that much of the population was functionally impaired by chronic lead poisoning originating from wine stored in lead-glazed vessels. Another theory attributes the same effect to the use of lead plumbing. Lead is also found in crystal glass, and can be extracted into wine stored in crystal decanters. Actually, lead contamination of alcoholic beverages of various kinds has occurred frequently through the use of lead components in presses, stills, or other equipment, and may still be a problem with illegal production such as "moonshine" whiskey.

From at least Roman times up to the eighteenth century, lead-containing materials were common additives used to sweeten sour wine,[27] and although the toxicity of lead in these applications had been recognized very early, this information was not generally understood. Lead compounds have also been used as food adulterants. White lead was added to flour to increase weight, and as late as 1994 many people in Hungary were poisoned when red lead, Pb_3O_4, was added to paprika. Even into the twentieth century, lead compounds were used in medicines.

Lead compounds (e.g., lead stearate, lead phthalate), have seen use as heat stabilizers in poly (vinyl chloride) materials (e.g., electrical wiring insulation,

[27]When wine is boiled down to a reduced volume with litharge, PbO, or grape juice is boiled down in a lead vessel, a sweet syrup containing lead is produced which was used in antiquity to sweeten and preserve wines. (Lead tetraacetate was called "sugar of lead.") Other alcoholic beverages have sometimes been produced with stills having lead components. For an interesting review, see J. Eisinger, Nat. Hist., 105 (7), 48 (1996).

water pipes). In the United States, current use is restricted to the first of these, and European use is expected to decrease as less toxic metal compounds such as organotin or calcium/zinc formulations take over. It is materials such as these that contribute to heavy metal release when "organic" materials such as plastics are incinerated.

Other sources of lead in the human environment have been lead arsenate and lead acetate used as insecticides. Lead also has widespread use in batteries, (discussed in Section 15.8.5.2), and solder in electrical equipment, and a significant amount enters the environment as spent ammunition from hunting. The problem here arises from the shotgun pellets used in duck hunting, which are ingested by the birds as they feed in sediments. Combustion of coal and oil releases trace amounts of lead. Castellino gives a concise history of the uses and toxicity of lead.[28]

There has been considerable controversy regarding the hazards associated with lead emissions into the environment. Various evaluations suggest that on the whole, body lead concentrations have not been increased very greatly over what is assumed to be a "normal" level by industrial and automotive emissions. On the other hand, there also are indications that symptoms of lead poisoning can be detected after only small increases. Children are particularly sensitive, and relatively small elevations in lead exposure can lead to learning disabilities.

10.6.11 Arsenic

Arsenic is a member of the same family as phosphorus. Although a classical poison, it is also an essential element. It occurs in the same phosphate rocks from which phosphorus chemicals are obtained, and in many industrial phosphates, arsenic remains as an impurity, and thus is found in small amounts in phosphate fertilizers and also among combustion products, mine tailings, and by-products from the metallurgical processing of copper and other metals. It is also used in pesticides—for example, sodium methanarsonate salts, dimethylarsinic acid, and inorganic arsenates and arsinites—although use of inorganic arsenical pesticides in agriculture is now prohibited in the United States. Earlier applications included pigments. Arsenic compounds have had considerable use as medicinals, especially against blood parasites, while other arsenic compounds, particularly salts of arsanilic acid [3-aminophenylarsonic acid, $NH_2C_6H_4AsO(OH)_2$], have been used as feed additives for poultry and pigs. The main current use of arsenic compounds is as a wood preservative, most frequently as "chromated copper arsenate," a mixture of chromic oxide,

[28]N. Castellino, in *Inorganic Lead Exposure*, N. Castellino, P. Castellino, and N. Sannolo, eds., Lewis Publishers, Boca Raton, FL, 1995, Chapter 2.

copper oxide, and arsenic pentoxide. Close to 20,000 metric tons of arsenic was used for this purpose in the United States in 1995, far more than in the rest of the world combined. There is growing concern about the release of arsenic from treated wood in sensitive areas such as playgrounds.

Arsenic, like phosphorus, is generally encountered as an oxoanion species, that is, arsenic acid, H_3AsO_4, an arsenate (AsO_4^{3-}), or an arsenite (AsO_3^{3-}) salt. The latter is considered to be more active biologically. In these forms its toxicity is well known, especially for soluble compounds from which the arsenic can readily be adsorbed. A lethal adult human dose of an arsenate is about 100 mg, but lower doses over a period of time can produce chronic poisoning. Effects include skin lesions and hardening of the skin (hyperkeratoses), which may lead to cancer. As of 1998, it was estimated that 200,000 people in West Bengal, India, were suffering toxic effects from arsenic-containing drinking water obtained from "tube wells" drilled to tap underground aquifers primarily to provide irrigation water, but also to eliminate the need to use contaminated surface water for drinking. Many more, potentially up to 70 million, are similarly exposed in Bangladesh.[29] The mechanism for what appears to be increasing arsenic release to the water is unclear at this point. The sediments that make up the aquifer are rich in iron sulfide, and the most probable explanation is that the arsenic (as arsenate) is largely adsorbed on $FeO(OH)$ particles produced from these sulfides. Anoxic conditions caused by a high organic content reduces the iron and releases the arsenic. Correlation of the arsenic concentrations with the anoxic character of the well and with the dissolved iron content in the water are evidence for this explanation, which suggests that the process might be reversed by aeration of the water and letting the contaminated ferric hydroxide produced settle. Less likely explanations propose an oxidative process connected to seasonal fluctuation of the water table that brings oxygen into the soils, along with flushing action from the water withdrawal, or displacement of arsenic by high levels of phosphorus from fertilizers.

In the United States, there is concern about low levels of arsenic occurring naturally in drinking water supplies. The limit has been 50 ppb, but there is evidence that at this level arsenic can contribute to bladder and lung cancer, and may also contribute to other health problems such as diabetes. The maximum level recommended by the World Health Organization is 10 ppb (also the level recommended in the United States by the EPA), but in 2001 a regulation setting this limit was withdrawn pending further stydy, largely for economic reasons connected to the costs of achieving it. As of this writing the 10 ppb limit had been reconfirmed.

---

[29] W. Lepkowski, Arsenic crisis in Bangladesh, *Chem. Eng. News*, p. 27; Nov. 16, 1998; also P. Bagla and J. Kaiser, *Science*, **274** (1996), and R. Nickson, J. McArthur, and W. Burgess, *Nature*, **395**, 338 (1998).

Aliphatic organic arsenic compounds have higher toxicity, and there is evidence that they can be produced from arsenates by the action of anaerobic bacteria, much as methylmercury is formed from inorganic Hg. The following sequence of steps can take place in sediments under the action of the appropriate bacteria:

$$
\underset{(a)}{\underset{\underset{O}{\|}}{\overset{\overset{OH}{|}}{HO-As-OH}}} \longrightarrow
\underset{(b)}{\underset{\underset{O}{\|}}{\overset{\overset{H}{|}}{HO-As-OH}}} \longrightarrow
\underset{(c)}{\underset{\underset{O}{\|}}{\overset{\overset{CH_3}{|}}{HO-As-OH}}} \longrightarrow
\underset{(d)}{\underset{\underset{O}{\|}}{\overset{\overset{CH_3}{|}}{HO-As-CH_3}}}
$$

where (a) is arsenic acid, arsenates; (b) is arsenious acid, arsenites; (c) is methylarsonic acid; and (d) is dimethylarsinic acid.

All these can be extracted into water. Further, additional biological processes can convert the methyl compounds to di- and trimethylarsine, $(CH_3)_2$ AsH and $(CH_3)_3As$, which are extremely toxic, volatile compounds. These are readily oxidized, and there is an arsine–arsenic acid biological cycle. As with methylmercury, these methylated arsenic compounds are fat soluble and may concentrate in the food chain. There are reports of illness and fatalities resulting from release of volatile organoarsenicals from the pigments in wallpaper under conditions of dampness and mildew; for example, it is claimed that Napoleon's death was due to this cause, and that an American ambassador to Italy became very ill for the same reason. Aromatic arsenic derivatives are less toxic than the aliphatic ones.

10.6.12 Selenium

Selenium falls below sulfur in the periodic table and resembles the latter element in many respects. Like S, in its inorganic environmental chemistry it exhibits four oxidation states; Se(VI) (selenates), Se(IV) (selenites), Se(0) (as the element), and Se(–II) (selenides). Metal selenides are found frequently with sulfides, and consequently selenium compounds may appear in the wastes from processing of sulfide ores. Reactions that produce SO_2, including combustion of fossil fuels, also produce SeO_2, but this compound is readily reduced by SO_2 to elemental selenium. In the elemental form, Se is insoluble and largely unavailable to organisms. Selenate resembles sulfate and is the most stable oxidized form of the element, although selenite is also comparatively stable and is believed to be the most active form biologically. In any event, interconversion of the two forms is slow, and selenite is a common environmental form of this element. Salts of both tend to be soluble. Selenides, like sulfides, are often very insoluble. Selenium and its compounds are employed in electronic and photoelectric devices, in glass for color, and in photocopying equipment.

The toxic nature of selenium is well known; effects vary with species, but in humans loss of hair and nails, skin lesions, and respiratory failure have been reported. Because of the acute toxicity, there has been a considerable emphasis on reducing selenium emissions. Selenium in the soil can be taken up by plants, which may be toxic to animals that consume them, as has occurred in some areas of the western United States. Ordinary grasses and cereals can absorb toxic levels, but some plant species can achieve very high degrees of concentration. Concentration of selenium compounds leached from soils by irrigation water and entering wetlands also has been problematic for wildlife (Section 11.7). Availability of soil selenium to plants varies with the total selenium content and soil acidity. The selenium in alkaline soils is chiefly present as soluble selenates that are readily taken up by plants (hence the early name "alkali disease" given to the symptoms of cattle that have eaten vegetation high in selenium), while under more acidic conditions the selenium seems to be present as selenites or elemental selenium. The latter is the predominant state under moderate and low pE conditions, and as already mentioned is insoluble and comparatively inert. Soil organisms have been shown to be able to convert soluble selenates to the insoluble elemental form.

Bacteria, perhaps other organisms in water sediments and sewage, and some plants can convert inorganic forms of selenium to volatile organic products such as dimethyl selenide, $(CH_3)_2Se$, and dimethyl diselenide, $(CH_3)_2Se_2$, much as inorganic mercury is converted to volatile methylmercury compounds. This may be a means by which these organisms protect themselves from toxic levels of selenium. Such conversion may, as well, set up a process of transport and cycling for this element. Other organoselenium compounds such as amino acids in which Se has substituted for S (e.g., selenocysteine) also are produced. Organoselenium compounds typically make up a significant fraction of the selenium in aqueous systems and sediments.

In spite of the toxicity of selenium, it is an essential trace element for a number of organisms, including humans. It is used in the glutathione peroxidase enzyme system to prevent the formation of peroxides and free radicals from the oxidation of unsaturated fats. Recently, evidence has been presented suggesting that selenium is effective in preventing the onset of cancer; that is, the incidence of cancer can be correlated inversely with selenium levels in the blood. Interestingly, this effect is counteracted by high levels of zinc. If the foregoing observation is valid, it illustrates the difficulty in dealing with trace elements. Eliminating selenium too thoroughly from food and water supplies may do harm, just as too high an intake certainly will.

Tellurium, the next element in the oxygen–sulfur–selenium family, is much less toxic than selenium because it is much more easily reduced and excreted as organotellurium compounds from organisms.

10.6.13 The Halogens

Chlorine is the most abundant halogen element, being widely distributed in water systems as the chloride ion. Seawater is near 0.5 M in chloride, but some chloride also occurs in fresh waters. The chlorides of some metals are insoluble (e.g., AgCl), but those commonly encountered in nature are soluble. Chlorine is a widely used oxidizing agent and is added to drinking water to oxidize organic matter and to destroy bacteria. In water, chlorine is hydrolyzed:

$$Cl_2 + 2H_2O \rightleftharpoons Cl^- + HOCl + H_3O^+ \qquad (10\text{-}34)$$

Hypochlorous acid, HOCl, is the effective agent in water treatment. It is a weak acid (pK_a 7.5), so there is a pH-dependent equilibrium between it and the hypochlorite ion, ClO^-, which is also commonly encountered as sodium or calcium hypochlorite used as an oxidizing agent or bleach. In addition to its oxidizing capabilities, hypochlorous acid is an effective chlorinating agent for ammonia and some organic molecules; the chemistry of these reactions is discussed in more detail in Section 11.5.

Chloride is not harmful, although the total salinity with which it is associated will, if high enough, render water unfit for use. NaCl (and to a lesser extent usually, KCl) are among the final salts to crystallize when freshwater evaporates. This leads to natural salt deposits in arid regions, and in some cases to a gradual buildup of salt in irrigated soils when irrigation waters containing small amounts of chloride evaporate in the absence of some periodic influx of freshwater to flush the soil. A good deal of chloride as solid NaCl or $CaCl_2$ is used to melt snow and ice, and this also contributes to locally high salinity of soils. The level of soil salinity is important for the growth of vegetation, in as much as many plants have limited salt tolerance. The Roman act of sowing salt on the ruins of Carthage reflected recognition, over 2000 years ago, that a high level of salinity would destroy soil productivity and at least symbolically was intended to prevent regrowth. Chloride ions will accelerate corrosion of metals such as iron by taking part in the mechanism by which the metal is attacked by oxygen.

Large amounts of chlorine in the form of HCl are released into the atmosphere by volcanic eruptions. For example, the Mount Pinatubo eruption of 1991 released an estimated 4.5 million metric tons of HCl. Most of this emitted HCl is removed from the atmosphere rapidly by dissolution in the water droplets that condense from the eruptions and subsequently precipitate. Claims that a large fraction of stratospheric chlorine comes from volcanic eruptions fail to recognize this fact. Chlorine also occurs naturally as organochlorine compounds produced by living organisms; some of these, including chloroform, carbon tetrachloride, and methylene chloride, are released to the atmosphere in natural processes.[30] Fires and volcanic activity also result in production and

---

[30]G. W. Gribble, Natural organohalogens, *J. Chem. Educ.*, **71**, 907 (1994).

release of organochlorine compounds. The importance of these natural releases of compounds with undesirable properties relative to releases from human activity has not been fully assessed, but the contribution of organochlorines to stratospheric ozone destruction, a prime concern with atmospheric chlorine compounds (Section 5.2.3), is not comparable to that of chlorofluorocarbons. In any event, there is no indication that their natural releases have changed in recent times to account for stratospheric increases in chlorine.

Chlorine and the heavier halogens form a number of oxo acids and their salts, of which hypochlorite, ClO^-, mentioned earlier is one. Another that has raised some concern is perchlorate, ClO_4^-. This has been found in groundwater in some western U.S. states, especially in areas where ammonium perchlorate was produced as a rocket fuel. It also receives use in fireworks and matches, and recently it has been found as a contaminant in a number of fertilizer brands in amounts up to 0.84 %. The health concerns are its potential for blocking iodide uptake and disrupting the function of the thyroid. The perchlorate ion is very inert under environmental soil conditions, persisting for decades. It is also poorly adsorbed by soils, hence is quite free to enter groundwater.

The heavier halogens bromine and iodine are much less common than chlorine in natural waters, but they are concentrated from seawater by seaweed. Iodine in particular is essential to thyroid function and is often added as an iodide to common salt, because water and foods are deficient in it in many areas, for example, the Great Lakes region of North America. Like chlorine, both these halogens are released as organic compounds from biological activity particularly from marine organisms. Methyl bromide, for example, is the chief bromine compound released upon burning biomass.[31] Like chlorine, both bromine and iodine are very effective in processes that destroy stratospheric ozone if they are carried high enough. Iodides in particular are readily decomposed by photochemical processes in the atmosphere at low altitudes, so that most of the release of either natural or technologically produced organoiodides never reaches stratospheric levels. However, it is possible that material released from polar marine sources during the winter could survive to reach the stratosphere, and that global warming could increase this possibility by increasing biological production.

The general chemical behavior of bromine is quite similar to that of chlorine. It is a weaker oxidizing agent, and chlorine treatment of water containing bromides will generate HOBr, which can result in bromination of organic molecules just as HOCl can chlorinate them. Iodine behaves somewhat differently. Hypoiodic acid, HOI, easily decomposes to I^- and iodate, IO_3^-, which is the stable form in water. The elemental form, I_2, is more stable in water than Cl_2 or Br_2. If the iodide ion I^- is present, iodine dissolves by forming the I_3^- ion.

---

[31] S. Manö and M. O. Andreae, *Science*, **263**, 1255 (1994).

The first member of the family, fluorine, is an essential element in trace amounts, entering into bones and teeth. It is best known to the public in the latter respect. Its action evidently stems from its ability to replace OH^- in the structures, for example:

$$Ca_{10}(PO_4)_6(OH)_2 + 2F^- \rightarrow Ca_{10}(PO_4)_6F_2 + 2OH^-$$
$$\text{Hydroxyapatite} \qquad\qquad \text{Fluoroapatite}$$

(10-35)

The OH^- and F^- ions are quite similar in size, with the result that the overall structure is not changed by a limited amount of such replacement, but evidently the hardness and crystallinity of the substances are improved. On the other hand, excess fluoride is harmful, with the most obvious effect being discoloration of the teeth (dental fluorosis). A significant portion of children who drink water at the EPA limit of 4 ppm display some evidence of this disease. Suggested links between fluoride and cancer, chromosome damage, and kidney problems have not been supported by conclusive studies. Fluoride is found in minerals, but only rarely does the level in waters become dangerously high from this source. It is added to water in fluoridation processes (in the range 0.7–1.2 ppm), from industrial processes, and as an impurity in phosphate products (e.g., fertilizers), where it often remains from fluorides present in the original phosphate mineral. Most simple fluoride salts are soluble and will enter the runoff water.

Recognition that low levels of fluoride ion in water are effective in reducing dental decay, especially among children, has led many communities to add fluoride to their water systems. Common fluoride materials that may be used in water fluoridation treatments are NaF, which may be obtained readily from reaction (10–36), or fluorosilicate salts (Na_2SiF_6 and others) obtained from phosphate purification.

$$CaF_2 + H_2SO_4 \rightarrow CaSO_4 + 2HF \xrightarrow{\text{NaOH}} 2NaF$$

(10-36)

The well-known toxicity of excess fluoride has always led some people to oppose such additions. This is a case in which small amounts of a material are beneficial while large amounts are harmful, and failure to recognize that this is a very common situation no doubt influences much of the fluoride opposition. At the controlled levels used in drinking water, fluoride has not been shown to be harmful, but legitimate questions may be raised about the ultimate environmental effects associated with a long-term, low-level injection of fluoride ions into freshwater systems. Total fluoride intake also involves sources other than drinking water, for example, food and fluoride-containing dentifrices. Other unforeseen effects of fluoride ions in water used for a multitude of purposes may also arise.

Fluorine has been widely used in chlorofluorocarbon compounds (freons) whose role in the destruction of the ozone layer was discussed in Section 5.2.3.

Small amounts of naturally occurring chlorofluorocarbons as well as HF have been detected in volcanic gases. Lest this be seized on by those reluctant to admit to human effects on the environment, it should be noted that amounts are small and are not correlated with the observed increase of the halogens in the stratosphere.[32] Organofluorine compounds also are emitted in aluminum refining (Section 12.2.5).

10.7 SUMMARY

In this chapter we have attempted to summarize the significant inorganic chemistry of the elements that have the greatest impact on humankind. As it turns out, many of these inorganic systems play vital organic/biochemical roles; sometimes essential for life, and sometimes toxic to it. Biological interactions with classical "inorganic" materials can have profound effects that have been recognized only comparatively recently. Examples include the methylation of mercury and the bacterial oxidation of sulfides. Other inorganic processes can render a benign material toxic—for example, increased solubility of metal salts through pH changes or complex formation. Many other interactions still await understanding. In the following chapters, we will discuss in more detail some of the applications of these aspects of inorganic chemistry as they pertain to specific segments of the environment.

Additional Reading

Alpers, C. N., and D. W. Blowes, eds., *Environmental Geochemistry of Sulfide Oxidation*, ACS Symposium Series 550. American Chemical Society, Washington, DC, 1994.

Butcher, S. S., R. J. Charlson, G. H. Orians, and G. V. Wolfe, eds., *Global Biogeochemical Cycles*. Academic Press, San Diego, CA, 1992.

Castellino, N., P. Castellino, and N. Sannolo, eds., *Inorganic Lead Exposure: Metabolism and Intoxication*. Lewis Publishers, Boca Raton, FL, 1995.

Chameides, W. L., and E. M. Perdue, *Biogeochemical Cycles*. Oxford University Press, Oxford, U.K., 1997.

Kaim, W., and B. Schwederski, *Bioinorganic Chemistry: Inorganic Elements in the Chemistry of Life*. Wiley, Chichester, U.K., 1994.

Laws, E. A., *Aquatic Pollution: An Introductory Text*, 2nd ed. Wiley, New York, 1993.

---

[32]Several letters in *Chemical and Engineering News* (Feb. 13, 1995, p. 4), discuss the natural vs anthropogenic generation of HF and organochlorine compounds. Reliable estimates of the amounts produced from these sources are not available, but the evidence for a major role for anthropogenic production comes from low and stable organohalide levels in sediments prior to about 1940 and from preindustrial samples of organic materials, and from increasing stratospheric concentrations that do not correlate with increased volcanic release and whose altitude profiles are not consistent with volcanic sources.

Oehme, F. W., ed., *Toxicity of Heavy Metals in the Enviromnment*, Parts 1 and 2. Dekker, New York, 1978.

Schlesinger, W. H., *Biogeochemistry: An Analysis of Global Change*. Academic Press, San Diego, CA, 1991.

Schnoor, J. L., *Environmental Modeling: Fate and Transport of Pollutants in Water, Air and Soil*. Wiley, New York, 1996.

Tessier, A., and D. R. Turner, *Metal Speciation and Bioavailability in Aquatic Systems*. Wiley, Chichester, U.K., 1995.

Thayer, J. S., *Environmental Chemistry of the Heavy Elements: Hydrido and Organo Compounds*. VCH Publishers, New York, 1995.

Trudinger, P. A., and D. J. Swaine, eds., *Biogeochemical Cycling of the Mineral-Forming Elements*. Elsevier, Amsterdam, 1979.

EXERCISES

10.1. (a) What are the main causes of imbalance in the carbon cycle?

   (b) How many tons of methane from natural gas would have to be burned to increase the carbon dioxide content of the atmosphere by 10% assuming that no competing removal processes occurred at the same time?

   (c) Discuss the possible means of sequestering carbon dioxide so that it does not enter the atmosphere.

10.2. Make a list of the atmospheric contaminants released by combustion of coal.

10.3. A particular 4000-MW power station burns 10 million tonnes of 3% sulfur coal a year. How many tonnes of limestone would be used up in scrubbing the SO_2 from the stack gases? If the final product is oxidized to gypsum, how many tonnes of gypsum would be produced? If this were disposed of as a solid in a landfill, what volume would it occupy? Is this a realistic estimate of the actual disposal volume?

10.4. Why is nitrogen gas inert?

10.5. What is meant by the terms nitrification and denitrification? Write some reactions that take place in these processes.

10.6. How are ammonia and nitric acid produced commercially? List the environmental problems associated with these processes.

10.7. What are the main sources of nitrogen oxides in the atmosphere?

10.8. What are the main sources of sulfur oxides in the atmosphere?

10.9. Describe the reactions that atmospheric nitrogen and sulfur oxides undergo that lead to their removal from the atmosphere.

10.10. How does fertilizer use contribute to atmospheric problems?

10.11. List the elements that are known to be essential for human health. List those believed to be essential whose role have not yet been definitely established.

10.12. Write the chemical reactions that are involved in the conversion of NO_x to HNO_3 in the environment.

10.13. A particular flow of water from a mine has a flow of 200 liters/min of water at pH 2. How many grams of sulfuric acid per day does this represent?

10.14. Compare the environmental chemistry of iron and manganese.

10.15. Discuss the pros and cons of using copper sulfate as an algicide in a small lake used for fishing and swimming.

10.16. Cr(VI) (as chromate) is a strong oxidizing agent that can oxidize most organic materials; nevertheless it can exist for considerable periods of time in biological systems. Discuss why this can be so.

10.17. Why is Cr(VI) as the chromate ion more readily transported into cells than Cr(III)?

10.18. Outline the reactions through which metallic mercury is converted into methylmercury in a body of water.

10.19. Mercury is found in fluorescent lights. What is its purpose?

10.20. Summarize the structures, uses, and hazards of organotin compounds.

10.21. What are the most common sources of ingested lead?

10.22. Explain how the amount of lead used over time can be inferred from analysis of Greenland ice. How does lead in the rest of the world get transported to Greenland?

10.23. Explain the relationship between arsenic poisoning and wallpaper.

10.24. Calculate the pE conditions under which selenite would become the stable form at pH 7. What pE would be required for Se(0) to be the stable form?

10.25. How many grams (ounces) of swordfish containing 1 ppm Hg would a person weighing 50 kg have to consume to reach the recommended daily maximum intake of 0.1 μg per kilogram of body weight?

11

WATER SYSTEMS AND
WATER TREATMENT

11.1 COMPOSITION OF WATER BODIES

The composition of natural water bodies depends on gain and loss of solutes through both chemical reactions and physical processes. For the most part, solutes undergo a geological cycle in which materials entering solution as products of weathering reactions of rocks, volcanism, and so on are carried to the oceans where they undergo further reaction, are deposited in sediments, and eventually are reincorporated into new rocks, which may repeat the cycle. Obviously, these processes have a long time scale. Volatile materials may enter the atmosphere for part of the cycle, and some elements may take part in biochemical as well as geochemical processes as described in Chapter 10. Physical transport of particulate material suspended in waters and as dust and sea spray in the atmosphere also has a part in these processes.

The composition of the ocean is not uniform, but variations from average values generally are not great. Dilution from freshwater sources is one means by which total concentrations can be altered locally. Relative concentrations of the dissolved materials are not changed by this, but they can be altered in regions of major chemical activity such as ocean vent systems. Fresh waters show much greater variation. Table 11-1 compares ocean and average river water compositions for some nonvolatile solute components. Oceans receive solutes from rivers, but the composition of the ocean is not simply that of a concentrated river. A major difference is in the Na/K ratios due to selective incorporation of K^+ in sediments. The natural freshwater sources of the materials in solution are chiefly weathering reactions of minerals. The most important reactions involve decomposition to other insoluble mineral species, with some decomposition products such as metallic cations entering solution; examples are given in Section 12.3.

If a fresh water containing typical amounts of dissolved Ca^{2+}, Na^+, and HCO_3^- in equilibrium with atmospheric CO_2 is allowed to evaporate, the pH will change from an initial slightly basic value, as could be calculated from the carbonate equilibria described in Section 9.2.2, toward greater basicity. The carbonate equilibria will shift to favor CO_3^{2-} as the predominant species. At some point, $CaCO_3$ will precipitate, providing some buffering action. However, the Na^+ will remain in solution, since it will form no insoluble salts, and the final solution will be a sodium carbonate solution. The actual behavior of a real freshwater system will be complicated by the other solute material present, but natural soda lakes, for example, Mono Lake in

TABLE 11-1

Typical Compositions of Rivers and Ocean Waters: Most Abundant Elements and Some Important Nutrients Only[a]

Element	Average river (mol/kg)	Average ocean (mol/kg)
Cl	2.3×10^{-4}	0.55
Na	2.2×10^{-4}	0.47
Ca	3.6×10^{-4}	0.01
K	3.4×10^{-5}	0.01
Mg	1.6×10^{-4}	0.053
S (as sulfate)	1.2×10^{-4}	0.028
C (mostly as HCO_3^-)	1×10^{-4}	2.3×10^{-3}
Br	2.5×10^{-7}	8.4×10^{-4}
B	1.7×10^{-6}	4.2×10^{-4}
Si	1.9×10^{-4}	1×10^{-4}
N (as nitrate)		3×10^{-5}
P (as phosphate)	1.3×10^{-6}	2.3×10^{-6}
Fe	7.2×10^{-7}	1×10^{-9}

[a]See S. M. Libes, *An Introduction to Marine Biogeochemistry*, Wiley, New York, 1992, p. 683, for a more complete list.

California,[1] approximate this situation. For most fresh waters, carbonate species are the predominant anions. Magnesium salts and silicates may also be among the precipitates, but F^-, Cl^-, and most SO_4^{2-} ions stay in solution as Na^+ and K^+ salts until late in the evaporation process.

11.2 MODEL OCEAN SYSTEMS

As was discussed in Section 9.3, the pH of the ocean corresponds to that of a solution of carbonic acid partially neutralized by reaction with basic minerals. That is, the means by which the pH of the oceans is established can be understood only if the interactions of these minerals with the aqueous components are considered; the carbonate system alone does not determine the natural pH of the oceans. (The carbonate system, along with the much lower concentration borate system, will provide short-term buffering action.) The average ocean pH is about 8.1; surface values are a bit higher (~ 8.3). The typical variation of pH with depth is shown in Figure 11-1. The high surface value may relate to biological factors, while variation at depth probably relates to pressure and temperature effects on equilibria, as discussed in Chapter 9.

Physicochemical models of the ocean have been set up to try to understand the pH-controlling reactions, among others. A model devised by Sillén[2] is a good example. In this and related models, it is necessary to consider not only the water and dissolved materials, but also the atmospheric and sediment components that could take part in environmental reactions. Equilibrium is assumed. The amounts of material to be considered per liter of seawater are shown in Table 11-2. Only the most abundant elements are listed here along with water, since these will have the most important effects.

One may consider the various chemical equilibria that could be set up among these major constituents. This is done most conveniently by imagining the successive addition of components to the water. The ions O^{2-}, OH^-, or H^+ may be associated with cations (or anions) in the following discussion.

After water, the most abundant material is silicon, which can be considered as being added to one liter of water as 6.06 mol of silica, SiO_2. Silica, which can be amorphous or crystalline, will undergo a variety of reactions with water. The solution process is as follows:

---

[1]Mono Lake is naturally highly alkaline because loss of the water flowing into it is largely by evaporation; the salts brought into it from the streams feeding it are concentrated and not flushed out. In recent years, the lake has evaporated to a fraction of its former self as the feeder water has been diverted to Los Angeles.

[2]L. S. Sillén, The physical chemistry of sea-water, in *Oceanography*, (M. Sears, ed.), p. 549. Publication No 67 of the American Society for the Advancement of Science, Washington, DC, 1961.

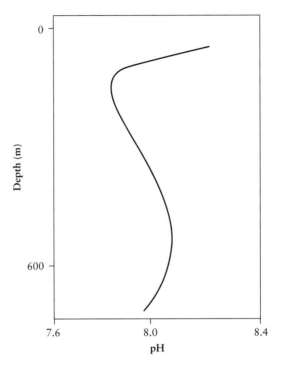

FIGURE 11-1 Variation of the ocean pH with depth.

$$SiO_2 + 2H_2O \rightleftarrows Si(OH)_4(aq)$$

$$K_{sp}(\text{crystal}) = 2 \times 10^{-4}; \ K_{sp}(\text{amorphous}) = 2 \times 10^{-3} \tag{11-1}$$

TABLE 11-2

The Major Components of the Ocean System, Relative to 1 Liter of Water

Component	Amount (mol)	Comments
H_2O	54.9	(1 liter)
Si	6.06	Mostly solid silicates and SiO_2
Al	1.85	Mostly solids
Na	0.76	0.48 mol in solution
Ca	0.56	0.01 mol in solution
Cl	0.55	Mostly in solution
C	0.55	See carbon cycle (Chapter 10)
Fe	0.55	Mostly solids
Mg	0.53	0.05 mol in solution
K	0.41	0.01 mol in solution

Solubility of amorphous SiO_2 is higher than that of the crystalline material, and consequently this form should dissolve with reprecipitation of the crystal. The fact that amorphous silica[3] is found in nature illustrates the problem of rate. Dissolution of the amorphous material is slow. Bottom ocean waters have about the expected dissolved silica concentration, but the surface concentration is influenced by biological (diatom[4]) action. The use of silica in the skeletons of these organisms is the reason for the existence of a biogeochemical cycle for silica.

The dissolved material undergoes several reactions. One such reaction is

$$Si(OH)_{4(aq)} + H_2O \rightleftarrows SiO(OH)_3^- + H_3O^+; \qquad K_a = 3.5 \times 10^{-10} \quad (11\text{-}2)$$

This means that $Si(OH)_4$ is a weak acid. There may be further dissociation steps, but with still smaller equilibrium constants. Another process is condensation, illustrated by the reaction

$$2Si(OH)_{4(aq)} \rightleftarrows Si_2O(OH)_{6(aq)} + H_2O \qquad (11\text{-}3)$$

Further steps may lead to higher polynuclear species.

The next most abundant element, aluminum, can be considered to be added as $Al(OH)_3$. The large affinity of this element for oxygen precludes the stable environmental existence of most aluminum species other than the hydroxide, oxide, and various minerals in which Al is associated with oxygen. The $Al(OH)_3$ can react

$$Al(OH)_{3(s)} + 2H_2O \rightleftarrows Al(OH)_{4(aq)}^- + H_3O^+ \qquad (11\text{-}4)$$

which can also be written as follows:

$$Al(OH)_{3(s)} + OH^- \rightleftarrows Al(OH)_{4(aq)}^-; \quad K = 10^{-1} \qquad (11\text{-}5)$$

There can also be reactions with silica to produce aluminosilicates. One such reaction is

$$Al(OH)_{4(aq)}^- + Si(OH)_{4(aq)} \rightleftarrows 1/2Al_2Si_2O_4(OH)_{4(s)} + 2OH^- + 2H_2O;$$
$$K = 10^{-6} \qquad (11\text{-}6)$$

This reaction can produce a basic solution.

Of the remaining elements in Table 11-2, the number of equivalents of the cations Na^+, K^+, Mg^{2+}, and Ca^{2+} exceeds the number of equivalents of the listed anions. Therefore, some of these cations must be added as the oxides or hydroxides, and if the calcium and magnesium associated with 0.55 mol of carbon as CO_3^{2-} were omitted, the basic materials would be in excess.

---

[3]Opal is amorphous SiO_2; it occurs widely in many forms besides the gemstone.
[4]Diatoms are members of the group of microscopic organisms called plankton, which are a major component of the food chain in the oceans. They have a skeleton made of silica.

However, the cations may further react with the excess SiO_2 and with the aluminosilicate minerals. One such process could be

$$3Al_2Si_2O_5(OH)_{4(s)} + 4SiO_{2(s)} + 2K^+ + 2Ca^{2+} + 15H_2O \rightleftarrows$$
$$2KCaAl_3Si_5O_{16}(H_2O)_{6(s)} + 6H_3O^+ \tag{11-7}$$

Such reactions produce H_3O^+ ions, and since the amounts of materials (alumina, silica, cations) are large, they can be the controlling reactions for establishing pH. Addition of carbonate still leaves the aluminosilicates in excess. Various reactions of this type involving known minerals can be proposed. Although accurate equilibrium calculations are difficult to make (partly because equilibrium constants for many of these reactions are not well known) and the hypothesis of equilibrium cannot be checked reliably, it is certain that such reactions can occur. The basic result of Sillén's model is that aluminosilicates control the pH, and the carbonate system merely serves as an indicator and a short-term buffer, since reactions involving sediments should reach equilibrium much more slowly. In any real application of buffering action, mixing times are important, as already mentioned.

Equilibria involving cation exchange reactions are important in mineral systems. Bonding of species held by predominantly Coulombic interactions will depend largely on packing efficiency, and for cations of the same charge, the affinity with which they are held in an aluminosilicate lattice will depend mostly on the relative sizes of the cations and the lattice positions available for them to occupy. For example, relative size (radius ratio) principles predict that a small cation is most stable in a site determined by four oxide ions that surround the cation in positions that correspond to the vertices of a tetrahedron (tetrahedral site) but fits less well and is less stable than a large cation if surrounded by six oxygens (octahedral site).[5] This simplification is often violated when factors other than simple electrostatic interactions become important.

In sediments, K^+ is more abundant than Na^+, and also Ca^{2+} is more abundant than Mg^{2+}. The molar concentration ratio K^+/Na^+ in sediments is 1.4, while in seawater it is 0.0026. That is, K^+ is preferentially bound in the sediments. If these processes are equilibrium reactions, it also follows that the sediments act to control the K^+/Na^+ (and Mg^{2+}/Ca^{2+}) ratios in seawater. Moreover, the K^+ and Ca^{2+} concentrations must appear in the equilibrium expression for the pH-controlling reactions. For the example given in reaction (11-7),

---

[5]The structures of many inorganic solids can be related to how the largest ions present can pack together geometrically. These are usually the anions. Packing of spheres naturally leaves spaces between them, and the cations occupy these positions, serving to hold the system together by Coulombic attraction. The usual geometries around these spaces are tetrahedral or octahedral; both occur in a given packing scheme.

$$K_{eq} = \frac{[H_3O^+]^6}{[K^+]^2[Ca^{2+}]^2} \qquad (11\text{-}8)$$

Thus, the pH of seawater is strongly related to these cation concentrations.

The chloride ions are not taken up significantly by the sediments, and nearly all chloride remains in solution. As a result, seawater is close to 0.5 M NaCl.

If the carbonates are now considered in Sillén's model, as 0.46 mol of $CaCO_3$ and 0.09 mol of $MgCO_3$, further reactions with the aluminosilicates become possible, but other processes involve precipitation of carbonate solid phases: (two crystalline modifications of $CaCO_3$ (calcite and aragonite), $MgCO_3$, and dolomite, $CaMg(CO_3)_2$. Carbonate will also enter into the pH-controlling reactions, but the significance of the model is that the amount of carbonate is small compared with the proton capacity available from aluminosilicates. For 1 liter of seawater in equilibrium with atmospheric CO_2, 10^{-3} mol of strong acid will change the pH from 8 to less than 6 (biological action may modify this). This is a small buffering capacity. However, when the minerals are included, the buffering capacity is about 1 mol/liter.

An additional element of great importance to the system is oxygen as O_2. Much of the oxygen remains in the atmosphere, but some is in solution according to Henry's law. This controls the pE of the ocean and determines the states of Fe and Mn to be found in the environment, as discussed in Section 10.6. The entire system is subject to the various physicochemical equilibrium equations, and equilibrium conditions can, at least in principle, be calculated.

11.3 RESIDENCE TIMES

For equilibrium models to have any validity, rates of the various reactions must be fast enough to approach equilibrium in the time period in question. Natural reaction rates are complicated by the mixing problem, which has been referred to before.

The length of time a species spends in a given phase (say, seawater) is important in terms of how rapidly appropriate cycles can be completed. This leads to steady-state models rather than equilibrium models of the type discussed earlier. A steady state assumes that input (for the ocean, weathering products brought in by rivers, volcanic activity, atmospheric input) is balanced by removal (sedimentation, ion exchange, biological production of inert material, etc.). That is,

$$\left(\frac{dC}{dt}\right)_{inflow} = \left(\frac{dC}{dt}\right)_{outflow} \qquad (11\text{-}9)$$

where dC/dt = rate of change in concentration. Inflow and outflow may include gain or loss through chemical reactions in addition to physical

TABLE 11-3

Residence Times for Some Elements in the Oceans[a]

Element	Residence time (yr)
Al	6.2×10^2
Ca	1.1×10^6
Cr	8.2×10^3
Fe	5.4×10^1
K	1.2×10^7
Mg	1.3×10^7
Mn	1.3×10^3
Na	8.3×10^7
Si	2×10^4
Zn	5.1×10^2

[a]See S. M. Libes, *An Introduction to Marine Biogeochemistry*, Wiley, New York, 1992, p. 638, for a more complete list.

processes. The residence time of a substance is defined as $T = C/(dC/dt)$. It is the average time a substance remains in a particular reservoir such as a lake or an ocean before it is removed by some transport process such as flow of the water to another system, or incorporation into sediments. Some examples of residence time for the ocean are given in Table 11-3.

In freshwater lakes, we have a similar situation, but output involves outflow as well as the other processes. An inert soluble material should have T about equal to that of the water itself, since it flows in and out of a given region with the water. Shorter T values imply rapid formation of insoluble products (e.g., those of Al), while larger ones suggest a cycling process that may be inorganic or biological. For example, in upper layers one may have $Fe^{3+} \rightarrow Fe(OH)_3$. However, if much organic material is present in the sediments, giving low pE, the precipitated ferric compound may be reduced to soluble Fe^{2+}, which is then reoxidized and recycled. Phosphorus and nitrogen are involved in biological cycles in both freshwater and oceans:

uptake by organisms → death and sedimentation → decomposition and solution

They often have long residence times.

11.4 ACID RAIN

We have frequently mentioned that oxides of sulfur and nitrogen entering the atmosphere from automobiles, other combustion sources, smelting, and some natural sources dissolve in raindrops and decrease the pH of the rain (Section 2.2). This activity has further effects on weathering reactions, solubility, and biological processes in lakes, rivers, and soils whose pH values are reduced as a consequence. Acidic deposition from the atmosphere may involve both wet deposition (rain, snow, fog, dew, frost) and dry deposition of acidic particulate and gaseous material on surfaces. Dry deposition is dependent on the nature of the surface and the relative humidity (there may be dissolution in a surface film of water), but is comparable to wet deposition in terms of amounts deposited. Vegetation is efficient for collecting dry deposition, which may be as damaging to leaves as the acid rain. Acidity also may come from ammonium sulfate that can be present in rain because of ammonia emitted to the atmosphere reacting with sulfuric acid; a solution of the salt of a weak base and a strong acid will be acidic. While sulfur oxides have been the major acidic component in acid rain, as major sulfur pollution sources have been cleaned up, nitrogen oxides are playing an increasing role. Typically, 60–70% of the acid comes from sulfur dioxide, and most of the rest from nitrogen oxides, but there may be a few percent HCl. Organic acids also may play a part.

The "normal" pH of rainwater in the absence of anthropogenic inputs is uncertain because natural sources of both acidic and basic materials in the atmosphere contribute. However, it is generally agreed that values significantly below 5 represent abnormal conditions, and precipitation with such values is called acid rain. The pH of precipitation over most of western and central North America averages 5 or slightly higher, while values for the eastern third of the continent are typically 4.6 to 4.4 (Figure 11-2) and sometimes less. Similar values less than 5 are found in areas of western Europe (Section 2.2).

Fog particles typically have higher concentrations of dissolved materials than raindrops because they have a higher proportion of the material that made up the condensation nuclei on which they formed. The drops will be diluted as they grow by condensation of more water vapor, although at the same time they can dissolve more gaseous pollutants. Scavenging of particulate and gaseous pollutants in clouds or as raindrops fall is referred to as washout or rainout.

Highly industrialized regions generate considerable amounts of sulfur and nitrogen oxides that are carried long distances by prevailing winds. Acid deposition in the northeastern United States and Canada has been blamed on releases in the industrialized Midwest and is believed to be responsible for acidification of lakes and die-off of forest trees in regions where they are already under stress from other causes. Loss of a large fraction of German forests is ascribed to acid rains from industry in Europe; much of northwest Europe is subject to such deposition as the prevailing winds carry pollutants from the industrialized areas to these regions. Figure 11-3 shows a correlation between forest damage and acid rain levels in Europe.

Part of the widespread distribution of the sulfur and nitrogen oxides that are the cause of the acidic precipitation comes from attempts to reduce local pollution from smelters, power plants, and other sources by employing very tall exhaust stacks. The concentration of pollutants at ground level is inversely proportional to the square of the effective stack height (which may be greater than the stack itself if the exhaust is released with appreciable upward velocity). Local effects thus can be reduced by tall stacks, but at the expense of injecting the pollutants into winds that can carry them long distances.

Ecological effects of acid deposition include disruption of species distribution and food chains in lakes as lower organisms, including plankton and benthic (sediment-dwelling) organisms, disappear, as well as through direct effects on pH-sensitive higher forms. In soils and sediments, aluminum may be solubilized, with potential toxic effects on vegetation. Heavy metals may also be solubilized from minerals, and less strongly tied up in complexes, as protons compete for the ligands or for adsorption sites; humic substances, which have complexing activity, are less soluble. Soil biota are affected. Direct effects on leaves have been reported.

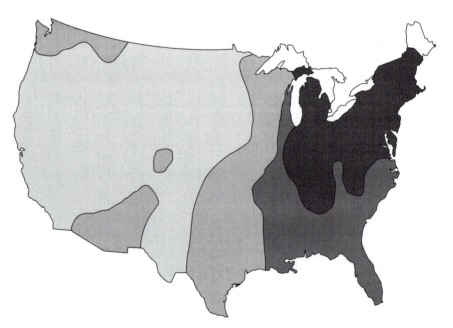

FIGURE 11-2 Approximate distribution of rainwater pH in the United States, based on U.S Geolological Survey data of 1996: light gray, pH 5.2–5.6; medium gray, pH 4.7–5.1; dark gray, pH 4.6–4.8; black, pH 4.3–4.6. Actual data can be found at the agency's Website: http://btdgs.usgs. gov/jgordon/acidrain.htm.

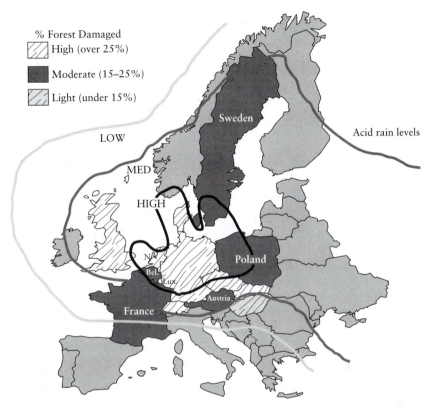

FIGURE 11-3 Acid rain and forest damage in Europe. From Database for Use in Schools Project, University of Southhampton: http://www.soton.ac.uk/~engenvir/environment/air/acid.how.big. problem.html.

11.5 WATER TREATMENT

Large-scale water treatment is necessary in two general circumstances; for water taken into distribution systems for household or industrial use, and for wastewater that must meet particular standards for pollution control. The details of the treatment processes depend on the initial quality of the water for domestic or industrial waters, and on the desired level of clean up of waste waters. We will discuss some general aspects here. Table 11-4 gives recommended maximum levels of some possible inorganic drinking water contaminants. Some, but not all, of these are mandated by regulation. The table is not complete; many (> 50) organic compounds that are specifically regulated are not included.

TABLE 11-4

Examples of Drinking Water Standards for Inorganic Substances

Substance	World Health Organization recommendations (mg/liter)[a]	U. S. Standards [(mg/liter), values in parentheses are recommendations][b]	Canadian standards (mg/liter)[c]	European Union Standards (mg/liter)[c]	Sources in drinking water[b]
Aluminum	0.2	(0.05–0.20)		0.2	
Ammonia	1.5			0.5	
Antimony	0.005	0.006	0.005	0.005	Fire retardants, ceramics, electronics, solder, fireworks
Arsenic	0.01	0.05	0.025	0.01	Smelters, glass, electronics, pesticides, natural deposits
Barium	0.7	2	1		Natural deposits, pigments, epoxy sealants
Beryllium		0.004			Electrical and aerospace industries, metal refineries, coal burning
Boron	0.3		5	1	
Bromate	25			0.01	
Cadmium	0.003	0.005	0.005	0.005	Galvanized pipe, batteries, paint, natural deposits
Chloride	250	(250)			
Chlorite	200	1.5	1.5		
Chromium	0.05	0.1	0.05	0.05	Mining, electro plating, pigments, natural deposits
Copper	1	1.3 (1)	1	2	Plumbing, wood preservatives, natural/industrial deposits
Iron	0.3	(0.3)	0.3	0.2	
Lead	0.01	0.015	0.01	0.01	Plumbing, solder, natural/industrial deposits
Manganese	0.1	(0.05)	0.05	0.05	
Mercury	0.001	0.002	0.001	0.001	Batteries, electrical equipment, crop runoff, natural deposits

(continues)

TABLE 11-4 (*continued*)

Substance	World Health Organization recommendations (mg/liter)[a]	U. S. Standards [(mg/liter), values in parentheses are recommendations][b]	Canadian standards (mg/liter)[c]	European Union Standards (mg/liter)[c]	Sources in drinking water[b]
Molybdenum	0.07				
Nickel	0.02			0.02	Batteries, natural deposits
Selenium	0.01	0.05	0.01	0.01	Mining, smelting, coal/oil combustion, natural deposits
Silver	0.1	(0.1)	0.05	0.01	
Thallium		0.002			Electronics, drugs, alloys, glass
Cyanide	0.07	0.2	0.2	0.05	Electroplating, plastics, mining, fertilizer
Fluoride	1.5	4 (2)	1.5	1	Fertilizer, aluminum industry, water additive, natural deposits
Nitrate	50 (as nitrate)	10 (as N)	10	50	Fertilizer, animal waste, sewage/ septic tanks, natural deposits
Nitrite	3 (as nitrite)	1 (as N)	3.2	0.5	As nitrate
Sulfate	250	(250)	500	250	Water processing, natural sources
Zinc	3	(5)	5		

[a]World Health Organization, *Guidelines for Drinking Water Quality Recommendations*, 2nd ed. WHO, Paris, 1993.

[b]U. S. Environmental Protection Agency, National Primary Drinking Water Standards; available at http://www.epa.gov/OGWDW/wot/appa.html.

[c]From Environmental Health Laboratories data as given at http://www.mastechnology.com/InfoPages/IntDW.html.

11.5.1 Domestic Water Treatment

The typical steps in treating water for household use in the United States and many other developed countries consist of (1) settling of suspended material, (2) aeration to aid in the oxidation of easily oxidized organic matter, (3) a preliminary chlorine treatment to remove bacteria and other organic material through oxidation, (4) use of a flocculating agent such as alum (aluminum sulfate) to carry down particulate matter, (5) addition of lime [$Ca(OH)_2$] to

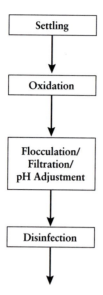

FIGURE 11-4 The common steps in the treatment of domestic water.

control pH and assist in the flocculation process, (6) filtration (through sand beds), and (7) a final chlorine treatment. Activated carbon may be used in some circumstances to absorb traces of material that give odor or taste, and a fluoride source (e.g., Na_2SiF_6) may be added as a dental health measure. Figure 11-4 shows a schematic block diagram of the process.

Suspensions of small particles in a liquid (colloids, Section 9.6) are often stabilized by adsorption of ions on the surface of the particles; if ions of a particular charge are adsorbed preferentially, the colloidal particles will repel each other and coagulation will be impeded. Various chemicals can be used that form insoluble precipitates that carry down the suspended materials. Examples are compounds of Fe(III) and Al(III), which on hydrolysis form the insoluble hydrated oxides discussed in Chapter 10. As mentioned earlier, aluminum sulfate, $Al_2(SO_4)_3$, is a widely used chemical for water treatment. Ferrous sulfate is another; it is oxidized by oxygen in the water to ferric ions. Hydrolysis of both Al(III) and Fe(III) produces polynuclear hydroxo species that are readily adsorbed on the surface of the suspended material, and as these hydroxo compounds further condense to insoluble forms, bind the particles with the floc as it settles. If the pH is near neutrality, both Al(III) and Fe(III) are completely hydrolyzed and the added materials are completely removed from the water (see Section 10.6.3). The pH may be adjusted to ensure this by the addition of a base such as $Ca(OH)_2$ (lime). The reaction can be summarized:

$$Al_2(SO_4)_3 + 3Ca(OH)_2 \rightarrow 3CaSO_4 + 2Al(OH)_3 \qquad (11\text{-}10)$$

The small amounts of calcium and sulfate ions left in solution are not harmful.

The best-known step in the treatment process for drinking water is chlorination. Chlorine is a strong oxidizing agent and in water produces hypochlorite, which is widely used in bleaches (Section 10.6.13):

$$Cl_2 + 2H_2O \rightarrow Cl^- + HOCl + H_3O^+ \tag{11-11}$$

This reaction is essentially complete at normal pH (the room temperature equilibrium constant is $4.5 \times 10^{-4}\,mol^2/liter^2$). HOCl is a weak acid ($K_a\ 3 \times 10^{-8}$) and will be present as both HOCl molecules and OCl^- ions, with relative amounts depending on the pH. (Concentrations will be equal at $pH = pK_a$; Chapter 9.) These species are referred to as free available chlorine. The HOCl molecule is most effective as a disinfectant, presumably because it can enter bacterial cells more easily than the ion.

The primary purpose of the chlorine treatment is to oxidize organic materials in the water, especially bacteria. The redox half-reaction of HOCl is

$$HOCl + H^+ + 2e^- \rightarrow Cl^- + H_2O; \quad E^\circ = 1.49\,V \tag{11-12}$$

Oxidizing conditions must be maintained throughout the entire distribution system to prevent redevelopment of septic conditions while ensuring that the chlorine content is not excessive near the addition point. The stability of the hypochlorite with respect to disproportionation to Cl^- and ClO_3^- is dependent on the pH, which may require adjustment for this reason, as well as to reduce the corrosive properties of the water.

Although chlorination oxidizes many organic materials, it does not oxidize all. It can result in chlorination of some molecules, with the possibility of producing chlorinated hydrocarbons with significant health hazards (Section 8.3.1). Chloroform is one of the potential by-products of water chlorination that is of concern. It may arise from the well-known haloform reaction, in which compounds that have structures $CH_3C(O)R$ or $CH_3CH(OH)R$, or can be oxidized to form these structures (e.g., olefins) undergo a reaction in which the hydrogens on the carbon adjacent to the one carrying the oxygen undergo a dissociation to form the carbanion intermediate $[CH_2C(O)R]^-$ followed by displacement of the Cl of HOCl to give the chlorinated product. This continues until all the hydrogens have been replaced, whereupon hydrolytic cleavage of the C—C bond releases chloroform, $CHCl_3$, as illustrated in the following reaction scheme.

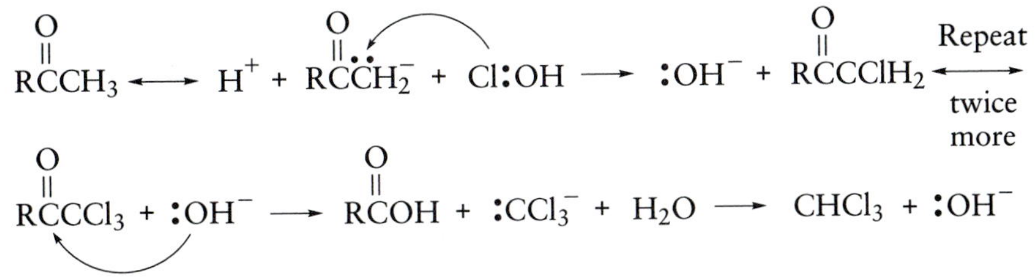

This reaction will occur with other halogens and is the basis of a qualitative test for the CH_3CO- group. Phenols are another class of organic compound that can be chlorinated by HOCl. Chlorinated phenols have very noticeable, antiseptic-like odors that are detectable at the parts-per-billion range.

These chlorination reactions are most expected in waters containing much industrial waste, where suitable organic molecules are present in significant amounts. Of course, such waters may already contain chlorinated hydrocarbon wastes, and the extent to which chlorination actually contributes to the total is not always clear in field situations. However, reactions with naturally occurring organic materials also are of concern. Humic substances (Section 9.5.7) in particular are suspected of being able to yield chlorinated products when they are present in large amounts in the water source. There are also concerns that harmful partially oxidized but not chlorinated by-products might be formed (e.g., epoxides).

In the presence of ammonia, HOCl will form chloramines (e.g., monochloramine) from the reaction

$$NH_3 + HOCl \rightarrow NH_2Cl + H_2O \qquad (11\text{-}13)$$

Under appropriate conditions, the reaction of the chlorine is fast and essentially complete, although at high chlorine/ammonia ratios or low pH values, di- and trichloramine can be formed. Analogous reactions take place with organic amines. Production of these materials may contribute to odor and taste problems in the treated water. On the other hand, combining ammonia with chlorine treatment can eliminate the formation of tastes or odors associated with the formation of chlorinated phenols, since the chloramine is a poor chlorinating agent and also gives a more persistent disinfectant action. This combined treatment (sometimes called chloramination) is used in a few public water supply treatment facilities in the United States. The chloramine slowly hydrolyzes to HOCl, prolonging the lifetime of the disinfectant, and producing it at a lower concentration so that chlorination reactions are reduced.

While bromine and iodine are not used to treat public water supplies, they have been used in disinfection of swimming pools. Iodine is used to treat water that is used for drinking in wilderness treks and so on, often as tetraglycine hydroperiodide: a complex of glycine, hydriodic acid and iodine having the composition $[NH_2CH_2COOH)_{16}(HI)_4(I_2)_5]$. The chemistry of bromine resembles that of chlorine closely, but in contrast to HOCl and HOBr, HOI is not stable and decomposes rapidly to I^- and IO_3^- so that there is no residual disinfection action.

While in many U.S. localities the law allows no substitute for chlorine treatment of drinking waters, an alternate method is available and receives some use in Europe and Canada, namely, treatment of the water with ozone, O_3. This method is as effective as treatment with chlorine, may have taste advantages, and avoids the risks of generating chlorinated products, although the question of whether it can produce partially oxidized toxic products has been raised. There is also some concern that ozone can react with inorganic bromide to produce substances that are effective in brominating organic compounds that have carcinogenic potentials as great as organochlorides. Ozonolysis is more expensive than chlorination, and because the ozone cannot be stored, it must be produced on site (by electric discharge in air). Ozone also does not provide a long-term disinfectant action in the distribution system, and so some chlorination remains necessary.

Another potential oxidizing agent for water purification is chlorine dioxide, also used in a few localities. This compound is unstable and cannot be stored; it must be prepared on site from sodium chlorite, usually by the reaction of chlorine,

$$NaClO_2 + 1/2Cl_2 \rightarrow ClO_2 + Na^+ + Cl^- \qquad (11\text{-}14)$$

although an alternative preparation method is to treat the chlorite with acid. In both cases, side reactions produce chlorate, a substance whose health effects do not seem to be fully studied. A third, more economical but more complex, method used in some pulp and paper mills involves reduction, in a strong acid solution of sodium chlorate, $NaClO_3$, with a reducing agent such as SO_2.

Because it acts as an oxidizer, chlorine dioxide is reduced to the chlorite ion:

$$ClO_2 + e^- \rightarrow ClO_2^- : E° + 0.954\,V \qquad (11\text{-}15)$$

It is not as effective a chlorinating agent as HOCl, and it does not produce $CHCl_3$ or chloramines, but there is some evidence that if organic material is in excess, ClO_2 can produce partially oxidized materials such as epoxides, quinones, and chlorinated quinones. It has no residual disinfecting action.

Ultraviolet irradiation is still another purification technique that has received limited use (e.g., in Switzerland). It is rapid and independent of pH and temperature, but less effective if suspended matter is present.

Passage of water through beds of activated carbon will remove organic contaminants that are not oxidized readily by the chlorine treatment and will also remove chlorinated organic molecules. This process has been used on a limited scale in water treatment plants for many years. It can also be employed by the householder in an attachment to the water tap. Activated carbon is produced by charring a variety of organic materials, followed by partial oxidation. This material has a high surface area and adsorbs most organic molecules effectively. The activity can be regenerated by oxidation of the adsorbed materials by heating in an atmosphere of air and steam. This method is perhaps the most practical means now available for removing chlorinated hydrocarbons, aromatics, and so on, which are among the most dangerous of the organic pollutants likely to be present in a domestic water source.

11.5.2 Wastewater Treatment

Clean water is a vital commodity, and usage is now so extensive that waste-waters must be repurified to avoid destruction of aquatic ecosystems and because, often, the water will be reused. Industrial wastewater may require specialized treatment that depends on the contaminants, while treatment of domestic waste (sewage) involves more general procedures.

A great deal of water is used for cooling purposes. Such water, if discharged back to freshwater systems, could lead to so-called thermal pollution by increasing lake and stream temperatures. Impurities are introduced only if additives are present to reduce fouling of heat exchangers (a common practice) or if substances are dissolved from them. Trace metals extracted as corrosion products would be possible examples of the latter case. The chemistry of thermal pollution lies chiefly in the effects of temperature on physicochemical and biological processes. For example, warm water will reduce the solubility of O_2, thus lowering its capacity to support life and to oxidize impurities. Large amounts of artificially warmed water, which will alter the relative volumes of the epilimnion and hypolimnion in a lake, may maintain the density gradient when normal conditions would lead to turnover (Section 9.1.2). Since fish and other aquatic organisms often prefer very restricted temperature conditions, the volumes of the habitats available for different species, and therefore the population balance of the water system, may be changed greatly. Dedicated cooling ponds from which the water may be recirculated or cooling towers which transfer the heat to the air, for example by evaporation as the hot water is sprayed through the tower, are alternatives to direct discharge. Use of the discharge water to heat buildings, greenhouses, and the like is attractive if such facilities are nearby.

Treatment of domestic wastewater (i.e., sewage) is important, and some steps used in common processes are described briefly below. Depending on the final water quality desired, treatment may be at the primary, secondary, or tertiary level. A typical treatment scheme is shown in Figure 11-5.

11.5.2.1 Primary Sewage Treatment

Domestic wastewater normally contains about 0.1 wt % impurities, composed mostly of a variety of organic materials. Many of these can be removed by filtration or sedimentation. Large particles (grit, paper, etc.) are removed by coarse mechanical screening followed by settling. Much suspended and fine particulate material remains, which is usually removed by sedimentation processes. Floating material such as grease can be skimmed off the top. In well-designed settling tanks, nearly 90% of the total solids and perhaps 40% of the organic matter can be removed. To remove the smaller particles, however,

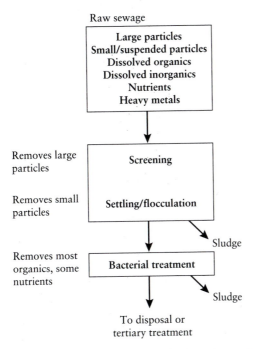

Raw sewage

FIGURE 11-5 Typical steps in primary and secondary sewage treatment.

assistance is needed and coagulation may be employed. This process, using aluminum sulfate, ferrous sulfate, or similar compounds, was discussed for drinking water purification in Section 11.5.1. The process is very similar when applied to wastewater.

In practice, sewage waters often contain considerable bicarbonate ion that can act as a base; $CaCO_3$ may be part of the precipitate. Lime, $Ca(OH)_2$, alone will cause $CaCO_3$ to precipitate and has some coagulant action. An added value to using calcium, iron, or aluminum in this process is that much of the phosphorus present can be removed as the insoluble phosphate salts. Up to 95% removal of phosphate is possible.

The solid produced in this process, which may contain toxic industrial wastes, grease, bacteria, and other substances, must be disposed of. It is often buried in landfill, or sometimes dumped at sea, although this mode of disposal is being phased out in many countries. The coarse materials removed in the preliminary steps must be disposed of in a similar way.

The treatment processes just described are called primary treatment, and a great deal of sewage receives no further processing. A large amount of dissolved material is left behind. Much of this is organic, but there are also nitrate, some phosphate, and metal ions. Secondary treatment removes the organic materials. This is important with respect to the overall quality of the discharged water, since a high organic content will use up the dissolved oxygen in the water, cause reducing conditions and the formation of various undesirable reduction products. Aquatic life also suffers. A practical measure of the reducing impurities in water is the so-called biological oxygen demand (BOD) (Section 7.2.5).

11.5.2.2 Secondary Sewage Treatment

The most widely used method of secondary treatment is a biochemical one, the activated sludge process. Bacterial action decomposes the organic matter in the sewage, producing a bacterial sludge that can be separated by sedimentation for disposal. This sludge is about 1% solids, and amounts to about 700 million gallons per day in the United States. Disposal is a major problem. As of 1993, over one-third (around 2.5 million tons dry weight) was used as fertilizer (often referred to as biosolids, which has a more attractive sound than sewage sludge), around 40% was landfilled, and under 20% incinerated. An even larger fraction of sewage sludge is used as fertilizer in Europe. Use as a fertilizer is attractive: there is significant N, P, and S content; but perhaps more significantly the practice builds up the organic content of soils, which in turn generally supports the natural soil biota and improves soil quality. Most is used in farmland, but other major uses are forest restoration and rejuvenation of mining areas. There are restrictions on the areas where sludge can be used to avoid excessive runoff, and other regulations limit the buildup of toxic materials from repeated applications. It is the possibility of introducing toxic materials that give rise to the greatest objections to this use. There are three categories of concern.

1. *Pathogens*. Various harmful organisms remain in the sludge, although they are maintained at acceptable levels and have not been directly implicated in disease. Various treatments are available to reduce these undesirable components to below detectable levels, but they are not used in most facilities.

2. *Undecomposed toxic organic compounds*. Highly stable materials such as dioxins and PCBs (Section 8.7) may be present, especially if the sewage contains industrial wastes. Numerous tests have shown that levels in most sewage sludge are small.

3. *Toxic heavy metals*. These known components of the wastes accumulate in soils in which they are used; take-up by plants can introduce them into the food supply, thus limiting long-term application of biosolids to agricultural land. These limits are indicated in Table 11-5. It is notable that European and U.S. standards are quite different. U.S. standards are based on risk estimates to humans based on various pathways for possible ingestion by humans, while

TABLE 11-5

Allowable Limits of Metals in Soils Treated with Sewage Sludge, (mg/kg)

Standards	Cd	Cu	Cr	Hg	Ni	Pb	Zn
United States	20	750	1500	8	210	150	1400
European	3	140	150	1.5	75	300	300

European standards are based on possible direct ecological effects on the biota of the soils.

Alternate sludge processing methods involve further bacterial digestion. Anaerobic digestion of the sludge converts carbon compounds chiefly to CH_4 and CO_2, while nitrate, sulfate, and phosphate are converted at least in part to NH_3, H_2S, and PH_3. About 70% of the gaseous products is methane. This can be burned to heat the reaction tanks, power pumps, and so on, and in fact such degradation of organic wastes to methane is a possible source of energy (Section 16.4). The other gaseous products (except CO_2) must be trapped or reacted to less noxious forms.

Aerobic digestion of the sludge with no additional source of nutrients results in autoxidation as the bacterial cells use up their own cell material. The organic materials are converted to CO_2 or carbonates, nitrogen, sulfur, and phosphorus to the oxo anions, and the solution ultimately produced is potentially useful as a liquid fertilizer. This approach to activated sludge disposal is comparatively new.

An alternative approach to sewage treatment is the use of lagoons, shallow open ponds in which the water can stand for several days to weeks while the biological activity of bacteria and algae under sunlight destroys most of the organic waste. Sludge must be removed from the lagoon periodically and disposed of as in other treatments. The final water still contains phosphorus and nitrogen nutrients. This approach depends on climate and available land areas.

11.5.2.3 Tertiary Sewage Treatment

Final cleanup of the water effluent from the secondary treatment is called tertiary treatment. Several types of materials must be considered in this respect. It should be kept in mind that while there are many tertiary treatment approaches, they are expensive and not widely used. Even secondary treatment is often not applied in many sewage disposal systems. The specific details of waste treatment vary considerably with different sewage handling facilities.

a. Colloidal Organic Matter

The remains of the activated sludge that do not settle out in the sedimentation tanks are returned to as colloidal organic matter. Different types of filter can be used, and it is common at this point to use an $Al_2(SO_4)_3$ coagulation process as described earlier as the key step in aiding the removal of these materials.

b. Phosphate

A phosphate concentration of 1 mg/liter is adequate to support extensive algal growth (blooms). Typical raw wastes contain about 25 mg of phosphate per liter. Thus, a large reduction is necessary to ensure that these effluents do not contribute to eutrophication of lakes. Removal of phosphorus by precipitation of an insoluble salt has been mentioned in connection with primary waste treatment. In fact, this precipitation may be done in the primary, secondary, or tertiary step. Lime is commonly used, with hydroxyapatite being the product:

$$5Ca(OH)_2 + 3HPO_4^{2-} \rightarrow Ca_5OH(PO_4)_3 + 3H_2O + 6OH^- \qquad (11\text{-}16)$$

Lime is cheap, and the efficiency of phosphate removal is theoretically very high, although colloid formation by the hydroxyapatite, slow precipitation, and other problems often reduce this efficiency in practice. Polyphosphates, if present, form soluble calcium complexes (Section 10.5). Other precipitants are $MgSO_4$ (giving $MgNH_4PO_4$), $FeCl_3$ (giving $FePO_4$), and $Al_2(SO_4)_3$ (giving $AlPO_4$). Phosphate precipitation may occur as part of a coagulation process, especially if iron or aluminum salts are used. An alternate process to precipitation is adsorption, using activated alumina (i.e., acid-washed and dried Al_2O_3). The alumina can be regenerated by washing with NaOH.

c. Nitrate

The activated sludge process converts some nitrogen to organic forms that are removed with the sludge; much, however, remains behind. Converting this to nitrate is possible if excess air is used, but this method of operation, more expensive than efficiently removing carbon compounds, is usually not used. The nitrogen remaining in the effluents then is largely some form of ammonia: NH_3 or NH_4^+. This can be stripped from solution by a stream of air if the solution is basic (the pH can be adjusted to be greater than 11 by the addition of lime, which can simultaneously remove phosphate). The chief problem lies in disposing of the ammonia in the air stream. It can be adsorbed by an acidic reagent, but at further cost. If allowed to escape, it is a potential local air pollutant, and eventually enters the runoff water through rain. Alternatively, aerobic bacterial nitrification of ammonia to nitrate, followed by denitrification by other bacteria in the presence of a reducing carbon compound (e.g., methanol) can be employed:

$$NH_3 + 2O_2 \xrightarrow[\text{bacteria}]{\text{nitrifying}} NO_3^- + H^+ + H_2O \qquad (11\text{-}17)$$

$$6NO_3^- + 5CH_3OH + 6H^+ \xrightarrow[\text{bacteria}]{\text{denitrifying}} 3N_2 + 5CO_2 + 13H_2O \quad (11\text{-}18)$$

Purely chemical processes for nitrogen removal are not available; suitable insoluble salts do not exist for a precipitation process, but reduction to N_2 or N_2O (which can be captured) by Fe(II) has been suggested for possible development.

d. Dissolved Organic Material

Comparatively little dissolved organic material remains after secondary treatment, but what remains is significant in terms of taste, odor, and toxicity. Chemical oxidation is a possibility, with a variety of oxidants being proposed, including chlorine, hydrogen peroxide, and ozone. Adsorption on activated carbon is effective and perhaps more practical.

e. Dissolved Inorganic Ions

Besides nitrate and phosphate ions, a significant quantity of other inorganic substances is present. Although many of these are not harmful, their accumulation eventually renders water unsuitable for reuse. In addition, toxic heavy metal ions may be present, although most of these precipitate under basic conditions. A variety of methods can be used to remove these substances, although in general the methods are expensive and rarely used in present tertiary wastewater treatments. When high-quality water is scarce, use of one or more of these methods may be practical as it is with brackish or seawater in arid areas, and for some of them the primary application is in desalinization plants to produce drinking water. Of main interest for desalination are distillation, electrodialysis, reverse osmosis, ion exchange, and freezing.

Distillation Distillation is an old technique, and quite simple if the waters do not contain volatile impurities such as ammonia. Although expensive, it has been used for many years to supply fresh water from seawater. Fossil fuel has normally served as the heat source, but solar distillation, the use of solar energy to evaporate the water has low cost potential in suitable areas.

Electrodialysis This is a newer technique. Water is passed between membranes that are permeable to positive ions on one side of the water stream and to negative ions on the other. An electrical potential is applied across the system so that the ions migrate, the positive ions passing out through the cation-permeable membrane (which prevents anions from flowing in), while the negative ions move out in the opposite direction through the anion-permeable, cation-impermeable membrane. The result is a stream of deionized water between the membranes, and this stream can then be separated as clean water. This process is critically dependent on the composition of the water; large ions, for example, do not pass through the membranes. It has been applied to tertiary treatment on a pilot-plant scale.

Reverse Osmosis   Sometimes called hyperfiltration, reverse osmosis is another newer process for water purification which is widely used. If pure water is separated from an aqueous solution by a membrane (e.g., cellulose acetate) that is permeable to water but not to the solute, then the solvent will flow from the pure water side into the solution. If the apparatus is designed appropriately, a hydrostatic pressure that will counteract the tendency of water to flow into the solution side of the membrane can be developed across the membrane. Without this hydrostatic pressure, liquid level in the solution compartment rises, corresponding to the osmotic pressure of the solution. However, if a hydrostatic pressure exceeding the value of the osmotic pressure is applied to the solution, water will flow from the solution to the clean water side of the membrane. This is the principle of reverse osmosis. The process is in use to obtain pure water from seawater in the Middle East, the Caribbean, on ships, and for some other pure water supplies. In practical systems, a very thin supported membrane is used; very small pores permit the water molecules but not the hydrated metal ions to pass.

Ion Exchange   Frequently used for small-scale water purification, ion exchange is often found in home water softening devices. Ion exchange materials are usually organic resins, but some inorganic materials can also function in this way. The bulk of the resin carries a charge (e.g., a group such as—OSO_3^- or a quaternary amine N^+), where an ion of opposite charge is held by Coulombic forces to produce electrical neutrality. An ion of a given charge can be replaced by another, and if those present initially are H^+ on the cation exchanger and OH^- on the anion exchanger, ionic impurities can be removed completely if water is passed through one and then the other. The exchangers are then regenerated by acid and base solutions, respectively. In household water softening, removing the Ca^{2+} ions, replacing them with Na^+ ions, is often adequate. Regeneration can then use an NaCl solution.

Partial Freezing   This is another method that can be used to produce useable water from a saline or contaminated source. Typically, when a dilute aqueous solution freezes, the first material to solidify is the solvent, water. Ice is formed, while the unfrozen solution becomes more concentrated. This ice can be separated and melted as clean water, and the remaining concentrated solution discarded.

11.6 ANAEROBIC SYSTEMS

The solubility of molecular oxygen in water is sufficient to support aerobic organisms and oxidizing conditions in much of the natural water systems that we encounter. However, inefficient mixing often permits bottom waters of lakes and oceans, and especially the waters in the sediments that underlie them, to become depleted in oxygen as biological processes use it up. This has been mentioned in several places (e.g., Section 9.1.2). Marshes and other systems with large amounts of decaying organic matter, soils with high organic contents and poor drainage, coal deposits, and landfill sites also lead to groundwaters that are anaerobic.

Normal respiration produces energy by the conversion of organic carbon compounds, the composition of which can be approximated by the empirical formula $(CH_2O)_n$, to CO_2 using oxygen as the oxidizing agent:

$$\frac{1}{n}(CH_2O)_n + O_2 \rightarrow CO_2 + H_2O \qquad (11\text{-}19)$$

This produces the maximum energy of any respiration process available to organisms; the free energy change involved in this process being about -500 kJ per "mole" of CH_2O. However, several alternative oxidizing agents are available to appropriate anaerobic organisms.

One of these alternatives is the nitrate ion:

$$\frac{5}{n}(CH_2O)_n + 4NO_3^- + 4H^+ \rightarrow 5CO_2 + 7H_2O + 2N_2 \qquad (11\text{-}20)$$

Use of nitrate in this way is a common process in soils, but nitrate is rarely abundant in sediments. In actuality, the microorganisms that carry out this process do so in several steps: nitrate $\rightarrow$ nitrite $\rightarrow$ nitrogen oxides $\rightarrow$ nitrogen. Some organisms utilize only some of these steps, and intermediates such as nitrous oxide may be released. The total free energy change is smaller than if oxygen were used, -480 kJ per "mole" of organic material, which means that nitrate-based metabolism is less efficient. As part of the decay process, nitrogen contained in the organic matter is converted to ammonia (recall that biological material contains about 16 nitrogen atoms for every 100 carbon atoms; Section 10.5). Under aerobic conditions, bacteria will oxidize the ammonia that is not assimilated, but under anaerobic conditions it too may be released. Reduction of nitrite to ammonia is somewhat more favorable thermodynamically than reduction to nitrogen, but does not seem to be an important step biochemically. Reduction of nitrogen to ammonia is of course an important biochemical process, but apparently not one that is used for energy production.

Sulfate is a second substrate available to anaerobic organisms, particularly in marine environments where sulfate is relatively abundant in seawater.

$$\frac{2}{n}(CH_2O)_n + SO_4^{2-} + 2H^+ \rightarrow 2CO_2 + 2H_2O + H_2S \qquad (11\text{-}21)$$

It is less energy efficient than the nitrate process, with a free energy change of only -103 kJ per "mole." The H_2S produced may be released to the atmosphere, but alternatively it may be precipitated as metal sulfides, since many transition metal and heavier metal sulfides are insoluble. For example, iron, which is relatively abundant, is likely to be present under reducing conditions as Fe^{2+} and can then generate ferrous sulfide deposits. Under other (aerobic) conditions, oxidation of sulfide or even of the Fe^{2+} can be energy sources for other organisms. In some cases, elemental sulfur can be a by-product.

Methane-producing organisms are likely to predominate when nitrate and sulfate concentrations are low. The process is

$$\frac{2}{n}(CH_2O)_n \rightarrow CO_2 + CH_4 \qquad (11\text{-}22)$$

with only 93 kJ of free energy available per "mole." Methane is frequently produced in marshy areas (swamp gas) and in landfills. Methane can also be used as an energy source for some bacteria. A similar process can generate hydrogen

$$\frac{1}{n}(CH_2O)_n + H_2O \rightarrow CO_2 + 2H_2 \qquad (11\text{-}23)$$

with an energy release of about 26.8 kJ.

11.7 IRRIGATION WATERS

Much agriculture depends on irrigation. Many of the most fertile soils occur in arid regions. Both surface water and groundwater sources are heavily used. Irrigation, however, is not simply adding enough water to the soil to maintain plant growth; it introduces some serious problems both for the water being used and for the soil itself. These problems relate to the mineral content of the water, which increases through the irrigation process. Inevitably, some water used for irrigation evaporates. The salt content, made up in part from the initial mineral content of the water, but increased by the salt leached from the soil as the water flows through the irrigation system, may reach a level at which further evaporation leads to deposition of additional salts in the soil rather than their removal. An essential feature of long-term irrigation is the need to use enough water to dissolve and carry away excess salts. In some areas, drainage systems are placed below the root level to facilitate this; an example is the San Joaquin Valley in California.

Irrigation water quality depends on at least four factors.

1. *Salinity,* the total salt content of the water. With a high soluble salt content, the soil may become unsuited to plants with low salt tolerance, and the returned irrigation water may be unfit for further use. Salt deposits from successive irrigations may carry deposited salt to lower depths, where it will accumulate in the root zone if adequate leaching does not take place.

2. *Sodium content relative to the magnesium and calcium ion content* (called the sodium absorption ratio, SAR). Because of ion exchange reactions with clay minerals in the soil, a high sodium ion content relative to the others can cause breakdown of the soil particles, leading to impermeability to water (waterlogging) and to hardening of the dry soil, even if the total salt concentration is low (Sections 9.7 and 2.3.1).

3. *Alkalinity,* the carbonate and bicarbonate content of the water (Section 9.2.2). This property influences soil pH, but also may result in precipitation of poorly soluble calcium and magnesium carbonates as the water evaporates, leading to an increased SAR for the remaining water.

4. *Toxic elements,* Some compounds that can be damaging to particular crops—for example, borates in parts of the western United States—may be present in the source irrigation water. These substances can accumulate in the soil over time, even if the initial levels are not harmful.

If enough excess water is used so that salts, carbonates, or toxic materials do not accumulate through evaporation, this excess mineral-laden water becomes runoff, entering either surface water systems or aquifers and perhaps degrading their qualities also. Buildup of toxic elements to hazardous levels can occur. Examples in the United States are boron and selenium (Sections 10.6.8 and 10.6.12), both of which have exceeded acceptable levels in some California irrigation waters. Selenium-rich water from irrigation uses in the San Joaquin Valley draining into the Kesterson Wildlife Refuge in California had serious effects on the aquatic life and wildfowl (it is teratogenic) until the input was stopped.

Additional Reading

Baker, L. A., ed., *Environmental Chemistry of Lakes and Reservoirs,* ACS Advances in Chemistry Series 237. American Chemical Society, Washington, DC, 1994.

Drever, J. L., *The Geochemistry of Natural Waters.* Prentice-Hall, Englewood Cliffs, NJ, 1982.

Forster, B. A., *The Acid Rain Debate.* Iowa State University Press, Ames, 1993.

Howells, G., *Acid Rain and Acid Waters,* 2nd ed. Ellis Horwood, New York, 1995.

Laws, E. A., *Aquatic Pollution: An Introductory Text,* 2nd ed. Wiley, New York, 1993.

Libes, S. M., *An Introduction to Marine Biogeochemistry.* Wiley, New York, 1992.

Sopper, W. E., *Municipal Sludge Use in Land Reclamation.* Lewis Publishers, Boca Raton, FL, 1993.

EXERCISES

11.1. Describe the changes in pH as typical river water evaporates. What solid products would you expect to deposit?

11.2. What are the five most abundant inorganic materials in seawater? How do these compare with elemental abundance in the environment as a whole?

11.3. Give equations for the reactions of silica in water; of alumina in water. Do these reactions occur to a significant extent in the environment?

11.4. Write some model equations to illustrate how the ocean pH might be determined by reactions involving alumina and silica.

11.5. What are the usual components of acid rain? What are the sources of these acids? What is the relative importance of the different sources?

11.6. Describe the usual geographic patterns of pH of precipitation in the United States and in Europe. Explain the reasons for these patterns.

11.7. We have mentioned acid precipitation in North America and Europe. Is it a problem in other parts of the world? Where else would you anticipate problems from acid precipitation? Explain why.

11.8. List the steps that are usually employed in the treatment of drinking water.

11.9. Use an equation to explain why the disinfectant power of chlorine is pH dependent.

11.10. What are the available practical methods for disinfecting drinking water? Give advantages and disadvantages of each.

11.11. Give equations showing the mechanism of formation of chloroform from acetophenone, $C_6H_5C(O)CH_3$, as a by-product of water chlorination.

11.12. Describe the chemistry involved in the process of flocculation in water treatment. Include equations.

11.13. Describe the processes of primary, secondary, and tertiary sewage treatment.

11.14. Indicate some practical methods for disposing of the sludge from sewage treatment. Compare the advantages and disadvantages of each.

11.15. Give the equations of the reactions that take place in anaerobic metabolic processes. Under what environmental conditions is each expected to be most important?

11.16. Compare the energy available from oxidation of one gram of organic material with O_2, NO_3^- and SO_4^{2-} as oxidizing agents.

11.17. Explain how irrigation can lead to deterioration of the agricultural quality of soils.

12

THE EARTH'S CRUST

12.1 ROCKS AND MINERALS

Current theories consider the earth to have formed from condensation of dust and gas in space. Gravitational and radiochemical heating melted the aggregated solids, and as the planet cooled, partial solidification and separation of materials took place. The present structure of the earth consists of a largely molten core composed chiefly of iron and nickel, surrounded by lighter rocks. The outer few miles, the crust, is the only portion of the earth that is accessible, and is the source of most of the substances used in a technological society. The crust has undergone a complex evolution as processes such as weathering and erosion have broken up the original rocks, and sedimentation and crystallization have produced new ones. High pressures and temperatures and volcanic activity have produced other changes. Indeed, these processes continue. These reactions, which fall into the realm of geology and geochemistry, are too extensive to be dealt with in detail here. We will discuss some of the aspects that most strongly impinge on human activity, in particular, those relating to some of the substances that we use extensively.

Many terms describing rocks and minerals are encountered in any discussion of the earth's crust and the processes that take place in it. The terms may describe a specific mineral, categorized by a specific structure and general composition [e.g., beryl, $Be_2Al_2(SiO_3)_6$] a class of related minerals (e. g., feldspars), or a conglomerate with particular properties (e.g., granite). Table 12-1 gives some of the most common of these terms; it is far from complete.

TABLE 12-1

Some Common Mineralogical Terms

Amphibole	A class of silicate minerals, containing chiefly Ca, Mg, and Fe, but also a wide variety of other cations, found in many igneous and metamorphic rocks. The approximate formula is $MSiO_3$, but the actual Si/O ratio is 4:11. Amphibole also refers to a specific mineral of the group, also called hornblende, $NaCa_2(Mg, Fe, Al)_5(Si, Al)_8O_{22}(OH)_2$. (Elements in parentheses mean any combination totaling the subscript number.) Many asbestos materials are amphiboles.
Anorthite	A common feldspar, $CaAl_2Si_2O_8$.
Apatite	A phosphate mineral, $Ca_5(OH, F, Cl)(PO_4)_3$.
Aragonite	One crystalline form of $CaCO_3$.
Asbestos	A term used to describe a fibrous form of some aluminosilicates, especially amphiboles but also a fibrous serpentine, chrysotile. Discussed in Section 12.4.5.
Basalt	A type of igneous rock formed on slow cooling; containing iron oxides, feldspar, and other species. The most common volcanic rock, underlies much of the ocean basins.
Bauxite	The primary ore used as a source of aluminum. It is a mixture of the minerals $Al(OH)_3$ (gibbsite) and two forms of $AlO(OH)$ (diaspore and boehmite).
Calcite	One crystalline form of $CaCO_3$, the most stable under ambient conditions.
Clay	Aluminosilicates with layer structures; weathering products of other aluminosilicate rocks and usually existing as small platelike particles. Also used to refer to any mineral of particle size under 2 μm.
Crocidolite	The most common amphibole asbestos, $Na_2Fe_4Al_2Si_8O_{22}(OH)_2$. It is the fibrous (asbestiform) version of the mineral riebeckite.
Cryolite	The mineral Na_3AlF_6.
Chrysotile	A serpentine (see below) asbestos, $Mg_3Si_2O_5(OH)_4$.
Dolomite	The mineral $CaMg(CO_3)_2$.

(continues)

TABLE 12.1 (*continued*)

Feldspar	A mineral group that is the most abundant in the earth's crust. Feldspars are the major constituents of igneous rocks and also important in most sedimentary and metamorphic rocks. The general composition is $KAlSi_3O_8$, but Na often substitutes for K, or another Al substitutes for a Si with the charge balance maintained by substitution of Ca or Ba for some or all of the alkali metal.
Granite	A common igneous rock formed on slow cooling; it is mostly quartz and feldspar.
Hematite	The mineral Fe_2O_3.
Igneous rocks	Rock formed on cooling of magma, the molten rock released from below the earth's surface by volcanic action. It consists of tightly interlocked crystallites of a variety of specific minerals.
Kaolinite	A common clay mineral, $Al_4Si_4O_{10}(OH)_8$.
Limestone	A sedimentary rock, usually deposited through biological processes, and mostly calcium carbonate.
Magnetite	Fe_3O_4; a combination of Fe(II) and Fe(III). It is an example of a spinel (a class of minerals based on the composition $M(II)M'(III)_2O_4$.
Marble	Metamorphosed limestone, $CaCO_3$.
Metamorphic rocks	A rock that has undergone change under high temperature and pressure, aided by chemical action (e.g., of water).
Mica	A group of aluminosilicate minerals with a sheetlike structure. Micas contain potassium and hydrogen, and often other elements. An example is muscovite.
Microcline	A common feldspar, KSi_3O_8 (the same formula as orthoclase but a different crystal structure).
Montmorillonite	A common smectic clay, $(Ca, Mg)Al_2Si_5O_{14} \cdot nH_2O$. (Smectic refers to the property of swelling when wet.)
Muscovite	The most common mica mineral, $KAl_3(OH)_2Si_3O_{10}$
Olivine	A group of abundant minerals in igneous rocks, with compositions running continuously from Mg_2SiO_4 (forsterite) to Fe_2SiO_4 (fayalite) and containing also Ca, Mg, and Mn as common bivalent cations.
Orthoclase	A common feldspar, KSi_3O_8 (the same formula as microcline, but a different crystal structure).
Plagioclase	Feldspars with Na or Ca as the primary cations (e.g., anorthite, $CaAl_2Si_2O_8$).
Pyrite	The mineral FeS_2 (fool's gold), which contains the S_2^{2-} ion. The term pyrite is sometimes used for related metal disulfides.
Pyroxene	A group of silicate minerals found in igneous and some metamorphic rocks. Their compositions are $MSiO_3$. An example is diopside, $CaMgSi_2O_6$.
Quartz	Crystalline SiO_2. Rocks composed mostly of quartz are called quartzite.

(*continues*)

TABLE 12.1 (*continued*)

Sandstone	A sedimentary rock composed of sand, mostly quartz but sometimes particles of feldspar or other minerals, cemented together with calcium carbonate, silica, an iron oxide, or a claylike mineral. The properties depend very much on the cementing material.
Sedimentary rocks	Rock formed from sediment deposits of material derived from the disintegration of another rock.
Serpentine	A group of layered-structure magnesium silicate minerals derived from the kaolinite type of aluminosilicates by replacement of Al by Mg. A fibrous form is the asbestos called chrysotile. Another example is talc, $Mg_3Si_4O_{10}(OH)_2$.
Shale	A sedimentary rock formed from mud, silt, or clay.
Slate	Rock formed by the metamorphosis of shale.
Zeolite	Aluminosilicate minerals with molecular-sized pores that can adsorb small molecules. Many synthetic zeolites have been made because of their adsorption and catalytic powers. Also used as "builders" in detergents in place of phosphates.

Table 12-2 gives the relative abundances of the most common elements in the earth's crust. As can be seen, by far the largest part of the crust is made up of silicon and oxygen, which form the basis of the abundant silicate minerals.

TABLE 12-2

Relative Abundance of Some Elements in the Earth's Crust

Element	Abundance (at. %)	Abundance (wt %)
Oxygen	63	47
Silicon	21	28
Aluminum	6.5	8
Iron	1.9	5
Calcium	1.9	3.6
Sodium	2.6	2.8
Potassium	1.4	2.6
Magnesium	1.8	2.1
Titanium		0.44
Hydrogen		0.14
Manganese		0.1
Phosphorus		0.1
Cu, Cr, Ni, Pb, Zn		$10^{-2}-10^{-3}$ each
Mn, Sn, U, W		$\sim 10^{-4}$ each
Ag, Hg		$\sim 10^{-6}$ each
Au, Pt		$\sim 10^{-7}$ each

The fundamental structure of all silicates consists of four oxygen atoms tetrahedrally arranged around a central silicon atom. In terms of oxidation numbers, these are considered to be $Si(+4)$ and $O(-2)$ (oxidation states $Si(IV)$ and $O(-II)$), but the Si—O bonding is covalent. However, an isolated SiO_4 tetrahedron would carry a charge of -4. Minerals based on the simple SiO_4^{4-} ion in association with cations (salts of silicic acid, Section 11.2) are unusual, and more typically the natural silicates are based on rings, sheets, and chains of linked tetrahedra in which the Si atoms share oxygens. The O:Si ratio is less than 4. Some typical structures are given in Figure 12-1.

In aluminosilicates, aluminum, the third most abundant element in the earth's crust, is also present, surrounded tetrahedrally or octahedrally by oxygen atoms.

In all these structures, negative charges on the silicate framework require the presence of cations for electrical neutrality. Within limits, the nature of the cationic species is of secondary importance. Particular structures are favored by particular cations to maximize crystal packing energies, Coulombic attractions, and other interactions leading to stabilization of the solid, but small amounts of many cations can be present in a given structure without influencing it in any major way. In particular, ions of similar size and charge may replace one another with little change in mineral structure (isomorphous substitution). If the similarity is great enough, complete replacement of one component by another can take place without significant structure change. As the similarity becomes less, then less replacement can take place in a given mineral before the structure changes. (Such behavior holds for many solid materials besides silicate minerals.)

From Table 12-2 it is seen that many of the elements essential for modern technology are present in very small amounts in the earth's crust compared with the dozen most abundant elements. The total amounts of these less abundant elements in the earth's crust are vast, however, and they are widely distributed. Most common rocks contain a few parts per million of many elements distributed randomly by isomorphous substitution of abundant elements (e.g., Pb replaces some K, and Zn replaces some Mg in most common rocks). For this reason, it has been suggested that humanity will never really run out of mineral resources, and as high-concentration ores are exhausted, new technology can be developed to tap the sources in the common rocks. However, the volume of material to be mined and processed, the energy requirements, and the waste disposal problems clearly set economic and environmental restrictions on the minimum concentrations that can be employed for large-scale use of any substance. Generally, society uses atypical, high-concentration sources for most of its mineral needs; little if any technological development would have been possible in the past if all elements were uniformly distributed in the earth's crust.

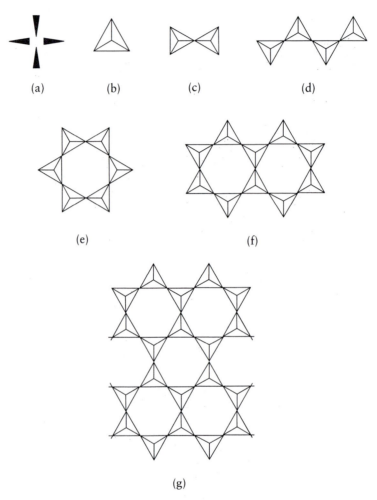

FIGURE 12-1 Examples of silicate structural units: (a) the SiO_4 tetrahedron; (b) a schematic view of a SiO_4 tetrahedron, where each apex represents an oxygen; (c) linked tetrahedra, the $Si_2O_7^{6-}$ ion; (d) a chain structure; (e) a cyclic silicate structure; (f) a double chain; and (g) a two-dimensional layered structure.

Imminent shortages of many metals were predicted during the third quarter of the twentieth century. These shortages have not materialized, partly because of improved methods of exploration, partly because of improved mining technology, and partly because of decreased demand and increased recycling. Current reserves of metal ores are adequate for many years.[1]

[1]C. A. Hodges, *Science*, **268**, 1305 (1995).

12.2 RECOVERY OF METALS FROM THEIR ORES

12.2.1 Sources

Minerals used as sources from which desired elements can be isolated are called ores. The ores of many elements are made up from particular compounds of the element in question, although these may be mixed with much inert rock. High-grade ores typically consist of veins of the compound; in low-grade ores the particles may be widely distributed in the rock. Mining may involve tunneling along the ore veins or open pit quarrying of large volumes of rock. The latter is particularly necessary when low-grade ores are used, as is becoming more common, and as modern earthmoving equipment makes this technique economical. Placer mining which involves recovery from riverbed gravels, is important in gold mining.

Mining has traditionally caused considerable environmental impact owing to the obvious damage to the land from the excavation itself, the waste rock deposits, and the disruption of rivers and streams. Less obvious but perhaps more critical are the effects from acid mine wastes (see Section 10.6.3) and solubilization of heavy metals. At least 50 abandoned mine sites in the United States have been designated as hazardous under the Superfund program.[2]

Ore bodies in nature were formed in a number of ways: for example, by selective crystallization as the molten material (magma) cooled to form rocks, and by hydrothermal processes, which have been important for many widely used metals. In a hydrothermal process, water or brine at high temperature and pressure dissolves metal ions from the bulk rock and redeposits them in concentrated form elsewhere. The extraction process is analogous to a weathering reaction. Exploitable deposits typically formed where the liquid flowed through a restricted channel in which a precipitation reaction could take place.

Formation of very insoluble sulfides and arsenides has provided many ore bodies. Oxides and carbonates are among the other most common ore minerals. Ongoing deposition at ocean hydrothermal vents may represent a present-day example of ore body formation. Hydrothermal circulation involves seawater penetrating into the ocean crust, where it is heated geothermally. Magnesium and sulfate are removed through interactions with the rocks, while at the same time components from the rocks are leached into the water. Conditions are normally reducing, resulting in formation of H_2S and sulfides, and metal ions such as Fe in reduced states. Under the high temperature and pressure of the crust, many of these are soluble. As the hot water rises, cooling allows deposition of secondary minerals such as metal sulfides and quartz as their solubility decreases. If by the time the water has reached the crustal surface it is still hot

---

[2]The Superfund program is a federal plan for funding the costs of cleanup of seriously polluted waste disposal and other sites.

enough to have retained a high mineral content, rapid mixing with the cold seawater results in formation of an immediate precipitate that contains materials such as FeS, ZnS, CuFeS$_2$, CaSO$_4$, and BaSO$_4$ (the minerals pyrrhotite, sphalerite, chalcopyrite, anhydrate, and barite, respectively). Since many of theses are black, the cloud of precipitating minerals around the release point gets the name *black smoker*; generally this cloud is accompanied by buildup of a "chimney" of precipitated minerals. The process is shown schematically in Figure 12-2. These systems are important not only for their inorganic chemistry, but also for the unusual biological activity that they support, based on the use of reduced mineral compounds as energy sources by certain bacteria that are capable of thriving under high-temperature conditions and of utilizing the energy released in biological oxidation of the reduced species to support their metabolism. These life-forms make up a class of organisms that are not dependent on solar radiation for their ultimate energy source.

Some specific ores and treatments used for individual metals are listed in Table 12-3.

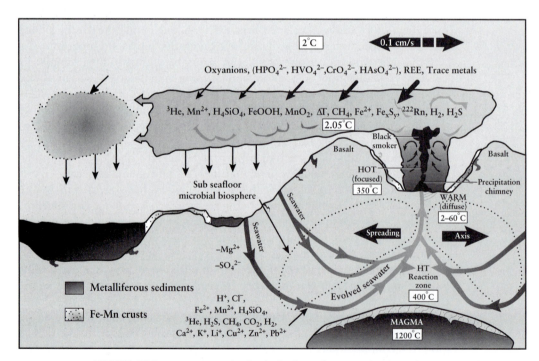

FIGURE 12-2   Processes involved in hydrothermal vent geochemistry. REE, rare earth elements. From U. S. Department of Commerce National Oceanic and Atmospheric Administration: www.pmel.noaa.gov/vents/chemistry/information.html

TABLE 12-3

Processes Used to Recover Some Metals from Their Ores

Metal	Chief ores	Common treatment processes	Comments
Aluminum	Bauxite, a combination of alumina clays	Purification with aqueous NaOH to give Al_2O_3 followed by electrolysis in molten cryolite, Na_3AlF_6.	Waste sludge of silicates and Fe_2O_3. Requires abundant electrical energy. Release of fluoride compounds.
Cadmium	Zinc ores	Volatilized from crude Zn and collected in flue dust, or precipitated from Zn solutions. Separated from Zn by distillation or electrolysis.	A by-product of Zn production.
Chromium	Chromite, $Fe(CrO_2)_2$	Reduction of ore with C produces ferrochrome alloy. This is dissolved in H_2SO_4, and Cr is then obtained by electrolysis. Alternatively, Cr_2O_3 can be formed and reduced with Al, C, or Si.	
Cobalt	Smaltite ($CoAS_2$), cobaltite ($CoAsS$), and other arsenides, sulfides, or oxides	Ore roasted to oxide (see Section 12.2.2.2). May be reduced to a mixture of metals, dissolved in H_2SO_4, other metals separated, and Co obtained by electrolysis or by precipitation as $Co(OH)_3$ with NaOCl; heating to oxide followed by reduction by C or H_2 gives Co.	Usually a by-product of the production of other metals (e.g., Cu, Pb, Ni).
Copper	Copper pyrite (CuFeS), cuprite (CuO), malachite $[Cu_2(OH)_2CO_3]$, and various arsenide, oxide, and carbonate ores	Ore concentrate is heated in air to remove some S, As, and Sb, then fused with silica; copper and iron sulfides and oxides separate (copper matte) from an iron silicate slag. The matte is oxidized in the presence of silica: $$2FeS + 3O_2 + SiO_2$$ $$\rightarrow \text{iron silicate slag}$$ $$3Cu_2S + 3O_2 \rightarrow 6Cu + 3SO_2$$ Crude Cu anode refined electrolytically; crude Cu anode dissolves in $H_2SO_4 - CuSO_4$ electrolyte and deposits as pure Cu cathode.	Much SO_2 released in the smelting process, along with As and Sb oxides; both gas and dust discharges involved. Also aqueous waste from the refining process. Anodic sludge is a source of Ag and Au. It also contains Se, Te, and heavy metals.

(continues)

TABLE 12.3 (*continued*)

Metal	Chief ores	Common treatment processes	Comments
Gold	Free metal, usually finely divided in rock.	Finely divided ore treated with aqueous cyanide solution and air; $4Au + 16NaCN + 3O_2 \rightarrow 4NaAu(CN)_4 + 12NaOH;$ then $2[Au(CN)_4]^- + 3Zn + 4CN^- \rightarrow 2Au + 3[Zn(CN)_4]^{2-}$ A lower technology process dissolves the Au in metallic mercury with recovery by distillation of the Hg.	Cyanide wastes can cause serious water pollution problems. Use of Hg releases metallic mercury to the environment; in primitive mining operations, the distilled Hg may not be condensed.
Iron	Magnetite (Fe_3O_3), hematite (Fe_2O_3), limonite ($2Fe_2O_3 \cdot 3H_2O$), siderite ($FeCO_3$).	Oxides reduced in a blast furnace with coke and limestone; carbonates are first roasted to oxide. Purification and steel making involve oxidation of C, Si, S, and P with O_2 and removal as CO_2 and as a slag formed with the limestone flux (see Section 12.2.2.2).	Large volumes of CO_2 gas produced, and Fe_2O_3 and Fe_3O_4 as dusts.
Lead	Galena (PbS), cerussite ($PbCO_3$), anglesite ($PbSO_4$, and others	Sulfide ores are oxidized: $3PbS + 3O_2 \rightarrow 3Pb + 3SO_2$	Much SO_2 is produced; As and Sb are often present and are released as oxides.
Manganese	Oxides (e.g., pyrolusite, MnO_2) and some carbonates and silicates.	Ferromanganese ($\sim 80\%$ Mn) produced in a process similar to that used for Fe; pure metal is produced electrolytically from MnO_2 dissolved in sulfate solution.	
Mercury	Cinnabar (HgS).	Roasted with air and the mercury vapors condensed: $HgS + O_2 \rightarrow Hg + SO_2$ Alternatively, roasted with Fe or CaO; metal sulfides and/or sulfates produced instead of SO_2	
Molybdenum	Molybdenite (MoS_2).	Concentrated and roasted to MoO_3, then reduced with H_2	SO_2 produced.

(*continues*)

TABLE 12.3 (*continued*)

Metal	Chief ores	Common treatment processes	Comments
Nickel	Pentlandite [(Ni,Fe)S], oxides, silicates, and arsenides.	Usually occurs with Fe and Cu. Fractionated by flotation, melted to a matte as with Cu and converted to NiO. Reduced with C or water gas, and purified by conversion to $Ni(CO)_4$, volatilized, and thermally decomposed back to the metal; alternatively, electrolysis.	Same as Cu.
Silver	Argentite (AgS), cerargyrite (AgCl) pyrargyrite (AgSSb), and other sulfide and antimonide ores.	By-product of Cu, Pb, and Zn ores; obtained during refining of the base metal. With Cu, the Ag is obtained from the anode slime and is recovered by roasting, leaching with H_2SO_4, and smelting to remove Se, Te, and other impurities.	
Tin	Cassiterite (SnO_2). Stannite ($Cu_2FeS_nS_4$)	Roasted as necessary to remove S as SO_2 and As as As_2O_3, and the tin oxide reduced by C.	
Titanium	Ilmenite ($FeTiO_3$), rutile (TiO_2).	The oxide is converted to the tetrachloride, which is reduced by magnesium: $$2TiO_2 + 3C + 4Cl_2 \rightarrow 2TiCl_4 + 2CO + CO_2;$$ then $$TiCl_4 + 2Mg \rightarrow Ti + 2MgCl_2$$	
Tungsten	Scheelite ($CaWO_4$), wolframite [(FeMn)WO_4].	The ore is roasted with Na_2CO_3 and $NaNO_3$, leached, and the tungstate precipitated with HCl as tungstic acid. This is calcined to WO_3 and reduced to metal with H_2.	
Zinc	Sphalerite (ZnS), carbonates.	The ore is roasted to ZnO, which is reduced with CO; the Zn vaporizes and is collected by condensation. It may be purified by distillation.	SO_2 is evolved in the roasting. Cd is a common associate; it has a higher volatility.

For all metals, it would seem economically preferable to recycle used metal, since this would eliminate the concentration and reduction steps necessary for virgin materials. Unfortunately, collection and sorting are also energy-intensive processes. Moreover, under cheap energy conditions, many processes have been developed without recycling as a major objective. No reprocessing can tolerate the presence of components (metallic or otherwise) that the refining steps are not designed to remove and that can produce deleterious properties in the final product. Thus, sorting and separation may have to be extensive. However, as waste disposal becomes more difficult, recycling must be increased for economic reasons as well as for conservation ones. Recycling is discussed further in Chapter 16.

The recovery of a metal from its ore, smelting, has historically been another major source of pollution. Sulfur oxides from sulfide ores have been a major emission, but dusts containing toxic elements such as selenium and heavy metals also have been of concern. Modern smelters can eliminate most of these problems. The details of the metal recovery process depend on the metal and often on the exact composition of the ore used. However, some steps employed are common enough to be discussed in general terms.

12.2.2 Ore Treatment

12.2.2.1 Concentration (Beneficiation)

Unless the ore is very rich, it is typically concentrated; that is, much of the worthless rock is separated physically from the desired phase. In all such processes the ore is first finely ground. In the common flotation method, the ore particles are suspended in water in the presence of flotation agents. The molecules of the flotation agent may attach themselves preferentially to the desired particles rather than to the worthless rock, rendering these particles hydrophobic, hence not wet table by water. They are then carried to the surface in foam when air bubbles are streamed through the water and can be skimmed off. Typical flotation agents used for CuS ores are alkali xanthate salts,

$$R-O-C\begin{array}{c}\diagup S \\ \diagdown S^{-}\end{array} \qquad \text{(R = alkyl group)}$$

which are surfactants analogous to detergents (Chapter 7), with an affinity for the metal sulfide particle. Other concentration methods involve, for example, density differences and magnetic separations.

In an alternate approach, the metals may be extracted in solution from the crushed rock by chemical or bacterial action (Section 12.2.1). A common chemical treatment uses cyanide, usually as NaCN, and is discussed in connection with gold mining in Section 12.2.6, although it is also used in silver mining and in the recovery of some common metals. Both the bacterial and chemical processes produce large pools of solution that are often acidic, with a high metal content. An example of the considerable potential for environmental damage if the containment fails is the release of several million gallons of acidic, metal-rich sludge and water into the Guadiamar River and surrounding areas in Spain in 1998 after a holding dam at a mining site failed. Other releases are referred to in Section 12.2.6. Even exposure of waste rock to air, water, and bacterial action may allow the release of residual heavy metals in soluble forms, which can enter water systems.

12.2.2.2 Roasting

Sulfide and arsenide ores, and some others, are heated with air at high temperature to convert them to oxides. This process may also involve the use of limestone or some other flux[3] that will permit removal of silicate materials, which react with the flux to form a molten slag. The slag can then be separated from the desired material that is insoluble in it. With sulfide ores, SO_2 is a major gaseous product and must be trapped to avoid air pollution problems, as discussed in Section 10.4. Arsenic oxide, As_2O_3, is another common product that volatilizes from the furnace but can be condensed to a solid at room temperature. Particulate materials are also given off to the atmosphere unless trapped or precipitated. Possible water pollution problems come from water used to scrub exhaust gases and to cool slag, and from aqueous treatment of the oxides in some processes.

12.2.2.3 Reduction

The oxides produced as just described, or oxide ores, must be reduced to the elemental metal. Carbon, as metallurgical coke made from low sulfur coal, is a common reducing agent. This reduction is carried out at high temperature, such as in blast furnaces, and offers the same waste control problems already mentioned. Large volumes of very hot gases are involved. Other reduction procedures are in use for some elements; electrochemical reduction of alumina is an obvious example. Some examples of these processes are described shortly.

12.2.2.4 Refining or Purification

Often the metal obtained initially is too impure for industrial uses. Impurities may be other metals or nonmetals such as S or P that can have very deleterious effects on the physical properties of the metal. Refining techniques may involve remelting of the metal and selective oxidation of the undesirable impurities, or electrolytic purification in which the impure metal is dissolved chemically or

---

[3]A flux is a material that will combine with some of the reaction products to form an easily melted phase, called a slag.

electrolytically and plated out in pure form on the cathode of an electrochemical cell.

Many ores are used as sources for several elements, so that actual flow schemes for metal recovery may be elaborate. In addition, unwanted elements are usually present, and if these include toxic materials (Chapter 10), considerable care must be taken in waste disposal practices. Many metal smelters and refineries in the past were noted for their air and water pollution.

12.2.3 Bacterial Mining

The discussion of acid mine drainage (Section 10.6.3) alluded to the ability of bacteria to use metal sulfides in their metabolism, and this capacity is the basis of applications of bacterial attack on sulfide ores to liberate the metals in a useful form. Such processes, which can be referred to as microbial mining,[4] permit recovery of metals from low-grade sources in an economical way. The most common bacterium involved in these processes is *Thiobacillus ferrooxidans*, which occurs naturally in some sulfide minerals. By oxidizing the sulfide to obtain energy, the microbe produces sulfate and releases the metal as soluble metal sulfate. In the copper recovery process, low-grade ore is simply treated with sulfuric acid to encourage growth of the acidophile bacteria, and the dilute copper sulfate solution leaching from the ore pile is collected in a catch basin. The copper can then be recovered by electrolysis and the acidic solution recycled. The normal high-temperature oxidation, with release of gaseous SO_2 and related air pollution or scrubbing problems, are thus avoided. About 25% of current worldwide copper production is based on this process.

Commercial gold mining is also beginning to apply this method to release gold from low grade sulfide mineral deposits so that it can be recovered by cyanide extraction. Even nonmetallic resources are being extracted in this way. A different bacterium is capable of solubilizing phosphate from its ores, thereby avoiding the usual sulfuric acid treatment and the production of large amounts of calcium sulfate by-product.

Bacteria commonly used in these processes have several limitations, such as sensitivity to poisoning by heavy metals and inability to function at the elevated temperatures that the biooxidation process can sometimes produce. Some strains are protected from heavy metals by special enzymes, however, and proper selection or genetic engineering of new strains may avoid this problem. Use of thermophilic bacteria from hot springs and deep sea vents shows promise of permitting higher temperature operation.

---

[4]A. S. Moffat, *Science*, **264**, 778 (1994).

12.2.4 Iron

Because iron is refined in greater amounts than any other metal, some processes used to obtain the metal will be discussed. The oxide (carbonate ores are first heated to decompose them to the oxide) is reduced directly in a blast furnace with coke, limestone being used as a flux. The solids pass down the furnace against a current of heated air that reacts with part of the coke to heat the ore. The liquid iron produced runs to the bottom of the furnace, while the liquid calcium silicate phase (slag) floats above it. The reactions that take place can be summarized as follows:

$$3Fe_2O_3 + CO \rightarrow 2Fe_3O_4 + CO_2 \tag{12-1}$$

$$Fe_3O_4 + 4CO \rightarrow 3Fe + 4CO_2 \tag{12-2}$$

$$Fe_2O_3 + CO \rightarrow 2FeO + CO_2 \tag{12-3}$$

$$FeO + C \rightarrow Fe + CO \tag{12-4}$$

The limestone decomposes and reacts with the silica:

$$CaCO_3 \rightarrow CaO + CO_2 \tag{12-5}$$

$$CaO + SiO_2 \rightarrow CaSiO_3 (slag) \tag{12-6}$$

Reaction (12-6) is oversimplified; the slag is not pure calcium silicate. The limestone also removes sulfides that may be present through the reaction

$$FeS + CaO + C \rightarrow CaS + Fe + CO \tag{12-7}$$

but excess sulfide is undesirable; hence the requirement that the coke be made from low-sulfur coal.

Some silica is reduced to Si, and phosphate to P. The product of the blast furnace, pig iron, ordinarily contains about 2% P, 2.5% Si, more than 0.1% S, 4% C, and 2.5% Mn by weight. The latter element is present in the iron ore and goes through reactions similar to those of Fe. Purification to give a steel of improved physical properties is necessary for most purposes. In particular, P, S, Si, and C must be removed or reduced in amount. In the now obsolete open hearth method, the pig iron with perhaps 40% scrap iron, limestone flux, and some Fe_2O_3 and Fe_3O_4 is melted in a furnace heated by gas and air. The P, C, Si, and S are oxidized, and the silicate and phosphate are removed as a slag by reaction with basic CaO.

Modern procedures use the basic oxygen process, in which the oxidation is carried out by a stream of oxygen gas. The process requires much less time (the order of half an hour compared with 12 h) and less energy than the open hearth process, but it uses less scrap than the latter and has had some depressive effects on recycling of scrap iron and steels.

Special steels, especially those requiring oxidizable components, are made in an electric furnace. This is expensive in terms of energy, but provided the overall composition is suitable for the steel being produced, a high proportion of scrap can be used.

Particulate matter in the form of iron oxides are the main emissions from the above processes. Although a great deal of carbon monoxide is involved [reactions (12-1)–(12-4)], most of it is recycled. Much water is used for cooling and scrubbing, but one of the most significant processes related to water pollution involves removal of oxide scale from the finished steel in a process called pickling. This is achieved in a H_2SO_4 or HCl bath:

$$FeO + 2H^+ \rightarrow Fe^{2+} + H_2O \qquad (12\text{-}8)$$

The waste liquors are highly acidic and contain ferrous salts; discharge to a water system is reminiscent of acid mine drainage (Section 10.6.3). Injection into deep wells is one means of disposal, although this could be dangerous with respect to groundwater contamination.

12.2.5 Aluminum

Aluminum is also smelted extensively, and the isolation of this metal illustrates the chemistry of an active metal for which chemical reduction is not economically favorable because of the need to use powerful, and consequently expensive, reducing agents. The aluminum ore used virtually exclusively is bauxite, essentially an alumina clay, $Al_2O_3 \cdot 2H_2O$, along with other minerals, often including iron oxide. Alumina, Al_2O_3, can be recovered from bauxite by treatment with NaOH solution. Solid impurities, including silica and Fe_2O_3, settle out, and $Al_2O_3 \cdot 3H_2O$ can then be crystallized. Heating produces Al_2O_3. Reduction is performed electrolytically on the Al_2O_3 dissolved in molten cryolite, Na_3AlF_6, in a cell with carbon electrodes; aluminum metal is produced at the cathode, while the anode is consumed by the oxygen that otherwise would be liberated at an inert anode in an electrochemical process. The net reaction is

$$2Al_2O_3 + 3C \rightarrow 4Al + 3CO_2 \qquad (12\text{-}9)$$

This requires less energy than electrolysis with inert electrodes to produce Al and O_2 because of the smaller free energy change involved in the overall process. It is still an extremely energy-intensive process because of the strong Al—O bond energy that must be overcome.

The electrolysis process produces some fluoride and other particulate emissions that must be trapped to avoid air pollution problems, but the most abundant waste products of aluminum mining are the insoluble materials from the NaOH treatment, the so-called red mud tailings, and, of course, any waste from the generation of the electricity employed. However, most of the electricity used for aluminum smelters is generated from water power for economic reasons.

Perfluorocarbons (e.g., CF_4) are produced when the Al content of the electrolysis bath is too low to maintain the reaction, and the voltage rises to cause secondary reactions to take place between the fluoride and the electrode material. It has been estimated that 30,000 metric tons of carbon tetrafluoride and 3000 tons of hexafluoroethane are released annually by worldwide Al metal production.[5] These are inert, nontoxic materials, but they have significant potential as greenhouse gases (Chapter 3). Because they are resistant to even the highly reactive atoms and free radicals of the upper atmosphere, breakdown is very slow (resulting mostly from photolysis by very short wave-length UV radiation). It is estimated that the lifetime of CF_4 in the atmosphere is as long as 50,000 years (Table 5-1).

12.2.6 Gold

The recovery of gold has some particular environmental consequences. Gold is usually found in the form of the free metal, but only rarely in the large nuggets of Gold Rush legend. Rather, the gold is in fine particles, often very intimately mixed with worthless rock. Because of gold's high density, it can be recovered in some cases by physical separation such as placer mining of river gravels, where weathering processes have released the metal from its original matrix. Such processes on a large scale have major physical consequences because the water system is disturbed through dredging and waste gravel deposition. However, the desire to use less tractable sources, and especially low-grade ores that can be profitable because of the high value of the metal, has resulted in reliance on some chemical processes that can have more serious environmental consequences.

One such process involves the dissolution of the metallic gold in mercury metal, in which it has high solubility. The liquid mercury solution (amalgam) is then easily separated from the debris, and the gold can be recovered by distilling the mercury away. This is a simple technology that can be used in relatively primitive mining operations: for example, the gold mines of the Amazon Basin of Brazil. Unfortunately, when carried out with simple technology, the mercury tends to be condensed relatively inefficiently, and there are losses to the environment both as the vapor and the liquid, with consequences that were discussed in Section 10.6.7.

A somewhat more sophisticated chemistry is involved in a second process that involves the use of cyanide salts. Although gold is a very nonreactive element, it does form strong complexes with some soft ligands (Section 9.5). One of these is cyanide. In the presence of a cyanide salt such as KCN, the gold

[5]P. Zurer, *Chem. Eng. News*, p. 16, Aug. 9, 1993.

can be slowly oxidized by air to the soluble $K_2Au(CN)_4$, which can be leached from the rock and concentrated for gold recovery (see Table 12-3). This process employs large volumes of rock and large amounts of cyanide, with considerable risk of release of toxic cyanide to local water systems. This risk has been realized several times. In late 1999, cyanide from a holding pond at a gold extraction operation in Romania was released into the Tisza River when a dike failed. This river carried the contamination into Hungary and into the Danube. Massive kills to fish and other fauna resulted. Six weeks later a dam broke at a similar installation in a tributary upstream of the original release. Pollution involves not only the highly toxic cyanide ion, which is fairly rapidly oxidized to less toxic cyanate, CNO^-, but also heavy metals, in this case especially copper and zinc, which will remain in the river sediments for a much longer time. This spill is just the latest of several similar examples; $3 \times 10^6 \, m^3$ of solution spilled into the Essequibo River in Guyana in 1995, and 2000 kg of KCN spilled into the Barskoon River in Kyrgyzstan in a 1998 mining truck accident. Cyanide wastes have contaminated several areas in the western United States.

12.3 SOILS

Soil, ordinary dirt that we pay little attention to, is a vital substance because it is the basis of much of our food supply. Unfortunately, human activity frequently causes degradation of soil in ways that range from loss of nutrients to changes in the physical character of the soil to contamination with toxic materials to loss of the soil itself. Many agriculturally productive soils are being washed or blown away through erosion as natural vegetation is destroyed. Others in dry areas are being converted to deserts through overgrazing and other activities that exceed the soils' carrying capacity. Much of the tropical land converted to agricultural use through forest clearing is abandoned after a few years. Soil degradation is of serious concern in terms of its impact on future food needs.[6]

12.3.1 Soil Formation and Composition

Surface rocks are subject to breakup and chemical change by a variety of processes called weathering. Some of these processes are listed briefly.

Physical disintegration of rocks caused by temperature effects, frost, wind, water, and ice erosion. Some of these factors are also responsible for transporting the weathering products away from their original sources. The small particles produced have a large surface area that facilitates other weathering processes.

[6]G. C. Daly, *Science*, **269**, 350 (1995).

Chemical reactions with water (e.g., hydration or hydrolysis). Hydration is illustrated by reaction (12-10); the change in crystal structure as water molecules are gained or lost will result in disintegration of the material. Hydrolysis involves a more extensive change in the material as shown, for example, in reaction (12-11). Reactions of this sort can liberate cations from the rocks [e.g., K^+ in (12-11)].

$$2Fe_2O_{3_{\text{hematite}}} + 3H_2O \rightarrow 2Fe_2O_3 \cdot 3H_2O_{\text{limonite}} \tag{12-10}$$

$$KAlSi_3O_{8_{\text{microcline}}} + H_2O \rightarrow HAlSi_3O_8 + K^+ + OH^- \tag{12-11}$$

Attack by acid, especially carbonic acid. Reactions (12-12) and (12-13) are examples of the weathering interactions of acid with minerals. Reaction (12-11) could also have been written as an example of acid attack.

$$CaCO_{3_{\text{insoluble}}} + H_2CO_3 \rightarrow Ca(HCO_3)_{2_{\text{soluble}}} \tag{12-12}$$

$$CaAl_2Si_2O_{8_{\text{anorthite}}} + 2H^+ + H_2O \rightarrow Al_2Si_2O_5(OH)_{4_{\text{kaolinite}}} + Ca^{2+} \tag{12-13}$$

Oxidation reactions involving substances in low oxidation states. Reaction (12-14) illustrates a hydrolysis reaction accompanied by oxidation of Fe(II):

$$12MgFeSiO_{4_{\text{olivine}}} + 8H_2O + 3O_2 \rightarrow$$
$$4Mg_3Si_2O_5(OH)_{4_{\text{serpentine}}} + 4SiO_2 + 6Fe_2O_3 \tag{12-14}$$

Biological effects. These may take the form of physical disruption by plant roots, chemical effects from substances excreted from roots, and bacterial action. For example, the reactions in the nitrogen cycle and the organic materials produced by biological action can determine the soil pH and consequently solubilities and leaching on a local level.

The weathering products that result from these various processes depend on the composition of the initial rock and the conditions. Rocks with high silica contents, that is, with more Si—O bonds, tend to have the greatest chemical stability simply because of the large Si—O bond energy, and igneous rocks (those produced from molten material), formed under the most extreme conditions of temperature and pressure, are most reactive to weathering decomposition. Igneous silicates high in magnesium and iron (e.g., olivines) produce soluble Mg^{2+} and SiO_4^{4-} and insoluble FeO(OH) under the action of H_2O and CO_2, while feldspars with magnesium, calcium, and aluminum contents give soluble Na^+, Ca^{2+}, and SiO_4^{4-}, and kaolinite as the insoluble products. However, at nearly neutral conditions (pH 5–8), further dissolution of the silicate component can take place, leading to a residue of $Al_2O_3 \cdot 3H_2O$, while under more acidic conditions, solid SiO_2 is left with soluble Al^{3+} species. This last process is of concern as a result of acid rain in view of the toxicity of Al^{3+} toward plants (Section 10.6.9).

Weathering patterns vary with climate zones. For example, in wet temperate forest regions, decay of organic material produces acidic leaching conditions, so that the upper levels are depleted in FeO(OH) and some alumina, while material with a high silica content is left behind. In tropical rain forest conditions, where temperature and moisture conditions lead to faster decay, conditions are less acidic, and leaching removes more of the silica components, leaving FeO(OH) and alumina clays. This is shown schematically in Figure 12-3.

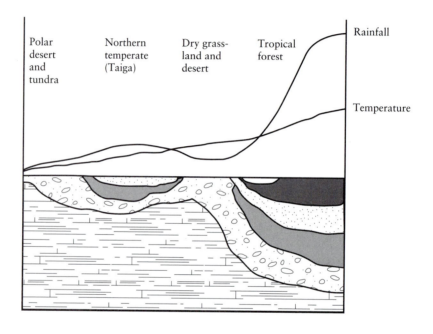

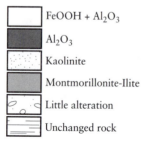

FeOOH + Al_2O_3

Al_2O_3

Kaolinite

Montmorillonite-Ilite

Little alteration

Unchanged rock

FIGURE 12-3 Weathering products from polar (left) to equatorial (right) regions, showing dependence on temperature and rainfall. Based on P. W. Birkeland, *Soils and Geomorphology*, Oxford University Press, New York, 1984.

The eventual result of the breakup and composition changes of weathering reactions is the formation of soil; this, of course, is the fraction of the earth's surface utilized for plant growth. Soil may be regarded as the upper layer of the regolith, which is the weathered and broken-up material overlying bedrock. Soil is the most heavily weathered portion of the regolith and is normally significantly influenced by plant materials. Soils play an important role in environmental chemistry besides providing a site for plant growth. Much of this role depends on the microorganisms that are important soil components. The absorption of CO has already been referred to, as have the carbonate equilibria (Section 9.2.2), which in soils are strongly dependent on CO_2 generated from bacterial degradation of organic materials. Other processes include rapid SO_2 absorption; soils have been shown to be effective in removing SO_2 from the air. Buffering of pH, and of metal ion concentrations, are also important aspects of soil chemistry.

A typical soil contains stones and gravel, coarse and medium-sized material such as sand and silt composed of silica and other primary minerals, and finer weathered material classed as clays. Most soils contain a few percent organic components such as humic substances (Section 9.5.7) and undecomposed plant materials, although a few organic soils (peats, mucks) may contain up to 95% organic matter. Typical soils exhibit several layers (horizons) of somewhat different composition that are important in soil classification. These horizons are the result of biological decay, weathering, and leaching processes. An idealized version is shown in Figure 12-4.

Soils play a significant role in the carbon cycle (Section 10.2), with about 60 billion tons of CO_2 released annually from soils. About the same amount is bound into land plants. Much of the CO_2 release comes from decomposing plant litter in the upper layer, but about two-thirds is released at deeper layers, from respiration in roots, fungi, and soil microorganisms. Carbon dioxide partial pressure in soils can be high. Factors that enhance microbial activity, which converts organic matter to CO_2, influence the amount of organic matter a particular soil will retain. Thus, temperate regions with cold winters allow a significant buildup of organic constituents, while tropical regions with year-round warmth promote microbial decay and often produce soils with little organic material below the upper layer of debris.

The size and the nature of the fine particles in the soil are the most important determinants of its chemical and physical properties, although the organic constituents can have a major influence as well. The voids or pores between the particles are important reservoirs for soil gases and soil moisture; large diameter pores allow water to drain, while small diameter pores retain water. Some clays swell when wet, as water molecules enter the crystal lattice. As a consequence of shrinkage on drying, such soils tend to crack. At the same time, the soil may cake and become impervious to moisture. Cohesion of the clay particles, which is also dependent on moisture and sodium ion content,

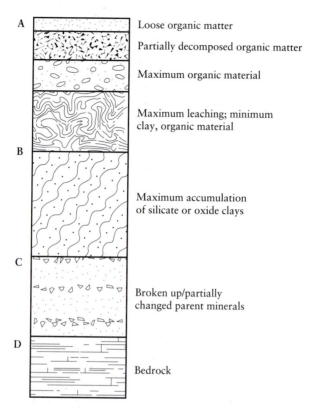

A — Loose organic matter

Partially decomposed organic matter

Maximum organic material

Maximum leaching; minimum clay, organic material

B

Maximum accumulation of silicate or oxide clays

C

Broken up/partially changed parent minerals

D

Bedrock

FIGURE 12-4 An idealized soil profile: A, B, C, and D are typical horizons (sometimes designated by other letters); they may be further subdivided, A_1, A_2, etc. An initial "O" horizon is often indicated.

determines the stickiness of the soil and its agricultural workability. The clays also are largely responsible for the very important ion exchange properties of soils, although humic materials contribute significantly.

Typical clays are aluminosilicates, present largely as colloidal particles; that is, the size is less than a micrometer. The particles are composed of layers or sheets somewhat analogous to the structure of graphite. The surface area of such small particles is very large and, in addition, internal area between the layers may be available to bind other materials. Thus, surface properties and adsorption are quite important. A variety of clays is recognized; two that are common components of soils are kaolinite and montmorillonite, the formulas for which are (idealized and simplified) $Al_2Si_2O_5(OH)_4$ and $NaMgAl_5Si_{12}O_{30}(OH)_6 \cdot nH_2O$, respectively.

The structure of aluminosilicate clays is based on tetrahedral SiO_4 units, each of which shares three of its oxygens with other tetrahedra to produce a planar sheet. If this layer is taken as infinite, the stoichiometry is $Si_2O_5^{2-}$. Associated with this is a layer of octahedral AlO_6 units that share the apical oxygens from the SiO_4 units in the tetrahedral layer. The octahedral unit can be written $AlO_n(OH)_m$, where n oxygens are shared with Si, and m OH^- groups bridge the aluminum atoms. Kaolinite (Figure 12-5) has an equal number of Si and Al atoms, and the combined layers are neutral. This combined layer is bound to others in the crystal by hydrogen bonds. In pyrophyllite, $AlSi_2O_5(OH)$, a second layer of SiO_4 tetrahedra is bound by sharing of the corner oxygens to the other side of the octahedral aluminum layer, giving a sandwich-type structure. These layers are held to others in the solid by van der Waals attractions. In both types of clay the layers are comparatively easily separated.

If some of the trivalent aluminum atoms are missing or replaced by bivalent ions such as Mg^{2+} or Fe^{2+}, the charge balance is no longer maintained. The layers carry a net negative charge, which is balanced by other cations between the layers. These additional cations also serve to hold the layers together by Coulombic attraction. In some layered structures, the interlayer cations, which may be Na^+, K^+, or Ca^{2+}, produce a fairly strong attraction between the layers, although they can still be separated readily. This class of material is known as mica; an example is muscovite, $NaAl_3Si_3O_{10}(OH)_2$. In other structures with lower charges on the layers, the interlayer cations hold the layers together more weakly; the ions themselves may retain their hydration sheaths, and there are water molecules between the layers. This is the case for montmorillonite (Figure 12-6). Because the water molecules between the layers can be removed or added easily, such clays show marked changes in volume upon wetting or drying. These clays are referred to generally as smectites.

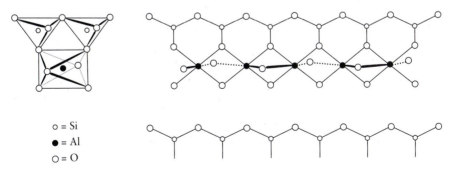

o = Si

● = Al

o = O

FIGURE 12-5 The structure of kaolinite. This is based on a two-dimensional layered structure of SiO_4 units (Figure 12-1g) perpendicular to the page; only three oxygens are shown around each Si for clarity. The geometric arrangement is shown schematically on the left.

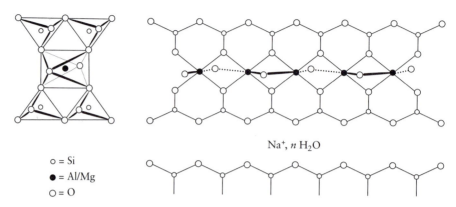

Na$^+$, n H$_2$O

o = Si
● = Al/Mg
o = O

FIGURE 12-6 The structure of montmorillonite. The silicate sheet structure similar to Figure 12-1g is perpendicular to the page; for clarity only three oxygens are shown around each Si. The geometric arrangement is shown schematically on the left.

Some soils, particularly in the tropics, contain colloidal metal hydroxides such as Al(OH)$_3$ and FeO(OH) that are also referred to as clays and have similar properties. The structure of the aluminum hydroxide particles is based on Al^{3+} ions occupying the octahedral holes between two close-packed layers of OH$^-$ ions, similar to the aluminum layer in the silicate clays described earlier. The layers are held together by van der Waals forces.

Clays are capable of entering into ion exchange reactions in two ways. First, H$^+$ ions can be lost from the OH groups, and replaced by other cations (which may later be exchanged for still other cations). This activity is pH dependent. Second, interlayer cations may be exchanged. Ion exchange is discussed in more detail in Section 9.7.

The colloidal-sized clay particles can pack together to give a low-porosity soil that has low permeability; that is, water has considerable difficulty flowing through it. A layer of clay can be quite effective in preventing the passage of groundwater, with the possibility of protecting a deep aquifer from contamination by polluted surface water, but also preventing its recharge. Impermeable clay layers are an important feature of a well-designed landfill (see Section 16.2), being used both as a cap to keep water out and as a liner to restrict contaminated leach water, which may enter the fill through imperfections in the cap, from seeping into the groundwater.

Hydrated metal oxide clays of Fe and Al that dominate some tropical soils and are also present in colloidal form in other soils have ion exchange properties similar to, but weaker than, those of the aluminosilicates. They tend not to be as sticky and adherent as the silicates, and soils with high proportions of hydrous oxide clays can be cultivated under moisture conditions that make siliceous clays unworkable. They can become concretelike when dry, however.

Every soil is assigned to one of a variety of classes. Although the details of soil classification are beyond the scope of this book, Table 12-4 gives some of their characteristics.

TABLE 12-4

Soil Types[a]

Name	Description
Entisols	No horizon formation, little organic matter; includes sandy areas, thin soils on rock, recently produced soil deposits. Mostly used as grazing land in dry areas; can be productive in humid areas. Have been called lithosols and regosols.
Vertisols	High content of swelling clays; crack extensively on drying. Large areas in India, Australia, Sudan; very difficult to farm. Have been called grumusols.
Inceptisols	Young soils without much weathering.
Aridsols	Found in arid regions, they have had little leaching. Upper layers tend to be neutral or basic; lower levels contain calcium carbonate and often soluble salts. Many deserts are of this type; with irrigation, they can be very productive.
Spodosols	Temperate region soils from humid, forested areas; they are highly weathered and leached. Found in large areas in northeastern North America, northern Europe, and Siberia. Productive if fertilized. Many have been called podzols.
Alfisols	Less weathered than spodosols, but also temperate, humid region soils formed under less acidic conditions because of different vegetation, especially deciduous forests. They are widely distributed worldwide and are generally excellent agricultural soils.
Mollisols	High nutrient and organic content, generally formed with grassland vegetation in temperate regions with moderate rainfall. Major areas are the Great Plains of North America, eastern Europe to the Urals, northern China, and South America. Very fertile.
Oxisols	Highly weathered and leached, with few nutrients; found in tropical forests, especially in South America and Africa. They have been called latosols and laterites.
Histosols	Very high organic content, water-saturated soils, found, for example, in bogs. When drained, they can be very productive.

[a]Soils are classified into a large number of types and subtypes. Classification systems have been based on how soils were formed (or thought to be formed) or on soil properties: texture, color, composition, moisture, and others. The presence or absence of certain horizons (layers) in the soil profile are important aspects of the classification scheme. This table gives the main soil types and some of the older equivalents.

Acid leaching, also called podzolization, one of the most important natural processes occurring in soils that determines their properties and agricultural characteristics. The process occurs most markedly in soils of moist, forested, temperate, and cold regions, such as the northeastern United States. The top layer of such soils consists of organic materials containing relatively few metal ions. As these materials decompose, acids are formed that are washed into the levels below. As a consequence, the soil beneath the organic layer is leached of most minerals except those with a high silica content. Aluminum and iron oxides and organic matter are precipitated in the lower layers, as soluble material and noncoagulated colloids are washed down. Such soils contain few nutrients except in the upper organic layers, the nutrients having been leached out of the levels beneath the surface. These soils are suitable for deep-rooted plants such as trees, but used agriculturally, they soon lose fertility. Besides true spodosols, many soils exhibit partial leaching (partial podzolization) when they have less organic matter on the surface and therefore, there is less acid leaching.

In wet, tropical forests, oxidation of the organic material is more rapid, and leaching is less acidic than in temperate regions, but very extensive. Silica and montmorillonite types of clay are removed under these less acid conditions, but iron and aluminum oxides remain, giving the soil a red or yellow color. Often deposits of these oxides form near the water table. Upon drying, these may form very hard materials that cannot be cultivated. Such levels are called laterites; the soils are called latosols or oxisols. They are common in much of South America and Africa. Modern farming techniques on such soils are effective only with extensive application of costly fertilizers. Because of this, much tropical land cleared for agriculture becomes converted to pasture after a few years.

Both spodosols and oxisols must receive large amounts of fertilization if they are to be useful for large-scale agriculture. With the latosols in particular, modern agricultural methods must be applied with great care to ensure that erosion does not expose the hydrous oxide layer, which could dry out, leading to irreversible loss of agricultural use of the soil. Indeed, primitive agricultural practices on many tropical forest lands, while not highly productive overall, display an excellent adaptation to the chemical nature of the soils. These methods have been the "slash and burn" type. Small fields are cleared of vegetation by burning. This converts the nutrient material in the growing vegetation to inorganic salts that provide a high level of fertilization for crops for a few growing seasons. When these nutrients are used up, the field is allowed to return to forestation for a few years, until the process can be repeated.

Soils with less abundant vegetation and less rainfall (called aridsols) have correspondingly less acidic leaching, and in fact may be neutral or basic. Calcium carbonate occurs at modest depths or even near the surface (calcareous soils), there may be soluble salts present, and the nutrient content is often very high. Some of these soils can be highly productive if irrigated.

Slightly acid to neutral soils are favored by many cultivated crops, and although some plants prefer strongly acid soils, it is often necessary to increase the pH of agricultural, garden, and lawn soils by the addition of lime. Agricultural limes are impure forms of calcium oxide, calcium hydroxide, or calcium carbonate, usually with considerable Mg present. Limestone and dolomite are common sources. On the other hand, if necessary, the acidity of the soil can be increased by the addition of sulfur, which is oxidized to H_2SO_4 by soil organisms. Many fertilizers (Chapter 10) increase soil acidity also, for example by reactions such as

$$NH_4^+ + 2O_2 \rightarrow 2H^+ + NO_3^- + H_2O \qquad (12\text{-}15)$$

The need for fertilization of soils has already been referred to, and several of the major fertilizers that are needed were discussed in Section 10.5. Phosphorus, nitrogen, and potassium are considered macronutrients, while many other elements are necessary micronutrients. Of the macronutrients, phosphate may be present in the soil as insoluble phosphate salts that form as a result of the reaction between the hydrogen phosphate ions with metal ions (chiefly, Fe, Al, Ca, either free or held in the clays) or with hydrous oxides. Reversible ion exchange reactions also may retain phosphate, for example, as in reaction (12-16):

$$Al(OH)_3 + H_2PO_4^- \rightarrow Al(OH)_2H_2PO_4 + OH^- \qquad (12\text{-}16)$$

Phosphate held in this way is more readily available than that of the insoluble phosphate salts.

Nitrogen may be present as nitrate, which is held only weakly by anion exchange, or as the ammonium ion, which can be held by cation exchange with the clays. The potassium ion is nearly the same size as the NH_4^+ ion, and is held in the same way. Weakly held nutrients are easily leached away, and much of the nitrogen applied as fertilizers can end up in lakes and streams rather than on the field. Micronutrients, especially metal ions, may be available in clays or other insoluble mineral forms. In some instances one or more such nutrients, although present, may be unavailable to plants because they are in insoluble forms, and some metal ion deficiencies have been relieved by treating the soil with complexing agents of the EDTA and NTA types that can solubilize these materials as complexes (Chapter 9). Some input of micronutrients such as chloride and sulfate comes from rainwater.

12.3.2 Contaminated Soils

Contamination of soils is an important concern. Contamination by pollutants can come from water leaching from landfills and waste dumps, spills and residues from industrial operations of many kinds, leaking storage tanks, and so forth. A contaminant may be mobile through solution in soil water, as a water-insoluble liquid, or to a lesser degree through volatilization in the soil. Contaminants in the form of dusts may also be spread by wind, but these remain on the surface unless other mechanisms are involved.

Water-soluble contaminants, which can be either inorganic or organic, will flow with the water in which they are dissolved, but generally not at the same rate. Adsorption to soil components will slow the motion of dissolved materials to an extent determined by the partitioning of the substance between the water phase and the solid. If a soluble pollutant is injected into the soil at a point source, as the adsorption capacity of the soil is exceeded the pollutant will follow the flow of groundwater (advection) but will also spread laterally through processes of dispersion. These can involve diffusion, but are largely the result of flow inhomogeneities. The concentration of the pollutant will decrease with distance from the source, as a result of adsorption and spreading. Because some of the contaminant will inevitably be left behind and will be released only very slowly, the contaminated region always extends to the source, even for a single spill event that seems to "sink in." Some of this may be taken up by later rainwater to provide a long-term input of pollutant into the groundwater.

A water-insoluble liquid, on the other hand, will flow as a separate phase. However, there will always be some solubility, even of organic liquids generally considered insoluble. The undissolved material will behave differently depending on whether it is heavier or lighter than water. In either event, a liquid spill will descend (and spread) until it reaches the water-saturated zone (Figure 12-7). The lighter liquid will "float" on the water layer—that is, it will occupy a zone above the saturated region. The heavier liquid, on the other hand, will pass through the water level until it reaches the impervious bed on which the water rests.

The details of this process are by no means simple. Globules of liquid may also remain in the water-saturated region. Through these globules and by pooling below the water level for the heavier liquid, or by floating on the water surface for a light one, a continuous reservoir is provided to contaminate the water through solubility. The heavier liquid may also enter cracks, pores, or capillaries in the soil or rock and remain strongly held. This too may be released slowly into the water phase through solution. A lighter liquid may also have some ability to diffuse as a vapor into soil above it.

It is frequently desirable to treat polluted soil to prevent the pollutants from spreading or to recover a site for safe use.[7] Remediation approaches may be removal of the pollutant, or its stabilization in a form that is immobile and

[7]John K. Borchardt, 'Dealing with soil contaminants', *Today's Chem. Work*, **4(3)**, 47 (March, 1995).

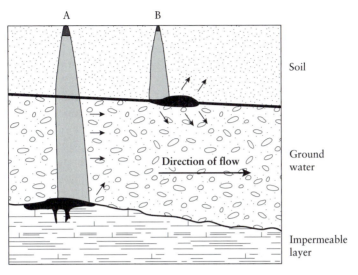

FIGURE 12-7 Schematic illustration of the behavior of spills of heavy (A) and light (B) insoluble liquids.

inert. One approach to soils in which the pollution is in the soil water is to drill wells to intercept the water and pump it out for treatment. This practice can be effective in reducing the water concentration of the pollutant while pumping is continued, but on cessation, water levels of the contaminant usually rise again as the more strongly held material is continually and slowly released. At any rate, this is a long-term treatment procedure. Injection of detergent solutions to help release organic materials may provide more efficient removal. Volatile contaminants may be removed by vapor stripping. Air is drawn from holes bored in the soil to carry vapors out where they are trapped. This can be enhanced by injecting hot air or steam. Such treatments are unlikely to remove all the contaminating materials, but they can reduce the concentration significantly, perhaps to safe levels, and can minimize the area of spread by reducing the concentration gradient that drives diffusion. (The remaining material will continue to spread in all directions by diffusion, subject to holdup by adsorption.)

Stabilization consists of using a chemical reaction to incorporate the contaminated soil into an inert solid from which the contaminant will not be released at a significant rate through leaching or other natural processes. As part of this, the contaminant may undergo a reaction, or it may simply be physically encapsulated, depending on its nature. This can be carried out directly on the soil *in situ* if contamination is close to the surface, or excavation may be used. The approach is also useful in dealing with sludges from stack gas scrubbing or other industrial processes to make landfilling safer. The soil (or sludge) is mixed with materials that react to solidify to a nonreactive solid. Most used are portland cement, alone or mixed with fly ash or silicates, lime–fly ash, and cement or lime kiln dust. Another solidification/stabilization technique that has been proposed is *in situ* vitrification, in which electrodes are inserted into the soil and a high current passed between them. Ohmic heating causes the soil to melt, pyrolizing some contaminants and trapping others in the glassy residue that forms on cooling.

Metals whose compounds are environmentally benign have been used with excellent results to reduce both organic and inorganic contaminants.[8] Iron is the usual choice, but zinc and tin are alternative possibilities. In this method, a porous barrier of metal filings is put in place to intercept the groundwater flow. Many organochlorine compounds undergo reaction:

$$Fe^\circ + RCl + H^+ \rightarrow Fe^{2+} + RH + Cl^- \qquad (12\text{-}17)$$

The result is a much less toxic product. Other inorganics such as dyes and pesticides can also be reacted in a similar manner. Inorganic materials such as $Cr(VI)$ can be reduced to less dangerous, immobile forms. The significant installation cost is offset by essentially zero operating costs. Metal reduction can also be used in aboveground treatments.

Excavated soils can also be cleaned by washing with appropriate reagents, or organic contaminants can be burned off by heating.

Bacterial degradation of organic pollutants is receiving more and more interest as a remedial process that can take place *in situ* as well as in excavated soil.[9] The maintenance of optimum conditions for bacterial growth, including nutrient supply, can be a problem. Bacteria can also be helpful in metal-contaminated soil by converting one chemical species to a less dangerous one, or concentrating the metal in the biological mass for recovery.

Another biological process, phytoremediation,[10] is applicable to metal ions in particular; this process uses plants that accumulate metal ions from soils in which they grow. Although plants generally take up metal ions from soils, some, called hyperaccumulators, can take up unusually large amounts of particular elements. A variety of metal ions, including copper, nickel, and lead, and some nonmetals such as selenium, can be taken up, in some cases by as much as 5% by weight of dry plant material. The metals can be concentrated in the ashes on incineration for recovery or safe disposal. This process appears to be applicable to shallow contamination (perhaps up to a meter) that can be reached by the plant roots. Development has just begun.

[8] E. K. Wilson, *Chem. Eng. News*, p. 19, July 3, 1995.
[9] D. R. Lovley, *Science*, **293**, 1444 (2001).
[10] A. S. Moffat, *Science*, **268**, 303 (1995); A. M. Rouhi, *Chem. Eng. News*, p. 21, Jan. 13, 1997.

Soil remediation is a rapidly growing field, with many approaches available. Clearly, the best will depend on the specific problem to be solved. As more experience is gained with these remediation methods, efficiency and cost-effectiveness should improve considerably.

12.4 NONMETALLIC MATERIALS FROM THE EARTH

Many nonmetallic inorganic materials are obtained from the rocks and minerals of the earth. Some of these are employed with little treatment, while others undergo considerable chemical modification before use. The environmental problems associated with these substances are chiefly those associated with mining or quarrying and crushing of large volumes of rock or clay: that is, with destruction of landscape and emissions of particulate materials into water and the air. The common minerals involved are not considered to be toxic, but the physical form in which the particulates are emitted can render them harmful to health. In the remainder of this chapter, we shall consider briefly a few aspects of some of the more important examples of substances of this class.

12.4.1 Stone

From the beginning of civilization, stone has been used as a material of construction. Marble, limestone (carbonates), sandstone, and granite (silicates) are examples of widely used building stones. Like other rock, the stone used for construction is subject to weathering processes and particularly to attack by acid constituents of urban airs. Sulfur and nitrogen oxides, which they form strong acids upon hydrolysis, are the primary acid components involved. Attack is particularly severe on carbonate stones. It is well known that many stone buildings and carvings that have existed for centuries have undergone rapid and extensive deterioration in modern times as air pollution from industrial and urban sources has increased and the acidity of the precipitation has led to faster deterioration. Attack can involve the acid rain itself, but dry deposition involving dissolution of gaseous pollutants in the film of moisture, which can condense on the surface of the stone may also do a great deal of damage.

While much of the weathering attack on building stone is due to acid attack from H_2SO_4 arising from SO_2 emissions and to carbonic acid, water itself can be destructive if it enters the pores in the surface layers of the stone and causes the surface layers to flake off through stresses imposed in freezing and thawing. The most serious cause of surface disintegration of this sort, however, is the strain imposed just beneath the surface by recrystallization of salts as water enters and evaporates from the pores. Any source of a salt having significant water solubility (e.g., sea spray) can contribute to this, but in general calcium and magnesium sulfates formed from the acid attack of industrial air [reaction (12–18)] are probably most important.

$$CaCO_3 + H_2SO_4 \rightarrow CaSO_4 + H_2O + CO_2 \qquad (12\text{-}18)$$

That is, air pollution does not just dissolve the stone from the surface, but can also cause layers a few millimeters thick to disintegrate. In many places, attempts are under way to protect stone monuments and historical buildings from deterioration from these causes.[11]

Protective measures were undertaken as long ago as the first century B.C. when workers rubbed wax on the warmed stone. Various waxes, resins, silicones, and other materials have been applied more recently. Such treatments rarely achieve deep enough penetration to be effective; water can get behind a thin film and eventually cause the treated layer to flake off. Some use is being made of polymers in processes where solutions of monomers of good penetrating power are applied to the stone, followed by *in situ* polymerization. Acrylics, epoxies, and alkoxysilanes are among the materials used. Another treatment for carbonate stones involves the deposition of $BaSO_4$ in outer layers of the stone. Much less soluble than $CaSO_4$ or $MgSO_4$, this material forms solid solutions with $CaSO_4$. Consequently, the strains introduced through the recrystallization process are much reduced.

12.4.2 Clay-Based Materials

Clays are among the oldest technological materials, used for bricks and ceramics from prehistoric times. Bricks for building purposes have been prepared simply by drying clays or muds. The hydrous oxide layers of lateritic soils are one example of soil that provides a good building material by simple drying, although generally aluminosilicate clays will produce a stronger and more permanent substance by firing. Brick, tile, and pottery are all made in this way. A nonswelling clay is required that can be formed into the desired shape as a paste with water and does not deform significantly on drying. The firing process causes changes in the chemical nature of the aluminosilicate and permits the individual particles of the clay to fuse together to produce a hard, strong structure. This is essentially irreversible; water will not reconvert the fired material to soft clay, as it will after simple drying. The fired clay is porous, but a nonporous surface can be achieved by a glaze of material that is melted to a glass.

---

[11]C. Wu, Consolidating the stone: New compounds may protect statues and buildings from weather damage, *Sci. News*, **151**, 56 (1997).

12.4.3 Glass

Glasses are rigid, noncrystalline materials that are often called supercooled liquids, but a glass is in fact a separate physical state. It is like a liquid in having a nonrepeating structural pattern, but it is a rigid structure without molecular mobility. In spite of the widely held belief, long standing does not cause glass to flow, and if old window panes are thicker at the bottom, it is because old glassmaking techniques produced glass that was not of uniform thickness and the panes were installed that way. There is a glass transition temperature, above which the material transforms into a true supercooled liquid of high viscosity.

Common glass (soda glass) is a silicate, but it does not correspond to a particular compound. Silica, SiO_2, is a three-dimensional network based on Si—O—Si linkages. If the atoms are arranged in a regularly repeating pattern, a crystalline form of quartz is produced, but they may be arranged irregularly, forming a glass. Addition of an oxide of a metal with a small tendency to form a covalent bond to oxygen (e.g., Na_2O) to SiO_2 results in breakage of the Si–O–Si links:

$$Na_2O + - Si-O-Si- \rightarrow - Si-O^- + - O-Si- + 2Na^+ \quad (12\text{-}19)$$

As a result, the properties of the network structure are changed and the softening temperature is lowered. These properties can be controlled by appropriate selection of the composition of the glass. Addition of B_2O_3 will result in Si—O—B linkages, since boron also has a strong affinity for oxygen. This is the basis of the borosilicate glasses of which Pyrex[12] is an example; they have higher softening temperatures and smaller coefficients of thermal expansion than the soda glasses. Lead oxide produces a high index of refraction and is used in "crystal" glassware. As discussed in Chapter 10, some of this lead can be leached out of the glass with use.

Common glass is made from SiO_2, Na_2O, and some CaO, which are fused at high temperatures. The source of the Na_2O and CaO may be the carbonates or other salts that decompose to oxide upon heating; SiO_2 is obtained as sand. There are restrictions on the purity of the starting materials, since many metal ions and colloidal metals, sulfides, and selenides produce colors in the glass. Indeed, this is how colored glasses are made. Scrap glass can easily be used in the melt (Section 16.17.1), but the need to sort the glass from contaminants and to separate the colored glasses causes problems in recycling. To produce a product with acceptable specifications, the starting materials must conform to appropriate compositions. For economic reasons, much recycled glass is used in products that do not depend very much on the glass properties, such as fillers in asphalt ("glasphalt").

[12]Trademark of the Corning Glass Company, Corning, New York.

12.4.4 Cements

Use of brick and stone for construction requires a cement or mortar to bind them together. A cement is the binding material, in this context a powder activated by water. If sand is added as a filler, the mixture is called a mortar. Mortars are used in bonding bricks, and other building units. Mixtures with coarse materials such as gravel or crushed rock are called concrete. Use of cements goes back to antiquity, with various substances having been used.

The earliest cements were based on gypsum, $CaSO_4 \cdot 2H_2O$, which can be dehydrated to plaster of paris at temperatures near $200°C$:

$$CaSO_4 \cdot 2H_2O \rightarrow 3/2H_2O + CaSO_4 \cdot 1/2H_2O \qquad (12\text{-}20)$$

Upon mixing with water, the reaction can be reversed, and the material will set to a hard mass. Plaster of paris has some uses even today. It evidently was used with sand as a mortar for masonry by the ancient Egyptians, but gypsum is slightly water soluble and unsuited to moist climates.

Lime mortars were also discovered quite early in the development of civilization. These are formed by heating (calcining) calcium carbonate, and slaking the quicklime (calcium oxide) produced with water to give slaked lime, calcium hydroxide:

$$CaCO_3 \rightarrow CO_2 + CaO \xrightarrow{\;H_2O\;} Ca(OH)_2 \qquad (12\text{-}21)$$

A paste of slaked lime, sand, and water can be used as a plaster on walls or as a mortar; it hardens by chemical reaction with atmospheric CO_2:

$$Ca(OH)_2 + CO_2 \rightarrow CaCO_3 + H_2O \qquad (12\text{-}22)$$

Lime mortars have two disadvantages. First, they depend on diffusion of CO_2 from the air into the interior of the mass, and consequently they cure slowly and not always completely. Second, they will not harden if contact with water occurs.

Hydraulic cements, those that will harden in the presence of water, first were made by mixing volcanic ash with lime. When mixed with water, these so-called lime–pozzolana cements, harden by a chemical reaction that produces calcium silicates and aluminosilicates from the calcium hydroxide, silica, and alumina. The last two are provided by the volcanic ash. Limestone containing considerable clay also produces hydraulic lime upon calcining; other similar reactions can take place. These materials were the forerunners of portland cement, the material now in general use for construction purposes.

Portland cement is chiefly composed of calcium silicates and calcium aluminosilicates; the primary compounds in it have the empirical formulas $3CaO \cdot SiO_2$, $2CaO \cdot SiO_2$, $3CaO \cdot Al_2O_3$, and $4CaO \cdot Fe_2O_3 \cdot Al_2O_3$. The CaO is ordinarily obtained from limestone and the SiO_2, Al_2O_3, and Fe_2O_3 from clay or shale, although blast furnace slag, fuel ash from power plants, and industrial calcium carbonate wastes are sometimes employed. The starting materials are ground and heated to incipient fusion. All water and CO_2 is thereby removed, and chemical reactions to form the silicate and aluminate compounds take place. After cooling, the solid residue, called clinker, is finely ground with gypsum. When water is mixed with the cement, complex reactions take place which involve hydration of the compounds present and the formation of new compounds. It is these chemical hydration reactions that are responsible for the hardening of the cement. These curing reactions are not completed for weeks, although they are fast enough to cause the paste to become solid in a matter of hours.

Cement is produced in kilns that are sloping, rotating cylinders, typically several hundred feet long, heated to 1500–2000°C by burning coal or oil at the lower end while the raw materials are added at the other. Normal operation releases carbon dioxide, nitrogen oxides, sulfur dioxide, and particulates. Trapping and disposal of the latter is in itself a problem. Because at the temperatures and long residence times involved, these kilns offer an attractive environment for the combustion of hazardous organic wastes, they are widely used for this; complete destruction of organic molecules is expected to take place. Combustion by-products such as heavy metals and chlorides, for example, are incorporated into the clinker and immobilized. This process has considerable economic advantage for the kiln operator in terms of fuel costs, but there is opposition to such use because of fear of incomplete combustion and possible leaching of toxic heavy metals from the cement, although leaching tests have indicated that the product is safe. Some of the concern stems from the regulations that govern the burning of hazardous waste in cement kilns, which are considerably less stringent than those imposed on regular hazardous waste incinerators (Section 16.5).

12.4.5 Asbestos

Several different aluminosilicate minerals are called asbestos (Table 12-5). The term refers to a crystal form or habit that many minerals can exhibit, but only three make up the deposits having important use as commercial asbestos materials. The characteristic of this crystal habit is its fibrous form. The crystals grow in a fine, whiskerlike structure with considerable strength and flexibility. All are based on shared SiO_4 tetrahedra, but some (amphiboles, Figure 12-8) are based on double chains of linked tetrahedra with the double chains linked to each other by interactions with cations. Others (chrysotiles) are based on double layers of SiO_4 units similar to the structure of kaolinite (Figure 12-5) but with Mg instead of Al. However, the layers are not flat but rolled into cylinders. Again, other ions link the units together. In the bulk, both

TABLE 12-5

Types of Asbestos

Name	Properties and comments	Formula
Chrysotile (white asbestos)	A serpentine; most widely used asbestos ($> 90\%$); strong but friable flexible white fibers, less acid resistant than other types	$Mg_3Si_2O_5(OH)_4$
Crocidolite (blue asbestos)	An amphibole; some commercial use ($<5\%$); long but thicker and more brittle fibers, with high chemical resistance to acids and other chemicals	$Na_2Mg_3(Fe, Al)_2Si_8O_{22}(OH)_2$
Amosite (brown asbestos); asbestos form of the minerals cummingtonite and grunerite	An amphibole; some commercial use, especially in plastic materials, sprayed insulation; even more brittle fibers	$(Fe, Mg)_7Si_8O_{22}(OH)_2$ (a crystalline structure different from that of anthophylite)
Anthophylite	An amphibole; some use as talclike material	$(Fe, Mg)_7Si_8O_{22}(OH)_2$
Actinolite	An amphibole	$Ca_2(Mg, Fe)_5Si_8O_{22}(OH)_2$
Tremolite	An amphibole	$Ca_2Mg_5Si_8O_{22}(OH)_2$

forms of asbestos are fibrous as they are built up of overlapping rods or cylinders. The most common form of asbestos, chrysotile (white asbestos) has long fibers that can be woven into fabric and has been widely used as

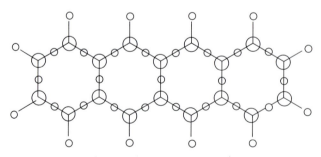

FIGURE 12-8 A portion of the amphibole chain. The large circles represent silicon overlaid by an o; the small circles are oxygen.

spray-on insulation in many buildings. An example of an amphibole is crocidolite (blue asbestos). This is less fibrous; the particles are of a more rigid, rodlike form than those of chrysotile.

Asbestos is noncombustible, and received many uses as a high-temperature insulating material for pipes, boilers, and other heating equipment, but also was widely used in other applications—asphalt roofing shingles, vinyl floor tiles and other plastic materials, laboratory bench tops, automobile brakes and gaskets, and in plaster, cements, and other construction materials where its low cost, chemical inertness, and fibrous nature were desirable. It is not uncommon to find one or another forms of asbestos as an impurity in other mineral-derived products, for example, talc and vermiculite.[13] The products themselves may not pose significant hazard, but the dusts involved in mining and processing may.[14]

All forms of asbestos are relatively inert chemically. However, asbestos is now recognized as having one environmental property that is of great importance for humankind: its dust is carcinogenic, producing lung cancer. Manipulation of the asbestos fibers invariably causes some breakdown and the release of asbestos particles; this can be quite extensive in asbestos mining and processing operations if precautions are not taken. Various industrial uses of asbestos, for example, insulation, also exposed workers who installed the materials to extensive dust. The carcinogenic character of the material evidently is due to the physical irritation caused by the tiny rodlike particles, particularly in the size range of 2–20 μm (Figure 12-9). The effect often has a long latency period, so that many workers were affected before the risk was noticed. Actually, three distinct diseases are related to exposure to inhalation of asbestos particles:

1. *Asbestosis*: basically a scarring of the lungs, referred to as fibrosis, that can also be caused by inhalation of other kinds of dust. It usually requires heavy exposure.
2. *Lung cancer*: appears to require exposure to high levels of asbestos dust, but the risk is strongly increased by smoking.
3. *Mesothelioma*: a rare cancer of the membranes of the chest cavity; most strongly related to blue asbestos. It has a very long latency period, and we have little knowledge of exposure–response relationships.

---

[13]Talc is magnesium silicate hydroxide, $Mg_3Si_4O_{10}(OH)_2$, most commonly encountered as a soft powder (talcum powder) but also in more massive forms such as stone bench tops; it is a major component of soapstone. Vermiculite is a form of mica that expands greatly upon heating, producing a low-density, porous material widely used for insulation, packing, and so on.

[14]Asbestos-related diseases have been significant for some exposed to heavy concentrations of dust from these materials as well as from asbestos itself.

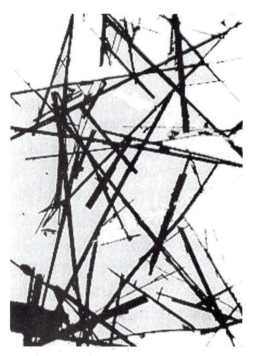

FIGURE 12-9   Electron micrograph of crocidolite fibers. Different forms of asbestos are identified microscopically as the fibers have different characteristics. From W. R. Parkes, *Occupational Lung Disorders*, 2nd ed., Butterworth-Heinemann, Oxford. Copyright © 1994. Used by permission of Arnold, as given by www.braytonlaw.com/asbestos/.

While all forms of asbestos are implicated in lung cancer, the actual degree of hazard associated with particular forms is in dispute.[15] The amphiboles are particularly dangerous because their more rodlike particles can penetrate the lining of the lungs more effectively and are less subject to further disintegration in the lungs. (The magnesium ions in chrysotile can be leached out in a kind of physiological weathering process.) Crocidolite also is associated with a particular form of tumor called mesothelioma, which can be triggered by smaller exposures than those typically associated with other lung cancers.

Although it is well established that inhalation of high concentrations of asbestos dust is harmful, there is more question about the effects of small amounts such as that to which the general population is exposed. Asbestos can be quite widespread in small amounts; normal building construction and destruction work, wear from vehicle brake linings, and asbestos contaminants

[15]B. T. Mossman, *Science*, *247*, 294 (1990).

in other mineral-derived products (e.g., talcum powder, plaster) are all sources of asbestos particles in the air, but whether they are general health hazards is undetermined. Most indications are that such casual exposure is not a cause for concern. Asbestos that is not disturbed or accessible is not a hazard while in place. There is very probably less risk in leaving asbestos insulation in well-maintained buildings than in removing it with consequent release of particles to the air.

There is also uncertainty, but cause for concern, over the presence of asbestos particles in drinking water; in some localities quite appreciable concentrations exist. Evidence is not conclusive about whether such particles are hazardous on ingestion; their interaction with the gastrointestinal tract need not be similar to their interaction with the lungs. Contamination may arise from discharge into lakes or rivers of wastewaters used in mining or mineral treatment, from waters used in scrubbers to remove dust from exhaust gases, or from dumping solid mineral wastes or sludges into water systems. It may also be noted that much inhaled asbestos dust eventually finds its way from the bronchial system into the alimentary system.

Because the physical, rather than the chemical, nature of asbestos particles is the cause of their carcinogenic properties, there is some concern that if substitutes—synthetic ceramic fibers, for example—break up into similar small particles, they could result in the same problems. This has yet to be confirmed.

Additional Reading

Allen, H. E., C. P. Huang, G. W. Bailey, and A. R. Bowers, eds., *Metal Speciation and Contamination of Soil*. Lewis Publishers, Boca Raton, FL, 1995.

Birkeland, P. W., *Soils and Geomorphology*. Oxford University Press, New York, 1984.

Boulding, J. R., *Practical Handbook of Soil, Vadose, and Ground-Water Contamination: Assessment, Prevention and Remediation*. Lewis Publishers, Boca Raton, FL, 1995.

Buckman, H. O., and N. C. Brady, *The Nature and Properties of Soils*, 7th ed. Macmillan, New York, 1969.

Cairney, T., ed., *Contaminated Land: Problems and Solutions*. Blackie Academic and Professional Publishers, Glasgow, 1993.

McBride, M. B., *Environmental Chemistry of Soils*. Oxford University Press, New York, 1994.

Schnoor, J. L., *Environmental Modeling: Fate and Transport of Pollutants in Water, Air and Soil*. Wiley, New York, 1996.

Stevenson, F. J., *Humus Chemistry: Genesis, Composition, Reactions*, 2nd ed. Wiley, New York, 1994.

Wilson D. J., and A. N. Clarke, *Hazardous Waste Site Soil Remediation*. Dekker, New York, 1994.

Yong, R. N., A. M. O. Mohamed, and B. P. Warkentin, *Principles of Contaminant Transport in Soils*. Elsevier, Amsterdam, 1992.

Zoltai, T., and J. H. Stout, *Mineralogy: Concepts and Principles*. Burgess Publishing, Minneapolis, MN, 1984.

EXERCISES

12.1. Give the formula for each of the following silicate ions that are based on linked SiO_4^{4-} tetrahedra:
(a) a cyclic structure of three SiO_4 groups
(b) a linear structure of three SiO_4 groups
(c) an infinite chain of SiO_4 groups

12.2. Zeolites are used as catalysts, in detergents, and as water softeners. What are the structural characteristics that enable them to perform these functions?

12.3. Discuss isomorphous substitution in minerals.

12.4. Some hydrothermal ocean vents are referred to as "black smokers." Describe the inorganic chemistry that is taking place at these vents. Include equations and explain why the term is used.

12.5. What are the typical steps used in the treatment of metallic ores?

12.6. Write the equations for the reactions that take place in obtaining metallic iron from its usual ores.

12.7. What are the three most heavily used nonferrous metals? Outline the processes used in recovering the metals from their ores and the environmental problems associated with each.

12.8. What are the two methods of treating gold ores that have the most serious environmental problems?

12.9. What types of reaction are involved in the weathering of minerals?

12.10. Explain how climate affects the weathering products of minerals.

12.11. Compare typical soils from moist temperate, wet tropical, and desert areas. How are they different in composition and properties, and why?

12.12. Many soil properties are determined by the relative proportions of the three main inorganic components; clay, sand, and silt. In this context, what is clay? Sand? Silt?

12.13. What are the structural differences between aluminosilicate clays that swell when wet and those that do not?

12.14. Summarize the methods available for decontaminating polluted soils.

12.15. Compare the behavior with respect to soil contamination of water-insoluble liquids that are lighter than water with those that are heavier than water.

12.16. Compare the main reactions of acid precipitation on a basic stone such as a carbonate and on a more acidic stone such as granite.

12.17. Kilns for the production of portland cement are sometimes used for the destruction of hazardous organic materials. Discuss the pros and cons of such use.

12.18. What are the main minerals whose asbestiforms have been used in commercial asbestos?

13

PROPERTIES AND REACTIONS OF ATOMIC NUCLEI, RADIOACTIVITY, AND IONIZING RADIATION

13.1 INTRODUCTION

The physical properties, chemical reactions, toxicity, carcinogenicity, and other properties of an element or compound present in the environment and discussed in the preceding chapters are determined by the electronic configuration of the atom, ion, or molecule of interest. Except for the fact that the number of electrons in an atom is determined by the number of protons in its nucleus, there is generally no need to give further consideration to the nucleus when dealing with the basic chemistry of an element having a stable nucleus. However, there are certain properties of the atomic nucleus that can have small but measurable effects on the physical properties and chemical behavior of an element and its compounds. We shall introduce those nuclear properties that contribute to the present environment and, perhaps more importantly, those that have application in the study of the paleoenvironment.

Atomic nuclei are either stable or unstable (radioactive). Because radioactivity is a major source of the ionizing radiation component of the environment, most of this chapter will be devoted to the quantitative aspects of radioactivity and to the nature, properties, and effects of ionizing radiation.

It is intended that this chapter supplement the coverage of nuclear chemistry in a general chemistry course and thereby provide background material for Chapter 14. Because of this limited objective, many topics contained in books on nuclear chemistry, radiochemistry, and radiation protection have been abbreviated or omitted.

In this introduction it is appropriate to note a few of the key scientific advances that prepared the way for the nuclear era. They are

1911 Postulation of the nuclear model of the atom by Ernest Rutherford based on studies of the scattering of α particles from a radioactive source by the atoms in a thin gold foil.

1896 Discovery of radioactivity by Antoine Henri Becquerel while using photographic plates to study x-ray-induced fluorescence of various substances, including uranium salts.

1895 Discovery of x rays by Wilhelm Conrad Roentgen while experimenting with Crookes tubes, which were the precursors of cathode ray tubes (CRTs) used in oscilloscopes, one type of TV display, and one type of computer monitor.

1830 Invention of the photographic plate by Louis Daguerre.

13.2 THE ATOMIC NUCLEUS

13.2.1 Composition, Types, and Selected Properties

An atomic species whose properties are of interest because of the composition and associated properties of its nucleus is called a nuclide. For the purposes of this book it will be adequate to identify and specify the composition of an atomic nucleus in terms of the number of nucleons, that is, neutrons (N) and protons (Z), that it contains.[1] In practice, the symbol for a nuclide consists of the chemical symbol, which determines Z, and the mass number (A), which is also the nucleon number ($N + Z$). Thus, for ^{6}Li the atomic number is 3 and the mass number is 6. Alternative forms of representation are $^6_3$Li, lithium-6, and Li-6.

Nuclides having the same atomic number are called isotopes; those having the same mass number are isobars; those with the same neutron number are isotones; and those with the same atomic number and the same mass number are called isomers. There are over 3100 known nuclides. Included are 266

---

[1] In terms of current theory, neutrons and protons are not fundamental particles but are composed of particles called quarks, three of which are bound in a nucleon by a strong interaction (nuclear force) involving the exchange of particles called gluons.

stable nuclides, about 515 isomer pairs and about 30 isomer triplets, and 21 primordial nuclides that undergo radioactive decay so slowly that they are treated chemically as if stable. All elements from $Z = 1$ (H) through $Z = 83$ (Bi) have at least one stable isotope except for the two manufactured elements, element 43 (technetium) and element 61 (promethium). All elements with $Z > 83$ are radioelements. Some of the radioisotopes of the elements polonium ($Z = 84$) through uranium ($Z = 92$) are members of three naturally occurring families or series discussed in Section 13.4.1. The transuranium elements, those beyond uranium in the periodic table, that have been synthesized at the time of writing are $_{93}$Np (neptunium), $_{94}$Pu (plutonium), $_{95}$Am (americium), $_{96}$Cm (curium), $_{97}$Bk (berkelium), $_{98}$Cf (californium), $_{99}$Es (einsteinium), $_{100}$Fm (fermium), $_{101}$Md (mendelevium), $_{102}$No (nobelium), $_{103}$Lr (lawrencium), $_{104}$Rf (rutherfordium), $_{105}$Db (dubnium), $_{106}$Sg (seaborgium), $_{107}$Bh (bohrium), $_{108}$Hs (hassium), $_{109}$Mt (meitnerium), $_{110}$El, $_{111}$El, and $_{112}$El, where El represents an unnamed element. Creation of elements 116 and 118 was announced in 1999, but the announcement was retracted in 2001 because repetition of the experiment in the same and in other heavy element research laboratories failed to confirm the initial results. The actinium series ends with lawrencium. Seaborgium has been shown to have the chemical properties of molybdenum and tungsten, bohrium those of technetium and rhenium, and hassium those of osmium. Neptunium and plutonium also are produced in trace amounts in uranium ores by the reaction of ^{238}U with neutrons. This reaction is discussed in Section 13.6.3.

Figure 13-1 shows the relationship between the number of neutrons and the number of protons in the stable nuclei. For the first 20 elements (through Ca), N is equal to or very close to Z and the data follow the line labeled ($Z = A - Z$). Beyond $_{20}$Ca, the stable nuclei contain an excess of neutrons and the excess increases with increasing Z. The number of stable isotopes varies from one for many elements to 11 for $_{50}$Sn.[2]

A radionuclide, A, that decays into a product nuclide B with a change of Z can be represented by a point that is either above or below the stability band in Figure 13-1, depending on the mode of decay. As A decays, the point will move toward the stability band and will be within the band if B is stable. If A is the first member of a radioactive series, [reaction (13-21) in Section 13.4.1], the point representing the last member will be in the band.

---

[2]Nuclides with 2, 8, 20, 50, or 82 protons or neutrons or 126 neutrons are characterized as having a magic number of nucleons. Such nuclides have several distinguishing properties (e.g., high abundance, a large number of stable isotopes, high stability). The correlation between magic numbers and nuclidic properties provides the basis for the shell model of nuclei. In this model, the magic numbers correspond to the number of protons in closed shells of protons or neutrons in closed shells of neutrons. A region of relative stability of superheavy radioelements has been predicted for nuclei close to those with 114 or 126 protons and those with 184 neutrons.

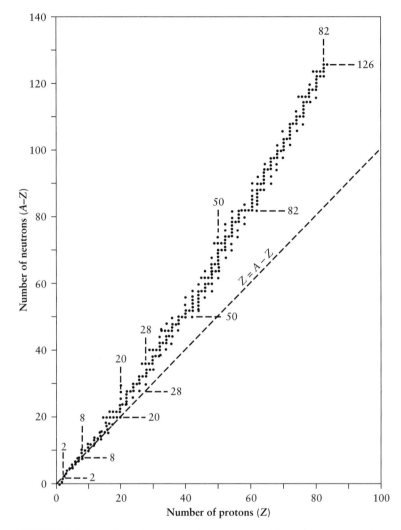

FIGURE 13-1 Numbers of protons and neutrons in stable nuclei. Magic numbers (see footnote 2) are indicated. Redrawn from S. Glasstone and A. Sesonske, *Nuclear Reactor Engineering*, 4th ed., Vol. 1, Chapman & Hall, New York. Copyright © 1994. Used by permission of Kluwer Academic Plenum Publishers.

At least 32 of the 91 naturally occurring elements (including the recently discussed plutonium, ^{244}Pu) have one or more naturally occurring radioisotopes and 10 have no stable isotopes.[3] Furthermore, all 91 elements have one or more (up to 36) anthropogenic radioisotopes.

The nuclidic mass (rest mass), M, is expressed on a relative scale based on the mass of an atom of the stable isotope of carbon ^{12}C, which is defined as having a mass of exactly 12.00000000 atomic mass units (amu).[4] For example, the values of M for stable ^{13}C and radioactive ^{238}U are 13.00335483 and 238.050785, respectively.[5] As discussed in the next section, the nuclidic mass is less than the sum of the rest masses of Z hydrogen atoms and N neutrons by an amount equal to the mass-equivalent of the binding energy that holds the nucleus together.

The two stable isotopes of carbon, ^{12}C and ^{13}C, have experimentally determined isotopic abundances of 98.90 and 1.10 at. %, respectively. By calculating the weighted average of the two isotopic masses, one obtains 12.011 for the atomic weight of carbon.

In addition to mass, the nucleons confer other properties on an atomic nucleus. One of the properties is nuclear spin, which is analogous to the spin assigned to an atom based on the orbital angular momentum and the intrinsic spin of the atom's electrons. (See Appendix A, Section A.2.) The properties of a nucleus are consistent with each nucleon having a quantized orbital angular momentum l, an integer value, and a quantized intrinsic spin angular momentum $s = 1/2$. When combined, the total nuclear angular momentum or nuclear spin j of a nucleon has a half-integral value. Nuclei with even mass number have a total nuclear spin I equal to 0 or an integer; those with odd mass number have $I =$ half-integer.

---

[3]The number of known naturally occurring radionuclides that decay exceedingly slowly increases as radiometric measurement techniques are improved. New, manufactured radionuclides also are being discovered as the products of nuclear reactions in which high-energy, heavy ions (e.g., ^{238}U ions) strike a target consisting of an element such as lead and as the products of reactions used to produce new heavy elements.

[4]The atomic mass unit is for the neutral atom, not for the nucleus alone. As defined, one amu (1 dalton) has a mass of 1/12 the mass of one atom of ^{12}C, that is,

$$(12.000000 \times 10^{-3}\,\text{kg}/12)\frac{1}{6.02 \times 10^{23}} = 1.661 \times 10^{-27}\,\text{kg}$$

[5]Values of M and abundances of naturally occurring isotopes and values of M for radioisotopes of the elements are readily available in handbooks. In some tabulations values of Δ, the mass excess, rather than M are given. The mass excess is defined in terms of the isotopic mass M and the mass number A, which is the whole number nearest M, by $\Delta = M - A$.

13.2.2 Physicochemical Effects of Nuclear Mass and Spin

13.2.2.1 Mass

The physical properties, such as density, viscosity, melting point, vapor pressure, spectra, and rate of diffusion, and the chemical properties, such as reaction rates and equilibrium states, of a given chemical substance are slightly different for isotopomers (chemical species differing in isotopic composition). For example, the normal melting and boiling points of D_2O (^2H_2O, heavy water) are 3.82°C and 101.42°C, respectively. This is an example of the isotope effect, which increases in magnitude with increasing $\Delta M/M$, the relative mass difference for two isotopes, and is, therefore, most pronounced for the light elements H, Li, Be, C, N, O, Si, S, and Cl. Even so, the isotope effect is sufficiently large to serve as the basis for methods of isotope separation (enrichment) for many elements, including uranium.

An important mass-dependent effect is isotope fractionation, which occurs in physical processes such as evaporation of a liquid and in chemical reactions. During the evaporation of water, which normally has a source-dependent deuterium concentration of about 0.015 atom percent of the hydrogen atoms, the liquid becomes slightly enriched in D_2O because of its lower vapor pressure.

Isotope fractionation can also occur in a chemical reaction so that the ratio of the atoms of one isotope to that of another in the reactant differs from the ratio in the product. An example is an isotope exchange reaction that reaches equilibrium. When $^{15}NH_3$ (g) is equilibrated with $^{14}NH_4^+$ (aq), the ratio of $^{14}N/^{15}N$ in the gas phase becomes about 1.023 times that in the aqueous phase. Another source of isotope fractionation is the kinetic isotope effect that occurs because the rate of a chemical reaction is affected by the mass of the reacting atoms in the reactant. Heavier isotopes form stronger bonds of a given type than lighter isotopes. Thus, the rate of a chemical reaction that involves the breaking of C-H or C-D bonds in isotopomers will be faster for the one with the C-H bonds. The kinetic isotope effect accounts for the enrichment of ^{12}C relative to ^{13}C in organic compounds formed in plants by photosynthesis.

As we shall see in Chapter 14, Sections 14.5.1, 14.5.2, and 14.5.4., isotope fractionation and its temperature coefficient have been used to study the mechanisms of a wide variety of biological, chemical, and physical processes in the environment, especially in the paleoenvironment.

13.2.2.2 Nuclear Spin

Using hydrogen as an example, we find that the total nuclear spin associated with an atom of hydrogen in any molecule such as benzene (C_6H_6) is that of a proton and is equal to 1/2. If we assume that the proton can be treated as a rotating charge, it will generate a magnetic field and have a magnetic moment that can interact with an external magnetic field. Thus, when a sample of a hydrogen-containing compound is placed in a strong, homogeneous, external magnetic field, the magnetic moment of the protons becomes oriented in one or the other of two allowed directions corresponding to two magnetic energy states. The energy difference between the two states is proportional to the strength of the external magnetic field and is the order of 10^{-26} J. If the sample is also irradiated with a pulse of radiofrequency (rf) radiation having energy equal to the energy difference, the sample will absorb the radiation, reduce the intensity of the radiation reaching a detector, and provide a signal that generates a nuclear magnetic resonance (NMR) spectrum. An NMR spectrum can be obtained by varying the rf energy for a fixed external magnetic field or by varying the magnetic field for a fixed rf pulse.

The electrons in the atoms of a molecule containing hydrogen shield the proton or protons from the external field and cause a small but measurable shift (chemical shift) of the NMR lines. As a result, the NMR spectrum of protons in benzene (C_6H_6), for example, is shifted relative to that for protons in ethanol (CH_3CH_2OH), because of the different electronic environment of the protons. Furthermore, there are three different electronic environments for protons in ethanol—one for CH_3, one for CH_2, and one for OH.

NMR spectroscopy provides information about the electronic environment within a molecule and is widely used to determine molecular structure and to study chemical bonding and chemical reactions. Other stable nuclides with $I = 1/2$ that are used in NMR studies are ^{13}C, ^{15}N, ^{19}F, and ^{31}P.

This very simplified and condensed introduction to NMR is intended to help in the understanding of an important application of stable nuclides and NMR in which the sample of interest is biological tissue and the purpose is to provide information for medical diagnosis. The application is discussed in Chapter 14, Section 14.6.1.1.

13.3 ENERGETICS OF NUCLEAR REACTIONS

Mass M and energy E are related through the Einstein equation

$$E = Mc^2 \qquad (13\text{-}1)$$

where c is the speed of light. The energy equivalent of a change in mass, ΔM, of 1 amu in a nuclear process corresponds to 149.2 pJ or 931.5 MeV.[6]

---

[6]The electron-volt (eV) was introduced as a unit of energy in Chapter 4, Section 4.1. One MeV is equal to 10^6 eV.

joules (J), on the magnitude of two electric charges q_1 and q_2 (in coulombs, C) and the distance of separation r (in meters) is given by

$$V = \frac{1}{4\pi\varepsilon_0} \frac{q_1 q_2}{r} \qquad (13\text{-}9)$$

In this equation ε_0 is the permittivity of a vacuum and is equal to $8.854 \times 10^{-12}\,\mathrm{J^{-1}\,C^2\,m^{-1}}$.

The quantity of interest in a nuclear reaction is V', the height of the Coulomb barrier in MeV, that is, the work required to bring the nucleus of the projectile species into contact with the target nucleus. Equation (13-9) becomes

$$V' = \frac{1.44 Z_1 Z_2}{R_1 + R_2}\,(\mathrm{MeV}) \qquad (13\text{-}10)$$

where the charges are Z_1 and Z_2 and r is replaced by the sum of the nuclear radii of the two reactants, R_1 and R_2. The radius of a nucleus in femtometers (10^{-15} m) is given by $R = 1.4 A^{1/3}$.[12] For reaction (13-7), the Coulomb barrier height is 4.35 MeV. When conservation of momentum is taken into account, the new threshold energy is 5.27 MeV.

Finally, it is necessary to point out that reaction (13-7) may occur for an incoming $^4\mathrm{He}^{2+}$ particle having a kinetic energy less than 5.27 MeV, but not less than 2.36 Mev. The value of 4.35 MeV for the Coulomb barrier height was calculated by means of classical mechanics [equation (13-10)]. When a nuclear reaction is examined in terms of quantum mechanics, an incoming charged particle with kinetic energy less than the Coulomb barrier height has a finite probability of penetrating (tunneling through) the barrier. This probability increases with increasing kinetic energy of the particle.

If an endoergic nuclear reaction involves a charged particle as a reactant or a product, any excess kinetic energy above Q that is given to the incoming particle because of the Coulomb barrier is distributed among the products. For an exoergic reaction with a Coulomb barrier, the energy available to the products is the sum of the kinetic energy of the incoming particle and the Q value.

Another type of energy associated with an atomic nucleus is the binding energy of its nucleons, which is usually expressed as the average binding energy per nucleon. The total binding energy is the energy required for the hypothetical reaction of pulling apart a nucleus into its constituent nucleons. Binding energy is, therefore, a measure of the stability of a nucleus. As an example, the total binding energy of a $^{23}\mathrm{Na}$ nucleus is the Q value based on the conservation of energy for the following hypothetical reaction:

$$^{23}\mathrm{Na} \rightarrow 11\,^1\mathrm{H} + 12\,^1\mathrm{n} + Q \qquad (13\text{-}11)$$

---

[12]When the particle in or out in a nuclear reaction is a proton, its radius is usually taken as zero except when the target or product nuclide has a very small mass number.

Then,

$$Q = [M(^{23}\text{Na})] - [11M(^1\text{H}) - 12M(^1\text{n})] \tag{13-12}$$

$$Q = 22.989769 - [(11 \times 1.007825) + (12 \times 1.008665)]$$

$$Q = -0.200288\,\text{amu} = -186.57\,\text{MeV}$$

and the average binding energy per nucleon is 8.11 MeV. Figure 13-2 illustrates the way average binding energy per nucleon varies with mass number for stable nuclides. The peak at 7 MeV is for ^{4}He. The maximum stability is found for elements of mid-range atomic number. A "rule of thumb" is to use a value in the range 8–8.5 MeV for the approximate average binding energy per nucleon for stable nuclides of elements between oxygen and tantalum, inclusive.

When a radionuclide decays into another nuclide, which may be stable or radioactive, there is a spontaneous decrease in the rest mass of the system. We discuss the energetics of radioactive decay in Section 13.5.

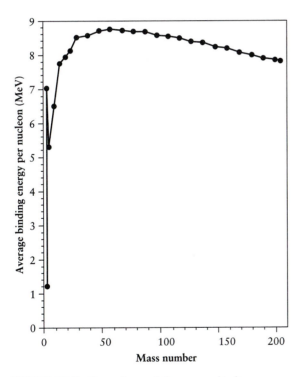

FIGURE 13-2 Dependence of the average binding energy per nucleon on mass number of a nuclide.

13.4 THE KINETICS OF NUCLEAR REACTIONS

13.4.1 Radioactive Decay

In the simplest case of radioactive decay, radionuclide A is transformed irreversibly (spontaneously) into a stable product B(s) according to an equation such as A → B(s).[13] The probability that the nucleus of a given atom of a radionuclide will decay does not depend on the age of the atom. Furthermore, characteristic of a first-order process, it is not necessary that such a nucleus interact with any other nuclei in a sample containing a large number nuclei in order to undergo radioactive decay. In a first-order process, the number of nuclei decaying per unit time at a given time in a sample containing a large number of a atoms of a given radionuclide is proportional to N, the number of such atoms present at that time. In other words, the fraction of radioactive nuclei decaying per unit time is a constant (λ). The differential rate equation for the process has the same form as that for a first-order chemical reaction

$$\frac{dN}{dt} = -\lambda N \qquad (13\text{-}13)$$

where N is used instead of concentration, λ is the traditional symbol for the rate constant ("decay constant," usually in reciprocal seconds), and the negative sign indicates that the number of radioactive atoms in the sample decreases with time.

Some radionuclides decay by branching, a more complex process in which decay may occur along two or more paths to give different products. The fraction of nuclei that decays by each mode is a characteristic property of the radionuclide that must be determined experimentally. Since the value of λ for a mode of decay is a measure of the probability of decay by that mode, the decay constant for the total rate of decay, $\lambda_{overall}$, is the sum of the λ values for each branch.

The disintegration rate λN of a radionuclide in a sample is the radioactivity or "activity" A of the radionuclide.[14] It is a measure of the amount of the radionuclide in a sample. The curie (Ci) was the accepted unit of radioactivity until 1975, when the new SI quantity of activity, the becquerel (Bq), was introduced.[15] The curie was originally equated to the number of disintegrations per second (dps) that occur in a fixed amount of ^{226}Ra. It is now defined

---

[13]When used to characterize a nuclide, (s) means "stable."

[14]By now the reader will have noted that two meanings have been assigned to N and three to A. Fortunately, this rather common practice should not create a serious problem because the intended meaning should be clear from the context in each case.

[15]In the late 1940s another unit, the rutherford (rd) was used in some publications. One rd was defined as 1×10^6 disintegrations per second.

as 3.700×10^{10} dps for any radionuclide. One Bq is equal to 1 dps for any radionuclide (see Table 13-1).

Another quantity that is often used, the specific activity, is the activity (in becquerels or curies) of a specified radionuclide per unit quantity of radioactive material (e.g., per millimole of a compound or per unit weight of an element).

When the differential rate equation (13-13) is integrated between the limits of $t = 0$ (the beginning of the time period of interest when the sample contains N^0 atoms of a radionuclide) and lapsed time $t = t$ (when the sample contains N atoms), the equation for the number of atoms remaining in the sample becomes

$$N = N^0 \exp(-\lambda t) \qquad (13\text{-}14)$$

In terms of natural logarithms,

$$\ln(N/N^0) = -\lambda t \qquad (13\text{-}15)$$

and, in terms of common logarithms,

$$\log(N/N^0) = \frac{-\lambda t}{2.303} \qquad (13\text{-}16)$$

After a sample has decayed so that $N/N^0 = 1/2$, the lapsed time is the half-life $t_{1/2}$ of the radionuclide. Then,

$$\ln(N/N^0) = \ln 0.5 = -\lambda t_{1/2} = -0.693 \qquad (13\text{-}17)$$

and the decay constant is related to the half-life by the equation

$$\lambda = \frac{0.693}{t_{1/2}} \qquad (13\text{-}18)$$

TABLE 13-1

Units of Radioactivity

Curies	Disintegrations per second (dps)	Becquerels
1 pCi	3.70×10^{-2}	37 mBq
27.0 pCi	1	1 Bq
1 nCi	37	37 Bq
1 µCi	3.70×10^{4}	37 kBq
1 mCi	3.70×10^{7}	37 MBq
1 Ci	3.70×10^{10}	37 GBq
1 kCi	3.70×10^{13}	37 TBq
1 MCi	3.70×10^{16}	37 PBq
1 GCi	3.70×10^{19}	37 EBq

Because it is the half-life of a radionuclide that is given in nuclear data tables,[16] it is more convenient to express equation (13-14) in terms of $t_{1/2}$. Thus we write

$$N = N^0 \exp\left(\frac{-0.693t}{t_{1/2}}\right) = N^0(1/2)^{t/t_{1/2}} \qquad (13\text{-}19)$$

In certain calculations the mean life τ is used. It is equal to $1/\lambda$ and $1.443t_{1/2}$.

For most purposes it is the change of the activity of a sample of radioactive material with time that is of interest rather than a change in the number of radioactive atoms. After multiplication of equation (13-14) by λ and replacement of λN and λN^0 by A and A^0, the equation for time dependence of activity becomes

$$A = A^0 \exp(-\lambda t) \qquad (13\text{-}20)$$

A typical semilog plot of the activity of a single radionuclide with a half-life of 2.0 days is shown in Figure 13-3 ($A^0 = 800$ Bq).

When the rate at which radiation is emitted by a radioactive source is measured in the laboratory or in the field, the observed counting rate R is proportional to the activity. If the proportionality factor does not drift during a series of measurements, the counting rate (e.g., counts per minute) will have the same time dependence as A. Then R and R^0 can be substituted for A and A^0, respectively, in equation (13-20). For a single radionuclide, the decay curve based on counting rates will resemble that in Figure 13-3 except that the data points will scatter about the "best" straight line because of statistical fluctuation of radioactive decay.

Most of the radionuclides that occur in nature and those produced in nuclear power reactors are components of a complex mixture. There are three kinds of mixture:

1. Those containing independent components, each decaying as just described
2. Those containing one series or several series, each having genetically related members (i.e., the first member in a series decays into the second, which decays into the third, etc., as discussed shortly)
3. Those containing a combination of types 1 and 2.

---

[16]In this text, half-lives are given in parentheses after the symbol for the radionuclide. The units are seconds (s), minutes (m), hours (h), days (d) and years (y). In some data compilations "a" is used for years. Nuclidic data in Chapters 13 and 14 were taken from Parrington *et al. Chart of the Nuclides* and the *CRC Handbook of Chemistry* listed at the end of the chapter under Additional Reading and Sources of Information.

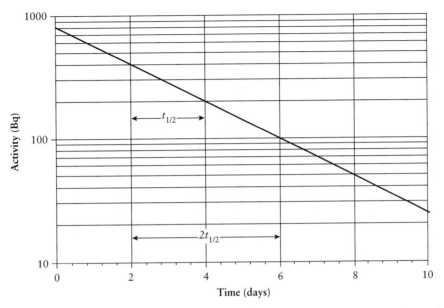

FIGURE 13-3 Semi log plot of the decay curve for a radionuclide with a half-life of 2.0 days and $A^0 = 800$ Bq.

A semilog plot of the decay curve for the total activity of a type 1 mixture will be concave upward corresponding to the sum of the straight-line decay curves for the activities of the components. An example is shown in Figure 13-4 for a mixture of two independent components, A and B, whose half-lives are 10.0 and 1.0 h, respectively, and $A^0 = 200$ Bq, $B^0 = 600$ Bq.

An example of the decay sequence for a type 2 mixture consisting of a chain with three radioactive members (without branching) and a stable end product is illustrated by

$$A \xrightarrow{\lambda_A} B \xrightarrow{\lambda_B} C \xrightarrow{\lambda_C} D(s) \qquad (13\text{-}21)$$

where radionuclide A is the parent of daughter B, which becomes the parent of C, which decays to the stable nuclide D.

The rate of decrease of the number of atoms of A [in reaction (13-21)] is independent of the radioactivity of the daughters and is given by the differential rate equation (13-13) with N changed to N_A and λ to λ_A. For atoms of daughter B, however, the total rate of change is equal to the rate of their formation by decay of A, the parent, minus the rate of their decay, so that the differential equation is

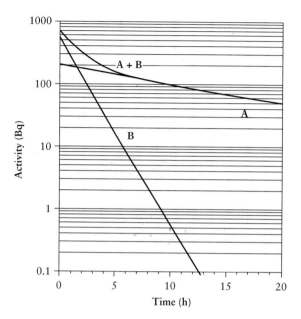

FIGURE 13-4   Semi log plot of the decay curves of the total activity and the activities of the two independent components, A $(10.0\,\text{h}, A_A^0 = 200\,\text{Bq})$ and B $(1.0\,\text{h}, A_B^0 = 600\,\text{Bq})$ in a mixture.

$$\frac{dN_B}{dt} = \lambda_A N_A - \lambda_B N_B \tag{13-22}$$

This equation is for the simplest case, in which the only removal process for nuclide B is radioactive decay. Other processes that can remove B include chemical separation, loss by volatilization, and conversion into another nuclide by nuclear reaction in an accelerator or nuclear reactor. After equation (13-22) has been integrated and the number of atoms changed to activity, the following equation is obtained:

$$A_B = A_A^0 \frac{\lambda_B}{\lambda_B - \lambda_A} [\exp(-\lambda_A t) - \exp(-\lambda_B t)] + A_B^0 \exp(-\lambda_B t) \tag{13-23}$$

where the last term accounts for the decay of any B that is present in the sample at time $t = 0$. General expressions (e.g., the Bateman equation),[17] are available for the equation obtained by integration of an equation like equation (13-22) for the nth member of a series.

[17]H. Bateman, Solution of a system of differential equations occurring in the theory of radioactive transformations, *Proc. Cambridge Philos. Soc.*, **15**, 423 (1910).

There are three naturally occurring, radioactive series, namely, the uranium, the thorium, and the actinium series. Each series is named after the longest-lived and, therefore, first member. The transitions that occur in the uranium series are represented in Figure 13-5. Data and names used in the older literature for the members of the uranium series are given in Table 13-2. Similar information for the thorium and actinium series is given in Appendix B.

The uranium series is also known as the $4n + 2$ series because the mass number of each member can be expressed as $(4n + 2)$, where n is an integer. Similar designations for the thorium series and the actinium series are $4n$ and $4n + 3$, respectively. The three series contain examples of branching, already mentioned, and the three types of parent–daughter system described next.

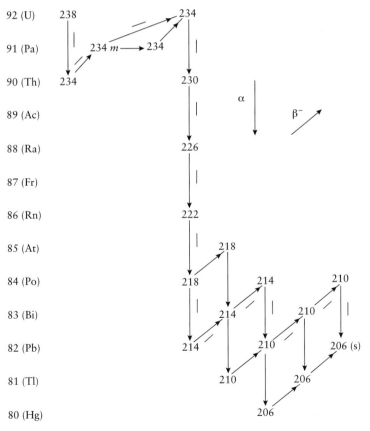

FIGURE 13-5 Uranium $(4n + 2)$ series. The main sequence is indicated by arrows and a dash.

TABLE 13-2

Uranium $(4n + 2)$ Series

Z	Nuclide[a]	Half-life	Radiation[b]
92	^{238}U (UI)	4.47×10^9 y	α
92	^{234}U (UII)	2.46×10^5 y	α
91	^{234m}Pa (UX$_2$)	1.17 m	IT, 0.13% β^-, 99.87%
91	^{234}Pa (UZ)	6.69 h	β^-, γ
90	^{234}Th (UX$_1$)	24.10 d	$\beta^-, (\gamma)$
90	^{230}Th (Io)	7.54×10^4 y	α
88	^{226}Ra (Ra)	1599 y	$\alpha, (\gamma)$
86	^{222}Rn (Rn)	3.8235 d	α
85	^{218}At	1.5 s	α
84	^{218}Po (RaA)	3.10 m	α, 99.98% β^-, 0.02%
84	^{214}Po (RaC′)	163.7 μs	α
84	^{210}Po (RaF)	138.38 d	α
83	^{214}Bi (RaC)	19.9 m	α, 0.02% β^-, 99.98%, γ
83	^{210}Bi (RaE)	5.01 d	$\alpha, 1.3 \times 10^{-4}$% β^-, 99$^+$ %
82	^{214}Pb (RaB)	27 m	β^-, γ
82	^{210}Pb (RaD)	22.6 y	$\alpha, 1.7 \times 10^{-6}$% β^-, 99$^+$%, (γ)
82	^{206}Pb (RaG)	Stable	
81	^{210}Tl (RaC″)	1.30 m	β^-, γ
81	^{206}Tl (RaE″)	4.20 m	β^-
80	^{206}Hg	8.2 m	β^-, γ

[a]Symbol used in the early studies of naturally occurring radioactivity is given in parentheses [(Io) = ionium].

[b](γ) indicates low intensity (1–5%); γ rays with intensity below 1% are not included. Where (IT and β^-) or (α and β^-) is given, branched decay occurs.

If nuclide A in equation (13-21) has a very long $t_{1/2}$ relative to that of B, corresponding to $\lambda_A \ll \lambda_B$, the system can attain a state in which the activity of B becomes equal (for all practical purposes) to A_A^0, the rate of production by its parent, and remains equal in the absence of physical or chemical separation. Daughter B is then said to be in secular equilibrium with its parent, A.[18] This condition arises because the parent activity decreases negligibly during the time required for the daughter to reach secular steady state. The equation is

$$A_B = A_A^0[1 - \exp(-\lambda_B t)] \tag{13-24}$$

[18]"Equilibrium" implies reversibility to a chemist. The term "secular steady state" would be better.

if B is absent when $t = 0$. For such a system the daughter activity reaches 50% of its steady-state limit after one half-life of the daughter, 75% after two half-lives, and so on. As a rule of thumb, steady state for practical purposes is attained in 10 half-lives. Thus, after 10 half-lives the daughter activity will be within $(1/2)$,^{10} that is, within 1 part in 1024 of the steady-state value. An example of two members of a series that can attain secular equilibrium is the formation of ^{222}Rn from ^{226}Ra in the uranium series according to the equation

$$^{226}\text{Ra}(1599\,\text{y}) \rightarrow {}^{222}\text{Rn}(3.8235\,\text{d}) \rightarrow {}^{218}\text{Po}(3.10\,\text{m}) \rightarrow \text{etc.} \qquad (13\text{-}25)$$

The growth curve for ^{222}Rn in a sample of pure ^{226}Ra ($A^0 = 660$ Bq) is shown in Figure 13-6. The activity of the equilibrium mixture will decrease very slowly with a half-life of 1599 y.

In the second type of system, $\lambda_A < \lambda_B$, so that A_A decreases significantly as A_B increases in a sample initially containing very little B or none at all. At first, A_B increases rapidly and then goes through a maximum before attaining a state of transient equilibrium (transient steady state) as shown in Figure 13-7 for A ($t_{1/2} = 10.0$ d, $A_A^0 = 600$ Bq) and B ($t_{1/2} = 3.0$ d). When the transient state is attained, the ratio of A_B to A_A becomes a constant and remains so as both A_A and A_B decrease with the half-life of the parent, A. Then, A_B is greater than A_A by the ratio $\lambda_B/(\lambda_B - \lambda_A)$.

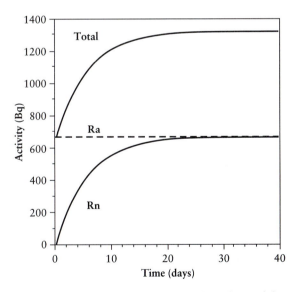

FIGURE 13-6 Linear plot of the time dependence of the total activity, the activity of ^{226}Ra (1599 y, $A^0 = 660$ Bq), and the activity of the ^{222}Rn (3.8235 d) daughter as it increases to secular equilibrium in a sample of initially pure ^{226}Ra.

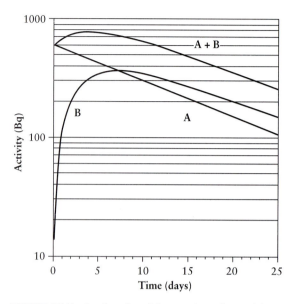

FIGURE 13-7 Semilog plot of the time dependence of the total activity, the activity of A (10.0 d, $A_A^0 = 600$ Bq), and the activity of the daughter B (3.0 d) as it increases and reaches transient equilibrium in a sample of initially pure A.

The third type of behavior ("no equilibrium") occurs when $\lambda_A > \lambda_B$. Steady state cannot be attained, as shown in Figure 13-8 for A ($t_{1/2} = 3.0$ d, $A_A^0 = 800$ Bq) and B ($t_{1/2} = 10.0$ d).

13.4.2 Induced Nuclear Reactions

Both stable isotopes and radioisotopes of the elements can be produced by nuclear reactions that occur when a target is irradiated with neutrons, charged particles (including radioactive nuclei), or high energy photons. In the production of synthetic radioisotopes of an element from its stable isotopes for use in nuclear medicine, for example, the target contains the element in the form of the pure element or in the form of a compound or an alloy. Neutron-induced nuclear fission (Section 13.6.3), the basis for nuclear power, is an example of the irradiation of a long-lived radionuclide such as ^{235}U (7.04×10^8 y). Nuclear fusion (Section 13.6.1.1), a potential source of energy in the future, may utilize a shorter-lived radionuclide, ^{3}H (12.32 y), as one of the reactants.

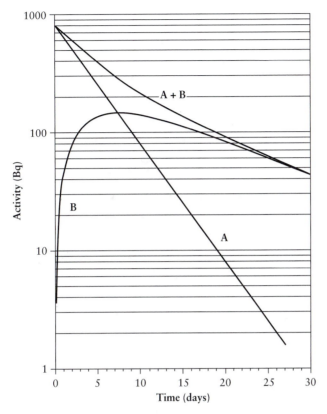

FIGURE 13-8 Semilog plot of the time dependence of the total activity, the activity of A (3.0 d, $A_A^0 = 800$ Bq), and the activity of the daughter B (10.0 d) in a sample of initially pure A.

The reaction for producing a single radionuclide, B, by irradiation of a target-containing nuclide, A (a specific isotope of the element reacting with neutrons or charged particles such as protons) can be written as follows:

$$A(\text{target}) \xrightarrow{\text{nuclear reaction}} B \xrightarrow{\text{decay}} C(s) \qquad (13\text{-}26)$$

If (1) the irradiation conditions are constant and (2) the target is thin and is not significantly depleted of target nuclei during the irradiation time t, then the rate of formation of nuclei of B, $(R_f)_B$, remains constant during the irradiation.[19] The reaction is pseudo-first order. Its rate depends only on the rate at which the irradiating species arrives and reacts with the target nuclei.

After about 10 half-lives of the product B, its activity A_B reaches the "saturation level," the point at which it becomes equal to $(R_f)_B$ for practical purposes. When written in terms of the half-life of B, the equation for the growth of A_B is

$$A_B = (R_f)_B \left[1 - \exp\left(\frac{-0.693t}{(t_{1/2})_B} \right) \right] \tag{13-27}$$

If R_f is constant and the half-life of B is short enough for the activity of B to reach saturation during the irradiation time, the process is analogous to the formation of a radioactive daughter by the decay rate of a parent that has a very long half-life. The growth curve for A_B is then similar to that shown for ^{222}Rn in Figure 13-6. Further irradiation, after saturation has been reached, does not increase the number of atoms of B in the target. After the irradiation has terminated, B decays exponentially. If the half-life of B is very long (e.g., $t_{1/2}$ in years) compared with the irradiation time, the increase of A_B is essentially linear (beginning of the growth curve) with irradiation time.

13.5 MODES AND ENERGETICS OF RADIOACTIVE DECAY

When a radionuclide decays, the radiation emitted usually, but not always, comes from the nucleus. From experimental measurements of the types, energies, and intensities of each type of radiation emitted by a radionuclide, a decay scheme (i.e., an energy level diagram) can be formulated. Such a diagram describes the mode or modes of decay in terms of transitions between energy levels of the two or more nuclei involved. Energy change (in MeV, not necessarily to scale) for the exoergic, spontaneous decay process is represented in the vertical direction and atomic number change in the horizontal direction.

For the decay of nuclide A into nuclide B, the energy change is the sum of the energies of the consecutive transition steps (modes of decay) required to take the system from the lowest nuclear energy level (ground state) in radionuclide A to the ground state in nuclide B. In the simplest case there is one mode of decay from the nuclear ground state of A to the nuclear ground state of a stable nuclide B. It is, however, more likely that for a given mode of

---

[19] $(R_f)_B$ for a target in a nuclear reactor is the mathematical product of three quantities: (1) φ, the neutron flux (neutrons per unit area per unit time), (2) σ, the cross section (reaction probability as an area) for the target nucleus for the specific reaction, and (3) N_A^0, the number of atoms of the specific isotope of the target element in the target. The cross section is expressed in barns, where $1\,b = 10^{-28}\,m^2$ ($10^{-24}\,cm^2$ in the older literature).

transition, nuclide A will decay to two or more nuclear energy levels of nuclide B. These may or may not include the ground state. An experimentally determined fraction of decays goes to each level of nuclide B that is populated by decay. Nuclei of B formed in nuclear excited states may de-excite to the ground state directly or stepwise (in cascade) by way of states with intermediate excitation energy. As discussed in Section 13.4.1, a radionuclide may decay to two different product nuclides, B and C, by two different modes of decay (branched decay) with a fraction of the decays going to each. The fraction for each mode is shown in the decay scheme.

Isomers are species of the same nuclide that differ because their nuclei are in different energy states. If there are two radioactive isomers, they will have different half-lives. In the discussion that follows, an isomeric pair is represented by A and A*, where A* has the higher nuclear energy state. Nuclide A may be stable ($t_{1/2} = \infty$) or radioactive. A* may decay (de-excite) by isomeric transition (IT) into A (discussed in Section 13.5.1), it may decay into B, or it may branch-decay—that is, a fraction may de-excite into A while the remainder decays into B. If A is radioactive, it may also decay into B, it may decay into another product C, or it may branch-decay into both B and C. The number of possible transitions by IT or the other modes of decay described in Section 13.5.1 is even greater for the members of an isomeric triplet A, A*, A**.

Examples of decay scheme diagrams illustrating the different modes of decay are given in the following sections.

13.5.1 Gamma-Ray Emission, Internal Conversion, and Isomeric Transition

When a nucleus is produced in an excited state by radioactive decay or an induced nuclear reaction, it may decay by a process that results in a change in atomic number or it may simply de-excite to a lower energy state or to the ground state by emitting a γ ray at once (i.e., within about 10^{-12} s) or after some delay. The γ rays emitted by a radionuclide are characteristic, monoenergetic photons having energies from about 0.1 MeV to a few MeV. They arise from transitions between nuclear energy states or levels that are unique for each nucleus and are analogous to the transitions between electronic energy levels associated with the inner electron shells of an atom that produce characteristic x rays. There is no change in atomic number, mass number, or neutron number.

The process by which radionuclide A decays to nuclide B* whose nucleus is in an excited energy state and de-excites immediately to the ground state, nuclide B, with the emission of γ rays is represented by the following equation:

$$A \xrightarrow{\lambda_A} B^* \longrightarrow B + \gamma \qquad (13\text{-}28)$$

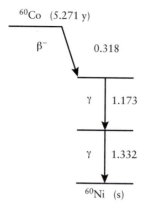

FIGURE 13-9   Decay scheme for ^{60}Co. Energy is in MeV.

There may be one or many monoenergetic γ rays of differing energy emitted per disintegration as B* de-excites. The number of different γ rays depends on how many intermediate excited states there are between the initial state and the ground state and how many transitions take place between the excited states before the ground state (B) is reached. As examples, see the decay schemes for ^{60}Co in Figure 13-9 and ^{131}I in Figure 13-10.[20] These illustrate γ-ray emission following negatron (β^{-}) emission (Section 13.5.2.1). In tabulations of nuclear data, the γ rays emitted in the de-excitation of ^{60}Ni are assigned to ^{60}Co and those of ^{131}Xe to ^{131}I as the radioactive sources of the γ rays.

Because a γ ray has momentum (E_γ/c) and momentum is conserved, the energy available for γ-ray emission is a small amount less than the Q value that corresponds to the decrease in rest mass of the system when the transition occurs. The difference is the recoil energy given to the product atom.[21] For a γ ray with energy E_γ (in MeV), E_{recoil} (in eV) is given[22] by

$$E_{recoil} = \frac{\left(537 E_\gamma^2\right)}{M_{recoil}} \qquad (13\text{-}29)$$

[20]Decay schemes are also drawn giving the energy (MeV) of each level relative to that of the ground state.

[21]When a radionuclide that is a constituent of a molecule decays by γ-ray emission or any of the other modes described in Section 13.5, the recoil energy may be distributed among the different types of internal energy of the molecule. The bond the product atom would have with the rest of the molecule may be broken. If the decay product atom becomes a free atom with kinetic energy above that for the ambient temperature (e.g., about 0.03 eV), it is called a "hot atom" and has properties that are described in discussions of hot atom chemistry.

[22]It is usually satisfactory to substitute A for M in equation (13-29).

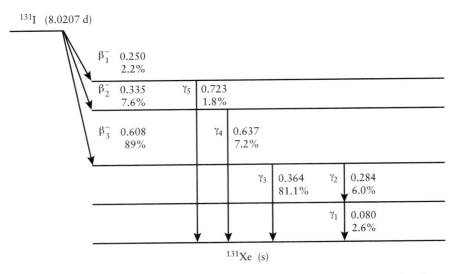

FIGURE 13-10 Simplified decay scheme for ^{131}I. Additional weak transitions ($2\beta^-$ and 4γ) have been omitted. Numbers assigned to the negatron transitions and γ-ray transitions are in the order of increasing energy (in MeV).

Thus a 1.5-MeV γ ray will transfer a recoil energy of 16 eV/atom (1.5×10^3 kJ/mol) to a product atom with $A = 75$.

For some radionuclides a process called internal conversion (IC) reduces the number of γ rays emitted when de-excitation occurs. Instead of emitting a γ ray, the nucleus transfers the energy available for emission of a γ ray to an orbital electron (usually a K electron or an L electron) in the same atom, causing the electron to be ejected from the atom. Conversion electrons are, therefore, monoenergetic, with an energy equal to the available energy minus the binding energy of the detached electron. The probability of IC increases as the energy available for γ-ray emission decreases.

The positively charged ion formed by IC is in an excited electronic state because of a missing inner orbital electron. Characteristic x rays are emitted as the vacancy is filled by an electron from a shell of less firmly bound electrons. Thus, a radioactive source can be a source of characteristic x rays of the product element.

The excited ion may emit Auger (pronounced "ohzhay") electrons instead of x rays. These electrons are monoenergetic. Each has a kinetic energy equal to the energy available for emission of a characteristic x ray minus the electron's binding energy. Auger electrons are emitted from shells having electrons that are less firmly bound than the conversion electron.

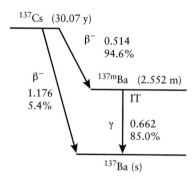

FIGURE 13-11   Decay schemes for ^{137}Cs and ^{137m}Ba. Energy is in MeV.

As described in the preceding section, an isomer A* can de-excite by isomeric transition (IT) with a measurable half-life and with the emission of one or more photons to isomer A. A* is written with its mass number followed by the letter "m" to indicate that the nuclide is in a metastable state. Isomeric transition can be represented by the equation[23]

$$^{Am}_{Z}\text{El} \rightarrow ^{A}_{Z}\text{El} + \gamma \qquad (13\text{-}30)$$

An example of isomeric transition to a stable nuclide is given in Figure 13-11 for the isomer ^{137m}Ba, an environmentally important radionuclide produced by decay of ^{137}Cs. The recoil energy given to the ^{137}Ba atom by the 0.662-MeV γ ray is 1.7 eV (1.6×10^2 kJ/mol). The extent to which the intensity of this γ ray is reduced by internal conversion is indicated by the difference between the percentage (94.6%) of ^{137}Cs that decays to ^{137}Ba and the percentage (85.0%) of such decays that are followed by γ ray emission. In tables of nuclear data, both ^{137}Cs and ^{137}Ba are given as sources of the 0.662-MeV γ radiation.

13.5.2 Beta Decay

13.5.2.1 Negatron (β⁻) Emission

A radioactive decay process in which there is no change in mass number but there is a change in both Z and N is called β decay. There are three such processes. Of the three, the one that was first called β decay after the discovery of radioactivity and α decay was the process in which a negatively charged electron (β^-) was emitted. The species emitted was called a β ray before it was

[23]El is used here as a general representation for any chemical element.

identified as a particle. The other two modes of β decay, positron ($β^+$) emission and electron capture (EC), are described shortly. The terms "β particle," "β decay," and "β emitter" are often used restrictively to mean emission of $β^-$ particles, and a nuclide that emits $β^-$ particles, respectively. For example, carbon-14, which emits only $β^-$ particles, is commonly referred to as a "pure β emitter." The meaning is usually clear from the context.

Although a $β^-$ particle becomes a common electron (e^-) after being emitted, the symbol $β^-$ is used to distinguish it from a common electron (e^-) because of its nuclear origin and its energy characteristics, both of which are described later. For the same reasons, "negatron" rather than "electron" is used interchangeably for $β^-$ particle in this text.

When a nucleus is unstable because it has an excess of neutrons relative to the number for a stable nucleus with the same Z, it emits a negatron and an electron antineutrino, ($\bar{v}_e$),[24] which has no charge. The product nuclide is an isotope of the neighboring element to the right in the periodic table. As N decreases by unity, Z increases by unity and A remains constant. A point representing the radionuclide moves from a position above the stability band toward the stability band in Figure 13-1. The two emitted particles are created and emitted when one of the neutrons in the nucleus is transformed into a proton. Thus, the negatron did not exist as an e^- prior to be emitted. The process is represented by the equation

$$n \rightarrow p^+ + β^- + \bar{v}_e \qquad (13\text{-}31)$$

In this equation and in equations (13-33) and (13-35) the proton is written with a (+) sign as a reminder of its charge, because charge must be balanced in the equation.

The nuclidic change associated with negatron emission is represented by the following process, where parentheses are used to indicate that γ rays may be emitted immediately following negatron emission:

$$^A_Z\text{El} \rightarrow \ _{Z+1}^{A}\text{El} + β^- + \bar{v}_e + (γ) \qquad (13\text{-}32)$$

Negatron emission occurs when there is a transition between a discrete energy level in the emitting nucleus ($_Z$El) and a discrete energy level in the product nucleus ($_{Z+1}$El). The difference between the two levels is fixed and determines the energy available to the negatrons. When many nuclei undergo a given transition, the negatrons are not monoenergetic but have energies between zero and the maximum available and, thereby, fail to conserve energy. They have a characteristic, continuous energy spectrum that goes through a broad maximum and then drops to zero at a finite maximum kinetic energy

---

[24] There are three types of neutrino. Each has its antiparticle (antineutrino), which differs from the neutrino with respect to the direction of its intrinsic spin. The electron neutrino ($\bar{v}_e$) is associated with positron emission (Section 13.5.2.2) and electron capture (Section 13.5.2.3).

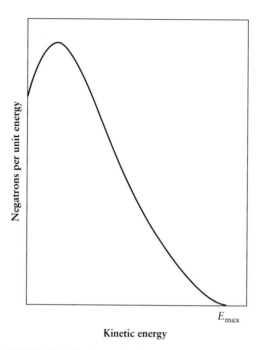

FIGURE 13-12   Example of a negatron spectrum.

E_{max} (the "end point energy") as shown in Figure 13-12. The energies listed for β particles in tables and charts of nuclear data are E_{max} values. Details of the spectral shape vary somewhat from one β emitter to another, depending on the changes in nuclear properties that occur in the transition. The average energy E_{ave}, which must be used in calculating the dose a person receives from exposure to β radiation (Section 13.12.2), is 30–40% of E_{max}, depending on the actual shape of the spectrum.

According to the generally accepted theory of β decay, the continuous energy spectrum of negatrons can be attributed to the simultaneous creation and emission of an electron antineutrino, which has no charge and a negligible rest mass (perhaps a very small fraction, e.g., 6×10^{-8} to 6×10^{-6}, of the rest mass of an electron.)[25] Both particles have continuous energy spectra and share the β-transition energy, keeping the total transition energy (sum of the kinetic energies of the two particles) constant. Thus, the law of conservation of energy is not violated.

Antineutrinos participate not only in the conservation of energy in negatron decay but also in the conservation of nuclear angular momentum, which

---

[25] The rest mass m_e of an electron is 5.485×10^{-4} amu, which is equivalent to 0.511 MeV.

requires that the nuclear spin I change by 0 or an integer in a nuclear reaction. Because a negatron (as an electron outside of the nucleus) has an intrinsic spin quantum number of $1/2$, its emission would result in a half-integer change in I. Simultaneous emission of an antineutrino, which also has an intrinsic spin quantum number of $1/2$, results in the required integer change in I.

When negatron decay occurs, the system loses m_e the rest mass of an electron (β^- particle emitted by nuclide A), but it also gains m_e because nuclide B contains one more orbital electron than nuclide A. The total energy change is equal to the sum of E_{max} for the negatron (the energy of the β^- particle plus that of the accompanying electron antineutrino) and the recoil energy acquired by nuclide B. If nuclide A decays to one or more excited states of the nucleus of nuclide B, each β^- transition is followed by one or more γ transitions as the nucleus de-excites to the ground state. The total transition energy ($\beta^- + \gamma$) is the sum of E_{max} for a given β^- transition and the energy of the photon(s) (or conversion electrons) emitted immediately after the β^- transition plus recoil energy. For example, the ($\beta^- + \gamma$) transition energy (0.970 MeV) for the decay of ^{131}I to ^{131}Xe (Figure 13-10) is equal to $\beta_1^- + \gamma_5$; $\beta_2^- + \gamma_4$; $\beta_3^- + \gamma_3$; and $\beta_3^- + \gamma_1 + \gamma_2$. Because of experimental errors in the β^- particle and γ-ray energies, the sums may differ slightly. The discrepancy between the frequency of emission of γ_2 and γ_1 results from the high degree of internal conversion of γ_1.

Negatron decay for three environmentally important, pure negatron emitters is illustrated in Figure 13-13 for ^{14}C, and the parent–daughter fission products, ^{90}Sr and ^{90}Y. The decay scheme for tritium (^{3}H) is like that for ^{14}C except that the maximum negatron energy is 0.0186 MeV. The more common mode in which negatron decay is followed by γ emission is illustrated by ^{60}Co and ^{131}I in Figures 13-9 and 13-10, respectively.

Calculation of the recoil energy given to the product atom following emission of a β^- particle is more complex than for γ-ray emission. Two examples of the maximum E_{recoil} from pure negatron emitters are as follows:

1. About 3.5 eV (3.4×10^2 kJ/mol) for ^{3}H (12.32 y, $E_{max} = 0.0186$ MeV), which decays to ^{3}He (s)
2. About 45 eV (4.3×10^3 kJ/mol) for ^{90}Y (Figure 13-13)

Although these recoil energies, like those from γ rays, are not large enough to significantly reduce the net energy of the radiation, they can cause disruptive chemical effects.

13.5.2.2 Positron Emission

A positron (a positively charged electron, e^+, written here as β^+), and a neutrino, ν_e, are created and emitted from a nucleus when a proton is transformed into a neutron. The positron is an antielectron, which is one of the

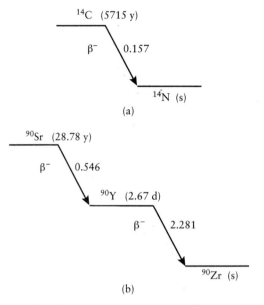

FIGURE 13-13   Decay schemes for (a) ^{14}C and (b) ^{90}Sr $-^{90}$ Y. Energy is in MeV.

antiparticles that make up antimatter. This decay process occurs in nuclei having a ratio of neutrons to protons that is too low for stability. In decay by positron emission Z decreases by unity, N increases by unity, and A remains constant as the nuclide moves from a position below the stability band in Figure 13-1 toward the stability band. The equations analogous to equations (13-31) and (13-32) are

$$p^+ \rightarrow n + \beta^+ + \nu_e \tag{13-33}$$

$$^A_Z El \rightarrow \, _{Z-1}^{A}El + \beta^+ + \nu_e + (\gamma) \tag{13-34}$$

Like negatrons, positrons have continuous spectra with an E_{max}, but with the broad maximum shifted toward higher energy. The role of the neutrino is analogous to that of the antineutrino in negatron emission. Because of their very small mass and the absence of a charge, neutrinos and antineutrinos have a negligible probability of interacting with matter and are, therefore, extremely difficult to detect. Since, however, the sun is a strong source of these particles, they are not in short supply.

In positron emission the rest mass of the system decreases by at least twice the rest mass of an electron. One electron is the β^+. The other is the extra-nuclear electron no longer needed to balance the nuclear charge of the product

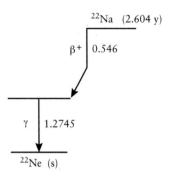

FIGURE 13-14   Decay scheme for ^{22}Na. Energy is in MeV.

nuclide. For positron emission to be energetically possible, the Q value must be at least 1.022 MeV (refer again to footnote 25) $E_{\max}$ for positrons for a given transition is equal to the total transition energy minus 1.022 MeV.

In the decay scheme for a positron emitter, the 1.022 MeV is indicated by the vertical section of the line for the β^+ transition to the decay product, as shown for ^{22}Na in Figure 13-14.

13.5.2.3 Electron Capture (EC)

When N/Z for a radionuclide is favorable for β^+ emission, but the Q value is less than 1.022 MeV, thus preventing positron emission, decay to the same product can occur by an alternative and competitive process called electron capture. The unstable nucleus can capture one of its surrounding, bound electrons—usually a K electron or, less likely, an L electron.[26] When the electron is captured, a proton is converted to a neutron in the nucleus and a monoenergetic electron neutrino, which conserves energy and nuclear spin, is emitted. The reactions are

$$p^+ + e^- \rightarrow n + \nu_e \tag{13-35}$$

and

$$^A_Z El + e^- \rightarrow \,_{Z-1}^{A}El + \nu_e + (\gamma) \tag{13-36}$$

To detect EC it would seem to be necessary to measure the E_{recoil} given to the product atom by the neutrino. Fortunately, electron capture is followed by electronic transitions in the product atom and sometimes by nuclear transitions in the product nucleus. These transitions are accompanied by the emission of detectable radiation. Very low intensity electromagnetic

---

[26]EC was called K-capture in the older literature.

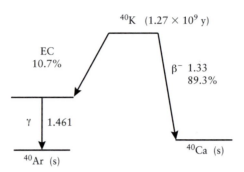

FIGURE 13-15   Decay scheme for ^{40}K. Energy is in MeV.

radiation (x radiation with a continuous spectrum) called inner bremsstrahlung is emitted when an electron (in this case the captured electron) experiences a rapid change in speed in the vicinity of an atomic nucleus.[27] The bremsstrahlen (bremsstrahlung photons) share the energy available to the monoenergetic neutrinos. Characteristic x rays of the product element are also emitted, because EC, like internal conversion, leaves a vacancy in, say, the K shell of the product atom, thereby establishing the conditions for electronic transitions that produce characteristic x rays. Auger electrons may be emitted instead of the x rays. If the product nucleus is formed in an excited state, γ rays characteristic of the product nucleus or conversion electrons also will be emitted.

The threshold energy for EC is the work required to detach the bound electron that is captured (i.e., the electron's binding energy). As examples, the K-electron binding energy is 8.98 keV for Cu and 88.00 keV for Pb. These are two of the many elements having radioisotopes that decay by EC. The decay scheme for ^{40}K (Figure 13-15), a primordial radionuclide of environmental importance, illustrates branched decay with β$^-$ particle emission and EC plus γ emission. EC often accompanies β$^+$ emission.

13.5.3 Alpha Decay

For some unstable nuclei, α particles, doubly charged ^{4}He nuclei, are emitted in monoenergetic groups corresponding to the finite differences between energy levels in the emitting and the product nucleus.

The process can be represented by the equation

$$_{Z}^{A}\text{El} \rightarrow {}_{Z-2}^{A-4}\text{El}^{2-} + {}_{2}^{4}\text{He}^{2+} + (\gamma) \tag{13-37}$$

[27]Section 13.7.1.3 contains a discussion of bremsstrahlung.

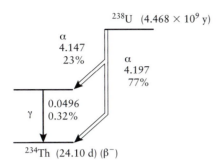

FIGURE 13-16   Decay scheme for ^{238}U. Energy is in MeV.

Thus, $\triangle Z = -2$, $\triangle N = -2$, and $\triangle A = -4$. Gamma rays are usually emitted also, as shown in the relatively simple decay scheme for α-emitting ^{238}U in Figure 13-16.

It is assumed that assemblies of nucleons containing two protons and two neutrons exist in nuclei and are emitted as α particles when reaction (13-37) is exoergic. Because of barrier penetration, the energies of α particles emitted in radioactive decay are generally much lower than would be expected on the basis of the classical Coulomb barrier. Thus, for ^{232}Th ($t_{1/2} = 1.40 \times 10^{10}$y) the barrier height [equation (13-10)] is 47 MeV, but the energy of the α particle is 4.01 MeV. The long half-life of ^{232}Th indicates that the probability of barrier penetration at the observed energy is very low.

If γ rays are emitted, the total disintegration energy is equal to the energy of each α transition plus the energy of the associated γ rays or conversion electrons plus the recoil energy of the product atom. The recoil energy E_{recoil} is relatively high for α particles. It is given approximately by $E_{recoil} = M_\alpha E_\alpha / M_{prod}$. For example, the recoil energy given to a ^{206}Pb atom formed when the nucleus of an atom of ^{210}Po (138.38d) emits a 5.3044-MeV α particle is 0.10 MeV (9.6×10^6 kJ/mol).

13.5.4 Other Modes of Decay

A few isotopes of the transuranic elements (elements beyond uranium in the periodic table) decay by spontaneous fission (SF) in which the nucleus splits into two unequal fragments.[28] An example is ^{252}Cf (2.646 y),

---

[28]Nuclear fission (discussed in Section 13.6.3) is normally induced by the nuclear reaction in which a fissionable nucleus (e.g., ^{235}U) captures a neutron to form a nucleus that typically splits into two different, smaller nuclei plus a few neutrons.

which decays by α emission (96.9%) and by SF (3.1%) with the release of an average of 3.76 neutrons per fission.

Neutrons, called delayed neutrons, are emitted by a few of the short-lived radioactive products formed by nuclear fission. These neutrons are emitted immediately after the emitting nuclide is formed in an excited nuclear energy state by a short-lived, β⁻-emitting precursor.

In addition to the environmentally significant modes of radioactive decay already examined, there are rare modes such as double β (2β⁻); double α; delayed proton, and double-proton emission; and emission of nuclei such as ^{3}H, ^{14}C, and ^{24}Ne. These do not warrant further consideration here.

13.6 SELECTED INDUCED NUCLEAR REACTIONS

13.6.1 Charged-Particle Reactions

Particle accelerators provide the kinetic energy needed for charged particles such as protons (^{1}H$^+$ or p), deuterons (^{2}H$^+$ or d), helium ions (^{4}He^{2+} or α) and heavier ions to react with nuclei of target atoms. The rate of reaction (Section 13.4.2) is proportional to the accelerator beam current (particles/s). As discussed in Section 13.3, charged incoming particles require energy to overcome the repulsion energy of the Coulomb potential barrier around a nucleus for both endoergic and exoergic reactions. The probability of barrier penetration increases and, therefore, the rate of reaction increases with increasing kinetic energy of the particle. However, when the particle energy is increased to achieve an increase in the rate of production of a particular product, other reactions with the target nuclei may become energetically feasible and contaminate the desired product.

A long-lived, α-emitting source combined with spectrometers in a portable, remotely controlled device can be used for chemical analysis by means of the characteristic energies of protons emitted by nuclei of the sample when the (α,p) reaction occurs and the energy of the x rays emitted after the α particles ionize the atoms of the sample.

13.6.1.1 Nuclear Fusion

Nuclear fusion is a thermonuclear reaction in which both reacting nuclei are given sufficient kinetic energy to penetrate the Coulomb repulsion barrier, thus allowing the nuclei to fuse. It has long been known that certain exoergic reactions can occur when a nucleus of deuterium combines with or fuses with another nucleus of deuterium or with one of tritium.

Among the various fusion reactions that deuterium and tritium can undergo are the following:

$$^2\text{H} + {}^3\text{H} \rightarrow {}^4\text{He} + {}^1\text{n} \quad (Q = 17.58\,\text{Mev}) \tag{13-38}$$

$$^2\text{H} + {}^2\text{H} \rightarrow {}^3\text{He} + {}^1\text{n} \quad (Q = 3.27\,\text{Mev}) \tag{13-39}$$

$$^2\text{H} + {}^2\text{H} \rightarrow {}^3\text{H} + {}^1\text{H} \quad (Q = 4.03\,\text{Mev}) \tag{13-40}$$

$$^2\text{H} + {}^2\text{H} \rightarrow {}^4\text{He} + \gamma \quad (Q = 24.85\,\text{Mev}) \tag{13-41}$$

D-D or (d-d) (deuteron–deuteron) reactions (13-39) and (13-40) are about equally probable; reaction (13-41) has a very low probability.

The mean kinetic energy required by the reactants depends on the specific reaction. For example, the D-T or (d-t) (deuteron–triton) reaction (13-38) requires about 15 keV (corresponding to a temperature of about 1.5×10^8 K), which is about one-tenth of that required for the D-D reactions. At these kinetic energies the electrons are stripped from the atoms to form a plasma, that is, a mixture of electrons and positive ions in which the fusion reaction takes place.

13.6.2 Photonuclear Reactions

Electrons that have been given energy (typically at least 10 MeV) in an electron accelerator can become an intense source of high-energy photons (bremsstrahlung).[29] These photons can be used for photonuclear reactions such as (γ, n), $(\gamma, 2n)$, and (γ, p).

High-energy γ rays emitted by radionuclides will undergo photonuclear reactions when absorbed by the nuclei of elements with very low atomic number (Section 13.6.4).

13.6.3 Nuclear Fission

When a ^{235}U nucleus captures a thermal (slow) neutron,[30] the resulting very short-lived compound nucleus $[^{236}\text{U}]$ may de-excite by emitting γ radiation to form the long-lived, α-emitting ^{236}U by the (n, γ) reaction, or it may deform

---

[29]Electrons accelerated in a linear electron accelerator (LINAC) emit bremsstrahlung when they are stopped in a water-cooled metallic target.

[30]See Section 13.7.3 for a discussion of fast and slow neutrons.

into the shape of a dumbbell and then undergo the induced fission (n,f) reaction and split into two fragments.[31] The two reactions are

$$^{235}_{92}U + ^1_0 n \rightarrow [^{236}_{92}U] \rightarrow ^{236}_{92}U(2.342 \times 10^7 y) + \gamma \qquad (13\text{-}42)$$

and

$$^{235}_{92}U + ^1_0 n \rightarrow [^{236}_{92}U] \rightarrow ^A_Z El + ^{236-A-x}_{92-Z}El + x^1_0 n + \gamma \qquad (13\text{-}43)$$

where Z is between 28 and 68 and x is the number of prompt neutrons i.e., neutrons emitted during the fission event). Fission is the more likely reaction by a factor of 5.91.

Equation (13-43) represents the *overall* nuclear fission reaction.[32] Prompt neutrons are fast neutrons having a continuous energy spectrum with an upper limit of about 3 MeV with an average of about 2 MeV. When averaged over a large number of fissions, the number of prompt neutrons released per thermal fission of ^{235}U is 2.46. Because more than one prompt neutron is released per neutron absorbed by ^{235}U, it is possible to establish the self-sustaining, fission chain reaction that takes place in a nuclear reactor.

Although fast neutrons also will induce fission in ^{235}U, the cross section (see footnote 19) for fission is much larger for thermal neutrons. The nuclides ^{235}U, ^{239}Pu, ^{233}U, and ^{241}Pu are fissionable with fast or thermal neutrons and are called fissile nuclides. Plutonium-239 is produced from ^{238}U ($4.47 \times 10^9 y$) by the following successive steps:

$$^{238}U(n, \gamma)\ ^{239}U\ (23.47\ m) \rightarrow \beta^- + ^{239}Np\ (2.355\ d) \rightarrow$$
$$\beta^- + ^{239}Pu\ (2.410 \times 10^4\ y) \rightarrow \alpha + ^{235}U \qquad (13\text{-}44)$$

and so on.[33] This is the basis for the production of ^{239}Pu for nuclear weapons and for a type of nuclear reactor fuel that contains ^{239}Pu.

Uranium-233 is made from ^{232}Th ($1.40 \times 10^{10} y$) by the following reaction and decay sequence:

$$^{232}Th(n, \gamma)^{233}Th\ (22.3\ m) \rightarrow \beta^- + ^{233}Pa\ (27.0\ d) \rightarrow$$
$$\beta^- + ^{233}U\ (1.592 \times 10^5\ y) \rightarrow \alpha + ^{229}Th \qquad (13\text{-}45)$$

and so on.

---

[31]Certain aspects of the fission process can be understood in terms of the liquid drop model of the nucleus. In this model, the binding of nucleons in an atomic nucleus and the surface properties of a nucleus are considered to be analogous to those of a drop of liquid composed of molecules.

[32]In less than 1% of the fission events a very light fragment such as ^3H or ^4He also may be emitted.

[33]The trace amounts of ^{239}Pu that have been found in uranium ores are produced mostly by reaction (13-44) with neutrons that are released by both ^{235}U and ^{238}U when they undergo spontaneous fission (with very low probability).

A number of heavy nuclides including ^{232}Th and ^{238}U are fissionable with fast neutrons. These two are also classified as fertile because they can be used to produce the fissile nuclides ^{233}U and ^{239}Pu, respectively, when they capture thermal neutrons.

After the highly unstable and very short-lived nuclei of the fission fragments have emitted negatrons and γ says, they become primary fission products, most of which are radioactive with measurable half-lives and characterizable radiation. They decay into secondary fission products by negatron emission because of the high neutron-to-proton ratio that they inherit from ^{235}U. Some of the fragments are long-lived and some are stable, and thus the observable fission products may be fragments or decay products. Commonly, an unstable fission product will decay into an unstable daughter, which in turn may decay into another radionuclide, and so on, as illustrated in reaction (13-21). Decay chains (increasing Z at constant A) are terminated with the formation of stable nuclides that are in the band of stable nuclides in Figure 13-1. Fission products with half-lives between a small fraction of a second and 10^{15} years are distributed among 97 chains, which have, on the average, three members.

Fission of ^{235}U with thermal neutrons can produce over 525 known radioactive fission products having mass numbers in the range 72 to 167. The large number of fission products reflects the relative rareness of symmetrical fission into two fragments of equal mass number when fission is induced by thermal neutrons. This gives rise to the double-peaked fission yield curve shown in Figure 13-17.[34] It is most likely that the mass numbers of the fission products will be in the range of 90 to 100 for the light fission product and 133 to 143 for the heavy fission product.[35] The fission yield for a given mass number is the total (maximum) yield for the chain of that mass number. It is the percent of fissions in which a fission product chain with a given mass number is formed. Individual fission products may be formed directly and, therefore, contribute to the cumulative yield along the chain. The total yield is 200%, since two fission product nuclei are produced by fission of one ^{235}U nucleus.

The fission products have a wide range of chemical behavior represented by the 41 elements from nickel through erbium. Twelve are members of the lanthanide series, and two, technetium ($Z = 43$) and promethium ($Z = 61$), are not only radioelements (i.e., all isotopes radioactive), but also have no naturally occurring isotopes. There are also the previously mentioned (Section 13.5.4), short-lived (seconds or less) delayed-neutron emitters among the

---

[34]Fission of ^{235}U becomes a more symmetrical process (single peak at the center of the yield curve) as the energy of the incident neutrons is increased. The yield curve for the fission of ^{239}Pu induced by thermal neutrons is similar to that for ^{235}U with the low-mass peak shifted toward slightly higher A values.

[35]At the time of fission, the atomic numbers of the two fragments must add up to 92, and when they become fission products, their mass numbers must add up to 236 minus the total number of prompt neutrons released.

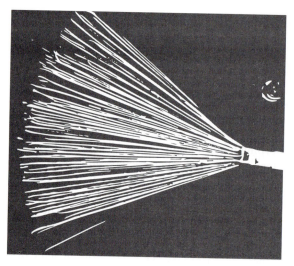

FIGURE 13-18   Cloud chamber photograph of tracks of α particles emitted by an α-emitting sample located at the right of the tracks. From I, Asimov and T. Dobzhansky, *The Genetic Effects of Radiation*, U.S. Atomic Energy Commission, Division of Technical Information, Oak Ridge, TN, 1966.

and short, nonlinear tracks caused by the secondary electrons branching from the main track. These short tracks are called delta rays. About two-thirds of the ionization produced by an α particle is secondary ionization.

In general, α particles emitted by radionuclides have energies between 4 and 11 MeV, with most between 5 and 9 MeV. Values for the linear range in air of α particles with energy between 2 and 11 MeV are given in Table 13-3.

As an illustration of the penetration of α particles and the use of the two units of absorber thickness, a 5.0-MeV α particle has the following linear ranges: 35,000 μm in air, 20 μm in aluminum, and 35 μm in water or tissue. The corresponding mass thicknesses are 4.5 mg/cm^{-2} in air, 5.5 mg/cm^{-2} in aluminum, and 3.5 mg/cm^{-2} in water or tissue. A piece of 20-lb copy paper typically has a linear thickness of 100 μm and a density thickness of 7.4 mg/cm^2. It will stop 6.0-MeV α particles.

13.7.1.3 Negatrons and Positrons

When the tracks of energetic electrons (e^-, β^-, e^-, β^+) in air are compared with those of α particles, three major differences are apparent. First, the specific ionization of electrons is much less than that for α particles. For example, the number of ion pairs increases from about 5 ion pairs per millimeter at the

TABLE 13-3

Mean Range of α Particles in Air[a]

Energy (MeV)	Range (cm air)
2	1.0
4	2.5
5	3.5
6	4.6
7	6.0
8	7.4
9	9.0
10	10.5
11	12.5

[a]Mean range takes into account straggling of α particles. That is, the track lengths for α particles with a given energy show a small random scattering about a mean value. Values are for dry air at 15°C and 1.013×10^5 Pa (1 atm).

beginning of the track of a 1-MeV electron to about 1000 ion pairs per millimeter near the end of the track. Second, the track of an energetic electron changes direction many times because an electron is easily scattered by interaction with atomic and molecular electrons. Third, consistent with the lower specific ionization, the total track length per MeV of particle energy is greater for electrons than for α particles in a given absorber.

The negatron tracks shown in the cloud chamber photograph in Figure 13-19 illustrate how easily an electron is scattered and how the specific ionization increases along the track and eventually approaches that of an α particle. The photograph does not show how an electron loses energy along its track also by transferral of an amount less than that needed to cause ionization of absorber atoms and molecules but sufficient to cause electronic excitation.

As energetic electrons slow down, they lose energy in an absorber not only by ionization and excitation, but also by producing bremsstrahlung (braking radiation). Emission of bremsstrahlung from the target (anode) of an x-ray tube or generator is well known. The radiation has a continuous spectrum and, therefore has been been called "white" radiation. Figure 13-20 shows bremsstrahlung with superimposed characteristic x rays emitted from a tungsten target that is irradiated with monoenergetic electrons for four values of applied voltage. The area under the curves represents the energy radiated. Wavelengths, λ, in angstrom units, can be converted to energy by the equation $E \text{ (keV)} = 12.4/\lambda$. Note that most of the energy radiated is in the form of

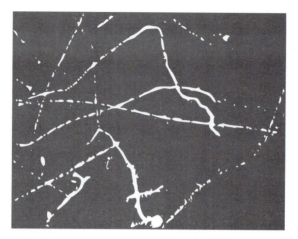

FIGURE 13-19   Cloud chamber photograph of tracks of negatrons emitted by a β^--emitting sample located at the left of the tracks. From I. Asimov and T. Dobzhansky, *The Genetic Effects of Radiation*, U.S. Atomic Energy Commission, Division of Technical Information, Oak Ridge, TN, 1966.

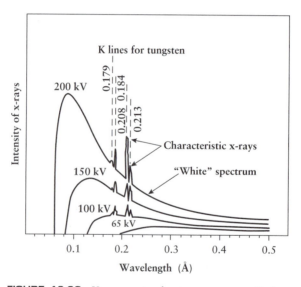

FIGURE 13-20   X-ray spectra for tungsten target. Redrawn from E. F. Gloyna and J O. Ledbetter, *Principles of Radiological Health* by courtesy of Marcel Dekker, Inc., New York. Copyright © 1969.

bremsstrahlung, the quantity increases with increasing energy of the electrons, and the increase is in the high-energy region of the spectrum.

When negatrons and positrons emitted by a radioactive source are stopped in an absorber, they emit bremsstrahlung having a spectrum peaked at a relatively lower energy than that produced by monoenergetic electrons in an x-ray tube because of the continuous spectrum of β particles. The ratio of the energy lost in bremsstrahlung production to the energy lost by ionization is given approximately by 1.25×10^{-3} EZ, where E is E_{max} and Z is the atomic number of the absorbing material. Thus, the relative importance of bremsstrahlung production increases with the increasing energy of the electrons and with increasing atomic number of the absorber. The energy range of the continuous bremsstrahlung spectrum is from zero up to E_{max}. As x rays, bremsstrahlen interact with matter in the manner described shortly (Section 13.7.2). Although only a small fraction of the β-particle energy is lost in an absorber (e.g., a container) by bremsstrahlung production, the intensity of this penetrating radiation emitted by the absorber can be important for high-activity sources. As an example, a source of ^{90}Sr $-^{90}$Y enclosed in aluminum would emit one bremsstrahlen photon for about 15 $β^-$ particles.

Positrons ($β^+$ or e^+) differ from negatrons ($β^-$) or electrons (e^-) as they approach thermal energy in the slowing-down process. At the end of the track, they may exist briefly as "positronium" (e^+e^-) before the two electrons are annihilated and are replaced by annihilation photons. Usually, the energy of the two annihilated electrons appears as two 0.511-MeV photons that are emitted at an angle of 180° to each other, but a single photon of 1.022 MeV may be emitted. Thus, an absorber of positrons is also a radiator of the 0.511-MeV annihilation photons that cause ionization by processes described in the next section.

Negatrons and positrons have finite ranges. They will travel up to several meters in air and several millimeters in tissue, as a function of the β-particle energy, E_{max}. Because of scattering, however, the track length of a negatron or positron in an absorber is longer than the depth of penetration. The range in an absorber is the finite thickness of absorbing material required to absorb (stop) the particles with the highest energy (E_{max}).

Because of the continuous spectra of negatrons and positrons and because of other factors, the absorption of negatrons and positrons is approximately exponential for absorbers having a thickness much less than the range, as just defined. The absorption becomes nonexponential as the thickness of the absorber approaches the finite range. Values for the range of negatrons and positrons in air, water, and aluminum are given in Table 13-4. A number of

TABLE 13-4

Range of β Particles for Selected Energies and Absorbers

Energy	Range			
(MeV)	Absorber X (mg/cm^{-2})^a	Air (m)^b	Water (mm)^c	Aluminum (mm)^d
0.05	4.0	0.034	0.040	0.015
0.1	14	0.12	0.14	0.05
0.3	76	0.64	0.76	0.28
0.5	155	1.31	1.55	0.57
0.7	250	2.11	2.50	0.93
1.0	400	3.38	4.00	1.48
2.0	950	8.02	9.50	3.52
3.0	1500	12.66	15.00	5.55

^{a}Absorber X is any absorber with low atomic number such as those in this table.
^{b}Air density $= 1.185\,mg/cm^3$ $(1.185\,kg/m^3)$ for dry air at 15°C, 1.013×10^5 Pa.
^{c}Water (or soft tissue) density $= 1.00\,g/cm^3 (1.00\,Mg/m^3)$.
^{d}Aluminum density $= 2.70\,g/cm^3 (2.70\,Mg/m^3)$.

empirical equations are available for relating R, the range, of β particles in aluminum, to E_{max}.[45] One such equation for E_{max} in MeV above 0.8 MeV is

$$R(g/cm^2) = 0.542E_{max} - 0.133 \qquad (13\text{-}46)$$

Another equation for absorption of β particles in aluminum for E_{max} between 0.15 and 0.8 MeV is

$$R(g/cm^2) = 0.407E_{max}^{1.38} \qquad (13\text{-}47)$$

A third equation for absorption in low-Z material (e.g., plastic) and for the energy range 0.01–2.5 MeV is

$$R(mg/cm^2) = 412E_{max}^{(1.265-0.0954\ln E_{max})} \qquad (13\text{-}48)$$

As an approximation based on equation (13-46), the range of β particles in an aluminum absorber expressed as grams per square centimeter is numerically equal to about $1/2E_{max}$ for β particles with energy above 0.8 MeV.

Although negatrons and positrons emitted by radionuclides are usually completely absorbed within a thickness of a few millimeters to a centimeter of a substance like glass, not all their energy remains in the absorber. As pointed out earlier, the absorber can become a source of high-energy photons that are more penetrating than the incident radiation.

---

[45]The equations were developed for the determination of E_{max} for β particles by measurement of their range in aluminum by using a graded series of aluminum foils differing in thickness.

13.7.2 X Rays and Gamma Rays

As photons, x rays and γ rays do not ionize atoms and molecules of an absorber by Coulomb interaction, but they do cause ionization indirectly. High-energy x rays and γ rays interact with matter in four ways: photoelectric absorption, Compton scattering, pair production, and photonuclear reaction. Only the first three of these modes are environmentally important. In each case the incident photon produces ion pairs within the absorbing medium, if it interacts at all. When ion pairs are formed, the electrons of the ion pairs cause further ionization as described for the interaction of charged particles.

13.7.2.1 Photoelectric Effect

In the photoelectric effect, the primary photon disappears in a single absorption event in which the photon transfers all its energy to a bound electron in an inner shell of an absorber atom. That electron, a photoelectron, is ejected from the atom with an energy equal to the energy of the absorbed photon minus the binding energy of the electron. A photoelectron may escape from the absorber intact, or it may transfer some or all of its energy to the absorber by causing ionization and electronic excitation of the absorber atoms. Because the residual positive ion has lost an inner electron rather than a weakly bound outer electron, it is in an excited electronic state of the type that leads to the emission of characteristic x rays of the absorber. These x rays may be absorbed by the photoelectric process or they may escape from the absorber. Auger electrons may be emitted instead of the x rays, as described in Section 13.5.1. Photoelectric absorption is relatively important for γ rays or x rays with energy less than about 100 keV. Its probability increases with the fifth power of the atomic number of the absorber and decreases with the 7/2 power of the photon energy.

Monoenergetic electrons that are emitted instead of γ rays when internal conversion (IT) occurs, and the monoenergetic Auger electrons that are emitted instead of characteristic x rays are essentially equivalent to electrons produced by photoelectric absorption. The difference is that photons are not involved in IC or Auger electron emission.

13.7.2.2 Compton Scattering

When the Compton process occurs, a portion of the energy of a primary (incident) photon is transferred to an electron (not a firmly bound one) during a scattering interaction. After the scattering event, the scattered photon has less energy relative to the primary photon by an amount equal to the energy transferred as kinetic energy to the Compton electron. The distribution of energy between the scattered photon and the Compton electron can be

where D_x is the absorbed dose of a reference radiation [e.g., x rays (200 keV)] required to produce the same biological effect as that produced by an absorbed dose D_R for radiation R. Because RBE is usually determined experimentally by irradiation of plants and laboratory animals at dose rates and levels high enough to permit observation of a particular biological effect, it was not considered suitable as a modifying factor in dose calculations for radiation protection of humans. The goal of radiation protection is to set dose limits *below* which the probability of occurrence of harmful effects of ionizing radiation at low doses and low dose rates will be negligible.

For the purpose of radiation protection, the importance of the quality of the ionizing radiation in determining the occurrence of stochastic effects was initially introduced by multiplying the absorbed dose(from external exposure) at a point by a weighting factor, the *quality factor*, Q or QF. The new quantity, the dose equivalent H was defined by

$$H = DQ \tag{13-63}$$

If D, the absorbed dose, is in rads, H has the same energy units as the rad and is called the rem (roentgen–equivalent–man). If D is in grays (Gy), the name of the unit of H is the sievert (Sv). In terms of SI units, $1 Sv = 1 J/kg$ and is, therefore, equivalent to 100 rems. The quality factor Q for a specified radiation is similar to RBE and many of the values are the same, but Q is evaluated as a function of the LET of the radiation in water. As discussed in Section 13.7.4, the LET of a radiation depends on its type and energy. Table 13-7 shows the correlation

TABLE 13-7

Quality Factors and Linear Energy Transfer (LET) Values for Various Types of Ionizing Radiation

Type of radiation	Quality factor Q^a	LET $(keV/\mu m)^{b-d}$
X, gamma, or beta (β^-, β^+)	1	≤ 3.5
Alpha, multiple-charged particles, fission fragments, heavy ions of unknown charge	20	≥ 175
Neutrons of unknown energy	10	53
High-energy protons	10	53

[a]Values in this column are used by the U.S. Nuclear Regulatory Commission, *Code of Federal Regulations*, Part 10, Title 20, 1004, January 2001.
[b]For water.
[c]Additional values are $Q = 1$–2 for LET 3.5–7.0; $Q = 2$–5 for LET 7.0–23; $Q = 5$–10 for LET 26–53; $Q = 10$–20 for LET 53–175.
[d]Values of Q can also be calculated from mathematical relationships between Q and LET for ranges of LET.

between Q and LET. If the radiation field contains a mixture of types, the values of H for each are summed.

From equation (13-63) and the values for Q in Table 13-7, we find that for external exposure, a given tissue or organ receiving an absorbed dose of only 0.05 Gy from α radiation would receive a dose equivalent to that from 1 Gy of γ (or x or β^- or β^+) radiation.

The International Commission on Radiological Protection (ICRP), which provides recommendations that nations use to establish regulations for protection against ionizing radiation, redefined some of the dose quantities and their method of calculation in 1977[62] and again with ICRP Publication 60 in 1990.[63] Changes in federal regulations on radiation protection (or any other safety regulation) affect the procedures used in industrial, government, and privately owned facilities and are usually made slowly after consideration of comments received from parties who may be affected by the proposed changes. Thus, the recommendations and regulations on radiation protection set by several agencies, commissions, and administrations of the U.S. government and published in the *Code of Federal Regulations* (CFR) are updated annually, but may not keep pace with recommendations of the ICRP and the U.S. National Council on Radiation Protection and Measurements (NCRP).

A new weighting factor, the radiation weighting factor w_R, was introduced by the ICRP to take into account the effect of the quality of the radiation (R) when the absorbed dose $(D_{T,R})$ was averaged over a tissue or organ (T) irradiated by an external or internal source. The new, weighted absorbed dose H_T, the equivalent dose (in sieverts), was defined by

$$H_T = \sum_R w_R D_{T,R} \qquad (13\text{-}64)$$

The summation applies when the radiation consists of components having different values of w_R. Values of w_R for different types of radiation and energy range are given in Table 13-8. They were selected by the ICRP to be representative of values of RBE of the radiation for causing stochastic effects at low doses and low dose rates. They are also compatible with the LET-related values of Q.

Because the equivalent dose does not take into account the dependence of the probability of occurrence of a stochastic effect on the particular tissue or organ (T) that is irradiated, the ICRP defined (in 1990) another weighting factor, w_T, the tissue weighting factor, for weighting the equivalent dose (H_T).

[62]*Recommendations of the ICRP*, ICRP Publication 26, Pergamon Press, New York, 1977. (Reprinted with additions in 1987.)

[63]*1990 Recommendations of the ICRP*, ICRP Publication 60, Pergamon Press, New York, 1991. Supersedes *ICRP Publication 26*.

TABLE 13-8

Radiation Weighting Factors

Radiation type and energy range	Radiation weighting factor, w_R
Photons, all energies	1
Electrons, all energies	1
Neutrons	
< 10 keV	5
10 keV to 100 keV	10
> 100 keV to 2 MeV	20
> 2 MeV to 20 MeV	10
> 20 MeV	5
Protons, other than recoil protons, > 2 MeV	5
Alpha particles, fission fragments, heavy nuclei	20

Source: Based on Table 1 of ICRP Publication 60, *1990 Recommendations of the International Commission on Radiological Protection*, Pergamon Press, New York, 1991.

This new weighting factor is a measure of the contribution of an organ or tissue T to the total of the health effects arising from whole body irradiation. The factor w_T is associated with the stochastic risk per sievert for the specified tissue or organ. The new dose, E, also in sieverts, is the effective dose and, when summed over all tissues, is defined by the equation

$$E = \sum_T w_T H_T. \tag{13-65}$$

Thus, the effective dose is defined by a double summation. Previously (1977) this dose was called the effective dose equivalent and the symbol was H_E.

The weighting factor w_T is independent of the quality of the radiation and is chosen so that for a uniform equivalent dose over the whole body, the sum of w_T values is unity. The value of w_T for a tissue or organ T is a measure of the stochastic risk from irradiation of that tissue relative to the risk from uniform whole body irradiation (Table 13-9). It is the relative contribution of the specified organ of the body. According to the table, the highest contribution to the total stochastic detriment for a uniform whole-body irradiation is associated with the gonads (hereditary effects) and that particular contribution is four times the contribution of the liver to the other stochastic effects. The radiation that delivers the absorbed dose can be from an external source or from a source within the body.

If the dose that an individual receives is from long-lived radioactive material located within the body [e.g., ^{90}Sr (28.78 y)] in bone, the dose is described as

TABLE 13-9

Tissue or Organ Weighting Factors

Tissue or organ	Weighting factor w_T[a, b]	Weighting factor w_T[c]
Gonads	0.20	0.25
Bone marrow (red)	0.12	0.12
Colon	0.12	
Lung	0.12	0.12
Stomach	0.12	
Bladder	0.05	
Breast	0.05	0.15
Liver	0.05	
Esophagus	0.05	
Thyroid	0.05	0.03
Skin	0.01	
Bone surface	0.01	0.03
Remainder[d]	0.05	0.3[e]
Total (whole body)	1.00	1.0[f]

[a]Based on Table 2 in ICRP Publication 60 (per source note, Table 13-8).

[b]The values apply to workers, to the whole population, and to either sex.

[c]U.S. Nuclear Regulatory Commission, *Code of Federal Regulations*, Title 10, Part 20, 1003, January 2001. (Based on reference given in footnote 62.)

[d]Includes adrenals, brain, upper large intestine, small intestine, kidney, muscle, pancreas, spleen, thymus, and uterus.

[e]0.30 results from 0.06 for each of 5 "remainder" organs (excluding the skin and lens of the eye) that receive the highest doses.

[f]For the purpose of weighting the external whole-body dose (for adding it to the internal dose), a single weighting factor, $w_T = 1$, has been specified. "The use of other weighting factors for external exposure will be approved on a case-by-case basis until such time as specific guidance is issued."

"committed." One commonly used period of commitment for a radiation worker is 50 years, the working life of an individual to age 70 years. The committed equivalent dose $H_{T, 50}$ is then the dose obtained by integrating the rate of change of H_T, dH_T/dt, from the date of intake ($t = 0$) to ($t = 50$ y). It includes all intakes of radioactive material in the time span. The absorbed dose D and, therefore, H_T are functions of the effective half-live (T_{eff}) of an internal radionuclide source. T_{eff} takes into account both the physical half-life (T_{rad}) and the biological half-life (T_{biol}) of a radionuclide (see Section 13.13.2). The committed effective dose is calculated in a similar manner.

At the time of writing, the U.S. Nuclear Regulatory Commission (NRC) used the following dose quantities and names:

H_T *Dose equivalent* (DQ × other modifying factors)

$H_{T, 50}$ Committed dose equivalent

For external whole-body irradiation, the most sensitive organs include the gonads (sterility), the blood-forming organs, and the lens of the eye. Doses of 5–6 Sv (500–600 rems) or more of any kind of ionizing radiation to the eyes can cause cataracts. The body extremities are less sensitive.[68] Some of the effects that result from an acute whole-body exposure (an exposure of short duration, i.e., seconds or minutes, as in an accident) to γ rays or x rays for the indicated values of dose equivalent or (equivalent dose) are summarized here.

- About 0.25 Sv (25 rems). This is the approximate lower limit for the detection of blood changes (e.g., a decrease in the leukocyte concentration in blood). For some individuals a dose of 0.50 Sv (50 rems) is required before the effect occurs. Although there are a number of possible causes for a decrease in leukocyte concentration, it can be the first indication of a significant exposure to ionizing radiation if a good preexposure, base line blood count is available. Blood cell count should return to normal after a few months as the body repairs or replaces damaged cells. When the repair mechanism of the cells is overwhelmed, the symptoms of permanent damage appear.

- About 2 Sv (200 rems). Nausea and fatigue are experienced within a few hours; epilation (loss of hair) occurs; and even death is possible within six weeks or so. Susceptibility to infection increases rapidly above this level.

- About 4 Sv (400 rems). The range of 4–6 Sv (400–600 rems) is known as the LD_{50} dose range (dose that is lethal for 50% of the exposed individuals). Higher doses are generally needed for LD_{50} when the exposure rate is low. Blood-forming organs (e.g., bone marrow and spleen) are damaged. To establish a reference time after exposure when the dose is counted as lethal (particularly in animal studies) a 30-day postexposure period is commonly used, and the dose is designated as $LD_{50/30}$.

- About 10 Sv (1000 rems) and above. Nausea and gastrointestinal problems begin shortly after the exposure. Death may occur in a few days. With increasing dose, the central nervous system is damaged. The exposed person becomes unconscious in a very short time, and death may follow within hours.

13.13.1.2 Stochastic Effects

Like chemical carcinogens, ionizing radiation causes body cells to lose their ability to control the rate at which they divide. Localized (not whole-body) doses of 7–15 Sv (700–1500 rems) from repeated exposure to x rays are known to have produced skin cancer on the hands of careless users of x-ray machines. Similarly, localized doses of x rays to the neck and head of children are known to cause thyroid disorders, including cancer.

---

[68]"Extremity" means hand, elbow, arm below the elbow, foot, knee, or leg below the knee.

Leukemia, also known as blood cancer, is the most common type of malignancy to affect children. The whole-body dose equivalent of γ radiation from an external source that will cause leukemia may be as low as about 0.3 Sv (30 rems). Although the symptoms may appear within months after an exposure, the appearance rate in an exposed population is usually at a maximum after about 7–10 years.

As a mutagen, ionizing radiation can cause gene mutations and chromosome aberrations. At the time of writing, the genetic effects are not considered to be distinguishable from effects produced by other mutagens.

13.13.2 Internal Sources

For internal sources of ionizing radiation, the most sensitive organs are the gastrointestinal tract, lung, bone, thyroid, kidney, spleen, and muscle tissue. The dose equivalent for a radioactive source located within the body as in the lungs, thyroid, or bones depends on the dose equivalent rate and the residence time in the body. In the case of an encapsulated source implanted in the body to provide radiation therapy, the residence time can be controlled precisely. However, when the radionuclide is incorporated in a body organ as the result of ingestion, inhalation, or injection into the bloodstream (as in a diagnostic procedure in nuclear medicine), it becomes a "body burden." Then, the exposure time or residence time is the effective half-life T_{eff} of the radionuclide in the organ or organs where it is located.

The effective half-life is defined in terms of the half-lives of the two independent spontaneous processes that remove a radionuclide from the body. These processes are radioactive decay, with a half-life T_{rad}, and biological elimination with a half-life T_{biol}, which is a characteristic property of the chemical element (or compound) whose radioisotope is involved. It is the time needed for the body to eliminate half the concentration by normal processes of elimination. The half-lives for the two competitive processes are analogous to two half-lives for branching in radioactive decay and are combined in the same way. Thus, we have

$$T_{eff} = \frac{T_{rad} \times T_{biol}}{T_{rad} + T_{biol}} \tag{13-69}$$

By inspection of equation (13-69), it can be seen that when T_{rad} and T_{biol} are very different, T_{eff} approaches the value of the shorter half-life. For example, T_{eff} in bone for ^{32}P ($T_{rad} = 14.28$ days and $T_{biol} = 3.16$ years) is 14.1 days. On the other hand, a relatively long-lived radioisotope of an element with a relatively long T_{biol} for an organ will have a long T_{eff} and will be potentially more hazardous. Values of T_{biol} for a few elements are given in Table 13-10. Chelating agents such as EDTA (Chapter 9, Section 9.5.6) have been used to

TABLE 13-10

Biological Half-Life of Selected Elements

Element	T_{biol}	Organ	Element	T_{biol}	Organ
H	12 d	Body water	Po	30 d	Total body
				70 d	Kidneys
				24 d	Bone
C	10 d	Total body	Ra	22 y	Total body
				44 y	Bone
Sr	35 y	Total body	Th	150 y	Total body
	50 y	Bone		200 y	Bone
				150 y	Liver
I	138 d	Total body	U	100 d	Total body
	138 d	Thyroid		300 d	Bone
Cs	70 d	Total body	Pu	180 y	Total body
	140 d	Muscle		200 y	Bone
				80 y	Kidneys
Rare earths	$\simeq 2$ y	Total body			

Source: *CRC Handbook of Radiation Measurement and Protection*, Section A, Vol. II: *Biological and Mathematical Information*, A. Brodsky, Ed. CRC Press, Boca Raton, FL, 1982.

reduce T_{biol} of certain elements (e.g., plutonium, in humans), but there are limitations to the amounts of such substances that can be used because of their chemical toxicity. Bentonite (a clay) (see Chapter 12, Section 12.3.1) and prussian blue [$K_4Fe(CN)_6$] have been used to reduce the contamination of ^{137}Cs (from fallout) in animals, especially meat-producing animals.

For a given radionuclide and for an initial absorbed dose rate $\dot{D}^0$ the absorbed dose at a time t is given by

$$D = \dot{D}^0 \frac{T_{eff}}{0.693} \left(1 - e^{-0.693t/T_{eff}}\right) \qquad (13\text{-}70)$$

13.14 CORRELATION BETWEEN DOSE AND EFFECT

13.14.1 High-Level Dose

There is no single value for absorbed dose that marks a sharp cutoff between a high-level and a low-level dose of ionizing radiation.[69] For each harmful health

---

[69]Sometimes an absorbed dose above about 1 Gy is referred to as a "high-level" dose. The U.S. Nuclear Regulatory Commission refers to a "High Radiation Area" as one in which an individual

effect, there is a range of values of dose within which the severity or the number of occurrences of the effect correlates with the dose in a statistically significant manner. Several types of relationship between dose and response have been observed. Examples are shown in Figure 13-21 for doses above background (the normal dose of ionizing radiation from natural radioactivity and cosmic radiation in the environment). The S-shaped curve in Figure 13-21a for a deterministic effect shows a threshold; that is, the response decreases to zero for a given type of dose greater than zero. In other words, with increasing dose starting at zero, there is no response until the dose has reached a minimum (threshold) value. At very high doses the curve becomes flat because the cells do not survive. Presumably, the mechanism for cell repair is effective below the threshold.

Most of the data used to establish the relationship between dose and response for a particular health effect have been obtained from laboratory experiments with animals. Because of the short lifetime and generation time for mice, for example, the health effects can be observed for a large population in experiments of reasonable duration. The key assumption is that the results of the animal experiments can be applied to human exposure.

In a number of instances, data on human exposure have been obtained directly from records of the dose received by patients exposed to ionizing

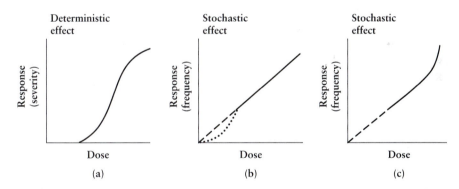

FIGURE 13-21 Examples of dose–response curves. (a) Curve showing threshold for a deterministic effect. (b) Linear curve for a stochastic effect with linear (dashed line) and nonlinear (dotted line) extrapolations to low-level dose (below ~ 1 Sv). (c) Linear–quadratic curve with linear extrapolation to low-level dose. In quadratic region, the response shows a marked increase with dose rate.

could receive (from external sources) "a dose equivalent in excess of 0.1 rem (1 mSv) at 30 centimeters from a radiation source or 30 centimeters from any surface that the radiation source penetrates." For a "Very High Radiation Area" an individual could receive from external sources "an absorbed dose in excess of 500 rads (5 grays) in 1 hour at 1 meter from a radioactive source or 1 meter from any surface that the radiation penetrates." These NRC definitions are given in the U.S. *Code of Federal Regulations* (10 CFR 20, 1003, January 2001).

radiation in medical procedures, radiologists exposed to x rays in their work, victims of accidents at nuclear facilities, and a population containing a statistically significant number of people exposed in known ways that can be documented. A few examples of populations studied in epidemiological investigations are workers who developed bone tumors, bone cancer, and anemia from ingesting luminous radium paint while painting clock and watch dials during the period from about 1915 to 1925; patients who developed malignancies after being given Thoratrast (a colloidal suspension of ThO_2) during the period 1930–1945 to improve the quality of diagnostic x-ray images of the liver; uranium miners who developed lung cancer; people exposed to ionizing radiation directly from nuclear weapons and from nuclear weapons tests; and people who were put at risk because of accidental release of radioactive material (e.g., the explosions at the Chernobyl nuclear power plant: Section 14.16.2.3).

Dose–response data for individuals in such groups who have received a dose high enough to cause a particular effect contribute to establishment of the solid-line section of the curves in Figures 13-21b and 13-21c.

13.14.2 Low-Level Dose

Unfortunately, the dose levels of interest for establishing regulatory dose limits for stochastic effects are in the low-level dose region, below the minimum dose for which the dose–response characteristics are reasonably well known. As the dose from ionizing radiation in a controlled-exposure experiment or the dose received by the cohort of a group or population is lowered, the frequency of occurrence of a particular effect decreases to the "normal" value associated with other causative agents. With decreasing dose, it becomes increasingly difficult to determine any statistically significant increase in the frequency of occurrence that can be attributed to ionizing radiation. The problem is similar to that of establishing the toxic level value for the chemical toxicity or carcinogenicity of a nonradioactive substance.

The usual procedure for predicting or estimating the value of a variable (response in this case) is to extrapolate the curve from the "known" region into the "unknown," low-level region. Dose–response predictions in the low-level dose region are customarily made by linear extrapolation (dashed straight lines in Figures 13-21b and 13-21c), which is based on what is variously referred to as the "linear theory," the "linear, no-threshold theory," the "linear hypothesis," or the "linear model."[70] It is the simplest extrapolation, and it is based on the assumption that for stochastic effects any dose above zero is harmful. In

---

[70]The U.S. Environmental Protection Agency uses linear extrapolation from large doses to zero to evaluate the effect of small doses of chemicals that can damage DNA.

practice, zero dose is relative to background, which provides a chronic, low-level dose.

Over the years, investigators studying the response for stochastic effects to low-level radiation have observed responses with high statistical certainty (e.g., > 90%) that are less (perhaps by a factor of 2 or more) than the value predicted by linear extrapolation. One interpretation of this discrepancy is that there really is a threshold dose below which the response is zero. Some investigators interpret the data for bone cancers among watch-dial painters using radium-containing paint as showing a threshold. Those who support the existence of thresholds for stochastic effects at low doses point out that the linear extrapolation will overestimate the risk for an effect where there is a threshold. On the other hand, those who support linear extrapolation believe that it is prudent to be conservative in estimating health risks and setting regulations. Resolution of the controversy is important and awaits researchers with an innovative approach to work in this field, and probably the availability of new technology.

Various nonlinear extrapolations have been proposed. One is shown as the arbitrarily drawn, nonlinear dotted line in Figure 13-21b. One of the goals of ongoing population studies is to resolve the question of the existence or nonexistence of a threshold dose of ionizing radiation (external or internal exposure) for cancer.

Whereas high-level doses of ionizing radiation clearly have harmful (negative) biological effects, low-level doses of low-LET radiation have been reported by numerous investigators to have stimulating (positive) effects. This phenomenon, called *radiation hormesis*, is similar to chemical hormesis. Examples of radiation hormesis for low-level doses have been observed for plants, in which early growth can be stimulated, and for animals, in which increased growth and proliferation rates, have been reported, as well as increased longevity and reduction in the incidence of cancer.

Additional Reading and Sources of Information

BEIR Reports. These are reports issued by the Committee on the Biological Effects of Ionizing Radiation, National Research Council, published by the National Academy Press, Washington, DC.

Cember, H., *Introduction to Health Physics*, 3rd ed. McGraw-Hill, New York, 1996.

Choppin, G. R., J.-O. Liljenzin, and J. Rydberg, *Radiochemistry and Nuclear Chemistry* (2nd edition of *Nuclear and Radiochemistry*). Butterworth-Heinemann, London, 1995.

CRC Handbook of Chemistry and Physics (Table of the Isotopes). Latest edition, CRC Press, Boca Rations FL. Revised annually.

Ehmann, W. D., and D. E. Vance, *Radiochemistry and Nuclear Methods of Analysis*, Vol. 116 of Series of Monographs on Analytical Chemistry and Its Applications. Wiley, New York, 1991.

Eisenbud, M. and T. Gesell, *Environmental Radioactivity from Natural, Industrial, and Military Sources*, 4th ed. Academic Press, San Diego, CA, 1997.

Firestone, R. B., and V. S. Shirley, eds., *Table of Isotopes*, 8th ed. Wiley, New York, 1997.

Friedlander, G., J. W. Kennedy, E. S. Macias, and J. M Miller, *Nuclear and Radiochemistry*, 3rd. ed. Wiley-Interscience, New York, 1981.

National Nuclear Data Center, Brookhaven National Laboratory, Upton, NY 11973–5000. http://www.nndc.bnl.gov/

Nuclear Regulatory Commission, *Code of Federal Regulations*, Title 10, Part 20. U.S. Government Printing Office, Washington, DC. Revised annually. One URL is http://www.access.gpo.gov/nara/cfr/cfr-table-search.html.

Parrington, J. R., H. D. Knox, S. L. Breneman, E. M. Baum, and F. Feiner, *Nuclides and Isotopes* (Chart of the Nuclides), 15th ed. General Electric Nuclear Energy, San Jose, CA, 1996.

Spinks, J. W. T., and R. J. Woods, *Introduction to Radiation Chemistry*, 3rd ed. Wiley, New York, 1990.

Turner, J. E., *Atoms, Radiation, and Radiation Protection*, 2nd ed. Wiley, New York, 1995.

EXERCISES

13.1. Calculate the average binding energy (MeV) per nucleon for
   (a) ^{19}F (18.9984033 amu)
   (b) ^{120}Sn (119.902198 amu)
   (c) ^{208}Pb (207.976636 amu)
   (d) ^{239}Pu (239.052156 amu)
   The isotopic masses of ^{1}H and ^{1}n are 1.007825032 and 1.008664924, respectively.

13.2. Calculate Q (MeV) for the following nuclear reactions:
   (a) ^{1}H$(n, \gamma)^2$H
   (b) ^{2}H$(\gamma, n)^1$H
   (c) ^{59}Co$(n, \gamma)^{60}$Co
   (d) ^{14}N$(n, p)^{14}$C
   (e) ^{27}Al$(n, \alpha)^{24}$Na
   The isotopic masses (amu) are ^{1}H, 1.007825032; ^{1}n, 1.008664924; ^{2}H, 2.014101778; ^{4}He, 4.002603250; ^{14}C, 14.0003241; ^{14}N, 14.003074007; ^{27}Al, 26.9815384; ^{24}Na, 23.990961; ^{59}Co, 58.933200; and ^{60}Co, 59.933819.

13.3. Calculate Q for the thermal neutron fission of ^{235}U into ^{100}Mo $+$ ^{134}Sn and 2 neutrons. The isotopic masses in amu are ^{1}n, 1.008664924; ^{100}Mo, 99.90748; ^{134}Sn, 133.92783; and ^{235}U, 235.043922.

13.4. A "rich" sample of pitchblende contains 60.0% U_3O_8 by weight. How many kilograms of ^{238}U, ^{235}U, and ^{234}U are present in one metric ton of the ore? How many becquerels and how many curies of ^{226}Ra are present in the same quantity of ore? How many kilograms of ^{226}Ra and of ^{222}Rn are present? Assume that all members of the uranium series are in secular steady state. Natural uranium contains 99.2745 at. % ^{238}U, 0.720 at. % ^{235}U, and 0.0055 at. % ^{234}U.

13.5. For the ore in Exercise 13.4, assume that all the ^{222}Rn could be removed quickly from one metric ton of ore, collected completely, and transferred to a 1.0-liter evacuated flask at 273 K. Calculate the initial pressure of the radon in pascals. Calculate the initial power output in watts for the α radiation (5.4895 MeV per disintegration).

13.6. A bottle with the original seal intact contains 1.0 kg of "chemically pure" U_3O_8 that is 2.0 years old. How many becquerels, microcuries, and micrograms of ^{234}Th and of ^{234}Pa does it contain?

13.7. Prepare a semilog plot of the following experimental data for the decay of a single radionuclide. (All data were obtained in a radiochemistry laboratory course. No measurement was made on the 28th day because of spring break.) Determine the half-life of the radionuclide. The data were obtained with a Geiger–Müller (GM) tube and have been corrected for the background counting rate in the absence of a sample and for coincidence losses at high counting rates.

Days	Counts/min	Days	Counts/min	Days	Counts/min
0	6297	35.0	1170	63.0	313
7.0	4502	42.0	852	70.0	221
14.0	3249	49.0	610	77.0	161
21.0	2298	56.0	441		

13.8. (a) Prepare a semilog plot of the following activity data in disintegrations per minute for a mixture of two independent radionuclides, A and B.

(b) Determine the half-life of A, the longer-lived component, extrapolate its decay curve back to $t = 0$, and determine A_A^0.

(c) Subtract values on the decay curve for component A from the total activity curve to obtain the decay curve for component B. Draw the decay curve for B and determine the values of $t_{1/2}$ and A_B^0.

Time (h)	Bq	Time (h)	Bq
0	1100	24	185
2	868	31	139
4	698	38	108
6	572	45	84
8	478	52	66
10	406	59	51
13	328	66	40
17	258		

13.9. The initial ratio of activities of ^{131}I (8.0207 d) to that of ^{24}Na (14.95 h) in a sample is 2.0. What is the ratio 7.0 days and 14.0 days later?

13.10. Iodine-131 (8.0207 d) emits a 0.608-MeV β ray in 89% of the decays and a 0.364-MeV γ ray in 81.1% of the decays (Figure 13-10). How many 0.608-MeV β rays and how many 0.364-MeV γ rays does a sample having an initial activity of 370 MBq (10.0 mCi) emit in 24 h?

13.11. Strontium-90 (28.78 y) decays into ^{90}Y (2.67 d) and Y-90 decays into stable ^{90}Zr. (see Figure 13-13). A sample of freshly separated ^{90}Sr has an activity of 3.7 GBq (100 mCi). Prepare a semilog plot showing the activity of ^{90}Sr and ^{90}Y and the total activity out to 30 days. What is the total β-ray emission rate per second in the sample after 30 days?

13.12. Transient equilibrium can be illustrated by the following fission product chain: ^{140}Ba (12.75 d) $\rightarrow$ ^{140}La (1.678 d) $\rightarrow$ ^{140}Ce(s). Prepare a semilog plot of the activities of ^{140}Ba, ^{140}La, and the total activity out to 30 days, assuming that at $t = 0$ the sample consists of 74 MBq (2.0 mCi) of radiochemically pure ^{140}Ba. When do the parent and daughter activities become equal? Show how the activity curves would change if the initial sample were not pure and contained 7.4 MBq (0.20 mCi) of ^{140}La.

13.13. Prepare a semilog plot of the total activity and the activities of ^{143}Ce and ^{143}Pr out to 30 days for the following fission product chain: ^{143}Ce (1.377 d) $\rightarrow$ ^{143}Pr (13.57 d) $\rightarrow$ ^{143}Nd(s) for a sample initially containing only 296 MBq (8.0 mCi) of pure ^{143}Ce.

13.14. Derive equation (13-23).

13.15. Derive equation (13-24) from equation (13-23).

13.16. From equation (13-23) with $A_B^0 = 0$, derive an equation for the time when A_B reaches a maximum. Write the resulting equation in terms of half-lives. What is the ratio of A_B to A_A when A_B is at its maximum?

13.17. A sample of NaCl weighing 0.100 g is placed in a nuclear reactor and is irradiated in a thermal neutron flux of 10^{13} neutrons cm^{-2}s^{-1} for 5.0 h. The thermal neutron activation cross sections (σ_γ, for the n, γ reaction) for the production of ^{24}Na (14.95 h) and ^{38}Cl (37.2 m) from ^{23}Na (100 at. %) and ^{37}Cl (24.23 at. %) are 0.53 and 0.43 barn, respectively. R_f for the activation reaction is $\varphi\sigma_\gamma N$, where φ is the thermal neutron flux and N is the number of atoms in the target having the σ_γ. Calculate the activity of ^{24}Na and ^{38}Cl in becquerels and millicuries at the end of the irradiation. Why is the activation of ^{35}Cl (75.77 at. %, $\sigma_\gamma = 43.6$ barns) to form ^{36}Cl (3.01×10^5 y) not important?

13.18. Calculate the weight of pure ^{137}CsCl (30.07 y) needed to make a 3.7-TBq (100-Ci) source.

13.19. Carbon-11 (20.3 m) decays to ^{11}B (s) by emission of a positron ($E_{max} = 0.960$ MeV) without γ-ray emission. The isotopic mass of ^{11}B is 11.0093055 amu. What is the isotopic mass of ^{11}C?

13.20. Zinc-65 (243.8 d) decays to ^{65}Cu (s) with the emission of a 1.115-MeV γ ray in only 50.75% of the disintegrations. What is the mode of β decay? Isotopic masses (amu): ^{65}Zn, 64.929243; ^{65}Cu, 64.927793 amu.

13.21. Beryllium-7 (53.28 d) decays to ^{7}Li (s) with the emission of a 0.477-MeV γ ray. What is the mode of decay? Isotopic masses (amu): ^{7}Be, 7.016928; ^{7}Li, 7.016003.

13.22. Calculate the energy (MeV) available for decay of ^{32}P by negatron emission into stable ^{32}S. Isotopic masses (amu): ^{32}P, 31.973907; ^{32}S, 31.972070.

13.23. The isotopic masses of ^{64}Ni, ^{64}Cu, and ^{64}Zn are 63.927968, 63.929765, and 63.929145 amu, respectively. Which nuclide is radioactive? Calculate the transition energy (MeV) available for it and identify the mode(s) of decay.

13.24. Samarium-147 (1.06×10^{11} y) decays by α-particle emission. Calculate Q and the Coulomb barrier height in MeV for the transition. Isotopic masses (amu): ^{147}Sm, 146.914894; ^{143}Nd, 142.909810; ^{4}He, 4.002603250.

13.25. Calculate the binding energy (MeV) of ^{4}He. (For M values, see Exercise 13-2.)

13.26. Calculate the power output in watts of a source containing 1 kg of ^{238}Pu. Plutonium-238 has a half-life of 87.7 years and emits α particles with energy 5.498 (71.1%) and 5.4565 (28.7%). Gamma-ray intensities are low and can be neglected.

13.27. A sample of a single radionuclide, A, has an activity of 9473 Bq 1.0 day after preparation. The activity is 7519 Bq 5.0 days later. The decay product, B, which was absent initially, is stable. How many atoms of B are present 21.0 days after preparation of A?

13.28. If 3.7 GBq (100 mCi) of ^{222}Rn is released into a sealed room that has the dimensions 10 ft $\times$ 10 ft $\times$ 8 ft (3.05 m $\times$ 3.05 m $\times$ 2.44 m) and if all of the ^{210}Pb daughter becomes uniformly adsorbed on the walls, ceiling, and floor of the room, what will be the surface concentration in atoms and the activity concentration in becquerels and picocuries per square meter 60 days after the radon release?

13.29. If the proton is unstable, its half-life must be exceedingly long. Assuming that the half-life is 10^{32} y, calculate the proton activity in becquerels per year in 1000 metric tons of water.

13.30. Describe with a sketch and explanation what you would expect to see in a cloud chamber photograph when a source emitting only γ rays is placed inside and near the wall (brass) of a cylindrical cloud chamber if $E_\gamma =$ (a) 0.050 MeV, (b) 0.50 MeV, and (c) 2.50 MeV.

13.31. Although the 4.147-MeV α transition in ^{238}U occurs in 23% of the disintegrations (Figure 13-16), the 49-keV γ ray is emitted in only 0.32% of the disintegrations. Explain.

13.32. Cobalt-60 (5.271 y) emits two γ rays per disintegration (1.173 and 1.332 MeV) (Figure 13-9). Estimate the exposure rate in coulombs per kilogram per hour and milliroentgens per hour at 1.0 m from an unshielded 7.4-GBq (200-mCi) point source of ^{60}Co. For a one-hour exposure, what would be the values of the absorbed dose in the units of grays and rads and the equivalent dose in terms of sieverts?

13.33. A collimated beam of 1.0-MeV γ rays passes through 2.0 m of air. What fraction of the energy of the beam is absorbed by the air at room temperature and pressure?

13.34. Phosphorus-32 emits negatrons ($E_{max} = 1.709$ MeV) without γ rays. Should the β-shielding container be made of lead, aluminum, or plastic (containing H, C, and O and having a density of 1.20 g/cm^3)? Why? What thickness (cm) of the correct shielding material would be required to stop all the negatrons?

13.35. A small 370-MBq (10-mCi) source of ^{60}Co (Figure 13-9) is placed in a lead cylinder having a 2.54-cm-thick wall and an outside diameter of 7.62 cm. Estimate the exposure rate in coulombs per kilogram per hour and milliroentgens per hour at the outer surface of the cylinder from the two γ rays. Estimate the fraction of the most energetic bremsstrahlen that will be absorbed in the cylinder wall.

13.36. While walking in a parking lot, you find what looks like a small pearl. You pick it up and decide to hold it tightly in your fist to avoid dropping it. The "pearl" is a 3.7-TBq (100-Ci) source of ^{137}Cs (Figure 13-11). Estimate the absorbed dose to your hand in grays if you hold the source for 30 minutes. Assume that all the β radiation is absorbed in your hand. The mass energy absorption coefficients for bone and muscle are 0.0315 and 0.0326 cm^2/g, respectively, for 0.60-MeV γ rays, and the value changes slowly with energy at this energy. (Note that for the same two types of tissue, the values are 19.0 and 4.96, respectively, for 10-keV photons.) The fraction of the incident γ-ray energy absorbed can be calculated from the decrease in γ-ray intensity in passing through your hand.

13.37. For the purposes of various radiological health calculations, a "reference man" (reference person) weighs 70 kg. A reference man also contains 140 g of potassium, which has a natural content of 0.0117 at. % of ^{40}K (1.27 $\times 10^9$ y). Potassium is distributed mainly in soft tissue (e.g., muscle).

    (a) Calculate the total becquerels and picocuries of ^{40}K in the reference man.

    (b) Calculate the number of negatrons emitted in the reference man over a period of 80 years, assuming a constant amount of potassium (see Figure 13-15).

13.38. Using the results of Exercise 13.37 and the decay scheme for ^{40}K (Figure 13-15), estimate the absorbed dose in grays per year and the equivalent dose in sieverts per year from the β^- radiation from ^{40}K in the soft tissue of a reference man.

13.39. The reference man contains 12.600 kg of carbon.

(a) Calculate the total activity in becquerels and the number of negatrons emitted per year by ^{14}C in the reference man. (Assume a pre-nuclear-weapons-testing equilibrium value of 0.255 Bq/g of carbon.)

(b) Estimate the absorbed dose in rads per year and in grays per year, and the equivalent dose in sieverts per year.

13.40. How long could you stand 10 ft from a 18.5-TBq (500-Ci) point source of ^{60}Co (Figure 13-9) before receiving a total-body dose corresponding LD_{50}? What would the time be at a distance of 30 ft?

13.41. How far (meters) from an unshielded 1.85-GBq (50-mCi) point source of ^{60}Co should you stand in an open area so that (a) the negatrons (see Figure 13-9) and (b) the γ rays do not reach you? How far should you stand from the source so that the γ-ray exposure rate is 2.58×10^{-6} C kg^{-1}h^{-1} (10 mR/h)? At this distance, what would the absorbed dose be in rads, and grays, for an exposure of one hour? Recalculate the values for a 185-GBq (5.0-Ci) source.

13.42. Calculate the absorbed dose rate (Gy/h) from the β^- radiation from 111 MBq (3.0 mCi) of ^{131}I uniformly distributed in a thyroid gland weighing 20 g. E_{ave} for β^- particles from ^{131}I is 0.19 MeV.

13.43. Derive equation (13-66) and the following equation for estimating exposure rate in roentgens per hour for a source with an activity of C curies: R/h at 1 ft) = $6nCE_\gamma$. (Use the absorption coefficient given in footnote 65.)

13.44. If the biological half-life of manganese is 17 days for the total body and the radioactive half-life of ^{54}Mn is 312.1 days, what is the effective total-body half-life of ^{54}Mn?

13.45. Prepare an essay, a class presentation, or a term paper on one of the following.

(a) A summary of the work that has been done to date to find a chemical substance ("protective agent") that could be ingested to protect a person from the effects of ionizing radiation. What type of substance (chemical properties) would be required?

(b) How much the absorbed dose received by patients from x rays in the dental office environment has been reduced, and how the reduction has been achieved for the period of 1930 to date.

(c) Results of research on methods to reduce the biological half-life of elements such as plutonium.

(d) A summary of the methods that have been used and are being used to study neutrinos and antineutrinos.

(e) The methods used by Madame Curie to isolate and purify radium. What connection was there between her laboratory work and her death?

(f) A summary of the properties of stable nuclides that led to the discovery of "magic numbers" and to the formulation of the Nobel Prize–winning shell model for the atomic nucleus.

(g) A brief summary of the chemistry of the hydrated electron.

(h) A summary of the repair mechanisms for damaged DNA.

(i) The status of the controversy on the existence of a threshold for stochastic health effects of ionizing radiation.

(j) The evidence, based on current literature, for and against radiation hormesis.

(k) The latest information about the flux of cosmic rays and solar neutrinos reaching the earth's surface.

(l) An update on any changes in recommended quantities and units for dosimetry of ionizing radiation and any changes in regulations involving changes in such quantities and units.

14

THE NUCLEAR ENVIRONMENT

14.1 INTRODUCTION

This chapter contains introductory discussions of the following topics:

- Natural and anthropogenic sources of ionizing radiation in the environment
- Ways in which both stable nuclides and radionuclides can be used as tools to study processes that occurred in the paleoenvironment, to determine the age of materials as old as the earth, and to serve as tracers to study physical, chemical, and biological processes
- Utilization of nuclear fission as an energy source that does not generate large quantities of CO_2, SO_2, or NO_x by the combustion of carbon-containing fuel, but does generate radioactive waste
- Production, testing, and proliferation of nuclear weapons
- Nature and management of radioactive waste
- Hazards, benefits, and regulation of sources of ionizing radiation
- Accidents that have resulted in the release of radionuclides into the environment

Because certain health effects of ionizing radiation (e.g., cancer) are characterized as having no threshold dose, so that their probability increases with cumulative dose of radiation, it is important to consider *all* sources, internal and external, that could contribute to the dose received by an individual member of the population.[1] In the course of daily living, a person may encounter many sources of ionizing radiation in both the indoor environment (e.g., in the home, workplace, and medical and dental facilities) and the outdoor environment (the atmosphere, hydrosphere, biosphere, and lithosphere). Therefore, in this chapter "environment" will mean the *total* or *extended* environment (outdoor plus indoor). The total environment is dynamic. Changes, both short-term and long-term, are made by nature with some assistance from the inhabitants of this planet. The inhabitants may thoughtfully or emotionally perceive the changes (especially those discussed in this chapter) as good or bad with respect to the quality of life at the present or in the future or they may simply find them acceptable without evaluation.

Inasmuch as the topics covered in this chapter require use of the concepts, definitions, terminology, and mathematical relationships specific for nuclear chemistry, a knowledge of the material in Chapter 13 is assumed.

14.2 COSMIC RADIATION

When cosmic radiation was discovered in 1912, it was believed to consist of penetrating, high-energy γ rays which were, therefore, called cosmic rays. The radiation is now known to consist mainly of particles having energies in the range of 10^8–10^{19} eV. Although much has been learned about cosmic radiation, especially in recent years from data obtained by means of satellites and spacecraft, many mysteries remain. It is known that the sun is a varying source of cosmic radiation at the low end of the cosmic energy range, that galactic sources (sources in the Milky Way, our galaxy) produce radiation having energy from 10^{10}–10^{19} eV, and that there are sources that produce very low intensity cosmic radiation with energy that may be as high as 3×10^{20} eV.

About 77% of the primaries are protons and about 20% are ^{4}He nuclei. The remainder consist mainly of nuclei of elements heavier than helium (up to Ni, $Z = 28$) plus a small contribution from electrons, positrons, and photons. There is evidence that these high-energy particles undergo a complex nuclear

---

[1]Although x rays such as those used for medical diagnosis are not nuclear in origin, they are equivalent to low-energy γ rays emitted by radionuclides and, in some cases, x rays are emitted as a consequence of nuclear processes. X rays contribute to the total dose of ionizing radiation that an individual receives and are, therefore, discussed in this chapter.

reaction (spallation)[2] with C or O nuclei in the interstellar medium to produce most of the Be and B and some of the Li in the universe.

Most of the primaries have energies below 10^{16} eV. Above this energy, the flux of primary cosmic particles drops rapidly, and very few of the highest energy primaries reach the earth's surface. Perhaps 0.05% of the primaries reach sea level. Almost all the cosmic radiation to which the earth's inhabitants are exposed reaches the surface of the earth as showers of secondary radiation produced in the earth's atmosphere (below an altitude of about 25 km) by the interaction of high-energy primary cosmic particles with nuclei of nitrogen, oxygen, argon, and the minor constituents of the atmosphere. Neutrons are released in these interactions. Cosmic radiation intensity near the earth's surface varies with magnetic latitude as the earth's magnetic field bends the trajectories of the primaries so that the intensity is highest near the poles. Cosmic radiation intensity increases with altitude and with decreased barometric pressure at a given altitude as a result of decreased absorption of the secondary radiation in the atmosphere. Because of the solar component of cosmic radiation, the intensity at an altitude of about 12 km can increase by as much a factor of 30 or more during a solar flare, which typically lasts one or two days.

Space travelers are exposed to the high-energy cosmic primaries and to high-energy protons and electrons that are trapped in the radiation belts in the earth's magnetosphere.[3] During the Apollo missions, the astronauts, who were shielded by the equivalent of about $2 \, g/cm^2$ of aluminum, received a dose between 1.6 and 11.4 mGy (0.16 and 1.14 rads).[4] In addition to being a health hazard for people traveling in spacecraft or living in space stations, cosmic radiation also creates problems with electronic components of instruments, especially computer chips in spacecraft and satellites. Thus, a high-energy proton can produce + and – charges (electron–hole pairs) by ionization in or near the depletion region of a junction and can cause a change in the logic state of the junction (see Figure 15-13a). Alternatively, a high-energy proton can undergo a nuclear reaction whose products would cause ionization.

The linear energy transfer (LET) for highly charged, high-energy primary cosmic particles can be 10 or more times that of α particles, making it difficult to evaluate Q, the quality factor, needed to calculate the dose equivalent in

[2]In spallation, the incident particle does not share its kinetic energy with the nucleons and become part of the target nucleus, which then emits a photon or some other reaction product or undergoes fission. Instead, a very high energy incident particle spalls or chips off small fragments (i.e., light nuclei) from the target nucleus.

[3]Also in the environment above the earth's atmosphere are x-rays and γ-rays emitted in pulses or "flashes" by sources unknown at the time of writing.

[4]J. E. Angelo, W. Quam, and R. G. Madonna, Radiation protection issues and techniques concerning extended manned space missions in *Radiation Protection in Nuclear Energy*, Vol. 2, Proceedings Series, International Atomic Energy Agency, Vienna, 1988.

sieverts from the absorbed dose in grays. At sea level, however, Q for the ionizing radiation can be taken as unity.

14.3 NATURALLY OCCURRING SOURCES OF RADIOACTIVITY IN THE ENVIRONMENT

The naturally occurring radionuclides in the environment can be divided into three types: (1) cosmogenic, those produced continuously by cosmic rays, (2) primordial, those that have been present since the earth was formed, and (3) progeny of the three primordial radionuclides, ^{232}Th, ^{235}U, and ^{238}U. Radionuclides of the third type make up the thorium, actinium, and uranium series.

Table 14-1 contains a list of cosmogenic radionuclides, that is, those produced by cosmic radiation in the atmosphere. Products of the primary reactions include high-energy protons and neutrons and radioactive nuclear debris, some of which are very short-lived. The protons and especially the neutrons, react with stable nuclei to form radionuclides mainly in the atmosphere, but also in soil and rocks. Cosmogenic radionuclides produced in the atmosphere have been collected and characterized in air samples, rainwater, snow, surface water, and seawater.

TABLE 14-1

Cosmogenic Radionuclides^a

Radionuclide (half-life)
^{3}H (12.32 y)
^{7}Be (52.28 d)
^{10}Be (1.5×10^6 y)
^{14}C (5715 y)
^{22}Na (2.604 y)
^{24}Na (14.95 h)
^{26}Al (7.1×10^5 y)
^{32}Si (1.5×10^2 y)
^{32}P (14.28 d)
^{33}P (25.3 d)
^{35}S (87.2 d)
^{36}Cl (3.01×10^5 y)
^{38}S (2.84 h)
^{38}Cl (37.2 m)
^{39}Cl (55.6 m)
^{39}Ar (269 y)
^{129}I (1.57×10^7 y)

^{a}All are β^- emitters except ^{7}Be (EC), ^{22}Na (EC, β^+), and ^{26}Al (EC, β^+).

Both ^{3}H (12.32 y) and ^{14}C (5715 y) are environmentally important radionuclides. Tritium is produced by the ^{14}N(n, t) ^{12}C reaction:

$$^{14}N + {}^1n \longrightarrow {}^{12}C + {}^3H \tag{14-1}$$

and ^{14}C by the ^{14}N (n, p) ^{14}C reaction:

$$^{14}N + {}^1n \longrightarrow {}^{14}C + {}^1H \tag{14-2}$$

About 0.9 PBq (24 kCi) of ^{14}C is produced annually in the atmosphere by cosmic radiation and the estimated global inventory is about 10 EBq (0.27 GCi), of which about 130 PBq (3.5 MCi) is in the atmosphere. The likely chemical species for ^{3}H and ^{14}C in the atmosphere are ^{1}H^{3}HO (or HTO) and ^{14}CO$_2$, respectively.

A radionuclide that has been receiving attention recently is ^{129}I (1.57×10^7 y), which decays into ^{129}Xe (s). On the basis of radioactivity measurements, it is considered to be an extinct primordial nuclide. It is present in the environment because it is made cosmogenically in the atmosphere, as a fission product of spontaneous nuclear fission of ^{238}U in the lithosphere, and also anthropogenically during the testing of nuclear weapons and as a fission product in nuclear power reactors.

Primordial radionuclides in the environment are given in Table 14-2. Except for ^{40}K, the independent primordial radionuclides with atomic number below 90 do not contribute significantly as sources of ionizing radiation because of a very low abundance in nature (i.e., a low abundance of the element or the radioisotope or both) or because of an exceedingly long half-life, which results in a very low specific activity. Nuclides ^{232}Th, ^{235}U, and ^{238}U are environmentally important as significant sources of ionizing radiation themselves, as parents of the three naturally occurring series, as fissile or fertile material for nuclear power reactors, and as fissionable material for nuclear weapons. The neptunium ($4n + 1$) series is not found in nature because the half-life of the longest-lived member, ^{237}Np (2.14×10^6 y), is too short for any of the original ^{237}Np to remain on the earth.

Isotopes of elements produced by decay of primordial radionuclides are characterized as "radiogenic." An example is the stable isotope of strontium, ^{87}Sr, the daughter of ^{87}Rb.

14.4 DISTRIBUTION OF NATURALLY OCCURRING RADIOACTIVITY IN THE ENVIRONMENT

14.4.1 Rocks and Soils

Granitic rocks in the upper part of the earth's crust contain average weight concentrations of 3–4 ppm of uranium, 10–15 ppm of thorium, and 1–5% of

TABLE 14-2

Primordial Radionuclides

Radionuclide	Abundance (at. %)	Half-life (years)	Mode of decay
^{40}K	0.0117	1.27×10^9	EC, β^-
^{50}V	0.250	1.4×10^{17}	EC
^{87}Rb	27.84	4.88×10^{10}	β^-
^{113}Cd	12.22	9×10^{15}	β^-
^{115}In	95.71	4.4×10^{14}	β^-
^{123}Te	0.908	$> 1.3 \times 10^{13}$	EC
^{138}La	0.090	1.05×10^{11}	EC, β^-
^{144}Nd	23.80	2.38×10^{15}	α
^{147}Sm	15.0	1.06×10^{11}	α
^{148}Sm	11.3	7×10^{15}	α
^{152}Gd	0.20	1.1×10^{14}	α
^{174}Hf	0.162	2.0×10^{15}	α
^{176}Lu	2.59	3.78×10^{10}	β^-
^{180}Ta	0.012	$> 1.2 \times 10^{15}$	EC, β^-
^{186}Os	1.58	2×10^{15}	α
^{187}Re	62.60	4.3×10^{10}	β^-
^{190}Pt	0.01	6.5×10^{11}	α
^{232}Th	100	1.40×10^{10}	α
^{235}U	0.720	7.04×10^8	α
^{238}U	99.2745	4.47×10^9	α
^{244}Pu[a]		8.0×10^7	α

[a] Detectable.

potassium [0.0117 at. % ^{40}K (1.27×10^9 y)]. The concentrations are generally less in basic igneous rocks, sedimentary rocks, and limestones. Based on the average concentrations in the crust, the ratio of Th to U is about 4.

Uranium in rocks and soils consists of three long-lived isotopes: ^{238}U (99.274%) (4.47×10^9 y), parent of the uranium series; ^{235}U (0.720%) (7.04×10^8 y), parent of the actinium series; and ^{234}U (0.0055%) (2.46×10^5 y), a member of the uranium series. In normal soil the radioactivity of ^{238}U is about 20 Bq/kg (0.54 nCi/kg) and that for ^{40}K about 400 Bq/kg (11 nCi/kg).

In rocks, soils, and sands containing the thorium series, the isotopes of thorium present are ^{232}Th (1.40×10^{10} y) and a daughter, ^{228}Th (1.913 y). The ^{238}Th activity in soils is about 40 Bq/kg (about 1 nCi/kg). If uranium is also present, the thorium isotopes will include those in the uranium series and the actinium series.

The γ-emitting members of the three series (Figure 13-5, Table 13-2, and Appendix B) and ^{40}K account for the above-normal radiation background often associated with granitic rock. In undisturbed rock, all the members of a given series are present in secular equilibrium and, therefore, have the same disintegration rate. However, each series contains a radioisotope of the inert gas radon, which can escape into the atmosphere to a limited extent from porous rock or rock with cracks but can escape much more readily from permeable soils. Usually the radon escapes into the atmosphere from the top meter or two of soil. In a given series, the members that follow the radon isotope have less than the equilibrium disintegration rate in rock or soil from which radon has escaped.

Of the three radon isotopes ^{219}Rn (3.96 s), ^{220}Rn (55.6 s), and ^{222}Rn (3.8235 d) that can escape into the atmosphere, ^{222}Rn in the uranium series has been linked to lung cancer and is the most important environmental contaminant. Its concentration in soil ranges from a few hundred becquerels per cubic meter to over $3.7\,MBq/m^3$ (tens of picocuries per liter to over 100 nCi/liter). Its nonvolatile parent, ^{226}Ra (1599 y), has a specific activity of 37–74 Bq/kg (1–2 pCi/g) in soil.[5]

Most coal used by power plants in the United States has a uranium concentration of 1–4 ppm and about the same range for thorium. When coal is compared with common rocks, the concentration is about the same for uranium but is lower for thorium. Eastern U.S. coal contains about 18 Bq/kg (0.5 pCi/g) of ^{226}Ra. The concentration in British coal is in the range of 1.8–11 Bq/kg (0.05–0.3 pCi/g). When coal is burned, the ^{222}Rn enters the atmosphere and its parent (^{226}Ra) remains in the ash, where its concentration is increased by a factor of perhaps 10 over that in the coal, depending on the composition of the coal.

14.4.2 Uranium and Thorium Ores

There are over 100 minerals that contain uranium. Only a few of these are found in deposits that are economically significant. The uranium content of an ore is expressed as a percentage of U_3O_8 and is commonly in the range of 0.1–0.5%, although ores with much higher uranium content have been found. Examples of specific ores are uraninite, which varies in composition from UO_2 to $UO_2 \cdot 2UO_3$ or U_3O_8 (known as pitchblende), and carnotite ($K_2O \cdot 2UO_3 \cdot V_2O_5 \cdot xH_2O$), which, in the past (before 1943), was mined in the United States mainly for its vanadium content. Commercial-grade uranium ores have also been mined in Australia, Canada, China, France, Gabon, Germany, Namibia, Portugal, Russia, South Africa, and the United Kingdom.

Thorium, like uranium, is widely distributed on the earth and is commonly found associated with uranium, but the concentrations of thorium in uranium ores are generally not high enough to be commercially interesting. The principal source of thorium is monazite, (Ce, La, Y, Th) PO_4, which occurs as beach and

[5]The density of soil is typically in the range of 1600–2600 kg/m^3.

river sands produced by the weathering of rocks. Significant sources of monazite occur in Australia, Brazil, China, India, South Africa, and the United States (Colorado, Florida, Idaho, Montana, North Carolina, and South Carolina). In the past, monazite was mainly of interest as a source of rare earths and ThO_2, which was used in gas mantles. The thorium content of monazite is usually in the range of 1–10% ThO_2.

14.4.3 The Atmosphere

14.4.3.1 Outdoor Air

The natural radioactive contaminants of the atmosphere are ^{222}Rn, its progeny, and, to a much lesser extent, the cosmogenic ^{3}H and ^{14}C. As far as is known, the latter two radionuclides have always been present in the atmosphere, but their rates of formation have not been constant over long periods of time (tens of thousands of years) owing to variations in the intensity of cosmic radiation, especially that of solar origin. Long-term changes in the $^{14}C/^{12}C$ ratio in the atmosphere resulting from changes in the earth's magnetic field are shown in Figure 14-1. Short-term variations are assumed to arise from variations in the solar component of cosmic radiation.

During the past hundred years, the $^{14}C:^{12}C$ ratio in the atmosphere has changed for two additional reasons. First, with the beginning of the industrial era at the end of the nineteenth century, the combustion of fossil fuels (high in ^{12}C, low in ^{14}C) led to a decrease in the ratio (the Suess effect). Second, beginning

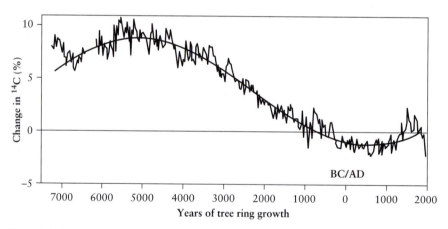

FIGURE 14-1 Long-term variation in atmospheric ^{14}C activity as determined from tree rings. Adapted from G. Choppin, J. O. Liljenzin, and J. Rydberg, *Radio-chemistry and Nuclear Chemistry*, 2nd ed. Copyright © 1995. Reprinted by permission of Butterworth-Heinemann, Oxford.

with the atmospheric testing of nuclear weapons in 1954, there was an increase in ^{14}C (5715 y) concentration caused by the ^{14}N(n, p) ^{14}C reaction [reaction (14-2)], with neutrons released in atmospheric testing of nuclear weapons, especially "H-bombs." (Section 14.11.1). Before testing, the ^{14}C content of plants and animals was 0.22 Bq/g of carbon (6 pCi/g). By the time atmospheric testing of nuclear weapons ended in 1963, the concentration of ^{14}C had risen to about double the natural value, as shown in Figure 14-2.

Tritium (12.32 y) in the atmosphere is present as HTO and is in equilibrium with HTO in water on the earth that is in contact with the atmosphere. Prior to 1954, the tritium content of the biosphere was estimated to be about 900 g or about 0.33 EBq (about 9 MCi). Its concentration is sometimes expressed relative to ^{1}H in tritium units, where 1 TU = 1 ^{3}H atom per 10^{18} atoms of ^{1}H. For water, 1 TU is 122 Bq/m^3 (3.3 pCi/liter). The level of ^{3}H in rainwater increased from the range of about 0.5 to 5.0 T U before 1954 to values as high as 500 T U after the testing of tritium-containing thermonuclear weapons. From the effects of weapons testing on ^{3}H concentration, the residence time of ^{3}H in the stratosphere has been estimated to be 2–4 years. Tritium is washed from the troposphere in about 2 months.

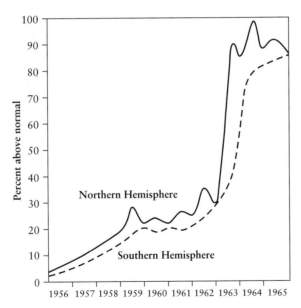

FIGURE 14-2   Increase in tropspheric ^{14}C from atmospheric testing of nuclear weapons. Normal air was taken to contain 74×10^5 atoms of ^{14}C per gram of air in 1954. Adapted from *Report of the United Nations Scientific Committee on the Effects of Atomic Radiation* (UNSCEAR), Supplement No. 14 (A/6314). United Nations, New York. Copyright © 1966. Used by permission of the United Nations.

Radon, predominantly ^{222}Rn (3.8235 d) in the uranium series, is present everywhere in the earth's atmosphere because it is a descendent of ^{238}U, and uranium is an ubiquitous element. Although the main source of radon in the atmosphere is radon in soil, radon dissolved in spring water or well water that has been in contact with rocks or soil or sediments containing uranium also escapes when the water is exposed to the atmosphere.

At any one location the concentration of radon in the atmosphere varies diurnally and seasonally and changes rapidly with atmospheric conditions. A typical value for ^{222}Rn in the United States is about 5.6 Bq/m^3 (0.15 pCi/liter). Worldwide concentrations have been reported to be from 4–18 Bq/m^3 (0.1–0.5 pCi/liter). For ^{220}Rn (55.6 s), a member of the thorium series, the value is about 0.7 Bq/m^3 (0.02 pCi/liter).

If decay products formed in very small amounts by branching are neglected, the important progeny of ^{222}Rn are given in the following equation:

$$^{222}\text{Rn} \xrightarrow{\alpha} {}^{218}\text{Po} \xrightarrow{\alpha} {}^{214}\text{Pb} \xrightarrow{\beta} {}^{214}\text{Bi} \xrightarrow{\beta}$$

$$^{214}\text{Po} \xrightarrow{\alpha} {}^{210}\text{Pb} \xrightarrow{\beta} {}^{210}\text{Bi} \xrightarrow{\beta} {}^{210}\text{Po} \xrightarrow{\alpha} {}^{206}\text{Pb (s)} \tag{14-3}$$

All are radioisotopes of nonvolatile elements, and except for ^{210}Pb (22.6 y) and ^{210}Po (138.38 d) they are short-lived. (See Figure 13-5 and Table 13-2.)

14.4.3.2 Radon in the Air in Mines and Buildings

When it was observed in the 1920s that uranium miners suffered from a relatively high incidence of lung cancer, the trend was at first attributed to radon in the air in the mines. Later it was shown that the effect is caused not by radon itself, but by its two short-lived α-emitting daughters, ^{218}Po (3.10 m) and ^{214}Po (163.7 μs).[6] Thus, radon is really the carrier for its nonvolatile progeny, and because of its half-life of almost 4 days it can, once airborne, travel a considerable distance from its point of origin. If the progeny are formed in air, a fraction becomes attached to aerosols (e.g., tiny droplets of water, dust, tobacco smoke, etc.), and the remainder remains unattached. As an inert gas, inhaled radon is readily exhaled. On the other hand, when the short-lived radon progeny are inhaled, they deposit in the lungs, where they remain and emit a particles that are absorbed in lung tissue.

In 1956 a unit of radioactivity, the working level (WL), was introduced in connection with the establishment of acceptable levels of radon daughters for the air in mines.[7] One WL, a unit of exposure, is associated with any combin-

---

[6]While airborne, the long-lived ^{210}Pb and ^{210}Po are not considered to contribute significantly to lung cancer. They are of interest, however, because they are washed out of the atmosphere by rain onto tobacco and other vegetation, where they become concentrated.

[7]The radon daughter hazard in underground mines is not limited to uranium mines, because uranium is a constituent of many rocks and ores.

ation of the short-lived radon daughters in one liter of air at ambient temperature that results in the ultimate release of 1.3×10^5 MeV of α-particle energy. One WL is, then, approximately equal to the α energy emitted per unit volume of air by the short-lived progeny in equilibrium with $3.7 \, \text{kBq/m}^3$ ($100 \, \text{pCi/liter}$) of ^{222}Rn. A related cumulative exposure unit, the working-level month (WLM), is defined for miners as the number of working levels of exposure in one working month (170 hours). One WLM is equal to 3.5×10^{-3} J/h per cubic meter of air. On the average, a miner with an exposure of 1 WLM receives a dose of about 8 mGy (0.8 rad). In 1961 almost 30% of the underground uranium mines in the United States had WLs greater than 10. Six years later, after improvements had been made in ventilation, only 1% of the mines had a WL exceeding 10.

Although a working-level month is defined as 170 hours, an actual month is 720 hours (30 days). Thus, for a person occupying a house for 16 hours a day for one actual month of 30 days at 1 WL, the cumulative exposure would be 2.82 WLM. The dose received by the average person for an exposure of 1 WLM would be about 4.5 mGy (0.45 rad).

Radon has been present in the air in our homes ever since people began living in caves, and there is extensive literature on both the sources and the health effects of radon. In recent years there have been numerous documented cases of elevated radon levels in houses and other buildings. The following are examples

- In 1983 it was discovered that radon levels were elevated in several houses in Glen Ridge, Montclair, and West Orange, New Jersey, because the soil used as fill around the basements and in the yards was contaminated with ^{226}Ra, radon's parent. Presumably, the soil came from a site where there had been a plant in which ^{226}Ra was separated from uranium ore. Some of the residents had to leave their homes temporarily. The contaminated soil was removed, packaged in drums, and stored on the site until, after several years, it was transported to a disposal site for low-level radioactive waste.

- In 1986 about 100 homes in Clinton, New Jersey, were found to have high radon levels [some $> 37 \, \text{kBq/m}^3$ ($1000 \, \text{pCi/liter}$)]. The homes were built on a cliff near an old quarry that contained limestone with an atypically high concentration of uranium. Remedial procedures described shortly were used to reduce the radon levels.

- Elevated radon concentrations have been found in the air in buildings in California located over shale that releases radon and in buildings whose construction involved the use of tailings from nearby uranium mines (e.g., in Colorado) and residues from phosphate mines (e.g., in Florida) (see Section 14.4.6).

The "radon hazard" in the home suddenly received nationwide attention early in 1985. In December 1984 an engineer who worked at a nuclear power plant in Boyertown, Pennsylvania, set off the alarm on a radiation monitor one day when he *entered* the plant. Such monitors are used to detect contaminated clothing on workers leaving a facility where radioactive material is used. In this incident, the person's clothing was contaminated with radon progeny present in the air in his house, where the radon level was found to be above $93\,\text{kBq}/\text{m}^3$ ($> 2.5\,\text{nCi/liter}$). In terms of the hazard of ionizing radiation, it was safer to be at work in a nuclear power plant than to be at home. The area where this event occurred is known as the Reading Prong (Figure 14-3), a geological formation that extends northeasterly from Reading, Pennsylvania, to the western edge of Connecticut. Radon levels found in some of the houses in the region are among the highest in the United States.

In 1985 the U.S. Environmental Protection Agency established a nationwide program to assist states and homeowners in reducing the risk of lung cancer from indoor radon. In the following year, EPA recommended that remedial action be taken if the radon level in a home exceeded $0.025\,\text{WL}$ or $148\,\text{Bq}/\text{m}^3$

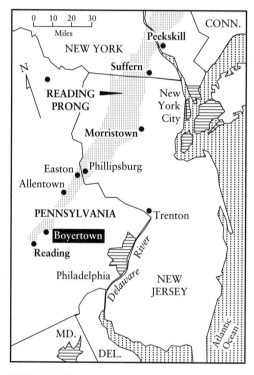

FIGURE 14-3 Uranium-rich Reading Prong. Redrawn from "Radiactive Gas Alters Lives of Pennsylvanians," by William R. Greer. *The New York Times*, October 28, 1985. p. A10. Used by permission of *The New York Times*.

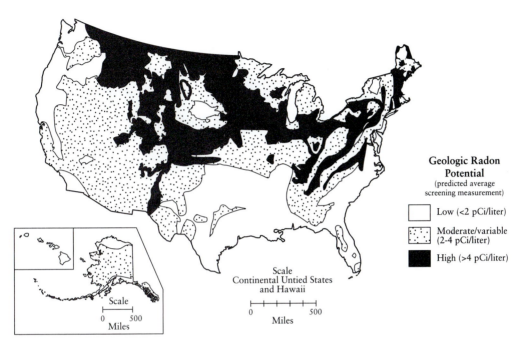

FIGURE 14-4 Generalized geologic radon potential of the united states by the U.S. Geological Survey. From EPA's map of radon zones (separate issue for each state), U.S. Environmental Protection Agency Document 402-R-93-052, 1993.

(4 pCi/liter).[8] In 1993 EPA issued reports containing maps of "radon zones" for each state. The reports contained two maps (Figures 14-4 and 14-5) of the continental United States and Hawaii. Figure 14-4, originally prepared by the U.S. Geological Survey, identifies regions where the *predicted* indoor radon potential is low, moderate, or high. Figure 14-5 shows three *predicted* zones of radon concentration in the air in the lowest livable area of a building: greater than 4 pCi/liter, equal to or greater than 2 pCi/liter, and less than 2 pCi/liter, for 3141 counties. The indicators used for radon potential included indoor radon measurements, geology, aerial radioactivity, soil parameters, and foundation types. It is very important to realize that the radon concentrations are predicted values and that actual values can differ significantly and unexpectedly.

---

[8]Because this value was obtained by using the linear no-threshold hypothesis and extrapolating from dose–effect data for miners, it has been criticized as being too high.

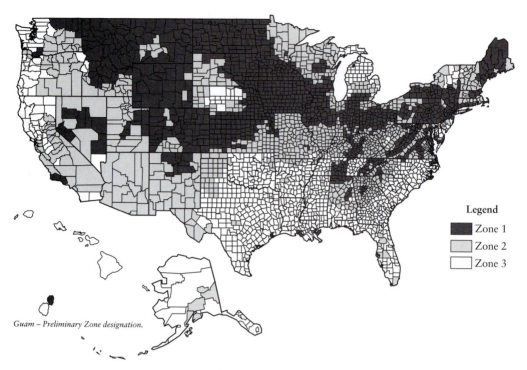

Guam – Preliminary Zone designation.

Legend
- Zone 1
- Zone 2
- Zone 3

FIGURE 14-5   EPA map of radon zones in the United States: Zone 1, $> 148 \, \text{Bq/m}^3 (> 4 \text{pCi/liter})$; Zone 2, $\geq 74 \, \text{Bq/m}^3 (\geq 2 \text{pCi/liter}) \leq 148 \, \text{Bq/m}^3$ ($\geq 4 \text{pCi/liter}$); Zone 3, $< 74 \, \text{Bq/m}^3 (< 2 \text{pCi/liter})$. The map is accompanied by the following statement from the EPA: "This map is not intended to be used to determine if a home in a given zone should be tested for radon. Homes with elevated levels of radon have been found in all zones. All homes should be tested, regardless of geographic location." From EPA's map of radon zones (separate issue for each state), U.S. Environmental Protection Agency Document 402-R-93-052, 1993.

A measured value for the concentration of radon in a building depends on the method of measurement used, the season of the year, the climatic conditions, the construction of the building, the type of soil surrounding and below the basement of the building, the source of water (e.g., well or reservoir), and so on.[9] As a very approximate guide, if the water used in a home has a radon

---

[9]There are several very different devices available for measuring the radioactivity that can be attributed to radon in the air in a home. Radon data require careful interpretation, and measurements should be made after consultation with the local health department. Radon measurements in a residence should be made at the living level and should be repeated several times. If the basement contains a bedroom or living room, it should be included.

concentration of 370 MBq/m^3 (10 μCi/liter), it could increase the radon content of the air by about 37 Bq/m^3 (1 pCi/liter).

In the United States, indoor radon levels of 3.7–7.4 Bq/m^3 (1–2 pCi/liter) are "normal." For levels above 148 Bq/m^3 (4 pCi/liter), the EPA recommends that follow-up measurements be made to establish a value that can be used to determine whether remedial action should be taken and, if so, how urgent it is to reduce the level to about 148 Bq/m^3 (4 pCi/liter) or less.[10] For levels between 148 and 740 Bq/m^3 (4 and 20 pCi/liter), the EPA recommends reducing the level within a few years; for levels between 740 Bq/m^3 and 7.40 kBq/m^3 (20 and 200 pCi/liter), within several months; for levels over 7.4 kBq/m^3 (200 pCi/liter), within several weeks. Estimates of the number of homes in the United States that have radon levels above 148 Bq/m^3 (4 pCi/liter) vary from region to region, but the country-wide value at the time of writing is about 6%.

The literature contains many publications on radon levels in buildings worldwide. For example, the national mean concentration of radon in indoor air in Finland is 120 Bq/m^3 (3.3 pCi/liter), but values as high as 370 Bq/m^3 (10 pCi/liter) have been measured. As another example, Figure 14-6 shows relative radon levels in homes in Great Britain, where the average level is 21 Bq/m^3 (0.54 pCi/liter), but values about 500 times greater have been measured. The two counties with the highest level, Cornwall at the southwestern tip and adjacent Devon, have an estimated 60,000 homes with radon levels above 200 Bq/m^3 (5.4 pCi/liter), the recommended action level.

A level of radon high enough to create a health problem in a building can generally be attributed to radon in the soil in contact with the walls and floor of the basement or under the slab in the absence of a basement. As a gas, radon can migrate most easily in coarse, gravelly soil and least easily in clay. It enters a basement by way of crawl spaces, gaps around pipes, cracks in a cement floor or in the basement walls, gaps between the basement floor and the walls, porous cinder blocks, the well of a sump pump, and so on. The maximum concentration of radon can be expected in the basement, with amounts decreasing as one goes to upper floors. Radon migrates to upper floors by way of stairways, gaps, around pipes, and so on.

Leakage of radon into a basement can be reduced somewhat by sealing all cracks and openings, but this approach has been shown to have limited benefit. There are two driving forces for the migration of radon from soil into a basement. One is diffusion from the higher concentraton in the soil to the lower concentration in the basement. The second and more important one, especially for permeable soil, is the pressure gradient that exists when the air

---

[10]The method used to set the value of 4 pCi/liter includes the assumption that there is *no* safe level; that is, there is not a cutoff value. Thus 3 pCi/liter is not safe. As a practical matter, it may be difficult to lower a very high level below 4 pCi/liter.

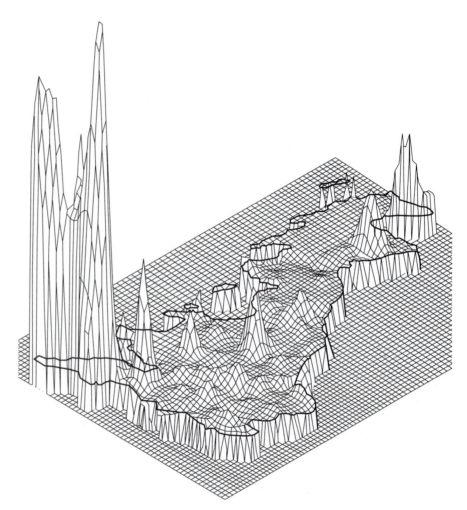

FIGURE 14-6 Radon levels in Britain showing the highest readings ($> 200 Bq/m^3$) in the counties of Cornwall and Devon in southwest England. From *Nucl. News*, 33, 56 (March 1990). Reproduced from a chart on radiation doses published by the National Radiological Protection Board.

pressure inside is less than that outside a building. It is very easy to create a slightly negative pressure in a house. A heating unit such as a furnace or fireplace that uses fossil fuel will reduce the air pressure as oxygen is removed from the air and vented as CO and CO_2. The heating units themselves, including radiators, and other sources of heat such as a clothes dryer and a kitchen range, cause the air to rise within the house, creating a pressure differential as the house itself behaves like a chimney. Even a strong wind can cause a reduced pressure in a building by the Venturi effect.

Ventilation may reduce or increase the radon concentration in a building. Window fans that exhaust air from a building reduce the air pressure in the building, thereby enhancing the flow of radon from the soil into the building. If the radon level is such that remedial action is advisable, the level can be lowered to a value approaching that of the air outside the building by pulling air in from the outside to provide a positive pressure in the building. This also can establish an airflow from the upper levels to the basement.

When the radon level in a house is very high, a more costly method of lowering the level may be necessary. An effective method is to pull the radon-laden air from the layer of gravel or crushed stone under the basement floor and/or outside the basement walls by a fan through piping, whereupon the air is exhausted into the atmosphere above the rooftop. For buildings located where the temperature can drop below freezing, a potential problem is that the cold air pulled under the basement floor can cause moisture condensation and structural damage to the floor.

If radon is known to be a potential problem before a house is built, features can be incorporated in the house to minimize the radon level. Thus, it would be expedient to provide a layer of crushed stone or gravel and to install pipes so to permit removal of the radon, and venting to the atmosphere, as just described.

The level of the important radon progeny (^{218}Po and ^{214}Po) in a building can also be lowered by methods that utilize the properties of the radon progeny. A ceiling fan, for example, will increase the rate at which these attached and unattached positively charged particles reach and become adsorbed by the walls, ceiling, and floor of a room. Because of the charge on the particles, positive ion generators assist in moving the particles toward surfaces of a room. Air cleaners that filter or electrostatically precipitate particulate matter in the air remove the progeny, but not the radon and, therefore, leave relatively little dust to which newly formed progeny can become attached. These smaller, unattached particles have the potential of producing a higher dose to the lungs.[11]

Another way by which the radon level in a home can become elevated is through the use of natural gas in space heaters and unvented kitchen ranges. Radon-222 in the natural gas is released into the air in homes that use such devices. Natural gas used for heating in homes and in industry and for power generation is also a direct source of ^{222}Rn in the atmosphere. The radon content of natural gas varies from source to source, but the average for most of the United States is about 740 Bq/m^3 (20 pCi/liter). Natural gas from gas fields in Kansas, Colorado, New Mexico, and the Texas Panhandle may contain higher levels of about 1.85 kBq/m^3 (50 pCi/liter).

---

[11]D. W. Moeller, *Nucl. News*, pp. 36–39, June 1989.

Radon can be removed effectively from water by filtering through granular activated charcoal. For water with very high levels of radon, care must be taken to change the charcoal filter before the adsorbed γ-emitting radio-nuclides become a radiation hazard. Several times beginning in 1997, the EPA has proposed regulations to limit the amount of radon in public water supplies and in schools and hospitals that use groundwater. At the time of this writing, the proposal was in the final review stage.

The main source of natural radioactivity in the ocean is ^{40}K $(1.27 \times 10^9 \, \text{y})$ $(11.2 \, \text{kBq/m}^3, \, 30 \, \text{pCi/liter})$. The second highest source is ^{87}Rb $(4.88 \times 10^{10} \, \text{y})$ $(1.1 \, \text{kBq/m}^3, \, 30 \, \text{pCi/liter})$. Secular equilibrium (Section 13.4.1) is not attained for the members of the uranium and thorium series in the ocean because of adsorption by sediment of some of the elements in the series, especially thorium (e.g., ^{230}Th, in the uranium series). Although the uranium content of seawater is very low (e.g., ~ 3 ppb), the uranium is extractable; hence, seawater is considered to be a future source of uranium for nuclear power plants. Uranium contained in the oceans represents a reserve of about 5 billion metric tons. Seawater also contains about $1 \times 10^{-10} \, \text{g}$ of thorium per liter.

14.4.5 Food

The naturally occurring α activity of food is mainly associated with ^{226}Ra (1599 y) in the uranium series, its α-emitting progeny [especially ^{210}Po (138.38 d)], and the α-emitting daughters of ^{228}Ra (5.76 y) in the thorium series. Both ^{210}Pb (22.6 y) and its daughter ^{210}Po can enter food not only from the soil, but also from the atmosphere, where they are present as the daughters of ^{222}Rn. Examples of atmospheric concentrations at ground level are 148 μBq/m^3 (4 aCi/liter) for ^{210}Pb and 37 μBq/m^3 (1 aCi/liter) for ^{210}Po. For leafy vegetables, the ^{210}Pb and ^{210}Po contents are determined more by "natural fallout and rainout" from the atmosphere than by the composition of the soil. Food accounts for 80–90% of the intake of ^{226}Ra, with the remainder coming from water. The average daily intake of ^{226}Ra is about 110 mBq (3 pCi) in food having 37–370 mBq/kg (1–10 pCi/kg). Cereals and other grain products, vegetables, and fruits are the main sources. Specific examples of *approximate* specific activity [in mBq/kg and (pCi/kg)] are whole-wheat bread 111 (3), milk 9 (0.24), potatoes 56 (1.5), fruit 48 (1.3), fresh vegetables 74 (2), fish 3.3 (0.9), meat 19 (0.5), and poultry 24 (0.8). For ^{228}Ra the intake is half or less than that of ^{226}Ra.

Most of the naturally occurring β^- activity in food can be attributed to ^{40}K $(1.27 \times 10^9 \, \text{y})$ (0.0117 at. %). A typical pair of approximate values [Bq/kg and (nCi/kg)] is 25–74 (0.7–2). A few of the foods with higher specific activities are bananas 130 (3.5), spinach 222 (6), and coconut 222 (6). For growing plants, the availability of potassium is subject to large variation. Depending on the nature of the soil, only a small percentage of the potassium in the soil may be exchangeable and, therefore, available.

In the United States the daily dietary intake of the β^- emitter ^{210}Pb (22.6 y) (uranium series) is about 37 mBq/kg (1 pCi/kg).

14.4.6 Materials Used for Building Construction

Buildings can provide some shielding from extraterrestrial radiation and from γ radiation from rocks and soil outside the buildings. However, certain materials of construction (e.g., stone, especially granite), containing uranium or a high potassium content, can raise the dose of ionizing radiation received by the occupants of a building.

In the 1950s it was shown that the concentration of radon could reach 1.1 kBq/m^3 (30 pCi/liter) in the air of a residence that was heated by radiant heating from concrete floors containing crushed Chattanooga shale having about 0.008% uranium. In Florida, phosphate rock (ore), which can contain about 0.004% uranium, is mined for the production of fertilizer. Conversion of the ore to fertilizer reduces the concentration of uranium and its decay product ^{226}Ra in the fertilizer. However, use of the mill tailings as fill around basements of buildings has resulted in elevated radon levels. In addition gypsum, CaSO$_4 \cdot$ 2H$_2$O, a by-product that may be contaminated with ^{226}Ra, is used to manufacture plaster of paris and wallboard for building construction. Finally, cinder blocks manufactured from coal ash containing ^{226}Ra also are a potential source of radon in houses and other buildings.

In the United States, following World War II, uranium was mined in 10 states: Arizona, Colorado, Idaho, New Mexico, North Dakota, Oregon, Pennsylvania, Texas, Utah, and Wyoming. Some of the approximately 24 million tons of tailings, a fine silt produced in the mining and milling of the ore, was used in several western states in concrete and as fill around the foundations of at least 5000 buildings that included private residences, schools, and public and commercial buildings. A similar situation occurred in Ontario Province in Canada. In the tailings that remain after the uranium has been leached from the ore, the ^{226}Ra activity can be as high as 3.7–37 MBq/kg (0.10–1.0 μCi/g). Concrete made from the tailings not only raised the level of γ radiation, but also provided a concentrated source of radon within the buildings. Remedial action has been very costly. Treatment of mill tailings as radioactive waste is discussed in Section 14.12.2.4.

14.5 APPLICATIONS OF NATURALLY OCCURRING STABLE NUCLIDES AND RADIO NUCLIDES TO THE STUDY OF THE EARTH AND ITS ENVIRONMENT

14.5.1 Types of Application

The many environmental applications of naturally occurring nuclides can be separated into three categories. The first category is the well-known use of isotopes as tracers. An example is measurement of the ratio of radiogenic, stable ^{87}Sr to stable ^{86}Sr to trace the source of groundwater and its path in an aquifer. Another is the use of cosmogenic ^{7}Be (53.28 d) as a tracer for river sediment and its resuspension.

The second category includes the study of chemical and physical processes associated with events and changes that are occurring or have occurred in and on the earth and in its atmosphere. Differences in chemical or physical properties associated with differences in the masses of the isotopes of an element result in isotope fractionation—a change in the relative abundance of isotopes of a given element. (See Chapter 13, Section 13.2.2.1.) Since isotope fractionation can be related to the environmental conditions on the earth when a change or event occurred in the past, it can be used as a research tool in atmospheric chemistry, geochemistry, paleogeology, paleoclimatology, and other fields related to the paleoenvironment.

The third category encompasses studies to determine the age of materials, date events that occurred in the past (e.g., for geochronometry), or measure the rates at which changes (e.g., climatic changes) have occurred, using radioactivity as a nuclear clock or chronometer. The length of time measured by radiometric dating may be from a few years to the age of the earth. Commonly, a study in the second category will also include the use of radionuclides to establish the time when the process or event being studied occurred.

Another method of determining the age of an ancient rock, sand, or objects buried in soil, for example, is based on a clock that measures the time interval since the undisturbed sample was last exposed to heat or sunlight. Time is proportional to the absorbed dose of ionizing radiation received by the sample from natural sources of ionizing radiation during the time interval. When the sample is heated or exposed to a laser, it emits light called thermoluminescence or optically stimulated luminescence, the intensity of which is proportional to the dose and, therefore, the exposure time (assuming the dose rate is constant). Luminescence occurs when electrons that were detached from atoms by the ionizing radiation and became trapped in imperfections in the crystal lattice of the material are given sufficient thermal energy to become free. The potential energy stored in the crystal is emitted as light. The concentration of trapped electrons also can be measured by electron spin resonance (ESR) techniques.

Data obtained by the various methods have been used in the formulation of models of naturally occurring processes (e.g., climatic changes) and in the testing of existing mathematical models and hypotheses. Such models are intended to enable prediction of future changes in the earth's environment.

14.5.2 Isotope Fractionation

Because isotope fractionation (Section 13.2.2.1) is very small, except for ^{1}H and ^{2}H, any effect on the normal chemical behavior of an element is usually ignored. However, when a change in the ratio of two isotopes occurs in some process in the environment, the change can be used to obtain very valuable information about the environment, especially the paleoenvironment. Extremely small changes in the abundance ratio of two isotopes of an element can be measured very accurately by mass spectrometric methods using a variety of ultrasensitive mass spectrometers.

The following are examples of such processes: photosynthesis of complex organic compounds, biological processes, and physical processes such as vaporization of a liquid, diffusion of a gas, and transport across the boundary of two phases. Isotope fractionation may have a temperature coefficient large enough to permit use of the coefficient as a paleothermometer to measure the temperature and temperature changes on the earth's surface (seawater and ice) over a very long period of time or at the time of a major environmental event in the past.

An emerging method for studying the sources and fate of organic pollutants in surface and groundwater systems utilizes compound-specific stable isotope measurements of isotope fractionation of H, C, N, and Cl in compounds accompanying transport, metabolism, and biodegradation. The method uses gas and liquid chromatography and high-precision stable isotope mass spectrometry.

The difference between the isotopic composition of an element in two samples of a material is commonly expressed in terms of the change in the ratio of numbers of atoms of two isotopes (e.g., $^{18}O/^{16}O$). When the isotope ratio for an element in a sample of interest is compared with that in a standard sample, the difference between the ratios is usually expressed in terms of delta, δ (values expressed in "per mil"). For example, $\delta^{18}O$ is defined by the equation

$$\delta^{18}O = \frac{(R_{sample})/(R_{std})}{R_{std}} \times 1000 \qquad (14\text{-}4)$$

where $R = (^{18}O/^{16}O)$.[13]

The following are a few examples of isotope fractionation:

1. Water in leaves of plants is enriched in ^{18}O relative to water in the soil where the plants are growing.

---

[13]Standard mean ocean water (SMOW) is an example of R_{std} for $\delta^{18}O$.

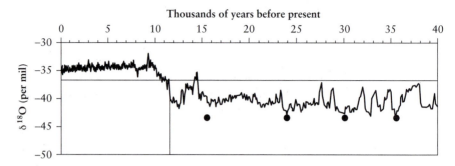

FIGURE 14-8   Variation of ^{18}O concentration in ice core samples from the Greenland Ice Sheet for the past 40,000 years. The abrupt change from a cold period to a warm period began about 11,650 years ago. The coldest periods in the North Atlantic are indicated by dots. From K. Taylor, *Am. Sci.* **87**, 320 (1999). Used by permission of Edward Taylor/*American Scientist*.

2. Photosynthesis in plants results in preferential uptake of ^{12}C so that biomass is depleted in ^{13}C relative to the atmospheric CO_2. In limestone, therefore, carbon-in-carbon inclusions formed by living organisms differ in their $\delta^{13}C$ from the inorganic carbon. The value of $\delta^{13}C$ varies with temperature. As little as 20 pg of carbon is required for mass spectrometric analysis.

3. Nitrous oxide emitted from soil is depleted in ^{15}N and ^{18}O relative to the values for atmospheric N_2O.

4. Stratospheric CO_2, O_3, and O_2 are enriched with respect to ^{18}O relative to their isotopic composition in the atmosphere.

5. The isotopic composition of phosphate in vertebrate bone depends on the temperature when the bone was formed. The ^{16}O content is higher at higher temperature.

6. When ice forms in the ocean, it is enriched in ^{16}O relative to water.

7. Both snow and rain are enriched in ^{18}O. The enrichment increases with increasing temperature at the time of precipitation.

8. Atmospheric CO_2 is enriched in ^{14}C relative to deep-water CO_2.

9. An example of the use of the temperature dependence of $\delta^{18}O$ to study the temperature profile of ice cores from the Greenland Ice Sheet over the past 40,000 years is shown in Figure 14-8. An abrupt change in climatic conditions occurred about 11,650 years ago. The time when this change began, which has been associated with a change in the behavior of the Gulf stream, can be pinpointed within about 20 years. It is not known whether anthropogenic activities can trigger a change in the way the Gulf stream transports warm water.

10. When atmospheric ozone is formed, the isotopic composition ($^{18}O/^{17}O$) is not that expected from the mass effect and is referred to as a "mass-independent" effect.[14]

[14]Y. Q. Gao and R. A. Marcus, *Science*, **293**, 259 (2001).

14.5.3 Radiometric Methods for Age Determination

Both primordial and cosmogenic naturally occurring radionuclides in the environment can be used as nuclear chronometers to determine the dates of a wide variety of processes and events in the earth, on the earth and in the earth's atmosphere. The radiometric method of determining the age of a material is based on equation (13-14), which is rewritten here as follows:

$$N_A = N_A^0 \exp(-\lambda_A t) \tag{14-5}$$

where N_A is the measured number of atoms of a radionuclide A in a sample at the unknown decay time t (the age of the sample), λ_A is the known disintegration constant of A, and N_A^0 is the number of atoms of A present when the sample became a closed system at time t_0.[15] Time t is then the length of time that the clock has been ticking since t_0.

If N_A^0 is known (as it is, e.g., for ^{14}C) the calculation is straightforward in principle, but not necessarily in practice. If N_A^0 is unknown (as is usually the case), there are several ways to evaluate it.

In general, N_A^0 can be calculated from N_B, the number of atoms of a single decay product, from N_C, the number of atoms of a long-lived decay product in a naturally occurring series, or from N_D, the stable end product ^{206}Pb, ^{207}Pb, or ^{208}Pb of a series. The corresponding equations are

$$A \longrightarrow B(s) \tag{14-6}$$

$$A \longrightarrow C \longrightarrow D(s) \tag{14-7}$$

Branched decay must be taken into account if it is significant. When the decay of A is represented by equation (14-6), N_A^0 can be calculated from

$$N_A^0 = N_A + N_B \tag{14-8}$$

where N_B is the measured number of atoms of B formed in the sample by decay of A.[16] Thus

$$N_B/N_A = \exp(\lambda_A t) - 1 \tag{14-9}$$

and

---

[15]Until recent years the procedures for measuring time by radioactive decay involved the measurement of the residual radioactivity of the decaying radionuclide. It is difficult to measure accurately the activity of very long-lived radionuclides with very low specific activity, by counting the nuclei that decay (a very small fraction of the total nuclei of the isotope). Mass spectrometric methods provide a direct measurement of the total number of atoms of an isotope in a sample and, therefore, require smaller sample size and can provide greater sensitivity.

[16]Equation (14-8) is based on the assumption that N_B^0 is zero; otherwise a correction must be made.

$$t = \frac{1}{\lambda_A} \ln \left(1 + \frac{N_B}{N_A} \right) \qquad (14\text{-}10)$$

Natural lead also contains a small amount of a fourth stable isotope, $^{204}Pb(s)$, which is not radiogenic. If a sample contains lead that is partially primordial (with a known isotopic composition measured in samples containing no parent radionuclide) and partially radiogenic, a correction for the primordial lead can be made by expressing the amounts of radiogenic lead relative to that of ^{204}Pb.

Samples containing uranium will contain the members of both the uranium and the ^{235}U (actinium) series. Both series can be used to measure time (as an internal consistency check). Similarly, the thorium series can be used separately or with one of the other series. The ratios of stable lead isotopes such as $^{207}Pb/^{206}Pb$ for the uranium and the actinium series vary with time and also can be used to determine the age of a sample.

A nuclear method of chemical analysis commonly used to determine the concentration of an element in a geological material is *isotope dilution analysis*. In general, the method is useful when methods of quantitative separation of an element are not suitable. A known quantity (a spike) of a stable isotope of the element is added to a sample of the material having a known isotopic ratio for the added isotope. From the change in the ratio in a pure sample separated from the mixture and containing only a fraction of the total amount of the element of interest in the mixture, the total amount can be calculated. For other types of samples (e.g., biological), radioisotopes can be used. In one procedure, for example, a known quantity of the appropriate chemical species of the element of interest that is labeled with a radioisotope and has a known specific activity is added to a sample to be analyzed. From the decrease in specific activity of a pure sample separated from the mixture, the total amount of the element initially present in the sample can be calculated.

Even before modern instrumental techniques became available and before all the members of the three naturally occurring series had been identified, radioactivity was used to estimate the age of the earth[17] from the ^{238}U and ^4He content of a uranium mineral. For a uranium-containing sample that is in secular equilibrium, eight α particles are emitted for each decaying ^{238}U. Unfortunately the $^{238}U/^4He$ method is not very accurate because of the ease with which helium can escape from a rock sample.

When the uranium series (Figure 13-5), for example, is not in secular equilibrium in a sample, the members of the series are said to be in disequilibrium. A solid sample with cracks and a porous solid are examples of materials that would very likely lose gaseous ^{222}Rn and cause disequilibrium. In natural waters such as ocean water, disequilibrium is normal and can be attributed to

---

[17]About 4.6×10^9 years, based on modern radiometric methods of dating. The age of the universe is about 13×10^9 years.

the aqueous chemistry of the daughters of ^{238}U. Specifically, ^{234}Th (24.10 d) and ^{230}Th (7.54×10^4 y) [both as Th(IV)] are readily removed from ocean water by adsorption on particulate matter such as ocean sediment. Similarly, ^{222}Rn can escape into the atmosphere from surface water, and ^{226}Ra can be removed by incorporation in $CaCO_3$ during formation of the shells of marine organisms. Much remains to be learned about the natural water chemistry of the elements that are members of the three natural series. Disequilibrium can be the basis for further study of the aqueous chemistry of these elements and for the study of chemical and physical processes in the environment.

Fission track counting is one of several methods that can be used instead of mass spectrometric analysis to determine the age of uranium-containing mineral specimens. The method is based on measuring the number of tracks of fission products per unit area of sample produced by the spontaneous fission of ^{238}U (Section 13.5.4). After the tracks have been counted, the sample is irradiated with thermal neutrons to fission some of the atoms of ^{235}U in the sample. From the number of new fission product tracks and the neutron irradiation conditions, the number of atoms of ^{235}U can be determined. Then, from the known ^{235}U/^{238}U ratio and the known λ for the spontaneous fission of ^{238}U the age of the sample can be obtained.

Age determination by the ^{40}K/^{40}Ar (^{40}Ar/^{39}Ar) method is useful for ages between a few thousand years and the age of the earth. The calculated age is relative to the last time the sample was heated and, therefore, lost some or all of the ^{40}Ar (s). Until 1966, the determination of the number of ^{40}K atoms in the sample depended on the difficult measurement of the very low specific activity of the long-lived ^{40}K (1.27×10^9 y). In the improved method, the sample is irradiated with fast neutrons to convert ^{39}K (s) into ^{39}Ar (269 y) according to the nuclear reaction[18]

$$^{39}\text{K}(n, p)^{39}\text{Ar} \qquad (14\text{-}11)$$

Next, the ^{40}Ar/^{39}Ar ratio is measured accurately by laser fusion (melting) mass spectrometrometry. From the number of ^{39}Ar atoms formed in the irradiation of the sample, the number of ^{39}K target atoms in the sample is calculated, and finally the number of ^{40}K atoms in the sample is calculated from the known ratio of ^{40}K to ^{39}K in normal potassium. Loss of ^{40}Ar before analysis is a potential source of error and varies with the type of sample. Correction for contamination by atmospheric Ar can be made based on the presence of ^{36}Ar (s) and the known ratio (295) of ^{40}Ar to ^{36}Ar in the atmosphere.

Equation (14.5) is used in a different way for the determination of the age of a sample by the decay of the cosmogenic radionuclides ^{14}C (5715 y) and ^{3}H (12.32 y): N_A is measured and N_A^0 is known. The age is the time since the

---

[18]See Section 13.4.2, equation (13.27) and related footnote 19. Because of the long half-life of ^{39}Ar (nuclide B), the number of atoms formed is $(R_f)_B$ multiplied by the irradiation time.

source of the sample ceased to be in isotopic exchange equilibrium with $^{14}CO_2$ and HTO in the atmosphere. For example, in the case of ^{14}C it could be the time when a tree is cut down. Thus, after a tree is cut down, the ratio of ^{14}C to stable ^{12}C in a sample of wood from the tree decreases with the half-life of ^{14}C. The age of carbonaceous material (e.g., wood and cloth) can be determined by comparing the ^{14}C content of the material with that which the material had when it was part of a living system in equilibrium with ^{14}C in the atmosphere.

In December 1949 Willard F. Libby introduced radiocarbon dating by using β-particle counting to measure the specific activity (disintegrations per second per gram of carbon) of the ^{14}C remaining in a sample. In 1977 a second method, in which the ratio $^{14}C/^{12}C$ is determined by accelerator mass spectrometry (AMS), became available. The measured ratio is N_A in equation (14-5). This method, in which the atoms of ^{14}C are counted, can extend the useful time range somewhat, but, as for any radionuclide used for dating, the useful time range is primarily limited by the half-life of the radionuclide. The AMS method does, however, make possible the use of smaller samples (e.g., 10^{-3} times smaller: $< 50\,\mu g$ to a few milligrams of carbon) than the counting method. Ages up to about 70,000 years can be measured using AMS. Beta-particle counting of larger samples can be used for calibration.

Radiocarbon dates are subject to many sources of error.[19] Three were discussed in Section 14.4.3.1. Additional errors can arise from fractionation[20] during the process whereby the ^{14}C was incorporated in the sample, and from contamination of the sample during preparation or analysis. Radiocarbon dates (or ages) have been calibrated by dendrochronology (dating of tree rings) combined with measurement of the $^{14}C/^{12}C$ ratio in the tree rings, by using other radionuclide dating systems, and by using historical data.

Radiocarbon dates are usually reported as years BP (before present). This practice is based on taking the concentration of ^{14}C in the atmosphere as constant since 1950 (the present). At that time the half-life of ^{14}C was taken to be 5568 years and, prior to the testing of nuclear weapons, the average value of the natural specific activity of ^{14}C in wood was 255 Bq per kilogram of carbon (6.9 pCi/g C). Dates BP are on the same scale and can be compared. Conversion of a BP date to a calendar date (e.g., BC) is not simple because the atmospheric concentration of ^{14}C has not remained constant and the half-life of ^{14}C has been determined more accurately. The conversion requires a cali-

---

[19]All applications of radiometric dating are subject to errors in addition to the usual experimental errors of measurement. Many errors result from assumptions that must be made about the system (e.g., that it has been a closed system and that it has not been contaminated). Additional errors may arise because the chemical behavior of the components is not fully understood.

[20]A correction for fractionation can be made using $\delta^{13}C$, defined in terms of measured $^{13}C/^{12}C$ ratios as for $\delta^{18}O$ in equation (14-4).

bration, as mentioned earlier. From the change in ^{14}C concentration in the atmosphere later, it was determined that the residence time for ^{14}C as CO_2 is about 4 years in the stratosphere and about 10 years in the atmosphere as a whole.

An application of tritium dating that is environmentally relevant is measurement of the age of water in aquifers. When water is no longer in contact with the atmosphere as surface water, the ^3H content as HTO decreases from its equilibrium level, with a half-life of 12.32 y.

Another example of the use of cosmogenic nuclides is to measure the buildup of ^3He, ^{10}Be, ^{21}Ne, ^{26}Al, and ^{36}Cl in the surface of rocks in a glacial moraine to determine how long the rocks have been exposed to cosmic rays.

14.5.4 Summary of Examples of Nuclides Used and Types of Application

Table 14-3 contains examples of nuclides that have been used in the two categories of environmental applications in recent years. The radionuclides used to measure time are cosmogenic, primordial, or a daughter of the latter. Table 14-4 gives examples of chemical and physical processes occurring in the environment that have been studied. For examples of the types of system that have been studied to determine the age of a substance or the date when a particular event occurred in the past, see Table 14-5.

14.6 SPECIAL-PURPOSE SOURCES OF IONIZING RADIATION IN THE ENVIRONMENT: TYPES AND APPLICATIONS

Special-purpose sources of ionizing radiation in the extended environment (not limited to the outdoor environment) range in size from low-level sources incorporated in consumer products used in the home to high-level sources located in industrial and medical facilities. Exposure to the radiation from such sources may be planned and based on risk–benefit considerations (e.g., exposure of a patient resulting from a procedure in a medical facility), or it may occur accidentally in the workplace (e.g., exposure of a worker in a facility where special-purpose sources are used). The dose of ionizing radiation received from any of these sources will contribute to the total dose that a person receives in the course of everyday living.

Other sources including nuclear reactors in central station power plants, nuclear weapons, facilities used to manufacture reactor fuel and weapons, and radioactive waste are discussed in Sections 14.7, 14.11, and 14.12, respectively.

TABLE 14-3

Examples of Naturally Occurring Nuclides Used for Studying Environmentally Related Reactions and Processes and for Radiometric Dating[a]

Reactions and processes[b]	Radiometric dating[b]
^{2}H/H or D/H	^{3}H*/^{3}He
^{3}He/^{4}He	^{10}Be*
^{10}Be*	^{14}C*
^{11}B/^{10}B	^{14}C*/^{12}C
^{13}C/^{12}C	^{14}C*/^{13}C
^{15}N/^{14}N	^{26}Al*
^{17}O/^{16}O	^{36}Cl*/^{35}Cl
^{18}O/^{16}O	^{40}K/^{40}Ar, ^{40}Ar/^{39}Ar*
^{20}Ne/^{22}Ne	^{87}Rb*/^{87}Sr
^{21}Ne/^{22}Ne	^{92}Nb*/^{92}Zr
^{34}S/^{32}S	^{129}I*/^{127}I
^{40}Ar/^{36}Ar	^{129}I*/^{129}Xe
^{44}Ca/^{40}Ca	^{146}Sm*/^{142}Nd
^{57}Fe/^{54}Fe	^{147}Sm*/^{143}Nd, ^{143}Nd/^{144}Nd*
^{56}Fe/^{54}Fe	^{176}Lu*/^{177}Hf
^{84}Kr/^{36}Ar	^{182}Hf*/^{182}W
^{87}Sr/^{86}Sr	^{187}Re*/^{187}Os
$^{124-136}$Xe/^{132}Xe	^{210}Pb*
^{230}Th*	^{230}Th*

Ratios of members of the U, Ac, and Th series:
^{238}U*/^{206}Pb, ^{238}U*/^{230}Th*^{c}, ^{235}U*/^{207}Pb
^{234}U*/^{238}U*^{c}, ^{232}Th*/^{208}Pb,
^{230}Th*/^{232}Th*, ^{230}Th*/^{234}U*,
^{226}Ra*/^{230}Th*^{c}, ^{231}Pa*/^{235}U*,
^{207}Pb/^{206}Pb, ^{208}Pb/^{204}Pb, ^{208}Pb/^{206}Pb
^{207}Pb/^{204}Pb, ^{206}Pb/^{204}Pb

[a]^{10}Be, ^{14}C, ^{26}Al, and ^{36}Cl are cosmogenic radionuclides.
[b]Radionuclides are marked with asterisks (*).
[c]Relative to secular equilibrium ratio.

TABLE 14-4

Examples of Studies of Chemical and Physical Processes[a]

Atmosphere
    Changes in the temperature at the earth's surface during the last three decades
    Changes in N_2O concentration in the atmosphere
    Levels of CO_2 in ancient times
    Rate of oceanic and terrestrial uptake of anthropogenic atmospheric CO_2
    Mixing of magmatic CO_2 with atmospheric CO_2
    Exchange of soil carbon and carbon in atmospheric CO_2
    Chemistry of $O_2, O_3,$ and CO_2

(continues)

TABLE 14-4 (*continued*)

Gravitational enrichment of Kr and Ar
Link between atmospheric CO_2 concentration and global warming in the past
Sources of sulfur and lead

Biosphere
Organic photosynthesis in sediments 2.8×10^9 years old
Precursors of organic compounds found in sedimentary rock
Body temperature of dinosaurs from $\delta^{18}O$ in bone phosphate
Living and dead carbon as sources of CH_4, CO, etc.
Change in type of vegetation
Uptake of atmospheric CO_2 on land
Isotope fractionation when polychlorinated biphenyls are dechlorinated by bacteria

Climate
Climatic conditions during the last glacial maximum
Rapid and slow paleoclimatic changes
History of glacial ice; glacial cycles
Validity of the Milankovich hypothesis
A warm period during relatively high solar radiance about A.D. 1000
Correlation between warming data from ice cores and climate changes in other regions of the earth

Hydrosphere
Groundwater: rate of mixing; transit time through an aquifer
Ocean sedimentation rates
Sources of wastewater
Rise and fall of sea level in the past
Changes in the isotopic composition of seawater in the past
Changes in sea surface temperature (SST)
Changes in subduction zones and mantle dynamics
Instability of the Antarctic Ice Sheet
Ocean circulation
Paleotemperatures of the oceans
Reduction of sulfate in seawater to form pyrite
Variation in paleosalinity in estuaries
Vertical mixing of ocean water (relevant to carbon cycle)
Stability of methane hydrate
El Niño in the past
Correlation between changes in the California Current and glacial maxima

Lithosphere
Earth's mantle
Geochemical processes: formation of igneous and metamorphic rocks; high-temperature metamorphism
Volcanism, plate tectonics
Weathering and erosion of the continental surface
Chemistry of thorium at low concentrations in seawater
Uranium series disequilibrium in basalts and magmas
Formation and growth of clays
Chemical exchange reactions between the earth's crust and its mantle
Magmatic degassing of CO_2
Continental weathering

Earth's orbit: eccentricity

[a]Some examples may apply to more than one category.

TABLE 14-5

Examples of Radiometric Age and Date Determination[a]

Biosphere
　　Pollen in ice cores
　　Evolution of fossil shells
　　Layers of peat identified with known paleoearthquakes
　　Pollen, spores, seeds, microfossils, wood, and charcoal in peat to detect prehistoric
　　　　earthquakes
　　Permian–Triassic extinctions

Climate
　　Date of last glacial maximum
　　Episodes of glaciation

Geomagnetic field
　　Lava flows that are studied to determine variations in geomagnetic paleointensity
　　　　(affects production of ^{14}C by cosmic rays)
　　Reversals of the geomagnetic field

Hydrosphere
　　Sea shells
　　Greenhouse gas (methane) hydrates in ocean sediments
　　Coral deposits
　　Sediments in freshwater lakes
　　Deep water in the Arctic Ocean
　　Well water

Lithosphere
　　Formation of the earth's crust
　　Sedimentary rocks
　　Geochronology of Yucca Mountain (proposed HLW repository)
　　Mantle recycling
　　Organic material trapped in glacial moraines
　　Potassium-containing minerals
　　Strata where fossils are found
　　Regions where asteroid impacts have occurred
　　Paleovolcanic events to determine suitability of sites for a geological repository
　　　　for high-level radioactive waste
　　Plate-tectonic events
　　Lava flow pulses separated by thousands of years
　　Sulfide minerals
　　Zircon crystals in volcanic rock and in sedimentary material
　　Mantle plumes near oceanic islands
　　Carbonates at archaeological sites
　　Rocks and glacial moraines to determine their "exposure age"
　　　　(i.e., length of exposure to cosmic rays)

[a]Some examples may apply to more than one category.

14.6.1 Sources Used for Structural Analysis of Solids, Radiography, Diagnostic Imaging, and Medical Therapy

14.6.1.1 Focused-Beam Sources

X-ray machines and neutron sources are used in research to determine the spatial arrangement of atoms by x-ray diffraction and neutron scattering in crystalline solids, respectively. X-ray machines for dental and medical radiography are commonplace special-purpose sources of ionizing radiation. Both the x-ray machines and the associated photographic film have been improved over the years to reduce patient exposure. X rays are also used for diagnostic imaging of soft tissue by computed tomography (CT) or CAT (computed axial tomography) scanning.[21] The x-ray source rotates completely around the patient's body, resulting in a series of cross-sectional views. Data for x-ray transmission are processed with a computer, which provides a display of an image of the region of the body being scanned.

Another type of CT scanner is the electron beam device (EBCT) in which pulsed beams of electrons strike tungsten targets arranged in a ring about the patient's body. The tungsten x rays (Figure 13-20) that are generated in the targets enable parts of the body (e.g., coronary arteries) to be observed in motion.

It is appropriate at this point to call the reader's attention to a nuclear diagnostic technique, magnetic resonance imaging (MRI), that uses computer analysis of nuclear magnetic resonance (NMR) measurements (see Section 13.2.2.2) to produce diagnostic tomographic images of internal body tissue without exposing the patient to ionizing radiation.[22] The most commonly used stable nuclide is ^{1}H.

A variation of MRI called functional MRI (fMRI) is becoming widely used to provide tomographic images of neural brain activity. Changes in blood flow in small regions of the brain can be identified when the patient carries out a mechanical task, responds by visual recognition, responds to music, and so on. In this technique a radiofrequency excitation pulse increases the magnetic energy of the nuclei (protons) in the sample. The measured time (relaxation time) required for the pulse to decay exponentially as these nuclei return to their initial energy states depends on the magnetic properties of nuclei in nearby molecules. Since bound oxygen ^{16}O (spin = 0) in blood affects the relaxation time, it can be used to measure blood flow.

---

[21]Tomography is the procedure in which a collimated beam of x rays or γ rays is focused on a specific plane or "slice" of an organ or region of the body. Variations in density of tissue cause variations in the attenuation of the photons, which in turn cause corresponding variations in the output of the radiation detectors that provide the data.

[22]MRI rather than NMRI is used because of the widespread fear of all things nuclear.

Because it is desirable that every effort be made to reduce a person's exposure to ionizing radiation, other procedures for noninvasive diagnosis such as those using MRI or ultrasound may replace or supplement the use of x rays and γ rays for certain types of medical diagnosis.

X-ray machines and γ-ray sources are also used for industrial radiography of metallic products to find flaws, defective welds, and so on. After nuclear reactors became available for large-scale production of γ-emitting radionuclides, ^{60}Co (5.271 y) and ^{137}Cs (30.07 y) replaced sealed radiographic sources containing ^{226}Ra (1599 y) and its daughters.

Both x rays and the γ rays from ^{226}Ra and its daughters have also been used as external sources of ionizing radiation for medical therapy. One or more highly collimated beams of γ rays are directed at the area to be irradiated. Common sources of γ rays in devices now used for cancer therapy include ^{60}Co and ^{137}Cs. As for radiographic sources, radiation therapy devices may contain terabecquerels (tens to hundreds of curies) of a radionuclide.

Boron neutron capture therapy (BNCT) has been used with some success to treat inoperable brain tumors. The patient is given a boronated pharmaceutical that concentrates in the tumor and then the tumor is exposed to a beam of neutrons from a nuclear reactor. The products of the ^{10}B(n, α)^{7}Li reaction, ^{4}He$^+$ and ^{7}Li$^+$, are ionizing and have a high linear energy transfer. Another procedure being investigated for cancer therapy is based on the use of an accelerator-produced beam of high-energy ions of elements such as neon, carbon, and silicon. The procedure requires that the beam of these high-LET ions be accurately focused on the site of the malignancy.

14.6.1.2 Implanted Sources Used in Medical Therapy

Ever since radium became available, small radioactive sources have been implanted in or near tumors to destroy them. Both ^{226}Ra (encapsulated in a needle) and ^{222}Rn (encapsulated in a glass or gold seed) have been used with limited success. Today, this type of radiation therapy, called brachytherapy, utilizes small rods, beads, seeds, or rice-sized pellets containing radionuclides emitting low-energy ionizing radiation such as ^{103}Pd (16.99 d, EC, e$^-$, x rays = 21 keV), ^{125}I (59.4 d, EC, e$^-$, E_γ = 35.49 keV), ^{90}Y (2.67 d, β$^-$), ^{192}Ir (73.83 d, EC, β$^-$, γ rays, mainly 0.316 MeV) and ^{252}Cf (2.646 y, α, e$^-$, low-intensity γ-rays and neutrons from spontaneous fission). Because of the limited depth of penetration of the radiations from these radionuclides, the risk of irradiation of nearby healthy tissue is low. The availability of imaging methods and improved delivery techniques facilitates precise positioning of the implants.

Examples of implant applications are ^{103}Pd and ^{125}I for prostate cancer and ^{90}Y for one type of liver cancer, hepatocellular carcinoma.

Californium-252 is useful in brachytherapy for tumors as a neutron source rather than photon source. Neutrons have a higher quality factor (Q) or radiation weighting factor (w_R) than photons (Tables 13-7 and 13-8).

A controversial application of ionizing radiation is intravascular therapy to prevent reclosing (restenosis) of coronary arteries after angioplasty. As one example of the techniques used, seeds containing a radionuclide such as ^{192}Ir or ^{90}Sr are placed in a catheter, which is positioned for a short time in the region from which the plaque was removed.

14.6.2 Sources Used in Nuclear Medicine

In nuclear medicine, a radiopharmaceutical, also called an imaging agent or a tracer and containing a particular radionuclide in a specific chemical form, is administered orally or intravenously to a patient. Diagnostic imaging is the usual purpose but, less frequently, radiation therapy is performed.[23] The radiation that is measured to obtain an image of an organ originates within the body rather than in an external source of x or γ rays. Radionuclides suitable for diagnostic imaging decay by the emission of γ rays that are detected by an array of detectors in a "γ-ray camera," which, in the simplest planar procedure, provides an image showing abnormalities in the region in which uptake of the radionuclide has occurred.[24] Radionuclides that decay by isomeric transition and electron capture have the advantage of not exposing the patient to high-energy particulate radiation. Suitable radionuclides must also have a relatively short half-life (minutes to days) and be readily incorporated into a nontoxic radiopharmaceutical.

A radiopharmaceutical contains a radionuclide in a chemical form that has been shown to concentrate in a specific organ (target organ) or region of the body. In a few radiopharmaceuticals, the radionuclide is present in a relatively simple chemical form. With few exceptions, the target organ of a radiopharmaceutical is not determined directly by the chemistry of the radionuclide, but by the chemistry of the organic or inorganic compound (e.g., an amino acid, a chelate, etc.) containing the radionuclide. Commonly used β^-- and γ-emitting radionuclides are ^{51}Cr (27.702 d), ^{67}Ga (3.261 d), ^{99m}Tc (6.01 h), ^{111}In (2.8049 d) ^{123}I (13.2 h), ^{125}I (59.4 d), ^{131}I (8.0207 d), ^{133}Xe (5.243 d), and ^{201}Tl (3.039 d). A few examples of specific radiopharmaceuticals and their

[23]An organic compound such as CH_4 in which a significant fraction of the stable ^{12}C atoms have been replaced by stable ^{13}C or by radioactive ^{14}C or ^{11}C is an isotopically labeled compound. A compound such as $Na^{99m}TcO_4$, sodium pertechnetate, is not labeled in the same sense because there is no stable isotope of technetium. In a more general usage, the word "labeled" indicates that the compound contains a stable nuclide or a radionuclide that makes the compound suitable for a particular use.

[24]The camera is a γ-ray detector with a specially configured shield of absorbing material to collimate the photons entering the detector.

diagnostic applications are listed in Table 14-6. The structure of one of these, technetium-99m sestamibi, is shown in Figure 14-9. The levels of activity needed for diagnostic imaging vary from about 200 kBq to 750 MBq (5.4 μCi to 20 mCi).

TABLE 14-6

Examples of Diagnostic Radiopharmaceuticals

Radiopharmaceutical[a]	Application
Gamma-ray emitters	
^{51}Cr-EDTA	Renal tracer
^{67}Ga-citrate	Tumor imaging
^{99m}TcO$_4^-$ (Na$^+$ salt)	Thyroid function, blood flow, brain imaging
^{99m}Tc-arcitumomab	Imaging recurrent and/or metastatic colorectal carcinomas
^{99m}Tc-DMSA	Renal function
^{99m}Tc-DTPA	Renal function
^{99m}Tc-MAA	Venous blood flow
^{99m}Tc-MAG$_3$	Renal function
^{99m}Tc-nofetumomab merpentan	Detection of small-cell lung carcinoma
^{99m}Tc-pyrophosphate	Bone imaging
^{99m}Tc-labeled red blood cells	Spleen imaging
^{99m}Tc-sestamibi	Myocardial perfusion imaging, imaging of breast lesions
^{99m}Tc-sulfur colloid	Liver imaging
^{111}In-capromab pendetide	Detection of recurrent and metastatic prostate cancer
^{111}In-imicromab pentetate	Imaging damaged myocardial tissue
^{111}In-labeled leukocytes	Imaging infections
^{111}In-DTPA	Flow of cerebrospinal fluid
^{123}I$^-$ (Na$^+$ salt)	Thyroid function
^{125}I-labeled antigen	Radioimmunoassay
^{131}I$^-$ (Na$^+$ salt)	Thyroid function
^{201}TlCl	Myocardial perfusion
^{133}Xe	Lung ventilation
Positron emitters (annihilation photon emitters)	
^{11}C-labeled compounds	Metabolic studies
^{11}C-palmitate	Long-chain fatty acid metabolism
^{13}N-labeled compounds	Metabolic studies
^{13}NH$_3$	Myocardial perfusion
^{15}O-labeled compounds	Metabolic studies
H$_2$[^{15}O]	Myocardial perfusion
^{18}F-labeled compounds	Variety of studies
^{18}F-deoxyglucose	Glucose metabolism
^{82}Rb (As Rb$^+$)	Myocardial perfusion

[a]EDTA, ethylenediaminetetraacetic acid; ^{99m}Tc-arcitumomab, radiolabeled monoclonal antibody; DMSA, dimercaptosuccinic acid; DTPA, diethylenetriaminepentaacetic acid; MAA, macroaggregated albumin; MAG$_3$, mercaptoacetylglycylglycine; nofetumomab merpentan, radiolabeled monoclonal antibody; sestamibi, 2-methoxyisobutyl isonitrile (Figure 14-9); ^{99m}Tc sulfur colloid, Tc$_2$S$_7$; capromab pendetide and imicromab pentetate, radiolabeled indium monoclonal antibodies.

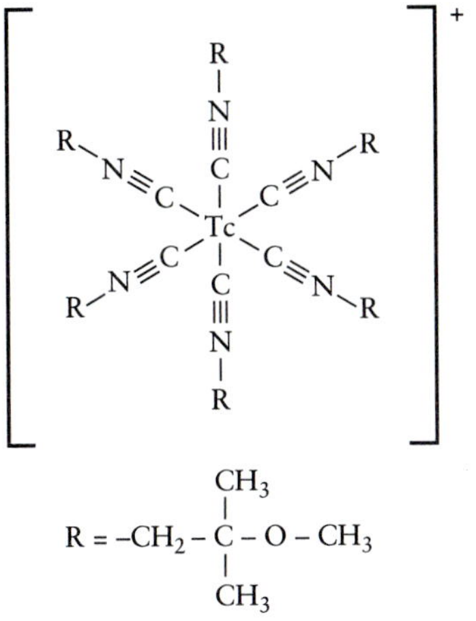

FIGURE 14-9 Sestamibi with ^{99m} Tc.

Two special diagnostic procedures are known as PET and SPECT. In positron emission tomography (PET), the radiopharmaceutical contains a positron emitter. The method is based on the fact that when a positron is annihilated, two photons (0.511 MeV each) are emitted at 180° to each other (see Section 13.7.1.3). The point of annihilation in three-dimensional space can be determined accurately by counting the photons that arrive in coincidence in a pair of detectors that are positioned at 180° relative to each other. The pair of detectors can be rotated around the patient in a circular or oval path, as for CAT and MRI scanning, or the patient may be surrounded by an array of pairs of detectors. Computer-assisted reconstruction generates the images. PET is also useful for studying metabolic kinetics and blood flow in the brain. Examples of positron emitters incorporated in radiopharmaceuticals are ^{11}C (20.3 m), ^{13}N (9.97 m), ^{15}O (122.2 s), ^{18}F (1.8295 h), ^{62}Cu (9.74 m), and ^{82}Rb (1.26 m). These radionuclides are produced with a dedicated particle accelerator, that is, apparatus located at or near the medical facility at which the positron emitters are used. Examples of production reactions are ^{12}C(d, n)^{13}N, ^{14}N(d, n)^{15}O, and ^{18}O(p, n)^{18}F. Table 14-6 gives only a few of the many compounds that are labeled with positron emitters.

One positron-emitting imaging agent, 2-[^{18}F]fluorodeoxyglucose (^{18}FDG or FDG), which has been used over the years for evaluation of glucose metabolism and myocardial viability, also has chemical properties that make it suitable for identifying cancerous tissue and detecting false indications of cancer obtained in standard tests. It, unlike a radiopharmaceutical designed to bind to cells of a particular organ, is nonspecific. That is, it will find any cancerous tissue and detect metastases. The rapidly growing cells in such tissue have a higher need for glucose than normal cells and accept FDG as glucose only to find that it is an unusable permanent resident. FDG is more effective for fast rather than slowly growing cancer cells.

In single-photon emission tomography (SPECT), the camera or array of cameras rotates around the patient's body and provides data for reconstructing images of planes. This procedure provides images with greater contrast than those obtained with the planar procedure for the same radionuclide.

Certain short-lived radionuclides used to prepare radiopharmaceuticals can be obtained from "generators" that are commercially available. These generators contain a relatively large amount of the longer-lived parent of the radionuclide of interest. The shorter-lived daughter grows [equation (13-23)] and is removed chemically as needed. An example is the separation by simple chromatographic elution of ^{99m}Tc as the pertechnetate ion, ^{99m}TcO$_4^-$, from the ^{99}Mo (2.7476 d) parent.[25] Other generator-produced radionuclides include ^{62}Cu from ^{62}Zn (9.22 h) and ^{82}Rb from ^{82}Sr (25.36 d).

Technetium-99m has been called the workhorse of nuclear medicine for several reasons: (1) there seem to be an endless number of ^{99m}Tc compounds that can be used to target virtually any organ of the body, (2) its half-life is quite suitable, (3) its γ-ray energy (142.7 keV) is ideal for measurement, and (4) it is readily available from a generator. The versatility of Tc arises from its electron configuration ($4d^5 5s^2$ for the outer electrons in the ground state) and, therefore, oxidation states from -1 to $+7$. Although ^{99m}TcO$_4^-$, the chemically stable form in which ^{99m}Tc is stored in aqueous solution, mimics I$^-$ and is itself used as a radiopharmaceutical, Tc(VII) is usually reduced to Tc(VI), Tc(IV), or Tc(III) before being combined with a ligand to form a radiopharmaceutical. In terms of the environment, it is of interest to note that ^{99m}Tc decays into ^{99}Tc, a fission product and a pure β emitter with a half-life of 2.13×10^5 y.

Iodine-131 (8.0207 d), which concentrates in the thyroid, is used not only diagnostically but also therapeutically to treat patients with hyperthyroidism and other thyroid diseases, including cancer, and after thyroidectomy. The quantities of ^{131}I (as iodide) used for hyperthyroidism are about 222–333 MBq (6–9 mCi). Another radionuclide used for therapy is ^{32}P (14.28 d) (as orthophosphate). It is used in the treatment of polycythemia vera, a condition in which the patient's blood has an abnormally high concentration of red cells. The quantity administered [up to about 185 MBq (5 mCi)] is limited by the extent of allowable irradiation of the bone marrow.

Additional examples of radionuclides that are used for radiotherapy (including radioimmunotherapy with labeled antigens to treat lymphoma and

---

[25]Molybdenum-99 can be made by neutron irradiation of molybdenum [i.e., ^{98}Mo(n, γ)^{99}Mo], or it can be separated from the other fission products produced by neutron irradiation of a sample of ^{235}U.

leukemia) are ^{89}Sr (50.52 d), ^{90}Y (2.67 d), ^{125}I (59.4 d), ^{153}Sm (1.928 d), ^{186}Re (3.718 d), ^{188}Re (16.94 h), ^{198}Au (2.6952 d), and ^{211}At (7.21 h). Yttrium-90 can be obtained as needed from a generator containing its parent, ^{90}Sr (28.78 y). The latter can be removed from fission product waste accumulated from reprocessed, neutron-irradiated ^{235}U or ^{239}Pu.

A radionuclide that is believed to show promise for destroying cancer by α-particle immunotherapy is bismuth-213 (45.6 m, β^- 97.8%, α 2.2%). The ^{213}Bi is attached to an antibody that in effect transports the radionuclide to the cancer cells. This therapy has been effective in destroying the cancer cells that remain after chemotherapy treatment of acute myeloid leukemia.

The availability of ^{213}Bi is rather interesting. It is a member of the ^{233}U $(1.592 \times 10^5$ y) decay series, which terminates in stable ^{209}Bi rather than a lead isotope. It is the granddaughter of actinium-225 (10.0 d, α), which can be separated from ^{233}U and can then be used as a ^{213}Bi generator. Uranium-233, which has potential use in nuclear weapons or in fuel for nuclear reactors, has been made in kilogram amounts by neutron irradiation of ^{232}Th and is in storage awaiting disposal.

It should be apparent that investigators with a good knowledge of the fundamentals and an interest in discovering new uses of radioisotopes for medical diagnosis and therapy, are limited only by ingenuity.

14.6.3 Multifarious Sources

The following are examples of other sources of ionizing radiation in the environment:

• For many years the oxides of uranium were used to produce orange or green glazes for tableware and other ceramic items that have become collectibles. The oxides are still used as pigments in cloisonné jewelry.

• Uranium has been used to color glassware, and thorium has been incorporated as an impurity in yellow glassware containing cerium.

• Radium is well known as the energy source in self-luminous devices. The radium content of the luminous paint used on the dials of men's wristwatches manufactured before 1960 varied between 3.7 kBq and 0.15 MBq (0.1–4 µCi). Exposure to the wearer of a wrist-watch with radium paint is reduced by absorption of the α and β^- particles and a fraction of the γ rays by the watch case and works.

• Tritium (12.32 y), which emits only low-energy negatrons ($E_{max} =$ 18.591 keV), and ^{147}Pm (2.6234 y), ($\beta^-, E_{max} = 224$ keV, low-intensity γ) are used as substitutes for radium in luminous paint. Up to 930 MBq (25 mCi) of ^{3}H and 7.4 MBq (200 µCi) of ^{147}Pm are used for watches and clocks. These two radionuclides are used in a variety of other products containing self-luminous paints or plastics. Although there is no risk of skin exposure to β radiation from tritium in a watch, a watch with a plastic back rather than a metal back can allow tritium to diffuse through the back and then to be absorbed through the skin. The glass capsules used in night sights for firearms contain about 1.85 GBq (50 mCi).

• Tritium is also used in sealed tubes in exit signs, miniature light sources, and so on in amounts up to about 1.1 TBq (30 Ci). Krypton-85 (10.76 y), $(\beta^-, E_{max} = 687\,keV$, low-intensity γ) is used in similar applications.

• Tritium, ^{85}Kr, ^{147}Pm, and ^{63}Ni (100 y) $(\beta^-$ only, $E_{max} = 66.9\,keV$) are used in electronic devices, fluorescent lamps, and so on in amounts from 0.37 kBq to 3.3 MBq (0.01–90 μCi).

• Alpha particles from ^{241}Am (432.7 y) are used to ionize the air in one type of smoke detector that is designed for use in any type of building but is especially suited for early detection of residential fires. Smoke particles attenuate the steady ionization current reaching the detector and trigger an audible alarm. These battery- or ac-operated devices are small, can be attached to a ceiling, and are not considered to be a radiation hazard. They may contain up to 925 kBq (25 μCi) of ^{241}Am.

• Polonium-210 (138.38 d), an α emitter, ^{3}H and ^{241}Am have been used in antistatic devices.

• A wide variety of radionuclides are routinely used as tracers in many fields of research. Radionuclides are available in pure form and in labeled compounds, some of which are very complex and are synthesized by animals. Hundreds of ^{14}C-labeled compounds and many compounds labeled with ^{3}H, ^{32}P, ^{35}S, and ^{125}I are commercially available. Included are products such as labeled antigens used in radioimmunoassay (RIA), mentioned in connection with nuclear medicine.

• Radioisotope thermoelectric generators (RTGs)[26] that utilize radioactive decay heat have been developed to supply power for space vehicles and satellites, remote terrestrial systems such as polar weather stations, seismic sensing stations, navigational buoys, and so on, and implanted cardiac pacemakers. The two radionuclides that have been used most often are ^{238}Pu (87.7 y) and ^{90}Sr (28.78 y). Plutonium-238, an α emitter with $E_\alpha = 5.4992\,MeV$ (71.1%) and 5.4565 MeV (28.7%) for the two important α particles and low-energy, very low intensity γ rays, is the preferred power source for space applications and pacemakers. It has a power density of 0.55 W (thermal) per gram of ^{238}Pu, requires very little shielding, and can be incorporated in generators providing tens of kilowatts of power. Strontium-90 sources require relatively heavy shielding, largely because of the relatively high-energy

---

[26]RTGs were originally called systems for nuclear auxiliary power (SNAP).

β^- radiation and associated bremsstrahlung from the ^{90}Y daughter (Figure 13-13). The power density for ^{90}Sr is 0.93 W per gram of ^{90}Sr.

• Small radioisotope heater units (RHUs) also containing ^{238}Pu as the dioxide are used to heat instruments and equipment used in experiments conducted in spacecraft. A unit contains about 2.7 g of the fuel, weighs about 40 g, and provides heat output of 1 watt.

• Direct energy conversion also has been used to obtain electrical energy from radioisotopes. Electrical power output of the order of microwatts is obtained by absorbing the ionizing radiation in a semiconductor material such as silicon, thereby generating a direct current for applications requiring polarity and negligible current.

• Over 30 small nuclear reactors with electrical power output in the order of kilowatts have been used in satellites placed in orbit by the former Soviet Union (FSU). Radiation from these space reactors has created serious interference with satellites designed to study γ rays in space, especially γ rays associated with solar flares.

• Thickness gauges using radionuclides are used in industry to measure and automatically control the thickness of products such as paper and sheet metal. As the material passes between the radioactive source and a radiation detector, it absorbs some of the radiation. Variations in thickness cause variations in the signal output of the detector. What appeared to be a malfunction of a thickness gauge containing ^{137}Cs in a plant making sheet steel led to the accidental discovery of ^{60}Co contamination in the plant's steel supply.

• Accelerators and nuclear reactors of several types are operated in research centers and in facilities used for the production of radioisotopes for use in research laboratories, medical facilities, and so on. Air that is used for ventilation in buildings containing accelerators that produce neutrons can contain short-lived radionuclides such as ^{16}N (7.13 s), ^{17}N (4.17 s), and ^{19}O (26.9 s), as well as the longer-lived ^{41}Ar (1.83 h). The neutron-activated air is generally exhausted through a high stack, so that the radioactivity is diluted to an acceptable level before it reaches the ground.

• Gamma-ray sources containing ^{60}Co or ^{137}Cs and particle accelerators are used commercially to produce radiation-induced changes in a variety of materials. Chemical effects that occur when a substance absorbs ionizing radiation constitute the branch of chemistry known as radiation chemistry. As in photochemistry, free radicals play an essential role in the radiation–chemical reactions that occur in an absorber after an initial interaction. Radiolysis, the decomposition of compounds by ionizing radiation, is discussed in Chapter 13 (Section 13.8) for the case of water. The three most common, cost-effective applications at this time are sterilization of packaged medical supplies, irradiation of food (Section 14.6.3.2), and modification of polymers (e.g., cross-linking of polyethylene). Cobalt-60 sources used for sterilization of medical supplies, for example, contain from a few peta becquerels to 2 EBq (a few tenths to about 5 MCi). Particle accelerators include positive ion accelerators, but most are electron accelerators having an electron energy range of 3–10 MeV. Radioisotopes or accelerators or both are in use as sources of ionizing radiation in over 40 countries.

• Electron accelerators are used to irradiate material with electrons or with photons (bremsstrahlen). Because electrons have a limited depth of penetration into an absorber, they are suitable for producing a radiation-induced effect in the surface region of the absorber. Small electron accelerators can be used to decontaminate biologically contaminated pieces of mail.

• A particular type of electron accelerator, the synchrotron, has become very popular as a research tool. A synchrotron accelerates electrons to very high energies (e.g., 1 GeV) as they travel along a circular path determined by a magnetic field. Unlike the electrons in a linear accelerator, which emit photons when suddenly stopped in a target, the electrons in a synchrotron continually emit low-energy photons that can have an energy up to about 25 keV and can be removed from the accelerator as a "bright," coherent beam of X rays. This type of accelerator is referred to as a "synchrotron light source." The photons are especially useful for protein crystallography, structural biology, structural genomics, studies of interacting molecules, surface chemistry, and environmental chemistry. There are over 40 synchrotron light sources in use worldwide, and about two dozen more are under construction or planned.

• The following are examples of processes that are in various stages of research and development for using ionizing radiation to remove environmental pollutants:

(1) Treatment of the gaseous effluents (flue gas) from fossil-fueled power plants to enhance removal of sulfur dioxide and oxides of nitrogen. Irradiation of a mixture consisting of the gas, water vapor, and ammonia leads to the formation of solid ammonium sulfate and ammonium nitrate that can be mechanically removed from the gas before it is discharged.

(2) Sterilization of sewage sludge used in agriculture.

(3) Degradation of a wide range of potential pollutants consisting of organic chemicals, including dichlorobiphenyl, contained in aqueous industrial waste.

(4) Purification of water drawn from municipal water supplies. These processes utilize the radiolysis products of water and require doses of tens of kilograys.

14.6.3.1 Neutron Activation Analysis and Thermal Neutron Analysis

Neutron activation analysis (NAA), a nondestructive method of quantitative chemical analysis, is based on equation (13-27). The unknown number of target atoms of an element of interest is N_A in Chapter 13 (Section 13.4.2, footnote 19). When a sample is irradiated with slow neutrons, the stable nuclei of some of the constituent elements will become radioactive, usually by the (n, γ) reaction. Ideally, the quantity of a radionuclide produced can be determined by quantitative measurement of its characteristic radiation, by γ-ray spectroscopy, without the need for chemical separation from radioisotopes of other elements.

NAA has been used for the determination of about 60 chemical elements in concentrations ranging from that of a bulk component to that of a trace component at the parts-per-million to parts-per-billion level. The sensitivity for a trace element depends on the neutron activation cross section of the target nuclide and on the properties of the product radionuclide (i.e., the half-life, and type, energy, and intensity of the radiation emitted).

Examples of determinations of trace concentrations in environmental samples are Cu and Mn in both raw and treated water supplies, Cl in pesticide residues, rare earth elements contained in cracking catalysts used in oil refineries and released into the atmosphere in particulate form, and a system of several elements (e.g., As, Sb, Se, Zn, In, Mn, V) present in atmospheric pollution aerosols and having the relative concentrations characteristic of their sources.

A variation of NAA is based on measurement of the characteristic energies of the prompt γ rays emitted by the target nuclei as the neutron capture reactions occur. It is advantgeous for the analysis of H, B, C, N, and Si. One application, known as thermal neutron analysis (TNA), was developed to detect chemical explosives in luggage at airports. Most chemical explosives have a relatively high nitrogen content. When nitrogen is irradiated with slow neutrons, the reaction $^{14}N(n, \gamma)^{15}N$ releases a 10.8-MeV γ ray—an easily detected signature for the reaction. The intensity of the γ ray is proportional to the nitrogen concentration. Several TNA detectors were built (using ^{252}Cf as the neutron source) and were tested with limited success (excessive false alarms). Changes in the design of devices using neutrons may eventually give them a significant security role.

14.6.3.2 Food Irradiation

Food irradiation, sometimes called "cold pasteurization" or "electronic pasteurization," is finding increasing worldwide use. This use of ionizing radiation has been endorsed by the UN Food and Agriculture Organization (FAO), the World Health Organization (WHO), and the International Atomic Energy Agency (IAEA). Its use has been approved in over 40 countries, about 30 of which now irradiate a variety of products up to a total of 60. Belgium, France, and the Netherlands are among the countries that have established markets for irradiated food. In the United States, however, food irradiation is a controversial topic (as was the pasteurization of milk for some 30 years). Because of fear, not cost, the process is not yet used to a significant extent even though millions of cases of foodborne illness might be ameliorated by irradiation, and indeed approximately 10,000 food-related deaths are reported annually, with the risk of such fatalities apparently increasing. Growing public concern about foodborne illness has raised the level of interest in the benefits of irradiation.

Examples of the types of food product for which processing with ionizing radiation has been developed include wheat, to disinfect insects; spices and seasonings, to kill microorganisms and insects; fresh fruits and vegetables, to delay maturation; carrots, potatoes, onions, and garlic, to inhibit sprouting; fresh fish, shellfish, and frozen fish, to extend shelf life; pork, to control trichinosis; poultry, to control *Salmonella* and *Campylobacter*; and red meat, especially ground beef, to control *Listeria* and *E. coli*. Worldwide disinfection of spices by irradiation increased from about 6000 metric tons per year in 1987 to about 80,000 metric tons in 1998. The U.S. Food and Drug Administration (FDA) approved irradiation of pork in 1985, poultry in 1990, and ground beef in 1997. Several of the examples just cited illustrate how irradiation of certain foods can be used to reduce the dependence on chemical pesticides and fumigants. Irradiated spices are produced commercially in at least 12 countries.

When food is irradiated, it is placed on a conveyor belt and moved past a source of ionizing radiation. The source may provide a beam of high-energy electrons (an e-beam source); x rays, including bremsstrahlen; or γ rays from a radioactive source sealed in a double-walled, stainless steel container. The source of γ rays is usually ^{60}Co but may be ^{137}Cs. Packaging material may require use of the more penetrating radiation from ^{60}Co.

The dose of ionizing radiation (γ radiation from radionuclides or bremmstrahlung from electron accelerators) used for food irradiation ranges from a few hundredths of a kilogray to about 50 kGy, depending on the objective. To inhibit sprouting and to kill insects and parasites, the required dose is usually less than 1 kGy. Extension of shelf life (by about 2–3 weeks) for normal methods of storage and removal of microorganisms that cause spoilage requires a higher dose (e.g., up to about 10 kGy). For sterilization of food for long-term storage at room temperature, a dose of about 50 kGy is used in combination with heat. Food consumed by astronauts receives a dose of 25 kGy or more.

An interesting radiation-resistant bacterium, *Deinococcus radiodurans* was found in 1956 in radiation-sterilized meat products. It is also found in soil, dust, and many other places. It has an extraordinary ability to repair DNA damage caused by high doses of ionizing radiation (1.5 million rads), especially in the absence of water. It is over 100 times more resistant to ionizing radiation than *E. coli*. It may have potential in radioactive waste treatment.

At high dose levels, radiation–chemical (radiolytic) changes in food not only can have an adverse effect on color and flavor (e.g., changing the flavor of dairy products and introducing rancidity in food with high fat content) but also can cause partial loss of vitamins. These deleterious effects can be reduced by vacuum-packing the food to remove oxygen and freezing it before irradiation, a practice that reduces the formation of substances by reaction of the radiolysis products of water (Section 13.8) with the chemical constituents of food. Radiolysis of the complex molecules in food also produces free radicals that can react to form a wide variety of molecules known to be mutagenic or carcinogenic. These are the same compounds that are introduced when food is processed without ionizing radiation (e.g., by cooking) and are, therefore, normally present in processed food.

Those opposed to food irradiation believe that irradiation makes the food unsafe. However, the Joint Expert Committee on the Wholesomeness of Irradiated Food of the World Health Organization (WHO), representing several international organizations, reached the conclusion that the irradiation of any food up to an overall dose of 10 kGy causes no toxicological hazard and introduces no special nutritional or microbiological problems. This dose will not produce sterile foods of the type needed by people who are diagnosed as having compromised immune systems. Apparently, ionizing radiation does not destroy food-borne viruses.

Irradiation of food with γ rays from radioactive sources does not make the food radioactive. As pointed out in Chapter 13, when ionizing radiation from such sources interacts with the atoms of an absorber, food in this case, it transfers a portion or all of its energy to the absorber, including the parasites or bacteria. The incident radiation is not stored in the absorber.

Accelerator-produced, high-energy bremsstrahlen (> 8 MeV) can be absorbed in photonuclear reactions [e.g., $(\gamma, n), (\gamma, p)$, and (γ, α)]. The cross sections (probabilities) for these reactions are relatively small except for photon energies corresponding to resonance absorption. For some of the isotopes of the light elements (isotopes with an odd number of neutrons, e.g., ^{2}H, ^{13}C, and ^{17}O), the binding energy of a neutron is less than 5 MeV; hence, the (γ, n) reaction is possible for bremsstrahlen used for food irradiation. Although the products of the (γ, n) reactions for these three nuclides are not radioactive, the neutron released in each case will be captured by some stable nuclide in the food and may produce a radionuclide by the (n, γ) reaction. Based on the chemical elements normally present in significant amounts in food, the only radionuclide that might reach a detectable level by neutron capture for an irradiation dose of 10 kGy is ^{24}Na (14.95 h).

In the United States, the Food and Drug Administration (FDA) and the Department of Agriculture (USDA) set the requirements for approval of new uses. The FDA, which establishes the rules for labeling food, requires that irradiated food be labeled and show the international logo, called a radura

FIGURE 14-10 The radura, the internationally recognized logo for irradiated food.

(Figure 14-10). Accompanying the radura must be a statement such as "Treated with Radiation" or "Treated by Irradiation." The size of the type used for labeling and the requirement itself have been part of the food irradiation controversy. The solid circle of the logo represents an energy source; the two petals (green, when colored) represent the food; and the five breaks in the circle represent radiation from the source. There is some concern that use of the logo and the accompanying statement could encourage careless handling during subsequent processing and preparation of the irradiated food, which has no protection against bacterial contamination and must be refrigerated after removal from its sealed container.

The FDA has placed regulatory limits on the energy and the dose of radiation that can be used for the irradiation of food for human consumption, animal feed, and pet food, and on the packaging materials for irradiated food.[27] Examples of the irradiation limits that apply to food for human consumption are given in Table 14-7.

14.7 NUCLEAR FISSION POWER PLANTS

Nuclear power reactors are of environmental concern for three major reasons. One is the release of radioactive nuclides during normal operation or accidentally. A second is the problem of disposal of spent fuel and other contaminated material. A third is the possibility of nuclear fuel being diverted to weapons. Before these can be discussed, it is necessary to discuss the reactors themselves with respect to types and the fuels they use.

[27]*Code of Federal Regulations*, Title 21, Part 179, Food and Drug Administration (HHS), U.S. Government Printing Office, Washington, DC (revised annually).

TABLE 14-7

Limitations on the Use of Ionizing Radiation for the Treatment of Food

Purpose	Dose limitations[a–c]
Control of *Trichinella spiralis* in pork	Min 0.3 kGy (30 krad) Not to exceed 1 kGy (100 krad)
Growth and maturation inhibition of fresh food	Not to exceed 1 kGy (100 krad)
Disinfestation of arthropod pests	Not to exceed 1 kGy (100 krad)
Microbial disinfection of dry or dehydrated enzyme preparations	Not to exceed 10 kGy (1 Mrad)
Microbial disinfection of specified dry or dehydrated vegetable substances used in small amounts for flavoring or aroma, (e.g., spices, herbs, etc.)	Not to exceed 30 kGy (3 Mrad)
Control of food-borne pathogens in fresh or frozen uncooked poultry products	Not to exceed 3 kGy (300 krad)
Sterilization of frozen, packaged meals used solely in the NASA space program	Min 44 kGy (4.4 Mrad)
Control of food-borne pathogens in and extension of shelf-life of refrigerated or frozen, uncooked meat or meat by-products or meat food products composed solely of intact or ground meat, meat by-products or both meat and meat by-products	Not to exceed 4.5 kGy for refrigerated products Not to exceed 7.0 kGy for frozen products
For control of *Salmonella* in fresh shell eggs	Not to exceed 3.0 kGy
For control of microbial pathogens on seeds for sprouting	Not to exceed 8.0 kGy

[a]The following radiation sources can be used for the inspection of food, for inspection of packaged food, and for controlling food processing: x-ray tubes with 500 kV peak or lower; sealed units producing radiations at energy levels of not more than 2.2 MeV from the isotopes: ^{241}Am, ^{137}Cs, ^{60}Co, ^{125}I, ^{85}Kr, ^{226}Ra, and ^{90}Sr. Californium-252 neutron sources used for moisture measurement in food have additional limitations.

[b]Conditions for using ionizing radiation for safe treatment of food: sources of ionizing radiation are limited to (1) γ rays from sealed units of the radionuclides ^{60}Co or ^{137}Cs, (2) electrons generated from machine sources at energies not to exceed 10 MeV, and (3) x rays generated from machine sources at energies not to exceed 5 MeV.

[c]For a food, any portion of which has been irradiated in accordance with regulations, the label and labeling and invoices or bills of lading must bear the statement "Treated with radiation (or by irradiation)—do not irradiate again."

Source: Based on *Code of Federal Regulations*, Title 21, Part 179, Food and Drug Administration (HHS), 2001.

14.7.1 Types of Nuclear Power Reactor

Several types of nuclear fission reactor have been used as heat sources to generate steam in power plants. A power reactor may be identified in different ways. For example, it may be a thermal or a fast reactor, depending on the energy of the neutrons causing fission (see Chapter 13, Section 13.7.3). A thermal reactor is usually identified according to the neutron moderator used. If the moderator and coolant are different, both are specified. Table 14-8 lists characteristics used to identify power reactors.

Almost all nuclear power reactors in operation worldwide at the time of writing are thermal reactors; that is, fission is induced when ^{235}U captures a thermal rather than a fast neutron. Thermal power reactors using H_2O as both

TABLE 14-8

Characteristics Commonly Used to Identify Nuclear Power Reactors

Characteristic	Examples
Fissile material	^{233}U, ^{235}U, ^{239}Pu
Moderator	H_2O, D_2O, C (graphite)
Coolant	H_2O, D_2O, molten Na, CO_2, He
Neutron spectrum	Slow (thermal), intermediate, fast
Fertile material	^{238}U, ^{232}Th

moderator and coolant are known as LWRs (light water reactors), those using D_2O as HWRs (heavy water reactors). LWRs, the most popular type in most countries, are further subdivided into PWRs (pressurized water reactors) and BWRs (boiling water reactors).

Nuclear reactors are also classified in terms of the way fissile material is used and formed. A reactor that uses highly enriched uranium as fuel without fertile material is a "burner." A nonmilitary research reactor is an example, although most of these devices have been modified to use fuel with no more than 20% ^{235}U. A "converter" is a reactor that burns one type of fissile nuclide (e.g., ^{235}U) and makes another (e.g., ^{239}Pu) but makes less fuel than it burns. LWRs are converters. Breeder reactors produce more of a given fissile material than they fission.

14.7.2 Nuclear Fuel

The methods of processing uranium ore and enriching the uranium for production of nuclear reactor fuel were developed in the nuclear weapons program and are discussed in Section 14.11.2. Most nuclear power reactors in operation at the time of writing use a fuel consisting of uranium that is slightly enriched in ^{235}U (e.g., 2–4% of the uranium vs 0.720% in natural uranium) and is known as low-enrichment fuel (LEU), fuel having less than 20% of the uranium as ^{235}U. As the ^{235}U is fissioned, some of the ^{238}U is converted to ^{239}Pu by the successive steps given in equation (13-44). A fraction of the ^{239}Pu fissions and contributes to the energy output of the fuel.[28] A fraction captures neutrons to form not only ^{240}Pu and higher isotopes of plutonium but also isotopes of transuranic elements with higher atomic number by a series of neutron capture and decay steps. The concentrations of these nuclides increase with increasing irradiation time.

---

[28]By the time the irradiated fuel has been removed, about half the fission energy is coming from plutonium.

Plutonium in the irradiated fuel can be chemically separated by reprocessing and then used as a reactor fuel. When used as a fuel, the plutonium is usually mixed with UO_2 to form a mixed oxide known as MOX and containing 3–5% Pu. The UO_2 is made from depleted uranium (DU), the uranium that remains in the ^{235}U enrichment process and has a ^{235}U content below the 0.720 at. % in natural uranium. Although MOX is required for fast reactors, it can also be used in LWRs and HWRs (described shortly) after some modification of the reactors. It has been used in Belgium since 1963. France, Great Britain, Japan, and Russia have programs involving MOX. MOX is being used in nuclear power plants to reduce the amount of weapons-grade plutonium now being stored (Section 14.11.6).

Another fuel (fissile nuclide) that has been used on a demonstration scale is ^{233}U. It can be made according to equation (13-45). By means of the ^{232}Th–^{233}U fuel cycle, the predominant, fertile isotope of thorium in nature could be used in the future to generate power. Either ^{235}U or ^{239}Pu would have to be used to make the initial supply of ^{233}U by neutron irradiation of thorium.

Almost all power reactors in operation in the 1990s contained the fuel in the form of cylindrical pellets of LEU UO_2 enclosed in a thin-walled metal tube (cladding) to form a fuel element or rod or pin (Figure 14-11a). Zircaloy, an alloy of zirconium, is commonly used for the cladding.[29] Zirconium has a low absorption cross section for thermal neutrons, but it does react with water and steam at high temperatures to produce hydrogen. Fortunately, a protective film of ZrO_2 prevents it from reacting with H_2O during normal reactor operation.

The rods are mounted in an assembly (Figure 14-11b) which may contain between 36 (6×6) and 306 (17×18) rods. The assembly has a width between 13 and 30 cm, a length between 4 and 4.5 m, and contains from about 180 to about 460 kg of uranium, depending on the type of reactor. A reactor core consists of an assembly of fuel assemblies in a structure that allows for the passage of control rods and the flow of coolant.

14.7.3 Nuclear Fuel Requirements

A fission rate of 3.1×10^{10} fissions per second will provide 1 W of power (thermal). For a power plant with a thermal output of 3000 MW_t as heat or an electrical output of about 1000 MW_e of electricity the number of ^{235}U nuclei that must be fissioned per second is 9.3×10^{19}.[30] This corresponds to the fissioning of 3.1 kg of ^{235}U per day. Uranium in spent fuel contains about 0.8 w % ^{235}U.

---

[29]Zircaloys are alloys of zirconium and tin with minor constituents.

[30]MW_e indicates the rate at which electrical energy is generated and MW_t indicates the rate at which thermal energy is generated. MW_e is about 30% of MW_t for nuclear power plants with water-moderated, water-cooled reactors. (Note: in much of the literature, the symbols MW_e and MW_t are used without subscripts.)

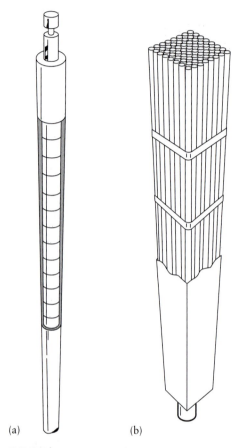

(a)   (b)

FIGURE 14-11   (a) Fuel rod containing uranium dioxide pellets. (b) Assembly of fuel rods. From *The Harnessed Atom*, U.S. Department of Energy, DOE/NE-0073.

The fuel rods must be replaced periodically, not only because of partial depletion of the ^{235}U, but also because of buildup of fission products that are neutron absorbers, distortion of the pellets as each ^{235}U atom is replaced by two atoms of other elements, and stress cracking of the cladding. The extent to which fuel can be irradiated before being replaced is expressed in terms of "burnup": MWd/MTHM (megawatt-days per metric ton of heavy metal as LEU uranium before irradiation). Variations in expressing burnup include the use of MWDT (megawatt-days thermal), MTIHM (metric tons of initial fuel loading), MWd/t, GWd/t (t = metric ton), and TJ/kgU.

For light water reactors (described in Section 14.7.4.1.1), typical burnup in the past has been 30,000–40,000 MWd/MTHM with refueling (replacing one-third or less of the fuel) every 12–18 months. Increased burnup (e.g., 55,000 MWd/MTHM), resulting in longer time period between refueling shutdowns, has been achieved in many nuclear power plants in various ways, including improved fuel element design. Fewer and shorter outages (reduced from 30–50 days to about 20 days) for refueling and maintenance have improved the capacity factor and the economics of plant operation. The capacity factor for a specified period of time is the percentage of design (maximum) power output actually achieved.

14.7.4 Types of Nuclear Power Plants

14.7.4.1 Pressurized Water Reactor Power Plants

14.7.4.1.1 LWR Plants

First developed for the propulsion of submarines and other naval vessels, PWRs are much more compact than the graphite-moderated reactors used in the nuclear weapons program during World War II.[31] The PWR type of reactor was also a logical choice for central power stations because its compactness made possible the use of relatively small containment buildings and because the electric utilities were experienced in the use of liquid water and steam in their fossil-fueled power stations.[32]

In pressurized water reactors the water that is pumped through the core containing the fuel rods is under pressure (about 15 MPa) to prevent its boiling. Hydrogen is injected into this primary loop to suppress the formation of oxygen by radiolysis of the water and thereby prevent the oxygen concentration in the core from reaching a value high enough to make an explosive mixture. The water (at about 316°C) in the primary loop is pumped through steam generators (e.g., four), where heat that has been removed from the core is transferred to water to generate steam in the secondary coolant loop as shown in the simplified schematic diagram of Figure 14-12.

Steam (at about 260°C and 5.5 MPa) goes to a steam turbine. Any radioactivity present in the primary coolant from activation of corrosion and erosion products or fission products escaping from defective fuel rods does not contaminate the steam unless leakage occurs from the primary coolant in the steam generators (SGs). Leakage through cracked tubes carrying the primary coolant in the steam generator has been a problem for PWRs. The number of such tubes is commonly over 3000 in a 350-ton SG that may be

---

[31]The first nuclear-powered (PWR) naval vessel was the submarine USS *Nautilus*, which was launched in 1955.

[32]The first central station, commercial nuclear power plant in the United States was built in Shippingport, Pennsylvania, and began operation in 1957. The reactor was a PWR and the electrical output was 90 MW$_e$.

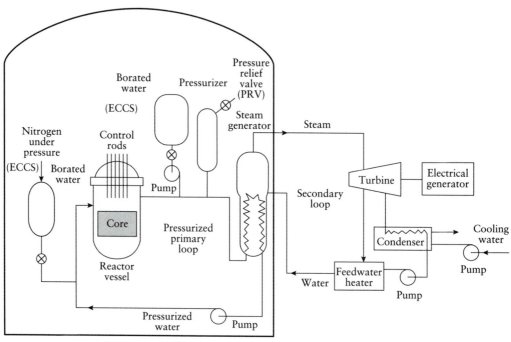

FIGURE 14-12 Simplified schematic for a pressurized water reactor (PWR) nuclear power plant.

about 60 ft long and about 15 ft in diameter. The entire reactor complex, with its control rods, primary loop, pressurizer that maintains pressure in that loop, emergency core cooling system (ECCS), and steam generators, is located in a specially designed containment building constructed of reinforced concrete (about 1 m thick) and having a steel liner (about 0.4 m thick). The purpose of the containment building is to prevent the release of radioactivity into the environment if the reactor is involved in a steam or chemical explosion.

The version of a PWR developed for use in submarines and in power plants in the former Soviet Union is known as the VVER (Russian acronym for water–water power reactor).

14.7.4.1.2 HWR Plants[33]

Pressurized heavy water power reactors using D_2O as both moderator and coolant were developed in Canada and are known as CANDU (**Cana**da

[33] The reactors are also referred to as PHWRs.

deuterium uranium) reactors. In HWR plants D_2O under pressure is circulated through pipes containing the fuel in the core and then through primary loops and heat exchangers, where it transfers heat from the core to H_2O in the steam generator. Because D_2O is a better neutron moderator than H_2O, the fuel is natural uranium rather than the LEU used in LWRs. Unlike the LWRs, the HWRs are designed so that they can be refueled without being shut down.

14.7.4.2 Boiling Water Reactor Power Plants

As the name implies, the water that removes the heat from the reactor core of the boiling water type of LWR is allowed to boil and generate steam directly within the core.[34] The steam thus generated (at about $288°C$ and $10\,MPa$) is used directly to drive the turbine. Steam is delivered to the turbine via a steam separator and a drier that removes the liquid from the water–steam mixture leaving the reactor. Volatile fission products that escape from any defective fuel rods are carried with the steam to the steam turbine. Devices such as catalytic recombiners are incorporated in a BWR system to prevent dangerous buildup of hydrogen and oxygen produced by the radiolysis of water in the core.

Two containment systems are used to prevent release of radioactivity in the event of an accident in a BWR plant. The primary containment is a steel pressure vessel that surrounds the cylindrical reactor vessel and is itself surrounded by reinforced concrete. The building that houses the reactor and the primary containment vessel is constructed of reinforced concrete and constitutes a secondary containment layer. Any radioactive material that escapes from the primary containment vessel is confined within the secondary containment space.

14.7.4.3 Water-Cooled, Graphite Moderated Reactor Power Plants

The first-ever nuclear power station ($5\,MW_e$) was built in the USSR and began operation in 1954. Its reactor was fueled with uranium metal (5% ^{235}U), moderated with graphite, and cooled with pressurized water that flowed through a primary loop and then through a heat exchanger to generate steam in a secondary loop.

Later (in the 1970s), large boiling water, pressure tube, graphite-moderated high-power boiling channel reactors were built for both the production of Pu and the generation of power. This type of reactor, called RBMK after the Russian acronym, has fuel rods consisting of UO_2 (enriched to about 2% ^{235}U) in Zircaloy cladding and control rods that move within channels in the graphite. Water flows around the fuel rods in vertical channels and is allowed

---

[34]The core of a BWR plant with an electrical capacity of $1000\,MW_e$ contains about 125 metric tons of LEU as UO_2.

to boil and generate steam in the core. In a 1000-MWe RBMK (RBMK-1000), the steam goes to two 500-MW$_e$ generators from two halves of the reactor that can be operated separately. The RBMK was the most popular type of civilian power reactor in the Soviet Union prior to the Chernobyl accident in 1986 (see Section 14.16.2.3). These reactors can be refueled while operating, and they allow the use of low-enrichment fuel.

Although large water-cooled, graphite-moderated reactors had been operated safely in the United States from the time of the Manhattan Project in the 1940s until 1994, they were used primarily for the production of Pu for nuclear weapons. Some also generated electricity, but as a type of reactor, they were never developed for use in U.S. commercial power plants. The last of these, the N Reactor at the Hanford Reservation in Washington State (near Richland), began operation in 1963 and was permanently shut down in 1986 after the Chernobyl accident. Although these reactors had many design features to make them safer than the RBMKs, they lacked a major safety feature, namely, a containment building.

14.7.4.4 Gas-Cooled, Graphite Moderated Reactor Power Plants

Gas-cooled (CO_2), graphite-moderated reactors (GCRs) using natural uranium and a magnesium alloy (Magnox) as cladding and advanced gas-cooled reactors (AGRs) using LEU fuel (UO_2) have been used in power plants in the United Kingdom for many years. These power plants, which generate steam, operate at higher temperatures than do LWRs. A different type of graphite-moderated, high-temperature, gas-cooled reactor, the HTGR, has been studied in Germany and in the United States for many years. Pressurized helium was the coolant. Uranium carbide (UC_2) or oxycarbide (UCO) initially containing ^{235}U in the form of coated microspheres was the fuel, and coated microspheres of ThO_2 served as fertile material for the production of fissile ^{233}U by reaction (13-45). Eventually, such a reactor would operate on the ^{233}U–^{232}Th fuel cycle. The helium transferred heat from the core to a steam generator at a temperature of about 700°C, providing a higher thermal efficiency than that for an LWR. Further improvement of efficiency was planned by going to a gas–turbine cycle.

The core of the HTGR was very different from that of an LWR. It consisted of an assembly of hexagonal graphite blocks containing vertical holes through which the helium could flow and holes for rods containing the microspheres, coated with a layer of graphite, a second coating of silicon carbide, and another coating of graphite. Fission products were retained in the microspheres by the coating.

A demonstration or prototype reactor (40 MW$_e$) was built in Peach Bottom, Pennsylvania, and a second commercial 330-MW$_e$ plant, built at Fort St. Vrain, Colorado, operated with many outages for mechanical problems for about 10 years beginning in 1979. Development of the HTGR in the United States was terminated in 1994. Two HTGRs operated in Germany have been shut down.

The reason for describing the HTGR is that the concept has been revived in South Africa, where the pebble bed modular reactor (PBMR) is being developed. As designed, the reactor is cooled by helium that leaves the reactor at about 900°C and generates electricity through a direct-cycle gas turbine. Each module has an output of only 110 MW$_e$. The core contains about 400,000 pebbles (about the size of tennis balls), each of which contains about 15,000 seeds of enriched uranium as UO_2 coated with layers of carbon and silicon carbide. The steel vessel containing the core is lined with graphite. The design allows spent fuel to be removed from the bottom of the core (into storage below the reactor) and new fuel to be added at the top without shutting down the reactor.

A small (10-MW$_t$), helium-cooled, pebble bed reactor began operation in China in 2000. It is designed as a source of process heat for industry and a source of electricity.

14.7.4.5 Sodium-Cooled, Fast Breeder Reactor Power Plants

Fast breeder reactors, known as liquid metal fast breeder reactors (LMFBRs), use ^{239}Pu as the fissile nuclide, ^{238}U as the fertile material, and fast rather than thermal neutrons. Reactors of this type have been built and operated in France, Japan, the former Soviet Union, the United Kingdom, and the United States to demonstrate their feasibility and to gain experience in their operation. The largest LMFBR power plant built up to mid-1998 is the 1240-MW$_e$ SuperPhénix located at Creys-Malville, France. These reactors are called breeders because they are designed to produce more ^{239}Pu than they "burn" by fission. They can breed with ^{239}Pu as the fissile material because ^{239}Pu has a larger cross section than ^{235}U for fission with *fast* neutrons and because the neutron absorption cross sections for fission products and for structural and other materials in the core usually decrease with increasing neutron energy, making a larger fraction of the neutrons released in fission available for converting ^{238}U to ^{239}Pu.

One fuel that has been used in fast reactors is MOX containing about 20% PuO_2 and about 80% UO_2. Stainless steel is a suitable cladding material for the fuel rods. Depleted uranium as UO_2 also is used as a blanket around the core so that ^{239}Pu is produced in both the core and the blanket.

Water cannot be used to cool the core of a fast reactor because it would function as a neutron moderator as well as the coolant. Sodium is not a good neutron moderator, and its physical properties make it a suitable coolant.[35]

---

[35]Lead has been proposed as a substitute for sodium.

It has a melting point of 98.7°C and a boiling point of 883°C. As liquid sodium passes through the core, ^{23}Na captures neutrons to form ^{24}Na (14.95 h) and ^{22}Na (2.604 y) by (n, γ) and (n, 2n) reactions, respectively. The coolant is, therefore, highly radioactive quite apart from any contamination by fission product leakage from fuel elements. Sodium is spontaneously flammable in air and must be handled in an inert atmosphere. Furthermore, it reacts violently with water to liberate hydrogen, which generally ignites. Although the technology for using molten sodium as a coolant is not new, sodium leakage problems have plagued operation of the few sodium-cooled power reactors that have been built.[36] Most, including the SuperPhénix and one in Japan following a sodium fire have been shut down at the time of writing.

In one type of fast reactor (pool type), the core is immersed in a pool of molten sodium, which is circulated through the core by pumps located in the pool of sodium. Hot sodium (about 500°C) also is pumped from the pool through a primary loop and an intermediate heat exchanger, where heat from the pool is transferred to nonradioactive sodium in a secondary loop. Specially designed heat exchangers having double-walled tubes to prevent accidental contact of the sodium with the water are then used to transfer heat from the nonradioactive, liquid sodium to water and generate steam.

In the second type of fast reactor (loop type), molten sodium is pumped through the core as for a PWR. The heat is transferred to a second loop of sodium and then to a steam generator as for the pool type reactor. The pool type design has been more popular than the loop type.

Breeding ^{239}Pu enables operation of nuclear power plants by utilizing ^{238}U, the most abundant isotope of uranium. Breeder reactors are seen as a means to energy independence by nations that must import their fossil and nuclear fuel. They are also seen as a potential source of material that could be used for nuclear weapons by terrorists and rogue nations.

14.7.4.6 Thermal Breeder Power Plants (LWBRs)

Breeding of nuclear fuel also can be achieved in an LWR by using ^{233}U as the fissile material and ^{232}Th as the fertile material. The nuclear power plant in Shippingport, Pennsylvania (footnote 32), was operated at a power level of 60 MW$_e$ (net) from 1977 to 1982 to demonstrate the feasibility of breeding of ^{233}U in an LWR. Much of the technology that will be needed in the distant future when the ^{233}U–^{232}Th fuel cycle becomes a competitive source of energy is now known.

---

[36]The second nuclear-powered submarine that was built in the United States, the USS *Seawolf*, was equipped with a sodium-cooled, beryllium-moderated reactor (not designed as a breeder). This submarine was launched in 1955.

TABLE 14-9

Number of Nuclear Power Reactors by Type That Existed and the Number Operational Worldwide at the End of 2000

Type	Total	Operational
Pressurized light water	289	256
Boiling light water	98	92
Gas cooled	32	32
Heavy water	52	43
Graphite moderated, light water cooled	14	13
Liquid metal, fast breeder	5	2
	490	438

Source: World list of nuclear power plants, *Nucl. News*, **44**(3), 61 (2001) (revised annually).

14.7.4.7 Other

Several additional concepts for nuclear plants have been studied since the 1950s. Some were developed to the prototype stage, but all were abandoned. Three examples are the liquid metal fuel reactor (LMFR), a graphite-moderated device that had fuel consisting of uranium metal dissolved in liquid bismuth; the molten salt reactor (MSR), in which the uranium as a fluoride was dissolved in a eutectic mixture of the fluorides of Be, Zr, and Li; and the organic-cooled reactor (OCR), which contained a mixture of polyphenyls (high boiling hydrocarbon compounds) as substitute for water as coolant and perhaps as moderator. The first two had materials problems related to containment of the fuel. Although polyphenyl compounds are relatively stable to radiolysis by ionizing radiation, they are not sufficiently stable to survive in the intense radiation field of a reactor core.

Table 14-9 lists the types of nuclear power plant that existed and the numbers that were operational worldwide at the end of 1999.

14.7.5 Safety Features of Nuclear Power Plants

14.7.5.1 The Reactor

This section is intended to provide background information and insight that should be helpful for understanding what went wrong in two of the nuclear reactor accidents described in Section 14.16.2. The material has been simplified (perhaps oversimplified for some readers) in order to avoid a quantitative discussion of nuclear reactor kinetics.

The fission rate and, therefore, the power level in a reactor, is controlled by rods that can be moved in or out of the reactor core. These control rods contain an element having at least one stable isotope that is a "neutron poison" (i.e., an isotope with a high absorption cross section for thermal neutrons). Examples are cadmium (^{113}Cd) and boron (^{10}B) as boron carbide. When the rods are fully inserted, the reactor is shut down. The reactor can be shut down by an operator or automatically by one of the engineered safety systems, if one of the various sensors in the core detects an unsafe condition (of neutron density, temperature, pressure, coolant flow, etc.).

Each type of power reactor has several intrinsic safety features. When a reactor is operating at a steady power level (steady fission rate), the rate of production of fission neutrons is equal to the rate at which they are absorbed in the core plus the rate at which they escape from the core. Absorption can be by H_2O,[37] control rods, fuel-rod cladding, fission products in the fuel, structural materials in the core, ^{235}U to cause fission or to form ^{236}U, ^{238}U to cause fission or to form ^{239}U (which is the source of the ^{239}Pu that accumulates in the fuel), and ^{239}Pu to cause fission or to form ^{240}Pu.

A safety-related question is, If the reactor is operating at full power and there is a temperature transient (e.g., a sudden increase in temperature of the water in an LWR, for example, because of a decrease in demand for steam or failure of a coolant pump and so on), how will the reactor respond? All the reactor parameters change with temperature. The cross sections for all the neutron reactions that can occur during the slowing-down process will change. Some will increase, others will decrease. Also, there will be density changes, especially for water. Analyzing this complex combination of changes shows that the overall temperature coefficient for changing the power level for LWRs is negative and the reactors are stable. Therefore, if there is an increase in sudden temperature, the fission rate will decrease until the temperature before the event is restored. A positive temperature coefficient would lead to a continuously increasing temperature and an unstable reactor.

Another question is, How will the reactor respond if a void is formed in the water of an LWR? Because a void is the absence of the neutron moderator, the number of thermalized neutrons will be insufficient to maintain a constant rate of fission with thermal neutrons. Thus, the void coefficient (again, for the power level in this discussion) is negative and the fission rate will decrease.

One of the factors that contributed to the Chernobyl accident (Section 14.16.2.3) was the void effect. The type of reactor involved, an RBMK-1000 (Section 14.7.4.3), contains graphite to thermalize the fission neutrons and

---

[37]A good neutron moderator is efficient in removing kinetic energy from a fast neutron in each elastic collision, *and* it has a low cross section for absorbing a thermalized neutron. The order for thermalizing is H > D > C, i.e., decreasing order with increasing mass. However, when absorption is taken into account, the ranking order of moderators is D_2O > C > H_2O.

boiling water as the coolant and minor moderator. A large void—one beyond that allowed for in the design (i.e., boiling water under pressure)—creates a positive void coefficient because it represents a removal of H_2O as an *absorber* of neutrons that have already been thermalized by graphite. This positive void coefficient is not overweighed by the negative coefficients for the other effects.

Even though nuclear power plants have engineered safety features designed to prevent accidents by automatically and rapidly adding neutron poisons to shut down the reactor if a dangerous condition is detected, they are vulnerable to a type of accident that is not a problem for a power plant using a fossil fuel. After the reactor has been shut down, the core is still a large heat source (5–7% of full-power heat output).[38] In fact, the decay heat generated in the fuel elements at the time of shutdown is more than sufficient to melt the reactor core if, after a normal or emergency shutdown, the heat is not removed by continued pumping of coolant through the core. To provide emergency cooling, the power plant contains an engineered safety feature called the emergency core cooling system (ECCS), which is turned on automatically to remove the decay heat if the normal cooling system fails. Such a failure is called a loss of coolant accident (LOCA). If power for the ECCS from outside the plant (from the power grid) fails, diesel-operated generators are supposed to turn on within a few seconds to supply emergency power.

We have not discussed the rate at which the power level of a reactor can rise. Fortunately, the delayed fission neutrons slow the rate so that within defined limits of safe operation, the time available for adjusting the power level is reasonable. For details, see the additional reading and sources of information at the end of the chapter.

14.7.5.2 Dose Received by Workers

Obviously a nuclear power plant is a very complex system, with many regions containing sources of ionizing radiation. During shutdowns for refueling and so on, and during normal operation and maintenance, workers must follow set procedures to limit their exposure to values that are, hopefully, below those specified in the operating license for the plant. During routine operation, workers must properly handle radioactive effluents of the type described in the next section.

The radiation dose received by workers at nuclear power plants in the United States has decreased steadily since 1980.[39] For PWR plants the median value for the man-rem dose per unit decreased from 417 in 1980 to 82 in 2000. The corresponding change in dose for BWR plants was from 859 to 150.

---

[38] As an example, for an LWR having a thermal output of $3000\,MW_t$, a burnup of $33,000\,MWd/MTHM$, and a fuel load of 82 MTHM, the decay heat immediately after shutdown is about $160\,MW_t$ or about 5.4% of the operating power.

[39] *Nuclear News*, **44** (6), 39 (May 2001).

14.7.6 Effluents from Nuclear Power Plants

A nuclear power plant is obviously not a source of greenhouse gases. If a 1000-MWe nuclear power plant is substituted for a coal-fired plant of the same size, the annual production of such gases is reduced by approximately the following amounts, depending on the composition of the coal and the type of pollution control equipment installed: about 7 million tons of CO_2, about 100,000 tons of SO_2, about 25,000 tons of NO_x, and about 1500 tons of particulates. In addition, there would be about 1 million tons less coal ash. On the other hand, a nuclear power plant has its own characteristic effluents that are of environmental concern.

During normal operation of a light water reactor (LWR), the cooling (and neutron-moderating) water will accumulate the following.

Gaseous radionuclides
Dissolved or colloidal, radioactivated corrosion products of the elements Fe, Co, and Mn
Any radioactive, nongaseous fission products and actinide elements that have escaped from the UO_2 fuel by way of imperfections in the cladding of the fuel rods
Fission products from the fission of traces of uranium contamination on the outside of the fuel rods

The amount of radioactivity a nuclear power plant is allowed to release into the environment by discharge as gas through a stack or as contaminated water is restricted by its license. The systems for preventing excess discharge of gases are complex, with holdup tanks for compressed gas containing short-lived radionuclides, filters to retain particulate matter, and charcoal absorbers to remove radioisotopes of Kr and Xe.

Table 14-10 shows the level of radioactivity for gaseous and liquid effluents and for solid waste released annually from BWRs and PWRs over a span of 20 years (1974–1993). One can discern a number of trends in this period. In interpreting such data or similar data about nuclear power plants, the following facts must be kept in mind.

• No two nuclear power plants of a given type are identical. As newer ones were built, the experiences gained in operating the older plants were incorporated in design changes and improved chemical treatment processes for radioactive waste. There are limits to how much an old ($\geq$ 20 years) plant can be upgraded.

• Fabrication techniques for fuel elements have improved fuel performance over the years.

• The data were not normalized to take into account the energy output of each plant each year. A given plant may have been shut down for weeks or

TABLE 14-10

Radioactive Materials (curies) Released from Nuclear Power Plants[a]

Year	Boiling water plant		Pressurized water plant	
	1974	1993[b]	1974	1993[b]
Airborne effluents				
Gases: fission and activation Products[c]				
Average[d]	324,000	861[e]	2624	372[e]
Number of nuclear power plants	20	36	25	75
[131]I and particulates				
Average	1.63	0.0187	0.207	0.0039
Number of nuclear power plants	20	40	25	75
Liquid effluents[f]				
Tritium				
Average	17.2	17.2	421	488
Number of nuclear power plants	18	31	24	73
Mixed fission and activation products[g]				
Average	18.3	0.339	1.69	0.476
Number of nuclear power plants	18	31	25	74
Solid waste[h]				
Average	728	12,820	494	1250
Number of nuclear power plants	21	39	20	70

[a]The reported releases are generally planned and in accordance with plant license. However, they include unplanned releases, perhaps from equipment failure, that are below "accident" level.

[b]The last year an annual report was prepared.

[c]BWR: Radionuclides. Major contributors: ^{41}Ar (1.83 h), ^{85m}Kr (4.48 h), ^{87}Kr (1.27 h), ^{88}Kr (2.84 h), ^{133}Xe (5.243 d), ^{135m}Xe (15.3 m), ^{135}Xe (9.10 h), ^{137}Xe (3.82 m), ^{138}Xe (14.1 m). Relative amounts vary from plant to plant. PWR: Radionuclides. Major contributors: ^{3}H (12.32 y), ^{41}Ar (1.83 h), ^{85}Kr (10.6 y), ^{85m}Kr (4.48 h), ^{87}Kr (1.27 h), ^{88}Kr (2.84 h), ^{131m}Xe (11.9 d), ^{133}Xe (5.243 d), ^{133m}Xe (2.19 d), ^{135}Xe (9.10 h). Relative amounts vary from plant to plant.

[d]Per plant, based on arithmetical average.

[e]Also, BWR: 25 Ci of tritium (data for 1993), PWR: 65 Ci of tritium (data for 1993).

[f]BWR: Aqueous waste that is diluted with water (perhaps by a factor of 10^2–10^5) before discharge. Final volume, 10^9–10^{12} liters (data for 1993). PWR: Aqueous waste that is diluted (perhaps by a factor of 10^2–10^3) before discharge. Final volume, 10^9–10^{12} liters (data for 1993).

[g]BWR: Radionuclides. Major contributors: ^{51}Cr (27.702 d), ^{54}Mn (312.1 d), ^{55}Fe (2.73 y), ^{60}Co (5.271 y), ^{65}Zn (243.8 d), ^{137}Cs (30.07 y). Relative amounts vary from plant to plant. PWR: Radionuclides. Major contributors: ^{51}Cr (27.702 d), ^{55}Fe (2.73 y), ^{58}Co (70.88 d), ^{60}Co (5.271 y), ^{124}Sb (60.20 d), ^{125}Sb (2.758 y), ^{133}Xe (5.243 d), ^{137}Cs (30.07 y). Relative amounts vary from plant to plant.

[h]BWR: Consists of contaminated ion exchange resin, filters, sludges, miscllaneous equipment, and irradiated components. Radionuclides. Major contributors: ^{51}Cr (27.702 d), ^{54}Mn (312.1 d), ^{55}Fe (2.73 y), ^{58}Co (70.88 d), ^{60}Co (5.271 y), ^{63}Ni (100 y), ^{65}Zn (243.8 d), ^{134}Cs (2.065 y), ^{137}Cs (30.07 y). Relative amounts vary from plant to plant. PWR: Radionuclides. Major contributors: ^{3}H (12.32 y), ^{14}C (5715 y), ^{51}Cr (27.702 d), ^{54}Mn (312.1 d), ^{55}Fe (2.73 y), ^{58}Co (70.88 d), ^{60}Co (5.271 y), ^{63}Ni (100 y), ^{95}Zr (64.02 d), ^{95}Nb (34.97 d), ^{125}Sb (2.758 y), ^{134}Cs (2.065 y), ^{137}Cs (30.07 y). Relative amounts vary from plant to plant.

Source: Based data in *Radioactive Materials Released from Nuclear Power Plants*, 1993, NUREG/CR-1907, BNL-NUREG-51581, Vol. 14. U.S. Nuclear Regulatory Commission, Washington, DC. December 1995.

many months for refueling, routine maintenance, replacement of equipment, and so on, or it may have operated year-round.

• There have been changes in NRC regulations requiring certain operational changes (e.g., the lowering of airborne radionuclides for BWRs).

• Changes in radionuclide release over the years also can reflect change in plant management.

Radionuclides in the primary coolant of a PWR can enter the secondary cooling system through imperfections in the many kilometers of tubing contained in the steam generators that are components of the cooling loops. Water is also withdrawn from the primary loops to remove corrosion products, and other contaminants. The aqueous effluents are passed through filters and demineralizers (inorganic or organic ion exchangers) to remove radionuclides. Such treatment does not, of course, remove two of the radionuclides produced in LWRs, namely, tritium (12.32 y) and ^{14}C (5715 y).

Tritium can be formed as a product of the rarely occurring ternary nuclear fission reaction in the fuel and escape through imperfections in the cladding. It can also be formed in LWRs by the ^{2}H(n, γ) ^{3}H reaction with deuterium in normal water.[40] In pressurized water reactors its major source is the ^{10}B(n, t) ^{4}He reaction in boric acid, which is added to the primary coolant as part of the reactor control system.

Carbon-14 is produced in an LWR mainly by the ^{14}N(n, p)^{14}C reaction on nitrogen or nitrogen-containing impurities in the coolant, moderator, or fuel. Carbon-14 is also produced to a lesser extent by the ^{17}O(n, α)^{14}C reaction.[41] When the ^{14}C escapes through leakage of the coolant, it may be as ^{14}CO$_2$, ^{14}CO, or ^{14}CH$_4$.

For a given type of power plant, PWR or BWR, the systems used to limit the amount of radioactivity in the effluents have become more complex and efficient over the years. Because the steam going to the turbines in a BWR plant is generated in the reactor core, the turbines become radioactively contaminated, the challenge of limiting the radioactivity in plant effluents is greater than for a PWR and, therefore, some of the effluents tend to have a higher level of radioactivity.

14.7.7 Advanced Nuclear Power Plants for the Future

The two major driving forces for improving nuclear power plants are economics and safety. Features that can be changed to improve the economics are lowering the cost of construction (dollars per kilowatt of capacity), and increasing the reliability, which will lower maintenance and operation costs

---

[40]The composition of normal hydrogen, is 99.985 at. % for ^{1}H and 0.015 at. % for ^{2}H.

[41]In a graphite-moderated reactor the ^{14}C is produced by the ^{13}C(n, γ) ^{14}C reaction.

and raise the capacity factor. Other improvements relate to minimization of radioactive waste and nonproliferation of nuclear weapons.

One class of advanced light water reactors (ALWRs), ABWRs and APWRs designed in the United States, have evolutionary improvements based on experience with existing LWRs. They are comparable in generating capacity to existing large LWR plants but are simplified (fewer welds, less piping) and have better instrumentation. The first ABWR [1315 MW$_e$ (net)] was built in Japan and began operation at the end of 1995. Canada has developed an evolutionary, standardized design AHWR, the CANDU 3. France and Germany have jointly developed an evolutionary design for the PWR, called the European pressurized water reactor (EPR), that is available for replacement of aging nuclear power plants. Advanced gas-cooled reactors (AGRs) have been built in the United Kingdom. These use enriched uranium instead of natural uranium as fuel and have concrete containment buildings.

Another class of advanced plants using LWRs is characterized as having passive cooling and passive safety features. Plants of this type differ markedly from existing plants while still providing safeguards against sabotage and diversion of plutonium. They require less human action, have fewer or no moving parts such as pumps and valves, have more safety features, and require no power to operate the safety systems. The latter include gravity feed of water stored above the reactor and greater use of convective and radiative heat transfer. Plants being designed with passive systems are generally mid sized (e.g., about 500 MW$_e$). Both types of ALWR are expected to benefit from the incorporation of standardized components, which should lower construction time and costs and expedite licensing.

Advanced gas-cooled and liquid-metal-cooled reactors also have been designed.

14.8 NATURAL FISSION REACTORS (THE OKLO PHENOMENON)

Natural fission, water-moderated reactors (at least 16) existed on the earth and were part of the environment about 1.96 billion years ago in deposits of uranium ore in the Oklo region near Franceville in the Gabonese Republic on the west coast of Africa. At that time the abundance of ^{235}U in the ore was about 3% (about that used in LWR reactors) instead of the current value of 0.720%. It is estimated that the reactors, which operated intermittently, had an operational life in the range of $5 \times 10^5 - 10^6$ years. Sufficient uranium remains so that there is an interest in mining at least one of the reactor sites.

The fossil reactors in the Oklo area have provided information that is helpful for understanding the probable behavior of high-level radioactive waste within a repository. Data have been obtained on the radiation damage of materials in and near the reactor zones and on long-term migration of ^{235}U, ^{239}Pu, and fission products. For example:

- About half the fission product elements remained in the UO_2 ore, which is analogous to spent reactor fuel.
- For five fission products that migrated from one of the reactor zones, the relative retentions were in the order Te > Ru > Pd > Tc > Mo. Some of these were retained in peripheral rocks.
- Most of the cadmium and tin escaped from the reactor zones.
- The heavy rare earth elements are more mobile than the light ones.
- The ^{239}Pu (2.410×10^4 y) present after the reactors ceased to operate remained immobilized long enough to decay into ^{235}U.

14.9 THERMONUCLEAR POWER PLANTS

Although power plants using controlled thermonuclear reactions (fusion reactions) as the heat source do not exist, progress is being made on generating and controlling fusion reactions and on developing concepts for fusion power plants. Because of the large quantity of deuterium (0.015 at. % of the hydrogen and present as HDO and D_2O) that is available in the earth's hydrosphere and because of the existence of well-established methods for increasing the deuterium content of water, nuclear fusion has the potential of providing an almost unlimited source of energy in the future.

The relatively low ignition temperature of the D-T reaction [equation (13-38)] has made it the reaction of choice in the facilities where fusion research and development are being conducted, and it will most likely be used in the first generation of fusion power plants. Ultimately, the two D-D reactions [equations (13-39) and (13-40)] would be used. Helium-3 is not an attractive fuel because its natural abundance is only 1.4×10^{-4} at. %, and it is more difficult to make by a nuclear reaction than ^{3}H.

In a controlled fusion reaction, it is necessary not only to create the plasma of reactant nuclei, but also to stabilize and confine it at the proper density and temperature ($\sim 10^8$ K). Furthermore, for power generation, the energy output must be greater than the energy input needed to produce and confine the plasma.

Two very different approaches to developing fusion power are being pursued. In the method that has been studied for the longer period of time, the plasma is contained in a vacuum chamber. The plasma can be produced by several methods—for example, passing a high current through the gaseous mixture of reactants (ohmic heating) followed by magnetic compression, or ionizing the reactants first and then injecting them into the device. Magnetic fields are used in a number of ways to confine the plasma within the reaction vessel and away from the wall, since there is no containment material that could survive the temperature of the plasma.

The tokamak, first built in the USSR, is an example of a device that uses magnetic confinement of the plasma in a torus (a single doughnut-shaped vessel). Magnetic confinement[42] has been investigated in the United States and in the former Soviet Union for many years. In the future, it is likely that major research and development projects based on magnetic confinement will be pursued abroad at the Joint European Torus (JET) and the International Thermonuclear Experimental Reactor (ITER) project, which is supported by Europe, Japan, and Russia and is scheduled to be completed and to demonstrate the feasibility of commercial fusion power by about the year 2013. The experimental facility is in Japan.

An early variation of the tokamak was the stellarator, in which the coils for generating the magnetic field are wound around the plasma chamber (torus) in the shape of a helix rather than a series of rings. Although study of the stellarator design was discontinued by most countries in the 1960s, a modern version of this device, known as the large helical device (LHD), is being developed in Japan.

In the second method, which uses inertial confinement, high-power lasers or sources of high-energy electrons or ions provide high-input energy in pulses lasting perhaps a microsecond or a nanosecond to irradiate and compress the fuel mixture ($^2H_2$ and $^3H_2$). Confinement is achieved by sealing the fuel in small pellets or microspheres that are about 1 mm or less in diameter and are made of glass, for example. Irradiation by converging, pulsed laser beams rapidly vaporizes the confining shell of the pellet, creating an implosion that increases the density of the fuel mixture and heats it to the ignition temperature needed for the reaction to become self-sustaining. This technology is being developed at the National Ignition Facility (NIF) at the Livermore National Laboratory in California, and in Japan. The fusion device, which will have about 190 converging lasers, will be used to study and develop inertial confinement fusion as a source of energy and to provide a means for maintaining the stockpile of nuclear weapons without testing.

For either method there are major materials problems to be solved, especially for the reactor vessel. At times articles have been written with misleading information, leading the reader to believe that a fusion reactor would be free of radioactivity. The first-generation reactors will very likely contain large amounts of radioactive tritium (12.32 y). The D-T reaction releases neutrons that will carry most of the energy released to the reactor vessel, where the resulting heat will be used to produce steam. The neutrons will also produce radionuclides by (n, γ) and (n, p) reactions with the various stable nuclides in

---

[42]Produced by the magnetic fields generated by current flowing in coils wound around the torus and current flowing through the plasma.

the reactor vessel and will cause embrittlement of the vessel, as is the case for fission reactor vessels. It is also likely that the heat will be removed by molten lithium and that the reactor will be surrounded by a lithium-containing blanket (possibly molten Li) where, as for the lithium coolant, neutrons will be used to produce tritium by the reactions

$$^6\text{Li} + {}^1\text{n} \longrightarrow {}^3\text{H} + {}^4\text{He} \tag{14-12}$$

and

$$^7\text{Li} + {}^1\text{n} \text{ (very fast)} \longrightarrow {}^1\text{n} \text{ (not so fast)} + {}^3\text{H} + {}^4\text{He} \tag{14-13}$$

If deuterium is eventually used alone as the fuel, the D-D reactions will produce both neutrons and tritium. In any event, thermonuclear power plants are not expected to become a reality until the distant future, perhaps by 2050.

14.10 COLD FUSION

In March 1989, a ripple of excitement swept the world when it was announced that the results of experiments were believed to show that the energy of nuclear fusion could be released as heat by simply electrolyzing a solution of LiOD in D_2O in a cell having a palladium cathode and a platinum anode. The energy output of the cell as heat was measured to be greater than the energy input. Since the experiments were carried out at room temperature, the process was referred to as "cold fusion." The implications of an inexpensive way of harnessing the energy of fusion were almost beyond imagination, especially for nations without energy resources.

The same experiments and variations of them have been carried out in laboratories in many parts of the world since 1989, but the original results have not been duplicated. Other effects that are not readily explained have been observed but they, too, cannot be repeated. When a small energy release is observed, neutrons, protons, tritons, helium nuclei, and γ rays are not observed in the quantities expected from reactions (13-39) and (13-40). It becomes necessary to assume that a new nuclear reaction mechanism can occur when deuterium is absorbed in palladium or some other metal.

In April 1989 a special 22-member Cold Fusion Panel was established by the Energy Research Advisory Board of the U.S. Department of Energy. The panel reached several conclusions, which can be summarized by the statement that there was no convincing evidence to support the reported experimental results.[43]

[43]John R. Huizenga, *"Cold Fusion: The Scientific Fiasco of the Century"*, University of Rochester Press, Rochester, NY, 1992.

14.11 NUCLEAR WEAPONS

The nuclear weapons programs that a number of nations have undertaken since 1945 have had significant effects on our nuclear environment.

14.11.1 Types

In the first generation of nuclear weapons ("A-bombs"), the explosive energy came from the fission of ^{235}U or ^{239}Pu. The main source of energy in a thermonuclear weapon ("H-bomb") is a reaction such as the D-T reaction (13-38), which is triggered by energy from the fission of ^{239}Pu. This type of weapon can be made to release much greater explosive energy than fission weapons (i.e., in the megaton energy range) with a smaller release of fission products per unit explosive energy.[44] On the other hand, if an outer layer of ^{238}U is added, the quantity of fission products increases, and the weapon has been called "dirty."

By the end of the Cold War, the arsenal of nuclear weapons contained a variety of types ranging from heads for intercontinental ballistic missiles to relatively small tactical weapons. Reduction in the size, weight, and amount of fissionable material was made possible by incorporating deuterium and tritium in the weapons. Tritium was also used in the so-called neutron bomb. Because 5.5% of the ^3H (12.32 y) decays per year, the tritium in stored weapons must be replenished by irradiation of ^6Li in rods containing lithium in a reactor [reaction (14-12)] or produced in an accelerator. The reactor route using commercial LWR power reactors has been chosen in the United States. Lithium-containing absorber rods are to be irradiated with neutrons at nuclear power plants operated by the Tennessee Valley Authority and shipped to a new processing plant at the Savannah River site.

14.11.2 Production

After the initial mining and milling of uranium ore, the uranium is leached from the ore, (e.g. with nitric acid),[45] purified by solvent extraction from uranyl nitrate $[UO_2(NO_3)_2]$ solution, converted to UO_3, reduced to UO_2 with hydrogen, converted to UF_4 by the reaction of UO_2 with hydrogen fluoride, and finally oxidized to UF_6 with fluorine. Uranium hexafluoride has a vapor pressure of 1 atm at 56.4°C and is suitable for increasing the ^{235}U

---

[44]A one-kiloton nuclear explosive has an explosive energy output of 4.184×10^9 kJ.

[45]Radon-222 is released from the ore when the uranium is extracted and from the radium-containing waste that is generated.

content above that in natural uranium by gaseous diffusion or centrifugation or magnetic separation.[46]

In the gaseous diffusion method of isotope separation, which is based on the kinetic theory of gases, the UF_6 diffuses through porous barriers. The theoretical separation factor is 1.0043, but in practice it is closer to 1.003. Over a thousand stages are required to increase the ^{235}U content from 0.72% to about 93% in highly enriched uranium (HEU). The depleted uranium (DU) that remains contains 0.2 w% or less ^{235}U.

Enrichment utilizing a gas centrifuge is more energy efficient. The $^{235}UF_6$ migrates toward the axis of rotation, where the enriched product is removed, while the $^{238}UF_6$ migrates toward the outside of the centrifuge, where it is removed.

Atomic vapor laser isotope separation (AVIS) has been studied in the United States and France. Another laser-based method, separation by laser excitation (SILEX), uses UF_6 as the feed material.

Although there should be negligible loss of radioactive material in enrichment processes, liquid waste is generated when the equipment is cleaned. Also, there can be leakage of UF_6 at various points in the processing. A medical monitoring program has been established for people who have worked at gaseous diffusion plants.

Next, the enriched uranium hexafluoride is converted to UF_4 followed by reduction with calcium to uranium metal, fabrication of weapon components, and assembly of the weapons. Workers at gaseous diffusion enrichment plants that also produce the metal may have been exposed to levels of ionizing radiation and toxic chemicals (e.g., beryllium) above current limits. There have been instances in which the metallic dust has escaped from air filters and has been carried into the environment outside the plant site.

Approximately 5.7×10^5 tonnes of depleted uranium as UF_6 were produced in the United States weapons program.[47] In metallic form (density of 19.07 g/cm^3), uranium is used for shielding in medical x-ray equipment and for making tank armor and conventional weapons (e.g., armor-piercing shells). In addition to the United States, nine other countries have developed such weapons. It is also used in the form of ballast and counterweights in civilian and military aircraft.

When ingested, ^{238}U (4.47×10^9 y), an α-emitter with a specific activity of 12.4 MBq/kg (335 μCi/kg), is generally considered to be a toxic heavy metal that is retained by bone and can cause kidney damage. Because metallic uranium is pyrophoric (ignites spontaneously) when finely divided, it can become airborne as smoke containing very fine particles of UO_2.

---

[46]Large mass spectrometers called *calutrons* also were used for enrichment at Oak Ridge during World War II. They later became the source of separated stable isotopes.

[47]After irradiation with neutrons to produce plutonium, the depleted uranium contains a small amount of α-emitting ^{236}U (2.342×10^7 y), which can contaminate the DU when it is put through the enrichment process to recover the ^{235}U.

External exposure to DU, with its very low specific activity, is not a radiation hazard. At the time of writing, inhalation of DU particles is not considered to present a risk of cancer arising from its being a source of ionizing radiation. The risk of cancer in organs of the body from DU, as a heavy metal, has not been ruled out. Extensive research on the health hazards of uranium has been carried out over the years. Additional research seems to be needed to settle the questions that have been raised about health effects such as leukemia for people exposed to DU during or after military operations in recent years. A complicating factor is that less is known about the health hazard of the traces of ^{236}U, ^{239}Pu, and ^{240}Pu that may be in the DU if its origin was irradiated uranium. See the additional reading at the end of the chapter for a report on an environmental assessment in Kosovo.

For a weapon based on ^{239}Pu, the plutonium is made according to reaction (13-44) by irradiation of ^{238}U in a nuclear production reactor. After irradiation, the fuel rods are chopped; the fuel (UO_2) is dissolved in acid (e.g., nitric acid), and the plutonium and uranium are chemically coseparated (e.g., by solvent extraction) from the fission products formed during the irradiation. The weapons-grade plutonium is chemically separated from the uranium, purified, converted to plutonium metal, fabricated into the required components, and assembled into weapons. This reprocessing of the irradiated uranium generates large quantities of liquid radioactive waste. It also releases volatile fission products. At the Hanford Reservation, for example, about 15 PBq (400,000 Ci) of ^{131}I (8.020 d) was released into the atmosphere between 1944 and 1947. The reactors used to produce the plutonium were cooled with water from the Columbia River until 1971. Radioactivated corrosion products [e.g., ^{60}Co (5.271 y)] and any fission products that escaped from the rods containing irradiated uranium contaminated the water that was returned to the river.

If a weapon contains tritium, there are the additional steps of removing the tritium from the irradiated target, purifying it, and fabricating a weapon component.

Because metallic plutonium, like metallic uranium, is pyrophoric, machining and other operations must be done with the fire hazard in mind. There have been cases of spontaneous combustion. The oxide smoke of the α-emitting ^{239}Pu is a major radiological (inhalation) hazard.

Until 1984 the government facilities located in some 32 U.S. states that produced nuclear materials and nuclear weapons were not required to operate in compliance with the applicable environmental laws and regulations that applied to commercial facilities such as nuclear power plants, factories, hospitals, and laboratories using or producing radioactive materials.[48] Weapons

---

[48]*Environmental Restoration and Waste Management Program*, U.S. Department of Energy, DOE/EM-001P (revised December 1992).

facilities of all types operated as they had during the Manhattan Project, with the urgency and secrecy that was part of wartime weapons production.

The nuclear weapons programs that flourished in the United States and the Soviet Union during the Cold War have left a legacy of environmental problems that include the following.

- Determination of health effects for workers in the weapons facilities and possibly for people living in the vicinity of the facilities
- Site assessment and selection of remedial method[49]
- Disposal of radioactive waste (Section 14.12) and nonradioactive toxic chemical waste stored at the facilities
- Disposal or storage of weapons-grade ^{235}U and ^{239}Pu
- Implementation of remedial action for facilities where groundwater has been contaminated by any hazardous substances
- Decontamination of the facilities with respect to hazardous chemical substances and radioactive materials
- Decontamination of the area around some of the facilities
- Decommissioning of the facilities to make them available for other use

14.11.3 Testing

Testing of nuclear weapons, especially atmospheric testing, has added to the radioactivity of the environment. At least 2419 tests have been carried out since 1945, and of these, 543 were atmospheric tests. In 1963 a Limited Test Ban Treaty banning nuclear tests in the atmosphere, in outer space, and under water was signed by the United States, the United Kingdom, and the Soviet Union. Testing continued, but the explosions were carried out underground with a limit of 150 kilotons under the 1974 Threshold Test Ban Treaty. The last such test in the United States occurred in September 1992.

In September 1996, 158 members of the United Nations signed the Comprehensive Test Ban Treaty (CTBT), which bans all testing. This treaty was signed by the five declared nuclear nations ("the Nuclear Club"), namely, China, France, Russia, the United Kingdom, and the United States. As of July 2001, 161 nations had signed the treaty and 77 had ratified it. The latter number includes only 31 of the 44 nations that have nuclear capability (i.e., possess nuclear reactors or conduct nuclear research). The treaty cannot go into effect, however, until it has been ratified by all the nuclear-capable nations. Ironically, experimental verification of the treaty (a signed agreement) is a critical aspect of the CTBT, and doubts about verification have delayed ratification. A network of over 300 stations equipped with sensors will be used

---

[49]A site may be suitable for a nonnuclear use, or it may be converted into a safe storage facility. Existing facilities may be entombed or may be completely dismantled.

for four types of monitoring: seismological, radionuclide, hydroacoustic, and infrasound [(sound waves with inaudible frequencies below 20 Hz) for atmospheric testing].

Nuclear weapons testing (underground) by the declared nuclear powers ended after announced tests by France and China in 1996. The world was taken by surprise when India conducted three unannounced, simultaneous, underground tests on May 11, 1998, and two more on May 13. The tests involved both fission and thermonuclear devices. India had previously detonated a "peaceful nuclear explosive" in 1974.

A little over two weeks later, on May 28, Pakistan conducted five underground tests. A sixth device was detonated on May 30. Thus, India and Pakistan moved from the group of probable (undeclared) nuclear nations to the group of declared nuclear nations.

14.11.4 Fallout and Rainout

The extent to which the fission products and residual fissionable material are deposited in the immediate area following detonation of a nuclear weapon in the atmosphere depends on the altitude at which detonation occurs and the type of weapon. The local effects of a large (kilotons to megatons) weapon detonated at an altitude of tens of thousands of feet are mainly those from the shock wave, from heat, and from irradiation by neutrons and γ rays. Weapons of a megaton or larger generally release most of the radioactive debris (atomic vapor) in the stratosphere, where it is dispersed and distributed worldwide. The long-lived components have a residence time in the stratosphere of one to five years, with an average of about two years. After the debris falls into the troposphere, it can reach the earth slowly as fallout of very fine particles, or it can be concentrated and carried to the earth's surface by rain (as rainout) or snow.

Although most of the fallout and rainout from atmospheric tests was expected to accumulate relatively near the test site, in some incidences the fission product cloud remained concentrated and traveled great distances before being brought to earth in rain. For example, after the Simon test, in Nevada on April 25, 1953, the highest fallout (rainout) recorded in the United States was in the region around Troy, New York. The weapons debris reached Troy in a jet stream 36 hours after the detonation and was deposited in a localized, exceedingly violent and destructive thunderstorm.[50]

In the case of detonations at or close to the earth, the bomb debris is adsorbed on soil, vegetable matter, and so on, which is carried upward as

---

[50]See M. Eisenbud and T. Gesell, *Environmental Radioactivity (From Natural, Industrial, and Military Sources)*, 4th ed., Academic Press, San Diego, CA, 1997, p. 292, and H.M. Clark, *Science*, **119 (3097)**, 619 (1954).

large particles and settles out in a matter of minutes. The fallout is deposited locally and includes radioactivity induced by neutron capture reactions in the soil (or seawater if the detonation is over an ocean).

When nuclear weapons are tested by underground detonations, most of the radioactivity remains underground, but there may be venting (tritium, noble gases, and other volatile fission products) and localized fallout. Such tests can be detected seismographically and have provided information about the inner regions of the earth.

Five weapons-produced fission products, namely, ^{90}Sr (28.78 y) and its daughter ^{90}Y (2.67 d), ^{131}I (8.0207 d), and ^{137}Cs (30.07 y) and its daughter ^{137m}Ba (2.552 m) are particularly hazardous when ingested (Section 14.13). More recently, ^{99}Tc (2.13×10^5 y) and ^{129}I (1.57×10^7 y) have also been receiving attention as long-term, internal radiological health hazards.

Most of the fission products from atmospheric weapons tests at the Nevada test site were released between 1952 and 1957. Beginning in 1983 and ending with a report in 1997, the National Cancer Institute (NCI) conducted a very challenging study to estimate the exposure to ^{131}I of the American people and to assess thyroid doses received by individuals across the country (3100 counties) from the Nevada tests.[51] The report contains maps showing the activities of ^{131}I deposited on the ground and dose estimates for all counties. A third part of the ^{131}I fallout study, namely, assessment of the risk for thyroid cancer from the estimated exposures, was completed by the Institute of Medicine (IOM).[52] Two conclusions in the IOM report are "first, that some people (who cannot be easily identified) were likely exposed to sufficient iodine-131 to raise their risk of cancer and, second, that there is no evidence that programs to screen for thyroid cancer are beneficial in detecting disease at a stage that would allow more effective treatment."

The potential climatic effects of nuclear warfare were discussed in Chapter 3, Section 3.3.4.

14.11.5 Proliferation

As more and more nations built nuclear power plants, the Soviet Union, the United Kingdom, and the United States became concerned that the plutonium in the irradiated fuel in these plants could be used to proliferate nuclear weapons. In 1970 the worldwide Nuclear Non-Proliferation Treaty (NPT)

[51]*Estimated Exposures and Thyroid Doses Received by the American People from I-131 in Fallout Following Nevada Atmospheric Nuclear Bomb Tests*, National Cancer Institute, U.S. Department of Health and Human Services, October 1997.

[52]*Exposure of the American People to Iodine-131 from Nevada Nuclear-Bomb Tests: Review of the National Cancer Institute Report and Public Health Implications*, Institute of Medicine, National Academy of Sciences, Washington, DC, 1998.

that had been signed by 82 nations during the preceding two years went into effect for a period of 25 years. Under this treaty a signatory nation agreed that in exchange for having access to the technology for peaceful uses of nuclear energy (i.e., generation of electricity) it would not develop nuclear weapons and would not assist other nations to develop such weapons. It further agreed to accept international safeguards that include monitoring of the acquisition, storage, transfer, and use of all fissionable materials and sources of such materials. Safeguards inspections are carried out by the International Atomic Energy Agency. In 1995, when the NPT was renewed for an indefinite period, there were over 162 signatories. Although the goal of the NPT is to prevent the spread of nuclear weapons, it also can be seen as protecting the environment from widespread radioactive contamination.

The type of proliferation just discussed is sometimes referred to as "horizontal proliferation." During the Cold War, the declared nuclear nations increased their stockpiles of nuclear weapons. This increase in number of weapons possessed by each nation is called "vertical proliferation."

A modern thermonuclear warhead is a complex system containing ^{235}U, ^{238}U, ^{239}Pu, Be, Li, ^{3}H, and ^{2}H as nuclear components. The uranium enrichment path to a weapon has a lower radiological health risk and may be more difficult to detect than the plutonium path, which requires chemical processing of the irradiated fuel with likely escape into the environment of detectable radioactive fission products. In one type of thermonuclear explosive the thermonuclear reactants, ^{2}H and ^{3}H, are enclosed within the fissile material, which compresses and ignites the fusion reaction. In a second type, the fissile material and the mixture of hydrogen isotopes are in separate compartments. Radiation from the fission reaction compresses the hydrogen mixture and ignites the thermonuclear reaction.

Various proposed schemes for preventing the proliferation of nuclear weapons by nations that have not signed the NPT have been studied and evaluated.[53] The general conclusion of the studies was that steps could be taken to make diversion of plutonium and enrichment of uranium more costly, more dangerous, and more time-consuming, but that there is no technical fix (including using the ^{233}U$-^{232}$Th fuel cycle) to prevent proliferation. At best, a degree of proliferation resistance could be created.[54] Also, it was concluded that fuel cycles involving reprocessing to remove the plutonium are more vulnerable than the once-through cycles in which the irradiated fuel is placed in a repository for long-term storage.

[53]Two in-depth studies were the International Fuel Cycle Evaluation (INFCE) and the Nonproliferation Alternative Systems Program (NASAP). The latter, which began in 1976, was a program of the U.S. Department of Energy and was in support of INFCE.

[54]For example, addition of fission products would provide a measure of deterrence. However, no matter what is done to the fuel before irradiation or what is done when the fuel is reprocessed, the plutonium can be recovered by chemical separation.

Prevention of the proliferation of nuclear weapons is dependent on nuclear safeguards, the measures that are taken to guard against the diversion of nuclear material from uses allowed by law or by treaty and to detect diversion or provide evidence that diversion has not occurred. The measures include physical security, material control (controlling access to and movement of nuclear material), and material accountancy.

When India and Pakistan increased the proliferation of nuclear weapons in May 1998, they followed different production routes. India used plutonium separated from irradiated nuclear reactor fuel. Tritium needed for the thermonuclear explosive was produced at a facility in India. Pakistan used highly enriched uranium obtained by the centrifuge separation process. However, Pakistan also has nuclear reactors and a facility for reprocessing irradiated fuel to separate plutonium.

Even though the end of the Cold War eliminated the need for large arsenals of nuclear weapons, the potential for diversion of fissionable material from nuclear power plants continues, and there is the added potential for theft of existing weapons or their components. Furthermore, the focus of nuclear deterrence moved to the "rogue states" and terrorist groups that might use nuclear or other weapons.

14.11.6 A Cold War Heritage: Weapons-Grade Plutonium and Weapons-Grade Uranium

When the Cold War ended, one of the obvious questions that arose was, What should be done with the surplus plutonium and HEU that have been produced in the weapons program? The risk of theft caused nations with these materials to examine and strengthen their safeguards programs. On paper, there are several long-term options available for disposal of each of the two fissile materials. An argument for not converting both the plutonium and the HEU into waste is that they represent two stockpiles ("treasures") of an energy source that were produced at great expense and have potential use as fuel in nuclear power plants in the future.

For HEU (about 93% ^{235}U), an obvious long-term solution for dealing with the surplus is to simply convert the metal to UO_2 and dilute the oxide with natural or depleted uranium oxide to obtain LEU (3–4% ^{235}U) for use as fuel for existing LWRs or for new LWRs in other countries. This dilution procedure is now well established and is being used. Since the material is no longer of interest for making a weapon after dilution below 20%, it can be stored. Among the objections to this dilution procedure is that plutonium will be produced from the ^{238}U in the LEU. However, because the plutonium will be sealed in irradiated fuel rods along with a large quantity of fission products, it is considered to be in a "theft-resistant" state.

The stewardship of the surplus weapons-grade plutonium is a much greater challenge than that for HEU. Isotope dilution is not an option. The estimated total amount of weapons plutonium in the world is 270 tonnes (metric tons). Perhaps about 200 tonnes will become surplus. Over 30 methods for guarding against use to make a nuclear explosive have been suggested. Five of these are as follows.

1. Store it as the metal in a criticality-safe manner and guard it for perpetuity for possible reuse in weapons or for future conversion to PuO_2 and dilution of the oxide with natural or depleted uranium UO_2 to obtain MOX that can be used in nuclear power reactors.
2. Convert it to MOX as soon as possible.
3. Treat it as waste by mixing it with fission products and use known technologies to dispose of it after vitrification (incorporation in glass).
4. "Destroy" it in an accelerator-driven, subcritical fission device in which the fission energy would be recovered
5. Bury it underground in boreholes more than 4 km deep.

Method 1 does not resolve the safeguards issue because a suitable, long-term storage facility does not exist, and ^{239}Pu (2.410×10^4 y) decays into the longer-lived ^{235}U (7.04×10^8 y). Method 2 requires that the existing commercial LWRs in the United States be modified to use MOX as fuel. MOX is not a weapons material. It is being used in some of the European reactors. Method 3 reduces the energy value of the plutonium to zero, but requires a geological repository. The fourth method would require extensive development.

In December 1996 the United States announced a decision to take a dual-track approach using two of the four methods just listed: vitrification and conversion to MOX, for disposal of about 50 tonnes of surplus plutonium over a period of about 30 years. Approximately four years later Russia and the United States agreed that each country would dispose of 34 tonnes of plutonium, either as MOX to be used as fuel for civilian reactors or by mixing the plutonium with fission product high-level waste and placing it in a waste repository.

14.12 RADIOACTIVE WASTE

14.12.1 Types and Sources

In the United States, radioactive waste (radwaste) is usually classified according to the following categories:[55]

[55]Acronyms HLRW, ILRW, and LLRW, where R designates radioactive, are also used.

High-level waste (HLW)

Spent nuclear fuel (SNF)

Transuranic waste (TRU) (waste containing elements beyond uranium in the periodic table)

Low-level waste (LLW)

Intermediate-level waste (ILW) (waste generated when water used to cool the core of a power reactor is decontaminated, and certain types of waste produced in reprocessing SNF)

Mixed waste (MW) (HLW or TRU or LLW containing other hazardous materials such as PCBs, mercury, organic solvents, heavy metals and materials that are carcinogenic, toxic, flammable corrosive, pyrophoric, or environmentally hazardous in some other way)

Uranium mill tailings

Contaminated soil

Contaminated groundwater

The disposal method used depends on the type of waste.

Radioactive waste of the various types that are now in storage has been generated mainly in facilities used in the nuclear weapons program and in nuclear power plants. Additional waste has been generated in research reactors and in facilities such as those in hospitals, universities, and industry. New waste is being generated at all these facilities. For the weapons facilities, the new waste will result from decontamination of the equipment and buildings as the facilities are decommissioned. The same holds for commercial nuclear power plants that cease operation. Plants continuing to operate will, of course, continue to produce spent fuel and LLW.

Normally, when a building (e.g., factory, apartment complex, etc.) is demolished, the debris includes concrete and scrap metal. The latter finds its way into the local metal supply. When a nuclear facility is decommissioned and leveled, both the concrete and the scrap metal may be radioactively contaminated, not simply on the surface where the unsafe areas are removable, but internally. Contaminated concrete is sent to a radwaste disposal site. Contaminated scrap metal, if detected, must not be recycled. Tens of thousands of tons of nickel and steel have been placed in storage for this reason. These metals may be of use in nuclear facilities.

Safety standards for the transportation of nuclear waste are set at the international level by the International Atomic Energy Agency. Domestic regulations based on relevant parts of the IAEA standards are set by the Department of Transportation, as well as the NRC and the EPA. Packages and containers of various types are tested to withstand transportation accidents. For high-level waste and SNF, the appropriate local authorities are notified of the route and schedule.

14.12.2 Methods of Disposal

Disposal is the process of isolating the waste from people and the environment. One criterion for a disposal site is location in an area of extremely low population density. A second criterion is that radionuclides in a permanent, underground disposal facility (e.g., a geological repository) not escape and contaminate the groundwater in the area for thousands of years. Predictions of how a repository will behave over extended periods of time are based on mathematical models that take into account the likelihood of catastrophic phenomena such as floods, earthquakes, and volcanism. In addition, the models provide predictions on how any radionuclides that enter the groundwater will be transported (e.g., as simple ions, as ions bound to complexing agents,[56] or depending on the pH, as colloids consisting of ions adsorbed on hydrous oxides such as those of iron and aluminum).

Despite the advances that have been made in chemistry to date, relatively little is known about the fundamental chemistry relevant to the migration of the elements on the earth's surface and below the surface. This area of research provides both a challenge and an opportunity. It includes investigation of the solubilities, complexation, redox reactions, colloid formation, and sorption and desorption mechanisms of the fission products and the actinide elements. In addition, there is need to develop more effective technologies that are both innovative and cost-effective to separate and convert hazardous wastes of all types into stable disposable forms.

An additional requirement for a long-term radwaste disposal site is that it be intrusion-proof. For a geological repository, "intrusion" could mean mining for plutonium or mining to reach mineral deposits in the area.

14.12.2.1 High-Level Waste (HLW) and Spent Nuclear Fuel (SNF)

HLW includes the mixture of fission products separated from irradiated reactor fuel when the fuel is reprocessed (for the nuclear weapons production) and often includes SNF that is stored without reprocessing. Transportation of HLW to sites for storage, predisposal treatment, and permanent disposal is a very important part of the management of such waste. Casks used to transport HLW by truck or rail are especially designed to provide radiation shielding and to remain intact in the event of an accident. The two major sources of HLW considered here are nuclear power reactors and facilities at which irradiated fuel and SNF have been reprocessed to remove the plutonium and recover ^{235}U.

For an LWR with a burnup of 33 GWd/MTHM and annual removal of the third of the fuel having the highest burnup, each MTHM of SNF would

---

[56]Complexing agents would include EDTA used as a decontaminating agent or naturally occurring substances such as humic and fulvic acids.

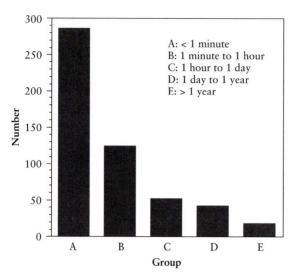

FIGURE 14-13 Distribution of fission products in spent LWR fuel according to half-life in specified time ranges.

contain about 35 kg of fission products and about 10 kg of transuranic elements (Np, Pu, Am, and Cm), of which about 9 kg would be plutonium.[57] The radioactivity of the fission products and that of the actinides at the time when the reactor is shut down decreases rapidly at first because of the high proportion of short-lived radionuclides. Figure 14-13 illustrates the half-life distribution for fission products in the SNF when it is removed from the reactor.

As spent nuclear fuel is stored, the fission product radioactivity will be determined increasingly by the radionuclides with long half-lives.[58] Fission products in the SNF having a half-life greater than one year are listed in Table 14-11. The four long-lived fission products that require long-term storage in a repository are ^{90}Sr (28.78 y), ^{99}Tc (2.13 × 10^5 y), ^{129}I (1.57 × 10^7 y), and ^{137}Cs (30.07 y).

[57]By the year 2008, burnup for LWRs in the United States is expected to increase to about 39 GWd/MTHM for BWRs and about 50 GWd/MTHM for PWRs. The concentrations of fission products and transuranic elements in the SNF will, therefore, increase.

[58]In an operating reactor, the short-lived fission products that are daughters of longer-lived fission products reach transient or secular equilibrium concentrations. Note that the short-lived daughters are counted according to their own half-lives in Figure 14-13, even though they decay with the half-life of the parent. As a good approximation, fission products with very long half-lives initially accumulate linearly with time.

TABLE 14-11

Fission Products and Actinides with Half-Life Greater than 1 Year That Are in LWR Spent Fuel

Fission products[a]		Actinides[a]	
Radionuclide	Half-life	Radionuclide	Half-life
^{3}H	12.32 y	^{234}U	2.46×10^5 y
^{85}Kr	10.76 y	^{235}U	7.04×10^8 y
^{90}Sr	28.78 y	^{236}U	2.342×10^7 y
^{90}Y*	2.67 d	^{238}U	4.47×10^9 y
^{93}Zr	1.5×10^6 y	^{237}Np	2.14×10^6 y
^{99}Tc	2.13×10^5 y	^{236}Pu	2.87 y
^{106}Ru	1.020 y	^{238}Pu	87.7 y
^{106}Rh*	29.9 s	^{239}Pu	2.410×10^4 y
^{107}Pd	6.5×10^6 y	^{240}Pu	6.56×10^3 y
^{121m}Sn	~55 y	^{241}Pu	14.4 y
^{125}Sb	2.758 y	^{242}Pu	3.75×10^5 y
^{125}Te*	58 d	^{241}Am	432.7 y
^{126}Sn	2.5×10^5 y	^{242m}Am	141 y
^{126m}Sb*	19.90 m	^{242}Am*	16.02 h
^{126}Sb*	12.4 d	^{243}Am	7.37×10^3 y
^{129}I	1.57×10^7 y	^{243}Cm	29.1 y
^{134}Cs	2.065 y	^{244}Cm	18.1 y
^{135}Cs	2.3×10^6 y	^{245}Cm	8.5×10^3 y
^{137}Cs	30.07 y	^{246}Cm	4.76×10^3 y
^{137m}Ba*	2.552 m		
^{147}Pm	2.6234 y		
^{151}Sm	90 y		
^{154}Eu	8.593 y		
^{155}Eu	4.75 y		

[a]Includes short-lived radioactive daughter (asterisk = *) present with the preceding long-lived parent.

The actinide radionuclides, including transuranic nuclides, having a half-life greater than one year that are contained in the fuel are also listed in Table 14-11. If SNF is reprocessed, the small amount of uranium and plutonium that is not recovered (perhaps 0.1–0.5% by weight) and the other TRU elements remain with the fission products in the HLW waste.

After 10 years of decay, the contribution of fission products to the long-term heat release consists of about 43% from ^{90}Sr–^{90}Y and about 49% from ^{137}Cs–^{137m}Ba. The isotopes of plutonium, americium, and curium contribute 52, 29, and 19%, respectively, of the heat released from actinides after 10 years.

When the nuclear power industry was established in the United States, it was intended that the spent nuclear fuel would be reprocessed after being stored for a "cooling" time of about 150 days under water in pools located at the power plant site. During the cooling time, the decay heat would be removed from the short-lived fission products. There are several reprocessing schemes for recovering the residual ^{235}U and ^{238}U and separating the reactor-grade ^{239}Pu (with higher concentrations Pu isotopes) from the fission products and then separating the uranium and plutonium. One such scheme was described in Section 14.11.2 for separating weapons-grade Pu. Reprocessing of spent fuel from commercial nuclear power plants was banned in the United States in 1976 because of concern about the possible proliferation of Pu-based nuclear weapons. After the ban went into effect, the spent nuclear fuel became part of the once-through fuel cycle (i.e., from reactor to a repository for permanent disposal as waste).[59] In 1982 (Nuclear Waste Policy Act) the U.S. Department of Energy agreed to begin accepting commercial SNF for storage or disposal by January 31, 1998. The electric utilities have been contributing to a Nuclear Waste Fund to pay for construction of a storage or disposal facility (repository). The Yucca Mountain geological storage facility (described shortly) will not be available until 2020 at the earliest.

As the storage pools for SNF became filled to capacity, some nuclear power plants built heavily shielded, above ground, dry, secure storage facilities to hold the older spent fuel. These facilities are passively air-cooled. Unfortunately, reliance on on-site, dry storage nuclear power plants keeps the SNF distributed about the country. Thus, spent fuel is stored in 35 states rather than in a single, isolated geological repository.

The projected annual rate of SNF discharge in the United States is about 1900 MTHM until the year 2014. In the year 2000, the total amount of spent fuel stored in various places amounted to 40,446 MTHM. If no new nuclear power plants are built, the rate of increase in spent fuel will drop to zero at a rate that depends on how many plants have their operating licenses extended (e.g., by 20 years) beyond the original period of 40 years.

Projections for cumulative discharges of spent nuclear fuel worldwide were made beginning with the year 1993 (when data for eastern Europe became available). According to one projection,[60] the annual discharge rate will decrease from 8513 MTHM in 2000 to 8179 MTHM in 2010. This change during the decade may be too small, since the increase in nuclear power generation in a country such as Japan may be more than offset by the rate of decrease resulting from the closing of nuclear power plants in countries that have a politically based nuclear power moratorium. In 2000, spent nuclear fuel was being generated in about 490 nuclear power plants in 33 nations. Not all of the spent fuel was being stored. Reprocessing is carried out in France, Japan,

[59]Although commercial reprocessing of spent fuel became legal again in the United States in 1980, it is not economically attractive for a number of reasons.
[60]Nuclear Energy Data 2000, Nuclear Energy Agency, OECD, Annapolis Junction, MD 20701.

Russia, and the United Kingdom. Some of these nations reprocess spent fuel shipped from other nations. Russia has indicated an interest in expanding the past practice of importing and storing nuclear waste (e.g., spent fuel) for a fee. Western Europe and Asia would be the source of the spent fuel, which would eventually be reprocessed.

Predisposal treatment of HLW that has been generated by reprocessing irradiated fuel in the weapons program in the United States for over 50 years presents a more difficult challenge. Most of the HLW is liquid (aqueous) waste that contains inorganic salts, organic solvents, complexing agents such as ethylenediaminetetraacetic acid, nonvolatile fission products, and small amounts of uranium, plutonium, and other transuranic elements.[61] At the Hanford Reservation, where plutonium was produced during the Manhattan Project and in subsequent years, the waste was stored in 177 large underground tanks having provisions for removing decay heat. Some tanks have vented gases (e.g., hydrogen), produced by radiolysis of water, and exothermic reactions involving nitrates have occurred in the stored waste. Most of the tanks (149) were single-wall tanks that were state-of-the-art vessels when built during World War II. HLW (about 140,000 gallons by 1996) had leaked from about 67 of these tanks. Transfer of liquid HLW from the old tanks to new, million-gallon, double-walled tanks (Figure 14-14) has been ongoing for several years.

Liquid HLW stored in tanks for a long time tends to separate into a stratified mixture of a supernatant solution, a salt cake consisting of compounds such as carbonates, phosphates, sulfates, and nitrates, and a sludge containing most of the radioactivity. After liquids have been removed to reduce the volume of the waste, the solid can be calcined to a granular solid to reduce its leachability, or the waste can be vitrified with a borosilicate glass having a low solubility in water.[62] Vitrification (at temperatures above 1000°C) reduces the volume of high-level liquid waste by about 75%. An additional "engineered barrier" to prevent escape of radionuclides from a repository into the environment (if water enters the repository) is introduced by sealing the glass "logs" in stainless steel canisters (3.0 m × 0.6 m).

Locations that have been proposed for permanent disposal of HLW include deep stable geological repositories, stable deep sea sediments consisting of clay and located in international waters, and outer space.[63] In the past, ocean disposal of various types of radioactive waste has been used by several nations.

[61]Strontium-90 and ^{137}Cs have been extracted from some of the waste for use in intense sources of ionizing radiation.

[62]Vitrification of HLW has been under development for about 40 years. It has been employed since 1978 in Europe, where there are at least three vitrification facilities. Such facilities began operation in the United States in 1997.

[63]Temporary disposal in outer space is a *de facto* method in the sense that there are nuclear-reactor-powered satellites parked in space.

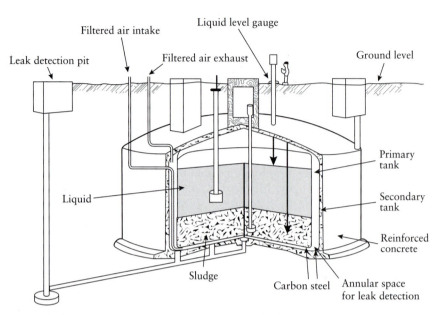

FIGURE 14-14 Double-shell tanks with improved leak detection systems and waste containment features built to replace the old single-shell tanks. From *Restoration and Waste Management (EM) Program. An Introduction*, U.S. Department of Energy. DOE/EM-0013P (revised December 1992).

Radwaste [estimated total of about 35 PBq (0.95 MCi)], including plutonium, from the Sellafield reprocessing plant in the United Kingdom has been dumped into the North Sea. The United States disposed of one submarine reactor, that of the *Seawolf* (see footnote 36), and lost two nuclear-powered submarines at sea. It also dumped sealed containers of radwaste into the Pacific Ocean about 50 miles from the coast of California, for an estimated total of about 3.5 PBq (95 kCi). The former Soviet Union disposed of 16 cores of reactors from submarines along with other containers of radioactive waste near the Novaya Zemlya archipelago in the Barents Sea. Additional waste was dumped into the Sea of Japan and the Kara Sea, so that the estimated total was about 93 PBq (2.5 MCi). At least 12 other countries are known to have used ocean disposal of radwaste in estimated amounts up to megacuries.

Although ocean dumping of radioactive waste is now prohibited by international law, disposal of radwaste at sea is still proposed. In the subseabed method, the waste would be placed in sealed canisters, which would be lowered into holes drilled hundreds of meters into the seabed of deep oceans. The canisters in a given hole would be separated by a layers of added sea sediment. Finally, the top section of the hole would be permanently sealed with sediment.

In the 1950s the USSR began to inject liquid HLW into the earth, 300 m or more below the surface, near the sites of three nuclear weapons facilities (Krasnoyarsk-26, Mayak, and Tomsk-7), where irradiated uranium was reprocessed to recover the plutonium. Some 55 EBq (1.5 GCi) of waste, mostly ^{90}Sr and ^{137}Cs, was disposed of in this manner. At Mayak, where plutonium was produced and the irradiated uranium was processed, liquid waste was released into the Techa River, which flows into a storage facility: Lake Karachai.

In the United States, a total of about 60 PBq (1.6 MCi) was injected underground or released into streams, soil, seepage basins, or ponds at the Oak Ridge, Savannah River, and Hanford facilities. LLW was injected into the ground at Hanford WA briefly at the beginning of the nuclear weapons program, and radwaste mixed with concrete was injected into shale formations for several years at the Oak Ridge nuclear facility. About 60 PBq of waste was injected underground or released into streams, soil, seepage basins, or ponds at Oak Ridge, Savannah River, and Hanford.

In the mid-1990s the strategy for disposal of HLW in the United States included construction of a temporary, above ground, monitored retrievable storage (MRS) facility, where spent nuclear fuel and other HLW would be stored, loaded into canisters, and eventually loaded into shipping casks for transfer to a repository. Difficulties in locating a politically acceptable site and fear that interim storage might delay construction of a repository left the future of an MRS in doubt, but did not rule out establishing several smaller interim storage sites.

Deep geological disposal of HLW and has been chosen by many countries, including Canada, Finland, France, Germany, Japan, Spain, Sweden, Switzerland and the United States.

In the United States, the Committee on Separations Technology and Transmutation Systems of the National Research Council examined the various methods of radioactive waste disposal and concluded the following, with respect to the health and safety issues for the disposal of nuclear waste: "For disposal, the lowest to highest risk alternatives appear to be—geologic repository—surface." (See Additional Reading and Sources of Information, *Nuclear Wastes* 1996.)

The rock formation surrounding the storage rooms serves as the natural and the main barrier against escape of the radioactive material into the environment. Most countries construct or plan to construct an experimental facility as a learning experience before selecting and developing a final repository site. A popular completion date is the year 2020.

A potential site for a repository in the United States is Yucca Mountain, Nevada, which is roughly 160 km (100 miles) northwest of Las Vegas. This site (Figure 14-15), which consists of volcanic tuff, has been undergoing characterization for several years. Work on the site began in the 1950s. A

FIGURE 14-15 Yucca Mountain, Nevada, candidate site for the first U.S. geologic repository. From *Managing the Nation's Nuclear Waste*, DOE/RW-0263P, U.S. Department of Energy, 1990.

few characterization projects include studies of redox reactions, radionuclide retention, hydrochemistry, and groundwater degassing.

The Yucca Mountain facility was initially scheduled to begin receiving 30,000 tonnes of spent fuel from commercial nuclear power plants and other HLW (defense waste, as glass cylinders in canisters), by January 31, 1998. Opponents of the use of this site point out that it is located in the state that is third, after California and Alaska, in frequency of earthquakes and that the mountain is within 70 km of several volcanoes. In addition, there is at least one fault within the mountain. For these reasons and others concerning the water table, the opponents argue that it is difficult to show that the HLW would remain undisturbed for the required 10,000 years. If the site is approved by the Environmental Protection Agency, disposal of about 70,000 tonnes of SNF (the design limit) and other HLW in rooms 300 m below the surface would begin in the year 2010 and proceed as shown in Figure 14-16. At a disposal rate of about 3000 metric tons of spent fuel per year, the design limit would be reached in 2033. Closure could be between 50 and 300 years after the first waste is accepted.

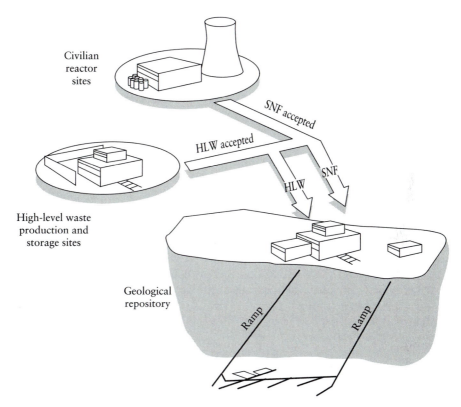

FIGURE 14-16 Civilian radioactive waste management system for spent nuclear fuel (SNF) and high-level waste (HLW). From U.S. Department of Energy, Office of Civilian Radioactive Waste Management, Report DOE/RW-0473.

Transmutation of long-lived fission products in HLW into shorter-lived radionuclides or even stable nuclides also has been proposed many times as a method for making HLW less radiotoxic and reducing the amount of HLW requiring storage in a geological repository. One technique, accelerator transmutation of waste (ATW), uses an accelerator rather than a dedicated reactor to change the isotopic composition of HLW.

An example of the transmutation of the long-lived fission product ^{129}I $(1.57 \times 10^7 \text{ y})$ by successive neutron capture and negatron decay is

$$^{129}\text{I} + {}^1\text{n} \rightarrow {}^{130}\text{I} \ (12.36 \, \text{h}) \ \beta^- + {}^{130}\text{Xe(s)} + {}^1\text{n} \rightarrow$$
$$^{131}\text{Xe(s)} + {}^1\text{n} \rightarrow {}^{132}\text{Xe(s)}$$

14.12.2.2 Transuranic (TRU) Waste

TRU waste contains α-emitting transuranic elements ($Z > 92$) with half-lives greater than 30 years and concentrations greater than 100 nCi per gram of waste. This category of radioactive waste consists mainly of paper, clothing, equipment, tools, and soil contaminated with plutonium; plutonium-containing liquid waste generated in decontaminating glassware, and so on; dust and chips of metallic plutonium used for weapons production; plutonium dioxide; and any other materials that may have been contaminated with plutonium or other transuranic elements. Perhaps 2% of the TRU waste requires remote handling.

A facility known as WIPP (Waste Isolation Pilot Plant) has been constructed for geological waste disposal of the TRU waste that has been generated in the weapons program. The waste must be stored in barrels, and there is a maximum allowable concentration of Pu that cannot be exceeded. WIPP (Figure 14-17), which includes facilities for research, is located in salt (rock salt, NaCl) deposits about 42 km (about 26 miles) east of Carlsbad, New Mexico. Salt beds or domes are considered to be suitable for geological disposal because large rooms can be built in the deposits without structural support. Also, salt has a relatively high thermal conductivity, is relatively impermeable to water,

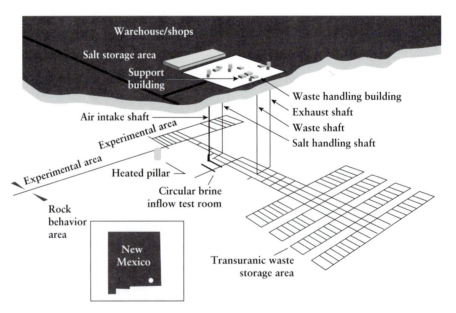

FIGURE 14-17 Cross-sectional view of the Waste Isolation Pilot Plant in Eddy County, New Mexico. From *Environmental Management 1995*, U.S. Department of Energy, DOE/EM-0228, February 1995.

and self-seals around voids so that the mined rooms, located 655 m (2150 ft) below the surface and containing the packaged waste, will eventually refill with salt and entomb the waste.

The facility was completed in 1988 but was not opened on schedule because of structural, management, and legal problems, as well as political opposition and questions that were raised about the retrievability of waste that would be stored in the salt deposit during a trial period. Technical approval for use of WIPP was given by EPA in May 1998, and the site was opened in March 1999. The initial permit from the New Mexico Environmental Department was for nonmixed TRU waste. Later a permit was issued for mixed TRU waste. It has been estimated that WIPP will be filled to capacity (175, 600 m^3) after about 35 years of operation and will be closed in about 2099. There are to be warning markers and barriers around the site.

Salt domes have been explored for radwaste disposal in Germany. An example is the Gorleben dome in northern Germany. Interim waste storage facilities have also been built at Gorleben.

14.12.2.3 Low-Level Waste (LLW)

In the United Sates, low-level waste consists of a broad spectrum of radio-actively contaminated material that does not qualify as HLW, SNF, TRU, or uranium mill tailings. Examples are contaminated clothing, gloves, shoes, tools, paper towels, pieces of equipment, ion exchange resin, and animal carcasses. Low-level waste is generated in nuclear power plants, government-owned facilities, industrial and academic research laboratories, particle accelerator facilities, hospitals, and plants manufacturing radiopharmaceuticals or devices containing radioactive material. Less than 5% of the LLW requires the use of shielding during handling and transportation.

It has generally been assumed that any LLW discharged (within legal limits) into public sewers by hospitals, laundries, and manufacturers is diluted to such a degree that it does not constitute a health hazard. Consistent with the known radiochemistry of radionuclides in very dilute aqueous solutions, however, certain radionuclides tend to reconcentrate on surfaces such as the particles of precipitates of hydrous oxides. For example, ^{60}Co (5.271 y) has been found in the sludge and ash produced in municipal waste treatment plants. Phytoremediation is a method for extracting radionuclides present at low concentration in soil and in aqueous solution. Sunflower plants, certain weeds, and willow trees, for example, have been shown to preferentially concentrate uranium, cesium, and strontium. If shown to be effective in a demonstration project, the method could be used on a large scale. Eventually the contaminated plants and trees would be incinerated to provide an ash containing the radionuclides at a higher concentration.

LLW is usually very bulky and, therefore, is also known as high-volume waste. The cost of disposal of LLW is determined by its volume; hence, there is a strong motivation to reduce the volume by compaction or incineration.

LLW is generally disposed of in licensed, near-surface (shallow land) facilities. These facilities must be designed to afford protection from releases of radioactivity to the general population and to the individuals who operate the site, as well as to inadvertent intruders. The design also must ensure stability of the site after closure. Waste (in sealed containers) is placed in trenches which are eventually covered with soil and other materials to restrict the inflow of water. In the United States, the burden of disposing of all civilian LLW rests with the individual states. Groups of neighboring states can form regional compacts to share a disposal site if they can find one that is politically acceptable. DOE has been disposing of LLW at the Nevada test site and at other major government sites that already have established LLW disposal facilities.

Since 1983 LLW has been classified further by the U.S. Nuclear Regulatory Commission. There are three classes, A, B, and C, based on the concentration of radionuclides and the length of time ($< 100\,y$, $100-500\,y$, and $> 500\,y$) required for the radioactive material to decay to a level such that an intruder would not receive a dose of more than 15 mrem/y.

In the United Kingdom, investigation of the geology near the Sellafield reprocessing facility was begun in 1997 for possible location of an underground repository for low-level and intermediate-level waste.

14.12.2.4 Mill Tailings

After acid leaching of the uranium ore, the residual sands contain 960–3700 Bq of ^{226}Ra (1599 y) and 2590–22.2 kBq of ^{230}Th (7.54×10^4 y) per kilogram (26–100 nCi of ^{226}Ra and 70–600 nCi of ^{230}Th per kilogram). These uranium mill tailings either are stabilized by a covering of soil and rock (Figure 14-18) or are removed for disposal at a more remote location on DOE-owned land and then covered. Figure 14-19 shows the location of most of the uranium mill tailings in the United States.

The EPA standard for the limit of release of ^{222}Rn from tailings with a permanent radon barrier is $740\,\mathrm{mBq\,m^{-2}\,s^{-1}}$ ($20\,\mathrm{pCi\,m^{-2}\,s^{-1}}$). Mill tailings that were used as fill around basements of buildings after World War II have been removed to a disposal site. Buildings that were significantly contaminated have been razed.

14.12.3 The Cold War Mortgage

The U.S. Department of Energy has referred to the environmental management problems associated with the facilities used for the production of nuclear

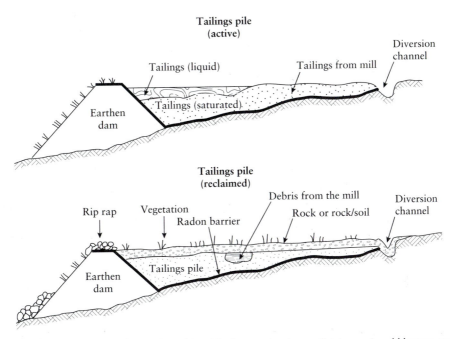

FIGURE 14-18 Reclaiming the tailings pile from a uranium mill (rip rap is cobblestone or coarsely broken rock used for protection against erosion of embankment or gully). From *Decommissioning of U.S. Uranium Production Facilities*, U.S. Department of Energy, DOE/EIA-0592, February 1995.

weapons as the "Cold War mortgage."[64] Another term is "the environmental legacy of nuclear weapons research, production and testing and DOE funded nuclear energy and basic science in the United States."[65] Actually, problems such as disposal of stored liquid HLW began during the Manhattan Project in World War II and increased in magnitude during the Cold War.[66]

Of the three major weapons production sites, Hanford, Oak Ridge, and Savannah River, the Hanford site is expected to be the most difficult and costly to stabilize and remediate.[67] Reprocessing irradiated uranium at this site to

[64]*Estimating the Cold War Mortgage*, DOE/EM-0232, U.S. Department of Energy, Office of Environmental Management, 1995 (updated annually).

[65]*Status Report on Paths to Closure*, DOE/EM-0526, U.S. Department of Energy, Office of Environmental Management, 2000.

[66]The actual date when the first Cold War ended is controversial. A commonly used range of dates is between 1989, when the Berlin Wall fell, and 1991, when the USSR ceased to exist.

[67]Estimated dates for completion of remediation are as follows: Hanford, 2046; Savannah River, 2038; and Oak Ridge, 2014. Remediation of the site of the Idaho National Engineering and Environmental Laboratory, in Idaho Falls, is scheduled for completion in 2050.

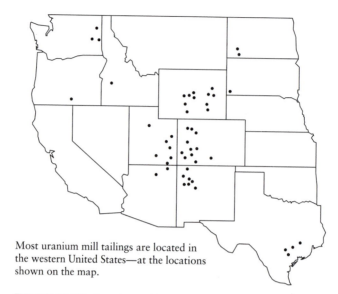

Most uranium mill tailings are located in the western United States—at the locations shown on the map.

FIGURE 14-19 Locations of uranium mill tailings. From *Radioactive Waste: Production, Storage, Disposal*, U.S. Nuclear Regulatory Commission, NUREG/BR-0216, July 1996.

recover the plutonium over the period from 1944 to the early 1980s generated about $2.1 \times 10^5 m^3$ (55×10^6 gal) of liquid, high-level radwaste with an activity of about 17 EBq (0.45 GCi). Also located at the Hanford Reservation are nine large reactors that were used to produce plutonium. It is likely that all or most of these will be buried on the site.

For several of the weapons production sites, the groundwater is contaminated with both radioactive and nonradioactive hazardous materials that are likely to migrate beyond the site boundaries. As an example, the groundwater beneath about 388 km^2 (150 mi^2) of the 1450 km^2 (560 mi^2) Hanford Reservation is contaminated and is believed to be moving toward the Columbia River.

The Idaho National Engineering Laboratory is another major site at which irradiated fuel (e.g., from the cores of nuclear submarine reactors) has been reprocessed and liquid HLW has been stored. Sections of hulls containing the reactors (minus the fuel) from decommissioned submarines are buried in disposal trenches at Hanford. At sites already mentioned, and at more than 100 additional sites, over $378 \times 10^3 m^3$ (100×10^6 gal) of radioactive and mixed liquid waste are stored in over 300 tanks. In addition, there are about $3 \times 10^6 m^3$ of buried radioactive or hazardous waste, about $250 \times 10^6 m^3$ of contaminated soil, and over $2.3 \times 10^9 m^3$ (600×10^9 gal) of contaminated groundwater.

Despite the large volume of the HLW stored at DOE sites, the total radio-activity of these wastes is less than that of wastes stored as spent fuel from commercial nuclear power plants. During the Cold War period, two plants for commercial reprocessing of spent fuel from nuclear power plants were built. Only one of the plants, that at West Valley, New York, was put into operation. It was operated from 1966 to 1972, when it was shut down for modifications to increase its capacity. It never resumed operation. Liquid HLW and unreprocessed spent nuclear fuel remained stored on the site until about 1985, when DOE initiated decontamination of the site. A facility (a melter) for vitrifying HLW was built and operated successfully, a result that led to the building of a second facility at the Savannah River site.

14.13 PATHWAYS FOR INTERNAL EXPOSURE TO IONIZING RADIATION

Unlike external exposure, which requires only that ionizing radiation from a source external to the body be absorbed in the body, the pathways for internal exposure can vary from direct injection of a radiopharmaceutical to a more complex pathway through the food chain.

The human body has evolved in an environment containing naturally radioactive materials in the atmosphere, rocks, soil, lakes and oceans, drinking water, and foods of all types. The pathways from these components of the environment into the human body are rather obvious and are usually of little concern to a person in the course of routine daily life. If, however, human activity leads to a significant increase in the concentration of radioactive material (naturally occurring or anthropogenic) in one or more of the many environmental components, questions about potential health hazards arise. Evaluation of the magnitude of a potential hazard requires an understanding of the chemistry of the material as it moves along an environmental pathway to a point where it may be ingested or inhaled by a human.

The information needed to evaluate a potential hazard includes the following:

- Identity and quantity of each radionuclide present in the environment and its key nuclear properties [half-life and type(s) of radiation emitted]
- Environmental chemistry of the elements represented by the radionuclides (e.g., chemical processes such as chelation) that could increase or decrease the concentration of the radionuclides as they move along a particular pathway
- Physiological chemistry of the elements including the biological half-life T_{biol} as a measure of retention by the human body, distribution among the organs of the body, chemical reactions (e.g., complexation), and physical processes such as isotope dilution with a stable isotope that could accelerate elimination from the body

- Biological concentration in plants and animals
- Particular radiological health hazard(s) associated with each radionuclide
- Identification of any segment of the population that might be especially vulnerable

Figure 14-20 is a simplified flow diagram that shows the major pathways by which radionuclides go from the atmosphere, surface water or groundwater, soil, and rocks into the human body.

The major pathway for ^{226}Ra (1599 y) and ^{228}Ra (5.76 y) is the same as that for calcium, that is, through agricultural produce and milk. Water used for drinking and cooking can also be an unsuspected source of radium. Potassium 40 (1.27×10^9 y) enters the body through fruits and vegetables that are correctly recommended as very good sources of potassium.

As discussed in Section 14.4.3.2, inhalation is the pathway into the body for ^{222}Rn and hazardous quantities of the two short-lived, α-emitting progeny of radon, ^{218}Po (3.10 m) and ^{214}Po (163.7 μs). Two relatively long-lived progeny of ^{222}Rn, the β emitter ^{210}Pb (22.6 y) and its α-emitting granddaughter ^{210}Po (138.38 d), also enter the body by inhalation as particulates contained in cigarette smoke. Polonium-210 has been implicated as a contributor to the carcinogenicity of cigarette smoke.

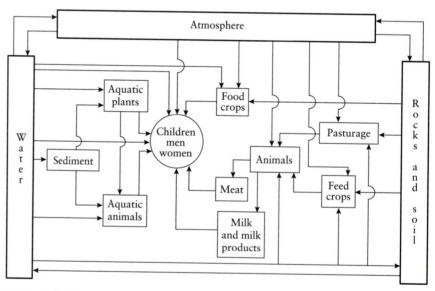

FIGURE 14-20   Major pathways for the uptake by humans of radionuclides in the environment.

Inhalation of fission product radionuclides now in the atmosphere from past weapons tests is not considered to be a significant source of exposure. However, inhalation of fission products and possibly plutonium could have serious consequences for someone a few miles downwind of the explosion of a nuclear weapon, an explosion at a storage site for high-level waste, or a chemical explosion at a large nuclear reactor.

The 17 radioisotopes of iodine that are produced in nuclear fission and are released into the atmosphere by a nuclear explosion or a nuclear accident return to the earth as fallout or rainout. They can contaminate local water supplies, leafy vegetables, grass in pastures and, therefore, milk supplies, if it is the season when cows are in the pasture. Because iodine concentrates in the thyroid when it is ingested, most of the absorbed dose from ionizing radiation emitted by a radioisotope of iodine is localized in the thyroid.

Except for ^{131}I (8.0207 d) and ^{129}I (1.57×10^7 y), the radioisotopes of iodine are short-lived and will probably decay to a very low level before they become an ingestion hazard. Because of the relatively long biological half-life (138 d) of iodine, ^{131}I (Figure 13-10) can deliver a large dose of ionizing radiation (β^-, γ) to the thyroid. Contaminated milk is the major pathway to the thyroid of children, who constitute the most vulnerable segment of the population because they drink a relatively large amount of milk and have relatively small thyroid glands (which leads to a higher radioiodine concentration than for an adult); moreover, since their thyroid glands are still growing, the cells are especially radiosensitive.

Relative to ^{131}I, ^{129}I has a very low specific activity and is considered to be a less hazardous component of "fresh" fallout or rainout. However, if released into the environment, it accumulates and could constitute a potential hazard in the future.

A compound such as KI, a "thyroid blocking agent," containing the stable isotope ^{127}I can be administered in tablet form prior to the arrival of ^{131}I or other radioisotopes of iodine, if there is advance warning. The stable iodide taken up by the thyroid reduces the uptake of radioiodide.

The fission products ^{137}Cs (30.07 y) and ^{90}Sr (28.78 y) also leave the atmosphere and enter the food chain or the water supply as fallout or rainout. Either they immediately contaminate any leafy vegetables in the fields before they can be harvested or they contaminate the soil in farms and dairies and, therefore, later and for many years, contaminate crops grown in the soil for human and animal consumption. As a soil contaminant, ^{90}Sr will have the chemistry of Sr^{2+} (i.e., similar to that of Ca^{2+}). The ions of both Sr and Ca will undergo cation exchange reactions.

Strontium, therefore, being like calcium, is a bone-seeking element. If ^{90}Sr is ingested, the ionizing radiation (β^-) from ^{90}Sr and its daughter, ^{90}Y, (2.67 d) (Figure 13-13) can cause bone tumors, bone cancer, and leukemia. Uptake of ^{90}Sr or ^{137}Cs, unlike that of ^{3}H and ^{14}C, is not reduced to a useful extent by natural isotope dilution with stable isotopes in food. Lime (CaO) has been used to treat ^{90}Sr-contaminated soil to compete with and reduce the uptake of ^{90}Sr by vegetables.

The main pathways for ^{137}Cs are from food grown in contaminated soil, from fish living in contaminated water, and from meat (e.g., that of sheep and reindeer). Fertilizers with high potassium content have been used to reduce the uptake of ^{137}Cs by plants from contaminated soil.

When ^{137}Cs is ingested, it becomes rather widely distributed in muscle tissue. As expected, the chemistry of cesium is similar to that of potassium, but cesium passes more slowly through cell membranes. Oral administration of prussian blue, $Fe_4[Fe(CN_6)]_3$, has been used since the mid-1980s to accelerate the clearance of ingested ^{137}Cs from the body. Prussian blue has low toxicity and is not absorbed from the intestinal tract, where it acts as an ion exchanger for monovalent cations such as ^{137}Cs$^+$ but does not interfere with the serum levels of K^+. This treatment has been found to reduce the biological half-life of ^{137}Cs (140 d) by a factor of 2–3. Prussian blue has been studied in laboratory animals and has been used to treat people and animals (as sources of meat) following the release of significant amounts of ^{137}Cs in accidents.

The actinide elements that have half-lives long enough to be of environmental concern are thorium, uranium, neptunium, plutonium, americium, and curium. A number of the isotopes of plutonium, americium, and curium (Table 14-11) are produced by multiple neutron capture reactions. Important oxidation states of these elements in the environment are Th(IV); U(IV), U(VI); Np(IV), Np(V); Pu(III), Pu(IV), Pu(V), Pu(VI); Am(III); and Cm(III). The extent to which these elements are mobile in soil and natural waters depends on a number of factors such as the degree of hydrolysis, the presence of adsorbents, and the presence of naturally occurring humic and fulvic acids (Chapter 9, Section 9.5.7), which can function as complexing agents and/or reducing agents.

In the environment, plutonium [^{238}Pu (87.7 y), ^{239}Pu (2,410 y), ^{240}Pu (6560 y), and ^{241}Pu (14.4 y) and its daughter ^{241}Am (432.7 y)] are strongly sorbed by components in soil (clay, hydrous oxides, organic matter. etc.). These radionuclides are, therefore, relatively immobile in soil and in sediments in natural waters. Most of the plutonium from nuclear weapons testing is concentrated in the upper 5 cm layer of soil and is very likely present in the form of the dioxide.

Plutonium, as the PuO_2^{2+} (plutonyl) ion, like calcium, strontium, and radium, is a bone-seeking element. Plutonium-239 resembles ^{226}Ra by being an α emitter with long biological and radioactive half-lives and is considered to be one of the most radiotoxic radionuclides associated with nuclear power and nuclear weapons. Once in the bloodstream, plutonium tends to concentrate mainly and about equally in the liver and in bone.

In addition to isotope dilution, nature provides another way of reducing the human uptake of certain radionuclides. Biological systems have an ability to discriminate against certain elements. The ratio of the ^{90}Sr concentration in milk to that in forage provides an illustration. Because of the chemical similarity between Sr and Ca, the concentration of ^{90}Sr in any material can be expressed in terms of a relative unit, the "strontium unit" (SU). One strontium unit is equal to 37 mBq (1 pCi) of ^{90}Sr per gram of calcium. The observed ratio (OR) for ^{90}Sr in milk relative to that in forage is given by the equation

$$OR = \frac{(^{90}Sr/Ca) \text{ in milk}}{(^{90}Sr/Ca) \text{ in forage}} \qquad (14\text{-}14)$$

Under steady-state conditions, the OR values are typically in the range of 0.1–0.2. The SU can be used to follow the dependence of the uptake of ^{90}Sr in human bone on diet.

Biological systems may also concentrate certain elements. Examples of such systems are plants growing in soil and organisms growing in a marine or a freshwater environment. A concentration factor (CF) for aquatic organisms, for example, is defined as C_O/C_W, where C_O is the concentration of the nuclear species of interest in the organism and C_W is the concentration of the same species in the ambient water. Equilibrium conditions are assumed. For the element zinc, as an example, CF values for plants, mollusks, crustacea, and fish in freshwater may be between 10 and 15,000. For similar marine organisms, the CF values for zinc may be between 100 and 300,000. Thus, for a given organism and a given element, the values vary widely with the local conditions.

14.14 DOSE OF IONIZING RADIATION RECEIVED BY THE U.S. POPULATION

14.14.1 Dose from External Exposure

External exposure is mainly from x rays (medical and dental), terrestrial γ rays (from soil and rocks), and cosmic rays. Most of the α and β particles emitted within a solid or liquid radioactive source are absorbed within the source or in the air, if they escape into the air. Similarly, for α and β radiation emitted by radionuclides in the atmosphere, only the fraction that is emitted close to a person will reach the area of that person's skin not covered with clothing.

14.14.2 Dose from Internal Exposure

14.14.2.1 Natural Radioactivity in Human Body Tissue

The "reference man" weighs 70 kg and contains 140 g of potassium (1.67×10^{-2} g ^{40}K) and between 10^{-10} and 10^{-11} g of radium. The latter is in the skeleton and may provide a dose of about 0.1 mSv/y (10 mrem/y). In soft tissue a dose of about 0.20 mSv/y (20 mrem/y) comes from the $\beta^-\gamma$ radiation of ^{40}K (1.27×10^9 y) and about 0.38 mSv/y (38 mrem/y) from the α radiation of ^{210}Po (138.38 d, in uranium series). The contribution from ^{14}C to the internal dose rate is negligible.

14.14.2.2 Radon

Estimation of the hazard of "normal" indoor domestic radon exposure is far more difficult than it is for underground miners. Except in rare instances, the level of domestic exposure is much lower and more uncertain; the number of hours of exposure per week is more variable; and the breathing rate is different. Furthermore, the incidence of lung cancer from radon progeny for miners or nonminers who are smokers is superimposed on the incidence of lung cancer attributed to cigarette smoking. Studies of the incidence of lung cancer strongly suggest that there is a synergistic interaction between radon exposure and cigarette smoking. That is, the number of cancers by radon in smokers is greater than that expected from the sum of the effects of smoking alone and radon alone. "The risk of lung cancer caused by smoking is much higher than the risk of lung cancer caused by indoor radon. Most of the radon-related deaths among smokers would not have occurred if the victims had not smoked."[68]

Factors that must be taken into account in the calculation of absorbed dose or dose equivalent include the size of the particles in the inhaled air and the fraction of the radon progeny attached to the particles, the breathing pattern of the individual, and the characteristics of the epithelial cells of the respiratory tract, which are the target cells for lung cancer.

Uncertainty in the estimation of the incidence of lung cancer that can be attributed to low-level dose of ionizing radiation from short-lived progeny of radon arises not only from uncertainty in the dose but also from the customary need to extrapolate dose–response curves for stochastic effects from higher level dose data (for underground miners, in this case). The recommended dose limit for radon in the home is based on linear extrapolation and is, therefore, subject to controversy. Doubling the exposure to radon doubles the risk of lung cancer and the Committee on Health Risks of Exposure to Radon, cited in footnote 68, states that "any exposure, even very low, to radon might pose some risk." Furthermore, mathematical models that have been developed to facilitate estimation of the risk of lung cancer from radon exposure necessarily use assumptions that also are controversial. The committee's best estimate, also found in the source cited in footnote 68, "is that among

[68]Committee on Health Risks of Exposure to Radon, *Health Effects of Exposure to Radon*, BEIR VI, National Academy Press, Washington, DC, 1999.

11,000 lung-cancer deaths each year in never-smokers, 2,100 or 2,900, depending on the model used, are radon-related lung cancers.... the 15,400 or 21,800 deaths attributed to radon in combination with cigarette-smoking and radon alone in never-smokers constitute an important public-health problem."

Radon's progeny in air are considered to be the second most important cause of lung cancer, and thus nonsmokers are not "safe." Radon and its progeny also contribute a small fraction of the total number of stomach cancers, as shown earlier in Figure 14-7.

14.14.3 Average Annual Dose

Table 14-12 contains a summary of values for the average annual effective dose equivalent H_E (Section 13.10.2) of ionizing radiation to a member of the U.S. population. Population density was taken into account in calculating the averages. Note that 82% of the annual dose is from natural sources.

It is important to understand that an individual can receive a higher annual effective dose equivalent than the average for the population given in the table. Consider a few examples:

TABLE 14-12

Average Annual Effective Dose Equivalent of Ionizing Radiation Received by a Member of the U.S. Population

Source of ionizing radiation	Effective dose equivalent
Cosmic rays	0.27 mSv (27 mrem)
Radon	2.00 mSv (200 mrem)
Terrestrial radioactivity (^{40}K, U series, Th series)	0.28 mSv (28 mrem)
Internal (mostly ^{40}K)	0.39 mSv (39 mrem)
X-ray machines (medical diagnostic)	0.39 mSv (39 mrem)
Nuclear medicine	0.14 mSv (14 mrem)
Consumer products	0.10 mSv (10 mrem)
Sleeping partner	1 μSv (0.1 mrem)[a]
Other[b]	< 0.01 mSv (<1 mrem)
Total	3.60 mSv (360 mrem)

[a]Included in "Other."
[b]Occupational, nuclear fuel cycle, fallout, and miscellaneous; each less than 0.01 mSv.
Source: Based in part on data in BEIR V, *Health Effects of Exposure to Low Levels of Ionizing Radiation*, Committee on the Biological Effects of Ionizing Radiation, National Research Council, 1990, and *Nevadan's Average Radiation Exposure*, U.S. Department of Energy, DOE/RW-0337P, 1992.

- The annual dose from cosmic radiation increases about 0.05 mSv (5 mrem) per 300-m ($\sim$ 1000-ft) increase in elevation. Thus the cosmic ray dose would about double for someone whose residence is 1.83 km (6000 ft) above sea level.

- During a flight from New York to Los Angeles the cosmic ray dose is about 1 mrem. At the altitudes reached by airplanes the radiation dose over the North pole is about twice the dose over the equator. Some European authorities have classified flight crews as radiation workers.

- In Denver for example, the terrestrial dose is about double the average.

- As discussed in Section 14.4.4, the concentration of naturally occurring radionuclides in water can vary over a wide range. Limits have been set by EPA, which also has set maximum contamination levels for "man-made radionuclides" in drinking water as follows: "The average annual concentration of beta particle and photon radioactivity from man-made radionuclides in drinking water shall not produce an annual dose equivalent to the total body or any internal organ greater than 4 mrem per year."[69] For drinking water containing tritium or strontium-90, the average annual concentrations assumed to produce a total-body or whole-organ dose of 4 mrem are 740 kBq/m^3 (20,000 pCi/L) of tritium (total-body dose) and 300 Bq/m^3 (8 pCi/L) of strontium-90 per liter (bone marrow).

- Medical x-ray examinations can add to the dose, depending on the region exposed. The dose a patient receives from x rays used for medical diagnosis and therapy is localized and is not a whole-body dose, because the procedures minimize the radiation dose to tissue outside the target tissue. The local dose to an arm, for example, would be much higher than what the whole body would receive. For an upper GI examination, not only would the local dose be relatively higher, but the fraction received by the whole body would also be higher.

- Radiopharmaceuticals used for diagnosis or therapy provide a localized dose as for x rays, but the γ rays are more penetrating and provide some dose to nearby organs of the body. In diagnostic nuclear medicine procedures, the equivalent dose to the target organ from a radiopharmaceutical (ideally containing a radionuclide that decays by IT or EC and emits low-energy γ radiation) is in the range of less than 1 mSv to 150 mSv, (0.1–15 rems) depending on the type of study and the radionuclide used. Radiopharmaceuticals are designed to provide maximum uptake by the tissue of the target organ and minimum uptake by normal tissue to minimize the effective dose equivalent for the whole body. The range of total-body, absorbed dose can be about 0.04–10 mSv (4 mrem to 1 rem) from a diagnostic radiopharmaceutical in a target organ. For ^{131}I therapy, absorbed dose to the thyroid can be 80 Gy (8 krad) or higher. Because of the need to limit the dose to tissue outside the

---

[69] *Code of Federal Regulations* Title 40, Part 141.16, Environmental Protection Agency, 2001.

target organ, especially radiosensitive cells such as those of the bone marrow, external beam radiotherapy is sometimes advantageous.

- The workplace (e.g., a nuclear power plant, an accelerator facility, a plant where radiopharmaceuticals are synthesized and packaged, a mine, a radiology facility, a nuclear medicine facility) is a potential source of added exposure to ionizing radiation.

A patient who is undergoing radiation therapy by means of either a radio-pharmaceutical or a radioactive implant (needle or pellet) will obviously raise the level of ionizing radiation in the home. Such patients must be advised on procedures to follow to minimize radiation exposure to family members. Those licensed to use the radiopharmaceutical or radioactive implant may authorize release of a patient if the total effective dose equivalent to any other individual from exposure to the released patient is not likely to exceed 5 mSv (0.5 rem).[70]

On November 13, 1789, long before x rays, radioactivity, and cosmic rays had been discovered, Benjamin Franklin wrote a letter to J. B. LeRoy in which he made the well-known statement, "In this world, nothing can be said to be certain except death and taxes." A third absolute certainty that could be added, with all due respect, is "ionizing radiation."

For a definitive 1220-page report having worldwide coverage of sources of ionizing radiation in the environment and the biological and health effects of such sources see UNCSEAR 2000, *Sources and Effects of Ionizing Radiation*, in the additional reading and sources of information at the end of this chapter.

14.15 REGULATORY LIMITS FOR RADIATION DOSE RECEIVED BY RADIATION WORKERS AND THE GENERAL POPULATION

The operator of a facility that is licensed by the Nuclear Regulatory Commission is required to make every reasonable effort to maintain the exposure and, therefore, the dose received by the workers (radiation workers) as far below each applicable dose limit [formerly called the maximum permissible dose (MPD)] as is practical consistent with the purpose of the work activity. This is known as the ALARA philosophy or concept. ALARA is the acronym for "as low as reasonably achievable." Setting the level involves balancing economic and social factors with radiation protection. An annual limit on intake (ALI) by inhalation or ingestion of radionuclides also is set for workers in licensed facilities. In addition, the operator is required to limit any radiation dose that the public might receive from the facility. Selected annual dose limits for occupational dose and dose to the public are given in Table 14-13. The limits do not include the dose from background or from medical procedures.

---

[70]*Code of Federal Regulations*, Title 10, Part 35.75, Nuclear Regulatory Commission, 2001.

TABLE 14-13

Annual Dose Limits for Individuals in the United States

Exposure classification	Dose
Occupational	
Adults	
Total effective dose equivalent	0.05 Sv (5 rems)
Sum of deep-dose equivalent and committed dose equivalent to any organ or tissue other than lens of the eye	0.5 Sv (50 rems)
Lens of the eye	0.15 Sv (15 rems)
Shallow-dose equivalent to skin or any extremity	0.50 Sv (50 rems)
Planned special exposure[a]	Not to exceed any of the foregoing limits in any year; or 5 times the foregoing limits during individual's lifetime
Minors	At a DOE facility, 1 mSv (0.1 rem) total effective dose limit in one year and 10% of the limits for adult workers
Pregnant women (embryo or fetus)	5 mSv (0.5 rem)
Nonoccupational	
Members of the public when exposed to radiation and/or radioactive material during access to a controlled area (area whose access is managed by or for DOE)	1 mSv (0.1 rem) in a year

[a]An approved exposure only in an exceptional situation when alternatives that might prevent a radiological worker from exceeding the specified limits (as given) are unavailable or impractical.

Source: *Code of Federal Regulations*, Title 10, Part 835, Energy, Department of Energy, 2001.

The operator of a facility that is licensed or has personnel who are licensed by local or state agencies for the use of sources of ionizing radiation (radionuclides, x-ray machines, accelerators, etc.) is required to follow regulations that are usually the same as the federal regulations but may have added restrictions. In any case, access to any area where the possibility of exposure to ionizing radiation exists must be restricted. Such restricted areas and the hazardous radiation sources in the areas must be identified with a warning sign or label that includes the standard symbol for ionizing radiation, shown in Figure 14-21. The symbol is accompanied by terms such as CAUTION or WARNING and then X RAYS, RADIATION AREA, RADIOACTIVE MATERIAL, or similar description.

For the population as a whole, the collective dose—that is, the total person-sievert (person-rem) dose—is an important quantity that should be kept to a minimum. The incidence of stochastic effects (e.g., cancer and genetic changes) would be expected to increase with population dose.

FIGURE 14-21   International symbol for ionizing radiation. (Background color: yellow; symbol color: magenta, purple, or black.)

The most important epidemiological study used to establish regulatory limits for the general population is that of the Hiroshima and Nagasaki survivors and their children.[71] One finding is that exposure to γ radiation produced by (n, γ) reactions of fission neutrons in the environment adds to that from radioactive fission products.

14.16 EXAMPLES OF ACCIDENTS AND INCIDENTS INVOLVING THE RELEASE OF RADIONUCLIDES INTO THE ENVIRONMENT

14.16.1 Therapy and Gauging Sources

There have been several incidents in which a radioactive source has been removed from its shield in a device designed for radiation therapy by persons who did not realize that the source was dangerous. Either the radiation warning symbol had been removed or its meaning was not understood.

For example, in November 1983 in Juárez, Mexico, a therapy device that had been stolen from a warehouse was opened in the back of the pickup truck being used to take it to a junkyard. About one-tenth of the 6010 pellets containing a total of about 17 TBq (450 Ci) of ^{60}Co (5.271 y) fell into the truck. Some of these metallic pellets were scattered along roads in the area. Most of the pellets were released in the junkyard, where they were either scattered or picked up by a magnet used to load scrap steel onto trucks. Pellets containing about 11 TBq (300 Ci) of the ^{60}Co were melted with other scrap metal. Among the contaminated steel products that were shipped to the United

---

[71]See reports of the Radiation Effects Research Foundation (RERF) and its predecessor, the Atomic Bomb Casualty Commission (ABCC).

States were office furniture and reinforcing rods. The radiation hazard was discovered by chance in January 1984, when a truck carrying contaminated steel took a wrong turn and drove near the Los Alamos National Laboratory in New Mexico, where it set off an alarm. Eventually items made with contaminated steel were traced and returned to Mexico.

Over 200 people, including children, in Juárez were exposed to γ radiation from ^{60}Co in the truck. At least one person received a "radiation burn" on one of his hands. Tests taken a few months after the incident showed that about a dozen individuals had suffered chromosome damage. The absorbed dose was estimated to be about 25 rads/h at 5 cm from a single pellet. Long-term health effects are expected to appear in the future, but the number of cases will probably be lost in "statistical noise."

Another case of contaminated steel occurred in County Cork in the Republic of Ireland in mid-1990.[72] An industrial gauge containing 3.7 GBq (0.10 Ci) of ^{137}Cs, which was being transferred from a firm in Scotland to one in England, arrived instead at an Irish steel mill in a consignment of scrap metal. The recipient of pelletized non-ferrous metals (zinc and lead) from the steel mill found it to be radioactive. Although the incident did not create any significant radiological health problems, decontamination of the steel plant was a major project. As a consequence of this incident and the one in Juárez, it is now common practice for facilities that receive scrap metal to monitor it for radioactive contamination.

A major incident that occurred in late September 1987 involved a radiation therapy device that contained about 52 TBq (1.4 kCi) of ^{137}Cs (30.07 y). It had been discarded by a cancer clinic and was dismantled in a junkyard in Goiânia, near Brasilia, Brazil. The junk dealer took the ^{137}CsCl source (about 4500 rads/h at 1 m) home, where he put it on display and scraped off small amounts to give to friends and relatives. Children played with the material (which exhibited a bluish glow in the dark), spread it about in their homes, and rubbed it on their skin. A bus and a hospital were contaminated, but two weeks passed before the presence of ^{137}Cs was discovered.

About 112,000 people out of a population of about 750,000 were checked for contamination. About 250 people had contaminated clothing or skin. Forty-four people were hospitalized, many with radiation damage to their hands. Seven of the eight people who received the largest dose (up to about 600 rads) died (four within 3 months). They were exposed to the source while it was on display in a house for five days. One of those who died was a 6-year-old girl who ingested about 0.9 GBq (about 25 mCi) of ^{137}Cs. Thirty-nine people, adults and children, were considered to have a high level of internal

---

[72]J. O'Grady, C. Hone, and F.J. Turvey, *Health Phys.*, **70**(4), 568 (1996).

contamination. Prussian blue treatment lowered the internal committed dose to levels below those from external exposure.

The ^{137}Cs was packaged in 6000 drums and buried. This radwaste disposal site has been made into a park that includes a visitor (tourist) information center, nature trails, and an auditorium. There are lessons to be learned from the social and economic impact of the accident on the residents of Goiânia.[73]

Another case of lethal exposure to a source used for radiotherapy occurred near Bangkok, Thailand, in February 2000. A cylinder containing ^{60}Co was stolen from a hospital storage area and eventually taken to a scrapyard, where the container was opened. Blood counts were measured for about 450 people living in the area. Ten people were hospitalized. Three died.

During the three years prior to the "accident" in Thailand, incidents involving uncontrolled access to large radioactive sources were reported in six other countries.

14.16.2 Nuclear Reactors

14.16.2.1 Windscale

In October 1957 a fire occurred in a plutonium production reactor at the Windscale site on the coast in the northwest of England.[74] Although the reactor was not a power reactor, it was similar to certain power reactors. The graphite-moderated, air-cooled Windscale reactor was fueled with aluminum-clad, natural uranium metal. During a routine annealing (heating) operation used to release the potential energy stored in the graphite,[75] regions of high local heating caused the graphite to burn and ignite the cladding, thereby exposing the uranium to air. Volatile fission products in the cooling air were released through a stack into the atmosphere. Fission products in particulate form were largely prevented from being released through the stack by filters. Essentially all of the core inventory of ^{85}Kr (10.76 y) [$\sim$ 59 TBq (1.6 kCi)] and ^{133}Xe (5.243 d) [$\sim$ 11.9 PBq (320 kCi)] was released to the atmosphere. About 10% of the ^{131}I (8.0207 d) [$\sim$ 590 TBq) (16 kCi)] and about 5–10% of the ^{137}Cs (30.07 y) [$\sim$ 24–48 TBq (600–1200 Ci)] were also released. Release of the less volatile ^{90}Sr (28.78 y) was about 74–222 GBq (2.0–6.0 Ci), that is, 2–5%. Also released were an estimated 6.7–10.7 TBq (180–290 Ci) of ^{210}Po (138.38 d).

---

[73]J. S. Petterson, *Nucl. News*, **31**, 84 (November 1988).

[74]The site of nuclear facilities in the area is now known as Sellafield.

[75]When the fast neutrons produced in fission are slowed down to thermal neutrons by graphite, some of the carbon atoms receive sufficient energy to leave their lattice positions and become interstitial atoms, with resultant increase in the potential energy of the graphite.

With respect to population exposure, the highest level of external exposure to γ radiation following the Windscale accident was about $1.0\,\mu Ckg^{-1}h^{-1}$ (4 mR/h) at a distance of 1 mile from the reactor. The concentration of ^{131}I (8.0207 d) in air samples varied widely, but the average for the surroundings corresponded to about 1.5 times the permissible exposure. Within hours after its release, ^{131}I began to appear in the milk from nearby dairies. Levels in milk varied from less than 2.2 kBq/liter at 80 km from Windscale to more than 19 kBq/liter (< 60 nCi/liter to > 500 nCi/liter) in isolated areas within 30 km. The maximum permissible concentration was set at 3.7 kBq/liter (100 nCi/liter) because of the potential risk of thyroid cancer for children. Milk distribution was restricted in certain areas. Even so, the maximum thyroid dose that a child might have received was estimated to be about 150 mSv (15 rems). Figure 14-22 shows the ^{131}I concentration in milk in the

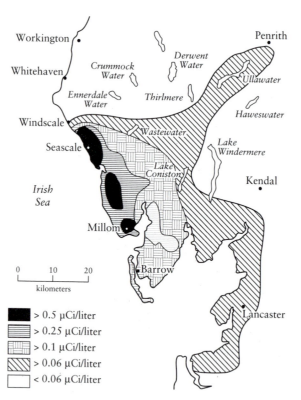

FIGURE 14-22 Geographical area surrounding Windscale showing the ^{131}I concentration in the milk from various districts 5 days after the accident in 1957. From M. Eisenbud and T. Gesell, *Environmental Radioactivity from Natural, Industrial, and Military Sources*, 4th ed. Copyright © 1997. Academic Press, San Diego, CA.

area around Windscale five days after the accident. Other fission products did not reach a sufficiently high concentration at ground level to require special action, although radioactivity released in the accident was detected outside England in Wales and in northern Europe.

14.16.2.2 Three Mile Island (TMI)

On March 28, 1979, the core of Three Mile Island Unit 2 (TMI-2), one of two units at a station located near Middletown, Pennsylvania, was partially destroyed by a loss-of-coolant accident. Figure 14-23 shows the site of the accident. The capacity of this PWR plant was 2700 MW$_t$ (906 MW$_e$).

The following sequence of events summarizes the accident.

• It began with a malfunction in the ion exchange system used to purify condensate water from the turbines before returning it as feedwater to the steam generator.

• An auxiliary feedwater system automatically turned on. Two valves in this system that should have been open were closed, preventing a supply

FIGURE 14-23 Three Mile Island nuclear power station near Middletown, Pennsylvania, after the accident at the TMI-2 plant, whose cylindrical (silo-type) reactor containment building and two cooling towers are in the foreground. From Pierce, William C., "A Nuclear Nightmare." *Time*, April 9, 1979. Used by permission of TimePix, a department of Time Inc.

of water from reaching the reactor core. The lights on the control panel indicated that the valves were closed. Unfortunately, the situation was not discovered until 8 minutes after the initiating event.

• Because heat was no longer being removed from the primary coolant loop in the steam generator, the temperature and pressure in that loop (normally about 315°C and 15 MPa) increased rapidly, causing the pressure relief valve (PRV) valve on the pressurizer to open (Figure 14-12). The pressure dropped to its normal value within about 15 seconds, and the valve should have closed. Although the indicator on the control panel for the PRV showed that it had closed, it remained open and allowed water from the pressurizer to flow onto the floor of the containment building for over 2 hours until the problem was discovered and the valve closed.

• In the meantime, the pressure in the primary loop continued to fall. The emergency core cooling system (ECCS) automatically began to inject water into the loop. Because of the reduced pressure, water in the core began to boil, forcing water into the pressurizer and leading the operators to believe that they should reduce the water level in the pressurizer by draining some of the water, reducing the flow of water into the loop, and finally turning off the emergency cooling system. The immediate result was increased generation of steam in the primary loop.

• When the main pumps that circulated water in the primary loop began to cavitate (because of the steam), the operators turned them off.

• The upper region of the core was no longer covered by water, and the temperature of the Zircaloy cladding rose above the temperature ($\sim 540°C$) at which zirconium reacts with steam to form hydrogen.[76] With increasing temperature, the gas pressure within the fuel rods increased and the cladding ruptured, releasing volatile fission products, which, together with hydrogen, entered the containment building through the open pressure relief valve. Sump pumps transferred the contaminated water to storage tanks.

• Nine and one-half hours after the initiating event, the hydrogen in the containment building ignited. The source of hydrogen included that normally injected into the primary loop to suppress the radiolysis of water plus that formed from the zirconium–steam reaction.

• Normal coolant circulation in the plant was restored about 16 hours after the initiating event.

Even though the containment building was sealed off, some radioactivity escaped into the environment by way of the ventilation systems in buildings containing equipment that was used to process the contaminated water. Between 0.11 and 0.48 EBq (3.0 and 13 MCi) of ^{133}Xe (5.243 d) ($\sim$ 2–8% of the inventory) and about 0.63 TBq (17 Ci) of ^{131}I (8.0207 d) ($\sim 2.7 \times 10^{-5}$% of

---

[76]The temperature of the core may have reached 2200°C.

the inventory) entered the environment. About 0.41 EBq (11 MCi) of ^{131}I remained in the containment building. A negligible amount of radiocesium and radiostrontium escaped from the containment building.[77]

Radiation dose levels inside the containment building were above 10 Sv/h (1 krem/h) but less than 1 mSv/h (100 mrem/h) near the perimeter of the site of the power plant. The highest dose was received by people who lived within 3 km of the plant. A few received 0.20–0.70 Sv (20–70 mrem), but the others received less than 0.20 Sv (< 20 mrem).[78]

Because there was concern about the possibility that a hydrogen explosion might occur and that such an explosion might release large quantities of fission products from the damaged core into the environment, people living in the vicinity were advised at first to stay indoors. Later, during the third day after the accident, temporary evacuation of young children and pregnant women living within 9 km of the plant was advised. Meanwhile, a large supply of potassium iodide tablets was acquired for distribution if a major release of fission products, especially ^{131}I, occurred. After the probability of an explosion was eventually judged to be negligible, distribution of the KI was deemed to be unwise because of the psychological effect it might have on the people. The level of ^{131}I was not high enough to require the use of KI.[79] Although the advisory on remaining indoors was withdrawn, those at greatest risk were advised to remain outside the evacuation area.

One of the many conclusions reached as a result of investigations following the accident at Three Mile Island was that it did not cause the release of sufficient radioactive material to create a public health threat. The only effects identified have been psychological (i.e., anxiety and fear). The accident also demonstrated the justification for requiring a containment building.

14.16.2.3 Chernobyl

The accident at the Chernobyl Unit 4 in the Soviet Union on April 26, 1986, was by far the most serious nuclear power plant accident that had occurred up

---

[77]Report to the American Physical Society of the Study Group on Radionuclide Release from Severe Accidents at Nuclear Power Plants, *Rev. Mod. Phys.*, **57**(3), Part II (July 1985). Staff Reports to the President's Commission on The Accident at Three Mile Island (the Kemeny Report), U.S Government Printing Office, Washington, DC 1979.

[78]T. M. Gerusky, Three Mile Island: Assessment of radiation exposures and environmental contamination, in The Three Mile Island nuclear accident: Lessons and implications, *Ann. N.Y. Acad. Sci.*, **365**, 54 (1981).

[79]In December 2001 the U.S. Food and Drug Administration issued the following guidelines for the use of KI to reduce the risk of radiation-induced thyroid cancer from radioisotopes of iodine in case of a nuclear accident: daily dose of 16 mg for infants less than 1 month old; 32 mg for children aged 1 month to 3 years; 65 mg for children and teenagers from 3 to 18 years old; and 130 mg for adults, including pregnant or lactating women, and adolescents over 150 lbs. [The KI should be taken 3 or 4 hours before exposure (assumes an early alert).]

to that time. It destroyed one of four graphite-moderated, water-cooled (boiling water) RBMK-1000 reactors (1000 MW$_e$) at the Chernobyl station located in a forested area in Ukraine, near the Kiev Reservoir and about 130 km north of Kiev. Unit 4 was among the second generation of 16 power reactors with similar design in the Soviet Union.

The accident occurred during an experiment designed to test the latest modifications that had been made to a turbogenerator to enable it to supply power to the coolant pumps as it coasted down after loss of steam following an emergency shutdown. Because the emergency cooling system would have been activated during the test, the operators disconnected it. They also turned off the automatic positioning system for the control rods and made an error in setting the control rods for the experiment so that the power level fell below the level considered to be safe and the reactor became unstable. In addition, a delay of over 10 hours in starting the experiment allowed a build up of the fission product ^{135}Xe in the fuel. Because the Xe-poison made it difficult to raise the power level up to that required for the experiment, the operators, in violation of safe operating procedures, withdrew more control rods, leaving fewer than needed for safe operation.

Next, the operators changed the flow of cooling water through the core to the extent that sensors in the protection systems would have responded to resulting changes in steam pressure and water levels in the plant and would have shut down the plant if the operators had not turned the systems off. Before starting the experiment, the operators made an additional adjustment of the flow of water to the core. This caused instrument readouts warning that the reactor was out of control and should be shut down. Unfortunately, it was too late.[80] The core had been damaged during the exponential increase in temperature and pressure so that the operators were unable to insert the slow-moving control rods before a steam explosion forced the rods out of the reactor, blew the cover plate off the reactor, and scattered parts of the core around the reactor building.

Shortly thereafter a second, larger explosion, believed to be caused by hydrogen and possibly carbon monoxide,[81] propelled chunks of burning graphite and fuel through the roof of the reactor building. Pieces landed on and set fire to the roof of the adjacent turbine generator building. There was partial meltdown of the upper part of the core. Clearly, there was need for a containment building such as those required for power reactors in the United States.

The fission reaction may have stopped at this point, but the decay heat from the fission products was capable of melting whatever fraction of the core

---

[80]Because of several design flaws, safety of the RMBK reactors at the time of the accident depended critically on control by the operators.

[81]Hydrogen is produced by the Zr-steam reaction and the reaction of graphite with steam, which also produces carbon monoxide. The first explosion exposed the core to air in the reactor hall.

FIGURE 14-24 Map of area within 560–965 km (350–600 mi) of Chernobyl. Adapted from "Once Again, Chernobyl Takes a Toll," by Francis X. Clines. *The New York Times*, September 30, 1989. First Section, p. 4. International and Noviye Gronyki Journal, used by permission of *The New York Times*.

remained. Unfortunately the fission product inventory in the reactor was high because three-fourths of the fuel had been in the core since the start-up about 27 months earlier.

The second explosion released about 0.74 EBq (20 MCi) of radionuclides, about half of which was in the form of core debris (perhaps 20% of the core) that fell to the ground within about a 3-km radius of the plant. Volatile fission products were carried into the atmosphere in a plume of smoke and hot gas, which drifted in a north-northwesterly direction (rather than the normal easterly direction) toward Pripyat, a city of about 45,000 inhabitants located 3 km from the Chernobyl station (Figure 14-24). In the reactor building, dose rates ranged from 1.0 to 5 Gy/h.

On the day after the explosions the graphite was still burning and the fuel was still releasing fission products in a rising column of hot gas. After efforts to extinguish the fire with water failed, sand, clay, boron compounds, lead, and dolomite were dropped onto or close to the core from a helicopter over a period of six days.[82]

---

[82]Some of the lead vaporized and contaminated the area around the plant with nonradioactive lead.

By May 1 the rate of release of radioactivity had dropped from about 0.40 EBq (11 MCi) on the day of the explosions to about 74 PBq/d (2 MCi/d).[83] Although the assorted materials that were dropped into the reactor building did not prevent the escape of fission products, they did provide some thermal insulation so that the temperature of the remains of the core increased to about 2500°C, causing additional meltdown of the core.[84] This caused a rise in the rate of release of fission products to about 0.30 EBq/d (8 MCi/d) on May 5.

The graphite fire was largely quenched shortly thereafter by the injection of liquid nitrogen, and by June 6 the release of radioactivity decreased to less than an estimated 7.4 TBq/d (200 Ci/d). In October the reactor was temporarily entombed in a concrete enclosure officially called a shelter, but dubbed "the sarcophagus" by the news media. The shelter was expected to have a life of approximately 25 years.[85]

The following are examples of the extent of contamination of the environment and radiation exposure to the population.

- Based on an early estimate, 1.9 EBq (50 MCi) of fission products [in addition to an estimated 1.9 EBq (50 MCi) of inert gas fission products] was released,[86] mostly during the 10 days following the explosions, not during the explosions. Revised estimates for the two values are 8.0 EBq (216 MCi) and 6.5 EBq (175 MCi), respectively.[87] The smoke and hot gases were carried away from the plant by winds that shifted several times, eventually going full circle. At various times the wind carried the airborne radioactivity into rainy weather so that rainout created areas of high radioactivity (hot spots) on the ground.

- By May 7 essentially all the core inventory of the inert gases (which were released first) had been released: 30–50% of the isotopes of the volatile elements, mostly ^{131}I (8.0207 d) with lesser amounts of ^{134}Cs (2.065 y) and ^{137}Cs (30.07 y) and perhaps 20% or less of ^{89}Sr (50.52 d), ^{90}Sr (28.78 y), ^{141}Ce (32.50 d), ^{144}Ce (284.6 d), and the isotopes of plutonium.

- The highest level of contamination [0.56–1.48 TBq/km^{-2} (15–40 Ci/km^2)] was within a 30-km radius of the plant (Figure 14-25). Evacuation of the inhabitants from the area, which became an exclusion zone, began 36 hours

[83]Unless stated otherwise, the estimated release rates do not include radioisotopes of inert gases and may be in error by ±50%.

[84]A few years later the additional meltdown of the core was observed to have produced a solidified lava like material, "coronium."

[85]Cracks in the roof of the enclosure were found less than 10 years after the accident. The roof was reinforced with steel beams in 1999. Some 27 countries are financing further stabilization and addition of a confinement cover.

[86]Report on the Accident at the Chernobyl Nuclear Power Station, U.S. Nuclear Regulatory Commission, NUREG-1250, Rev. 1, December 1987.

[87]R. F. Mould, *Chernobyl Record*, Institute of Physics Publishing, Bristol, UK, 2000.

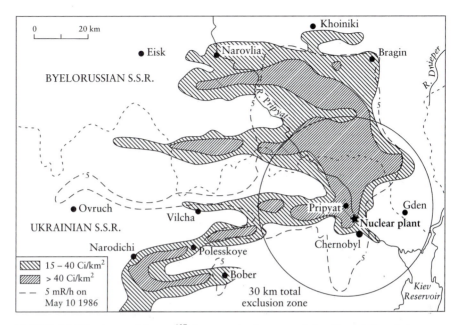

FIGURE 14-25 Areas of heavy ^{137}Cs contamination around the Chernobyl exclusion zone (circle) as measured during 1988. The contour marked by isolines indicates the territory where the exposure rate for γ radiation was above 5 mR/h on May 10, 1986. From Z. A. Medvedev, *The Legacy of Chernobyl*, "map p. 87." Copyright © 1990 by Zhores Medvedev. Used by permission of W.W. Norton & Company, Inc.

after the explosions. The zone contained Pripyat, many villages and small settlements, and Chernobyl, a town of about 10,000 inhabitants, located about 18 km southeast of the plant.

• About a month after the accident, some soil samples taken in the exclusion zone were found to have 18.9 TBq/km^2 (510 Ci/km^2) of ^{131}I and 10.0 TBq/km^2 (270 Ci/km^2) of ^{137}Cs. Plutonium contamination was as high as 0.37 TBq/km^2 (10 Ci/km^2) near the reactor but did not extend very far except for a few hot spots. The total (worldwide) deposition of ^{137}Cs was estimated to be about 100 PBq (2.7 MCi).[88]

• About 0.25 million square kilometers of land in an area lying in Belarus, Russia, and Ukraine remained contaminated with ^{137}Cs at a level of about 37 GBq/km^2 (1 Ci/km^2) some 10 years after the accident.

• Two of the workers in the Chernobyl nuclear power plant at the time of the accident on April 26, 1986, died at once—one from steam burns and one from falling debris. Two other workers who were sent into the reactor

[88]L. R. Anspaugh, R. J. Catlin, and M. Goldman, *Science*, **242**, 1513 (1988).

building to investigate the accident died of radiation exposure soon after the accident, as did several of the firemen who received total-body exposure rates over $0.258\,C\,kg^{-1}\,h^{-1}$ (1000 R/h) as they tried to extinguish the burning graphite. Between 197 and 300 of the people who were on the scene at the time of the explosions or soon thereafter as emergency personnel were hospitalized. They received extensive thermal burns and high doses of radiation from external exposure [e.g., up to 16 Sv (1600 rems)] and internal exposure [e.g., 4 Sv (400 rems)] from inhalation of radioactive dust and smoke. Many became incapacitated within a few hours. At least 10 of the patients were given bone marrow transplants; others received liver tissue transplants.

• In the town of Pripyat, the exposure rate on the streets was reported to have reached a level as high as about $2.58 \times 10^{-4}\,C\,kg^{-1}h^{-1}$ (1 R/h).

• By July 19, 1986, within days or weeks of the accident, 28 of 35 people who received between 4 Gy (400 rads) and 16 Gy (1.6 krads) had died. In October 1993 the reported death total was 31.

• Over 600,000–800,000 people were exposed sufficiently, either externally or internally, to be registered in a study program set up to follow their health until they die. About 200,000 of these were "liquidators" (workers who participated in the cleanup after the accident). The liquidators handled radioactive debris (including graphite from the reactor core) on the plant site or were involved in the construction of the sarcophagus in 1986–1987. Some worked for only 90 seconds. The average dose was reported to be 0.25 Sv (25 rems) during the first year.

• Until the residents of Pripyat and about 90,000 other residents of the zone were evacuated from the exclusion zone, they carried out normal outdoor activities (cultivating gardens, etc.), unaware or unafraid of the fallout. Estimates of external doses for individuals living in Pripyat are as high as 0.50 Gy (50 rads) from γ radiation and 0.20 Gy (20 rads) from β radiation plus internal doses up to 0.25 Gy (25 rads) to the thyroid.

• In Ukraine, ^{137}Cs contaminated about 6.5 million hectares (about 25,000 mi^2) of forests and farmland.

• Potassium iodide as tablets or as an iodide solution was distributed in Pripyat and nearby areas, in eastern Poland, and in Yugoslavia to decrease radioiodine uptake.[89]

• Because the variable winds at Chernobyl carried the fission products, especially ^{131}I and ^{137}Cs, across both eastern and western Europe, some of the milk from dairies in Austria, Hungary, Poland, and Sweden was contaminated (Figure 14-26). Fresh fruit and vegetables, fresh meat, freshwater fish, milk, and milk products from countries in eastern Europe[90] within

---

[89]There is some evidence that high doses of KI may be harmful.
[90]Bulgaria, Czechoslovakia, Hungary, Poland, Romania, the Soviet Union, and Yugoslavia.

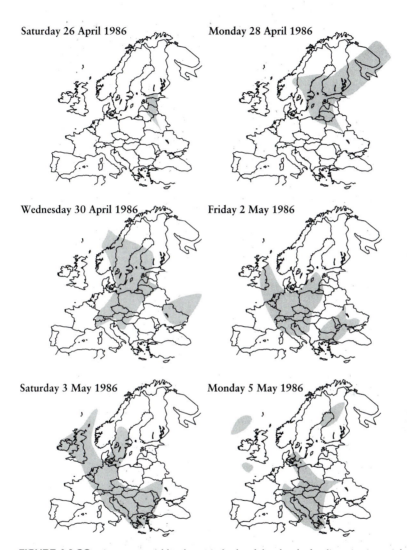

Saturday 26 April 1986   Monday 28 April 1986

Wednesday 30 April 1986   Friday 2 May 1986

Saturday 3 May 1986   Monday 5 May 1986

FIGURE 14-26 Areas covered by the main body of the cloud of radioactive material on various days during the release from the Chernobyl accident. From *The Radiological Impact of the Chernobyl Accident in OECD Countries*, Nuclear Energy Agency (NEA), Organization for Economic Co-operation and Development, Paris, 1987.

1000 km of Chernobyl were banned for several weeks by the 12-nation European Economic Community.[91] Vegetables grown in parts of France, Greece, and Italy that spring could not be consumed locally or exported as would normally have occurred. Later it was found that in the northern parts of Finland, Norway, and Sweden, where the Lapps raise reindeer for their livelihood, samples of the reindeer meat contained over 8.0 kBq/kg (216nCi/kg) of radiocesium [^{134}Cs (2.065 y) and ^{137}Cs (30.07 y)]. Maximum values for ^{137}Cs were about 150,000 Bq/kg for reindeer meat and about 40,000 Bq/kg for lamb. In Norway, bentonite was supplied to farmers as a Cs binder that could be added to animal feed to lower the ^{137}Cs content of meat and milk. Prussian blue incorporated in salt licks was also used in mountain areas where reindeer, sheep, and goats grazed. Both methods were effective (i.e., a 50–90% reduction in radiocesium contamination was achieved).

• In areas of northwest England, north Wales, Northern Ireland, and Scotland, where it was raining when the cloud of radioactivity arrived about a week after the accident, the increase in dose level was about the same as normal background. A few weeks later the cumulative level of ^{131}I, ^{103}Ru (39.27 d), ^{134}Cs, and ^{137}Cs in sheep exceeded the maximum permissible level of 1.0 kBq/kg (27 nCi/kg), and the animals were banned as a source of lamb and mutton. Sheep from over 500 farms were still banned some 18 months after the accident.

• Dose levels in the then Federal Republic of Germany were probably the highest in western Europe. Again, there were hot spots from rainout in the southern part of the country. Iodine-131 and ^{137}Cs contamination forced the destruction of leafy vegetables in the areas of highest contamination.

• In northern Italy the level of contamination from the same two radionuclides required restrictions on the consumption of milk and vegetables. In France there was normal fallout, requiring restrictions on the use of food until the ^{131}I level decreased by decay. Fallout was also detected on crops in Greece and Turkey.

• Eventually, radionuclides from the accident were detected in the atmosphere over Canada, Japan, and the United States. Evidence for their arrival in Troy, NY is given in Figure 14-27. The spectra show the γ-ray activity of particulates in air samples taken over 24-hour periods by standard air-sampling methods in which the particulates were captured on fiberglass filters on a rooftop at Rensselaer Polytechnic Institute. The spectra were obtained with a sodium iodide scintillation detector and a γ-ray spectrometer set to divide γ-ray energies up to 2 Mev into 200 channels (10 keV/channel).

[91]After the ban was lifted at the end of May, imports were still monitored during most of 1987. Intervention limits for ^{137}Cs were set at 600 Bq/kg for foodstuffs in general and 370 Bq/liter for milk and baby foods.

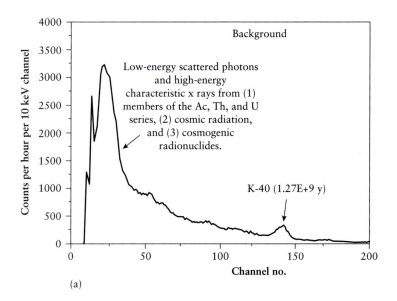

(a)

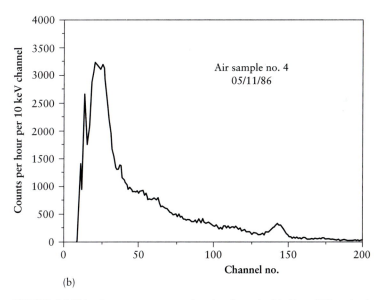

(b)

FIGURE 14-27 Gamma-ray spectra showing the arrival in Troy, NY, and subsequent decay of fission products from the Chernobyl accident on April 26, 1986. Dates given are when the spectrum was measured. Data provided by H. M. Clark.

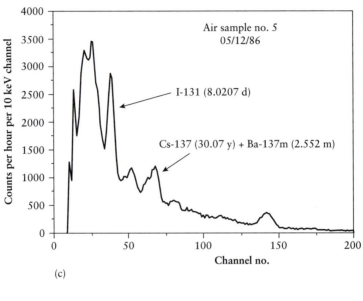

(c)

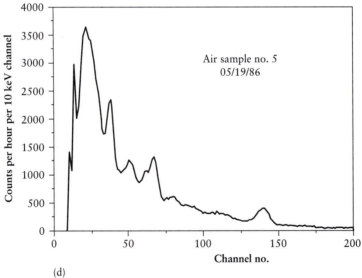

(d)

FIGURE 14-27   *(Continued)*

• Spectrum (a) is the background spectrum for the detector, while (b) is that of an air sample taken May 7–8, 1986, before arrival of fission products from Chernobyl. The spectrum (c), for a sample collected May 9–10, shows peaks

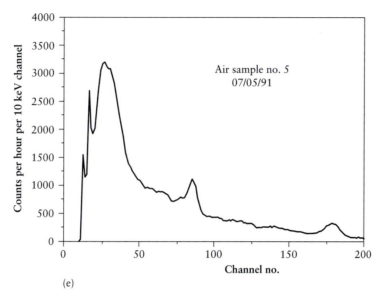

(e)

FIGURE 14-27 *(Continued)*

for ^{131}I (8.020 d) and ^{137}Cs (30.07 y). When the same sample was counted 7 days later, spectrum (d) was obtained. Spectrum (e) for the same sample a little over 5 years later shows the continuing presence of ^{137}Cs in secular equilibrium with its daughter ^{137m}Ba (2.552 m). Fission product activity in air samples became undetectable by May 21.

• By the end of 1995, the number of cases of thyroid cancer among children in Ukraine that were reported after the accident was about 600 rather than an estimated 50 based on the preaccident rate.

• Thyroid doses for 57% of the children who were under one year of age and were living in the exclusion zone were over 2 Gy.

• In the heavily contaminated Gomel region of Belarus, an increase in thyroid cancer (presumably from ^{131}I)[92] was observed 4 years after the accident instead of the expected 8–10 years. In 1991 the rate was almost 20 times that of 2 years earlier, and by 1996 the rate was about 200 times the normal rate. The increase in incidence of thyroid cancer in children up to age 15 is the long-term health effect that is accepted as being linked to the Chernobyl

---

[92]Although it is generally believed that ^{131}I causes thyroid cancer, proof may be found in studies of children in the Chernobyl area. Other shorter-lived iodine isotopes, ^{132m}I (1.39 h), ^{132}I (2.28 h), ^{133}I (20.8 h), and ^{135}I (6.57 h) that were released in the Chernobyl accident are also suspect.

accident and is beyond the increase that might be expected from intensified levels of examination.[93]

- UNSCEAR 2000[94] reports that there have been about 1800 cases of thyroid cancer in children who were exposed to fission products (radioisotopes of iodine) at the time of the accident. Additional cases are expected during the next decades.

- Initially, the important source of ionizing radiation to the population was ^{131}I via inhalation or contaminated milk. Later, the hazard shifted to ^{137}Cs.

- A pine forest (500–600 hectares) in the exclusion zone died, as did some of the plants and wildlife.[95] Plants and wildlife showed little permanent radiation damage and recovered within about 3 years.

- The number of cases of leukemia found up to 10 years after the accident seemed to be less than expected (based on data for Japanese bomb survivors)[96] for the exposed population as a whole. UNSCEAR reported that the expected elevation in cases of leukemia was not found 14 years after the accident.

- Monitoring of ^{131}I and ^{137}Cs provided data indicating that existing models for the movement of these two fission products from the atmosphere into the food chain and for ground migration need to be improved.

- The worldwide annual average effective dose from the Chernobyl accident 14 years afterward is estimated to be 0.002 mSv.[94]

- The health and environmental effects of the Chernobyl accident are being studied by a number of international organizations. An International Chernobyl Center (ICC) has been established at the site itself and in the exclusion zone to study radioecology, nuclear safety, and radioactive waste handling. Epidemiological studies are expected to add to our understanding of several health effects, including dose–response relationships associated with ionizing radiation (e.g., thyroid cancer and leukemia). However, uncertainty in dose will be a limiting factor, but less so than for the Japanese survivors.

14.16.3 Radioactive Waste

An accident that released significant amounts of radioactive material into the environment occurred in September 1957 at the Chelyabinsk-65 facilities

---

[93]The increase in thyroid cancer was observed sooner than expected (latency of about 10 years). Some critics believe the observed number of cases is too high and is the result of more intensive screening since the accident.

[94]United Nations Scientific Committee on the Effects of Atomic Radiation (UNSCEAR), UNSCEAR 2000 Report to the General Assembly, United Nations, NY, 2000.

[95]Conifers are particularly sensitive to ionizing radiation. These trees received 80–100 Gy. The forest remains contaminated with ^{90}Sr and ^{137}Cs. Trees receiving up to about 10 Gy recovered.

[96]See reports of the Radiation Effects Research Foundation (RERF), which monitors the health of some 120,000 people who are survivors of the Hiroshima and Nagasaki bombings.

located near Kyshtym in the southern Urals in what was then the Soviet Union. The event is variously known as the Urals accident, the Kyshtym disaster, and the Chelyabinsk tragedy. Details of the accident did not become available until 1989. The Chelyabinsk-65 site contained storage and reprocessing facilities for spent reactor fuel, fuel fabrication plants, and facilities for storage and treatment of radioactive waste as part of a complex initially dedicated to the Soviet weapons program. Liquid waste containing fission products as a mixture of nitrates and acetates was stored in several large tanks, which were continuously cooled to remove the fission product heat. Because of problems with the cooling system of one tank, the waste was allowed to evaporate to dryness. The ^{137}Cs but not the ^{90}Sr had been removed from the waste. Eventually the temperature of the solid reached about 500°C and the mixture exploded with a force estimated to be that from up to 100 tons of TNT.

Most of the approximately 740 PBq (20 MCi) of radioactivity ejected from the tank fell to earth near the site of the explosion. About 74 PBq (2 MCi) moved away from the site as a cloud and contaminated over 200 towns and villages having a total population of about 0.3 million people. In the contaminated areas the ^{90}Sr activity varied from 3.7 GBq/km^2 to 18.5 TBq/km^2 (0.1–500 Ci/km^2).

Over 10,000 people were evacuated from the more highly contaminated regions during a period beginning a week after the accident and continuing for several months. Prior to evacuation, many of the people cultivated their gardens as usual and ate the crops. Estimated doses received were up to 1 Gy (100 rads) for individuals at the accident site and up to 3 Gy (300 rads) for people who accumulated their dose in the most contaminated villages during the first month after the accident. The average dose of about 10,000 evacuees was about 50 rems.

14.16.4 Nuclear Fuel Processing Plant

On September 30, 1999, a criticality accident occurred in a nuclear fuel processing plant in Tokaimura, Japan. Workers in facilities where fissionable materials are handled are given special training to prevent the materials from achieving critical mass.

In the Tokaimura accident, the workers used short-cut procedures instead of the licensed, safe procedures in preparing and handling a uranyl nitrate solution containing enriched (18.8%) uranium. Although the uranium oxide was dissolved in buckets instead of the proper vessel, the critical mass needed for the chain reaction was not reached until the workers had transferred 15.8 kg of uranium (seven times the normal amount of uranium) into a tank that had a relatively large diameter and a jacket to allow circulation of cooling water. The water in the solution served as a neutron moderator and that in the jacket served as a neutron reflector, which scattered neutrons that would have escaped from the container back into the solution and reduced the critical mass requirement. The solution became a nuclear reactor. Normally, the heat generated by the fissioning uranium would have caused the volume of the solution to increase, reducing the effectiveness of water as a moderator and reducing the fission rate. Unfortunately, but predictably, the cooling system enabled the reaction to continue for about 17 hours until the cooling water was drained and boron was added to the solution as boric acid.

At the moment when the chain reaction began, the solution became an intense source of neutrons and γ rays. Volatile fission product radioisotopes of iodine, krypton, and xenon escaped into the room and the ventilating system. Some of the stable ^{15}N in the air would have been converted to ^{16}N (7.13 s) by neutron capture.

There were three workers in the room at the time of the accident. One who received an estimated dose equivalent of about 17 Sv died 82 days later. The second, who received an estimated dose equivalent of about 10 Sv, died 207 days later. The third received an estimated dose equivalent of 1–5 Sv and was discharged from the hospital after about 90 days. Other workers (about 148) in the plant received dose equivalents from 1 to 50 mSv. Some 300,000 residents in the area near the plant were warned to stay indoors overnight.

This was by no means the first criticality accident. Several resulting in fatalities to workers occurred in the United States during a period of about 20 years beginning with the Manhattan Project in the 1940s. A criticality accident in 1970 at the Kurchatov Institute of Atomic Energy in Moscow resulted in the death of two technicians and very high-level exposure to two others at an experimental critical assembly.

Additional Reading and Sources of Information

General

Choppin, G. R., J.-O. Liljenzin, and J. Rydberg, *Radiochemistry and Nuclear Chemistry* (2nd edition of Nuclear and Radiochemistry) Butterworth-Heinemann, London, 1995.

Eisenbud, M., and T. Gesell, *Environmental Radioactivity from Natural, Industrial, and Military Sources*, 4th ed. Academic Press, San Diego, CA, 1997.

Various authors, *Reports* of the (1) Committee on the Biological Effects of Ionizing Radiation, National Research Council, National Academy of Sciences (BEIR reports), (2) Energy Information Administration (EIA), U.S. Department of Energy, (4) International Atomic Energy Agency (IAEA), (4) International Commission on Radiological Protection (ICRP), (5) National Academy of Science-National Research Council (NAS-NRC), (6) National Council on Radiation Protection and Measurements (NCRP), (7) Organization for Economic Cooperation and Development (OECD), (8) Radiation Effects Research Foundation (RERF), (9) United Nations Scientific Committee on the Effects of Atomic Radiation (UNSCEAR), (10) U.S. Department of

Energy (DOE), (11) U.S. Environmental Protection Agency (EPA) and (12) U.S. Nuclear Regulatory Commission (NRC).

Code of Federal Regulations pertaining to this chapter: Title 10 (Energy), Title 21 (Food and Drugs), Title 29 [Labor (OSHA)], Title 40 (Protection of the Environment), and Title 49 (Transportation).

Internet websites. Information on topics covered in this chapter is available from websites such as the following: www.nap.edu (National Academy of Science Press), www.doe.gov (U.S. Department of Energy), www.nrc.gov (U.S. Regulatory Commission), www.epa.gov (U.S. Environmental Protection Agency), www.gpo.gov (U.S. Government Printing Office), and www.nsrb.org (National Radon Safety Review Board).

Radon

A Citizen's Guide to Radon, 3rd ed. U.S. Environmental Protection Agency, Document 402-K-92-001 (9/92), U.S. EPA, National Center for Environmental Publications (NSCEP), P.O. Box 42419, Cincinnati, OH. (One of many documents available.)

Commission on Life Sciences, Committee on Risk Assessment of Exposure to Radon in Drinking Water, *Risk Assessment of Radon in Drinking Water*, National Research Council, National Academy Press, Washington, DC, 1999.

Cothern, R. C., and P. A. Rebers, eds., *Radon, Radium and Uranium in Drinking Water*, Lewis Publishers, Chelsea, MI, 1990.

Applications of Isotopes

Attendorn, H.-G., and R. N. C. Bowen, *Radioactive and Stable Isotope Geology*, Chapman & Hall, London, 1997.

Beck, J. W., *et al.*, Extremely Large Variations of Atmospheric ^{14}C Concentration During the Last Glacial Period, *Science*, **292**, 2453 (2001).

Dickin, A. P., *Radiogenic Isotope Geology*, Cambridge Univ. Press, Cambridge, U.K., 1995.

Geyh, M. A., and H. Schleicher, *Absolute Age Determination*, Springer-Verlag, Berlin, 1990.

Taylor, R. E., Fifty Years of Radiation Dating, *Am. Sci.*, **88** (1), 60 (2000).

Journals that publish research on isotope fractionation and isotope dating include *Applied Geochemistry, Chemical Geology, Earth and Planetary Letters, Environmental Science and Technology, Geochimica et Cosmochimica Acta, Geology, Journal of Climate, Journal of Geophysical Research, Journal of Paleoliminology, Nature, Paleoceanography, Paleogeography Paleoclimatology and Paleoecology, Phytochemistry, Pure and Applied Geophysics, Radiocarbon, Science.*

Nuclear Medicine

Wagner, H. N., Jr., Z. Szabo, and J. W. Buchanan, eds., *Principles of Nuclear Medicine*, 2nd ed., W. B. Saunders, Philadelphia, 1995.

Journal of Nuclear Medicine (Society of Nuclear Medicine)

Nuclear Reactors and Nuclear Power Plants

Bodansky, D., *Nuclear Energy: Principles and Practices*, American Institute of Physics Press, New York, 1996.

Cochran R. G., and N. Tsoulfanidis, *The Nuclear Fuel Cycle: Analysis and Management*, 2nd ed., American Nuclear Society, La Grange Park, IL, 1999.

Cowan, G. A., A Natural Fission Reactor, *Sci. Am.*, **235**, 36, (1976).

Glasstone, S., and W. H. Jordan, *Nuclear Power and Its Environmental Effects*, American Nuclear Society, La Grange Park, Il, 1980.

Glasstone, S., and A. Sesonske, *Nuclear Reactor Engineering*, Vols. 1 and 2, 4th ed., Chapman & Hall, New York, 1994.

Murray, R. L., *Nuclear Energy: An Introduction to the Concepts, Systems, and Applications of Nuclear Processes*, 5th ed. Butterworth-Heinemann, Boston, MA, 2001.

Quinn, E. L., New Nuclear Generation—In Our Lifetime, *Nuclear News*, **44**, 52–59 (October 2001).

The New Reactors, *Nuclear News*, **35**, 66–90 (September 1992).

Report to Congress on *Small Modular Nuclear Reactors*, Office of Nuclear Energy, Science and Technology, U. S. Department of Energy, Washington, DC, May 2001.

U. S. Nuclear Regulatory Commission, *Reactor Safety Study, An Assessment of Accident Risks in the U. S. Commercial Nuclear Power Plants*, WASH-1400 (NUREG 75/014), 1975.

Cold Fusion

Close, F., *Too Hot to Handle: The Race for Cold Fusion*, Princeton Univ. Press, Princeton, New Jersey, 1991.

Taubes, G., *Bad Science: The Short Life and Weird Times of Cold Fusion*, Random House, New York, 1993.

Nuclear Weapons and Depleted Uranium

Carter, L. J., and T. H. Pigford, "Confronting the Paradox in Plutonium Policies", *Issues* **XVI (2)**, 29, (Winter 1999–2000). Also letters to the editor on this paper *Issues* **XVI (3)** (Spring 2000), **XVI (4)** (Summer 2000).

Depleted Uranium in Kosovo: Post-Conflict Environmental Assessment, United Nations Environmental Programme (UNEP), United Nations, New York, 2001.

Jeanloz, R., Science-Based Stockpile Stewardship, *Phys. Today*, **53**, 44 (2000).

Office of Technical Assessment, Congress of the United States, *Nuclear Proliferation and Safeguards*, Praeger, New York, 1977.

Sullivan, J. D., The Comprehensive Test Ban Treaty, *Phys. Today*, **51**, 24–29 (1998).

Radioactive Waste

Committee on Separations Technology and Transmutation Systems, *Nuclear Wastes: Technologies for Separations and Transmutation*, National Academy Press, Washington, DC, 1996.

Macfarlane, A., Interim Storage of Spent Fuel in the United States, *Annu. Rev. Energy Environ.*, **26**, 201–235 (2001).

Murray, R. L., *Understanding Radioactive Waste*, 4th ed., Battelle Press, Columbus, OH, 1994.

Radioactive Waste, *Phys. Today*, **50**, 22–62 (1997).

Wiltshire, S. D., *The Nuclear Waste Primer*, rev. ed., League of Women Voters Educational Fund, Washington, DC, 1993.

Sources and Effects of Ionizing Radiation

Committee on the Biological Effects of Ionizing Radiation (BEIR), *The Health Risks of Radon and Other Internally Deposited Alpha-Emitters*, BEIR IV, 1998; *Comparative Dosimetry of Radon* in Mines and Homes (companion to BEIR IV), 1991; Health Effects of Exposure to Low Levels of Ionizing Radiation, BEIR V, 1990; Health Effects of Exposure to Radon, BEIR VI, 1999, National Research Council, National Academy Press, Washington, DC.

Hendee, W. R., and F. M. Edwards, eds., Health Effects of Exposure to Low-Level Ionizing Radiation, Institute of Physics, London, 1996.

Moeller, D. W., Environmental Health, Harvard University Press, Cambridge, MA, 1992.

Peterson, L. E., and S. Abrahamson, eds., Effects of Ionizing Radiation: Atomic Bomb Survivors and Their Children (1945–1995), Joseph Henry Press, Washington, DC, 1998.

UNSCEAR (United Nations Scientific Committee on the Effects of Atomic Radiation), Reports to the General Assembly of the United Nations, United Nations, NY:

UNSCEAR 1988 (includes Chernobyl)

UNSCEAR 1993

UNSCEAR 2000 Vol. I: Sources, Vol. II: Effects (Both volumes include Chernobyl update)

Chernobyl

Kryshev, I. I., ed., Radioecological Consequences of the Chernobyl Accident, Nuclear Society International, Moscow, 1992.

Medvedev, Z., The Legacy of Chernobyl, Norton, New York, 1992.

Mould, R. F., Chernobyl Record, Institute of Physics, London, 2000. (Contains updated data on levels of radioactivity and effects on the population.)

Vargo, G. J., ed., The Chernobyl Accident: A Comprehensive Risk Assessment, Battell Press, Columbus, OH, 2000. (Contains details about the accident and its effects written by a group of Ukrainian scientists.)

EXERCISES

14.1. What is the heat output (watts) associated with the α decay of a sample containing 1 kg of (a) ^{235}U $(t_{1/2} = 7.04 \times 10^8 \, \text{y})$, $E_\alpha = 4.4$ MeV; (b) ^{239}Pu $(t_{1/2} = 2.410 \times 10^4 \, \text{y})$, E_α (ave) $= 5.13$ MeV; (c) ^{238}Pu $(t_{1/2} = 87.7 \, \text{y})$, $E_\alpha = 5.5$ MeV?

14.2. Calculate $\dot{D}$ (Gy/h) in 100 g of soft tissue containing 2.1×10^6 Bq of (a) ^{14}C, (b) ^{3}H. [See Equation (13.68).]

14.3. Show that the theoretical separation factor for enriching ^{235}U in natural uranium by gaseous diffusion is 1.0043.

14.4. How many kilograms of D_2O would be required per year to replace the D_2O consumed by the (d,d) reaction in a thermonuclear power plant with an output of 4000 MWt?

14.5. How many grams of ^{3}H would be "burned" if the D-T reaction is the source of energy for the power plant in Exercise 14.4?

14.6. Show that the power density of a sample of ^{90}Sr that is a month old is 0.93 W/g. (See Figure 13.13.)

14.7. Look up the composition of granites, choose one and calculate the heat output in watts from the ^{40}K for a metric ton of the granite. The abundance of ^{40}K is 0.0117 at. % and the decay scheme is given in Figure 13-15.

14.8. Look up and summarize any changes that have been made during the past 12 months in the *Code of Federal Regulations* (CFR) for DOE, NRC, EPA, and FDA that apply to the topics covered in this chapter.

14.9. In a poll taken in 1992, it was found that 36% of the U.S. population believed that milk containing radioactive contaminants could be made safe by boiling it. Comment on the validity of the belief. How do you account for so considerable a percentage of the population having such a belief?

14.10. At a time prior to 1952, when the specific activity of ^{14}C in new wood was 15.3 disintegrations per minute per gram of carbon, how old was a sample of wood having a specific activity of 9.7 dpm?

14.11. If the forest vegetation and soils on the earth contain approximately 1.14×10^{18} g of carbon, what is the corresponding ^{14}C activity in on becquerels and curies? Assume the specific activity in given in Exercise 14.10.

14.12. If the total global inventory of ^{14}C is about 8.3×10^9 GBq and the rate of production in the atmosphere is about 9.2×10^5 GBq per year, is the inventory increasing or decreasing? At what rate?

14.13. If the standard 70-kg man contains 12.6 kg of carbon and the specific activity given in Exercise 14.10 is assumed, how many β particles are emitted in his body per year?

14.14. Boiled codfish contains about 200 mg of potassium per 100 g of fish. If you eat 200 g of codfish, how many becquerels and curies of ^{40}K will you ingest? (See Exercise 14.7.)

14.15. Explain how fission products could appear in the steam turbines of a PWR power plant.

14.16. A sample containing ^{235}U is irradiated with slow neutrons for one year. After a decay period of 10^9 seconds, it is found that 99.6% of the decay heat is produced by four fission products, Which are they? Why are ^{99}Tc and ^{129}I not among the four?

14.17. If the fuel in an LWR contains 7.4 PBq (0.20 MCi) of ^{131}I per metric ton of heavy metal after an irradiation of 200 days, what will be the activity of ^{131}I be if the irradiation is continued for another 50 days? If the fuel that has been irradiated for 200 days is "cooled" for 150 days, what will be the ^{131}I activity?

14.18. If a mixture of spent fuels stored in a repository contains 100 g each of ^{90}Sr (28.78 y), ^{137}Cs (30.07 y), ^{239}Pu (2.410×10^4 y), and ^{241}Am (432.7 y), what will be the activity of each in becquerels and curies after 10,000 years?

14.19. Assume that over a period of time the total quantity of ^{99m}Tc (6.01 h) given to many patients is 370 PBq (10 MCi). In becquerels and curies, how much ^{99}Tc (2.13×10^5 y) is added to the environment?

14.20. It is found that δ^{18}O values remain fairly constant as measurements are made for samples taken deeper in an ice core until a certain depth is reached, at which point the value begins to drop until it reaches a new average value about 15% lower. What does this reveal about the environment?

14.21. Much has been learned about salt domes from centuries of salt mining. Describe typical salt domes in some detail. What are the desirable and undesirable properties of salt domes as receptacles for storage of radioactive waste?

14.22. What is the ratio of ^{235}U/^{238}U today? Calculate what the ratio would have been 2 billion years ago when the all-natural Oklo reactors were operating. What is it for fresh fuel in an LWR? Why was the presence of water important at the Oklo sites?

14.23. Is the weight of fission products in spent LWR fuel greater or less than the weight of ^{235}U fissioned? Explain.

14.24. Prepare an essay, a class presentation, or a term paper on one of the following.
   (a) The methods currently being funded in the United States, in Europe, and in Asia to develop thermonuclear energy.
   (b) Current concepts of thermonuclear power reactors in the future.
   (c) The design features of a nuclear power plant under construction or placed in operation in any country during the past 12 months that qualify it as having an "advanced" design.
   (d) Latest values for concentration factors for plants, animals, fish, etc. Comment on the ranges of values for a given species.
   (e) The latest radiopharmaceuticals available for diagnosis.
   (f) The latest radiopharmaceuticals available for therapy.
   (g) Your conviction about food irradiation. Give reasons.
   (h) The types and locations (worldwide) of nuclear power plants under construction.
   (i) The status of accelerator disposal of waste from spent fuel from nuclear power reactors.
   (j) Current thinking about the health hazard posed by ^{222}Rn.
   (k) Recent examples of studies using isotope fractionation.
   (l) Recent examples of age determination that are relevant to the environment.
   (m) Recent accidents involving high-level sources (e.g., industrial or therapy).

(n) Recent accidents involving nuclear reactors of all types.

(o) Recent changes in federal and/or state regulatory limits for dose of ionizing radiation.

(p) Recent changes in the half-life values for the radionuclides ^{3}H, ^{7}Be, ^{14}C, ^{40}K, ^{90}Sr, ^{137}Cs, and ^{239}Pu.

14.25. Prepare an essay, a class presentation, or a term paper on the current status of one of the following.

(a) Oklo reactors and mining of uranium in the region

(b) WIPP

(c) Yucca Mountain repository and geologic repositories in other nations.

(d) Health of people exposed to ionizing radiation tracable to the Chernobyl accident

(e) Environmental effects of the Chernobyl accident

(f) ^{99m}Tc radiopharmaceuticals

(g) New uses for radiopharmaceuticals

(h) New radiopharmaceuticals other than those of ^{99}Tc

(i) Radiotherapy

(j) Use of particle accelerators for radiation therapy

(k) Radioactive waste at Hanford or another facility

(l) Radioactive waste disposal in Europe and in Asia

(m) Food irradiation in the United States

(n) Food irradiation in Europe and in Asia

(o) Low-level waste disposal (especially for medical facilities) in your home state or country

(p) Studies of the survivors of Hiroshima and Nagasaki

(q) High-temperature, gas-cooled reactors

(r) Use of MOX to reduce the stockpile of weapons-grade plutonium

(s) Proliferation of nuclear weapons

(t) CTBT: Which nations have signed the treaty? Which nations have ratified it? Which nations are eligible, but have not yet signed it?

(u) Reduction of stockpiles of nuclear weapons worldwide

(v) "The nuclear deterrent"

(w) Cold fusion

(x) Origin and health effects of cosmic radiation

(y) Status of breeder reactors

(z) Advances in the applications of MRI and fMRI and radiopharmaceuticals for medical diagnosis

15

ENERGY

15.1 INTRODUCTION

Human societies have always required energy—for cooking and heating, for transportation, and for industrial activities. For most of human existence, wood has provided the main fuel for cooking, heating, and such high-temperature techniques as pottery and metallurgy. Human and animal power have provided the mechanical energy, supplemented from some ancient time by wind and water power as windmills and water wheels were developed. But it was only with the invention of the steam engine and later the internal combustion engine and electrical generators, all with a need for an energy supply, that power for technological development became freely available. Transportation similarly was by muscle or wind power until these other power sources were developed. Use of wood as a major energy source resulted in deforestation and large environmental changes when population density became large; some third-world nations have serious problems today from demands for wood as cooking fuel that outstrip regeneration. For other uses, wood is a relatively inefficient and variable fuel. Even primitive technologies attempted to improve on it by the use of charcoal: partially burned wood from which the volatile components have been driven off and the solid converted to nearly pure carbon.

Industrialization required the use of a superior fuel, coal, which became the main fuel for industrial and mechanized transportation applications. Coal is abundant, but much less convenient than liquid fuels such as petroleum. Widespread availability of the latter permitted it to replace coal in most transportation and some stationary power generation and heating applications by the mid-twentieth century. More recently, natural gas has been playing a larger role. After 1945, nuclear energy from fission began to be considered as a major component of power generation, and in some nations (e.g., France) fills that role today, although in other nations public concern about safety and general fear of radiation make its future questionable. Still more recently, attention has shifted to liquid fuels that are readily prepared from biological sources (e.g., alcohols), mainly for environmental reasons.

All these sources of energy are finite, and in the case of petroleum in particular, reserves are projected to have limited lifetimes at present rates of use, although predictions of when we will exhaust them have repeatedly been shown to be wrong as more reserves have been discovered. The need to conserve these limited resources is one reason to use them with more efficiency and to find alternative, preferably renewable sources. Another reason, and one that has driven a great deal of research and development in recent years, has been the realization that availability of much of these resources is controlled by a small number of nations, which consequently wield considerable power to influence world events. This was brought out in the "oil crisis" of the early 1970s, when the large Middle East producers cut production to drive up prices. Not only did prices for petroleum products and materials dependent on petroleum for their production go up, but severe shortages developed that resulted in restrictions on driving and, in some places, gasoline being available only on alternate days of the week. Still another reason to look for alternatives to the traditional fossil fuels is concern about environmental degradation that accompanies their use, as we have discussed many times. Human need for energy has had and will have major impacts on the environment.

Much of our energy is based on combustion of carbon-based fuels. The energy released upon combustion is essentially the difference between the energy of the bonds that are formed in the product molecules (C=O and O—H) and those in the reactants that are broken (O=O in dioxygen, and C—C, C—H, C—O, etc., depending on the fuel). Some illustrative values of heats of combustion (heating values) are given in Table 15-1 for equal masses of material. Wood and some coals can be highly variable because of moisture content. Note that heating values for oxygenated fuels are lower than for others because some of the bonds that appear in the products (O—H) are already present in the reactants.

TABLE 15-1

Heating Value for Various Fuels

Fuel	Heating value (kJ/g)
Methane	55.6
Propane	50.3
n-Butane	49.5
Gasoline, kerosene	41–48
Fuel oil	44.4–49.4
Coal	
anthracite	32.5
Bituminous	25.5–34.8
Lignite	11.6–17.4
Wood (12% moisture)	17.5–18.3
Methanol	22.7
Ethanol	29.7

In this chapter, we shall discuss some thermodynamic considerations underlying efficiency, and then consider the various sources of energy that are in use or that are reasonable proposals for future use. Not all these are chemical, but we shall include them for completeness. We shall also comment on some predictions for future energy needs.

15.2 THERMODYNAMIC CONSIDERATIONS

15.2.1 The First and Second Laws of Thermodynamics

Energy flow and energy use are governed by the law of conservation of energy, which states that energy can neither be created nor destroyed, although it can be changed from one form to another. This law appears sometimes as the first law of thermodynamics:

$$\Delta U = q + w \qquad (15\text{-}1)$$

where ΔU is the change in internal energy of the system being studied when heat q flows in or out, and work w is done on or by the system. The *system* being studied can be any part of the universe—a person, a city, a forest, a gasoline engine, a nuclear power plant—within certain limits that will not be discussed further here.[1] All the rest of the universe is called the surroundings. When ΔU is positive, the internal energy of the system is increasing; when it

[1]For rigorous application of the laws of thermodynamics, the system in question must have either isothermal or adiabatic boundaries at any time.

is negative, the internal energy of the system is decreasing. When q is positive, heat is being absorbed *by* the system *from* the surroundings; when it is negative, heat is being evolved *by* the system *to* the surroundings. When w is positive, work is being done *on* the system *by* the surroundings; when it is negative, work is being done *by* the system *on* the surroundings.

There are many forms of work. For example, mechanical work can be done on the system by compressing it, or the system can do mechanical work on the surroundings by expanding against a resisting pressure. Electrical work can be done on a charged particle by moving it through a potential difference. Most of the time we will be concerned here only with the mechanical work of expansion or compression of a system from volume V_i to V_f against a resisting pressure P, in which case

$$w = - \int_{V_i}^{V_f} P dV \qquad (15\text{-}2)$$

If the resisting pressure P is constant, $w = -P\Delta V$, where $\Delta V \ (= V_f - V_i)$ is the change in volume of the system when it expands ($\Delta V > 0, w < 0$) or is compressed ($\Delta V < 0, w > 0$).

We shall be discussing the heat released during various chemical reactions—for example, the combustion of coal or oil. The enthalpy H of the system is defined as $H = U + PV$, so that the change in enthalpy for a process is $\Delta H = \Delta U + \Delta(PV)$. If we consider only the mechanical work of expansion or compression, it is easily shown from equation (15-2) that *at constant pressure* the heat absorbed for a process is equal to the change in enthalpy:

$$q_p = \Delta H \qquad (15\text{-}3)$$

A heat engine is a device that generates mechanical work from heat. An example is the automobile internal combustion engine, in which chemical energy is released as heat at about 1000 K when gasoline is burned in the combustion chamber above the pistons. Some of this heat appears as work as the pistons move and drive the transmission, generator, and so on through a complex series of linkages, and much heat is "rejected," some at the temperature of the exhaust, and some at the temperature of the cooling system. When a car is cruising down a highway at about 50 mph (80 km/h), about 75% of the chemical energy is lost as "rejected" heat, and only about 25% of the chemical energy is eventually converted to useful mechanical work to move the automobile. The automobile internal combustion engine operates through a six-process cycle called the Otto cycle.[2] In the following discussion, however, we

---

[2]M. W. Zemansky and R. H. Dittman, *Heat and Thermodynamics: An Intermediate Textbook*, 7th ed., McGraw-Hill, New York, 1997.

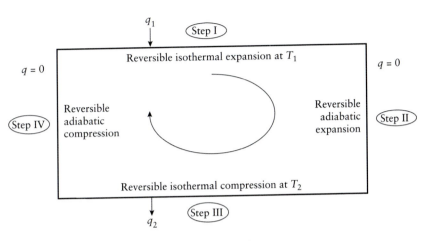

FIGURE 15-1 Schematic diagram of the Carnot cycle.

will use a simpler hypothetical heat engine that operates reversibly[3] through a four-step cycle that is known as the Carnot cycle.

We have seen that the first law of thermodynamics states that energy can be changed from one form to another but that it cannot be created or destroyed; it says nothing about how much one form can be changed to another. The second law of thermodynamics puts a limit on the amount of heat that can be converted into work. Specifically, one statement of the second law is that no process is possible in which the sole effect is the conversion of heat into work.[4] A corollary of this is that a heat engine operating through a cyclic process must undergo heat transfers with at least two heat reservoirs at different temperatures. Consider now the Carnot cycle (Figure 15-1). It operates reversibly between two temperatures, a high temperature T_1 and a low temperature T_2. (The Otto cycle requires more than two reservoirs.) A quantity of heat q_1 is absorbed isothermally at T_1 from the high-temperature reservoir by reversibly expanding a gas (step I), and work is done on the surroundings as the gas is cooled by reversible adiabatic (no heat absorbed) expansion from T_1 to T_2 (step II). To complete the cycle (return to the starting conditions), the gas is

---

[3]A reversible process by a system is carried out in such a manner that at any stage, the direction of the process may be reversed by an infinitesimal change in the conditions of the surroundings. It is required by the second law that all naturally occurring (spontaneous) processes be irreversible.

[4]Of course, many processes are possible in which heat is converted completely into work. An example is the isothermal expansion of an ideal gas from V_i to V_f. There is no change in internal energy, so that all the heat absorbed by the gas is converted to work performed by the gas on the surroundings. However, this is not the *sole* result of the process (i.e., the gas has also expanded) and therefore the second law of thermodynamics is not violated.

compressed isothermally at T_2, evolving heat q_2 to the low-temperature reservoir (step III), and finally is adiabatically compressed back to the high temperature T_1 with some work being done on the gas (step IV). The efficiency of a heat engine η is defined as the net work output (work done on the surroundings) divided by the heat input at the high temperature:

$$\eta = \frac{-w_{net}}{q_1} \qquad (15\text{-}4)$$

Thus, for the reversible Carnot cycle operating with no frictional or other losses (i.e., operating under maximum efficiency conditions), it can be shown[5] that

$$\eta_{max} = \frac{T_1 - T_2}{T_1} \qquad (15\text{-}5)$$

where T_1 and T_2 must be in kelvin. It turns out, in fact, that η_{max} is the maximum efficiency for any heat engine operating between two temperatures. Thus, the maximum efficiency depends on the difference between the two temperatures, $T_2 - T_1$, at which the engine operates; it becomes more efficient as the temperatures T_1 and T_2 move farther apart. Since, for most purposes, room temperature (300 K) is usually the lowest practical value for T_2, much effort has been expended to make T_1 as high as possible for various engines. For example, superheated steam, rather than ordinary steam has been used in steam engines (high pressures are necessary for this). The important thing to note here is that there *is* a theoretical maximum efficiency for engines that do useful work [equation (15-5)]; an actual engine running between these two temperatures generally will have an efficiency considerably less than this maximum possible efficiency because it operates irreversibly and has various other losses. We have seen, for example, that an engine in an automobile being driven under highway conditions has an efficiency of approximately 25%. If this engine could be assumed to be operating as a reversible Carnot cycle engine between 1000 and 300 K, its efficiency would be 70%. The second-law limitation becomes even more important when the high-temperature heat source is at a much lower temperature, such as geothermal, or nonfocused solar sources, or bodies of water. It is pointed out in Section 15.5.3.3, for example, that solar ponds have at most a temperature gradient of about 50°C, giving a maximum second-law efficiency of the order of 15%.

If a Carnot cycle heat engine is reversed so that a net amount of work is done on the system accompanied by withdrawal of an amount of heat q_2 from the low-temperature reservoir, the net effect is the transport of an amount of heat from the low-temperature reservoir to the high-temperature reservoir by the expenditure of work. This is the principle behind the heat pump that is used

---

[5]See any elementary physical chemistry textbook: for example, R. A. Alberty and R. J. Silbey, *Physical Chemistry*, 3rd ed., Wiley, New York, 2001.

in many areas for domestic heating of buildings. (It is also the principle of an electric compressor refrigerator.) The outside of the building is the low-temperature reservoir at T_2 and the inside is the high-temperature reservoir at T_1. The minimum amount of work that would have to be done (e.g., electric work to run a compressor) to pump q_2 heat from the outside to the warmer inside can also be shown to be

$$w_{min} = q_2 \frac{T_1 - T_2}{T_2} \tag{15-6}$$

Thus, consider a building being maintained internally at 22°C (295 K) with an outside ambient temperature of −5°C (268 K). From equation (15-6), only a minimum of about 100 J of electric work would have to be expended in order to pump 1000 J of heat into the building, or about 10% of the energy required if the 1000 J of heat was supplied to the building at maximum efficiency by electric resistive heating.

It is important to emphasize that the maximum efficiency limitation given by equation (15-5) is a second-law limitation restricted to the conversion of heat into work. Other parts of the overall conversion from chemical to electrical energy are generally much more efficient than the heat-to-work component. Combustion efficiency can be as high as 98%, boiler heat transfer efficiencies are typically 80–90%, and mechanical-to-electrical is often greater than 90%. If heat is not involved, the second-law limitation does not apply, although the first law conservation of energy limitation will still be applicable. The direct conversion of solar energy to electrical energy (see Sections 15.5.4 and 15.5.5) and of chemical energy to electrical energy as in fuel cells (Section 15.9) are examples of such processes, as are conversions of wind or wave energy to electrical energy. However, other limitations on the efficiencies of some of these processes—for example, restrictions on the efficiencies of solar photovoltaic cells (Section 15.5.4)—may be even more restrictive than the second law. Indeed, at the present time virtually all the chemical energy of fossil fuels and the nuclear energy of fissile fuels is first converted to heat before being converted to mechanical (and then on to electrical) energy.

The waste heat from our various engines is usually rejected into the environment and there becomes "thermal pollution," although in some cases this "rejected" heat is used as a by-product for domestic or industrial heating. We saw in Chapter 3 that cities are always warmer than their surrounding countryside partly because of the rejected heat from all the machines and engines used in the city. Large electric power plants, whether of the steam or nuclear type, reject their waste heat into a very small volume of the environment. When these power plants are water cooled, the streams and lakes into which this water is usually released may become much warmer and change the environment for aquatic life. Large populations of fish may be killed, or, alternatively, attracted by the new, warmer environment.

15.2.2 Units of Energy and Power

In this book we generally use the SI system of units, in which the unit of energy is the joule (J) and the unit of power (energy per unit time) is the watt (W): $1\,W = 1\,J/s = 3600\,J/h$. However, so many other units of energy and power are in common use that some useful conversion factors are given here:

$$1\,J = 10^{-3}\,kJ = 10^{-6}\,MJ = 10^{-9}\,GJ = 10^{-12}\,TJ = 10^{-15}\,PJ = 10^{-18}\,EJ$$
$$(k = kilo;\ M = mega;\ G = giga;\ T = tera;\ P = peta;\ E = exa)$$
$$1\,J = 10^{7}\,ergs = 0.239\,cal = 9.48 \times 10^{-4}\,Btu = 2.778 \times 10^{-4}\,Wh$$
$$1\,cal = 4.184\,J$$
$$1\,Btu = 1054\,J = 1.054\,kJ$$
$$1\,W = 44.27\,ft\text{-}lb/min = 1.341 \times 10^{-3}\,HP$$
$$1\,kWh = 1000\,Wh = 3.6 \times 10^{6}\,J = 3600\,kJ$$
$$1\,quad(Q) = 10^{15}\,Btu = 1.054 \times 10^{18}\,J = 2.93 \times 10^{11}\,kWh$$

Since the conversion of thermal energy (heat) to mechanical and ultimately to electrical energy is limited by the second law, thermal energy (designated by subscript "th") is often referred to as low-grade energy and electrical energy (subscript "e") as high-grade energy. It always takes more than $1\,kWh_{th}$ heat from the combustion of coal or from any other fuel to produce $1\,kWh_e$ of electricity from a fuel-fired electric power plant, not only because of the second-law limitation but also because no real process is as efficient as a hypothetical reversible cycle.

15.3 FOSSIL FUELS

15.3.1 Introduction

Fossil fuels in the form of coal, petroleum, and natural gas account for 90% of all the energy used in the United States (Section 15.10.1). The growth in the use of fossil fuels as an energy source has been very rapid, as evidenced by the fact that wood supplied 90% and coal only 10% of the energy used in the United States in 1850. Coal and wood supplied equal proportions of energy in 1885. Undoubtedly this change from wood to coal reflected the greater ease in obtaining large amounts of coal required for the new industries developing in the United States. However, some of the incentive for the change may have also been the "energy crisis" resulting from cutting down most of the readily available trees in the eastern United States.

Petroleum, coal, and natural gas are expected to continue to be the dominant energy sources in the United States and the rest of the world well into the twenty-first century (Figure 15-2). An optimistic estimate of the world's oil

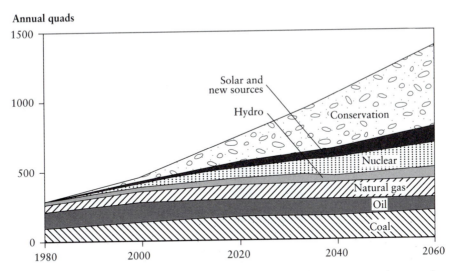

Annual quads

FIGURE 15-2 Projected global primary energy production by energy type (conservative case). Coal remains the primary fossil fuel source, with a rise and eventual decline of oil and growing levels of energy from natural gas, uranium, solar, and other (e.g., biomass) sources. From C. Starr, M. F. Searl, and S. Alpert, Energy sources: A relative outlook, *Science*, **256** 981–987. Copyright © 1992 American Association for the Advancement of Science.

reserves was published by the U.S. Geological Survey in 2000 which indicates that there is 20% more oil to be discovered in the world than was estimated in 1994 (Figure 15-3). In this estimate the world's oil production will peak in 2050, while the more pessimistic 1994 estimate is 2015. The pessimists note that global oil consumption is 27 billion barrels a year and is increasing by 1.5–2% each year. The postulated and undiscovered reserves are estimated to amount to about 2300 million barrels, with the United States having about 8% of that. The projected U.S. reserves were doubled in previous estimates, but the new estimates note no further increases. The extraction of oil from these U.S. reserves will be more costly because more expensive technologies will be required to retrieve the remaining oil. There was little change in the estimates of coal reserves in this time period, and the current supply of coal is sufficient to meet all our energy needs for the next 200 years. In 1999 coal-powered generators accounted for 56% of the electricity produced in the United States and 36% in the world. Energy conservation is projected to be the biggest "energy source" from about 2020 on into the future. That is, projected future energy needs, based on current practice, will be reduced by greater efficiency.

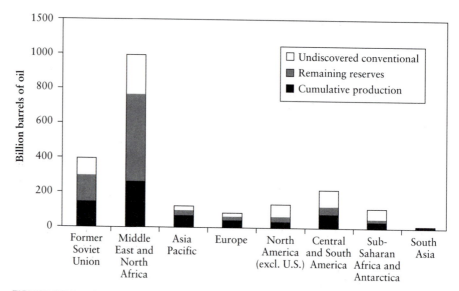

FIGURE 15-3 The amounts of oil already produced, known to be in the ground, and yet to be discovered in the world. Estimates for the United States are not included in this figure but can be found in earlier estimates. Redrawn from R. A. Kerr, *Science*, **289**, 237 (2000).

15.3.2 Liquid Hydrocarbon Fuels

15.3.2.1 Petroleum

Oil is a convenient energy source because it can be obtained cheaply, it is easy to transport, and, when burned properly, it produces low levels of environmental pollutants. The technology for the separation and conversion of crude oil into a variety of hydrocarbon fractions is well worked out (Section 6.2). Unfortunately the tremendous use of petroleum in the United States has resulted in the consumption of over 50% of our known reserves and, as a consequence, the rate of U.S. crude oil production is decreasing. The country has imported over 50% of its crude oil since 1997 to supplement the diminished domestic supply.

15.3.2.2 Tar Sands and Oil Shales

Products very similar to those obtained from crude oil may be obtained from tar sands and oil shales. The potential for petroleum production from these sources is estimated to be equal to that of all the crude oil reserves in the world. However, in most instances, the cost of the recovery of this petroleum in 1998 is not competitive with the simple extraction of oil and gas. Oil shales and tar sands may be an important source of petroleum by the middle of the twenty-first century, when the cost of crude oil increases because of its diminishing supply.

Tar sands are a mixture of approximately 85% sand and 15% hydrocarbons, chiefly the heavier fractions of crude oil. The hydrocarbons have the consistency of paving asphalt. There are large deposits of this black sand in Alberta, Canada, and in Venezuela, and smaller ones in the western United States. Some of the tar sands can be obtained using open-pit mining techniques. The hydrocarbon–tar mixture is then stirred with NaOH and surfactant to give a frothy foam that contains the hydrocarbons. This foam scraped off the top and the sand are sent to settling ponds. The tar is then cracked and refined to give oil (Section 6.2). Suncor Energy of Fort McMurray, Alberta, Canada, produces and profitably sells 28 million barrels of oil annually using this procedure.[6] However, most of the tar sands deposits are so deep that other methods must be used to extract them. The approach looked on with favor at the present time is to warm the tar by injecting steam into the bed; the hot oil forms pools in the vicinity of the injected steam and can be pumped to the surface and processed like crude oil.

The largest oil shale deposits (sedimentary rock containing hydrocarbon material) in the world are located in Utah, Colorado, and Wyoming. The organic portion of shale oil is a highly cross-linked material called kerogen. Consequently, the shale must be heated to crack this hydrocarbon so that the organic products may be removed from the shale. When shale is heated, gaseous hydrocarbons, an oil, and a cokelike residue are obtained. The oil can be processed in the same fashion as natural petroleum (Section 6.2), while the residual coke is used as an energy source to thermally crack the kerogen in more shale rock.

15.3.2.3 Synthetic Petroleum Products from Coal

The conversion of coal to petroleum has been investigated sporadically for over 60 years. Germany obtained some of its petroleum by this means during World War II. South Africa has been preparing petroleum this way since the mid-twentieth century. At present the United States is only testing potential coal liquefaction techniques at the pilot-plant level. This process is discussed in more detail in Section 6.9.3.

15.3.3 Coal

About 25% (Figure 15.2) of the energy produced in the United States and in the world is generated from coal. Unfortunately severe environmental problems

[6]R. L. George, Mining for oil, *Sci. Am.* pp. 84–85, March 1998.

may result from using greater amounts of coal in energy production. One major problem is the release of oxides of sulfur when coal is burned. Bituminous coal from the eastern and midcontinent United States usually has a high sulfur content (3–6%). About half the sulfur in coal is in the form of pyrite (FeS_2), of which about 90% can be removed by mechanical cleaning; but the remainder of the sulfur is covalently bound to the carbon and can be removed only by chemical reaction. If this high-sulfur coal is used, then the sulfur oxides (mainly SO_2) must be removed from the stack gases after the coal is burned, or the coal must be chemically changed (e.g., coal gasification or liquefaction) in such a way that the covalently bound sulfur can be removed.

The removal of SO_2 from stack gases has been studied for many years, but the technology that has been developed is still controversial (Section 10.4). The method that is generally considered to be the most efficient is the use of limestone to absorb the SO_2. The main drawback is that up to 350 lb (159 kg) of limestone is required to absorb the SO_2 emissions from a ton of high-sulfur bituminous coal. As a consequence, $8–9\,ft^3$ ($0.2–0.3\,m^3$) of sludge, which must be disposed of, is produced per ton of coal processed.

A second problem associated with coal combustion is the formation of nitrogen oxides. This formation is enhanced by the need to carry out the combustion at near 1500°C to get high thermal efficiency.

A third problem is the release of toxic heavy metals, e.g., Hg, Pb, and Cd present at low levels in coal. There are calls for restrictions on Hg emissions from coal burning plants (Section 10.6.6).

15.3.4 Natural Gas

Natural gas is a nearly ideal fuel (Table 15-1). Gas is virtually pollution free; it is easy to transport by pipeline, and it requires little or no processing after it comes from the gas well (although some sources contain H_2S that must be removed). Natural gas consists mainly of methane together with some other low molecular weight hydrocarbons. In 1997 it supplied about 23% of the energy requirements of the United States. In 2000 the worldwide reserves of natural gas were estimated by the U.S. Geological Survey to be about 85% of those of petroleum in energy content, and the United States had about 8% of this total amount. These reserves are postulated to last longer because the rate of their consumption is slower than that of oil.

The use of natural gas in place of coal or oil has many environmental benefits. It usually contains virtually no sulfur or sulfur compounds, so its combustion does not contribute to the emissions of SO_2 and its use contributes only 13% of the U.S. NO_x emissions, while coal and oil contribute 84%. Being the least carbon intensive of the fossil fuels, upon combustion it contributes the least amount of carbon dioxide per unit of energy.

15.3.5 Synthesis of Methane from Coal

A commercial-sized plant for the conversion of coal to methane, the Great Plains Gasification Facility in Beulah, North Dakota (Figure 15-4), has been in operation successfully since 1984. It is the outcome of five U.S. pipeline companies' desire to have an alternative supply of methane gas and the U.S. Department of Energy's interest in having a domestic synthetic fuels demonstration facility built. After default on the loan by the original owners, the project reverted to the Department of Energy for three years. It was purchased by the Basin Electric Cooperative in 1988 and is run as both a commercial source of SNG (substitute natural gas, or methane) and as a demonstration facility to investigate the development of various by-products.

The SNG is made from lignite coal that is strip-mined nearby, and the facility processes about 32,000 tons of lignite a day. The first step is to crush and screen the coal. The larger pieces (17,000 tons/day) are used for methane formation, while the fines (up to 15,000 tons/day) are sent to a nearby power plant for the generation of electricity. The coal is gasified in 14 Lurgi reactors [Section 6.9.2, equations (6-9)–(6-11)] in the presence of oxygen and steam. The oxygen is separated from other atmospheric constituents in a cryogenic air separation unit on site, and the steam is partially produced as a by-product of cooling processes and partially from boilers at the facility. In addition to the formation of carbon monoxide and hydrogen, carbon dioxide, hydrocarbon

FIGURE 15-4 The Great Plains Gasification Facility in Beulah, North Dakota, for the generation of methane from coal. The gasification facility is in the foreground. Electricity is generated from the coal fines, which are not suitable for methane synthesis, in the power plant shown in the background next to the large smokestack. Photograph provided by Fred Stern at GPGF.

gases, H_2S, NH_3, and ash are formed. The heat from this process volatilizes phenols, tars, and oil from the coal, which are collected as separate fractions. A portion of the carbon monoxide is further reacted with water to produce additional hydrogen [("shift reaction": equation (6-10)] so that there is sufficient hydrogen to convert the remaining carbon monoxide to methane in the final synthesis reaction [equation (6-12)]. After cleanup of this raw gas to remove carbon dioxide, sulfur compounds, and light hydrocarbons, the final synthesis of methane is carried out. Finally, water is removed from the gas to yield methane suitable for pipeline transportation and use as an energy source. The methane from this facility, 160 million cubic feet a day, is shipped by pipeline to the Midwest.

The cost of producing methane in this facility is significantly higher than the cost of obtaining methane from gas wells. Therefore the current focus is on the sale of the by-products to ensure that the overall process remains economically viable. Originally about 10% of the income for the facility came from the sale of these by-products, and the goal is to eventually have these compounds provide more income than the sale of methane. The hydrogen sulfide together with the waste light hydrocarbons is burned in the boilers and combined with the ammonia in a flue gas desulfurization unit to produce ammonium sulfate, which is sold as DakSul 45 fertilizer. The remainder of the ammonia is stored in liquefied form and sold. Some of the hydrocarbons and oils formed are used as an energy source at the facility, while the carbon dioxide is used for the enhanced recovery of oil from oil wells. Liquid carbon dioxide dissolves in some petroleum, thus increasing its volume and reducing its viscosity so that it flows more readily. Carbon dioxide is not soluble in heavy oil so its main function is to displace the heavy oil to a location near the pipe used to remove the oil from the ground.

Analysis of the phenolic fraction distilled from the lignite during gasification revealed the presence of phenol and alkylated phenols. Currently the gasification facility in Beulah is exploring methods for the separation of these compounds from tars to obtain catechols, which can be sold as starting materials for the synthesis of pesticides, dyes, and pharmaceuticals.

In addition, the by-products from the cryogenic separation of oxygen have been found to have commercial value. These include nitrogen, argon, krypton, and xenon. Some of the liquid nitrogen is now being sold and some is used at the site. Krypton and xenon are now being sold while the economics for other gases, such as argon, is being explored.

The direction of the research at this facility suggests that the key to the utilization of coal as an energy source may be to focus on the valuable by-products formed in addition to the production of a clean liquid or gaseous fuel. It probably will not be possible to come up with one process for all facilities because the by-products will be strongly dependent on the source of the coal used and the type of gasifier employed. For example, some of the features the

Great Plains facility tried to adopt from the SASOL process of South Africa did not work as anticipated, presumably because of the different grade of coal used.

15.4 NUCLEAR ENERGY

15.4.1 Nuclear Power and the Environment

The discussion of nuclear energy in this chapter is limited to the established use of nuclear fission to generate electricity in central station power plants. Weapons-related use is summarized in Chapter 14. Although nuclear power plants (NPPs) do not release the atmospheric pollutants that are emitted by fossil fuel power plants and are associated with the "greenhouse effect" and acid rain, some of their effluents contain radionuclides and are, therefore, of environmental concern. Nuclear power plants are designed to be very nearly closed systems with respect to release of radioactivity into the environment during normal operation. They may release small amounts of fission products and neutron-activated radionuclides. In the United States, the quantities released into the environment are legally limited by the regulations of the Nuclear Regulatory Commission (NRC), which licenses NPPs.

To prevent release of radionuclides into the environment if there is an accidental loss of coolant (e.g., pump failure or break in the piping) in an NPP with an LWR[7] steam generator, there is an emergency cooling system designed to prevent melting of the fuel rods and release of fission products, uranium and plutonium. If the emergency system fails, escape of radionuclides is prevented by a physical barrier. The reactor and plant components that could contain radioactivity originating in the core of the reactor plant are enclosed in a steel-lined, reinforced concrete, single or double containment vessel (or structure or enclosure or building). The special enclosure also protects the reactor from damage from external forces.

Both nuclear and fossil fuel power plants cause thermal pollution by transferring unused heat into the environment. Nuclear plants reject about 66% of the available heat; fossil fuel plants using steam turbines, about 60%.

15.4.2 Status of Nuclear Power

15.4.2.1 Worldwide

The first central station power plant for the generation of electricity with fission energy began operation in 1954 in the former Soviet Union. The

---

[7]A light water reactor (LWR) is one that is moderated and cooled by H_2O.

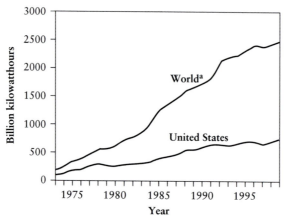

^aEastern Europe and the former USSR are included beginning in 1992.

FIGURE 15-5 World and U.S. nuclear electricity gross generation, 1973–1999. Adapted from *Monthly Energy Review*, Section 10, January 2001. Energy Information Administration, U. S. Department of Energy, Washington, DC. http://www.eia.doe.gov/mer.

world nuclear electricity gross generation 1973 through 1999 is shown in Figure 15-5. The U.S. gross generation is shown for comparison. Countries that had one or more nuclear power plants at the end of 2000 are listed together with the total number of plants in Table 15-2. About 435 were in operation. Table 15-3 shows the nuclear share of total net electricity generated at the end of 1999 and estimated values for the year 2010 for countries that are members of the Organization for Economic Co-operation and Development (OECD).[8]

Prediction of the worldwide nuclear generation of electricity in, say, 20 years cannot be made simply by predicting the demand for electricity in each country, because nuclear power has become a politically divisive issue in many countries, including the United States. For example, Italy has phased out nuclear power, and Belgium, Germany, the Netherlands, and Sweden have made political decisions to phase out nuclear power. When this decision was made in Germany in June 2001, it was assumed that energy conservation, renewable energy sources, and importation of electrical energy from other countries would replace nuclear energy by the time all nuclear power plants are phased out—about 2021.

In the United States it is likely that several nuclear power plants will be retired even though their licenses (for 40 years) might be renewable for another 20 years. However, the number of new plants to be built in France, South Korea,

---

[8]Net generation is the electricity available for distribution (i.e., gross minus that needed to operate the plant).

and Japan may approximately balance the number to be phased out world-wide. In countries where the nuclear power industry is growing, there is an increasing interest in small ($< 300\,MW_e$) and medium-sized (300–$700\,MW_e$) plants. There is also an interest in using small nuclear plants to eventually replace gas turbines in cogeneration plants and diesel engines on ships.

TABLE 15-2

Countries with Nuclear Power Plants at the End of 2000

Country	Total number
Argentina	3
Armenia	1
Belgium	7
Brazil	3
Bulgaria	6
Canada	22
China	11
Czech Republic	6
Finland	4
France	59
Germany	20
Hungary	4
India	18
Iran	1
Japan	58
Lithuania	2
Mexico	2
Netherlands	1
North Korea	2
Pakistan	2
Romania	5
Russia	31
Slovakia	8
Slovenia	1
South Africa	2
South Korea	20
Spain	9
Sweden	11
Switzerland	5
Taiwan	8
Ukraine	18
United Kingdom	33
United States	107

Source: World list of nuclear power plants, *Nucl. News*, (3) **44**, 61 (2001). (Revised annually.)

TABLE 15-3

Nuclear Share of Total Electricity Generated (net TWh) in OECD Countries at the End of 1999 and Estimated Share for the Year 2010

Country	Share (%)	
	At end of 1999	Estimated for 2010
Belgium	57.7	58.3
Canada	12.4	11.3
Czech Republic	18.2	34.4
Finland	32.8	23.3
France	75.0	75.3
Germany	34.8	34.3
Hungary	37.9	32.0
Japan	37.5	45.4
Korea (South)	43.1	39.9
Mexico	5.8	2.9
Netherlands	4.0	0.0
Spain	29.0	23.9
Sweden	45.5	42.8
Switzerland	36.0	30.0
Turkey	0.0	3.6
United Kingdom	25.4	13.1
United States	19.2	14.6

Source: *Nuclear Energy Data 2000*, Nuclear Energy Agency, Organization for Economic Co-operation and Development, OECD Publications, 2 rue André-Pascal, 75775 Paris Cedex 16, France (OECD Washington Center, 2001 L Street N W., Suite 650, Washington, DC 20036–4922).

15.4.2.2 United States

In the 1950s and 1960s, nuclear power was seen as the least expensive, most reliable way for the country's electric utilities to meet the projected annual increase in demand for electricity of about 7.5% in the 1970s, 1980s, and beyond. However, an abrupt change in projected demand for electricity in the United States followed the oil crises in 1973 and 1978. Energy utilization not only of oil but also of all energy sources was lowered significantly by conservation and by improved efficiency of automobiles, appliances, and so on. As a result, the annual rate of increase in demand for electricity in the subsequent years dropped to less than 3.0%. It is expected that the annual increase in demand and replacement of aging central station power plants in the future will be met by gas-turbine plants fueled with natural gas.

A full explanation of why no new nuclear power plants have been ordered in the United States since 1978 must go beyond the unexpectedly low increase in demand for electricity in the future. Factors contributing to this situation include safety, waste disposal, and the cost of generating electricity. Before a

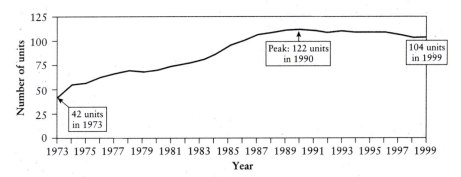

FIGURE 15-6 Number of operating nuclear power plants in the United States 1973–1999. From *Monthly Energy Review*, Section 8, January 2001. Energy Information Administration, U. S. Department of Energy, Washington, DC. http://www.eia.doe.gov/mer.

supplier of electric power will be willing to order a new nuclear power plant, the third of these factors must be favorable relative to that for new fossil-fueled plants that are being built and are scheduled to be built. The majority of these are relatively efficient plants that burn natural gas. Even if generating costs are favorable, a proposed NPP might never be built because of public opposition based on the first two factors.

In discussions of new NPPs in the United States it is usually assumed that such plants would be advanced BWRs or PWRs. However, a very different type of nuclear power plant that uses a gas-cooled, pebble bed modular reactor (PBMR) (see Section 14.7.4.4) has been receiving publicity as an alternative. If a proposed prototype (demonstration) plant is built (in South Africa) and is operated successfully, the safety characteristics of such a plant can be tested and the cost of generating electricity can then be estimated.

Growth and decline of the number of NPP units operating in the United States from 1973 through 1999 are represented in Figure 15-6. Figure 15-7 illustrates the relative contribution of nuclear power to the total for the same period of time. A comparison of energy production from various sources is presented later (Figure 15-18).

15.4.3 Factors Affecting the Choice of Nuclear Energy for the Generation of Electricity in the United States

15.4.3.1 Availability of Other Energy Sources

The United States is fortunate in having fossil fuel resources and well-established technologies for their use. A number of industrialized nations (e.g., France, Japan, and South Korea) that lack such resources depend heavily on

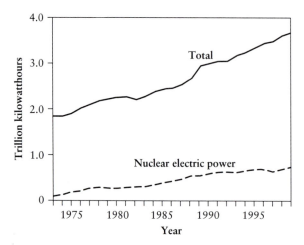

FIGURE 15-7 Total and nuclear net electricity generation in the United States, 1973–1999. Adapted from *Monthly Energy Review*, Section 8, January 2001. Energy Information Administration, U. S. Department of Energy, Washington, DC. http://www.eia.doe.gov/mer.

nuclear energy to meet their present energy needs. They have, in fact, supported the development of fast breeder reactors, because such reactors convert the relatively abundant, nonfissile ^{238}U into fissile ^{239}Pu and have the potential of providing long-term energy independence. However, the availability of ^{235}U (after dilution with ^{238}U) and ^{239}Pu (after dilution with ^{238}U to form MOX (see Section 14.7.2) contained in the post–Cold War nuclear weapons stockpiles as fuel for NPPs has reduced somewhat the urgency for developing fast, liquid-metal-cooled breeder reactors. On the other hand, India, which has an abundant supply of monazite sand containing thorium but little uranium, is developing power reactors, including fast breeder reactors, which use the thorium fuel cycle (Sections 13.6.3 and 14.7.2) with ^{233}U as the fuel.

15.4.3.2 Safety

Prior to the Three Mile Island accident in March 1979 (Section 14.16.2.2), there had been numerous "close calls" or near accidents at NPPs. The TMI accident, together with the well-established association of nuclear power with the "atom bomb," escalated uneasiness to fear, which generated an enduring negative reaction to nuclear power among the population in the United States and elsewhere. This fear of anything nuclear, exacerbated by the Chernobyl accident, was an expression of lack of confidence in those responsible for safe operation of NPPs. However, since the TMI accident a steady effort has been made in the United States by the nuclear power industry, the Nuclear Regulatory Commission through its regulations and inspections, and other safety organizations providing education and training to improve the safety record of nuclear power plants. In general, the effort has been successful and is perceived to have made nuclear power somewhat more acceptable to the public.

15.4.3.3 Fuel and Spent Fuel

Table 15-4 shows the unit energy content of coal, natural gas, oil, and uranium-235, as well as the quantities of fossil fuels that are equivalent in thermal energy content to one kilogram of uranium-235.

Nonnuclear power plants that burn oil or natural gas have no spent fuel disposal problem. Spent fuel from a NPP is highly radioactive and requires safe storage at the plant site or at a temporary storage site until it can be taken to a geological repository for disposal as high-level radioactive waste. Although ash from a coal-fired plant contains chemical components of environmental concern (Section 6.9.1), its disposal is less challenging than that of nuclear fuel. However, spent nuclear fuel is discharged from the reactor less frequently (e.g., once in 12 months to once in 24 months).

15.4.4 Costs

When the nuclear power industry was created in the United States, it was predicted that the cost of electricity to the consumer would be negligible. This seemed reasonable based on the cost of nuclear fuel. In reality, of course, the cost per kilowatt-hour of electricity delivered to the power transmission system

TABLE 15-4

Unit Energy Values for Fuels and Quantities of Fossil Fuels with Thermal Energy Content Equal to That of One Kilogram of ^{235}U

Fuel	Unit energy content[a]		Equivalent energy content
	SI units	Btu	
^{235}U[b]	82 TJ/kg	35 GBtu/lb	1 kg
Coal[c]	30 MJ/kg	12.9 kBtu/lb	2.73×10^6 metric tons
Oil	39 GJ/m^3	5.8 MBtu/bbl	2.13×10^3 m^3
Natural gas	38 MJ/m^3	1.02×10^3 Btu/ft^3	2.16×10^6 m^3

[a]1 Btu = 1British thermal unit = 1.054 kJ; 1 bbl = 1 barrel = 42 gal(U.S.) = 0.159 m^3.
[b]Values for ^{239}Pu and ^{233}U are essentially the same as that for ^{235}U.
[c]Unit energy varies from 14 MJ/kg for lignite to 35 MJ/kg for bituminous coal.

by any type of central station power plant involves many costs in addition to that for fuel.

15.4.4.1 Capital

Nuclear power plants are more complex and, therefore, take longer and are more expensive to build than other plants having the same generating capacity. In addition to the reactor, safety-related construction costs such as those for the reactor vessel, the containment structure, the emergency cooling system, and temporary storage facilities for spent fuel are among the many costs that are unique for NPPs. Capital carrying charges for construction of a plant increase as the construction time increases. Construction time for nuclear power plants in the United States increased from an average of about 6 years prior to 1979 to an average of 11 years (with an upper value of 19 years) by 1989. By contrast, construction times have been about 6 years in France and South Korea and as low as about 4 years in Japan. One reason for increased construction time in the United States was delay caused by litigation[9] brought by antinuclear groups and by others with the NIMBY (not in my backyard) philosophy. A second reason was delay resulting from design changes that were made during construction to ensure that the plant would meet new licensing criteria that reflected new safety features required by the NRC. Similarly, retrofitting new safety features into existing plants raised capital costs.

Construction costs (adjusted for inflation) for a NPP in the United States increased from an average of about $800/kWe of generating capacity in the early 1970s to as high as about $3000/kWe for a plant at the end of the growth period 20 years later. It is expected that the new modular LWR plants that are precertified by the NRC will take 5 or 6 years to build and will have construction costs estimated to be between $1000/kWe and $1700/kWe. It is very likely that if modular reactors are built, they would be placed on the sites of existing NPPs. Capital charges are estimated to be about 81% of the total cost for generating electricity for a new NPP, about 73% for a coal-fired plant, and about 28% for gas turbine plants.

15.4.4.2 Fuel

The contribution of fuel to the total cost is about 8% for NPPs, about 18% for plants burning coal, and about 67% for gas turbine plants.

15.4.4.3 Operation and Maintenance (O & M)

Compared with a fossil-fueled plant, a nuclear plant has very different control and operating characteristics and requires specially trained operating and maintenance personnel. Furthermore, the decay heat that is released after

---

[9]The cost of the litigation itself adds to the total cost.

shutdown makes operation of a power reactor uniquely different from any other steam supply. The percentage of total generating cost stemming from O & M is about 11% for NPPs, about 10% for coal-burning plants, and about 5% for gas turbine plants.

When an LWR power plant requires refueling, it must be shut down. During the scheduled shutdown (outage), maintenance work is performed. By 1999 the outages for some plants had been reduced from about 120 days to less than 20 days. In addition, the frequency of unplanned shutdowns has been reduced so that most plants are operating above 88% of annual capacity and some are at 100%. As a means of improving operation and lowering operating and maintenance costs for existing plants, a number of utilities have arranged to have their nuclear power plants either operated by or purchased by companies that specialize in the management and operation of nuclear power plants. Some utilities have merged and others have combined to form joint operating companies for their nuclear power plants in order to make them more competitive. These efforts to reduce O & M costs for well-operated NPPs in the United States have succeeded in bringing the average cost of electricity down sufficiently to be competitive with plants fueled with coal.

The presence of ionizing radiation makes maintenance of certain regions of a nuclear power plant difficult and expensive. The requisite radiation safety program for monitoring plant workers adds to the operating cost of a nuclear power plant. A number of utilities have been fined heavily by the NRC because of operational safety problems. In general, safety of operation has greatly improved over the years. For example, the collective exposure to ionizing radiation of plant workers is about 80% lower than it was two decades ago.

15.4.4.4 Decommissioning

When an NPP ceases to operate and is decommissioned, it can be mothballed temporarily or it can be dismantled immediately, decontaminated, and the buildings put to other use or demolished to provide a "green field" site. Whenever an NPP is dismantled, a large volume of radioactively contaminated structural material and equipment is generated. The rate that the owner of an NPP charges its customers for electricity includes an amount that is accumulated in a decommissioning fund during the lifetime of each of its nuclear power plants.

15.4.4.5 Other

Also included in the cost of operation of a nuclear power plant are (1) a security program to protect the plant from intruders including terrorists and (2) a safeguard and accountability program to assure that the plutonium produced in the plant is not accessible for use as a weapons material or as a radiotoxic contaminant.

15.5 SOLAR ENERGY

15.5.1 Introduction

Although only about 2×10^{-9} of the radiation from the sun reaches the earth, sunlight represents a tremendous source of renewable, greenhouse-gas-free energy. We saw in Chapter 3 that the solar constant, which is the average solar energy flux (i.e., solar power) of all wavelengths at the top of the atmosphere, is about $1.37\,kW/m^2$. This corresponds in one second to $1.37\,kJ/m^2$ of energy. Commercially, energy is often expressed in terms of kilowatt-hours, or kWh, which is the amount of power in kilowatts provided or produced over a time period of one hour. Thus, the amount of energy from the sun at the upper edge of the earth's atmosphere is $1.37\,kWh/m^2$ in one hour, or in one year (8760 hours), $12.0\,MWh/m^2$ ($1\,MW = 1000$ kW). Not all this energy is available at all times at a given site on the earth's surface, however. For one thing, it varies diurnally—from day to night over a 24-hour period—as the earth rotates. We have also seen that much of the sun's radiation (particularly in the ultraviolet and infrared spectral regions) is absorbed in the atmosphere, so that most of the radiation reaching the biosphere is in the visible region. Of this, about 30% (the earth's albedo) is reflected back to space. The amount impinging on one square meter of the earth's surface also varies seasonally and with the latitude. Figure 15-8 shows the annual average available solar energy across the United States, using a collector that can be rotated to keep it perpendicular to the sun's rays, ranging from $3.6\,MWhm^{-2}y^{-1}$ at Las Vegas to $1.8\,MWhm^{-2}y^{-2}$ at Seattle. Assuming a national average of $2.1\,MWhm^{-2}\,y^{-1}$, this amounts to approximately $2.1 \times 10^{13}\,MWh/y$ of solar energy for the whole country ($10^{13}m^2$), or about 1000 times the energy consumed by the United States.[10]

There are several important ways that solar energy may be directly utilized, and these are covered in this section. It should also be pointed out that although wind and water are reviewed in a separate section (Section 15.7), the ultimate source of these energies is also of course solar energy. These energy sources are renewable, in contrast to fossil and even nuclear fuel sources. Table 15.5 at the end of Section 15.7.3.3 gives a summary comparison of most of them.

Available sunlight is both *direct* and *diffuse*; that is, some of the sun's radiation comes directly from the sun, while the rest of it reaches the earth's surface indirectly by atmospheric scattering or refracting. (The total of the two is called global.) On very clear, cloudless days the direct sunlight may make up

---

[10]In 1990 the United States consumed 2.4×10^{13} kWh (82 quads) of energy, which was 24% of the whole world's use of about 10^{14} kWh that year. By the year 2026 the annual world energy use is predicted to increase by about 50%.

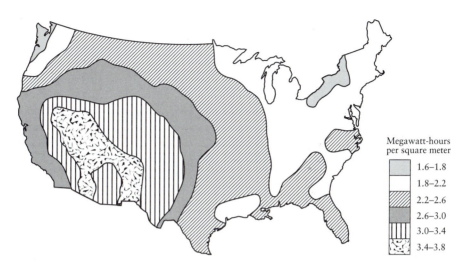

FIGURE 15-8 Average annual solar radiation in the United States. From K. Zweibel, "Harnessing Solar Energy," p. 220. Plenum Press, NY, 1990.

Megawatt-hours per square meter

	1.6–1.8
	1.8–2.2
	2.2–2.6
	2.6–3.0
	3.0–3.4
	3.4–3.8

as much as 90% of the total solar radiation, whereas on very cloudy days 100% of the available sunlight may be diffuse. Utilization of the two components of sunlight depends on the technique being employed. Diffuse light produces "low-grade" energy, which can be used, for example, for solar heating. However, production of "high-grade" energy, such as conversion directly to electrical energy with photovoltaic cells, in general requires direct sunlight. Probably the most effective uses of solar energy will involve simultaneous integration of technologies, such as the hybrid units of photovoltaic and solar thermal energy cells currently being developed.

15.5.2 Biological Utilization of Solar Energy: Biomass

In its broadest sense, biomass (living matter) is the earth's vegetation—that is, plant material produced by photosynthesis, with the result that solar energy is stored.[11] This encompasses agricultural crops, forests, grasslands and savannas, and so on. The overall equation of oxygen-producing photosynthesis is

[11]Photosynthesis is the process by which green plants, algae, and some bacteria convert light energy to chemical energy to sustain their life. It has existed in its present form for at least 3.5 billion years, and undoubtedly it is the most important photoprocess occurring in nature. Since all the earth's coal and oil deposits come from past photosynthesis, all our food and most of our fuels are products of this process.

$$CO_2 + H_2O \xrightarrow{\text{light}} \frac{1}{n}(CH_2O)_n + O_2 \qquad (15\text{-}7)$$

in which $(CH_2O)_n$ represents any carbohydrate, including sugars, starches, and lignocellulose. Biomass has been used for many centuries for heating, lighting, and cooking. It is still the primary source of energy for half the world's population and produces 13% of the world's energy, but in the developed industrial countries fossil fuels have been much more widely used for many years.[12] The use of biomass for the generation of electric power, which historically has been used primarily by the paper and wood products industries utilizing waste wood fuel, is expected to increase in the United States from approximately 10 GW in 1997 (about 2% of the total U.S. generating capacity) to more than 100 GW by the year 2030. In addition, there are well-established methods for producing liquid (biofuels) and gaseous (biogas) fuels from biomass (Chapter 6). Brazil, for example, has developed an extensive fuel program for production of ethanol from sugarcane: in 1995, 35% of the passenger cars in Brazil were fueled with pure ethanol, the remainder using blended fuels containing 22% ethanol. An 85% ethanol (15% gasoline) fuel is being marketed in the United States for use in automobiles designed to run on it and on 100% gasoline.

We see by reaction (15-7) that photosynthesis uses up CO_2; therefore, biomass is an important "sink" for this primary "greenhouse gas" (see Chapter 3). It follows, then, that the ongoing destruction of tropical forests, in which photosynthesis is a major naturally occurring reaction, is one of the major contributors to enhanced global warming. One way of moderating the current increase in atmospheric CO_2 might be to develop extensive new forests. However, uptake and storage of CO_2 by trees greatly decreases as the trees mature, and even at that most of the trees will die and decay within a century or so if not harvested, releasing again the stored CO_2. In the long run it may be better to develop fast-growing biomass plantations for eventual CO_2-neutral energy replacement of fossil fuels rather than for limited-time CO_2 reduction given by permanent forests. An example of such a process is short-rotation coppice (SRC). In this technique fast-growing young growth from woody biomass is harvested, and resprouting (coppicing) growth from the left-behind stumps is cultivated and reharvested. SRC can produce high yields of biomass in a relatively short time—up to about 10 years—on a sustainable basis. It needs to be emphasized, however, that biomass fuel will be completely CO_2 neutral only if no nonrenewable energy sources are used in all stages of production, processing, and distribution. This will not be the case in the foreseeable future. For example, fossil fuels are extensively used in the manufacture of nitrogen fertilizers and in transportation. Furthermore, biomass use

---

[12]However, in the United States biomass is nevertheless the second largest *renewable* energy source (after hydroelectric power: see Section 15.7.3.1).

will make sense only if there is an overall net positive energy balance,[13] and processes such as the production of nitrogen fertilizers are very energy intensive. It is estimated, for example, that the net energy balance for the production of ethanol from wheat, a cultivated crop grown from seed, either is negative or has a low positive energy balance (between 1.0 and 1.2). On the other hand, energy balances for SRC may be positive by at least an order of magnitude.

It is also worth noting that biomass is generally lower in sulfur than most coals. This may be an important benefit of biomass in view of recent regulations in some countries reducing SO_2 emissions by electric power utilities.

There are, however, several serious environmental concerns with use of biomass as a major renewable energy source. Intensive agricultural practices often lead to nitrification of water supplies and pesticide and herbicide residues. Major production of biomass for energy usage will also require large land area commitments. For example, it is estimated by the U.S. Department of Energy that about $100 \, \text{mi}^2$ (25,000 hectares) of land will be required for an advanced 150-MW biomass power plant, which means that approximately $70,000 \, \text{mi}^2$ (about 2% of the total land area of the country) will be needed to meet the above estimate of 100 GW of electrical power from biomass by 2030 if there are no major increases in biomass productivity. Some major agricultural areas in developed countries can produce far more food than is needed by their residents, and biomass growth for energy purposes may lead to more effective use of productive but uncultivated land. For some developing countries, however, emphasis on biomass production could lead to intense competition between land and water use for food and for biomass energy production.

15.5.3 Physical Utilization of Solar Energy

15.5.3.1 Solar Heating and Cooling

About 20% of all the energy consumed in the United States goes for domestic hot water and for heating and cooling of buildings. Solar heaters to provide domestic hot water for homes and residential buildings have been commercially available in countries such as Japan and Israel for many years. Following the 1973 oil embargo (Section 15-1), a large number of electric utilities initiated major research and development programs in solar heating and cooling; these have involved literally thousands of experimental and developmental installations demonstrating energy conservation as well as utilization of

---

[13]Net energy balance is a measure of the difference between the energy contained in the biomass fuel and the total energy required to grow, harvest, and process the crops used in its production. Often the net energy balance is expressed (as we will do here) as a ratio of these two quantities (i.e., the energy contained in the biomass fuel divided by the total energy used in its production).

solar energy. The renewal of an abundance of low-cost oil led to a decrease in this activity by utilities since the early 1980s. Nevertheless, at this time the space heating and cooling of buildings is probably the most developed direct usage of solar energy, although hydroelectric power, which is a form of solar energy utilization (see Section 15.7.3.1), generates much more energy than is converted to thermal energy by solar energy collectors. Since this is low-temperature energy usage, it readily lends itself to utilization of solar energy, and the basic technology is available now. Homes and buildings with roof solar collectors for domestic hot water are now common sights, particularly in regions of high and extensive sunshine such as the southwestern United States.

Solar heating and cooling may be divided into two categories: passive and active. Passive systems use only natural conduction, convection, radiation, and evaporation, while active (nonpassive or hybrid) systems employ external energy to run fans, solar circulating pumps, water pumps, and so on. The early active systems were quite complex and unreliable, and the trend in recent years has been toward simplicity in passive and hybrid systems that use minimal nonrenewable energy sources. In most locations the installation and maintenance costs of active systems still substantially exceed the energy cost savings, but passive systems are cost-effective in many areas.

A typical procedure is to use a solar collector to absorb the solar energy, converting it to thermal energy, which is transferred by heat pipes or pumped fluids directly for low-temperature ($< 100°C$) heating or for storage. The collector should therefore have high thermal conductivity and low thermal capacity, such as metals (copper, steel, and aluminum) and some thermal-conducting plastics. The most common collectors are flat blackened plates, since they convert both direct and diffuse (cloudy) solar radiation into heat. Heat losses from collectors result mainly from reflection, thermal (blackbody) emission (Section 3.1.1), and convection. Reflection and thermal emission losses can be reduced by selective surface coatings, and convection losses are reduced by covering, or glazing, the front of the collector with a glass that transmits most of the radiation but traps the convection heat between the absorber and the glazing, and by insulating the back side of the collector. (Further reduction in convection loss—almost to zero—is attainable by evacuating the space between the collector and the glazing, but such collectors can be very expensive to manufacture and maintain.) In many cases (up to 100°C), water is the heat transfer fluid. Liquids with high specific heats (such as mineral oils and glycols) are employed at higher temperatures, however, and at very high temperatures molten salts or liquid sodium are used.

Although it is best to use the converted solar energy directly, sunlight is intermittent. Thus storage of the converted thermal energy is a very important factor in its efficient and economical utilization. Several large-scale energy storage techniques are covered in Section 15.8. However, heat can be stored for several days by sensible (specific) heat storage using water or other fluids in insulated tanks or rocks (pebble bed) for gas, and by latent heat storage with phase-change materials.[14] Materials used for latent heat storage are generally more expensive than those used for sensible heat storage but have a much higher density for energy storage. For example, calcium chloride hexahydrate ($CaCl_2 \cdot 6H_2O$) is extensively used as a phase-change material involving its latent heat of fusion (melting); it melts at about 27°C, has high thermal stability, and can store $2.8 \times 10^5 \text{ kJ/m}^3$. By comparison, water would have to be heated about 67°C to store as sensible heat an equivalent amount of thermal energy stored as latent heat by $CaCl_2 \cdot 6H_2O$ for the same volume.

Direct (but only a small fraction of diffuse) solar radiation can be concentrated by reflecting light with an array of mirrors (a heliostat) onto a common collector. Heat losses are minimized because the surface area of the absorber can be relatively small, and very high temperatures are possible. An extreme example is the experimental High Flux Solar Furnace built in 1989 at the National Renewable Energy Laboratory (NREL) in Colorado, in which the intensity absorbed by the collector furnace, using a series of focused heliostats, is equivalent to 50,000 times the intensity of direct (unfocused) sunlight, generating temperatures up to 3000°C. We will see in Section 15.5.3.2 that this technique is extensively used in solar thermal electric power systems.

Solar air conditioning, although not as well developed as solar heating, has the advantage that it is most needed on hot sunny days when solar radiation is also plentiful, particularly in commercial buildings, where less air conditioning may be needed at night. There are two main solar air conditioning systems currently in use involving cycles based on the very old technique of evaporative cooling; these are the absorption cooling system and the desiccant cooling system. Figure 15-9 is a very simplified schematic drawing of an open-cycle absorption cooling system. Liquid water, the refrigerant,[15] is added to the system at the evaporator, where the pressure is reduced and the water is flash-vaporized. This results in the absorption of heat—the latent heat of vaporization of the water—hence removal of heat (via a heat-exchanger coolant) from the space to be cooled. The water vapor is fed to the absorber, where the concentrated absorbent solution (often a lithium bromide solution) absorbs the water vapor and the heat evolved in this dilution process is discharged to the atmosphere. The diluted absorbent solution is pumped to the regenerator, where it is heated by solar light; this results in desorption of water and thus regeneration of the concentrated absorbent solution, completing the cycle.

[14]Sensible heat storage involves heating up the material (with no phase change), the stored energy being released as the material is cooled. Latent heat storage involves a phase change that absorbs energy (e.g., melting a solid) and releases the energy when the phase-change process is reversed (solidification).

[15]In the open-cycle system, water is almost always used as the refrigerant; it is continuously fed into the system and discharged to the atmosphere after being used for cooling. In a closed-cycle absorption cooling system, the refrigerant is reused in subsequent cycles.

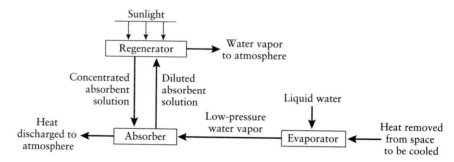

FIGURE 15-9 Simplified schematic of an open-cycle absorption cooling system using water as the refrigerant. Modified from M. Wahlig, "Absorption Systems and Components," in *Active Solar Systems*, G. Löf, ed. The MIT Press, Cambridge, MA, 1993.

Similarly, in the open-cycle desiccant cooling system ambient air is dried by blowing it over a solid desiccant such as silica gel or a molecular sieve, or a liquid desiccant such as glycol is circulated through the airstream; in either case the air heats up because heat is given off when water is removed by a desiccant. The hot, dry air is then cooled by passing it through a heat exchanger and by evaporating water into it similar to the absorption cooling system, with the desiccant being regenerated with outside air warmed by solar heating. Both systems permit independent control of humidity and temperature.

In general, the initial cost of either a solar heating or cooling system alone is so great that they are not economically feasible and cannot be justified for seasonal operation only. Since much of the solar energy collecting and recycling equipment can be common to both systems, however, combined heating–cooling systems may make this form of solar utilization viable; even greater economies may be realized if the solar units can also be used to supply domestic hot water to the buildings being heated or cooled.

15.5.3.2 Solar Thermal Electric Power

Solar thermal electric power is generated by converting solar energy first into heat and then into electric power, generally by steam-generating techniques similar to those used in conventional power plants where the source of heat is a gas-, coal-, oil-, or nuclear-fired furnace. As with solar heating and cooling, major solar thermal electric power research and development programs evolved after the oil shortage of the early 1970s, only to decline in the 1980s after world oil prices dropped and long-term, low-cost electricity seemed assured. Nevertheless, this technology has the potential to provide an important renewable energy supply in the future.

Conversion to thermal energy is an intermediate step in the conversion of solar energy to electrical energy, in contrast to direct conversion as with photovoltaic or photoelectric utilization (see Sections 15.5.4 and 15.5.5). Therefore the fundamental thermodynamic second-law limitation (Section 15.2.1) on the efficiency of the process applies. We see by equation (15-5) that the greater the temperature difference, the greater the efficiency of the energy conversion, hence the need in most cases of solar thermal electric technology to use a high-temperature solar heat source. (One exception to this is the use of solar ponds for electric generation; see Section 15.5.3.3.) This requires focusing direct sunlight onto a collector, similar in principle to the technique just described in connection with solar heating (Section 15.5.3.1). There are three basic methods used for this: the central receiver, the parabolic dish, and the parabolic trough. All three systems eventually convert reflected sunlight into thermal energy, which is transported by a heat transport fluid (water/steam, molten salts, or liquid sodium) to a central location for thermal-to-electric energy conversion or for energy storage as sensible heat. The central receiver system uses a single tower-mounted receiver to collect focused, reflected sunlight from a large field of assembled sun-tracking mirrors.[16] Parabolic dish systems use a collection of point-focusing reflectors; each reflector tracks the sun in two axes and focuses the light onto a receiver at its focal point. Parabolic troughs are U-shaped (line-focusing) reflectors that reflect the sunlight onto a linear receiver located along the focal line of the trough. The parabolic trough system is the most fully developed of the three, and probably because of its lower operating temperatures, it is the most useful for industrial process heat applications. Several hybrid parabolic trough plants (also using natural gas) are now in commercial operation in California. However, the potential for future economic utilization of solar energy is greater for either the central receiver or the parabolic dish technology, although neither is technically feasible for commercial deployment now.

15.5.3.3 Solar Ponds

Hot water normally rises by convection and mixes with cooler water in a pond because above 4°C the density of water decreases with increasing temperature. However, a temperature inversion can result if the pond is made nonconvective by some convection barrier. In a salt gradient solar pond, this is accomplished by establishing an increase in salt concentration (hence an increase in density) from top to bottom, so that the water is denser at the bottom than at the top even if the water at the bottom is heated by transmitted solar light to a

---

[16]For example, a 10-MW central receiver demonstration solar power plant in the Mojave Desert (California) uses 2000 motorized mirrors to focus sunlight onto a receiver, heating to 565°C a molten nitrate salt that boils water to drive a steam electric turbine.

relatively high temperature. Thus, solar ponds are a means for collecting direct solar energy and storing it as heat for low-temperature applications. Naturally occurring salt gradient ponds in Transylvania were first studied in 1902, and this technique is currently being investigated in many countries for economically feasible collection and storage of solar energy.

Considering only buoyancy, a salt gradient pond will be stable—that is, nonconvective—as long as the salt concentration is high enough to compensate for thermal expansion. Since water in this state is not convective, it behaves as a thermal insulator, the only loss being by conduction. The largest temperature difference is between the surface and the bottom (which may be a difference of 40–50°C), and thus most of the useful energy is in the lower regions of the pond. Actually most ponds do not have a continuous salt gradient, but rather at least three distinct zones or layers, as shown in Figure 15-10. The top (surface) and bottom (storage) zones frequently are of uniform concentration and therefore are convective within each one separately, but the middle (gradient) zone must have a concentration gradient and therefore be nonconvective.[17] The concentration gradient is established by filling the solar pond with concentrated salt solution to a depth of the storage zone plus about half the gradient zone; fresh water is added to the solution at successively higher levels, creating the gradient, followed finally by adding a layer of fresh water on

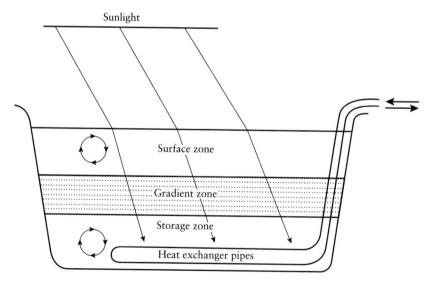

FIGURE 15-10 Schematic diagram of a three-zone salt-gradient solar pond.

[17]Advanced solar ponds have an additional stratified thermal layer consisting of several sub-layers of different densities between the gradient and storage zones.

top, forming the surface zone. The ponds are lined with clay or a plastic membrane buried in clay to minimize heat loss to the earth and to prevent salt loss. Heat is extracted by placing heat exchanger pipes in the storage zone or by pumping brine from the storage zone to an external heat exchanger. (Heat can also be removed from the stratified thermal zone in an advanced solar pond.)

As with most other solar energy utilization systems, solar ponds do not pose major environmental problems, although leaks or waste disposal procedures can lead to serious salt contamination. The energy gained per unit area of a solar pond is roughly the same as for a flat plate collector, and the pond has the additional advantage of having its own built-in energy storage. If reasonable sunlight and adequate low-cost supplies of water and salt are available, and if the land site can tolerate a large open pool of water, then a solar pond can provide low-temperature energy much more favorably than flat-plate collectors for many uses such as thermal heating and cooling, domestic hot water, desalination (distillation or electrodialysis), crop drying, and energy-intensive processes in hydrometallurgical mining (that is, the recovery of metals from ores by leaching with aqueous solutions). Also, although this is low-temperature thermal energy and therefore low efficiency for solar energy utilization, since it involves indirect conversion to electrical energy via thermal energy, it may turn out to be economical for electric power generation in areas where nonrenewable fuel costs are very high. An example is the electric power station at Beit Ha'Arava, Israel, at the north shore of the Dead Sea. This plant has been in operation since 1984, and uses a 40,000-m^2 pond and a 210,000-m^2 pond to operate a 5-MW power station.

15.5.3.4 Ocean Thermal Energy Conversion

The process called ocean thermal energy conversion (OTEC) converts thermal to mechanical to electrical energy by using the temperature difference between ocean surface waters that are warmed by absorption of solar energy and the colder waters at depths below 1000 m that come from Arctic regions. Generally the temperature difference ΔT even in the tropical and subtropical oceans is at best only about 20°C, and it is estimated that this gives an overall operating energy efficiency of the order of 2%.

Three systems have been studied for OTEC. In the closed-cycle system, Figure 15-11a, a low boiling working fluid such as ammonia is vaporized by the warm surface water and the working fluid vapor drives a turbine electric generator. The vapor is then condensed by the cool deep ocean water and the cycle is repeated. The working fluid in the open-cycle system (Figure 15-11b) is the warm surface ocean water itself. It is vaporized in a low-pressure flash evaporator, and the vapor (steam) drives a low-pressure turbine. The steam is condensed by the cool water via a heat exchanger, producing desalinated

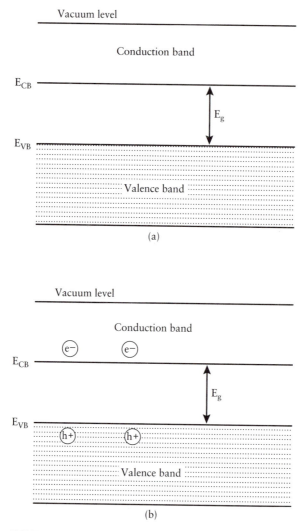

FIGURE 15-12 Schematic energy diagrams for a pure crystalline semiconductor. (a) 0 K; (b) room temperature.

If now a p-type semiconductor and an n-type semiconductor are intimately joined, a pn-junction semiconductor is formed. Some of the conduction band electrons in the n side diffuse a very short distance into the p side, where they combine with holes, resulting in a layer of negatively charged acceptor atoms; at the same time some of the extra positive holes diffuse from the p side into the n side and combine with electrons, giving a layer of positively charged donor atoms. This produces a double layer of charge, hence a built-in internal potential at the junction, the junction potential (Figure 15-13a).[21] This permanent potential barrier, penetrable only by highly energetic electrons, is necessary to maintain the separation between the remaining weakly bound excess electrons in the n-type side and the excess holes in the p-type side.[22]

Most solar cells are pn-junction semiconductors. Generally they consist of a p-doped base coated on the front with a thin layer of n-doped material, producing the pn junction. On top of this is a metal grid for electrical contact that also serves as a window to allow a high percentage of the incident radiation to pass into the cell. The entire bottom is covered with a metal to provide the other electrical contact. When a photon with energy at least greater than the band gap E_g is absorbed by the semiconductor, an electron–hole pair can be formed with the electron excited into the conduction band, leaving behind the positive hole in the valence band, as shown in Figure 15-13b. The photoexcited electrons migrate preferentially to the n-type side and the photo-produced positive holes move toward the p-type side, regardless of the side of the pn junction on which the electronic transition takes place, making the n-type side negative with respect to the p-type side. This set of migrations produces the photopotential, so that if a wire is connected to the two sides, electrons will flow in the external circuit from the n-type side to the p-type side.[23]

The efficiency of a solar cell is the percent of the incident sunlight power that is converted to electrical output power. In general the major factor limiting the efficiency is the amount of the solar spectrum utilized in the conversion, although recombination processes between electrons and holes that reduce the photocurrent may also be important depending on the types of semiconductors involved.

For power usage, many individual monocrystalline solar cells are connected together to give a module, or amorphous or polycrystalline layers are deposited on a large surface such as glass or plastic or another semiconductor. These can be flat-plate modules to absorb low-intensity diffuse sunlight, or lenses and a tracking mechanism can be used to focus direct sunlight onto the cell.[24] A group of connected modules is called an array.

---

[21]It is assumed as an approximation that there are no charge carriers (neither electrons nor holes) in the depletion region.

[22]There are other ways of generating a potential barrier in a semiconductor (e.g., the Schottky diode), which are, however, beyond the scope of this book.

[23]However, only the photoproduced electrons and holes that cross the depletion region before they recombine contribute to this external current.

[24]Concentrating the light also tends to increase the efficiency of solar cells at lower light intensities, up to concentrating ratios of from 20-fold to several hundredfold, but the cells must be kept cool because high temperatures *reduce* the efficiency.

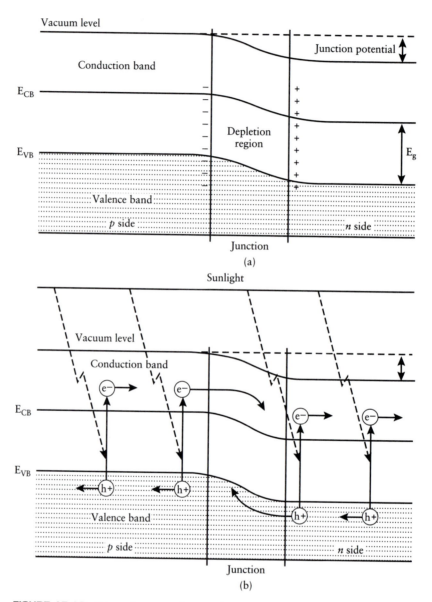

FIGURE 15-13 Schematic energy diagrams for an *np*-junction semiconductor. (Thermally excited electron-hole pairs not shown.) (a) Dark; (b) illuminated.

Many thousands of small stand-alone photovoltaic or hybrid (photovoltaic plus diesel) power systems are in use throughout the world. These provide power for applications such as water pumping, lighting and home power, telecommunications, and vaccine refrigeration, particularly in remote rural areas where in many cases this has become the least expensive method of electric power generation. Presumably, though, the most important impact of photovoltaic devices in addressing the energy requirements of the United States will be with large (up to gigawatt) plants for utility grid distribution. Many experimental and demonstration units for bulk power generation have been built around the world by public and private utilities. For example, the largest ever constructed to date was the 6.5 MW facility at Carissa Plains, California; completed in 1985, it has recently been dismantled. The largest rooftop solar power plant at this date is the 1-MW installation built in 1997 at the New Munich Trade Fair Center. The first photovoltaic project attached directly to a utility power grid near where the electricity will be used (rather than centrally located and distributed to outlying areas as is traditionally done) is the 500 kW plant at Kerman, California that became operational in 1993. Experience has shown that utility-sized photovoltaic plants made up of many single-crystal silicon solar cell arrays providing megawatt power can operate reliably and with satisfactory performance. Several techniques appear to be close to economical feasibility, and when the relatively benign environmental impacts and the appreciable social benefits are considered, photovoltaic-generated utility-scale power is probably competitive with fossil fuel electricity even at the time of this writing.

15.5.5 Photoelectrochemical Utilization of Solar Energy

15.5.5.1 Electrochemical Photovoltaic Cells

We have seen that photovoltaic solar cells are essentially light-sensitive semiconductor rectifiers constructed of solid-state materials. Electrochemical photovoltaic cells, or photoelectrochemical cells, are similar to ordinary electrolytic cells except that one of the electrodes is a photosensitive semiconductor instead of a conducting metal. Irradiation of the semiconductor electrode converts solar energy to electric energy, generating a photoelectric current and potential in the external circuit. This may be the only effect if the oxidation–reduction (redox) reaction at one electrode is the opposite of that at the other electrode so that no overall reaction occurs; this is called a regenerative cell. On the other hand, solar energy may be stored as chemical energy (fuel) if electrolyte products are produced at one or both of the electrodes (a nonregenerative cell). An example of this latter effect, the photoelectrochemical production of hydrogen, is covered in Section 15.5.5.2.

When a semiconductor electrode is immersed in an electrolyte, charge-transfer equilibration between the semiconductor and the electrolyte results in formation of a thin space-charge layer in the semiconductor near the surface that is in contact with the electrolyte. For n-type semiconductors with an excess of electrons (see Section 15.5.4), the equilibration reduces the concentration of the electrons within the space-charge layer, whereas for p-type semiconductors with a shortage of electrons (an excess of positive holes) the concentration of the positive holes is reduced. This produces a potential barrier analogous to the internal potential at a pn junction in a photovoltaic cell (Figure 15-13a); however, in a photoelectrochemical device the barrier is within the semiconductor just beneath the interface with the electrolyte rather than in the interface between the two solids as in a solid-state photovoltaic device. Shining light of photon energy equal to or greater than the band gap energy E_g of a photosensitive semiconductor again produces electron–hole pairs, exciting electrons from the valence band to the conduction band and leaving positively charged holes in the valence band. Thus, if an illuminated n-type semiconductor electrode is connected by the external circuit to an inert metal counter electrode and both are immersed in an appropriate reversible redox electrolyte, the holes migrate to the semiconductor surface, where they oxidize the reduced form of the redox electrolyte, and the electrons flow through the external circuit to the counter electrode where the oxidized component of the redox electrolyte is reduced. Conversely, for p-type semiconductor electrodes the photoexcited electrons migrate to the redox electrolyte at the semiconductor surface, reducing the oxidized form, and the positive holes oxidize the reduced part of the electrolyte at the counter electrode.

The efficiencies for conversion of solar to electrical energy can be quite high for high-quality single-crystal or thin-film semiconductor electrodes. However, in general these devices are very expensive to construct. Polycrystalline semiconductor electrodes, on the other hand, are much less expensive but usually their efficiencies are determined by processes occurring at grain boundaries (such as recombination of photoexcited electrons with positive holes) rather than by inherent characteristics of the semiconductor materials.

A major problem with solar photoelectrochemical cells is the corrosion and degradation of the semiconductor surface.[25] It is possible to minimize this problem by using an n-type semiconductor electrode with sufficiently high band gap ($> 3\,eV$, such as TiO_2)[26] to make it stable, but this has a negative result: namely, the photon energy must be so high to be absorbed that only a

---

[25]A silicon electrode, for example, lasts only a matter of minutes in most aqueous electrolytes.

[26]Titanium dioxide (TiO_2) exists in two crystalline forms with different band gaps: anatase ($E_g = 3.23\,eV$, corresponding to $384\,nm$); and rutile ($E_g = 3.02\,eV$, corresponding to $411\,nm$). It is quite stable in photoelectrochemical processes, but it does corrode slowly in $1\,M$ acids such as H_2SO_4 or $HClO_4$. It is nontoxic and is one of the cheapest and most widely available metal oxides, currently being primarily used in house paints.

small fraction of the sun's spectrum can be utilized. One method of overcoming this disadvantage is by using a visible light-sensitive dye adsorbed to the high-band-gap stable electrode.[27] This approach also allows the exciting photon to be photoabsorbed by the adsorbed dye molecule, not by the semiconductor, which means that a photogenerated positive hole is not produced, and therefore efficiency-reducing electron–hole recombination that takes place in solid-state photovoltaics cannot occur.

15.5.5.2 Direct Solar Energy Production of Chemical Fuels: Hydrogen

We shall show in Section 15.8.3.1 that electrolytic dissociation of water, producing molecular hydrogen and oxygen, is one of the most promising methods of chemical energy storage. Photoelectrolysis of water (i.e., direct solar-to-chemical energy conversion) can also be carried out in a non regenerative photoelectrochemical cell if the band gap of the photosemiconductor is at least 1.229 eV (1000 nm), which is the difference between the H_2O/H_2 and H_2O/O_2 electrode potentials.[28] If the semiconductor electrode is an n-type (see Section 15.5.4), the valence band positive holes react with the water to produce O_2 at the semiconductor surface, the overall electrode reaction in acidic solution being[29]

$$2h^+ + H_2O \rightarrow \frac{1}{2}O_2(g) + 2H^+ \tag{15-8}$$

(h^+ is a positive hole) and electrons reduce H_2O at the counter electrode:

$$2e^- + 2H^+ \rightarrow H_2(g) \tag{15-9}$$

On the other hand, with p-type semiconductor photocathode electrodes, the roles are reversed: the photoexcited electrons reduce water to H_2 at the semiconductor surface and water is oxidized to O_2 at the counter electrode by the positive holes.

---

[27] B. O'Regan and M. Grätzel, *Nature*, 353, 737–740 (1991); M. K. Nazeeruddin, A. Kaye, I. Rodicio, R. Humphrey-Baker, E. Müller, P. Liska, N. Vlachopoulos, and M. Grätzel, *J. Am. Chem. Soc.*, 115, 6382–6390 (1994); A. J. McEvoy and M. Grätzel, *Solar Energy Mater. Solar Cells*, 32, 221–227 (1994) and following papers in Issue No. 3; M. Grätzel, *Renewable Energy*, 5, 118–133 (1994); M. Grätzel, *Prog. Photovoltaics*, 8, 171–185 (2000).

[28] This difference does not depend on whether the electrolyte is acidic or basic, although the individual standard electrode potentials do. Thus, at 298 K in acidic solution $E°(H_2O/H_2) = 0.0$ and $E°(H_2O/O_2) = 1.229$ V, whereas in basic solution $E°(H_2O/H_2) = -0.828$ V and $E°(H_2O/O_2) = 0.401$ V. See Section 9.4.

[29] Note that this is equivalent to

$$H_2O \rightarrow \frac{1}{2}O_2(g) + 2H^+ + 2e^-$$

The first demonstration of this direct method of solar dissociation of water was carried out with an n-TiO_2 semiconductor electrode.[30] The efficiency of hydrogen production was low, however (1%), since as pointed out in Section 15.5.5.1, the energy gap of TiO_2 is over 3.0 eV, resulting in very poor utilization of the total solar radiation reaching the earth's surface. The efficiency can be improved by using semiconductors for both electrodes—an n type for one and a p type for the other. When both electrodes are simultaneously illuminated, the sum of the two energy gaps contributes to the energy needed in the dissociation. An example of such a two-semiconductor photoelectrode cell is one containing an n-GaAs electrode coated with manganese dioxide and a platinum-coated p-Si electrode.

It is worth noting that the space-charge layer exists near the semiconductor–electrolyte interface even if the semiconductor is not part of an electrochemical cell (i.e., even if it is not connected to an inert counter electrode by an external circuit). Therefore, dissociation of water can occur simply on the surface of illuminated semiconductor particles suspended in the electrolyte. Advantages of this method include that the particles can have very high surface-to-volume ratios (if colloidal, e.g.), and that very expensive crystal growth or thin-film deposition processes are unnecessary. The obvious major disadvantage of this technique is that the products of water dissociation (hydrogen and oxygen) are both gases and therefore are evolved together, making it difficult to recover hydrogen as a fuel. On the other hand, if dissolved hydrogen sulfide, H_2S, rather than water is dissociated in a sulfide electrolyte, the oxidized product sulfur forms a variety of polysulfides in solution, the overall reaction being

$$(n-1)H_2S(aq) + S^{2-}(aq) \rightarrow (n-1)H_2(g) + S_n^{2-}(aq) \qquad (15\text{-}10)$$

Elemental sulfur can subsequently be separated from the sulfide solution as a solid. Hydrogen sulfide is an abundant by-product of crude oil desulfurization and low-purity natural gas. An appropriate semiconductor has been found to be platinum-catalyzed CdS; although this undergoes photoanodic degradation, it is stabilized in a high concentration of sulfide ions.

The efficiencies of several of the foregoing photoelectrochemical cells for the production of hydrogen have been shown to be reasonable and have encouraged ongoing research and feasibility studies.[31] However, to date there have been no studies carried out utilizing them in pilot-scale facilities.

15.6 GEOTHERMAL ENERGY

Hot springs and geysers occur in many parts of the world and are visible evidence for the reserves of thermal energy present in the earth. Two applica-

---

[30]A. Fujishima and K. Honda, *Nature*, **238**, 37–38 (1972).
[31]J. R. Bolton, 'Solar photoproduction of hydrogen: A review', *Solar Energy*, **57**, 37–50 (1996).

tions of this geothermal energy are possible: local or district heating, as is widely practiced in Iceland and parts of France and Hungary, for example, and generation of electricity. Principal geothermal electric generation sites are more localized; they include California, Italy, New Zealand, and the Philippines. In 1997 twenty-seven countries were using over 11,000 MW of geothermal energy for various heating purposes, while twenty-one countries generated 7000 MW of electricity (over a third of it in California) from these sources. One estimate suggests that there is as much energy available in accessible thermal reservoirs worldwide as in all the coal deposits. Further development of such sources is technologically feasible, and in appropriate locations they may be expected to contribute significantly to future energy supplies.

The 2000 MW$_{th}$ Nesjavellir power plant in Iceland illustrates a dual use facility. Steam and water are taken from boreholes 1 to 2 km deep and are used in part to provide steam at 190°C and a pressure of 12 bar to generate 60 MW$_e$ of electricity, and in part to provide 1100 l/s water at 100°C for heating purposes in Reykjavik and surroundings (1999 values). This water is transported via an insulated pipeline 25 km long with a temperature drop of about 1°C.

Although steam and hot water issue naturally from geothermal sources, practical utilization of these involves drilling wells into the heated strata. Three types of system are possible.

1. *Dry steam*, that is, steam accompanied by very little water, is the simplest system and is produced by some reservoirs. The steam can be used directly to power turbines for the generation of electric power.
2. *Wet steam*, which is steam mixed with a considerable amount of hot liquid water, is the most common geothermal energy source. Indeed, the liquid usually predominates. These waters, which have been in contact with hot rock under pressure, contain large concentrations of dissolved minerals that give rise to severe problems of corrosion and scale deposition from precipitation of dissolved substances. Three approaches to the use of such energy sources are as follows:
   (a) Separation of the liquid from the steam, and use of only the latter for power generation
   (b) Use of turbines capable of operating with a two-phase mixture
   (c) Use of the mixture or the hot water to vaporize a second, low boiling liquid such as isobutane to operate a turbine in a so-called binary system (a closed cycle is required for the working fluid, which must be condensed and reused)
3. *Hot, dry rock* is a potential energy source, although thermal reserves of this sort are not in use at present. For use, holes would be drilled into the porous or fractured rock, water pumped into some of them, and steam collected from others.

The environmental effects of geothermal power plants have been discussed for one particular example by Axtman.[32] Many of the problems are expected to be common to geothermal sources in general. Most obvious is the hot water waste from condensed steam, geothermal water, and condensed cooling water, although some hot water would be released naturally, regardless of whether it is used, from most reservoirs. Since the temperature of geothermal water and steam is comparatively low (of the order of 100–400°C), the thermal efficiency of power generation from these sources is low in comparison to a modern fossil fuel plant, and a large fraction of the thermal energy of such a system is waste unless used for secondary heating. In addition to the thermal pollution problem, the high mineral content of geothermal waters can be highly corrosive to equipment and can be a hazard, particularly since significant amounts of toxic materials such as arsenic and mercury may be present. This, of course, depends on the composition of the rocks at the source. Injection of waste geothermal waters back into the system could avoid the pollution problem and also prolong the useful life of the source.

A second pollution potential arises from gases released with the steam and water. The CO_2 released may exceed that from an equivalent fossil fuel plant, and hydrogen sulfide is a common constituent. Small amounts of many other gases may be released, depending on the source, but one of major concern is radioactive radon, which occurs in some geothermal sources. Although there are other possible problems, such as earth subsidence, for the most part the environmental problems associated with geothermal energy do not seem to be great.

15.7 WIND AND WATER

15.7.1 Introduction

The winds and waters of the earth offer abundant renewable energy, if their energy can be harnessed economically. Since few, if any, chemical reactions are involved in its use, it is free of most of the chemical pollution problems associated with fossil, nuclear, and even geothermal sources.

15.7.2 Wind Energy Conversion

Wind was one of mankind's earliest sources of energy. Windmills have been used to pump water and perform other kinds of mechanical work for centuries,

[32]R. C. Axtman, 'Environmental impact of a geothermal power plant,' *Science*, **187**, 795 (1975).

but they have been used only since the latter part of the 1800s to produce electric power. After the oil shortages of the 1970s (Section 15.1) and the pressures for developing environmentally benign energy sources, however, major efforts were made to harness wind energy. The United States, through the use of incentives provided by the 1978 Public Utility Regulatory Policies Act, state and federal tax credits, and a guaranteed utility electricity market, has until recently been the leading country in the development of wind turbines and wind farms. The early 1980s expectations for utilization of wind power were overly optimistic, partly owing to poorly designed wind turbines, high capital costs of installation, and little 'benefit' to public utilities to improve the equipment. However, great advances in new turbine performance and reliability have been made, and it appears that there are now no technological barriers to significantly greater usage of wind power.

In fact, wind power is now the second fastest growing energy source in the world. In 2000 total global power capacity of wind turbines was approximately 17.3 GW, with roughly 15% of this in the United States, and the amount is growing about 22% per year. Almost all the existing wind farms in the United States are located in California in three major mountain passes. The wind farm at San Gorgonio Pass, between Los Angeles and Palm Springs, consists of 3500 turbines (Figure 15-14). But we see in Figure 15-15 that

FIGURE 15-14 A typical wind farm in California.

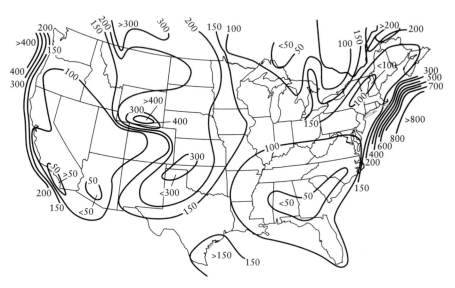

FIGURE 15-15 Average annual wind power available in the United States. The constant-power contours (*isodyns*) are in W/m^2 of rotor-swept area ($100\,W/m^2$ corresponds to an average annual energy of $0.88\,MWh/m^2$). From J. O'M. Bockris, B. Dandapani, and J. C. Wass, "A Solar Hydrogen Energy System," in *Advances in Solar Energy*, Vol. 5, K. W. Böer, ed., Plenum, New York, 1989.

there are at least 14 states, primarily in the Great Plains from Texas to Canada, with equal or greater wind power potential.[33] Two problems with this region for utilization of wind energy are that it also has large coal deposits, making for potentially much cheaper coal-fired electric generation, and that the desirable wind sites are generally in remote areas far from large-load areas, requiring major transmission distribution facilities. Even at that, it is estimated that eventually about 20% of U.S. electrical power may come from the wind. Significant increases in wind power generation are also planned in other countries. Some estimate, for example, that 100 GW of wind power will be in operation in Europe by the year 2030—well above the projection for the United States.

Wind power does not emit polluting substances, nor does it produce unwanted material that requires disposal. There appear to be only minor environmental impacts associated with the installation of rural wind farms for the production of electric energy, aside from the possible disturbance of

[33]Thirty-seven states have been identified with sufficient wind resources to support utility-sized wind farms.

wildlife habitat[34] and farming or ranching[35] and the negative visual impacts of large wind farms on the natural beauty of an area. Wind turbines are not considered to be noisy machines; however, some noise is generated in their operation, and this has led to negative reactions of the public in some areas where construction of wind farms are proposed. The primary sources of noise are aerodynamic noise from the rotating blades sweeping through the air and the machinery noise from the electrical generator equipment and other mechanical components (gearbox, bearings, etc.). Although there are not established noise standards for wind farms, noise greater than 10 dB (decibels) above ambient may be considered to be undesirable. Wind turbines may also interfere with telecommunications, but not if the rotor blades are nonmetallic.

15.7.3 Water Energy Conversion

15.7.3.1 Hydroelectric Power

Hydroelectric power, or simply hydro power, is the power generated when the potential energy of water at a higher level is converted eventually to electrical energy by the water flowing through a turbine to a lower level. In a sense it is a massive utilization of solar energy, since 23% of the solar radiation illuminating the earth is involved in evaporating the ocean waters that condense and form the sources of hydro power. First used to generate electricity commercially in 1880, hydro power is now the most developed form of renewable energy, producing 24% of the world's electricity. In the United States it provided about 9% of the electricity generated in 1999. In Norway, a very mountainous country, 99% of the electrical energy is produced by hydro power.

The main sources of harnessable (i.e., developable) hydro power are the earth's large river systems. Whether the power potential of a river system can be harnessed depends on several factors, including topography, climate, rainfall, and the altitude gradient. Mountainous regions have larger altitude gradients than plains, but in many cases the greater surface areas of the plains may provide more energy potential than the mountains. Although the harnessed potential of the Columbia River in the northwestern United States and southwestern Canada is currently the greatest in the world at over 90 TWh, the Zaire (Congo) River in Africa has the largest harnessable potential at 700 TWh. In 1985 the harnessed potential of the world was 2200 TWh, or 22.4% of the harnessable power. The United States has 39.6% harnessed, while the

---

[34]Some of the wind farms in the California passes are in bird migration paths, and significant bird kills have been reported.

[35]The actual physical structures occupy only about 1% of the land devoted to a wind farm power plant. Thus the remaining 99% is available for agricultural and natural habitat uses.

harnessed power of France is close to 100%. The world's largest hydro power plant (12.6 GW) is the Itaipú Dam between Paraguay and Brazil. However, it will be surpassed by the massive Three Gorges Dam under construction in China. This dam, expected to be completed by 2009, will have 18.2 GW of installed power generation.

There are many advantages to the utilization of hydro power. It is nonpolluting and is automatically self-replenishing. The water is unchanged chemically after its potential energy is converted to electrical energy, and therefore is available "downstream" for further power generation, and so on. Unlimited by the second law of thermodynamics, hydro power plants have high efficiencies (potentially > 80%); their usable lifetimes are long, and in general they seldom experience power outages and maintenance shutdowns. They emit no atmospheric pollutants. Construction of dams to form storage reservoirs often leads to other advantageous uses such as flood control, irrigation, and recreation, although the concomitant submergence of land areas may have adverse environmental, economical, and recreational effects, particularly in agricultural or wilderness areas.

The damming of rivers can affect wildlife habitat, particularly where wetlands used by migratory birds and shallow pools needed for spawning of fish are submerged below layers of sand in deep water. Damming can also have a negative influence on the migration of fish, although this can be substantially mitigated in some cases by installing appropriate fish ladders or elevators, and by controlling operations to allow young fish to bypass turbines. In a dramatic policy reversal, at the time of this writing the government is considering the removal of several small power dams in the eastern and western parts of the United States, particularly where the loss of power would be minimal compared to the impact the dams continue to have on migratory fish.

The gradual deposition of silt (siltation) may affect the capacities of storage reservoirs behind dams. Generally this effect is negligible in forest-covered areas, but it may be enormous in regions such as desert areas that receive violent rainstorms and are not protected by vegetation or where there is concurrent deforestation of the watershed.

15.7.3.2 Tidal Power

Energy can be extracted from the rise and fall of ocean tides—an enormous source of power. Extraction of tidal power is quite simple in principle. A tidal dam or "barrage" across an estuary creates an enclosed basin for storage of water at high tide. Turbines in the barrage are used to convert the potential energy resulting from a difference in water level ultimately into electrical energy.

There are five ways that tidal barrages can operate to generate electricity. The simplest (and probably the most economical) is ebb generation, where flow is toward the sea at low tide. Reverse flow—from sea to basin at high tide—is flood generation. A modification of ebb generation is ebb generation plus pumping: the turbines are operated in reverse soon after high tide to pump water into the basin (to give a net energy gain, the additional water is released at a higher head than when it was pumped). Here the advantage is that electrical generation can start sooner than with ebb generation alone. Two-way generation utilizes flow in both directions. In the two-basin generation scheme, which provides smoother power output, two interconnected water reservoirs are used.

There are several environmental concerns with tidal power. Obviously, construction of a tidal barrage across an estuary changes the natural scenery. It also affects water quality (salinity, turbidity, dissolved oxygen, bacteria, sediment), fish and ocean fowl, and surface travel, since strong currents are produced and tides are modified within the estuary.

The oldest large tidal power scheme operating in the world (commissioned in 1966) is the 320-MW barrage at La Rance, near Saint-Malo, France (site of a twelfth-century tidal mill), where the mean tidal range is about 8 m but is 13.5 m during the spring equinox. The other currently operating scheme is a 20-MW pilot program at Annapolis Royal in the Bay of Fundy, Nova Scotia, Canada, commissioned in 1985. (The Bay of Fundy has the largest listed mean tidal range in the world, 11.28 m.) It is now estimated that tides of at least 5 m are required for tidal power to be economically feasible in competition with other energy sources. In the United States, only the southern part of the Bay of Fundy (Passamaquoddy Bay), with a mean tidal range of about 5.4 m, meets this criterion. Work was started there on a 110-MW two-basin scheme in 1935 but was abandoned the following year when it was determined that normal hydroelectric sources in Maine were more economical. Other sites in the world at which extensive economic feasibility and engineering studies have been carried out over the past 20 years, include the Severn and Mersey estuaries on the west coast of England and Wales, and sites in Russia and China.

Tidal power is a truly renewable and predictable energy source. Carefully designed and constructed tidal barrages, while requiring several years for construction, are expected to last many years. Unfortunately, the economics are such that electricity to be produced many years in the future has little present value. Its real value, though, may be very large when accessible oil reserves are depleted or when the discharge of atmospheric pollutants from combustion of fossil fuels becomes a major concern.

15.7.3.3 Wave Power

Wave energy comes from the winds as they blow across the oceans. As such, the wind energy is concentrated in the water near its surface and is attenuated only over large distances.

The first patent describing a wave energy device was issued about 1800 in France, and an apparatus constructed near Bordeaux, France, in 1910 used the oscillations of seawater in a vertical bore hole to pump air that ran a turbine, producing 1 kW of electrical power. Small wave power generators have been in use for over 25 years for navigational aids. However, much research and development has been undertaken, particularly since the energy crises of the 1970s, in several of the developed maritime countries. Viable commercial facilities are now actively being developed in Japan, Norway, and the United Kingdom. (It is estimated that if the entire coastlines of these three countries were to be used to convert wave to electrical energy, the percentages of the 1985 national electricity demand would have been 15, 50, and 15%, respectively; for the United States, it would have been 7%.)

There are many ways to convert wave energy to more usable forms; as with other forms of renewable power, though, it is important that the systems be rugged and simple. The most practical method now appears to be the pneumatic technique. Basically this system, shown in Figure 15–16, consists of a capture chamber (in which wave energy is trapped), which produces an oscillating water column that pumps an air turbine mounted above the water surface. Generally the capture chamber is designed to resonate with the wave frequency, but chambers can also be constructed that have two resonance frequencies, one for the chamber and one for the column, which in effect

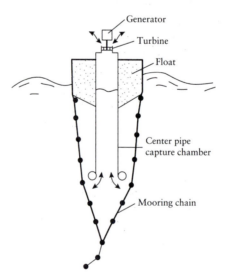

FIGURE 15-16 Wave energy conversion chamber. From M. E. McCormick and Y. C. Kim, eds., "Utilization of Ocean Waves—Wave to Energy Conversion," American Society of Engineers, New York, (1987).

makes it possible to double the energy production. In practice, many air chambers that match the frequency of the incident wave are combined. Oscillating water column systems can be floating, as on a ship, or shore-fixed. A 10-year study using a floating system known as the KAIME I was undertaken by Japan in 1976. This system generated approximately 620 kW of instantaneous maximum power and an annual energy output of 190 MWh. Norway has constructed a shore-fixed multiresonant prototype power station in the 500-kW range on its west coast at Tofteshallen.

Wave power stations attenuate wave motion and thus make waves less steep, which may tend somewhat to reduce beach erosion. The greatest environmental effect of wave energy converters probably will be on non-bottom-feeding fish. There is also the possibility of interference with shipping, fishing, and ocean recreational activities. In general, though, environmental impact problems appear to be much less severe than those of virtually any other energy conversion process.

The various renewable energy sources discussed in Sections 15.5–15-7 are summarized in Table 15-5. Many of the quantitative estimates given in that table may be expected to change with further technological development.

15.8 ENERGY STORAGE

15.8.1 Introduction

If large thermal or nuclear power stations are operated below a fairly narrow range of their rated output capacity, fuel consumption does not decrease in proportion to the decrease in electricity output. This leads to a waste of fuel and increased unit power cost. On the other hand, power demand is not constant over the seasons or even over a single day, being the greatest usually in midmorning and evening and the lowest after midnight to early morning. Thus, appreciable savings are possible if such plants can be operated full time near maximum capacity, with the excess electricity stored during low-demand periods and released quickly during peak demands.

Similarly, solar energy is intermittent. Direct usage during sunlight hours may be adequate in some applications, but it is obvious that for increasing long-term utilization of solar energy (as well as other fluctuating renewable energy sources such as wind power), some auxiliary storage is required to provide effective continuous energy supply. We have already seen in Section 15.5.3 that water (e.g., solar ponds), rocks, molten salts, and liquid sodium are frequently used for solar energy storage as sensible and/or latent heat. These are short-term and limited-capacity storage means, however, and generally not sufficient for large-scale usage.

TABLE 15-5

Comparison of Renewable Energy Sources[a]

Energy source	Capital cost ($U.S./kW)	Anticipated change in capital cost over next decade	Delivered cost ($U.S./kWh)	Advantages	Disadvantages
Biomass	2500–3500	30–50% decrease	0.05–0.08	CO_2 emission-neutral; usable for base and peak energy loads; large potential	Water pollution; requires large land area (competes with food production); may lead to some biodiversity loss; energy input may be a large fraction of output
Solar thermal	2000–4000	25% decrease	0.10–0.25	Negligible environmental pollution; potentially usable over $\pm 40°$ latitude; plant sizes from small (10 kW) to large (hundreds of megawatts); usable with other energy sources (hybrid)	Considered to be a high-risk technology
OTEC (ocean thermal energy conversion)	5000–25,000	25–35% decrease	0.06–0.32	Negligible environmental pollution; desalinated water and mariculture desirable by-products	Probably will need to use by-products to be economically viable; very low efficiencies
Photovoltaic	8000–35,000	40–50% decrease	0.25–1.50	Negligible environmental pollution; modular; no moving parts; eventually should have low operating and maintenance expenses	Intermittent; very high capital and production costs; large land areas needed for big plants; very dependent on location
Geothermal	~ 2000		0.015–0.05	Commercially feasible; capable of large-scale base load electrical power generation	Noxious emissions (e.g., H_2S) and water pollution unless reinjected; available only in discrete locations
Wind	800–3500	20–35% decrease	0.04–0.10	No emissions; can range from few hundred watts to many-megawatt stand-alone applications; already cost-effective in some applications	Intermittent; possible impact on wildlife and bird migration; undesirable visual and sound effects

(continues)

TABLE 15-5 (*continued*)

Energy source	Capital cost ($U.S./kW)	Anticipated change in capital cost over next decade	Delivered cost ($U.S./kWh)	Advantages	Disadvantages
Hydro	1000–2000	Slight increase	0.01–0.03	Most developed form of renewable energy; provides electric generating and "pumped storage" capabilities; automatic self-replenishing; large dams can control flooding and provide irrigation and recreation	May occupy large land areas; disruption of water flow pattern may seriously impact the ecology downriver and in adjacent areas; affects fish migration
Tidal	~ 1500			No emissions; predictable source of energy; practical from small to mega scale	Very limited sites available; may have major impact on water quality, fish, and marine mammals; produces strong currents; modifies tides; may lead to sediment buildup or erosion
Wave	1600–7000		0.04–0.33	Negligible environmental pollution; potential for decreasing beach erosion; possible simultaneous mechanical water desalination)	Minor effects on marine environment; variable with weather; low efficiency, available only near coast (potential for negative visual and ecological impact on coastal areas)

[a]Cost estimates are those available in 1998 and are subject to change
Sources: *Key Issues in Developing Renewables. Report of the International Energy Agency*, OECD, Paris, 1997. *The Future for Renewable Energy: Prospects and Directions*, EUREC Agency, James & James, London, 1996. N. Cole and P. J. Skerrett, *Renewables Are Ready: People Creating Renewable Energy Solutions*, Chelsea Green, White River Junction, VT, 1995.
R. J. Seymour, ed., *Ocean Energy Recovery: The State of the Art*, American Society of Civil Engineers, New York, 1992. U.S. Department of Energy website: http//geothermal.id.doc.gov/

A key parameter for evaluating and comparing various means of storing energy, particularly for powering mobile vehicles where weight may be a controlling factor, is energy density (or, perhaps more correctly, specific energy).

This is the energy stored in the carrier, in watt-hours or kilowatt-hours, per kilogram of material (Wh/kg or kWh/kg). For example, we have seen (Section 15.5.3.1) that calcium chloride hexahydrate can store 2.8×10^5 kJ/m^3; this is equivalent to an energy density of 0.045 kWh/kg. A typical 12-V lead–acid storage battery (Section 15.8.3.2) has a useful energy density of about 0.03 kWh/kg. Advanced thermochemical heat storage technologies have energy densities as high as 1.1 kWh/kg.

15.8.2 Physical Energy Storage

15.8.2.1 Pumped Storage

Pumped storage is an extension of hydroelectric power. Electricity is in effect stored as potential energy by pumping water up to a high-level reservoir during off-peak load periods and released to a lower level (usually at least 100 m) through turbine power plants during peak power demands. This is an established flexible and proven technology. Maximum output can be attained in less than one minute, and the overall efficiency of such a pump-discharge cycle can be as high as 65–75%. In the United States alone, total pumped-storage electricity capacity in 1999 was approximately 19 GW.

Instead of pumped water, compressed air can also be used for energy storage. Off-peak electricity is used to pump air into underground caverns, apparently at considerably less cost if the appropriate cavern (either natural or excavated from rock or salt) is already available. Electricity is generated by withdrawing air and heating and expanding it through gas turbines. This technique was pioneered by Germany with a 290-MW plant at Huntdorf, and in Alabama a 500,000-m^3 salt dome reservoir now being used for stored pumped air is capable of generating 110 MW of electric power for 26 hours.

15.8.2.2 Flywheels

An electrically run spinning flywheel stores energy by converting electrical energy into mechanical energy of rotation. Ideally, in the absence of a load, it runs freely without friction by rotating on noncontacting magnetic bearings in an evacuated chamber. (But even with bearings of this type, flywheels will lose on the order of 0.1% of stored energy per hour.) In essence it acts both as a motor (the electric current spins the wheel) and as a generator (magnets mounted in the flywheel produce an electric current). These devices are particularly attractive for storing solar energy because a direct-current motor can be used to power them and an alternating-current motor can be used to remove the stored energy; thus they can serve as a dc-ac converter. Although they can be very efficient ($> 70\%$) and may have an energy density as high as 0.15 kWh/kg, rotational velocities as high as 100,000–200,000 rpm may be necessary for practical applications, requiring high-strength composite material. To date a fully integrated flywheel system (sometimes called an electromechanical flywheel battery) has not been demonstrated, although such units are being developed for stationary energy storage (where large units can be buried for safety purposes in the event of a flywheel "explosion") and for automotive propulsion.

15.8.2.3 Superconductivity

Superconductors have zero resistance, and therefore virtually no energy loss in the absence of an applied load. This means that a coil made out of a superconducting material can serve as an energy storage device: electric energy, fed into the coil, is conducted almost indefinitely until it is withdrawn as needed. One problem, however, is that the known superconducting materials exhibit superconductivity only at low temperatures. The critical temperature T_c is the temperature at which the resistance suddenly drops to zero. Before 1986 the highest known T_c was about 23 K. Since then new materials, such as the nonstoichometric oxide of yttrium, barium, and copper, $YBa_2Cu_3O_{7-x}$ (where $x \leq 0.1$), and mercury-based cuprates have led to T_c values well above the boiling point of nitrogen (77 K, $-196°C$). This means that liquid nitrogen, which is plentiful and cheap, can be used as coolant instead of the much more expensive liquid helium. However, the new "high-T_c" superconducting materials are very expensive, and construction costs will undoubtedly be quite high. For example, the storage units probably will have to be underground to control twisting forces from the high electric currents in the coil and to allow for possible coolant failure and the resultant loss of superconductivity, which would lead to violent release of the stored electric energy. Other engineering and fabrication problems exist, but the technology looks promising, and several prototype systems are currently being developed.

15.8.3 Chemical Energy Storage

15.8.3.1 Hydrogen

Hydrogen can be produced from the dissociation of water (an inexhaustible source) by several means, and it can be stored and transported over long distances by well-established techniques. It has excellent combustion properties and can be used in virtually all applications now primarily fueled by fossil sources such as power stations, internal combustion engines, and many other household and industrial areas. Its stored chemical energy can be converted directly to electrical energy by a fuel cell.[36] Based on energy utilization by combustion to water, it has an energy density of about 40 kWh/kg of hydrogen,[37] and catalytic burning of H_2 gives high heat value with an efficiency greater that 95%. It has, however, a wider flammable range, lower ignition energy, and higher flame velocity when mixed with air than do most gas and liquid hydrocarbon fuels, making it a more hazardous material than most fossil fuels.

We saw in Section 15.5.5.2 that experimental techniques using photosensitive semiconductor electrodes are being developed for the photoelectrolysis of water by direct solar-to-chemical energy conversion. However, electrolysis of water has been commercially available for over 80 years. It is one of the simplest methods for producing hydrogen, and in fact is the only process virtually free of pollutants when carried out with electricity produced from renewable primary energy sources such as solar energy (i.e., solar to electrical to chemical energy). Electrolysis of water is up to 95% efficient and has the advantage that the H_2 and O_2 products are physically separated when produced.[38] Conventionally the electrolysis is carried out ($\leq 90°C$) at atmospheric pressure in aqueous 25–30% KOH solution with nickel-plated steel or steel gauze electrodes, with the electrodes separated by a thin membrane. Hydrogen is evolved at the cathode

$$2H_2O + 2e^- \rightarrow 2OH^- + H_2(g) \qquad (15\text{-}11)$$

and oxygen at the anode

$$2OH^- \rightarrow H_2O + \frac{1}{2}O_2(g) + 2e^- \qquad (15\text{-}12)$$

with the sum of the two giving the net two-electron decomposition

$$H_2O \rightarrow H_2(g) + \frac{1}{2}O_2(g) \qquad (15\text{-}13)$$

---

[36]Fuel cells are electrochemical cells that convert chemical energy, in the form of a continuous supply of fuel, directly into electrical energy. See Section 15.9.

[37]Chemicals generally have energy densities from 1 to 10 kWh/kg. Hydrogen is exceptionally large because of its low molecular weight.

[38]It should be pointed out that at the present time electrolysis is too expensive (primarily because of the cost of electricity) to be used on a large scale for the production of hydrogen. Industrial hydrogen is almost exclusively produced by the catalytic steam reforming of hydrocarbons

$$C_nH_m + nH_2O \rightleftharpoons nCO + \left(n + \frac{m}{2}\right)H_2$$

and the water-shift reaction

$$CO + H_2O \rightleftharpoons CO_2 + H_2$$

However, this method involves a nonrenewable feed source (hydrocarbon), and therefore it is not functioning as an energy storage procedure. It also generates CO_2, a greenhouse gas (see Section 3.1.2).

The theoretical cell potential for the dissociation of water is 1.229 V at 25°C and 1 bar; in practice, this is higher because of electrode surface overvoltages and resistive losses. (Most of the energy loss is at the oxygen electrode.) Processes are being developed to carry out the electrolysis of water at temperatures as high as 900°C. One of the major advantages of operating at a high temperature is that the electrical energy required is decreased with increasing temperature.[39] However, there are yet-to-be-solved problems in adequate high-temperature materials and construction techniques.

Water can be dissociated in combination (hybrid) thermochemical–electrochemical processes or by purely thermochemical means. As an example, the Mark processes involve as a first step the endothermic high-temperature ($\leq 1000°C$) dissociation of sulfuric acid:

$$H_2SO_4 \xrightarrow{\text{heat}} H_2O + SO_2 + \frac{1}{2}O_2 \qquad (15\text{-}14)$$

The thermal energy can come from combustion, nuclear, or solar sources. In the hybrid process, this is followed by an electrochemical step in which the SO_2 is oxidized back to H_2SO_4

$$SO_2 + 2H_2O \rightarrow H_2SO_4 + 2H^+ + 2e^- \qquad (15\text{-}15)$$

and hydrogen gas is evolved by the reduction of H^+:

$$2H^+ + 2e^- \rightarrow H_2 \qquad (15\text{-}16)$$

giving the overall reaction

$$SO_2 + 2H_2O \rightarrow H_2SO_4 + H_2 \qquad (15\text{-}17)$$

In the purely thermochemical process, the oxidation of SO_2 and formation of H_2 occurs in the presence of I_2 catalyst via a complex cyclic process involving polyiodides and sulfuric acid intermediaries as well as other catalytic decomposition reactions. The advantage of the purely thermochemical method is that thermal energy is converted into chemical energy (the H_2 and O_2 products of the dissociation) without first converting the heat into mechanical energy and then into electrical energy. Since, however, these are multistage processes with some stages that involve highly corrosive materials, equipment and apparatus costs become an important factor in comparing these techniques to conventional water electrolysis. Several demonstration pilot plants are currently in use.

---

[39]This is shown as follows. The electrical energy required is equal to ΔG, the change in the Gibbs function. At a constant temperature T, $\Delta G = \Delta H - T\Delta S$. For the dissociation of liquid water, both the change in enthalpy and the change in entropy are positive: $\Delta H = 285.8$ kJ/mol and $\Delta S = 163.3 \, \text{JK}^{-1}\text{mol}^{-1}$ at 25°C. Therefore, assuming ΔH and ΔS are independent of temperature (a reasonable first approximation), ΔG decreases with increasing temperature.

Hydrogen can be stored as a gas or as a liquid, or chemically, or cryoadsorbed (adsorption at low temperatures). It is the least expensive long-term energy storage medium among electrochemical systems. Because of its low volume density, however, gaseous hydrogen generally has to be compressed for feasible storage. The triple point of hydrogen (the temperature at which solid, liquid, and gas are in equilibrium) is 14 K, its normal boiling point is 20.4 K, and its critical temperature is 33.3 K. Although liquefying hydrogen at 20 K increases its density by a factor of 845, the costs of cooling and storage at this low temperature (e.g., about 30% of its combustion energy is required to cool and condense it) make liquid storage economically unrealistic for large-scale storage or transport over long distances.

An example of solid chemical storage of hydrogen is the exothermic formation of regenerable metal hydrides

$$M + xH_2 \rightleftharpoons MH_{2x} \qquad (15\text{-}18)$$

where M can be a metal such as magnesium, titanium, zirconium, or lanthanum, or intermetallic compounds or mixtures of metals (alloys). Nickel and its alloys are particularly promising, and metals such as magnesium doped with 5–10 wt% nickel show greater storage capacity and better cycling (hydrogenation/dehydrogenation) stability than the metals alone. The theoretical maximum energy density of nickel-doped magnesium hydride as a hydrogen storage material is about 3 kWh/kg. Hydrogen can also be converted to other molecules such as ammonia or methanol for easier storage and transportation, but at the cost of energy for its subsequent release. It is also cryoadsorbed below 77 K (the boiling point of liquid nitrogen) on a variety of active surfaces of different materials, such as active carbon, which increases the energy density of the hydrogen. However, both metal hydrides and cryoadsorbed hydrogen materials are affected by repetitive cycling. Another technique being studied for hydrogen storage is the use of carbon nanotubes with diameters the size of several hydrogen molecules; tubes are bundled together and hydrogen is drawn into them, making essentially a lightweight hydrogen sponge.

15.8.3.2 Batteries

A battery is an electrochemical cell in which stored chemical energy is spontaneously converted to electrical energy. This produces an electric potential across the cell's two electrodes, which are separated in the cell by an electrolyte solution (or sometimes a solid ionic conductor), and generates a current through a connected external load. Electrochemical reactions occur at the electrodes: reduction, a gain of electrons, at the cathode; and oxidation, a loss of electrons, at the anode. Many commonly used consumer batteries are discarded when most of the chemical energy has been converted (the cell has been discharged). For a battery to function as an intermediate energy storage device, however, it must also be capable of being recharged—that is, of restoring its chemical energy from electrical energy, generated, for example, by photovoltaic solar cells. To be an effective storage battery a device also must be able to cycle through this charge/discharge process many times.

The most commonly used rechargeable storage battery for photovoltaic solar energy storage and for mobile vehicles[40] is the lead–acid battery, made up of one or more series-connected cells. Each cell in the charged state consists of a solid (sponge) lead (Pb) electrode, a solid lead dioxide (PbO_2) electrode, and an aqueous sulfuric acid electrolyte. Upon spontaneous discharge, the Pb is oxidized to solid $PbSO_4$ at the negative electrode,

$$Pb(s) + H_2SO_4(aq) \rightarrow PbSO_4(s) + 2H^+(aq) + 2e^- \qquad (15\text{-}19)$$

and the PbO_2 is reduced to solid $PbSO_4$ at the positive electrode:

$$2H^+(aq) + H_2SO_4(aq) + PbO_2(s) + 2e^- \rightarrow PbSO_4(s) + 2H_2O(l) \quad (15\text{-}20)$$

with the net cell discharging reaction being

$$Pb(s) + PbO_2 + 2H_2SO_4(aq) \rightarrow 2PbSO_4(s) + 2H_2O(l) \qquad (15\text{-}21)$$

The reverse reactions occur on charging. Maintenance-free, hermetically sealed lead–acid units are now being produced. A major disadvantage of these for applications requiring portability, however, is their relatively low energy storage capacity of 0.03 kWh/kg (see Section 15.8.1). They also have a limited cycle life, they contain toxic and environmentally damaging materials, and they are damaged by complete discharge and overcharging (deep cycling), which removes active solid material from the electrodes, as well as by high operating temperatures. Environmental contamination from lead in manufacturing and in disposal or recycling processes is also of concern (see Section 10.6.10).

The nickel–cadmium (NiCd) battery is also used commercially in solar cell storage applications and in some experimental electric vehicles. When charged it consists of a cadmium negative electrode, a positive electrode of nickel oxohydroxide $NiOOH \cdot H_2O$ (with some graphite for conduction), and an aqueous potassium hydroxide electrolyte. When being discharged, cadmium is oxidized to $Cd(OH)_2$

$$Cd(s) + 2OH^-(aq) \rightarrow Cd(OH)_2(s) + 2e^- \qquad (15\text{-}22)$$

and the oxohydroxide is reduced to $Ni(OH)_2$

$$2NiOOH \cdot H_2O(s) + 2e^- \rightarrow 2Ni(OH)_2(s) + 2OH^- \qquad (15\text{-}23)$$

---

[40]At the time of this writing, commercially available electric vehicles require roughly 1 kWh to travel 10 km.

giving the net spontaneous cell reaction

$$Cd(s) + 2NiOOH \cdot H_2O(s) \rightarrow Cd(OH)_2(s) + 2Ni(OH)_2(s) \qquad (15\text{-}24)$$

Initially NiCd batteries are more expensive than the comparable lead–acid systems and have only slightly higher energy density (0.05 kWh/kg). However, they can tolerate more extreme operating temperatures and greater cycling depths—for example, they can be completely discharged and/or overcharged without permanent harm—hence generally have a longer design life ($\leq$ 20 years). They also can be hermetically sealed. Since the hydroxide ion is neither a reactant nor a product in the overall reaction, the density of the electrolyte solution, potassium hydroxide, does not change during charging or discharging, in contrast to the lead–acid battery. However, cadmium is a toxic, environmentally undesirable material (Section 10.6.7), and proposals have been made in the European Union to phase out (ban) the use of these batteries.

The battery that appears at this time to be the most probable replacement for the lead–acid battery in future electric vehicles is the rechargeable nickel–metal hydride (NiMH) battery.[41] We have seen that hydrogen can be stored in a solid metal hydride phase (Section 15.8.3.1). In the nickel–metal hydride battery, the negative electrode is an appropriate metal hydride, MH,[42] and the positive electrode is solid nickel oxohydroxide (as in the NiCd battery), with a concentrated KOH aqueous electrolyte. Upon discharging (the reverse reactions occur during charge), the hydrogen in MH is oxidized to water,

$$MH + OH^- \rightarrow M + H_2O + e^- \qquad (15\text{-}25)$$

and nickel oxohydroxide is reduced to nickel hydroxide [reaction (15-23)], giving the overall discharge reaction

$$NiOOH \cdot H_2O + MH \rightarrow Ni(OH)_2 + M + H_2O \qquad (15\text{-}26)$$

This battery has a nominal voltage of 1.2 V and an acceptable energy density of approximately 0.07 kWh/kg. It can be sealed, and it has an appreciably longer lifetime than a comparable lead–acid battery. However, the nickel metal hydrides are very expensive to prepare and in the long run may be environmentally unacceptable.

Another system that may turn out to have a reasonable energy density is the lithium ion cell. (The name comes from the exchange of lithium ions between

---

[41]S. R. Ovshinsky, M. A. Fetcenko, and J. Ross, A nickel metal hydride battery for electric vehicles, *Science*, 260, 176–181 (1993). It is worth noting that a NiMH battery is now being used in the hybrid Toyota Prius automobile.

[42]M is usually a metal alloy, optimized to give the most desirable physical and chemical properties. A conventional alloy for metal hydride batteries is LaNi$_5$ mixed with rare earth elements such as Ce, La, Nd, and Pr. The proprietary MH electrode of the Ovonic battery (Ovonic Battery Co., a subsidiary of Energy Conversion Devices, Inc.) consists typically of a V, Ti, Zr, Ni, and Cr alloy.

the positive and negative electrodes during charging and discharging.) Electrochemical cells using lithium electrodes can theoretically produce a higher voltage than lead–acid or NiCd cells, and this in principle can lead to lighter weight storage batteries and therefore higher energy densities (0.15 kWh/kg). However, lithium metal reacts violently with water, forming LiOH and H_2. This means either that a nonaqueous electrolyte solution (such as lithium ion salts dissolved in organic compounds) must be used or that the lithium atoms need to be "inactivated"—for example, by intercalating[43] them in an appropriate solid that binds them so tightly that they do not react with water. An example of the former is the Li—MnO_2 battery, in which Li is the negative electrode, solid MnO_2 is the positive electrode, and the electrolyte is lithium perchlorate ($LiClO_4$) dissolved in a mixed organic solvent such as propylene carbonate and 1,2-dimethoxyethane. The spontaneous discharging reactions are

$$Li \rightarrow Li^+ + e^- \tag{15-27}$$

and

$$Mn^{IV}O_2 + Li^+ + e^- \rightarrow Mn^{III}(Li^+)O_2 \tag{15-28}$$

with the overall net reaction

$$Li + Mn^{IV}O_2 \rightarrow Mn^{III}(Li^+)O_2 \tag{15-29}$$

One advantage of using cells of the latter type (where the lithium atoms are "inactivated") is that a relatively cheap aqueous electrolyte can be used. A promising experimental rechargeable cell has been developed[44] with an aqueous 5M LiOH electrolyte that utilizes solid intercalation compounds such as $LiMn_2O_4$ and VO_2 as the electrodes in the completely discharged cell. During the charging process, some lithium is oxidized and removed from $LiMn_2O_4$ at the positive electrode

$$xLiMn_2O_4 \rightarrow xLi_{(1-1/x)}Mn_2O_4 + Li^+ + e^- \tag{15-30}$$

and the lithium ion is reduced and intercalated into vanadium oxide at the negative electrode,

$$yVO_2 + Li^+ + e^- \rightarrow yLi_{(1/y)}VO_2 \tag{15-31}$$

so that the overall charging reaction is

$$xLiMn_2O_4 + yVO_2 \rightarrow xLi_{(1-1/x)}Mn_2O_4 + yLi_{(1/y)}VO_2 \tag{15-32}$$

---

[43]Intercalation is the reversible insertion of a substance into the bulk of a solid material.

[44]W. Li, J. R. Dahn, and D. S. Wainwright, "Rechargeable lithium batteries with aqueous electrolytes," *Science*, **264**, 1115–1118 (1994).

Thus, the spontaneous discharge of the cell, which is the reverse of reaction (15-32), represents the removal of intercalated lithium from $Li_{(1/y)}VO_2$ and its intercalation into $Li_{(1-1/x)}Mn_2O_4$. This cell has a working voltage of about 1.5 V, and it is estimated that a practical battery constructed of these materials would have an energy density of the order of 0.055 kWh/kg. However, safety is a major problem in scaling up to the battery sizes that would be necessary for economic use in electric vehicles. Another rechargeable lithium ion battery is the lithium polymer system, in which the problem of reaction between lithium and the solvent is eliminated by using a polymer matrix membrane, such as lithium perchlorate in polyethylene oxide, to separate the negative lithium metal electrode from the positive electrode. The negative electrode is made of polymer-bonded graphite attached to metal foil; passive films that form on the surface prevent reaction with the electrolyte (a solution of lithium hexafluorophosphate in propylene carbonate and other solvents), yet allow relatively free passage of Li^+ ions during current flow. The positive electrodes are made from metal (nickel, manganese, or vanadium) oxides.

The sodium–sulfur battery has two molten electrodes (a molten sodium anode and a molten mixture of sulfur and sodium polysulfides cathode) with a solid β- alumina—nominally $Na_2O \cdot 11Al_2O_3$—electrolyte that allows transport of sodium ions at elevated temperatures. When discharging, sodium is oxidized at the negative electrode

$$Na(l) \rightarrow Na^+(solv) + e^- \tag{15-33}$$

and sulfur is reduced to polysulfides at the positive electrode:

$$2Na^+(solv) + xS(solv) + 2e^- \rightarrow Na_2S_x(solv) \tag{15-34}$$

where $x = 3$–5. The net cell reaction is

$$2Na(l) + xS(solv) \rightarrow Na_2S_x(solv) \tag{15-35}$$

Sodium and sulfur are cheap and readily available. They are light elements and have a large reaction energy (the battery has a theoretical energy density of about 0.15 kWh/kg, more than four times that of the lead–acid battery, and an energy efficiency of approximately 85%), so it can be light and compact—an important property for propulsion of electric vehicles. The major disadvantage is that the battery must be operated between 300 and 350°C to keep the sodium, sulfur, and polysulfides liquefied, as necessary to obtain sufficient electrolyte conductivity.[45] These liquids can be very corrosive and hazardous.

The zinc–air battery also has a high energy density (0.15–0.2 kWh/kg), and the materials used in its construction are relatively cheap compared with most of the other batteries that have been described. The oxidizing agent (oxygen)

---

[45] β-Alumina is still a solid at these temperatures, but even as a solid it is as conductive at 300°C as is the liquid sulfuric acid electrolyte in the lead–acid battery at room temperature.

comes from the surrounding air rather than from material that is part of the battery, and this characteristic is largely responsible for the high energy density. During spontaneous discharge, zinc is oxidized at the negative electrode in a highly alkaline electrolyte to the zincate ion,

$$Zn + 4OH^- \rightarrow [Zn(OH)_4]^{-2} + 2e^- \tag{15-36}$$

and oxygen from atmospheric air is reduced at a catalyst-activated carbon electrode, represented by the net reaction

$$O_2 + 2H_2O + 4e^- \rightarrow 4OH^- \tag{15-37}$$

to give the overall reaction

$$2Zn + 4OH^- + O_2 + 2H_2O \rightarrow 2[Zn(OH)_4]^{-2} \tag{15-38}$$

In contrast to the electrical recharging already described for other rechargeable battery systems, the zinc–air battery being developed for possible large-scale utility energy storage and electric vehicle uses is recharged by recovering the spent zinc from the electrolyte and the metallic zinc electrode is mechanically replaced.

Although we have focused attention on heavy-duty applications such as electric vehicles or power storage, many of these batteries already see much use on a much smaller scale in portable electronic devices and other consumer goods. A great deal of work is under way to improve performance in this area.[46] Nonrechargeable batteries also see widespread use for these applications. They are also chemical energy storage devices in which energy is used to produce the reactive chemicals that go into them; but because the discharge reactions are not easily reversed, the batteries are used once and then discarded. (If recharging is possible at all with these types, it is usually to a small degree.)

The most familiar nonrechargeable battery is the zinc–carbon dry cell, which consists of a carbon cathode (the positive pole), an aqueous paste of MnO_2 and NH_4Cl, and an anode of zinc (which also acts as the container). The active elements are the zinc and MnO_2, which react to give zinc ions (as an ammonia complex) and $MnO(OH)$. In practice, other additives may be present to enhance performance; in some designs these have included small amounts of mercury compounds. The alkaline manganese cell is similar, but the cathode is a compressed MnO_2–graphite mixture, and the anode a powdered zinc–KOH paste; an absorbent material impregnated with KOH solution is the electrolyte. The reaction produces ZnO and Mn_2O_3. The mercury battery is another type of dry cell that was used extensively in cameras, medical equipment, and other devices requiring a more constant potential than is provided by conventional

---

[46]M. Jacoby, *Chem. Engi. News*, **76** (31), 37 (1998).

dry cells. It consists of a zinc anode and a HgO–graphite (for conductivity) cathode and a concentrated KOH electrolyte. The cell reaction produces ZnO and Hg. Because of the hazard associated with mercury disposal, this type of cell is being phased out.

15.9 DIRECT GENERATION OF ELECTRICITY FROM FUELS

All important electrical power generation from fossil fuels (and also from nuclear fuels) takes place by allowing the heat from burning fuel to vaporize a liquid, which in turn drives a mechanical device such as a turbine to operate a generator. The heat is rejected at some lower temperature determined by the water used for cooling the working fluid. The Carnot cycle efficiency limitations discussed earlier (Section 15.2.1) and other losses result in the waste of much of the energy of the fuel. A number of direct heat-to-electricity conversion processes have been proposed to eliminate the inefficiencies associated with the mechanical steps, but they do not affect the heat cycle losses.

Three methods of converting heat directly to electricity are the thermoelectric, thermionic, and magnetohydrodynamic conversions. Thermoelectric conversion operates on the familiar principle of the thermocouple: if two different conductors are made to form two junctions that are maintained at different temperatures, a potential difference and a flow of electricity is set up between them. The thermionic process involves the emission of electrons from a heated cathode. These electrons are collected by an anode at lower temperature, and electricity can flow in the external circuit between the two. In a magnetohydrodynamic generator, the fuel is used to heat and ionize a gas to produce a plasma. Passage of the plasma through a magnetic field generates an electrical potential and current at right angles to the flow of plasma. None of these has yet proven practical for commerical power generation use, although thermoelectric power units utilizing the heat from the decay of radioactive nuclei (e.g., ^{238}Pu) have been used for some time in special-purpose applications such as spacecraft.

Direct conversion of chemical to electrical energy would also eliminate the Carnot cycle restriction, and may be done in fuel cells. A hydrogen fuel cell, for example, would involve the reaction[47]

$$2H_2 \rightarrow 4H^+ + 4e^- \tag{15-39}$$

at one electrode, and the reaction

---

[47]As we have noted elsewhere (footnote 28), redox reactions involving hydrogen, oxygen, and water can be written with H^+ and H_2O or H_2O and OH^- depending on pH, but these are equivalent versions of the same reaction. Standard potentials are different but are related through the ion product of water (Chapter 9, Section 9.4).

$$O_2 + 4H^+ + 4e^- \rightarrow 2H_2O \qquad (15\text{-}40)$$

at the other. The overall reaction is the combination of hydrogen with oxygen to produce water, but if the two processes of reactions (15-39) and (15-40) take place at the surfaces of separate electrodes connected by an external conductor, the flow of electrons necessary for the reactions can be used to do work. This work is equal to that required to transport the required number of electrons through the potential difference E of the cell. On a molar basis, this work is nFE, where n is the number of the electrons transferred in the reaction [4 in reaction (15-40)], and F is the value of the Faraday constant. If the process is carried out reversibly, the total energy available is equal to the free energy change of the process, that is,

$$\Delta G = -nFE \qquad (15\text{-}41)$$

Under these conditions, all the chemical energy is available to do work, and the theoretical efficiency is 100%.

To obtain energy at a practical rate from an electrochemical cell, the cell would not in fact operate at equilibrium, and some energy would be lost. The actual potential V would be smaller than E, and the voltage efficiency can be expressed as

$$\varepsilon_v = \frac{V}{E} \qquad (15\text{-}42)$$

Because a working cell would not operate under equilibrium conditions, it is more realistic to base an efficiency comparison on the enthalpy change of the reaction than on free energy. In this case the maximum intrinsic efficiency of an electrochemical energy-producing process is

$$\varepsilon = \frac{\Delta G}{\Delta H} \qquad (15\text{-}43)$$

In most cases of interest, the calculated efficiency from equation (15-43) is greater than 80% (e.g., it is 83% for the H_2-O_2 cell). Some reactions have apparent efficiencies greater than 100% when calculated in this way; this happens if the entropy change is positive, making $|\Delta G| > |\Delta H|$. Such a cell must convert heat energy into power and would cool unless heat were provided to it from the surroundings. In this case both the energy of the fuel and the additional heat energy provided to the cell are converted to electricity. The foregoing theoretical efficiency does not allow for the losses in a real operating cell. The actual fuel efficiency of most fuel cells in practice is more typically in the 30–50% range, depending on the type; one must allow for energy that must be expended to supply the fuel, cool the system, and provide the required oxidant flow, all of which may be significant in a large cell. Nevertheless, efficiencies can be far greater than in a mechanical generator.

The cell electrodes must be separated by an ionic conductor (electrolyte) to permit transport of the hydrogen ions. This must be of low resistance to minimize internal voltage drop and heating, while preventing H_2 and O_2 from coming into direct contact with each other.

Operating fuel cells generally employ hydrogen as the electroactive fuel, although this often is produced through reaction of coal, hydrocarbons or alcohols and the overall products in these cases are H_2O and CO_2. For large-scale, economical operation, hydrogen derived from coal or natural gas is likely to be the preferred choice in the near future. In any event, a fuel processing unit is a necessary component in fuel cell power generation (although the solid oxide cell may be able to use natural gas without a re-former), and at the time of writing, a system for on-board conversion of gasoline to hydrogen for use in automotive fuel cells is in the development stage. Direct use of organic fuels, particularly methanol, has been studied extensively and would provide the most efficient operation, but such cells are not yet practical. However, studies are continuing. In any event, CO_2 is a product unless H_2 can be generated by nuclear, solar, or other non–fossil fuel means.

Air is the most economical source of oxygen for these cells. Small amounts of NO_x are also produced, but much less than from a traditional combustion source.

Much research has been carried out on fuel cells over the past quarter-century to overcome the limitation of the small currents that can be transferred per unit area by most electrode materials. In many cases, expensive electro-catalytic materials such as platinum have been employed to obtain satisfactory high current densities at normal operating temperatures, although modern designs can use very small amounts or other, less expensive metals. High-temperature cells avoid this problem, but materials become more expensive and corrosion becomes a major problem. The major cell types currently considered for large-scale use,[48] categorized by electrolyte material (Figure 15-17), are as follows.

1. *Alkaline cells*, with an aqueous KOH electrolyte and using purified hydrogen and oxygen. These cells have been widely used in the space program and are most developed. They are not expected to be economically suitable for electric utility generation, however.
2. *Phosphoric acid cells*, with hydrogen produced by steam re-forming of hydrocarbons or alcohols. These are being used in commercially available power generators with capacities of several megawatts.

---

[48]J. Haggin, (*Chem. Eng. News*, p. 29, Aug. 7, 1995) summarizes demonstration applications of fuel cells for electric power generation as of mid-1995.

	Anode	Electrolyte	Cathode	
Anode exhaust, H_2O (CO_2) ←				Cathode exhaust →H_2O
Alkaline, 80°C $H_2 + 2OH^- \longrightarrow H_2O + 2e^-$	Platinum on graphite →H_2 ←H_2O	Aqueous NaOH ←OH^-	Platinum on graphite O_2 ←	$1/2 O_2 + H_2O + 2e^- \longrightarrow 2OH^-$
Phosphoric acid, 200°C $H_2 \longrightarrow 2H^+ + 2e^-$	Platinum on graphite →H_2	Phosphoric acid H^+ →	Platinum on graphite O_2 ← H_2O →	$1/2 O_2 + 2H^+ + 2e^- \longrightarrow H_2O$
Polymer membrane, 80°C $H_2 \longrightarrow 2H^+ + 2e^-$	Platinum on graphite →H_2	H^+-conductive polymer membrane H^+ →	Platinum on graphite O_2 ← H_2O →	$1/2 O_2 + 2H^+ + 2e^- \longrightarrow H_2O$
Molten carbonate, 650°C $H_2 + CO_3^{2-} \longrightarrow CO_2 + H_2O + 2e^-$	Nickel →H_2 ←H_2O ←CO_2	Molten Li_2CO_3 - Na_2CO_3 ←CO_3^{2-}	Nickel O_2 ← CO_2 ←	$1/2 O_2 + CO_2 + 2e^- \longrightarrow CO_3^{2-}$
Solid oxide, 1000°C $H_2 + O^{2-} \longrightarrow H_2O + 2e^-$	Ceramic →H_2 ←H_2O	Zirconium oxide ←O^{2-}	Ceramic O_2 ←	$1/2 O_2 + 2e^- \longrightarrow O^{2-}$
Fuel, H_2 →				Oxidant, O_2 or ← air (CO_2)

Direct   Current

FIGURE 15-17 Schematic illustration of the different types of fuel cells. (CO_2 added to the cathode of the carbonate fuel cell may be recycled from the anode.) Exhaust will contain any unreacted fuel or oxidant. Cathode and anode half reactions are given.

3. *Molten carbonate cells*, using a molten alkali carbonate electrolyte at about 650°C. These have low internal losses, and at the time of this writing megawatt-sized plants are planned. High-temperature cells such as these are less dependent on high-purity reactants than the lower temperature types.

4. *Solid oxide electrolytes at high temperatures* give cells with high efficiencies, and several test units for power generation have been demonstrated. Natural gas may be used directly as the fuel.

5. *Solid polymer electrolyte (proton exchange membrane) cells*, which can produce high power densities because of their light weight, are beginning

to be used in transportation and other applications requiring portability. Stationary power generation units are under test.

High-temperature cells (i.e., the carbonate and solid oxide types) have practical efficiencies in the 50–60% range. They also produce waste heat at high temperatures that can be used in a steam cogeneration system. Phosphoric acid cells, which operate in the 40–45% efficiency range, with waste heat produced at low temperatures, are less attractive for supplemental energy generation use.

Fuel cells have been used in specialized applications such as space vehicles since the 1960s and in demonstration power generation applications since 1992. Large-scale applications in electric utilities and in transportation are expected as further development brings costs down.

15.10 THE ENERGY FUTURE

Any discussion of the energy future for either the United States alone or for the whole world is very difficult because of assumptions that must be made about population growth, industrial growth, changes in transportation, and political changes. Unforeseen political changes may occur at any time. For example, the political and economic collapse of the Soviet Union was not widely predicted before it happened, neither was the present trend to minimize emission of sulfur dioxide and greenhouse gases, if possible, in the developed countries. Nevertheless, a few comments will be made here.

15.10.1 The United States of America

Before looking at the future, let us take a brief look at the past half-century. Energy production in the United States from 1949 through 1999 is shown in Figure 15-18. Because Figure 15-18 does not include imports, it shows considerably less petroleum and liquefied natural gas products than we actually use. A complete energy flow diagram that includes imports and exports for the United States in 1999 is shown in Figure 15-19. Both figures use the same unit of energy, the quad; $Q1Q = 10^{15} Btu = 1.054 \times 10^{18} J$.

Figure 15-20 shows the total energy consumption, the energy consumption per capita, and the energy consumption per 1996 dollar of gross domestic product in the United States from 1949 through 1999. Total energy consumption increased by about a factor of 3 in that time period, while per capita energy consumption increased by less than a factor of 2, and energy consumption per dollar of gross domestic product actually decreased. The U.S. population was about 150 million in 1949, increasing to 273 million in 1999. That

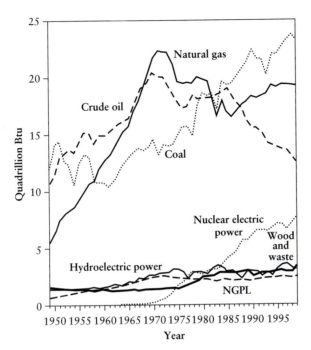

FIGURE 15-18 Energy production by major source in the United States, 1949–1999. NGPL refers to natural gas plant liquids. From the U.S. Department of Energy, Information Administration website, http://www.eia.doe.gov.

the population increased much faster than the per-capita energy consumption and the energy consumption per 1996 dollar of gross domestic product actually decreased may be attributed to the energy conservation that followed the 1973–74 interruption of petroleum exports from the oil-producing countries in the Middle East and the concomitant rise in oil prices; this is sometimes referred to as the "energy crisis" or "oil shock" of 1973 and 1974. Figure 15-20 shows that nevertheless, this "energy crisis" resulted in only a short-term decrease in total and per-capita energy consumption in the United States. The post-1973 low point in energy use occurred in 1983 after a period of high oil prices.

Energy demand forecasts for the United States have been made by the Office of Integrated Analysis and Forecasting, Energy Information Administration, U.S. Department of Energy, using the National Energy Modeling System (NEMS) up to 2020, a long time before nonrenewable resources like oil, natural gas, and coal run out. (It is very difficult to predict how long these nonrenewable resources will last because new reserves are still being discovered.) The modelers feel that the economy and the nature of energy

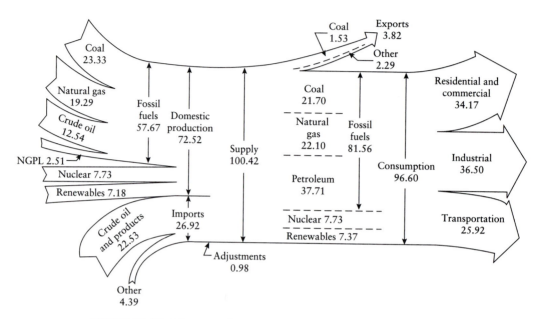

FIGURE 15-19 The energy flow pattern in the United States in 1999. Units are quads (Q), and NGPL refers to natural gas plant liquids. From the U.S. Department of Energy, Information Administration Website, http://www.eia.doe.gov.

markets are reasonably well understood in this "middle term" time horizon. Figure 15-21 shows these forecasts of the energy use per capita and per dollar of gross domestic product (GDP) up to 2020 using real data through 1999. The projected future decrease in energy use per dollar of GDP, which follows the real decrease already mentioned, is partly the result of the Energy Policy Act of 1992 and the National Appliance Energy Conservation Act of 1987, which mandated energy efficiency standards for new energy-using equipment in residential and commercial buildings and for motors in industry. With the information available up to 1999, the projected energy use per capita was expected to increase slightly up to the year 2020.

15.10.2 The Rest of the World

Prediction of global energy trends is much more difficult than forecasting for the United States, but, again, projections have also been made up to 2020 in the International Energy Outlook 2000 (IEO2000) from the U.S. Department of Energy. Figure 15-22 plots the known world energy consumption from

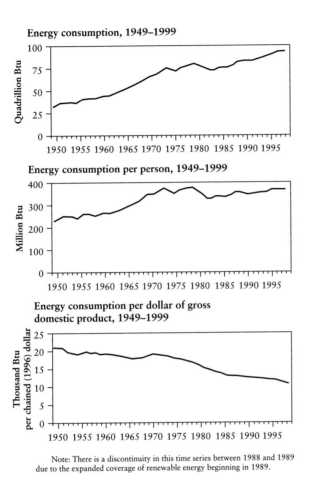

Energy consumption, 1949–1999

Energy consumption per person, 1949–1999

Energy consumption per dollar of gross domestic product, 1949–1999

Note: There is a discontinuity in this time series between 1988 and 1989 due to the expanded coverage of renewable energy beginning in 1989.

FIGURE 15-20 Total energy consumption, energy consumption per person, and energy consumption per dollar of gross domestic product in the United States, 1949–1999. From the U.S. Department of Energy, Information Administration website, http://www.eia.doe.gov.

1970 through 1997 and the projected world energy consumption through the year 2020, using three different predictions for world economic growth. In the case considered to be most likely at present, called the Reference case in Figure 15-22, the world is expected to consume 608 Q in 2020, an increase of 228 Q over 1995, with 66% of the increase attributed to the developing world, especially Asia and Central and South America (see Figure 15-23). The world gross domestic product (WGDP) grew by 4.3 trillion (1997 U.S. dollars) between 1970 and 1997 and is expected to almost double between 1997 and

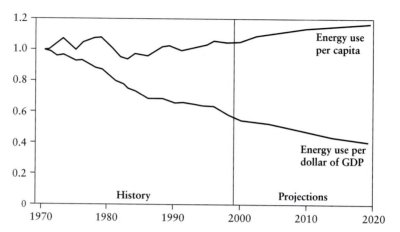

FIGURE 15-21 Energy use per capita and per dollar of gross domestic product, 1970–2020 (index, 1970 = 1). From the U.S. Department of Energy, Information Administration Website, http://www.eia.doe.gov.

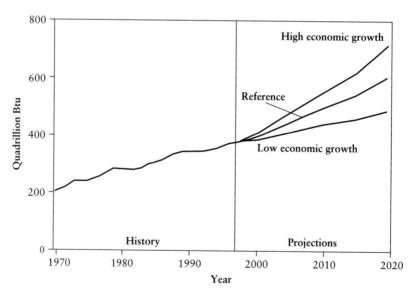

FIGURE 15-22 World energy consumption by economic growth case, 1970–2020. From the U.S. Department of Energy, Information Administration website. Historical data from Energy Information Administration (EIA), Office of Energy Markets and End Use, International Statistics Database and *International Energy Annual 1997*, DOE/EIA-0219(97), Washington, DC, April 1999. Projections from E/A, World Energy Projection System, 2000. Website, http://www.eia.doe.gov.

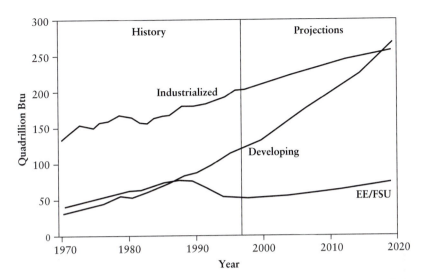

FIGURE 15-23 World energy consumption by region, 1970–2020. EE/FSU stands for eastern Europe/Former Soviet Union. From the U.S. Department of Energy, Information Administration Historical data from Energy Information Administration (EIA), Office of Energy Markets and End Use, International Statistics Database and *International Energy Annual 1997*, DOE/EIA-0219(97), Washington, DC, April 1999. Projections from EIA, World Energy Projection System, 2000. Website, http://www.eia.doe.gov.

2020, reaching 54 trillion by 2020. The IEO2000 projection makes many assumptions about oil prices and economic development in many parts of the world that may or may not be borne out. It is predicted that the "energy intensity," usually shown in thousands of Btu per dollar of WGDP, will decrease in all parts of the world (see Figure 15-24). As one may imagine, energy forecasts change every year. As this volume is being written, they can be accessed on the World-Wide Web, starting with the U.S. Department of Energy, Energy Information Administration home page (http://www.eia.doe.gov). World energy use forecasts, which give very similar results, are also made at intervals by the International Energy Agency (IEA) home page (http://www.iea.org). Other world energy forecasts, also giving similar results, are made by Petroleum Economics Limited (PEL), and Petroleum Industry Research Associates (PIRA).

Figures 15-22 and 15-23 both show that the total energy used over the whole world is expected to increase in the future. The projections shown in these figures assume that the energy will come from fossil fuels, mostly because these projections were made with no regard for the Kyoto Protocol of December 1997, which could forestall much of the growth in energy

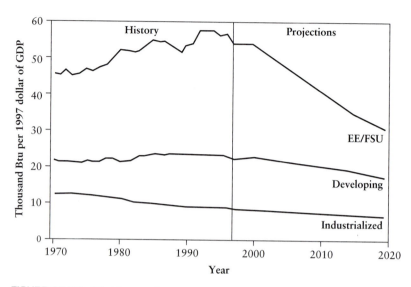

FIGURE 15-24 World energy intensity (energy consumption per 1997 dollar of gross domestic product) by region, 1970–2020. EE/FSU stands for eastern Europe/former Soviet Union. From the U.S. Department of Energy, Information Administration. Historical data from Energy Information Administration (EIA), Office of Energy Markets and End Use, International Statistics Database and *International Energy Annual 1997*, DOE/EIA-0219(97), Washington, DC, April 1999. Projections from EIA World Energy Projection System, 2000. Website at http://www.eia.doe.gov.

demand from industrialized countries.[49] If the CO_2 emission targets written into the protocol were to be achieved by reducing the use of fossil fuels, energy consumption would have to be reduced by 40–60 Q, about 10% of the total expected for the year 2020. Replacement of fossil fuels by biomass would recycle rather than increase CO_2 emission, but the agricultural capabilities of the earth as a whole seem unlikely to be able to support the large-scale biomass program that would be necessary to make a major contribution while still producing food. Nuclear energy is the one obvious energy source that could substitute for fossil fuels without contributing CO_2 to the atmosphere, but with the significant problem of disposing of the accompanying radioactive waste. Large-scale nuclear power plants in some of the less stable underdeveloped countries are a disconcerting idea, since some of these facilities could be used to make nuclear weapons. Large hydroelectric potential is still available in some parts of the world, but only at the expense of major alterations in river

[49]As of January 2000, the date used for the energy use projections, too few countries had ratified the Kyoto Protocol to make it effective.

flow and enormous dislocation of human beings; the Three Gorges project in China[50] is an example. Economics is a limiting factor in any case, since most new energy sources are more expensive than fossil fuels. The "bottom line" in the energy future is that we really do not know what that future will be, but, if it is simply an extrapolation of the past, the consequences could be severe.

15.10.3 Conclusions

It is virtually certain that energy use worldwide will increase because of increasing populations and continued development efforts in the presently underdeveloped nations. Continued reliance on fossil fuels causes problems quite apart from continued availability; these include possible effects on climate due to further increases of atmospheric carbon dioxide. Stabilizing CO_2 emissions at somewhere near present values, or decreasing them, will be difficult to achieve as developing nations attempt to advance their economies. Although improvements can be made through conservation and increased efficiency, other energy sources seem essential. Many other energy sources yielding no or little pollution, as discussed in this chapter, can play a role, but most are limited with respect to location or amount. Table 15-5 summarized the current status of renewable energy sources. Cost is a critical factor. As long as fossil fuels remain inexpensive, other energy sources such as solar power, hydrogen fuels, and biomass will be unable to compete unless subsidized or their use mandated by regulation. In any event, costs will go up. Nuclear energy is the one tested source that has the potential for replacing carbon dioxide producing fossil fuels, but it has its own set of problems. These include waste disposal and the potential for radiation release, possible use of nuclear material by terrorist groups, and the risks of putting nuclear reactors in the hands of unstable or irrational governments whose leaders might divert nuclear fuels into weapons. In the United States, public fears of reactors and radiation are so strong that it is difficult to foresee much development in this field. Even if built, reactors would be of high cost because of safety and licensing requirements. Other parts of the world have a different view. France, for example, generates most of its electricity from nuclear energy. Controlled nuclear fusion, highly touted as a future energy source, is too uncertain to be included in serious projections.

---

[50]This project, to be completed in 2009, is forcing the resettlement of about 1.2 million people. The dam, across the Yangtze River, will be a 1.2-mile-long stretch of concrete with a reservoir about 400 miles long behind it.

Additional Reading

15.1 Introduction

Berger, J. J., *Charging Ahead: The Business of Renewable Energy and What It Means for America.* Henry Holt., New York, 1997.

Schipper, L., and S. Meyers, *Energy Efficiency and Human Activity: Past Trends, Future Prospects.* Cambridge University Press, Cambridge, U.K., 1992.

15.3 Fossil Fuels

Starr, C., M. F. Searl, and S. Alpert, Energy resources: A realistic outlook, *Science*, **256**, 981 (1992).

Preventing the next oil crunch. A series of articles by several authors. *Sci. Am.*, pp. 75–95, March 1998.

U.S. Geological Survey Circular 1118, 1995 National Assessment of United States Oil and Gas Resources.

U.S. Geological Survey, U.S. Crude Oil, Natural Gas and Natural Gas Liquids Reserves, 1998, Annual Report. For a discussion of this report, see R. A. Kerr, USCS optimistic on world oil prospects, *Science* **289**, 237 (July 14, 2000).

15.4 Nuclear

Ahearne, J. F., The future of nuclear power, *Am. Sci.* **81**, 24 (1993).

Bodansky, D., *Nuclear Energy, Principles, Practice and Prospects.* American Institute of Physics Press, New York, 1996.

Committee on Future Nuclear Power Development, Energy Engineering Board, Commission on Engineering and Technical Systems, National Research Council, *Nuclear Power: Technical and Institutional Options for the Future.* National Academy Press, Washington, DC, 1992.

Darmstadter, J., and R. W. Fri, Interconnections between energy and the environment: Global challenges, *Annu. Rev. Energy Environ.*, **17**, 45 (1992).

Levine, M. D., F. Liu, and J. E. Sinton, China's energy system: Historical evolution, current issues, and prospects, *Annu. Rev. Energy Environ.*, **17**, 405 (1992).

Pace University Center for Environmental Legal Studies, *Environmental Costs of Electricity.* Oceana Publications, 1990.

Rhodes, R., *Nuclear Renewal: Common Sense about Energy.* Viking Penguin, New York, 1993.

Worledge, D. H., Role of human performance in energy systems management, *Annu. Rev. Energy Environ.*, **17**, 285 (1992).

15.5 Solar Energy

15.5.1 Introduction

Scheer, H., The economy of solar energy, in *Advances in Solar Energy*, Vol. 9, K. W. Böer, ed., pp 307–337. American Solar Energy Society, Boulder, CO, 1994.

Wieder, S., *An Introduction to Solar Energy for Scientists and Engineers.* Krieger, Malabar, FL, 1992.

Zweibel, K., *Harnessing Solar Power.* Plenum Press, New York, 1990.

15.5.2 Biological Utilization of Solar Energy: Biomass

Abrol, Y. P., P. Mohanty, and Govindjee, eds., *Photosynthesis: Photoreactions to Plant Productivity*. Kluwer Academic Publishers, Dordrecht, 1993.

Bain, R. L., and R. P. Overend, Biomass electric technologies: Status and future development, in *Advances in Solar Energy*, Vol. 7, K. W. Böer, ed., pp. 449–494. American Solar Energy Society, Boulder, CO, 1992.

Beadle, C. L., S. P. Long, S. K. Imbamba, D. O. Hall, and R. J. Olembo, *Photosynthesis in Relation to Plant Production in Terrestrial Environments*. Tycooly, Oxford, U.K., 1985.

Biofuels. Energy and Environment Policy Analysis Series, International Energy Agency, Paris, 1994.

Klass, D. L., *Biomass for Renewable Energy, Fuel and Chemicals*. Academic Press, San Diego, CA, 1998.

Patterson, W., *Power from Plants. The Global Implications of New Technologies for Electricity from Biomass*. Earthscan Publications, London, 1994.

15.5.3 Physical Utilization of Solar Energy

Avery, W. H., and C. Wu, *Renewable Energy from the Ocean. A Guide to OTEC*. Oxford University Press, New York, 1994.

Cavrot, D. E., Economics of ocean thermal energy conversion, *Renewable Energy*, **3**, 891–896 (1993).

DeMeo, E. A., and P. Steitz, The U.S. electric utility industry's activities in solar and wind energy: Survey and perspective, in *Advances in Solar Energy*, Vol. 6, K. W. Böer, ed., pp 1–218. Plenum Press, New York, 1990.

Hurdes, J. V., and B. Lachal, *Industrial Solar Heat*. Delta Energy, Schaffhausen, Switzerland, 1986.

Kreider, J. F., and F. Kreith, *Solar Heating and Cooling: Active and Passive Design*, 2nd ed. McGraw-Hill, New York, 1982.

Löf, G., ed., *Active Solar Systems*. MIT Press, Cambridge, MA, 1993.

Nielsen, C. E., Salinity-gradient solar ponds, in *Advances in Solar Energy*, Vol. 4, K. W. Böer, ed., pp. 445–498. Plenum Press, New York, 1988.

Takahashi, P. K., and A. Trenka, Ocean thermal energy conversion: Its promise as a total resource system, *Energy*, **17**, 657–668 (1992).

15.5.4 Photovoltaic Utilization of Solar Energy

Coutts, T. J., and J. D. Meakin, *Current Topics in Photovoltaics*. Academic Press, London, 1985.

Fahrenbruch, A. L., and Bube, R. H., *Fundamentals of Solar Cells. Photovoltaic Solar Energy Conversion*. Academic Press, New York, 1983.

Foraser, D. A., *The Physics of Semiconductor Devices*, 4th ed. Clarendon Press, Oxford, U.K., 1986.

Fuhs, W., and R. Klenk, Thin-film solar cells, in *Advances in Solar Energy*, Vol. 13, D. Y. Gosami and K. W. Böer, eds., pp. 409–443. American Solar Energy Society, Boulder, CO, 1999.

Green, M. A., *Solar Cells: Operating Principles, Technology, and System Applications*. Prentice-Hall, Englewood Cliffs, NJ, 1982.

Holl, R. J., and E. A. DeMeo, The status of solar thermal electric technology, in *Advances in Solar Energy*, Vol. 6, K. W. Böer, ed., pp. 219–394. Plenum Press, New York, 1990.

Lasnier, F., and T. G. Ang, *Photovoltaic Engineering Handbook*. Hilger, Bristol, U.K., 1990.

Maycock, P. D., and E. N. Stirewalt, *A Guide to the Photovoltaic Revolution*. Rodale Press, Emmaus, PA, 1985.

Möller, H. J., *Semiconductors for Solar Cells*. Artech House, Boston, 1993.

Neville, R. C., *Solar Energy Conversion: The Solar Cell*. Elsevier, Amsterdam, 1978.

Treble, F. C., ed., *Generating Electricity from the Sun*. Pergamon Press, Oxford, U.K., 1991.

15.5.5 Photoelectrochemical Utilization of Solar Energy

Aruchamy, A., ed., *Photoelectrochemistry and Photovoltaics of Layered Semiconductors*, Vol. 14 of *Physics and Chemistry of Materials with Low-Dimensional Structures*, F. Levy, editor-in-chief. Kluwer Academic Publishers, Dordrecht, 1992.

Grätzel, M., ed., *Energy Resources through Photochemistry and Catalysis*. Academic Press, New York, 1983.

Koval, C. A., and J. N. Howard, Electron transfer at semiconductor electrode–liquid electrolyte interfaces, *Chem. Rev.*, **92**, 411–433 (1992).

McEvoy, A. J., and M. Grätzel, Sensitization in photochemistry and photovoltaics, *Solar Energy Mater, Solar Cells*, **32**, 221–227 (1994).

Memming, R., Photoelectrochemical utilization of solar energy, in *Photochemistry and Photophysics*, Vol. II, J. F. Rabek, ed., pp. 143–189. CRC Press, Boca Raton, FL, 1990.

Norris, J. R., Jr., and D. Meisel, ed., Photochemical energy conversion, in *Proceedings of the Seventh International Conference on Photochemical Conversion and Storage of Solar Energy*, July 31–August 5, 1988. Elsevier, New York, 1989.

Santhanam, K. S. V., and M. Sharon, eds., *Photoelectrochemical Solar Cells*. Elsevier, New York, 1988.

15.7 Wind and Water

Asmus, P., *Reaping the Wind*, Island Press, Washington, D.C., 2001.

Baker, A. C., *Tidal Power*. Peter Peregrinus, London, 1991.

Dodge, D. M., and R. W. Thresher, Wind technology today, in *Advances in Solar Energy*, Vol. 5, K. W. Böer, ed., pp. 306–359. Plenum Press, New York, 1989.

Jog, M. G., *Hydro-Electric and Pumped Storage Plants*. Wiley, New York, 1989.

McCormick, M. E., and Y. C. Kim, eds., *Utilization of Ocean Waves—Wave to Energy Conversion*. American Society of Engineers, New York, 1987.

Quarton, D. C., and V. C. Fenton, eds., *Wind Energy Conversion 1991*. Mechanical Engineering Publications, London, 1991.

Raabe, J., *Hydro Power*. VDI-Verlag, Düsseldorf, 1985.

Sesto, E., and C. Casale, Wind power systems for power utility grid connection, in *Advances in Solar Energy*, Vol. 9, K. W. Böer, ed., pp. 71–159. American Solar Energy Society, Boulder, CO, 1994.

Shaw, R., *Wave Energy. A Design Challenge*. Ellis Howood, Chichester, U.K., 1982.

15.8 Energy Storage

Bockris, J. O'M., B. Dandapani, and J. C. Wass, A solar hydrogen energy system, in *Advances in Solar Energy*, Vol. 5, K. W. Böer, ed., pp. 171–305. Plenum Press, New York, 1989.

Bockris, J. O'M., and T. N. Veziroglu, with D. Smith, *Solar Hydrogen Energy—The Power to Save the Earth*. MacDonald Optima, London, 1991.

Bolton, J. R., Solar photoproduction of hydrogen: A review, *Solar Energy*, **57**, 37–50 (1996).

Chaurey, A., and S. Deambi, Battery storage for PV power systems: An overview, *Renewable Energy*, **2**, 227–235 (1992).

Douglas, T. H., ed., *Pumped Storage*. Thomas Telford, London, 1990.

Gretz, J., Solar hydrogen, *Renewable Energy*, **1**, 413–417 (1991).

Mantell, C. L., *Batteries and Energy Systems*, 2nd ed., McGraw-Hill, New York, 1983.

Pichat, P., *Photochemical Conversion and Storage of Solar Energy*. Kluwer Academic Publishers, Dordrecht, 1991.

Serpone, N., D. Lawless, and R. Terzian, Solar fuels: Status and perspectives, *Solar Energy*, **49**, 221–234 (1992).

Yürüm, Y., ed., *Hydrogen Energy Systems*. Kluwer Academic Publishers, Dordrecht, 1995.

15.9 Fuel Cells

Blomen, L. J. M. J., and N. M. Mugerwa, eds., *Fuel Cell Systems*. Plenum Press, New York, 1993.

15.10 Resources

Starr, C., M. F. Searl, and S. Alpert, Energy resources: A realistic outlook, *Science*, **256**, 981–987 (1992).

U.S. Geological Survey Circular 1118, 1995 National Assessment of United States Oil and Gas Resources.

EXERCISES

15.1. Coal is combusted at 1500°C to obtain the maximum amount of heat (Section 15.3.3). One ton (2000 lb) of bituminous coal ($\Delta H_{combustion} = 35$ kJ/g) is combusted in a Carnot cycle heat engine operating at maximum thermodynamic efficiency, with the excess heat discharged to a water reservoir at 68°F (20°C). The specific heat and density of water are 1.0 cal/g and 1.0 g/cm^3, respectively. Calculate the following:

(a) The maximum thermodynamic efficiency of this engine

(b) The amount of heat discharged to the river, assuming that the temperature of the river is unchanged

(c) The amount of water in the reservoir if the amount of heat calculated in part b actually causes the temperature to rise to 71°F.

15.2. The potential supply of hydrocarbons in tar sands and oil shale is said to be equal to all the crude oil reserves in the world. List the economic and environmental reasons why neither of these sources is used as a source of oil at the present time. When do authorities predict they will be used? Why?

15.3. (a) List the environmental problems resulting from the combustion of coal for the generation of electricity. Why is coal still being used on a large scale for the generation of electricity in the United States? (b) What are the economic and environmental advantages and disadvantages for the conversion of coal to methane (syngas). Why is this not done extensively today? Write the equation for the conversion of coal to methane. In writing your answer, remember that most coal contains the following elements: C, H, N, O, S.

(c) Both nuclear and coal-fired power plants are used extensively for the generation of electricity. List the environmental advantages and disadvantages of each and then decide which is the more cost-effective and the lesser source of environmental problems. Finally, assuming that you are the leader of the world, which would you decree should be used for power generation world wide. Explain your answer.

15.4. A basic problem with the use of fossil fuels as an energy source is the emission into the atmosphere of the green-house gas carbon dioxide. Suggest how to burn carbon compounds without releasing carbon dioxide into the atmosphere (this is not discussed in this book). What are the environmental and economic costs of your solution(s) to the carbon dioxide emission problem? If you cannot come up with an idea to solve this problem look, for a possible solution in the scientific literature or on the World-Wide Web.

15.5. Calculate the maximum (ideal thermodynamic) efficiency for (a) a PWR plant with steam to the turbine at $330°C$ and condenser temperature at $30°C$ and (b) a coal-fired plant with steam at $530°C$ and the same condenser temperature. The net efficiencies are about 30 and 50%, respectively. Why are these less than the maximum?

15.6. A certain nuclear power plant (an LWR with 3% enriched fuel) has a net electrical output of $1000\,MW_e$ and operates for 320 days per year. Assuming as an approximation that half the energy comes from the fission of ^{235}U and the remainder from ^{239}Pu, how many kilograms of ^{235}U are fissioned? How many kilograms of fission products are produced? How many gallons of fuel oil and how many tons of coal would be required to provide the same energy per year?

15.7. Compare the relationship between energy utilization and the environment in the days of the Industrial Revolution and in the present.

15.8. An electrical utility has asked for proposals to replace an aging $1000\text{-}MW_e$ nuclear power plant located in northeastern New York State. Presentations are to be made for replacing the NPP with (a) solar panels, (b) an array of windmills, and (c) a coal-fired plant. As a consultant, you have been asked to evaluate the three proposals. Select one of the options and prepare a statement containing the various reasons for and against selecting that option. Include a detailed statement on how the environment would be affected. As a variation, select a different U.S. location for the NPP that is to be replaced.

15.9. In regions of the United States where deregulation of the electric utility industry has been partially or completely achieved, what consequences can be characterized as (a) "expected" and (b) "unexpected"? How is deregulation affecting the environment?

15.10. How do you account for the fact that in different countries the attitude toward nuclear power ranges from strong opposition to indifference to opposition to the closing of an NPP?

15.11. Prepare an essay, a class presentation, or a term paper on one of the following.

    (a) The current status and future prospects for nuclear power in the United States, Europe, and Asia. This amounts to updating Tables 15-3 and 15-4 and adding comments.

    (b) The options available to the United States for decreasing the emission of greenhouse gases. Is lowering energy consumption a viable option?

    (c) Your personal attitude toward nuclear power and the basis for it. Discuss the extent that your attitude is influenced by facts and by emotions.

    (d) What do you believe to be (not wish to be) the future of nuclear power in the United States? Why? How much of your belief is based on risk? On environmental impact? On political and other factors (to be identified)?

15.12. Silicon is the most used semiconductor material for solar cells (Section 15.5.4). It has a band gap $E_g = 1.107\,eV$.

    (a) What is the wavelength, in nanometers of a photon of energy $1.107\,eV$? What spectral region is this?

    (b) Assuming a Boltzmann distribution of thermal energy [equation (4-2)], calculate the fraction of Si atoms that possess an energy equal to $1.107\,eV$ at $25°C$.

15.13. Make a list of the regions of the world that are most favorable for geothermal power generation.

15.14. What are the most likely environmental problems associated with geothermal power sources?

15.15. List the sources of renewable energy and summarize the advantages and disadvantages of each. In so far as you can, include economic considerations.

15.16. Write the electrode reactions that take place when an ordinary carbon–zinc dry cell is discharged.

15.17. When lead–acid batteries "wear out," they must be disposed of or recycled. Describe the processes and hazards that would be involved in recycling them. Compare this to the hazards of disposal in a land-fill.

15.18. What are the electrode reactions (anodic and cathodic) in the various types of fuel cells?

15.19. Calculate the theoretical efficiency of a hydrogen–oxygen fuel cell operating at room temperature.

15.20. What are the primary types of electric battery under consideration for electric power storage? For electric vehicle propulsion? Summarize the advantages and disadvantages in each case.

15.21. Search the Internet for the most recent predictions of energy use and compare them with those given in this chapter. If there are significant changes, discuss the reasons for them.

16

SOLID WASTE DISPOSAL AND RECYCLING

16.1 INTRODUCTION

Safe disposal of solid wastes is a serious problem. With our culture, which generates ever larger amounts of disposable materials and an increasing population density, we can no longer simply "throw things away." If we discard them on land, they must be buried for aesthetic, safety, and health reasons. Even this is not enough, because toxic materials can be dissolved and enter the groundwater. Consequently, disposal site construction that minimizes leaching and includes elaborate leachate recovery systems is required; sites must be chosen carefully to ensure that the inevitable accidental breaches of the system used to seal the completed landfill will have minimal impact. Remote sites increase transportation requirements, but of course, sites must not be in our backyards (NIMBY). Ocean disposal has potentially serious effects on ocean life and thus on a vital food supply, as well as having safety and aesthetic consequences when refuse materials wash up on beaches that are heavily used for recreation (see Chapter 1 for other comments on this).

Some waste materials can be disposed of by burning. Unfortunately, combustion of many substances can generate toxic products that are released to the atmosphere and widely dispersed. Special incinerators and scrubbers may be necessary, and even these may not satisfy the concerns of those who live downwind.

Recycling will reduce the amount of material to be disposed of. It has its own problems of collection, sorting, and cost. Composting of degradable organic materials also reduces disposal while producing a useful product, while anaerobic digestion is a possible source of methane fuel. A mix of these processes will be needed in future solid waste handling techniques.

Municipal wastes, that is, those not including industrial wastes or hazardous materials, vary widely in composition. Sometimes they include demolition debris, industrial wastes, or water treatment sludge, for example. Most of the total municipal solid waste generated in the United States is disposed of in landfills; according to 1998 EPA estimates, of the total 220 million tons of waste produced, 55% went to landfills, 17% was incinerated, and 28% recycled. Waste generation is projected to continue to increase, while recycling is expected to take up a significantly larger fraction. Disposal of hazardous wastes is a separate problem.

Estimates of municipal waste production and recycling in the United States in 1998 are given in Tables 16-1 and 16-2 and recent trends are shown in Figures 16-1 and 16-2.

TABLE 16–1

Waste Production and Recycling of Different Types of Material in the United States in 1998

Material	Weight generated ($lb \times 10^6$)	Percent recycled
Paper/paperboard	84.1	41.6
Plastics	22.4	5.4
Glass	12.5	25.5
Metals:		
Iron/steel	12.4	35.1
Aluminum	3.1	27.9
Other nonferrous	1.4	67.4
Rubber and leather	6.9	12.5
Textiles	8.6	12.8
Wood	11.9	6.0
Yard wastes	27.7	45.3 (composting)
Food wastes	22.1	2.6
Miscellaneous materials	3.9	23.1
Miscellaneous inorganic wastes	3.3	Negligible

Source: EPA Environmental Fact Sheet: Municipal Solid Waste Generation, Recycling and Disposal in the United States: Facts and Figures for 1998. http://www.epa.gov/osw.

TABLE 16-2

Examples of Industries Using Recycled Materials as Feedstocks in 1989

Recycled material	Type of industry	Virgin materials	Number of manufacturers	Consumption (1989) (tons $\times 10^6$)	Benefits
Paper	Paper and paperboard mills	Wood pulp, other plant libers	700	85	Less costly feedstock Reduced water consumption Lower capital requirement
Glass	Container manufacturers	Sand, limestone, soda ash	16	11	Energy savings Cleaner furnace firing Potential furnace life extension
Tin cans	Steel mills and detinning plants	Iron ore, lime, coke	50–150	80	Reduced capital requirements Reduced raw material requirements Higher quality input (if purchased from detinners)
Aluminum cans	Primary and secondary aluminum producers	Refined bauxite, carbon	> 60	8	Avoided capital cost
Plastic	Plastic resin, film, and fiber manufacturers and processors	Products of petroleum and natural gas distillation and refinement	>14,000	30	Reduced energy consumption Cost savings Availability

Source: L. F. Diaz, G. M. Savage, L. L. Eggerth and C. G. Golueke, *Composting and Recycling Municiple Solid Waste*, Lewis Publishers, Boca Raton, FL, 1993.

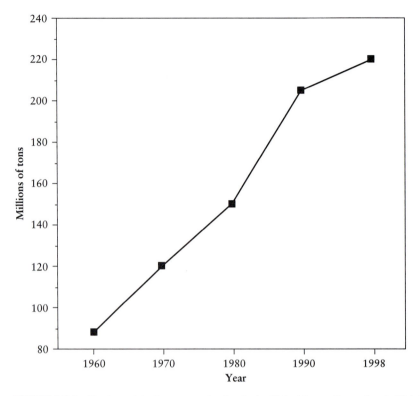

FIGURE 16-1 Total municipal waste production in the United States. From data in EPA Environmental Fact Sheet: Municipal Solid Waste Generation, Recycling and Disposal in the United Stated: Facts and Figures for 1998. http://www.epa.gov/osw.

Containers and packaging, including cans and bottles, made up 72.4 million tons of waste in the United States in 1998, of which 40% overall was recycled; over half the steel and about 44% of the aluminum packaging (mostly cans) and over half the paper and paperboard packaging was recycled, but only 29% of the glass packaging materials. Nondurable goods provided 60.3 million tons of waste, with paper and paper products contributing 40 million tons in this category. Durable goods (appliances, etc.,) amounted to 34.4 million tons.

16.2 LANDFILLS

Originally, landfills were simply places where waste could be dumped where it did not disturb too many people. As problems associated with this form of

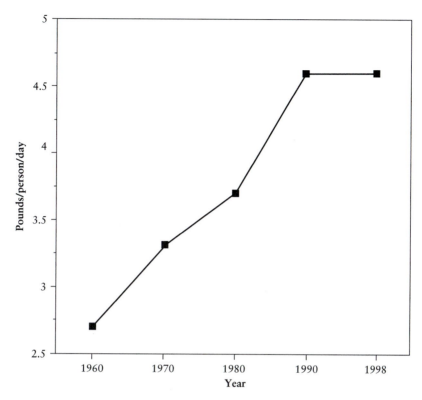

FIGURE 16-2 Municipal waste production in the United States. From data in EPA Environmental Fact Sheet: Municipal Solid Waste Generation, Recycling and Disposal in the United Stated: Facts and Figures for 1998. http://www.epa.gov/osw.

disposal became to be recognized, more elaborate arrangements became necessary. Many old landfills that did not meet modern standards, or that were full, have been closed, and it is not a simple matter to open new ones. In the United States, operating landfills went from about 8000 in 1988 to about 2300 in 1998.

A modern landfill is sealed underneath by a flexible membrane liner such as various forms of polyethylene, rubber, or poly(vinyl chloride) that resist degradation, and/or a layer of impermeable clay. Some of the properties of clays for this purpose were discussed in Section 12.3.1. A system for leachate collection also is necessary. The waste is placed in compacted, earth-covered layers, and, when the landfill is full, closed by a water-impermeable top layer. Figure 16-3 is a schematic illustration of a landfill with a double liner and double leachate collection system. Not all landfills will have these redundant features.

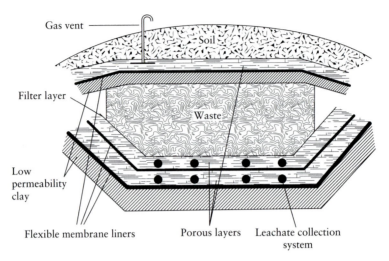

FIGURE 16-3 Schematic model of a modern landfill.

The leachate collection system typically consists of plastic pipes leading to a sump with a pumping system to remove the liquid. The pipes are embedded in a porous layer that may be an appropriate soil or an artificial material. A filter layer screens out fine particles of earth that could clog the drainage system. Gases, chiefly methane, are released as the organic materials in the waste decay. These gases must be controlled to avoid the buildup of potentially explosive concentrations. In some cases, the methane can be collected for use as a fuel.

Although a typical landfill is a pit, some are constructed as mounds above the surface. These above-grade deposits can be large (a popular name is Mount Trashmore); the largest is the Fresh Kills landfill site on Staten Island, New York, which is expected to top out as a mound more than 150 m high, along with several smaller mounds, by 2005. This type of landfill avoids excavation and makes leachate collection easier, but more earth must be brought into cover the trash layers, and the mound is more subject to erosion. When completed, the earth-covered mound can be used as a recreational area, as has been done at Virginia Beach and other locations, but any use must avoid penetration of the cap.

Although biodegradable materials do decompose in landfills, the process is not rapid. Conditions are anaerobic and not conducive to rapid biological action. Even after decades, newspapers have been recovered that are readable, and food items identifiable. In some sense, a landfill can be regarded not as a disposal site but as a storage site.

On a volume basis, plastics are more important than their low weight percentage implies. They have low densities, often do not crush well, and have a tendency to work their way to the surface. Discarded automobile tires have a similar tendency, which is one reason many landfills will not accept them.

Although not nominally considered to contain hazardous materials, toxic substances from household chemicals, batteries, paint solvents, and so on inevitably make up some fraction of household wastes. However, EPA regulations prohibit disposal of hazardous wastes in an ordinary municipal landfill. Hazardous wastes, which include flammable, volatile, toxic, and pathological wastes, include wastes from many industrial operations. Radioactive wastes are a special class of hazardous material, discussed in Chapter 14. Hazardous wastes are often disposed of in landfill facilities specifically intended for the purpose, and with extensive documentation. In the past, there was considerable mixing of hazardous and municipal waste and few records of landfill composition or even location were kept; landfill construction was much more casual. Accidental breaching of these disused landfills has had serious effects, as illustrated by the notorious Love Canal[1] site near Niagara Falls in New York State.

Love Canal was a disposal site used by a number of firms for dumping of chemical wastes over a long period beginning in the 1930s. Sealed in the 1950s, the site was later used for school and housing development. Such use breached the clay cover that had sealed the site and allowed entry of water. Leaching of chemicals such as benzene and chlorinated hydrocarbons into basements and other sites of exposure caused considerable health risks requiring abandonment of the housing and the expenditure of millions of dollars in cleanup programs. This was the beginning of the program to identify potentially hazardous waste sites in the United States and the setup of the "Superfund" program to monitor them and clean them up. Many such sites exist. They are by no means confined to the United States; a situation very similar to Love Canal occurred in the Netherlands, where a new village (Lekkerkirk) was built on a waste dump in 1970 but had to be abandoned in the 1980s with large costs for cleanup. There are thousands of identified abandoned and potentially hazardous waste sites worldwide, and probably many more that have not been identified.

16.3 COMPOSTING

Biodegradable organic materials have beneficial effects when applied to agricultural land, in part as fertilizer because of their N, P, and K content (although this is typically relatively low), but mostly in terms of improving soil quality

---

[1]Love Canal was an uncompleted navigation and electrical generation canal intended to bypass Niagara Falls. It was started in the late 1800s but abandoned with comparatively little excavated. The "ditch" left behind was later used as a site for waste disposal.

by increasing its organic content (see Section 12.3.1). Raw organic waste, such as food processing residues, garden wastes, or sewage sludge (if sterilized) can be applied directly, but composting provides important advantages. (The problem of heavy metal contamination of sewage sludge was referred to in Section 11.5.2; the same considerations apply if this sludge is used in composting.) Composting consists of breakdown of the organic material through microbial activity to derivatives of the lignins, proteins, and celluloses that resist further reaction. It produces a material with the characteristics of humus. Pathogens are destroyed during the process through the action of the heat that is generated. Temperatures of 55–60°C for a day or two will kill most pathogens of concern, and such conditions are produced in normal composting behavior. Obviously, adequate mixing is essential, to avoid cold spots and to ensure that all pathogens spend adequate time in the heated regions. Composting of wastes that contain potentially large concentrations of human pathogens, (e.g., sewage sludge) must be done with careful control of conditions.

Composting involves the interaction of the organic substrate with the organisms in the presence of water and oxygen to produce heat, carbon dioxide, and the decomposed organic materials. Conditions such as substrate composition, aeration, and moisture content affect the process and need to be well controlled to give a good quality product and to ensure operation of a large-scale plant. As with recycling, sorting is necessary to avoid contamination with nondegradable materials. Size reduction, particularly of waste wood, branches in yard waste, and so on, is necessary. The C/N ratio of the substrate strongly affects the rate, and nitrogen-rich materials may have to be added to act as fertilizer. The consistency of the material must allow air to circulate. Therefore the mix must have reasonable particle size and cannot be excessively wet. Materials such as sewage sludge need to be mixed with coarser materials, for example. The process is not rapid. Two weeks and usually more is required between the start of the process and its completion, so that a large-scale composting plant will contain a large volume of material.

One process is static; the organic waste is placed in an appropriate pile over a pipe that passes air into it, or applies suction to draw air through it. Odors can be trapped by a layer of finished compost, which has good adsorption properties. A second process involves the construction of windrows (elongated piles) of the waste, and mechanically turning and mixing them on a regular basis. These piles may be 6 or 7 ft high, 10 to 13 ft wide (narrower with manual turning), and as long as necessary. Shelter is needed for both these systems to keep out excessive moisture from rain, and for temperature protection in cold climates. Odor control is necessary for the windrow method.

An alternate process, mechanical composting, involves the use of rotating drums, or tanks equipped with mixing devices along with aeration, to provide optimal aeration and mixing. These afford more rapid reactions in the initial steps of the process when the greatest generation of odor is likely, but still require considerable time either in the reactor or in a windrow stage for complete reaction to a usable material. Control of possible leachate from the wastes waiting to be processed, and from the composting materials, is necessary.

As indicated, odor is one of the main concerns raised in objections to composting sites. Although the process is aerobic, some anaerobic decomposition can occur locally in the composting mass, generating typical anaerobic decay products such as ammonia, hydrogen sulfide, and organic sulfides such as mercaptans. Aerobic products such as low molecular weight fatty acids (e.g., acetic and butyric acids), aldehydes, ketones, esters, terpenes, and others with objectionable odors also may be released. A properly designed composting facility must have provision for capturing these materials, usually by passing the exhaust air through absorbers or scrubbers of some kind to either trap or chemically (sometimes biologically) destroy the obnoxious materials.

16.4 ANAEROBIC DIGESTION OF BIOLOGICAL WASTES

It was pointed out in Section 11.5.2 that the volume of sewage sludge that must be disposed of can be minimized by anaerobic decomposition to produce methane, a useful by-product. This process can be extended to other organic waste materials, including animal and human wastes, domestic wastes and crop residues. The wastes are made into a slurry and digested by anaerobic bacteria over a period of weeks to produce a gas (called biogas) that is approximately 60% methane. Most of the remainder is carbon dioxide; undesirable components such as H_2S are usually quite low. The process can be quite simple and has applications as an energy source in undeveloped areas, since the biogas can be used for heating, cooking, or running engines.

The digestion process uses a number of complementary bacterial species that convert protein, carbohydrate, and lipid material to simpler compounds— amino acids, simple sugars, fatty acids—and finally into methane as the bacteria use these compounds to grow. Except for some chemicals that might be found in industrial wastes, composition of the waste is not critical. Operating conditions, such as temperature and pH, must be kept within certain limits for optimum gas production. The final output from such digesters is a sludge containing half to two-thirds of the original organic content. This can be used as a fertilizer, although there may be concern about the complete destruction of pathogenic bacteria.

16.5 INCINERATION

Combustible wastes have long been disposed of by burning. As is the case with burial, this solution is not simple when carried out on a large scale. Three major problems must be dealt with in burning of waste: smoke, soot, and ash released to the atmosphere; highly toxic compounds produced through incomplete combustion and also released to the atmosphere; and toxic residues, chiefly heavy metals, that contaminate the ash and may enter groundwater from ash disposal. On the other hand, the volume of solid waste that must be disposed of in landfills is markedly reduced, and the energy released in the combustion process may be recovered, either to produce steam for heating or to generate electricity.

In all incineration processes, several possible modes exist for release of undesirable materials in the exhaust. These include material that escapes combustion (e.g., if the residence time is too short, or if the waste, is poorly mixed and does not experience sufficiently high temperature), material that is only partially oxidized to other, perhaps more hazardous compounds, fragments from thermal decomposition (pyrolysis) of larger molecules (e.g., benzene rings, small chlorinated molecules), and new molecules that may form from partially decomposed material in temperature regions that are high enough to promote reaction but not high enough for decomposition to be complete. Dioxins are an example; their formation was discussed in Section 8.10.3. The presence of chlorine- or bromine-containing materials can lead to the formation of hazardous chlorinated or brominated organics, including halogenated dibenzofurans (Section 8.10.3) and biphenyls (Section 8.7), as well as HCl, while sulfur compounds will produce SO_2, and the combustion process itself will generate nitrogen oxides. Such incomplete combustion can release hydrocarbons, particularly polycyclic compounds such as benzo-(a)-pyrene,

as well as carbon monoxide. Other toxic elements present at trace concentrations in the waste, such as Se, As, and heavy metals, can also escape as fine particulates; fly ash is generally higher in heavy metal concentration than the coarse ash left in the incinerator, but even the latter may have enough heavy metal content to require special disposal.

Various designs of incinerators are available for dealing with municipal wastes. They must allow thorough mixing of the waste with air and maintain a temperature high enough (750–1000°C), along with a residence time of the combustible material in the hot zone long enough (at least 3–4 s) to permit complete degradation of all organic molecules to carbon dioxide. This may involve a two-stage combustion process. Exhaust gases must be scrubbed to remove HCl and other components, and particulate materials must be removed by electrostatic precipitators and filters.

For incineration of municipal waste, two approaches are in use. In mass burning, the waste is burned with only minimal preseparation (e.g., of oversize noncombustibles). Shredding or shearing of large combustibles is necessary to provide reasonable handling. Such incinerators may be used solely for waste disposal, or the heat generated may also be used as a source of energy. The heat value of municipal solid waste ranges from about 7,000–15,000 kJ/kg, depending on the composition (cf. anthracite coal, 30,000–37,000 kJ/kg). Plastic waste has the greatest heating value per unit weight, while paper and other combustble components are considerably lower. Combustion is a way of recovering some of the energy value of plastic and other wastes if recycling is not practical.[2] Although incineration of large amounts of plastics generates more emissions of chlorine and of lead and cadmium from compounds that are still used as stabilizers, studies have shown that properly operating modern plants can maintain emissions well within safe limits.

Complete combustion will preclude release of smoke containing soot, but fine ash particles and gaseous pollutants must be removed by precipitation and/or scrubbing of the stack gases as already mentioned. The waste normally contains considerable water, and under some conditions this will lead to a visible steam plume as the vapor condenses. A plume will form if, upon mixing of the warm stack gas with the ambient air, the temperature and relative humidity of the mix reach a value in which the saturation level of the air is exceeded. This is often interpreted by neighbors as harmful, although in itself it is not unless it hides particulate smoke.

Ash, both the fine fly ash recovered from the gas stream and the coarser material left in the incinerator, must be disposed of, usually in landfills. If the heavy metal content is low enough, however, the ash may be used as a fill or aggregate. Some material, especially ferrous metal, can be recovered from the coarse ash before disposal. If scrubbers are used to treat the stack gases, solid or sludge residues from these will also be placed in landfills.

While incineration as a municipal solid waste disposal method can be safe and effective, the technology is not easy to control because of the variable nature of the material to be burned. Even with careful selection, one can never be sure that a householder has not put some highly hazardous material into the garbage, and thus explosions when the waste enters the incinerator are possible. Care in maintaining the monitoring systems and in controlling the operating parameters is essential if release of highly hazardous organic by-products is to be avoided.

Municipal incinerators have been criticized not only because of their potential for release of materials to the atmosphere, but also because of their effect on recycling. The cost of construction, particularly if the facility is also used to generate energy, makes it necessary from an economic point of view to operate

[2]B. Piasecki, D. Rainey and K. Fletcher, *Am. Sci.*, **86**, 364 (1998).

an incinerator as continuously as possible. Consequently, removal of combustible materials from the waste stream for recycling may be discouraged. Also, efficient plants need to be large and consequently need large amounts of waste; transportation becomes a factor.

The second approach to incineration of municipal waste is the use of refuse-derived fuels for energy generation. In this approach, the waste is first processed for the recovery of recyclables and to assure a more uniform material that can be handled and burned in a more controlled fashion. The refuse-derived fuel is co-burned with another fuel, usually coal. Typically, in these co-burning systems, 30% of the fuel, producing 15% of the energy, is refuse derived. This is the limit for regulations applying to coal-burning generators. If more refuse is used, more stringent rules requiring more monitoring of potential toxic releases go into effect.

Various special purpose waste incinerators are in use—for example, to burn sewage sludge or hospital wastes. Hazardous wastes that are not highly combustible and must be burned at an appropriately high temperature for safety reasons are treated in incinerators with auxiliary gas or oil fueling. A large use of incineration is for the disposal of liquid industrial hazardous wastes.

Wastes designated as hazardous may be disposed of in incinerators intended specifically for that purpose (e.g., Figure 16-4), but much is disposed of in industrial boilers or furnaces, such as the cement kilns discussed in Section 12.4.4. In fact, over half the hazardous wastes incineration in the United States is done in cement kilns. Incineration of hazardous wastes in the United States is governed by the fairly elaborate EPA Boilers and Industrial Furnace (BIF) regulations, which in summary call for minimum destruction of 99.99% (99.9999% for certain materials like dioxins) destruction of the waste. Notice that this is not a limit on the amount of unburned material that can be released in total, just a limit on the fraction of the input that can be released. There are other restrictions such as on CO, metal, and particle release.

A well-designed combustion chamber can minimize stagnant regions that may not reach proper temperatures and maximize mixing, but operation under less than optimum conditions can defeat this. Because hazardous wastes are likely to be of variable composition (except for burners dedicated to incineration of a specific process stream), maintaining optimum conditions may not be easy. Monitoring release is also a problem because of the large number of possible products involved. Carbon monoxide monitoring can be used to indicate deviations from proper conditions but is not a direct measure of what is being released.

Ocean incineration, carried out aboard special incinerator ships, has had some use. Here, the theory is that the incineration occurs far from any populated area, and any emissions will be absorbed by the ocean before they can be carried to land. This practice would permit incineration of very toxic materials,

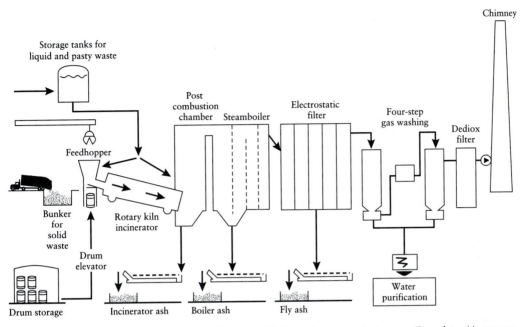

FIGURE 16-4 Schematic of a rotary kiln hazardous waste incinerator. From http://www.eurtits.org/reports/9702.htm. Used by permission of the European Union for Responsible Incineration and Treatment of Special Waste and Indaver.

but if carried out on a large scale raises questions about effects of toxic releases on the ocean food chain if combustion were not complete.

16.6 THE TIRE PROBLEM

About 250 million worn-out automobile tires are discarded annually in the United States, and many more globally. Disposal of these waste tires has long been a problem. Landfilling is difficult because tires tend to work their way to the surface, as mentioned earlier. Open dumps, of which there are many, are breeding grounds for mosquitoes because of the stagnant water that readily collects in the tires, and they are subject to fires that are both extremely smoky and notoriously difficult to put out. These fires release toxic products and leave toxic residues. One tire fire, in Winchester, Virginia, burned for nine months. A number of states have enacted legislation dealing with scrap tire disposal.

However, nearly two-thirds of the tires now being discarded are not disposed of in any productive way.[3]

Useful rubber products are produced by vulcanization, a process that uses sulfur to cross-link the polymer chains of the raw rubber to give a material with physical and chemical properties that make it tough, stable, and nonbiodegradable. Fillers such as carbon black are added to improve properties, while tires also contain steel or fiber reinforcing belts. Depolymerization of the rubber, while chemically feasible, is not economically practical. A number of processes can be used to pyrolyze the scrap tires to produce gaseous and liquid hydrocarbons useful as fuels or as chemical feedstocks, along with carbon black and recovered steel. Although these have promise, at the time of writing no commercial plants are operating, again because such enterprises are not economically viable. A couple of plants, one in Wind Gap, Pennsylvania, and another in Centralia, Washington, have operated intermittently, and a demonstration pilot plant has been built near Leipzig, Germany, in the town of Grimma.

The major use of scrap tires at present is as tire-derived fuel. Whole or shredded tires, which have very high heat values, are used to supplement traditional fuels in cement kilns, pulp and paper mill boilers, and coal-fired power generating facilities. Growing use as fuel is anticipated as it becomes more commonly recognized that controlled combustion does not involve the smoke emissions associated with free burning.

The properties of scrap tires make them directly useful for some purposes. They can be used to construct artificial reefs in which the tires are tied together and anchored in coastal waters, where they are attractive to barnacles and other marine organisms and for many types of fish. Tires can also be used to construct floating breakwaters to protect harbors and beaches. Stacks of tires have been used as highway crash barriers at bridge piers and other obstructions and for the construction of retaining walls for steep-sided areas next to highways. In a number of states including Florida, West Virginia, Ohio, and Pennsylvania, scrap tires are used constructively in landfills, either as a daily landfill cover or as part of the leachate collection system. In some cases, the tires are shredded before use in landfills.

Shredded tires have many other applications. For example, the shreds have been used to make floor mats and sandals and have been used as fill and insulation under road beds. Very small shreds, known as ground rubber, have been used for running tracks on athletic fields, railroad crossing beds, carpet underlay, and as an asphalt additive.

In some cases, rubber strips cut from the tires are used to make dock bumpers, floor mats, conveyor belts, and so on. In 1998, 23% of U.S. tires were recycled, excluding those used for fuel or recovered for retreading.

---

[3]K. Reese, *Today's Chem. Work*, 4(#2), 75 (February 1995), provides a discussion of the 1994 situation.

16.7 RECYCLING

Recycling is an alternative to the waste disposal methods discussed in the preceding sections of this chapter. The present discussion will be mostly about "municipal solid waste," which comes from households, commerce, industry, and government. The approximate composition of this municipal solid waste in 1998 and the percent of the different materials that were recycled were shown in Table 16-1. Some solid waste can be recycled in a variety of ways to make useful objects. Before any of the various materials that make up this solid waste can be recycled, however, they must be separated, at least in part. Separation at the point of origin provides cleaner, higher quality material than that separated from mixed solid waste, and many municipalities now encourage such separation, but even separated waste does not provide pure materials for recycling. Sometimes the separated but not pure solid wastes can be recycled as such, but often further separation is necessary. For example, as will be seen shortly, glass waste and beverage cans must be further separated before the materials can be reused.

It is difficult to treat the economics and cost-effectiveness of recycling in isolation. In general, the improvement of the quality of the environment due to recycling has not been factored into the economics. In many cases, this makes many types of recycling appear to be cost-ineffective on purely economic grounds. In some cases, however, the use of recycled materials can be cost-effective even without considering these intangible factors. Table 16-2 gave some examples of industries in the developed world that used recycled materials as feedstocks in 1989.

16.7.1 Glass

The municipal solid waste stream contains about 10 wt % glass, most of which is container glass, the glass used to make jars and bottles: mayonnaise and pickle jars, beer bottles, wine and liquor bottles, some soft drink bottles, and so on. Container glass is the only glass that is being recycled in large quantities at present. Glass used in such items as light bulbs, drinking glasses, mirrors, and cookware is different in composition from container glass and is not accepted for recycling by the container industry. Glass pieces that can be recycled are called "cullet" and must often be color-sorted before reuse. Much cullet, called primary cullet, is generated during the manufacture of glass. Traditionally, about 15% of the material used in the manufacture of glass containers has been cullet, but a goal of 30–50% has been established within the container industry. It is actually advantageous to use cullet because it liquefies at a lower temperature than the raw materials: sand (SiO_2), lime (CaO), and soda ash (Na_2CO_3) used in the manufacture of glass. This results in energy savings (for every 1% of recycled glass used, there is a 0.5% drop in energy use) and the extension of the useful life of furnace linings, among other things. If the cullet is not color-separated, it is used for the production of glass beads, roadway materials, and building materials such as fiberglass insulation.

16.7.2 Paper

Paper and paper products are made from wood and consist mostly of cellulose, as discussed in Section 7.3.2. Most paper products contain various additives—fillers such as clay or titanium dioxide, resins to improve wet strength, sizings, coloring, and so on. These differ depending on the type of paper, adding to the complications of paper recycling. Separation of paper to be recycled into various grades is required. As shown in Table 16-1, nearly 42% of the paper produced in the United States was recycled in 1998. The types of paper that are recycled are newsprint, corrugated cardboard, magazines, and so-called high-grade and mixed papers. Domestic demand for reclaimed papers has been rather erratic over the years, and a large amount has been exported: 6.5 million tons in 1990, for example, mostly to the Far East. In 1995, about 20% of all paper reclaimed in the United States was exported. Corrugated cardboard is the main reclaimed paper product that is exported.

Recycled paper is used mainly by four different industries. The first is the paper products industry, to make newsprint, various other writing and printing papers, bags, towels, and tissues. To use recycled newsprint for fresh newspapers, it must first be de-inked in special de-inking mills that were in short supply until the mid-1990s. New de-inking plants were built only after a number of states required that a minimum percentage of recycled paper be used in newsprint. De-inking consists of repulping the paper in a solution of sodium hydroxide and hydrogen peroxide, along with surfactants to release the ink, which is separated by flotation. Other contaminants such as staples and glue must also be removed. The process breaks up the fibers to some degree, reducing the quality of the product and limiting the number of times a given sample of paper can be recycled. To maintain quality, some percentage of virgin fibers or higher quality paper (typically 30% coated paper—magazines, etc.) must be added to recycled newsprint. The second industry is the paperboard products industry, to make corrugated boxes, shoe boxes, and file folders. The third is the construction paper industry, to make roofing materials and acoustic tile. The fourth is the molded paper products industry, to make egg cartons and layers for packing fruit. Recycled paper products are also used to a small extent to make cellulose insulation, bedding for animals, mulch, and liquid and solid fuels.

16.7.3 Metal

The metal content of municipal solid waste consists mostly of aluminum or ferrous metals and these are generally recycled separately.

16.7.3.1 Aluminum

Aluminum scrap that is produced during the manufacture of aluminum products (new scrap) is almost completely reused and recycled. Old scrap comes from discarded aluminum products including cans, aluminum siding, lawn furniture, automobiles, pots and pans, and venetian blinds. Recycling of aluminum cans began in the early 1960s, encouraged by the major aluminum companies, and was viewed mostly in terms of public relations. By 1995, 65% of the aluminum from used beverage cans was recycled back to aluminum can manufacturers, who processed it into so-called can sheet to produce new beverage containers. Each of these newly produced aluminum cans contained about 51% recycled metal. Other types of old scrap aluminum often contain other alloyed metals and are thus not used to make beverage containers or aluminum foil but may be sold to aluminum mills for production of other products.

16.7.3.2 Ferrous Metals

There are three types of ferrous metal scrap. *Home scrap* is produced in steel mills during production of the steel and is recycled internally. *Prompt industrial scrap* is produced during machining, stamping, and other fabrication processes during the production of iron and steel items. *Obsolete scrap* consists of iron and steel products that have served a useful purpose and are now being discarded. Railroad cars and tracks, automobiles, and trucks are major sources of obsolete scrap; other sources include manhole covers, old water pipes, lawn mowers, pots and pans, and steel cans. Steel cans can be collected along with aluminum cans because the two types of can can be easily sorted by means of magnetic separation (the steel cans can be magnetized, while the aluminum cans cannot). Steel mills use scrap iron and steel in the manufacture of new steel (Section 12.2.4).

The obsolete scrap most often recycled by consumers consists of steel cans that have been used as containers for food, beverages, paint, or aerosols. To protect the contents of the cans from corrosion, these steel cans are usually lined with a very thin coating of tin, about 3×10^{-5} in. thick. These cans, often called "tin" cans, have the tin recovered by detinning companies during recycling. Detinning companies use either a chemical and electrolytic process or heat treatment (tin melts at about 232°C) to remove the tin. The chemical process consists of dissolving the tin as sodium stannate with a sodium hydroxide–sodium nitrate solution. In the electrolytic process, the scrap steel is made anodic, and the tin electrolytically dissolved and redeposited on the cathode. Some food cans, including tuna cans, are made with tin-free steel, while some others have an aluminum lid and a steel body (bimetal cans) whose parts can be separated magnetically after shredding. In 1995, almost 56% of steel and bimetal cans in the United States were recycled.

16.7.4 Plastics

All plastics are made of polymers, which were discussed in Section 7.3. Although most plastics waste in the United States has been placed in landfills, Switzerland, Japan, and some other countries, have disposed of most of their plastics waste by incineration. Although the incinerators used include excellent antipollution devices, they have not been accepted by the populations of many countries. Most plastics wastes are not being recycled at this time. For example, the plastics used in some durable goods such as automobiles may consist of mixtures of as many as 50 or 60 different plastics that are not currently being separated for recycling. A number of ways are being found to get around this problem:

1. It is possible to use fewer different plastics in the manufacture of auto-mobiles.
2. Some companies—for example, General Electric—have made long-term agreements with scrap dealers who buy junked automobiles to obtain the plastic parts from these automobiles for recycling.
3. Manufacturers may be asked to put identifying bar codes on each poly-mer-containing component.
4. Uses can be found for the mixed polymers—for example, tiny beads of mixed plastic can be used by the aircraft industry in a process analogous to sand blasting for removing paint from aluminum.

Many manufacturers of plastics now attempt as much recycling of plastics wastes produced during production as possible. In the rest of this section, however, we shall be concerned only with the postconsumer plastics waste that has been labeled with the Society of the Plastics Industry (SPI) system shown in Figure 16-5. The six main polymers now labeled for recycling make up most of the postconsumer plastics wastes (numbers 2 through 6 make up about 85% of these). Some municipalities collect mixed plastics wastes, while others collect only plastics wastes that have been separated by number. In 1993 about 7% of the plastics labeled with numbers 1 through 6 in Figure 16-5 were recycled in the United States. In western Europe, in the same year, about 21% of municipal plastics waste was either reused or incinerated in a way that produced useful energy (see Section 16.5).

Recycling symbol	Polymer	Examples of use
1 PETE*	Poly(ethylene terephthalate) 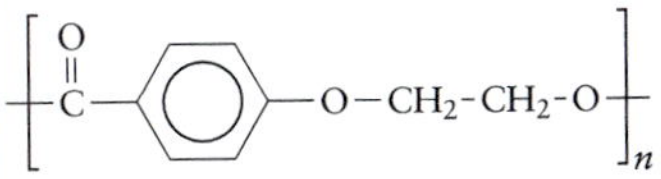 $\left[\!\!\begin{array}{c} \overset{O}{\underset{\parallel}{C}}-\bigcirc\!\!\bigcirc-O-CH_2\text{-}CH_2\text{-}O\end{array}\!\!\right]_n$	Soft drink bottles, carpets, fiberfill, rope, scouring pads, fabrics, Mylar tape(cassette and computer)
2 HDPE	High density polyethylene $-\!\!\left[CH_2\text{-}CH_2\right]_n$	Milk jugs, detergent bottles, bags, plastic lumber, garden furniture, flowerpots, trash cans, signs
3 V	Poly(vinyl chloride) $-\!\!\left[CH_2\text{-}CHCl\right]_n$	Cooking oil bottles, drainage and sewer pipes, tile, institutional furniture, credit cards
4 LDPE	Low density polyethylene $-\!\!\left[CH_2\text{-}CH_2\right]_n$	Bags, squeeze bottles, wrapping films, container lids
5 PP	Polypropylene $-\!\!\left[\begin{array}{c}CH_2\text{-}CH-\\ \ \ \ \ \ \ \ \ \ \ \ CH_3\end{array}\right]_n$	Yogurt containers, automobile batteries, bottles, carpets, rope, wrapping films
6 PS	Polystyrene 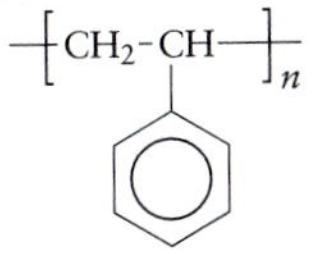 $-\!\!\left[\begin{array}{c}CH_2\text{-}CH-\\ \ \ \ \ \ \ \ \ \bigcirc\end{array}\right]_n$	Disposable cups and utensils, toys, lighting and signs, construction, foam containers, and insulation
7 Other	All other polymers	Various food containers, hand cream, toothpaste, and cosmetic containers

* PETE is used as an abbreviation for poly(ethyleneterephthalate) in recycling codes, but most chemists use the abbreviation PET.

FIGURE 16-5 Recycling codes for plastics.

16.7.4.1 Mixed Postconsumer Plastics Waste

Mixed plastics, after cleaning and shredding, can be melted and extruded under pressure to make "plastic wood." Since most plastics, including those in postconsumer plastics waste, are not miscible with each other, the particles in the "plastic wood" do not adhere well to each other, and the material contains voids and often pieces of newspaper, aluminum foil, and anything else that was not removed from the plastics waste before extrusion. The material is thus not very strong and cannot be used under tension; also, the extruded pieces must have a large cross section, several inches or more. Plastic wood is used in agriculture (fences, animal pens, tree supports), marine engineering (seawalls, boat docks, boardwalks, lobster traps), recreational equipment (park benches, picnic tables, stadium seating), gardening (compost bins, garden furniture), civil engineering and construction (signposts, siding insulation, shutters, roof tiles), and for industrial uses (highway construction, pipe racks, traffic barriers).

It is generally more useful to separate the postconsumer plastic waste into its constituent plastics (Figure 16-5), either at the source or after collection. There are many ways to do this; some of these are separation by hand, various flotation methods that separate the (shredded) plastics by density, and many other methods that are applicable to particular plastic mixtures. For example, bottles made of poly(vinyl chloride), PVC, can generally be separated from other bottles by using various devices based on the infrared absorption or the fluorescence emission of these bottles as contrasted with that of bottles made from other plastics. The bottles are on a conveyor belt; when the sensor detects a PVC bottle, this bottle is ejected from the belt into a special container. Methods based on the different solubility of the various plastics in different solvents or in the same solvent (xylene) at different temperatures are also being considered; in these cases, the polymers must be recovered from the solvents separately.

16.7.4.2 Reuse of Pure Plastics

Most 1- and 2-liter soda bottles are made primarily of poly(ethylene terephthalate), PET, and, in states such as New York State, are recycled separately from other plastic waste; this may be done by imposing a bottle deposit, which is returned to the consumer when the empty bottle is turned in at a collection center. These bottles can be used to produce polyester fiber, which is used in carpets (in 1995, it was used in 50% of all polyester carpeting sold in the United States), outdoor clothing, insulation, furniture stuffing, and so on. Waste nylon is also processed into fiber to make carpets, tennis ball felt, and other items.

High-density polyethylene (HDPE), used for milk and other bottles when first produced, has been recycled for use in non food containers for such products as oil, antifreeze, and laundry detergent. It can also be used for drainage pipes, flower pots, trash cans, kitchen drain boards, and so on. Recycled milk jugs have been made into a superior type of plastic wood that is hard to split, easy to saw, colorable while being produced, and stands up better than treated lumber to weather and insects. Although it costs much more than treated lumber, its durability makes it less expensive in the long run.

16.7.4.3 Tertiary Recycling of Plastics

Tertiary recycling is the production of basic chemicals and fuels from plastic waste, as defined by the American Society for Testing and Materials (ASTM). The polymers may be chemically decomposed by various methods, or they may be pyrolyzed. Tertiary recycling is used even for polymers that may be recycled as such (see Section 16.7.4.2) because food and drug regulations in the United States forbid the use of melt-process recycled plastics for anything that will contact food.

16.7.4.3.1 Chemical Decomposition of Polymers

Many polymers can be decomposed back into their monomers by using thermal or chemical treatments. These include many more plastics than those few used as illustrations in this section.

A number of different chemical companies have treated PET with ethylene glycol, methanol, or water at elevated temperatures under pressure to produce the chemicals from which PET may be resynthesized, dimethyl terephthalate and ethylene glycol. Equation (16-1) shows the methanolysis reaction, which is widely used because it is relatively insensitive to the additives and contaminants that may be included in the recycled polymer.

$$
\left[\overset{\displaystyle O}{\underset{\displaystyle \|}{C}} - \!\!\!\bigcirc\!\!\! - \overset{\displaystyle O}{\underset{\displaystyle \|}{C}} - O - CH_2CH_2\text{-}O \right]_n + 2n\,CH_3OH \longrightarrow
$$

PET

$$(16\text{-}1)$$

$$
n\,CH_3O - \overset{\displaystyle O}{\underset{\displaystyle \|}{C}} - \!\!\!\bigcirc\!\!\! - \overset{\displaystyle O}{\underset{\displaystyle \|}{C}} - OCH_3 + n\,HO - CH_2CH_2\text{-}OH
$$

Polyurethanes, used in foam mattresses, foam insulation, and proprietary elastic fibers such as Spandex, are various polymeric materials that contain urethane, —NHCO—, groups formed from isocyanates, often aromatic ones, and polyols. When polyurethane foam products are mixed with superheated steam, they liquefy and hydrolyze to the amine corresponding to the starting isocyanate, usually 2,5-diaminotoluene, propylene glycol, and carbon dioxide.

$$\left[\begin{array}{c} \overset{O}{\overset{\|}{C}}HN \underset{}{\overset{CH_3}{\bigcirc}} NH\overset{O}{\overset{\|}{C}}OCH_2CH_2CH_2O \end{array} \right]_n + 2n\ H_2O \longrightarrow$$

[Simplified polyurethane structure]

(16-2)

$$n \quad H_2N \underset{}{\overset{CH_3}{\bigcirc}} NH_2 \quad + \quad 2n\ CO_2 \quad + \quad n\ HOCH_2CH_2OH$$

Equation (16-2) shows a very simplified structure of a polyurethane. More than one glycol and more than one aromatic diisocyanate may have been used to make the polyurethane and, in addition, parts of the polymer chains may consist of polyesters, polyethers, and polyamides. Therefore, additional low molecular weight compounds are usually produced by hydrolysis. Alcoholysis using a short-chain alcohol has also been used to decompose polyurethanes; in this case, no carbon dioxide is formed. The low molecular weight compounds produced can be mixed with diisocyanates and new foamed polyurethanes can be produced.

Nylon 6 can be depolymerized at high temperature or by using various hydrolysis reactions to form the monomer, which can be purified and repolymerized:

$$\left[NH-(CH_2)_5-CO \right]_n \longrightarrow \begin{array}{c} CH_2-CH_2-CH_2 \\ | \\ CH_2-CH_2-CO \end{array} \hspace{-0.5em}\searrow NH$$

(16-3)

Nylon 6   Caprolactam

16.7.4.3.2 Pyrolysis

Pyrolysis is the thermal fragmentation of plastics into small molecules, either in the absence of oxygen or in an oxygen-deficient atmosphere, at temperatures as high as 500–1000°C. Waste items made of a single polymer, a mixture of polymers, or, for that matter, mixed household wastes, may be and have been pyrolyzed. Gases, liquids, and solids are obtained. The gases obtained may include CO, CO_2, H_2, CH_4, ethylene, propylene, and others. Some of these gases are used to heat the pyrolysis plant so that no external energy source is needed for this purpose. The liquids that are formed include benzene, toluene, naphthalene (this is dissolved in the other liquids), fuel oil, and kerosene, depending on the plastics that are pyrolyzed. The solids obtained are mostly carbon black and waxes (these are hydrocarbons). Figure 16-6 shows a schematic of a pyrolysis plant using a fluidized bed of sand.

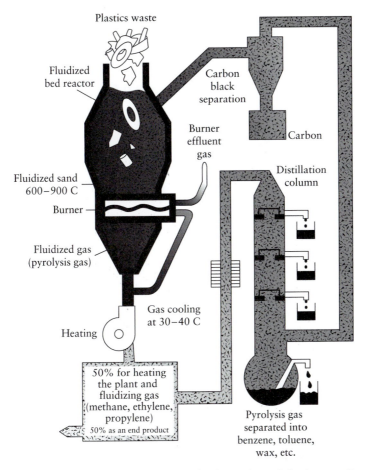

Plastics waste

Fluidized bed reactor

Carbon black separation

Burner effluent gas

Carbon

Distillation column

Fluidized sand 600–900 C

Burner

Fluidized gas (pyrolysis gas)

Gas cooling at 30–40 C

Heating

50% for heating the plant and fluidizing gas (methane, ethylene, propylene)
50% as an end product

Pyrolysis gas separated into benzene, toluene, wax, etc.

FIGURE 16-6 A fluidized-bed reactor for the pyrolysis of plastic waste. From N. Mustafa, ed., *Plastic Waste Management: Disposal, Recycling, and Reuse* by courtesy of Marcel Dekker, Inc. Copyright © 1993.

Additional Reading

Andrews, G. D., and P. M. Subramanian, eds., *Emerging Technologies in Plastics Recycling*. American Chemical Society, Washington, DC, 1992.

Bagchi, A., *Design, Construction and Monitoring of Landfills*, 2nd ed. Wiley, New York, 1994.

Brandrup, J., M. Bittner, W. Michaeli, and G. Menges, eds., *Recycling and Recovery of Plastics*. Hanser/Gardner, Cincinnati, OH, 1995.

Diaz, L. F., G. M. Savage, L. L. Eggerth, and C. G. Golueke, *Composting and Recycling Municipal Solid Waste*. Lewis Publishers, Boca Raton, FL, 1993.

Haug, R. T., *The Practical Handbook of Compost Engineering*. Lewis Publishers, Boca Raton, FL, 1993.

Kreith, F., ed., *Handbook of Solid Waste Management*. McGraw-Hill, New York, 1994.

Lund, H. F., ed., *The McGraw-Hill Recycling Handbook*. McGraw-Hill, New York, 1993.

Manser, A. G. R., and A. A. Keeling, *Practical Handbook of Processing and Recycling Municipal Solid Waste*. Lewis Publishers, Boca Raton, FL, 1996.

Mustafa, N., ed., *Plastics Waste Management: Disposal, Recycling, and Reuse*. Dekker, New York, 1993.

Niessen, W. R., *Combustion and Incineration Processes*, 2nd ed. Dekker, New York, 1995.

Polprasert, C., *Organic Waste Recycling*, 2nd. ed. Wiley, Chichester, 1996.

Strong, D. L., *Recycling in America*. ABC-CLIO, Santa Barbara, CA, 1997.

Wilson, D. J., and A. N. Clarke, *Hazardous Waste Site Soil Remediation*. Dekker, New York, 1994.

APPENDIX A

Designation of Spectroscopic States

A.1 INTRODUCTION

The recipes briefly summarized here are used in arriving at atomic and molecular state (energy) descriptions. The reader is referred to standard textbooks of quantum mechanics and spectroscopy, some of which are given in the Additional Reading, for the mathematical formulations of these principles.

A.2 ATOMS

The state of the electron in the hydrogen atom or hydrogen-like ion (e.g., He^+ and Li^{2+}, each containing only one electron) is specified by the three quantum numbers n, l, and m_l. The principal quantum number n can have the values 1, 2, 3, ...; it defines the shell of the atom occupied by the electron, and therefore the total energy of the single bound electron (neglecting relativistic effects) is specified completely by n. The azimuthal or orbital angular momentum quantum number l can have the values $0, 1, 2 \ldots (n-1)$; it determines the total angular momentum of the electron, which is quantized and can be represented by a vector having magnitudes $(h/2\pi)[l(l+1)]^{1/2}$ (where h is Planck's constant). The states (or atomic orbitals) with $l = 0, 1, 2, 3, \ldots$ are designated, respectively, s, p, d, f, ... orbitals, so that the electron in the lowest electronic state of the hydrogen atom ($n = 1$, $l = 0$) is called a 1s electron. The quantum number m_l is the magnetic quantum number (so named because it is related to the effects of a magnetic field on atomic spectra), and can have the values $-l, -(l-1), \ldots, 0, \ldots (l-1), l$; it determines the allowed components of the angular momentum in a definite z direction, which are also quantized and can only have values of $m_l h/2\pi$.

In addition to the three quantum numbers given, another factor, electron spin, must be considered to explain the fine structure observed in atomic spectra. This fourth parameter is also a natural consequence of quantum mechanics if relativistic effects are included in the detailed wave mechanical treatment. In addition to orbital angular momentum described by the quantum number l, an electron has an intrinsic magnetic moment as if (classically) it were spinning about its own axis in addition to orbiting the nucleus. Associated with this moment is a quantum number s, such that the total spin angular momentum vector has a magnitude $(h/2\pi)[s(s+1)]^{1/2}$; however, s can have only the single value of $1/2$. Analogous to the orbital angular momentum, the z component of the spin angular momentum is also quantized with allowed values of magnitude $m_s h/2\pi$, where $m_s = \pm 1/2$ (so that there are only two possible orientations of the spin angular momentum). The electron spin and orbital angular momentum magnetic moments interact, so that l and s are coupled to give a total electronic angular momentum vector of magnitude $h/2\pi[j(j+1)]^{1/2}$ where $j = l \pm s = l \pm 1/2$.

With polyelectronic atoms, the complexities of the mathematics involved in describing the system as a result of electrostatic interactions among the electrons as well as between the electrons and the nucleus are such that only approximate solutions of the quantum mechanical treatment are possible. The concept of hydrogen-like orbitals is retained with the principal quantum number n designating the electronic shells of the atom, but now the energy of the state is a function of the total angular momentum of the electrons as well as the quantum number n. The number of electrons in a given orbital is governed by the Pauli exclusion principle, which states that no two bound electrons can exist in the same quantum state; that is, no two electrons in the atom can have the same values of n, l, m_l, and m_s. Thus, an atomic orbital characterized by a specific set of n, l, and m_l values can be occupied by two electrons only if their spins are opposed with $m_s = +1/2$ and $-1/2$. Subject to this constraint, the orbitals are filled in the order of increasing energy (the aufbau principle). If two or more orbitals have the same energy (i.e., are degenerate), then the electrons go into different orbitals as much as possible, thus reducing electron–electron repulsion by keeping them apart, with spins parallel and in the same direction rather than opposed (Hund's rule).

For light atoms of the type we shall be concerned with here (nuclear charge < 40), the orbital angular momenta of all the electrons strongly couple together to give a total, or resultant, orbital angular momentum. This is designated by the quantum number L, which is obtained by the combination of the moments of the individual electrons. Similarly, the spin angular momenta combine to give a resultant spin, designated by the quantum number S. It follows that the total orbital angular momentum in the z direction is characterized by a quantum number $M_L = \Sigma m_l$, and the total spin angular momentum in the z direction has associated with it a quantum number $M_S = \Sigma m_s$.

Allowed values of M_L are $-L, -(L-1), \ldots, 0, \ldots, (L-1), L$ (i.e., $|M_L| \leq L$) and allowed values of M_S are $-S, -(S-1), \ldots, 0, \ldots, (S-1), S$ (or $|M_S| \leq S$). Closed (completely filled) shells and subshells have to a good approximation zero net resultant orbital and spin angular momenta, so that the combination need be carried out over only the electrons in partially filled subshells. Examples of this process are given later. The L and S momenta then couple together (Russell–Saunders coupling) to form a total atomic angular momentum J. Possible values of J are

$$J = L + S, \ L + S - 1, \ L + S - 2, \ \ldots, \ |L - S + 1|, \ |L - S| \qquad \text{(A-1)}$$

where $|L - S|$ is equal to $L - S$, if $L \geq S$, and equals $S - L$ if $S \geq, L$ (i.e., J can only have positive values). The complete term symbol, which designates the electronic state of the atom, is written

$$n \, {}^{2S+1}L_J \qquad \text{(A-2)}$$

Frequently, n is omitted. The quantity $(2S + 1)$ is the multiplicity of the atom (called singlet, doublet, triplet, etc., respectively, for values of $1, 2, 3, \ldots$); it gives the total number of possible spin orientations. Often the value of J is omitted from the term symbol, particularly if the energy differences among the different J states are so small that they are not a factor in photochemical considerations. Analogous to hydrogen or hydrogen-like atoms, the atomic states of $L = 0, 1, 2, 3$, are designated, respectively, S, P, D, F, states.

As a specific example, consider the oxygen atom with eight electrons in an electronic configuration $1s^2 2s^2 2p^4$. The 1s shell and the 2s subshell are completely filled, hence only the four 2p ($l = 1$) electrons need be considered. The state of lowest energy, hence the most stable, is called the ground state. According to Hund's rules, this state is the state of maximum multiplicity; for a given multiplicity, the state of maximum L; and for given L and S, the state of minimum J if the partially filled subshell is less than half-occupied, or conversely, the state of maximum J if the partially filled subshell is more than half-occupied. Thus the ground state of oxygen is given by the four electrons occupying the three 2p orbitals in the following manner:

The quantum numbers for the four electrons in this unfilled p subshell are as follows:

Electron	n	l	m_l	m_s
1	2	1	$+1$	$+1/2$
2	2	1	$+1$	$-1/2$
3	2	1	0	$+1/2$
4	2	1	-1	$+1/2$

For this configuration M_S is $\Sigma m_s = 1/2 - 1/2 + 1/2 + 1/2 = 1$, which is the maximum possible for 4 electrons in 3 orbitals to be consistent with the Pauli exclusion principle, and therefore $S = 1$ (since $|M_S| \leq S$) and the multiplicity $(2S + 1) = 3$. Similarly, $M_L = \Sigma m_l = 1$, which is also the maximum possible value; hence L must also be 1, since $|M_L| \leq L$. The quantities S and L are maximum allowed values, and therefore they give the state of lowest energy which is a 2^3P (a "triplet-P") state. Possible J values are 2, 1, and 0; since the subshell is more than half-filled, maximum J gives the most stable state, and therefore the complete term symbol for the oxygen atom in its lowest energy state is 2^3P_2.

Electronic excitation may result in change of J (generally unimportant in photochemistry, an exception being the rather large separations between the halogen $^2P_{3/2}$ and $^2P_{1/2}$ states), transition to another orbital within the subshell, or transition to a higher energy subshell. For example, two possible excited states of the oxygen atom within the same subshell (same $n = 2$ and $l = 1$) are as follows:

	↑↓	↑↓	—
m_l	+1, +1	0, 0	
m_s	+1/2, − 1/2	+1/2, − 1/2	

and

	↑↓	—	↑↓
m_l	+1, +1		−1, − 1
m_s	+1/2, − 1/2		+1/2, − 1/2

The term symbols are 2^1D_2 and 2^1S_0, respectively, the former being of lower energy because of maximum L.

Similarly, it can readily be verified using the same procedure that the total momenta for nitrogen with seven electrons $(1s^2 2s^2 2p^3)$ in the ground-state configuration is $S = 3/2$ (multiplicity $= 4$), $L = 0$, and $J = 3/2$; hence the term symbol is $2^4S_{3/2}$. A possible excited state is $2^2D_{3/2}$.

It should be pointed out that some excited electronic configurations are excluded by the Pauli exclusion principle for atoms with electrons with the same values of n and l (called equivalent electrons), because simply interchanging the order in the designation does not lead to different states.

A.3 DIATOMIC MOLECULES

For diatomic (or linear polyatomic) molecules, it is possible to designate the electronic states by a set of quantum numbers analogous to atoms. Thus, electrons with atomic orbital angular momentum quantum number l can have quantized molecular values λ from 0 to l, and these combine in a manner analogous to polyelectronic atoms to give a quantum number Λ representing the total orbital angular momentum along the internuclear axis. It is possible for Λ to have values $0, 1, 2, \ldots, L$ (where L is the quantum number for the resultant orbital angular momentum for all the electrons in the molecule), and these are designated, respectively, $\Sigma, \Pi, \Delta, \ldots$ states, analogous to s, p, d, $\ldots$, for the hydrogen atom and to S, P, D, $\ldots$ for polyelectronic atoms. Similarly, the total molecular spin quantum number is S, and the component of S along the internuclear axis is Σ with possible values $S, S-1, \ldots, 0, -(S-1), -S$. The sum of these two quantum numbers is a quantum number analogous to J, $\Omega = |\Lambda + \Sigma|$, and the total term symbol for a specific diatomic or linear polyatomic electronic state is

$$^{(2S+1)}\Lambda_\Omega \tag{A-3}$$

In addition to these quantum numbers $\Lambda, S,$ and Ω, there are two other properties of homonuclear diatomic species dealing with the symmetry of its wave function:

1. If inversion of all electrons through a center of symmetry of the molecule leads to no change in sign of the electronic wave function, then the state is *even*, or gerade (g); if a change of sign results, the state is *odd*, or ungerade (u). The symbol g or u is included in the molecular term symbol as a second subscript to Λ.

2. If reflection in a plane of symmetry passing through the nuclei (containing the internuclear axis of symmetry) leads to no change in sign of the electronic wave function; then the state is *positive* (+); if it changes sign, the state is *negative* (–) . The symbol + or – is written as a second superscript to Λ. (This element of symmetry applies only to $\Lambda = 0$ or Σ states.)

Molecular oxygen provides a practical atmospheric photochemical example for illustrating these state designation factors. The electronic configuration for ground-state O_2 (16 electrons) is

$$(\sigma_g 1s)^2 \quad (\sigma_u^* 1s)^2 \quad (\sigma_g 2s)^2 \quad (\sigma_u^* 2s)^2 \quad (\sigma_g 2p_z)^2 \quad (\pi_u 2p_x)^2$$
$$(\pi_u 2p_y)^2 \quad (\pi_g^* 2p_x)^1 \quad (\pi_g^* 2p_y)^1$$

The $(\sigma_g 1s)$, and so on designate molecular orbitals, the asterisk (*) refers to an antibonding orbital (i.e., the density of electrons between the two oxygen atoms in this type of orbital is less than that for the two free atoms, and therefore there is a repulsive force between them), and the superscript gives the number of electrons in the molecular orbital. Since there are two unpaired electrons in the equivalent (degenerate) $(\pi_g^* 2p_x)$ and $(\pi_g^* 2p_y)$ orbitals, by Hund's rule of maximum multiplicity the total spin is 1 and the ground state is a triplet. The quantum number Λ can be 2 or 0; however, if $\Lambda = 2$, both electrons would have the same $\lambda = \pm 1$ value, which is forbidden by the Pauli exclusion principle for electrons in equivalent orbitals also with the same spins; therefore only the $\Lambda = 0$, or Σ, state is allowed for $S = 1$. Figure A-1 shows the wave functions for one of the two unpaired electrons—the one in the $(\pi_g^* 2p_x)$ orbital—for the lowest energy configuration, from which it is seen that inter-change through the center of symmetry does not lead to a change in sign, whereas interchange through the plane of symmetry does. The ground state of oxygen is therefore gerade and negative, and the term symbol is $^3\Sigma_g^-$.

While the ground electronic state of O_2 is the $^3\Sigma_g^-$ state, there are two other possible states (of somewhat higher energies) for the electrons in these same two equivalent orbitals in which the total spin is zero (multiplicity = 1). In this case, both electrons can now have either the same λ values (each +1 or −1), leading to $\Lambda = 2$ and a $^1\Delta$ state, or different λ values (+1 and −1) giving a net angular momentum component $\Lambda = 0$ and hence a $^1\Sigma$ state. From the same symmetry considerations based on Figure A-1 that were used to obtain the term symbol for the ground state, it follows that the term symbols for these two electronically excited states are $^1\Delta_g$ and $^1\Sigma_g^+$. On the other hand, if an electron in a lower energy orbital [such as the $(\pi_u 2p_y)$ state] is excited to the

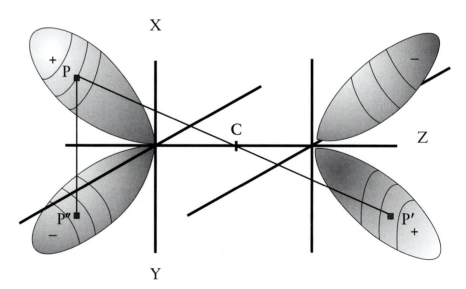

FIGURE A-1 Schematic drawing of the $(\pi_g^* 2p_x)$ orbital of O_2 in the $X-Z$ plane, the plane of the page. [The $(\pi_g^* 2p_y)$ orbital, not shown, is in the $y-z$ plane perpendicular to the plane of the page.] Inversion through the center of symmetry C from P to P′ does not change sign, whereas inversion through a plane of symmetry passing through the two nuclei such as the $y-z$ plane (from P to P″) does lead to a change in sign.

$(\pi_g^*2p_y)$ orbital, then two nonequivalent orbitals, (π_u2p_y) and $(\pi_g^*2p_x)$, are singly occupied; the Pauli exclusion principle no longer applies, since the two unpaired electrons occupy separate orbitals, and six states are possible: $^1\Sigma_u^+$, $^1\Sigma_u^-$, $^1\Delta_u$, $^3\Sigma_u^+$, $^3\Sigma_u^-$, and $^3\Delta_u$.

A.4 POLYATOMIC MOLECULES

Although the concepts developed for atoms and diatomic species can be extended to polyatomic molecules, the designation of electronic states for more complex molecules can become quite complex if the maximum spectral information is to be retained. However, many of the more subtle points that are especially important for fine-structure spectroscopic characterizations in most cases simply do not affect understanding of photochemical transformations involving complex species, and therefore for our purposes we can get by with relatively simple terms involving orbitals for only a single optical electron. The optical electron is the one electron promoted in the light absorption process.

Of importance for these orbitals are their symmetry and multiplicity characteristics, plus their involvement in bonding within the molecule. These may be of the following types involving both bonding and antibonding characteristics:

1. σ *and* σ^**orbitals*: these are associated with two atoms and involve two tightly bound electrons
2. π *and* π^**orbitals*: for example, the contribution of p electrons to the double bond in ethylene, or the conjugated electrons in benzene
3. *n orbitals*: these are nonbonding orbitals occupied by lone pair electrons in heteroatomic molecules

Excited states are then designated by the initial and final orbitals associated with the transition of the optical electron. Thus, the first excited state of ethylene is formed by promoting an electron from a bonding π orbital to an antibonding π^* orbital, leading to two unpaired electrons. Hund's rule of maximum multiplicity again suggests the spins of these electrons should be in the same direction, so that $S = 1$ and the multiplicity is 3. The designation of this state is $^3(\pi, \pi^*)$. For a molecule involving a carbonyl group,

$$\begin{array}{c} R \\ {}^{\diagdown}C{=}O \\ R' \end{array}$$

an additional low-lying state is the $^3(n, \pi^*)$ state formed by exciting an electron from the nonbonding n orbital to the antibonding π^* orbital. The state of lower energy, whether the $^3(n, \pi^*)$ or the $^3(\pi, \pi^*)$ state, will depend somewhat on substituent (R and R') groups and on the physical environment of the molecule.

Additional Reading

Harmony, M. D., *Introduction to Molecular Energies and Spectra*. Holt, Rinehart, & Winston, New York, 1972.

Herzberg, G., *Atomic Spectra and Atomic Structure*, 2nd ed. Dover, New York, 1974.

Herzberg, G., *Spectra of Diatomic Molecules*. Van Nostrand, Princeton, NJ, 1950.

Herzberg, G., *Electronic Spectra and Electronic Structure of Polyatomic Molecules*. Van Nostrand, Princeton, NJ, 1966.

Karplus, M., and Porter, R. N., *Atoms and Molecules*. W. A. Benjamin, New York, 1970.

McQuarrie, D. A., and Simon, J. D., *Physical Chemistry: A Molecular Approach*. University Science Books, Sausalito, CA, 1997.

APPENDIX B

Thorium and Actinium Series

TABLE B-1

Thorium ($4n$) Series

Z	Nuclide[a]	Half-life	Radiation[b]
90	^{232}Th (Th)	1.405×10^{10} y	α
90	^{228}Th (RdTh)	1.913 y	α, (γ)
89	^{228}Ac (MsTh$_2$)	6.15 h	β^-, γ
88	^{228}Ra (MsTh$_1$)	5.76 y	β^-
88	^{224}Ra (ThX)	3.66 d	α, (γ)
86	^{220}Rn (Tn)	55.6 s	α
84	^{216}Po (ThA)	0.145 s	α
84	^{212}Po (ThC$'$)	298 ns	α
83	^{212}Bi (ThC)	1.009 h	α, 35.9%; β^-, 64.1%, γ
82	^{212}Pb (ThB)	10.64 h	β^-, γ
82	^{208}Pb (ThD)	Stable	
81	^{208}Tl (ThC$''$)	3.053 m	β^-, γ

[a]Symbol used in the early studies of naturally occurring radioactivity is given in parentheses: (MsTh$_1$), mesothorium I; (RdTh), radiothorium; (Tn), thoron.

[b](γ) indicates low intensity (1–5%); γ rays with intensity below 1% are not included. Where both α and β^- are given, branched decay occurs.

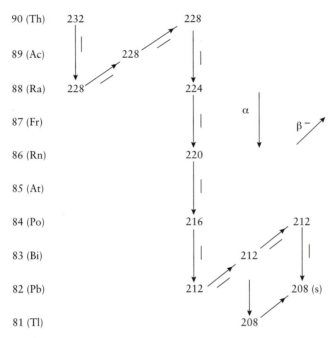

FIGURE B-1   Thorium ($4n$) series. The main sequence is indicated by arrows and a dash.

TABLE B-2

Actinium ($4n + 3$) Series

Z	Nuclide[a]	Half-life	Radiation[b]
92	^{235}U (AcU)	7.04×10^8 y	α, γ
91	^{231}Pa (Pa)	3.28×10^4 y	$\alpha, (\gamma)$
90	^{231}Th (UY)	1.063 d	β^-, γ
90	^{227}Th (RdAc)	18.72 d	α, γ
89	^{227}Ac (Ac)	21.774 y	β^-, 98.62%; α, 1.38%
88	^{223}Ra (AcX)	11.435 d	α, γ
87	^{223}Fr (AcK)	21.8 m	β^-, 99+%, γ; α, ~ 0.005%
86	^{219}Rn (An)	3.96 s	α, γ
85	^{219}At (At)	56 s	α, ~ 97%; β^-, ~ 3%
85	^{215}At (At)	0.10 ms	α
84	^{215}Po (AcA)	1.780 ms	α, 99%; β^-, 3.3×10^{-4}%
84	^{211}Po (AcC')	0.516 s	α
83	^{215}Bi (Bi)	7.6 m	β^-

(*continues*)

TABLE B-2 (*continued*)

Z	Nuclide[a]	Half-life	Radiation[b]
83	^{211}Bi (AcC)	2.14 m	α, 99.72%, γ
82	^{211}Pb (AcB)	36.1 m	β⁻, γ
82	^{207}Pb (AcD)	Stable	
81	^{207}Tl (AcC″)	4.77 m	β⁻

[a]Symbol used in the early studies of naturally occurring radioactivity is given in parentheses. (AcU), actinouranium; (RdAc), radioactinium; (An), actinon.

[b](γ) indicates low intensity (1–5%); γ rays with intensity below 1% are not included. Where both α and β⁻ are given, branched decay occurs.

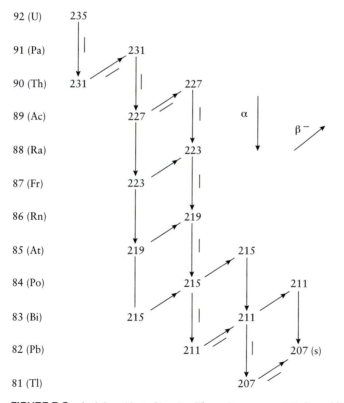

FIGURE B-2 Actinium (4*n* + 3) series. The main sequence is indicated by arrows and a dash.

APPENDIX C

Units

SI Units

Quantity	Unit	Symbol
Amount of substance	mole	mol
Electric current	ampere	a
Energy	joule	J
Length	meter	m
Mass	kilogram	kg
Power	watt (1 J/s)	W
Pressure	pascal	Pa
Radioactive decay	becqerel	Bq
Temperature	kelvin	K
Time	second	s

Prefixes

Name	Symbol	Multiplying factor
atto	a	10^{-18}
femto	f	10^{-15}
pico	p	10^{-12}
nano	n	10^{-9}
micro	μ	10^{-6}
milli	m	10^{-3}
centi	c	10^{-2}
deci	d	10^{-1}
deka	da	10
hecto	h	10^{2}
kilo	k	10^{3}
mega	M	10^{6}
giga	G	10^{9}
tera	T	10^{12}
peta	P	10^{15}
exa	E	10^{18}

Derived Units

Name and symbol	Value
acre	0.4047 hectare
angstrom (Å)	10^{-10} m or 0.1 nm
atmosphere (atm)	1.01325×10^5 Pa
bar	10^5 Pa
barrel (petroleum) = 42 gal = 306 lb	0.159 m^3
calorie (cal)	4.184 J
curie (Ci)	3.70×10^{10} Bq
electron-volt (eV)	1.602×10^{-19} J
foot (ft)	0.3048 m
gallon (gal)	3.786×10^{-3} m^3
hectare	10^4 m^2
metric ton—see tonne	
mile (mi)	1609.344 m
ton (U.S. long); 2240 pounds	1.016×10^3 kg
ton (U.S. short); 2000 pounds	9.072×10^2 kg
tonne (metric ton); 2205 pounds	1×10^3 kg

APPENDIX D

Symbols and Abbreviations

Note: Some of these symbols have other meanings when used in specific contexts; other symbols that are used only in specific contexts may not be included.

α	Earth's albedo (reflectivity); also Beer–Lambert proportionality constant; also alpha particle
β	Beta particle
γ	Gamma rays
Δ	Difference or change in a quantity
ε	Extinction coefficient; also efficiency of a fuel cell
ε_0	Permittivity in vacuum
η	Efficiency
λ	Wavelength; also nuclear decay constant
Λ	Orbital angular momentum quantum number (along internuclear axis)
μ_{at}	Attenuation coefficient for γ rays
ν_e	Frequency; also kinetic rate; also symbol for electron neutrino
$\bar{\nu}_e$	Electron, antineutrino
σ	Absorption cross section, also nuclear reaction cross section
τ	Mean-life
T	Residence time
ϕ	Primary quantum yield
Φ	Overall quantum yield
2,4-D	2,4-dichlorophenoxyacetic acid
2,4,5-T	2,4,5-trichlorophenoxyacetic acid
A	Ampere; electric current
A	Activity (in radioactive decay); also mass number; also Arrhenius preexponential factor
Å	Ångstrom
ABS	Alkylbenzenesulfonates
amu	Atomic mass unit
atm	Atmosphere

BOD	Biological oxygen demand
Bq	Becquerel (1 disintegration per second)
Btu	British thermal unit (1 054 J)
BWR	Boiling water reactor
c	Speed of light in vacuum (2.998×10^8 m/s)
C	Coulomb; also Celsius (as in °C)
C	Concentration
CAFE	Corporate average fuel economy
cal	Calorie
CFC	Chlorofluorocarbon
Ci	Curie; 3.7×10^{10} disintegrations per second
COD	Chemical oxygen demand
C_p	Heat capacity at constant pressure
CTBT	Comprehensive Test Ban Treaty
D	Dose of ionizing radiation
DDE	Dichlorodiphenylethylene; 1,1-dichloro-2,2-bis(p- chlorophenyl)ethylene
DDT	Dichlorodiphenyltrichloroethane; 1,1,1-trichloro-2,2- bis(p-chlorophenyl)ethane
DNA	Deoxyribonucleic acid
DOC	Dissolved organic carbon
DOE	U.S. Department of Energy
DU	Dobson unit
E	Energy; potential
E_a	Arrhenius activation energy
EC	Electron capture (nuclear process)
ECCS	Emergency core cooling system (nuclear reactor)
EDTA	Ethylenediaminetetraacetic acid
E_g	Band gap
E_H	Cell potential on the hydrogen scale
$E°$	Standard potential
EPA	U.S. Environmental Protection Agency
eV	Electron-volt
exp	Exponential; $\exp(y) = e^y$
F	Faraday constant, 96,485 coulombs
ft	Foot
g	Gram; also molecular symmetry designation, gerade (even)
G	Free energy; Gibbs function
gal	Gallon
GWP	Global warming potential
Gy	Gray = 100 rads
h	Planck's constant (6.626×10^{-34} J · s)
H	Enthalpy (not to be confused with the symbol for hydrogen, roman H)
HC	Hydrocarbon
HCFC	Hydrochlorofluorocarbon
HEU	High enrichment uranium
HFC	Hydrofluorocarbon
HFE	Hydrofluoroether
HLW	High-level waste (radioactive)
HWR	Heavy water reactor
I	Radiation intensity
IC	Internal conversion (nuclear process)
ILW	Intermediate-level waste (radioactive)

IT	Isomeric transition (nuclear process)
J	Joule
J	Total angular momentum quantum number of an atom
k	Boltzmann constant (1.381×10^{-23} J molecule^{-1}K^{-1}); also rate constant
K	Kelvin
K	Equilibrium constant
K_a	Acid dissociation (ionization) constant
K_b	Base dissociation (ionization) constant
keV	Thousand electron-volts
kg	Kilogram
K_{sp}	Solubility product constant
K_w	Ion product of water
L	Net electronic orbital angular momentum quantum number
LAS	Linear alkylsulfonates
LD$_{50}$	Lethal dose for 50% of the population
LET	Linear energy transfer
LEU	Low-enrichment uranium
LLW	Low-level waste (radioactive)
LMFBR	Liquid metal fast breeder reactor
LOCA	Loss of coolant accident (nuclear reactor)
LWBR	Light water breeder reactor
LWR	Light water reactor
m	Meter, mass
M	Molarity (mol/dm^3)
MeV	Million electron volts
MH	Metal hydride
mi	Mile
mol	Mole of substance
MON	Motor octane number
MOX	Mixed oxide (Pu and U oxides)
MRI	Magnetic resonance imaging
MTBE	Methyl t-butyl ether
MTHM	Metric ton of heavy metal
MWd	Megawatt-day
MW$_e$	Megawatts of electricity
MW$_t$	Megawatts of heat (t = thermal)
N	Neutron number, also number of atoms
N_A	Avogadro's number; 6.023×10^{23} particles
NO$_x$	Nitrogen oxides
NPT	Non-Proliferation Treaty
NRC	U.S. Nuclear Regulatory Commission; also National Research Council
ODP	Ozone depletion potential
p	Proton
P	Pressure
Pa	Pascal
PBB	Polybrominated biphenyl
PCB	Polychlorinated biphenyl
pE	Negative log of electron activity (a measure of redox potential)
PET	Poly(ethylene terephthalate); also position emission tomography
PFC	Perfluorocarbon

pH	Negative log of H^+ activity (approx. concentration)
pK	$-\log K$
POC	Particulate organic carbon
ppb	Parts per billion
ppbv	Parts per billion by volume
ppm	Parts per million
ppt	Parts per trillion
pptv	Parts per trillion by volume
PSC	Polar stratospheric cloud
PVC	Poly(vinyl chloride)
PWR	Pressurized water reactor
q	Heat; also electric charge
Q	Quad (1.054×10^{18} J)
Q	Energy in nuclear reactions; also quality factor (for ionizing radiation)
r	Rate of reaction
R	Roentgen
R	Universal gas constant ($8.314 \, \text{J} \cdot \text{K}^{-1} \, \text{mol}^{-1}$)
rad	Radiation absorbed dose (10.0 mJ/kg)
RBE	Relative biological effectiveness
RBMK	Reactor high-power boiling channel nuclear reactor (Soviet)
rem	Roentgen-equivalent man (a measure of radiation dose)
RON	Research octane number
s	Second; also spin quantum number for an electron
S	The solar constant; also used for entropy and net spin angular momentum of an atom
SNF	Spent nuclear fuel
Sv	Sievert, dose equivalent (ionizing radiation)
T	Absolute temperature (K)
$t_{1/2}$	Half-life
T_c	Critical temperature
TCDD	2,3,7,8-Tetrachlorodibenzodioxin
T_{eff}	Effective half-life
TMI	Three Mile Island
TOC	Total organic carbon
TRU	Transuranic waste
u	Molecular symmetry designation—ungerade (odd)
U	Internal energy
UV	Ultraviolet radiation; < 400 nm
UV-A	Radiation in the range 400–320 nm
UV-B	Radiation in the range 320–290 nm
UV-C	Radiation in the range 290–200 nm
V	Volts
V	Volume, potential energy
VOC	Volatile organic compound
w	Work
W	Watts
WL	Working level (for exposure to α particles from radon progeny)
w_R	Radiation weighting factor
w_T	Tissue weighting factor
Z	Collision frequency; also atomic number
Z'	Specific collision frequency

INDEX